The chemistry of

the hydrazo, azo and azoxy groups

Part 1

THE CHEMISTRY OF FUNCTIONAL GROUPS

A series of advanced treatises under the general editorship of
Professor Saul Patai

The chemistry of alkenes (published in 2 volumes)
The chemistry of the carbonyl group (published in 2 volumes)
The chemistry of the ether linkage (published)
The chemistry of the amino group (published)
The chemistry of the nitro and nitroso group (published in 2 parts)
The chemistry of carboxylic acids and esters (published)
The chemistry of the carbon–nitrogen double bond (published)
The chemistry of amides (published)
The chemistry of the cyano group (published)
The chemistry of the hydroxyl group (published in 2 parts)
The chemistry of the azido group (published)
The chemistry of acyl halides (published)
The chemistry of the carbon–halogen bond (published in 2 parts)
The chemistry of the quinonoid compounds (published in 2 parts)
The chemistry of the thiol group (published in 2 parts)
The chemistry of the hydrazo, azo and azoxy groups (published in two parts)

$$-NH-NH- \qquad -N{=}N- \qquad -N{=}\underset{\substack{\downarrow \\ O}}{N}-$$

The chemistry of
the hydrazo, azo and azoxy groups

Part 1

Edited by

SAUL PATAI

The Hebrew University, Jerusalem

1975

JOHN WILEY & SONS

LONDON — NEW YORK — SYDNEY — TORONTO

An Interscience ® Publication

Library of Congress Cataloging in Publication Data:

Patai, Saul
    The chemistry of the hydrazo, azo, and azoxy groups.
    (The Chemistry of functional groups)
    'An Interscience publication.'
    Includes indexes.
    1. Azo compounds. 2. Azoxy compounds. I. Title. II. Series.
QD341.A9P36   547′.043   75–2194
ISBN 0 471 66926 1 (Pt. 1)
ISBN 0 471 66927 X (Pt. 2)
ISBN 0 471 66924 5 (Set)

Printed in Great Britain by J W Arrowsmith Ltd,
Bristol, BS3 2NT.

Contributing Authors

R. Allmann	Philipps University, Lahnberge, 355 Marburg, Germany
J.-P. Anselme	University of Massachusetts at Boston, Boston, Massachusetts 02125, U.S.A.
K. G. Boto	James Cook University of North Queensland, Townsville, Australia 4811
M. I. Bruce	University of Adelaide, South Australia 5001
E. Buncel	Queen's University, Kingston, Ontario, Canada
R. A. Cox	Queen's University, Kingston, Ontario, Canada
A. J. Dolenko	Eastern Forest Products Laboratory, Canadian Forestry Service Environment Canada, Ottawa, Ontario, Canada
R. J. Drewer	Avondale College, Cooranbong, N.S.W. 2265, Australia
B. L. Goodall	The University, Bristol, England
D. A. R. Happer	University of Canterbury, Christchurch, New Zealand
A. F. Hegarty	University College, Cork, Ireland
A. Horowitz	Soreq Nuclear Research Centre, Yavne, Israel
G. Koga	Ibaraki University, Mito, Japan
N. Koga	Ibaraki University, Mito, Japan
P. J. Krueger	University of Calgary, Calgary, Alberta, Canada
K. Mackenzie	University of Bristol, Bristol, England
T. Miyadera	Central Research Laboratories, Sankyo Co., Ltd., Tokyo, Japan
B. T. Newbold	University of Moncton, Moncton, New Brunswick, Canada
L. A. Rajbenbach	Soreq Nuclear Research Centre, Yavne, Israel
M. B. Robin	Bell Laboratories, Murray Hill, New Jersey, U.S.A.
R. Shaw	Stanford Research Institute, Menlo Park, California, U.S.A.
Y. Shvo	Tel-Aviv University, Tel-Aviv, Israel
F. Snatzke	University of the Ruhr, Bochum, Germany
G. Snatzke	University of the Ruhr, Bochum, Germany

J. C. Stowell	Louisiana State University in New Orleans, Louisiana, U.S.A.
S. W. Tam	The Chinese University of Hong Kong, Shatin, N.T., Hong Kong
F. G. Thomas	James Cook University of North Queensland, Townsville, Australia 4811
J. W. Timberlake	Louisiana State University in New Orleans, Louisiana, U.S.A.
J. Vaughan	University of Canterbury, Christchurch, New Zealand

Foreword

The present volume, 'The Chemistry of the Hydrazo, Azo and Azoxy Groups' is organized and presented on the same general lines as the other volumes of the series and which are described in the 'Preface to the series' printed on the following pages.

The subject matter of this volume does not include either diazonium salts or diazo-alkanes. These groups will be treated in a separate forthcoming volume of the series. In addition it was decided not to treat azo dyes as such extensively since this would have enlarged the volume to unreasonable proportions while recent and satisfactory treatises exist on this subject.

The plan of the present volume included also a chapter on 'Detection and Determination of Hydrazo, Azo and Azoxy Groups' which however owing to the illness of the author, did not materialize.

Some additional material on the Azo and Azoxy Groups will also be available soon in the first supplementary volume of the series now in the press which will include material on double-bonded groups. While this supplementary volume will treat mainly the C═C, C═O and C═N groups it will also treat some aspects of the N═N groups in several of its chapters.

Jerusalem, January 1975 SAUL PATAI

The Chemistry of Functional Groups
Preface to the series

The series 'The Chemistry of Functional Groups' is planned to cover in each volume all aspects of the chemistry of one of the important functional groups in organic chemistry. The emphasis is laid on the functional group treated and on the effects which it exerts on the chemical and physical properties, primarily in the immediate vicinity of the group in question, and secondarily on the behaviour of the whole molecule. For instance, the volume *The Chemistry of the Ether Linkage* deals with reactions in which the C—O—C group is involved, as well as with the effects of the C—O—C group on the reactions of alkyl or aryl groups connected to the ether oxygen. It is the purpose of the volume to give a complete coverage of all properties and reactions of ethers in as far as these depend on the presence of the ether group, but the primary subject matter is not the whole molecule, but the C—O—C functional group.

A further restriction in the treatment of the various functional groups in these volumes is that material included in easily and generally available secondary or tertiary sources, such as Chemical Reviews, Quarterly Reviews, Organic Reactions, various 'Advances' and 'Progress' series as well as textbooks (i.e. in books which are usually found in the chemical libraries of universities and research institutes) should not, as a rule, be repeated in detail, unless it is necessary for the balanced treatment of the subject. Therefore each of the authors is asked *not* to give an encyclopaedic coverage of his subject, but to concentrate on the most important recent developments and mainly on material that has not been adequately covered by reviews or other secondary sources by the time of writing of the chapter, and to address himself to a reader who is assumed to be at a fairly advanced post-graduate level.

With these restrictions, it is realized that no plan can be devised for a volume that would give a *complete* coverage of the subject with *no* overlap between chapters, while at the same time preserving the readability of the text. The Editor set himself the goal of attaining *reasonable* coverage with *moderate* overlap, with a minimum of cross-references between the chapters of each volume. In this manner, sufficient freedom is given to each author to produce readable quazi-monographic chapters.

The general plan of each volume includes the following main sections:

(a) An introductory chapter dealing with the general and theoretical aspects of the group.

(b) One or more chapters dealing with the formation of the functional group in question, either from groups present in the molecule, or by introducing the new group directly or indirectly.

(c) Chapters describing the characterization and characteristics of the functional groups, i.e. a chapter dealing with qualitative and quantitative methods of determination including chemical and physical methods, ultraviolet, infrared, nuclear magnetic resonance and mass spectra: a chapter dealing with activating and directive effects exerted by the group and/or a chapter on the basicity, acidity or complex-forming ability of the group (if applicable).

(d) Chapters on the reactions, transformations and rearrangements which the functional group can undergo, either alone or in conjunction with other reagents

(e) Special topics which do not fit any of the above sections, such as photochemistry, radiation chemistry, biochemical formations and reactions. Depending on the nature of each functional group treated, these special topics may include short monographs on related functional groups on which no separate volume is planned (e.g. a chapter on 'Thioketones' is included in the volume *The Chemistry of the Carbonyl Group*, and a chapter on 'Ketenes' is included in the volume *The Chemistry of Alkenes*). In other cases, certain compounds, though containing only the functional group of the title, may have special features so as to be best treated in a separate chapter, as e.g. 'Polyethers' in *The Chemistry of the Ether Linkage*, or 'Tetraaminoethylenes' in *The Chemistry of the Amino Group*.

This plan entails that the breadth, depth and thought-provoking nature of each chapter will differ with the views and inclinations of the author and the presentation will necessarily be somewhat uneven. Moreover, a serious problem is caused by authors who deliver their manuscript late or not at all. In order to overcome this problem at least to some extent, it was decided to publish certain volumes in several parts, without giving consideration to the originally planned logical order of the chapters. If after the appearance of the originally planned parts of a volume it is found that either owing to non-delivery of chapters, or to new developments in the subject, sufficient material has accumulated for publication of an additional part, this will be done as soon as possible.

The overall plan of the volumes in the series 'The Chemistry of Functional Groups' includes the titles listed below:

The Chemistry of Alkenes (*published in two volumes*)
The Chemistry of the Carbonyl Group (*published in two volumes*)
The Chemistry of the Ether Linkage (*published*)
The Chemistry of the Amino Group (*published*)
The Chemistry of the Nitro and the Nitroso Group (*published in two parts*)
The Chemistry of Carboxylic Acids and Esters (*published*)
The Chemistry of the Carbon–Nitrogen Double Bond (*published*)
The Chemistry of the Cyano Group (*published*)
The Chemistry of Amides (*published*)
The Chemistry of the Hydroxyl Group (*published in two parts*)
The Chemistry of the Azido Group (*published*)
The Chemistry of Acyl Halides (*published*)
The Chemistry of the Carbon–Halogen Bond (*published in two parts*)
The Chemistry of the Quinonoid Compounds (*published in two parts*)
The Chemistry of the Thiol Group (*published in two parts*)
The Chemistry of the Carbon–Carbon Triple Bond
The Chemistry of Amidines and Imidates (*in press*)
The Chemistry of the Hydrazo, Azo and Azoxy Groups (*published*)
The Chemistry of the Cyanates and their Thio-derivatives (*in preparation*)
The Chemistry of the Diazonium and Diazo Groups (*in preparation*)
Supplementary Volume on the Chemistry of Double-bonded Groups (*in press*)

Advice or criticism regarding the plan and execution of this series will be welcomed by the Editor.

The publication of this series would never have started, let alone continued, without the support of many persons. First and foremost among these is Dr. Arnold Weissberger, whose reassurance and trust encouraged me to tackle this task, and who continues to help and advise me. The efficient and patient cooperation of several staff-members of the Publisher also rendered me invaluable aid (but unfortunately their code of ethics does not allow me to thank them by name). Many of my friends and colleagues in Israel and overseas helped me in the solution of various major and minor matters, and my thanks are due to all of them, especially to Professor Z. Rappoport. Carrying out such a long-range project would be quite impossible without the non-professional but none the less essential participation and partnership of my wife.

The Hebrew University,
Jerusalem, ISRAEL

SAUL PATAI

Contents

CHAPTER **1**

Electronic structures of the azo, azoxy and hydrazo groups

M. B. ROBIN

Bell Laboratories, Murray Hill, New Jersey, U.S.A.

I. INTRODUCTION

Our intentions are to present a unified discussion of the basic chemical groups ⟍N=N⟍ (azo), ⟍N=N(O)⟍ (azoxy) and >N—N< (hydrazo), focusing on their electronic structures as derived from semiempirical and/or *ab initio* calculations, and also to compare the predicted structures with the results of spectroscopic experiments where they are available. This effort is thwarted in part by the fact that in certain cases the calculations are untested because of a lack of spectroscopic data and, in others, the spectra have no calculation for comparison; in the case of the azoxy group there seem to be neither spectra nor calculations available. The neglect of these compounds by most chemists working in the field of large-molecule electronic structure is rather odd, since if the N atoms of hydrazine, diimide, nitrogen and diimide monoxide are replaced by CH groups, then the molecular series ethane, ethylene, acetylene, ethylene oxide is generated, and each of these has received more than adequate attention.

Probably the most interesting aspect of the azo, azoxy and hydrazo groups taken together is that each has two pairs of nonbonding electrons which interact with one another to an extent governed by geometric considerations. Certainly, the extent of lone pair–lone pair interactions is a major factor governing the electronic spectra of these chromophores, and must likewise be an important factor in deciding their paths of chemical reaction.

With respect to the calculations discussed here, most are *ab initio* calculations performed in a double-zeta Gaussian orbital basis. They are quite reliable when applied to the interpretation of photoelectron spectra via Koopmans' theorem, but do an uneven job of predicting bound optical excitations. Interpretations of the optical spectra are further complicated by the frequent presence of Rydberg transitions, which are not accounted for in valence shell calculations of whatever sophistication. The *ab initio* calculations are also very useful and instructive when the wave functions have been analysed for the atomic and overlap populations. In the case of large molecules, such Gaussian orbital calculations become too expensive, and it is more practical to use the semi-empirical CNDO-type techniques.

Where appropriate, we also compare the electronic structures and properties of these groups among themselves and with the nitrogen molecule, for they all can be thought of as various oxidized and reduced forms of N_2.

II. THE AZO GROUP

The parent substance for this class of compounds is diimide, H—N═N—H, a fugitive material[1] about which relatively little is known, compared to the isoelectronic systems ethylene and formaldehyde. Because data on the parent are so scanty, we will supplement them with data for the related derivatives *trans*-azomethane, diazirine and difluorodiazine.

A. Theoretical Calculations

The electronic structure of planar *trans*-diimide has been calculated several times[2–4] with different purposes in mind. The calculation of Snyder and Basch is best suited for review purposes, since the population analyses have been reported, Table 1. The numbering of the atoms and the disposition of the local axes (the same on each atom) are shown in Figure 1. The three highest MOs are of the greatest interest in this molecule. The highest, $4a_g$, is represented very nicely by the two lone pairs on the nitrogen atoms, taken out-of-phase so as to give an N—N overlap population of −0·5291. (The upper component formed by the interaction of two lone-pair orbitals will be called n_- throughout this work. The lower component is

TABLE 1. Atomic and overlap populations for diimide[2]

	$4a_g$	$1a_u$	$3b_u$	$3a_g$	$2b_u$	$2a_g$
H(1s)	0·1072	0·0000	0·0335	0·2819	0·2560	0·0266
N(2s)	0·1435	0·0000	0·1951	0·0398	0·5189	0·8242
$N(2p_x)$	0·1793	0·0000	0·0965	0·5846	0·0728	0·1426
$N(2p_y)$	0·5700	0·0000	0·6748	0·0936	0·1521	0·0065
$N(2p_z)$	0·0000	1·0000	0·0000	0·0000	0·0000	0·0000
N_1—H_1	0·0112	0·0000	0·0054	0·2534	0·3374	0·0351
N_1—H_2	−0·0705	0·0000	0·0214	−0·0680	−0·0444	0·0019
N_1—N_2	−0·5291	0·4604	0·2771	0·1328	−0·3732	0·6332
H_1—H_2	0·0077	0·0000	−0·0008	0·0094	−0·0028	0·0000

referred to as n_+.) The corresponding in-phase combination (n_+) correlates with the third lowest MO, $3b_u$, where the N—N overlap is +0·2771. As might be expected from the large values of the overlap populations, the $4a_g$–$3b_u$ splitting is appreciable; 6·15 eV as reckoned by Koopmans' theorem. The third orbital, $1a_u$, falls between the two combinations of lone-pair AOs, and constitutes the π bond between the two nitrogen atoms (overlap equals +0·4604). The nitrogen 1s ionization potentials of diimide are also of some interest; as might be expected, those of N_2H_2 are predicted to be intermediate between those of N_2 and N_2H_4.

The most remarkable aspect of the diimide electronic structure is the calculated splitting of 6·15 eV between the lone-pair MOs, implying an extraordinarily large interaction between the lone-pair AOs. A quantitative test of the energy level structure calculated for diimide is at present not practical since the photoelectron spectrum has not been determined.

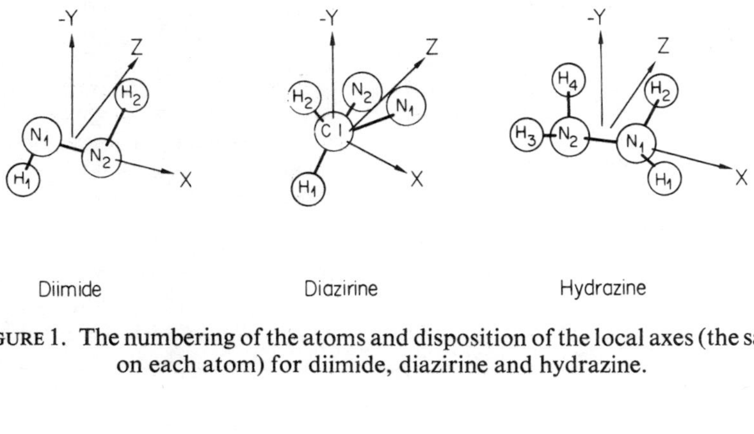

FIGURE 1. The numbering of the atoms and disposition of the local axes (the same on each atom) for diimide, diazirine and hydrazine.

However, we can turn to certain derivatives of diimide for this type of information. In *trans*-azomethane[5–7] the n_-, π, n_+ pattern is observed with an overall splitting of 3·3 eV, while in diazirine and dimethyl diazirine it is 3·5 and 3·55 eV, respectively, and 3·7 eV in diazabicycloheptene. Inasmuch as the calculated split for diazirine is 4·0 eV, it seems that the calculations are reliable and that the n_--n_+ split in both *cis*- and *trans*-azoalkanes is as large as 3·5–4·0 eV; as for diimide itself, the generally high reliability of the Gaussian orbital calculations leads us to expect an n_--n_+ split substantially larger than 4·0 eV. The lone-pair splitting in *cis*-diimide is calculated to be about 3 eV smaller than in the *trans*-isomer[4].

Let us imagine diimide produced by pulling two protons from the nuclei of the oxygen molecule. If these protons are pulled out along the N—N internuclear line, then the diimide will have the same electronic symmetry as the ground state of oxygen, $^3\Sigma_g^-$. Orgel[8] shows that as these protons are brought off the N—N line, the $^1\Delta$ upper state splits, and with increasing angle, a component of this finally drops below $^3\Sigma$ to become the ground state. Thus the ground-state spin multiplicity of R—N=N—R azo compounds in general will depend upon the angular displacements of the R groups from the N—N line. Though all derivatives of diimide are known to be diamagnetic, implying the crossover described above, a spectroscopic study of N_2H_2 led Trombetti[9] to conclude that the ground state of this molecule is ^3B_g. Though later e.p.r. work in matrices has failed to confirm

TABLE 2. Atomic and overlap populations for diazirine[2]

	$3b_1$	$2b_2$	$6a_1$	$5a_1$	$1b_2$	$2b_1$	$4a_1$
H(1*s*)	0·0000	0·2286	0·0227	0·1699	0·2108	0·0000	0·1610
C(2*s*)	0·0000	0·0000	0·0257	0·0460	0·0000	0·0000	1·1165
C($2p_x$)	0·7842	0·0000	0·0000	0·0000	0·0000	0·2208	0·0000
C($2p_y$)	0·0000	0·2568	0·0000	0·0000	0·9301	0·0000	0·0000
C($2p_z$)	0·0000	0·0000	0·1073	0·5896	0·0000	0·0000	0·0084
N(2*s*)	0·0771	0·0000	0·1918	0·0605	0·0000	0·7802	0·0507
N($2p_x$)	0·0819	0·0000	0·7146	0·0059	0·0000	0·0856	0·0444
N($2p_y$)	0·0000	0·6430	0·0000	0·0000	0·3241	0·0000	0·0000
N($2p_z$)	0·4730	0·0000	0·0044	0·4458	0·0000	0·0232	0·1814
H_1—H_2	0·0000	−0·0781	0·0067	0·0315	−0·0220	0·0000	0·0073
C_1—H_1	0·0000	0·1825	0·0039	0·1444	0·2101	0·0000	0·2049
N_1—H_1	0·0000	−0·0450	−0·0031	−0·0209	0·0114	0·0000	−0·0025
N_1—C_1	0·0568	−0·1501	0·0569	0·0634	0·1304	0·1174	−0·0954
N_1—N_2	−0·2281	0·3520	−0·0013	0·1857	−0·0034	−0·8095	0·2048

this triplet ground state, it still must be considered as a possibility in the gas phase. The photoelectron spectrum of this molecule in the gas phase would readily settle the question of its spin multiplicity.

Let us take a closer look at diazirine as a prototype for the azo group, for abundant theoretical and experimental data are available for it[2,10,11]. The transition from diimide to diazirine is predicted to considerably decrease the lone pair–lone pair interaction. As seen from Table 2, the N—N overlap population for the n_- orbital ($3b_1$) drops to –0·2281 in diazirine and the calculated n_--n_+ split decreases to 4·0 eV. However, in diimide the n_+ MO is correspondingly N—N bonding (overlap equal to +0·2771) whereas in diazirine this decreases to –0·0013 while taking on a certain small amount of C—N σ character. Since the two-centre π bond of diimide is spread over three centres in diazirine with almost equal weight at each centre, the N—N π orbital population understandably drops (0·4604 to 0·3520). There is a lower π MO in diazirine ($1b_2$) which is also somewhat N—N bonding, and adding its contribution to that of $2b_2$ results in an overall π population of 0·4486, which is very close to that of diimide. The remaining MOs of diazirine are involved in the σ bonding within the ring and between the C—H atoms.

Various ground-state properties of the diazirine molecule have been calculated with the Gaussian orbital wave function[10]. A dipole moment of 2·34D is predicted, while 1·59D is observed. It is not unusual with such calculations that the predicted moment is too large by 0·5–1·0D. The calculated values of the second moments of the molecular charge distribution appear to be reliable, for the anisotropies $\langle z^2\rangle - \langle x^2\rangle = 4{\cdot}19$, $\langle z^2\rangle - \langle y^2\rangle = 11{\cdot}98$ and $\langle x^2\rangle - \langle y^2\rangle = 7{\cdot}79 \times 10^{-16}$ cm² compare well with Flygare's experimental values $4{\cdot}2 \pm 0{\cdot}2$, $12{\cdot}0 \pm 0{\cdot}2$ and $7{\cdot}7 \pm 0{\cdot}2 \times 10^{-16}$ cm², respectively[12].

B. Photoelectron Spectra

These predictions of the ground-state electronic structure of diazirine can be most readily tested by comparison with the photoelectron spectrum, Figure 2. First, the calculated orbital energies should agree quantitatively with the successive ionization potentials, and second, the width and vibrational structure of a photoelectron band should be related to the overlap populations involved, with essentially nonbonding MOs giving narrow lines and the others showing the excitation of many vibrations in the regions of large overlap population.

Actually, the first ionization potential of diimide has been measured by mass spectrometry[13] and the value found, $9{\cdot}85 \pm 0{\cdot}1$ eV, is remarkably close to that predicted by Koopmans' theorem, 9·90 eV. However, since

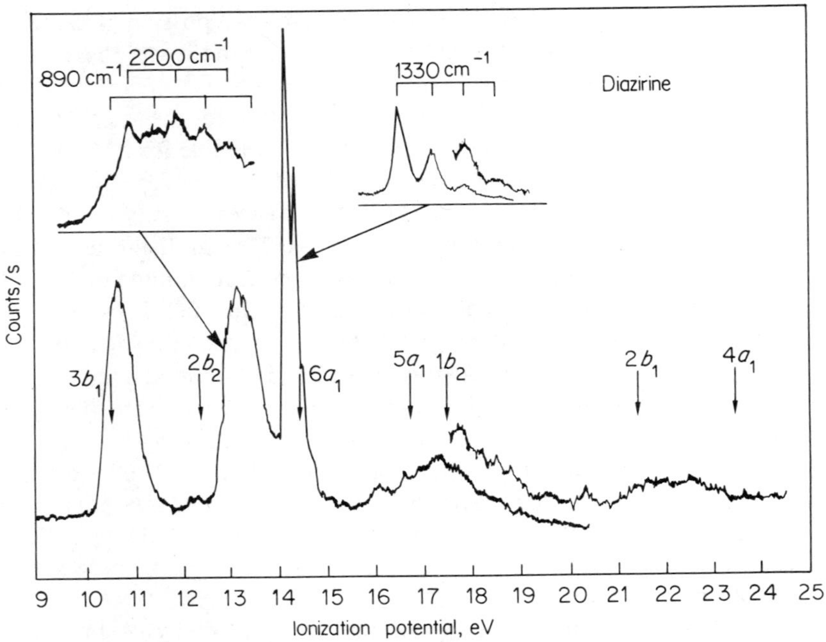

FIGURE 2. The photoelectron spectrum of diazirine, with vertical arrows at the predicted ionization potentials[11].

further ionization potentials are unknown for diimide, it is best to turn to diazirine for comparison with experiment.

The agreement between calculated and observed ionization potentials in diazirine is quite good (see Table 3) and so supports the calculated wave

TABLE 3. Observed and calculated ionization potentials of diazirine[11]

Orbital	Vertical ionization potential	0·92 K.T.[a]
$3b_1$	10·75	10·48
$2b_2$	13·25	12·32
$6a_1$	14·15	14·48
$5a_1$	16·5	16·74
$1b_2$	17·5	17·48
$2b_1$	21·5	21·40
$4a_1$	22·5	23·48

[a] 92% of the Koopmans' theorem value.

functions. Vertical ionizations from n_- ($3b_1$) and n_+ ($6a_1$) are observed at 10·75 and 14·15 eV, Figure 2, with the second of these considerably more narrow and vertical than the first. The $6a_1$ band in fact looks very much like an ionization from a true lone-pair AO, as in an alcohol, for example. The difference between the $6a_1$ and $3b_1$ band shapes is readily understood since the $6a_1$ MO is indeed calculated to have the smallest overlap components of any MO in the molecule. This assignment of the narrow feature in the diazirine photoelectron spectrum as originating with the n_+ lone-pair MO rather than with the π MO is supported by the work of Haselbach and co-workers[14], on dimethyl diazirine, where the n_-, π, n_+ ordering of levels is preferred. The 1330 cm^{-1} vibration excited in the $6a_1$ band of diazirine is thought to be the N=N stretch, ν_3'. Vibrational structure is seen as well in the π $2b_2$ ionization, but it is much less vertical owing to the strongly bonding nature of this MO. Ionization from the deeper π MO $1b_2$ (an orbital which is largely C—H bonding, Table 2) is predicted to come at 17·48 eV, and is to be associated with the broad ionization centred at 17·3 eV, Figure 2.

C. Optical Spectra

From our earlier experiences with the spectra of the better-known olefin and keto groups, we can readily guess what sort of optical spectra to expect for diimide and the azoalkanes. First, excitations from the uppermost lone-pair orbital, n_-, to π^* will be low-lying as in the ketones. Further, this $n_- \rightarrow \pi^*$ transition is of $a_g \rightarrow b_g$ symmetry in *trans*-azo compounds, and so is electronically forbidden whereas it is allowed in *cis*-azo compounds where the symmetry change is $b_1 \rightarrow a_2$. Even though formally allowed, the $n_- \rightarrow \pi^*$ band of *cis*-compounds will not be very intense due to the lack of spatial overlap of the n_- and π^* MOs[4].

A second $n \rightarrow \pi^*$ excitation is also expected in azo compounds, this being allowed in *trans*-compounds ($b_u \rightarrow b_g$), but forbidden in *cis* ($a_1 \rightarrow a_2$). The very large separation of the n_- and n_+ MOs in azo compounds (ca. 3·5 eV according to the photoelectron spectra) means that the $n_+ \rightarrow \pi^*$ band will come at ca. 2000 Å in these compounds, again with modest intensity.

The most prominent feature in the olefin spectra is the intense $\pi \rightarrow \pi^*$ excitation in the 1600–1900 Å region. In ketones, the identification of the corresponding band is still very much in doubt, though it seems likely that it is at much shorter wavelengths and far less intense than the $\pi \rightarrow \pi^*$ band of the olefins. With these two contrasting examples, it is very hard to know what to expect for azo compounds. If an intense band is observed in the 1200–1900 Å region of azo compounds ($\varepsilon \sim 6000$; $f \sim 0{\cdot}3$), then it is probably $\pi \rightarrow \pi^*$, but the absence of such a band should be no surprise. Some confusion will undoubtedly arise due to transitions within the alkyl groups of the azoalkanes, for such transitions are also expected to fall in the 1200–1600 Å region.

Finally, transitions originating at n_- and terminating at various Rydberg orbitals will fall in the vacuum u.v. region. Usually, in strongly alkylated chromophores only the transitions to 3*s*, 3*p* and 3*d* have sufficient intensity to be observed, and empirical rules are known whereby their absorption

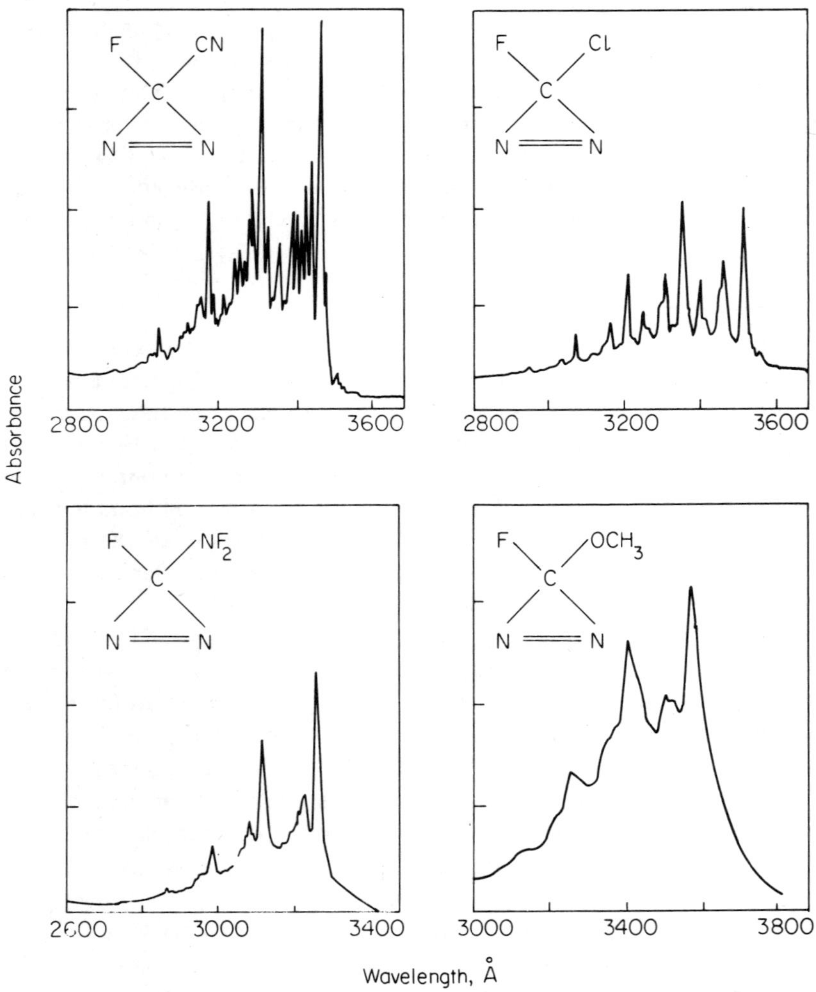

FIGURE 3. The $n_- \rightarrow \pi^*$ bands in several substituted diazirines[19].

wavelengths can be estimated once the corresponding ionization potential is known[15].

Let us now compare our expectations with reality. In *trans*-azoalkanes, the $n_- \rightarrow \pi^*$ band comes at ca. 3600 Å with a molar extinction coefficient ε of 5–10, whereas in cyclic *cis*-azo compounds the absorption wavelength is much the same, but ε is in the 100–200 range, as expected. In the *cis*-compounds, the transition is allowed with out-of-plane polarization. Single-crystal polarized absorption by the forbidden $n_- \rightarrow \pi^*$ band of *trans*-azobenzene was interpreted as showing a vibronic borrowing of in-plane-polarized intensity from the $\pi \rightarrow \pi^*$ band in the near u.v.[16]. Because the $n \rightarrow \pi^*$ band can readily borrow intensity from low-lying $\pi \rightarrow \pi^*$ transitions, the formally forbidden transition readily achieves an ε of ca. 500 when aromatic groups are substituted adjacent to the azo group. Aromatic groups also move the wavelength of maximum absorption downward to 4500 Å, where it is observed in a wide variety of compounds regardless of substitution. Robin and Simpson[16] explain this in the following way: the removal of an electron from the n_- MO leaves a positive hole tied to the nitrogen atoms which strongly attracts the optical electron in the π^* MO. The attraction is so strong in fact that the optical electron moves no further away than the adjacent carbon atom. Thus, on replacing one alkyl group of an azoalkane with a π-electron substituent, the $n \rightarrow \pi^*$ band moves to 4000 Å, and thence to 4500 Å upon replacing the second alkyl group with a π-electron substituent. Once at that point, it makes no difference whether the substituent is phenyl, naphthyl, etc., as long as it has a π orbital in the position α to the azo group. The substitution of silyl or pentavalent phosphorus groups adjacent to the azo group will move the $n_- \rightarrow \pi^*$ excitation to 5000–6000 Å[17]. Baird and co-workers[18], have calculated that as the NNC angle of *cis*-azomethane is opened from 105° to 135° the $n_- \rightarrow \pi^*$ absorption wavelength will increase smoothly from 2850 to 5200 Å.

Trombetti has observed the $n_- \rightarrow \pi^*$ band of diimide in the gas phase at ca. 3500 Å as a weak, broad continuum[9]. Only in the diazirines and a few other cyclic *cis*-azoalkanes, Figure 3, is the $n_- \rightarrow \pi^*$ band structured. Several of the diazirines have been studied by Merritt and co-workers[20,21] and others[22,23], and it is found that in diazirine the most prominent vibration excited is the totally symmetric C—N stretch, but with no appreciable N—N stretch excited, whereas in the difluoro, chloromethyl and bromomethyl derivatives, the $n_- \rightarrow \pi^*$ absorption shows a long progression of N—N stretch, but no appreciable excitation of C—N stretching.

Careful vibronic analysis of the $n_- \rightarrow \pi^*$ bands of several diazirines shows what is apparently a second electronic origin within a few hundred cm^{-1} of the more intense origin. Tentatively, this second band has been assigned as the $\pi \rightarrow \pi^*$ singlet–triplet excitation made allowed by spin-orbit mixing with the adjacent $n_- \rightarrow \pi^*$ band[23]. Rau also postulates a triplet state within the $n_- \rightarrow \pi^*$ band envelope in order to explain the lack of luminescence in azo compounds and the general lack of vibronic structure

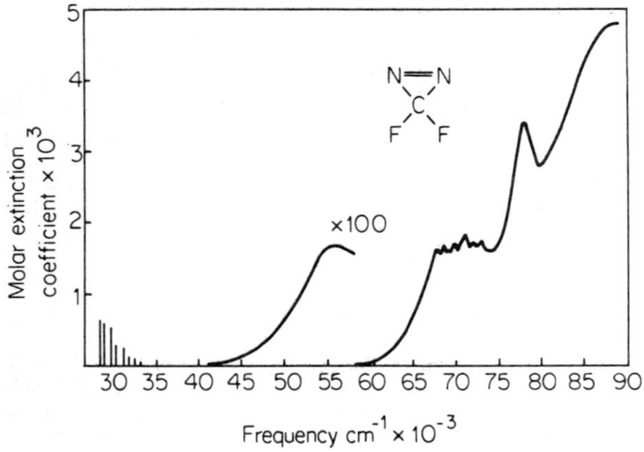

FIGURE 4. Optical absorption spectrum of difluorodiazirine[10].

in their $n_- \rightarrow \pi^*$ bands[24]. In any case, the second band is clearly *not* the second $n \rightarrow \pi^*$ excitation, as has been suggested by some.

Locating the second $n \rightarrow \pi^*$ band in azo compounds is difficult indeed. It will be allowed in *trans*-compounds, but forbidden in *cis*-compounds. Though many suggestions have been made, the only real possibility at present is the 1800-Å band of difluorodiazirine, Figure 4, which is appropriately weak ($\varepsilon = 18$) and falls in the expected region. As judged by the term values, Rydberg excitations would not fall in this region of difluorodiazirine.

Each of the *trans*-azoalkanes shows a pattern of three bands in the vacuum u.v., Figure 5, with that labelled band I having been considered repeatedly as a possible $n_+ \rightarrow \pi^*$ excitation. Trombetti[9] reports on the spectrum of diimide and describes an intense, structured band with origin at 1726·52 Å which no doubt corresponds to band I in the alkylated species. Since Trombetti's rotational analysis of this band resulted in an out-of-plane polarization, the out-of-plane polarized $n_+ \rightarrow \pi^*$ assignment is a possibility for band I of the azo group.

According to Robin[15], the $n_- \rightarrow 3s$ Rydberg transitions in the *trans*-azoalkanes will be forbidden, but $n_- \rightarrow 3p$ will be allowed and should come at ca. 20,000 cm^{-1} below the respective n_- ionization potentials. For diimide

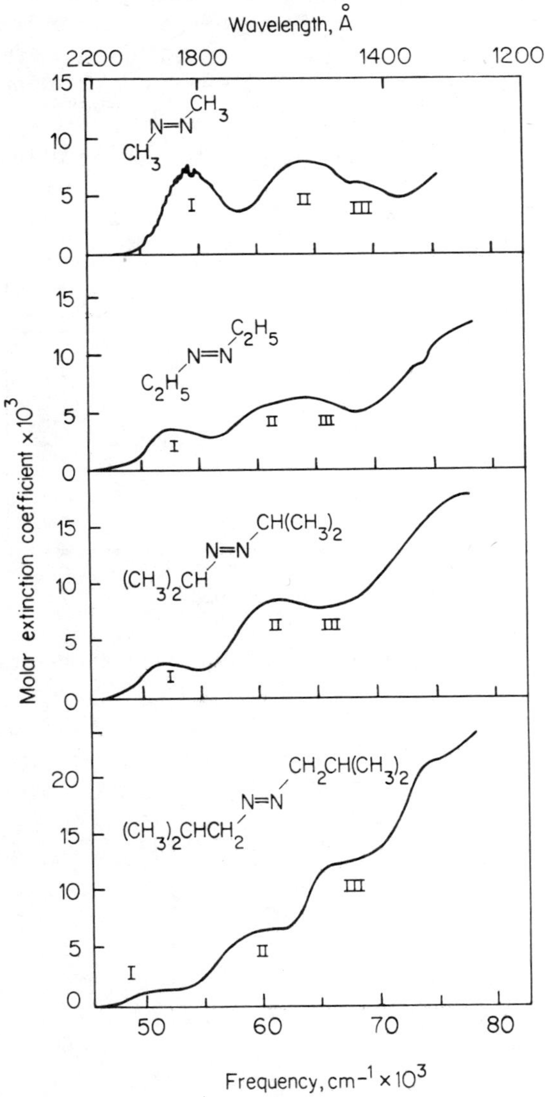

FIGURE 5. Optical absorption spectra of various azoalkanes[4].

and azomethane where the n_- ionization potentials are known, this places the $n_- \rightarrow 3p$ transitions at 1683 and 1910 Å, which agree quite well with the observed band I wavelengths. Band I of azomethane shows some discrete vibrational structure and this can be used to test further the Rydberg character of this band, for it has been demonstrated[25] that the vibronic

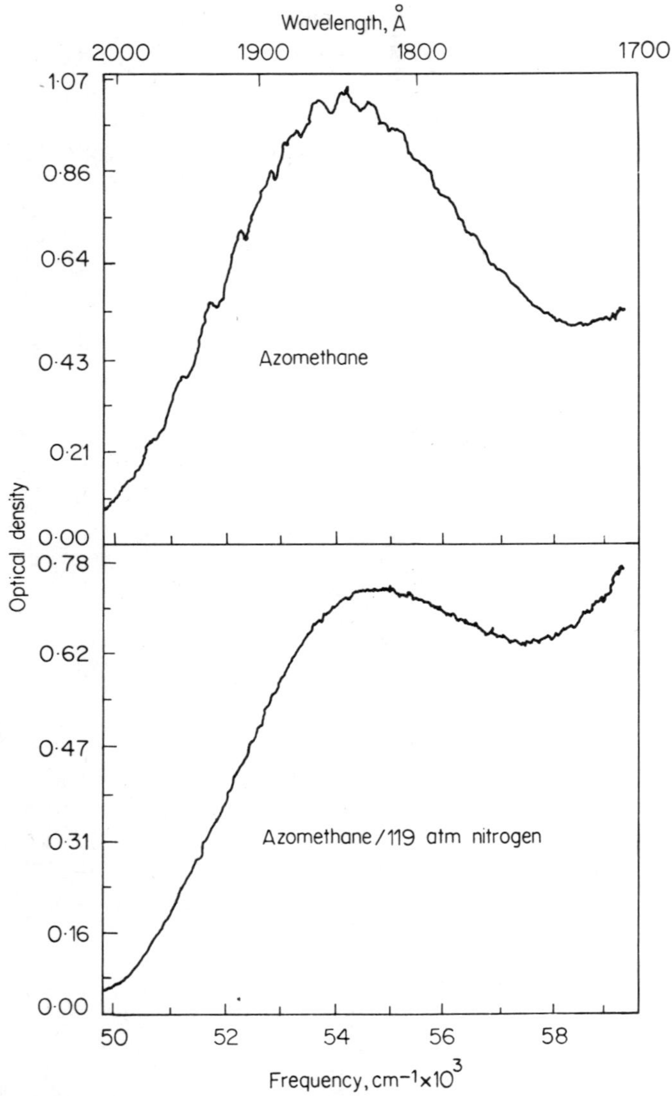

FIGURE 6. Optical absorption spectrum of band I of azomethane before and after pressurization with 119 atm of nitrogen.

structure of transitions to big-orbit Rydberg states is effectively washed out upon pressurization with a second gas, whereas valence shell excitations are unaffected by this perturbation. Actually, the Rydberg response is a broadening to the high-frequency side of each vibronic component, but this often results in washing out the vibrational structure. In Figure 6, the result of applying this test to band I of azomethane is shown; the typical Rydberg orbital response is observed upon applying the N_2 gas perturbation. Band II of the azoalkanes is at the proper frequency for the $n_- \rightarrow 4p$ Rydberg excitation, but such an assignment must be considered tentative.

In the spectrum of difluorodiazirine, Figure 4, the Rydberg transitions to $3s$ and $3p$ are both allowed by symmetry from n_-, and should be located about 26,000 and 20,000 cm^{-1}, respectively, below the n_- ionization potential at 95,700 cm^{-1} (11·78 eV). These expectations are met by the two strong bands in the spectrum at 1430 and 1320 Å, respectively.

Where then is the $\pi \rightarrow \pi^*$ transition of the azo group? This must remain as one of the more interesting unanswered questions of molecular spectroscopy, for there are no compelling arguments for its assignment to any of the observed bands. For the moment, one might take the somewhat unsatisfying position that it appears in these spectra as a broad underlying continuum in the 1650–1250 Å region, but this is tantamount to admitting total ignorance. Haselbach and Schmelzer[26] have calculated the optical spectrum of azomethane, using the observed ionization potentials together with theoretically determined J and K integrals, and conclude that $\pi \rightarrow \pi^*$ comes in the vicinity of 1100 Å, but this is certain to be at too short a wavelength due to the neglect of correlation in the π,π^* state.

The position of the lowest triplet state of the azo group is only slightly more secure. Since phosphorescence is not observed for the simple azo-containing molecules, the only evidence in this regard comes from photosensitization and quenching experiments, which show that the lowest triplet of azomethane is longwards of 4000 Å[27] and that of azobenzene is longwards of 4750 Å[28].

III. THE AZOXY GROUP

Experimental and/or theoretical studies aimed at exploring the electronic structure of the azoxy group are relatively infrequent, and the few studies which have been done do not mesh with one another very well. Hopefully others will serve this neglected area.

According to the semiempirical calculations of Kuhn and co-workers[29], the three uppermost MOs of azoxymethane involve two lone-pair MOs and a π MO, as in azomethane. However, where the ordering was n_-, π, n_+ in *trans*-azomethane, in *trans*-azoxymethane it is predicted to be π, n_O, n_N. Here, n_O is the $2p$ lone-pair on the oxygen atom and n_N is an approximately sp^2 lone-pair hybrid on the nitrogen; the splitting between the two lone-pairs amounts to 0·98 eV. The π MO must now span three centres, but its exact composition is unknown.

The net charges on the atoms of azoxymethane have been calculated also, and it was found that the oxygen bears a negative charge of –0·538 *e*, while the nitrogen to which it is bonded has a charge of +0·536 *e* and the second nitrogen is –0·352 *e*. The extreme differences in the charges of the two nitrogen atoms of the azoxy group are readily obvious in the N(1*s*) binding energies, for a very large split of 4·2 eV is observed in the e.s.c.a. spectrum of azoxybenzene[30]. The photoelectron spectrum of the only azoxyalkane run to date[31] is shown in Figure 7. In azoxy-*t*-butane, the only bands in the photoelectron spectrum which are not part of the massive $(CH_3)_3C$ group ionization are those at 9·30 and 10·00 eV. If the azoxymethane calculation is a good one (and appropriate), then these two bands are the π and n_O ionizations.

As regards the optical spectrum of the azoxy group, this is still waiting to be determined. At this point, all one can say is that the $n_O \rightarrow \pi^*$ and $n_N \rightarrow \pi^*$

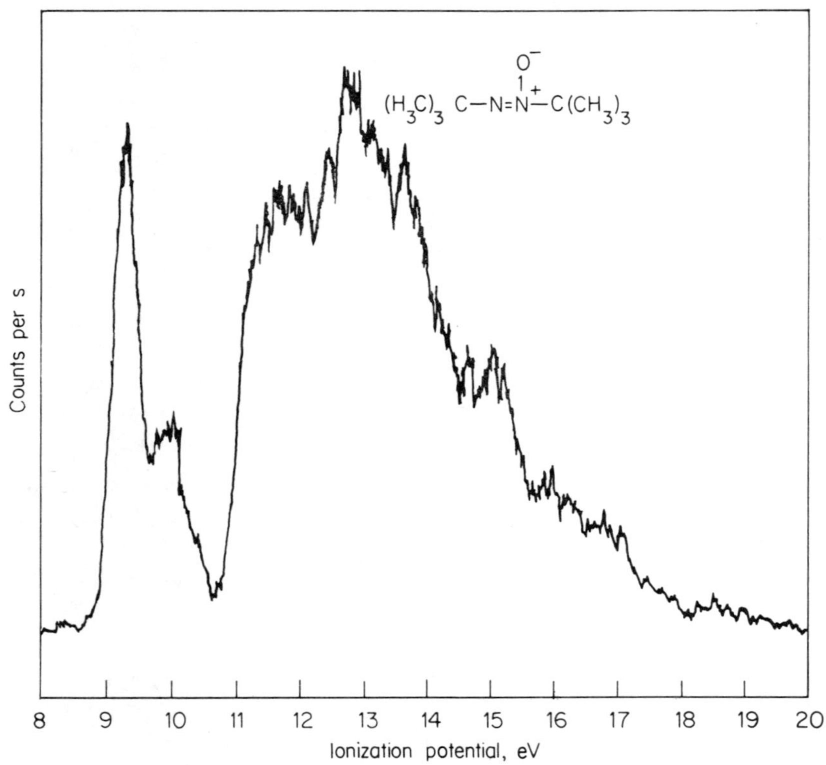

FIGURE 7. Photoelectron spectrum of azoxy-*t*-butane[31].

transitions must be at shorter wavelengths than those in the comparable azo compounds, since azoxy-*t*-butane and azoxybenzene are both colourless.

IV. THE HYDRAZO GROUP

The reduction of diimide by hydrogen leads to the related system hydrazine, which can be thought of as resulting from the addition of H_2 across the double bond of diimide, leaving the σ structure relatively unchanged. However, once the π bond is destroyed, there is no other strong factor working for planarity, and the molecule is free to distort in such a way that the very intense N—N antibonding character of the $4a_g$ MO is relieved. In hydrazine, this is accomplished by the relative rotation of the $—NH_2$ groups by approximately 90°, thereby placing the lone pairs in nearly orthogonal planes, while in the cyclic systems such as the diaziridines, the lone pairs assume a *trans* configuration[32] which again minimizes the lone pair–lone pair interaction. Because of this disadvantageous geometrical factor, it is expected that the lone-pair splitting in hydrazine and its derivatives will be considerably smaller than in diimide and the azoalkanes.

A. Theoretical Calculations

The *ab initio* calculation of the electronic structure of the hydrazine molecule in its ground state has been performed several times[2, 33–35]. By the criterion of lowest total energy, that of Snyder and Basch[2] is best, a total

TABLE 4. Atomic and overlap populations for hydrazine[2]

	5*a*	4*b*	4*a*	3*b*	3*a*	2*b*	2*a*
H_1 (1*s*)	0·0502	0·0013	0·1913	0·1894	0·0958	0·1225	0·0417
H_2 (1*s*)	0·0091	0·1125	0·0239	0·1552	0·2762	0·1099	0·0523
N_1 (2*s*)	0·0807	0·0847	−0·0033	0·0135	0·0153	0·6713	0·8376
N_1 ($2p_x$)	0·2053	0·0513	0·6406	0·0042	0·0234	0·0367	0·0433
N_1 ($2p_y$)	0·0307	0·0516	0·0586	0·5333	0·4714	−0·0003	0·0009
N_1 ($2p_z$)	0·6231	0·6984	0·0890	0·1048	0·1175	0·0600	0·0240
H_1—H_2	−0·0147	−0·0189	0·0261	−0·0436	−0·0460	0·0074	0·0003
H_1—H_3	0·0022	−0·0010	0·0089	−0·0047	0·0036	−0·0006	0·0000
H_2—H_3	−0·0040	0·0053	0·0069	0·0105	−0·0115	−0·0012	−0·0001
H_2—H_4	0·0074	−0·0340	0·0062	−0·0275	0·0448	−0·0032	0·0000
N_1—H_1	0·0392	0·0105	0·1446	0·1582	0·0951	0·1747	0·0555
N_1—H_2	0·0239	0·0913	0·0023	0·1404	0·1754	0·1649	0·0674
N_1—H_3	−0·0198	−0·0117	−0·1103	0·0278	0·0153	−0·0221	0·0045
N_1—H_4	−0·0288	−0·0712	−0·0340	−0·0069	0·0102	−0·0271	0·0082
N_1—N_2	−0·2038	−0·0624	0·3060	0·1024	0·1300	−0·3449	0·4030

energy of $-111 \cdot 1261$ a.u. having been achieved with a double-zeta Gaussian orbital basis set. The orbital wave functions from this calculation have been analysed, and the populations are given in Table 4, while the numbering of the atoms and the disposition of the local axes (the same at each atom) are shown in Figure 1. From this table one sees, for example, that MOs $5a$ and $4b$ are constructed largely of the $2p$ AOs on nitrogen with orientations appropriate to lone pairs, that $4a$ is the N—N σ bond made largely of $2p\sigma$ AOs, etc. As always happens in such calculations, these localized descriptions of the bonding are only approximate. Since the geometry chosen for the hydrazine calculation has the two halves of the molecule rotated by ca. 90° with respect to each other, the lone-pair orbitals on each centre are also nearly perpendicular.

Osafune and co-workers[36] have performed semiempirical MO calculations on hydrazine and give a simple analysis of the orbital structure. The local orbitals of concern to us are the following: the lone-pair orbitals on each nitrogen atom, the N—N σ bond orbital, and four MOs involved in N—H σ bonding. According to Osafune and co-workers[36] the N—N σ orbital of symmetry a mixes strongly with the a combination of the lone-pair orbitals pushing up this latter MO, whereas the corresponding b combination of N—N σ orbitals is at far higher energy (a virtual MO) and so has a much smaller effect on the lone-pair orbital of b symmetry, pushing it down slightly. Thus the through-bond interactions act to split the lone-pair orbitals with a above b. The semiempirical calculation shows this splitting to be very sensitive to the relative angular orientation of the two $—NH_2$ groups. The results of *ab initio* calculations[37] on the cyclic hydrazine derivative diaziridine $H_2N_2CH_2$, in which the imino protons are locked into either the *cis*- or *trans*-arrangement confirm our intuitive reasoning, for the *trans*-isomer was found to be the more stable one, and the lone-pair splitting in the *cis*-isomer (2·40 eV) approached that in diazirine itself (3·5 eV), while the splitting in the *trans*-isomer (0·92 eV) was predicted to be much smaller.

The more extensive Gaussian orbital calculation[2] suggests however, that the lone-pair interactions are more complicated than given above, for $4a$ is N—N bonding, and if it simply mixed into $5a$, then this latter also would be N—N bonding, whereas the overall N—N overlap population for $5a$ is antibonding (−0·2038), Table 4. Since this is also the sense of the direct overlap interaction (i.e., $5a$ is N—N antibonding) both the through-space and through-bond interactions act to place $5a$ above $4b$[32].

A dipole moment of 2·87D is calculated for hydrazine, while an experimental value of $1 \cdot 75 \pm 0 \cdot 08$D is reported by Kasuya[38]. The agreement is only fair, but it quite often happens that the dipole moment calculated in a

Gaussian orbital basis is 0·5–1·0D larger than the experimental value. The atomic population analysis predicts a net charge of –0·5688 e at each nitrogen and charges of +0·3069 e and +0·2609 e at H_1 and H_2, respectively. The quadrupole coupling constant, eqQ, at the ^{14}N nuclei of hydrazine has been measured (–4·09 ± 0·05 Mc/s), but has not yet been calculated with the GTO wavefunction. For comparison, values of –3·6 and –4·084 Mc/s are observed for methyl amine and ammonia, respectively.

B. Photoelectron and Optical Spectra

With respect to the calculated electronic structure of hydrazine, the photoelectron spectrum is of prime relevance since, by Koopmans' theorem, the negatives of the orbital energies calculated for the molecule in its ground state are the successive ionization potentials measured in the photoelectron spectrum. As shown in Table 5 and Figure 8, the agreement between the

TABLE 5. Observed and calculated ionization potentials in the hydrazine molecule

Orbital	Observed ionization potential	0·92 K.T.[a]
$5a$	9·91	10·32
$4b$	10·64	10·97
$4a$	15·61	14·50
$3b$	16·66	16·59
$3a$		16·98
$2b$	24·5	25·78
$2a$	31	31·71

[a] The Koopmans' theorem value multiplied by 0·92.

observed vertical ionization potentials[36] and the Koopmans' values (reduced empirically by 8 %) is very good and supports the calculated wavefunctions. It is clear from the calculations that there are two levels in the 16–17 eV region, though only one is apparent experimentally. The $5a$ and $4b$ lone-pair levels discussed above correspond to the two bands found at 9·91 and 10·64 eV in the photoelectron spectrum.

Dewar and co-workers[39] have studied the photoelectron spectra of several alkylated and phenylated hydrazines using a spectrometer of low resolution. Consequently, the most accurate values which they can report

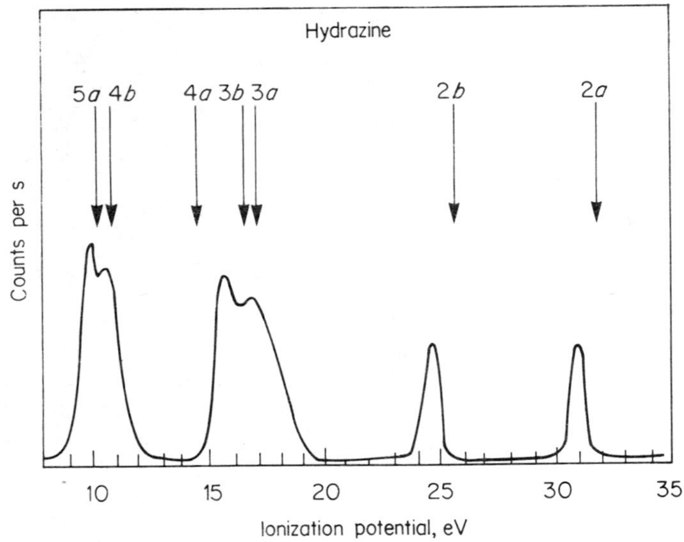

FIGURE 8. Photoelectron spectrum of hydrazine with predicted ionization potentials shown by the vertical arrows.

are the adiabatic (threshold) values for the first ionization potentials in each compound, and the splittings between the first and second ionizations are recognizable but not accurately measureable. This work was recently expanded upon by Haselbach and co-workers[32], using a high-resolution photoelectron spectrometer to measure the lone-pair splittings in several diaziridine derivatives. As expected, methylation decreases the first ionization potential of hydrazine, the decrease amounting to 1·0 eV on going from hydrazine to tetramethyl hydrazine, and the largest lone-pair splitting (1·40 eV) was found in that compound in which the lone pairs are thought to be in a *cis*-like arrangement.

The nitrogen 1*s* binding energies of N_2H_4 and N_2F_4 have been measured by Finn and co-workers relative to that of nitrogen gas[40]. They found that this ionization required 2·4 eV more energy in N_2F_4 than in N_2, but 3·8 eV less for N_2H_4 than it did in N_2. These chemical shifts of the binding energies relate to the net charges on the nitrogen atoms and indicate that in hydrazine the bond polarity is H^+—N^-, as is also apparent from the population analysis, whereas in tetrafluorohydrazine it is F^-—N^+. Quantitatively, the Gaussian orbital calculations predict a 1*s* energy difference of 3·8 eV between N_2 and N_2H_4, as observed.

Once the photoelectron spectrum is at hand and can be interpreted fairly readily via the Gaussian orbital calculation, then one is prepared to investigate the optical spectrum in the vacuum u.v. region. The connexion here is that many of the bound excitations in the u.v. spectrum of a saturated molecule bearing lone-pair electrons are Rydberg excitations which are the leading members of series converging to the ionization potentials found in the photoelectron spectrum. The lowest members of the various series are the most prominent, and it has been shown that their term values (ionization potential – absorption frequency) can be estimated readily[15]. For a molecule such as hydrazine, the 3*s* term value will be about 28,000 cm^{-1}, that to 3*p* will be about 21,000 cm^{-1} and that to 3*d* will be 13,000 cm^{-1}. Thus, given these values and the known ionization potentials, one quickly sees where the lowest Rydberg excitations will fall. Of course, such an approach says nothing about the valence shell part of the spectrum, but in saturated molecules such as ammonia, hydrazine, etc., the valence shell bands are very broad and only serve as a continuum upon which the Rydberg excitations rest.

The optical spectrum of hydrazine is almost featureless[41], see Figure 9, but one can make some sense of it. The first band centred at 1900 Å (52,600 cm^{-1}) has a term value of 27,400 cm^{-1} with respect to the 5*a* ionization potential and is most reasonably assigned as $5a \rightarrow 3s$. The broad band peaked at 1710 Å (58,500 cm^{-1}) has a term value of 21,500 cm^{-1} with respect to the 5*a* ionization potential ($5a \rightarrow 3p$), but also a term value of 27,500 cm^{-1} with respect to the 4*b* ionization potential ($4b \rightarrow 3s$), and so probably contains both of these excitations. The problem will be further complicated by the splitting of the 3*p* manifold into three components (unresolved) in hydrazine. Similarly, the band at 1510 Å (66,300 cm^{-1}) would be a composite of the components of the $5a \rightarrow 3d$, $5a \rightarrow 4s$ and $4b \rightarrow 3p$ excitations. The more intense absorption commencing at 1400 Å is most likely a valence shell excitation ($\sigma \rightarrow \sigma^*$?).

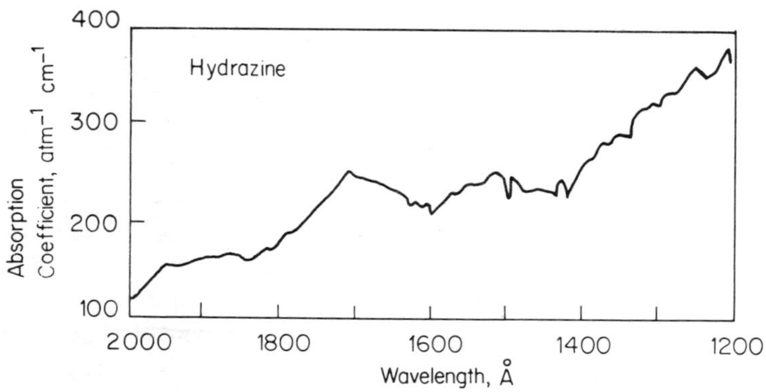

FIGURE 9. Optical absorption spectrum of hydrazine[41].

ADDENDUM

Several significant papers have appeared since the text of this article was submitted to the Editor. The contents of these recent papers are briefly described in this section.

With respect to the interesting question of the spin-multiplicity of the diimide ground state, Wagnière has published the results of *ab initio* calculations which predict closed-shell singlet ground states for *cis*- and *trans*-diimide, but a triplet ground state derived from the (n_-, π^*) configuration for the 1,1-compound, $H_2\overset{+}{N}{=}\overset{-}{N}$ (G. Wagnière, *Theoret. Chim. Acta* **31**, 269 (1973)). Its total energy however, is higher than those of the other two isomers.

Boyd and co-workers have presented the photoelectron spectra of the series of azo compounds

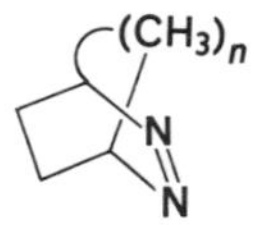

having $n = 1$–4 (R. J. Boyd, J. C. Bunzli, J. P. Snyder and M. L. Heyman, *J. Amer. Chem. Soc.* **95**, 6478 (1973)). Once again, the n_-, π, n_+ orbital ordering was observed in each member of the series, and it was further found that the n_--π split was constant at $2{\cdot}5 \pm 0{\cdot}2$ eV, whereas the n_--n_+ split ranged from 2·9 to 3·5 eV, depending upon the NNC angle. This changing NNC angle also has the anticipated hypsochromic effect on the $n_- \rightarrow \pi^*$ absorption wavelength.

The effects of altering the NNC angle on the lone-pair splittings in several tetraalkyl hydrazines have been charted by Nelsen and Buschek, who find the splitting varies from 2·45 eV in

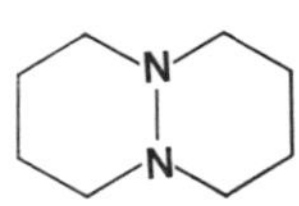

where the lone pairs are constrained to be coplanar, to 0·40 eV in tetramethyl hydrazine, where they are perpendicular (S. F. Nelsen and J. M. Buschek, *J. Amer. Chem. Soc.* **95**, 2011 (1973)). The preferred perpendicular orientation of the lone pairs in unconstrained hydrazines is said to involve considerable stabilization due to charge transfer from the lone-pair on one atom into the (N—X) σ^* MO on the adjacent atom (L. Radom, W. J. Hehre and J. A. Pople, *J. Amer. Chem. Soc.* **94**, 2371 (1972)). This strong dependence of the n_{-}-n_{+} splitting on the NNC angle has been used to good advantage in determining the configurations of several cyclic hydrazines from their photoelectron spectra (S. F. Nelsen, J. M. Buschek and P. J. Hintz, *J. Amer. Chem. Soc.* **95**, 2013 (1973)).

V. REFERENCES

1. C. Willis and R. A. Back, *Nature*, **241,** 43 (1973).
2. L. C. Snyder and H. Basch, *Molecular Wave Functions and Properties: Tabulated from SCF Calculations in a Gaussian Basis Set*, Wiley–Interscience, New York, 1972.
3. D. P. Wong, W. H. Fink and L. C. Allen, *J. Chem. Phys.*, **52,** 6291 (1970).
4. M. B. Robin, R. R. Hart and N. A. Kuebler, *J. Amer. Chem. Soc.*, **89,** 1564 (1967).
5. E. Haselbach and E. Heilbronner, *Helv. Chim. Acta*, **53,** 684 (1970).
6. C. R. Brundle, M. B. Robin, N. A. Kuebler and H. Basch, *J. Amer. Chem. Soc.*, **94,** 1451 (1972).
7. E. Haselbach, J. A. Hashmall, E. Heilbronner and V Hornung, *Angew. Chem. Intern. Ed.*, **8,** 878 (1969).
8. L. E. Orgel, *J. Chem. Soc.*, 1276 (1953).
9. A. Trombetti, *Can. J. Phys.*, **46,** 1005 (1968).
10. M. B. Robin, H. Basch, N. A. Kuebler, K. B. Wiberg and G. B. Ellison, *J. Chem. Phys.*, **51,** 45 (1969).
11. M. B. Robin, C. R. Brundle, N. A. Kuebler, G. B. Ellison and K. B. Wiberg, *J. Chem. Phys.*, **57,** 1758 (1972).
12. Personal communication from W. Flygare.
13. S. N. Foner and R. L. Hudson, *J. Chem. Phys.*, **28,** 719 (1958).
14. E. Haselbach, E. Heilbronner, A. Mannschreck and W. Seitz, *Angew Chem. Intern. Ed.*, **9,** 902 (1970).
15. M. B. Robin, *Higher Excited States of Polyatomic Molecules*, Vol. I, Academic Press, New York, 1974; *Intern. J. Quantum Chem.*, **6,** 257 (1972).
16. M. B. Robin and W. T. Simpson, *J. Chem. Phys.*, **36,** 580 (1962).
17. H. Bock, *Angew. Chem.*, **77,** 469 (1965).
18. N. C. Baird, P. de Mayo, J. R. Swenson and M. C. Usselman, *J. Chem. Soc. Chem. Comm.*, 314 (1973).
19. R. A. Mitsch, E. W. Neuvar, R. J. Koshar and D. H. Dybvig, *J. Heterocyclic Chem.*, **2,** 371 (1965).
20. L. C. Robertson and J. A. Merritt, *J. Mol. Spectrosc.*, **19,** 372 (1966); **24,** 44 (1967).
21. J. A. Merritt, *Can. J. Phys.*, **40,** 1683 (1962).
22. J. D. Simmons, I. R. Bartky and A. M. Bass, *J. Mol. Spectrosc.*, **17,** 48 (1965).
23. J. R. Lombardi, W. Klemperer, M. B. Robin, H. Basch and N. A. Kuebler, *J. Chem. Phys.*, **51,** 33 (1969).
24. H. Rau, *Angew. Chem. Intern. Ed.*, **12,** 224 (1973).
25. M. B. Robin and N. A. Kuebler, *J. Mol. Spectrosc.*, **33,** 274 (1970).
26. E. Haselbach and A. Schmelzer, *Helv. Chim. Acta*, **54,** 1575 (1971).
27. R. E. Rebbert and P. Ausloos, *J. Amer. Chem. Soc.*, **87,** 1847 (1965).
28. E. Fischer, *J. Amer. Chem. Soc.*, **90,** 796 (1968).

29. J. Kuhn, W. Hug, R. Geiger and G. Wagniere, *Helv. Chim. Acta*, **54,** 2260 (1971).
30. K. Siegbahn, C. Nordling, A. Fahlman, R. Nordberg, K. Hamrin, J. Hedman, G. Johansson, T. Bergmark, S.-E. Karlsson, I. Lindgren and B. Lindberg, *ESCA*, Almqvist and Wiksells Boktryckeri Ab, Uppsala, 1967.
31. M. B. Robin and N. A. Kuebler, unpublished work.
32. E. Haselbach, A. Mannschreck and W. Seitz, *Helv. Chim. Acta*, **56,** 1614 (1973).
33. A. Veillard, *Theoret. Chim. Acta*, **5,** 413 (1966).
34. L. Pedersen and K. Morokuma, *J. Chem. Phys.*, **46,** 3941 (1967).
35. W. H. Fink, D. C. Pan and L. C. Allen, *J. Chem. Phys.*, **47,** 895 (1967).
36. K. Osafune, S. Katsumata and K. Kimura, *Chem. Phys. Lett.*, **19,** 369 (1973).
37. R. Bonaccorsi, E. Scrocco and J. Tomasi, *Theoret. Chim. Acta*, **21,** 17 (1971).
38. T. Kasuya, *Sci. Papers Instit. Phys. Chem. Sci.*, **56,** 1 (1962).
39. N. Bodor, M. J. S. Dewar, W. B. Jennings and S. D. Worley, *Tetrahedron*, **26,** 4109 (1970).
40. P. Finn, R. K. Pearson, J. M. Hollander and W. L. Jolly, *Inorg. Chem.*, **10,** 378 (1971).
41. M. Schürgers and K. H. Welge, *Z. Naturforsch.*, **23A,** 1508 (1968).

CHAPTER 2

Structural chemistry

R. ALLMANN

Philipps-Universität, Lahnberge, 355 Marburg, Germany

I. INTRODUCTION

This chapter ranges a little wider than the actual functional groups treated in this volume and it also comprises some related compounds and general aspects. Structural chemistry is understood here mainly as a discussion of bond lengths and angles of the groups concerned and the dependence of these values on their environment, i.e. on the molecules into which the groups are incorporated.

As long as molecules are packed in crystals by van der Waals' forces only, they exhibit little difference in their configuration as compared with the free molecules in the gaseous state. Especially, there are no significant differences in bond lengths. Packing effects may influence some bond angles and certain twisting or torsion angles, so as to allow the densest possible packing. In the crystalline state the free rotation of methyl or other terminal groups is hindered by adjacent molecules but this is not the case in the gaseous state. In the crystalline state one important molecular interaction may be the formation of hydrogen bonds X—H···Y by which the other bonds to X become somewhat strengthened and those to Y weakened.

II. GENERAL REMARKS ABOUT BOND LENGTHS

A. Determination and Accuracy of Bond Lengths

There are several methods in existence to elucidate the configuration of molecules. Smaller ones are often examined in the gaseous state, larger ones in the crystalline state. As the various methods use different properties of the molecule to be measured and different models to transform the measured data into statements about bond lengths etc., the results obtained by different methods can only be compared with caution and after applying certain corrections, which are mostly in the range of 0·01–0·02 Å. The most common kinds of distances are shown below.

1. r_e: The minimum of the potential energy function. This is an ideal distance within a rigid molecule without thermal vibrations. It cannot be measured directly because of the zero-point vibrations, but it may be the result of *theoretical calculations*.

2. r_o: The nuclear distance within a molecule in the state of zero-point vibration as calculated from average rotational constants obtained by *several spectroscopic methods*, e.g. i.r., Raman, microwave or molecular beam resonance, on molecules in the gaseous state. The results are dependent on the vibrational state of the molecule. If one knows the rotation–vibration coupling constants, r_o can be extrapolated to r_e. Because of the asymmetry of the potential trough, $r_o > r_e$ holds for most cases. To give an impression of the differences between r_o and r_e, some results are compared in Table 1.

TABLE 1. Comparison of some r_o and r_e values

Molecule		r_o (Å)	r_e (Å)	Method
H_2		0·75105	0·74130	Raman[1]
N_2		1·1000	1·0976	Raman[2]
HCN	HC	1·0637	1·0630	microwave[3]
	C≡N	1·1563	1·1538	

For non-hydrogen atoms, the difference $r_o - r_e$ is within ±0·005 Å. Costain[3] has introduced a new parameter, r_s, for polyatomic molecules which need isotopic substitution to get sufficient rotational constants; r_s is about $(r_e + r_o)/2$, mostly nearer to r_e than to r_o; $|r_s - r_e|$ is in the range of 0·003–0·005 Å.

3. r_g: The average value of an instantaneous internuclear distance. The r_g is obtained from the *electron-diffraction* examination of gases as the

centre of gravity of a maximum in the nuclear radial distribution curve[4]. If a Morse potential function is assumed, r_e may be calculated from r_g by extrapolation[5] (e.g., for CCl_4: $r_g = 1{\cdot}771(5)$, $r_e = 1{\cdot}760(4)$ Å). Other authors quote r_m, the position of the maximum itself, and several corrections and modifications may be applied to the evaluation of r-values. By this, the results of electron-diffraction examinations of several authors may differ somewhat according to their definition of r. The results of spectroscopic methods and of electron-diffraction often show astonishingly good agreement, that means $r_o - r_e$ and $r_g - r_e$ are almost equal (e.g., N—N in hydrazine: $r_o = 1{\cdot}453 \pm 5$ Å[6], $r_g = 1{\cdot}449 \pm 4$ Å[7]). Both methods allow the calculation of the average vibration amplitudes, which are in the range of 0·05 Å for bound atoms. The amplitudes vary rather slowly with temperature (e.g., for C—Cl in CCl_4: $\sqrt{u^2} = 0{\cdot}051$ Å at 0 K and 0·056 Å at 373 K[8]).

4. r_z: The distance between average atomic positions in a molecular coordinate system. This is the uncorrected value obtained by *neutron diffraction* of crystals. It may be the result of an *X-ray diffraction* examination of crystals if the centre of gravity of the electronic charge coincides with the site of the nucleus, which holds for atoms heavier than fluorine within the recent limits of error (see later). The r_z is always less than r_g by up to 0·01–0·02 Å, depending on the vibrational amplitudes of the molecule. The intermolecular vibration amplitudes are mainly in the range of 0·1–0·3 Å and are greater than the bond vibrations (see above). Of both kinds of molecular libration (as a rigid body), only the torsional oscillations cause a shortening of measured distances, whereas the translational oscillations only increase the uncertainties in deriving bond lengths (for corrections see references 9 and 10.) For instance, for pyrene[11], with translational amplitudes of 0·17–0·26 Å and torsional amplitudes of 3·2–4·75°, the C—C bond lengths increased by 0·004–0·007 Å after applying the rigid body oscillation correction. For nonrigid molecules the distinction between intra- and intermolecular vibrations breaks down and the problem of correction is then formidable. Most affected are bonds to terminal atoms, for which the riding motion model[12] yields fairly reliable correction terms. (The higher experimental thermal vibration values of a terminal atom are interpreted by a 'riding' motion of the terminal on the rather rigid, but also vibrating, rest of the molecule. See later quoted examples with uncorrected and corrected r values, e.g. **8** or **10**.) These corrections are also in the range of 0·01–0·02 Å. The shrinkage effect of molecular vibrations may be decreased by measuring at low temperatures. But, even in this case, r_z values can only be compared with r_g or r_o values after applying the necessary corrections.

5. r_x: The *uncorrected* value from *X-ray crystallographic examination*. This is the most complex value which is reported in the majority of structural

papers and also in many of the examples given here if not mentioned otherwise. It not only needs corrections for thermal vibration as r_z does, but also for the asperity of the electronic charge density around the nuclei of the lighter atoms. As X-rays are diffracted at electrons and not at the nuclei (as neutrons are), by the usual least-squares methods using spherical atomic form factors, one gets the time-averaged centroids of the several atomic electron clouds. For bonded H-atoms this centroid is shifted by about 0·1 Å in direction to the atom bound to H as compared with the site of the proton[13]. For atoms of the first period (Li to F) the effect is smaller, but may reach 0·01 and even 0·02 Å, which is several times the statistical standard deviation. For *s*-triazine (**3**) Coppens[14] found the following atomic shifts as compared with neutron data: N: 0·009 Å towards the lone electron pair, C: 0·015 Å in the direction of the ring centre, H: 0·12 Å towards C. As long as X-ray data are calculated with spherical atom form factors, the results for organic molecules cannot be trusted to be better than ±0·01 Å, even if the statistical standard deviation is 0·002 Å and the bond lengths are corrected for thermal vibration (for recent approaches to overcome these difficulties see for instance reference 15). In the later reported structures determined with X-rays, r_x-values are given, if not otherwise mentioned, together with the conventional *R*-value

$$R = \sum ||Fo| - |Fc|| / \sum |Fo|$$

and the estimated standard deviation σ. The smaller the residue *R*, the less is the discrepancy between observed (*Fo*) and calculated (*Fc*) structure factors. Per nonhydrogen atom of a molecule about 100 *Fo*-values are to be measured. For the calculation of *Fc* the atomic coordinates and thermal vibration constants are used. For recent structure determinations, *R* should be less than 0·1. *R*-values less than 0·05 are excellent but do need an appreciable amount of refining effort.

B. *Standard Bond Lengths (for* NN, NC, CC, NO *and* CO*)*

As long as no conjugation is present in molecules, chemically similar bonds have about the same bond lengths, which can be fairly well approximated by a sum of covalent radii[16]. When structure determinations increased in accuracy, it became obvious that the length of a single σ-bond is quite sensitive to the hybridization states of the two atoms involved. *Tetrahedron*, **17**, 125–266 (1962) contains several papers which discuss the influence of hybridization and resonance on the C—C bond lengths. The general conclusion is that the shortening of the central bond in butadiene (1·483 Å) as compared with ethane (1·534 Å) is mainly due to the different state of hybridization (sp^2 instead of sp^3). Dewar and Schmeising[17] give the following set of C—C single distances (in Å):

$$sp^3\text{-}sp^3 = 1{\cdot}544 \quad sp^3\text{-}sp^2 = 1{\cdot}515 \quad sp^3\text{-}sp = 1{\cdot}459$$

$$sp^2\text{-}sp^2 = 1{\cdot}482 \quad sp^2\text{-}sp = 1{\cdot}439 \quad sp\text{-}sp = 1{\cdot}379$$

These values agree within 0·01–0·02 Å with structural data. (For sp^3-sp^3, the value for ethane = 1·534 Å may be preferred.) For carbon atoms it is rather easy to postulate sp^3, sp^2 or sp σ-bonding orbitals if the C atom is bound to four, three or two atoms with about normal bond angles (109·5°, 120° or 180° respectively). For nitrogen atoms the case is more complicated. Because of the lone-pair electrons the hybridization state is more difficult to estimate. N bound to three other atoms may be either sp^3 (e.g., NH_3) or sp^2 (e.g., NO_3^-). Therefore, until now no sets of N—N and N—C single bonds dependent on the hybridization state have been reported.

One may estimate this effect to be as great for N as for C. For instance the C—N bond in $CH_3—\overset{+}{N}\equiv\overset{-}{C}|$ (sp^3-sp = 1·424 Å[3]) is shorter than in $CH_3—NH_2$ (sp^3-sp^3 = 1·474 Å[18]) by 0·05 Å; or N—N in hydrazine (sp^3-sp^3 = 1·453[6], 1·449[7]) is longer by 0·05 Å than in tetrakis(trifluoromethyl)hydrazine (**23**), $(CF_3)_2N—N(CF_3)_2$, ($\sim sp^2$-sp^2 = 1·402 Å[19]) or 0·06 Å longer than in diformylhydrazine (sp^2-sp^2 = 1·392 Å[20]).

In conjugated systems, the bond lengths (mostly sp^2-sp^2) are between those for single and double bonds. When comparing with calculated bond orders '*no unique bond length–bond order curve exists* for pairs of atoms such as (C—C) or (C—N), *unless the state of hybridization of the two atoms remains the same* for a series of compounds' (Orville-Thomas[21], p. 85). Even if the state of hybridization is known for an atom, e.g. sp^2 for N with three planar bonds, the hydridizations of the bond orbitals are different if the bond angles are not equal, e.g. 120° for sp^2 (for instance: in formaldehyde, $H_2C{=}O$, and difluoroformaldehyde, $F_2C{=}O$, the s-characters of the carbon hybrid orbitals to O are calculated[21] from the angles H—C—H = 125·8° and F—C—F = 112·5° as 26 and 44% respectively, in accordance with the found C=O bond lengths of 1·23 and 1·17 Å).

For C—C atoms the π-bond order p as estimated by the simple valance bond method (i.e. by averaging over all possible Kekulé formulae, e.g. $p = \frac{1}{2}$ for benzene) yields quite good estimations of bond lengths r using Cruickshank's formula[22] $r = 1{\cdot}477 - (1{\cdot}477 - 1{\cdot}337) \times 4p/(p+3)$, but for C—N and N—N bonds polar limiting formulae must also be considered as, for instance, in the case of the peptide bond with C—N = 1·32 Å[16]:

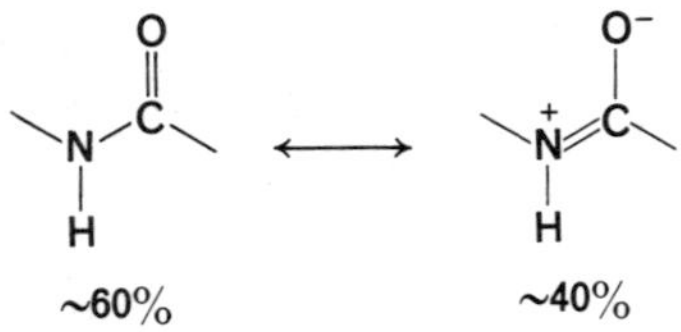

In Table 2 bond lengths are estimated for atoms in sp^2-state for a given total bond order n (from simple VB, $n = 1 + p$) using Pauling's[16] formula $r(n) = r(1) - k \times \log n$. The constant k (in Å) is fixed to fit the assumed values for $n = 1{\cdot}0$ and $n = 2{\cdot}0$ (i.e., $k = (r(1) - r(2))/\log 2$). These values differ somewhat from those reported in the volumes on C=N and N_3 in this series. The single bonds are shorter than the usual values because of the omission of sp^3-sp^3 bonds. For C=N double bonds recent structure determinations yield values of 1·26 Å and even shorter (e.g., (**8**)), N=N bonds may be as short as 1·21 Å[23]. Most uncertain are the N—O bonds, for which the limiting values are taken from reference 21. If the deviation of the bond angles from the normal values is neglected, the given values should be good within ±0·03 Å. For bond orders calculated by other methods (e.g., M.O.) Table 2 must be modified.

TABLE 2. Estimated bond lengths in Å for atoms in sp^2-state[a] dependent on the total bond order n

n	N—N	N—C	C—C	C—O	N—O
1·0	1·41	1·43	1·48	1·40	1·41
1·1	1·385	1·41	1·46	1·375	1·37
1·2	1·365	1·39	1·44	1·35	1·34
1·3	1·345	1·37	1·425	1·33	1·31
1·4	1·325	1·35	1·41	1·31	1·28
1·5	1·31	1·335	1·40	1·295	1·25
1·6	1·295	1·32	1·385	1·28	1·225
1·7	1·28	1·305	1·37	1·26	1·20
1·8	1·265	1·29	1·36	1·245	1·18
1·9	1·25	1·28	1·35	1·23	1·16
2·0	1·24	1·27	1·34	1·22	1·14

[a] If N or C is in sp^3-state in single bonds, 0·03 Å must be added for each sp^3-atom; for atoms in sp-state, about 0·04 Å must be subtracted.

The bond lengths thus calculated for $n = 1{\cdot}5$ may be compared with the lengths reported for some benzene-like heterocycles, e.g. N—N = 1·321 in *s*-tetrazine (**1**) and 1·314 Å in diphenyl-*s*-tetrazine (**2**); N—C = 1·316 in *s*-triazine (**3**), = 1·335 in *s*-triphenyl-triazine (**4**) and 1·34 in melamine (**5**);

s-Tetrazine, $C_2H_2N_4$
F. Bertinotti, G. Giacomello and A. M. Liquori, *Acta Cryst.*, **9**, 510 (1956).
$\sigma \sim 0{\cdot}01$, $R_{h0l} = 0{\cdot}077$, $R_{0kl} = 0{\cdot}061$

a	= 1·321	ab^1	116·3°
b^1	= 1·345	$a'b^2$	115·6
b^2	= 1·323	b^1b^2	127·4

(1)

3,6-Diphenyl-*s*-tetrazine, $C_{14}H_{10}N_4$
N. A. Ahmed and A. I. Kitaigorodsky, *Acta Cryst.*, *B*, **28**, 739 (1972).
R = 0·13, ~planar (±0·025 Å)

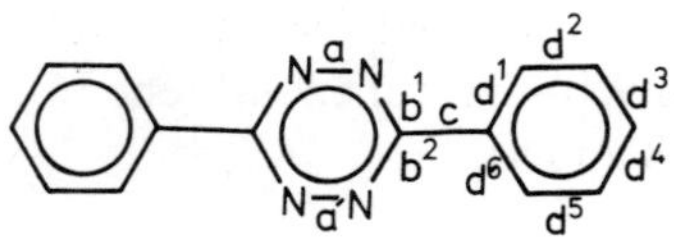

a	= 1·314	ab^1	117·5°
b^1	= 1·353	$a'b^2$	120·6
b^2	= 1·338	b^1b^2	121·8
c	= 1·454	⟨bc⟩	119·1
⟨d⟩	= 1·409	⟨cd⟩	120·2

(2)

s-Triazine[14], $C_3H_3N_3$
R = 0·05 (X-ray), 0·073 (neutron) (length and angles calculated from given coordinates)

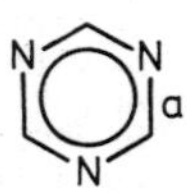

	X-Ray	Neutron
a =	1·315(4)	1·317(7)
C—N—C	113·0°	114·8
N—C—N	127·0	125·2

(3)

s-Triphenyltriazine, $C_{21}H_{15}N_3$
A. Damiani, E. Giglio and A. Ripamonti, *Acta Cryst.*, **19**, 161 (1965).
$\sigma \sim 0{\cdot}007$, R = 0·114, twisting angles 7·6, 10·9 and 6·9°

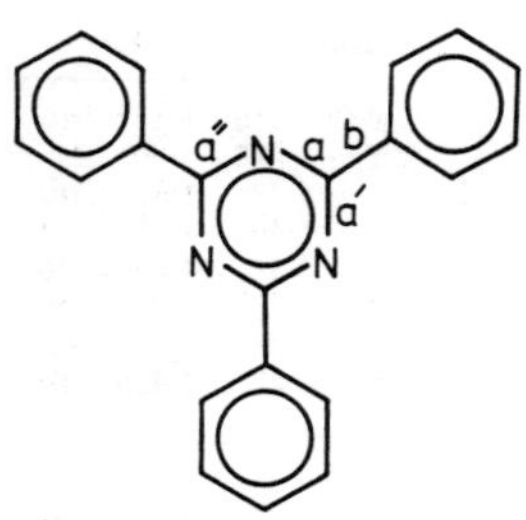

Average			
a = 1·335		aa″	115·5°
b = 1·475		aa′	124·5
		ab	117·8

(4)

Melamine, $C_3H_6N_6$
Y. Akimoto, *Bull. Chem. Soc. Japan*, **28**, 1 (1955).

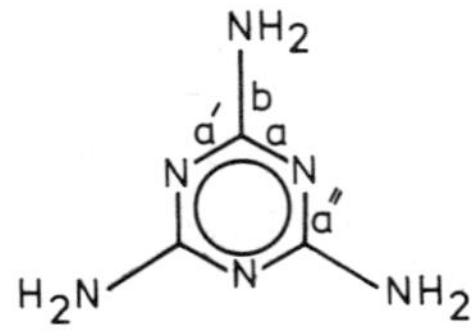

Electron diffraction*	
a =	1·34(1)
b =	1·37(3)
aa″	123°(±3°)

(5)

and C—C = 1·397 in benzene[24] (X-ray examination yields only 1·392[25] because of reasons discussed under r_x) as compared with the calculated values of 1·31, 1·335 and 1·40 Å respectively in Table 2.

In some heterocyclic compounds the reported distances can be explained without polar limiting forms, as in 9,10-diazaphenanthrene (**6**), in dichinoxalylene (**7**) or in 1,2-diazepines (**8** and **9**).

9,10-Diazaphenanthrene, $C_{12}H_8N_2$
H. van der Meer, *Acta Cryst. B*, **28**, 367 (1972).
$\sigma = 0{\cdot}003–0{\cdot}005$, R = 0·05, average over two independent halves

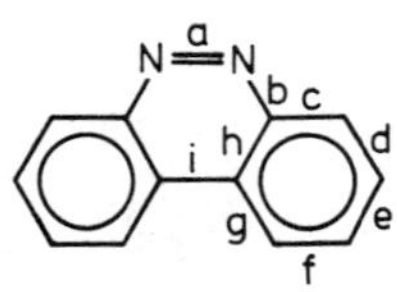

a = 1·292		ab	120·3°
b = 1·396		bc	116·4
c = 1·415		bh	123·1
d = 1·373		hc	120·5
e = 1·340		cd	119·5
f = 1·377		de	120·4
g = 1·417		ef	121·1
h = 1·411		fg	120·0
i = 1·436		gh	118·5
		gi	124·9
		hi	116·6

(6)

Dichinoxalylene, $C_{16}H_8N_4$
R. Allmann, *Cryst. Struct. Comm.*, **3**, 57 (1974).
$\sigma = 0{\cdot}006$, R = 0·049, planar, average over chemically equivalent bonds

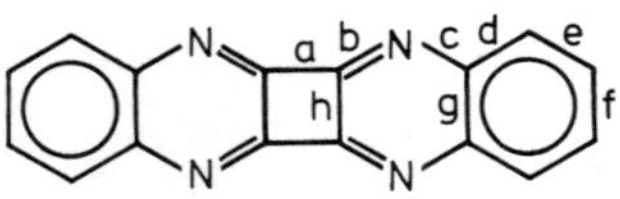

	Uncorr.	Corr.		
a =	1·515	1·525	ah	90°
b =	1·295	1·30	bh	127
c =	1·42	1·42	bc	110
d =	1·40	1·41	cg	123
e =	1·34	1·35	dg	119
f =	1·385	1·39	de	120
g =	1·44	1·44	ef	121
h =	1·42	1·425	ab	143

(7)

* In the structure tables all lengths are given in Å; σ (in Å) is the standard deviation of the reported bond lengths (the probability for a length to be within $\pm\sigma$ is 68·3%, for $\pm 2\sigma$: 95·5% and for $\pm 3\sigma$: 99·7%). For single values, if at all, σ is added in parentheses.

1-Tosyl-1,2-diazepine, $C_{12}H_{12}N_2O_2S$
R. Allmann, A. Frankowski and J. Streith, *Tetrahedron*, **8**, 581 (1972).
$\sigma = 0{\cdot}003–0{\cdot}006$, $R = 0{\cdot}038$

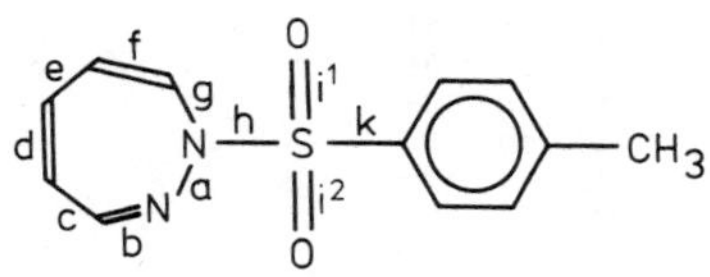

	Uncorr.	Corr.	Torsion ∢		
a =	1·447	1·458	74·0°	ab	112·2°
b =	1·255	1·262	−6·0	bc	129·4
c =	1·460	1·463	−38·8	cd	123·5
d =	1·326	1·326	7·5	de	125·1
e =	1·436	1·439	28·9	ef	123·1
f =	1·333	1·343	2·9	fg	119·1
g =	1·428	1·434	−71·8	ga	113·2
h =	1·649	1·651		gh	117·1
i¹ =	1·433	1·447		ah	110·7
i² =	1·419	1·440		⟨hi⟩	105·3
k =	1·759	1·759		⟨ik⟩	108·3
				i¹i²	120·0

(8)

1-Isopropoxycarbonyl-1,2-diazepine,$Fe(CO)_3$, $C_9H_{12}N_2O_2$—$Fe(CO)_3$
R. Allmann, *Angew. Chemie*, **82**, 982 (1970); *Int. Ed.*, **9**, 958 (1970).
$\sigma = 0{\cdot}01$, $R = 0{\cdot}045$, planar except for atoms of bond d and methyl groups

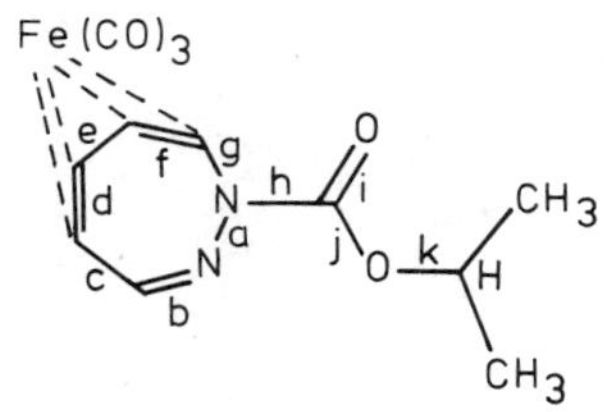

a	= 1·37	ag	126°
b	= 1·27	ah	116
c	= 1·47	gh	117
d	= 1·41	ab	118
e	= 1·38	bc	133
f	= 1·43	cd	127
g	= 1·42	de	119
h	= 1·40	ef	118
i	= 1·20	fg	125
j	= 1·32	hi	122
k	= 1·47	hj	113
⟨Fe—C⟩	= 2·07	jk	115
⟨Fe—CO⟩	= 1·79		

(9)

But in most compounds limiting forms with N^+ or N^- are involved as in **10**, **11** or **12**. In 1,2,4-triazole (**13**) the bond lengths are found to be fairly uniform (N—N ~ 1·36, C—N ~ 1·34Å).

3-Acetyl-4-methyl-*as*-triazino[4,3-b]-2*H*′)-indazol, $C_{12}H_{10}N_4O$
R. Allmann, T. Debaerdemaeker, W. Grahn and Ch. Reichardt, *Chem. Ber.*, **107**, 1555 (1974).
$\sigma \sim 0{\cdot}003$, $R = 0{\cdot}041$, planar

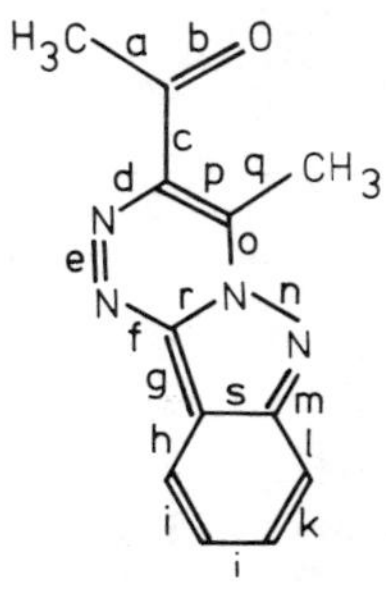

	Uncorr.	Corr.		Uncorr.	Corr.		
a =	1·488	1·515	l =	1·409	1·421	ro	120·6°
b =	1·211	1·233	m =	1·372	1·377	op	114·2
c =	1·504	1·511	n =	1·338	1·345	pd	124·2
d =	1·362	1·372	o =	1·356	1·357	rg	105·2
e =	1·321	1·321	p =	1·381	1·384	gs	104·5
f =	1·328	1·340	g =	1·487	1·503	sm	113·1
g =	1·405	1·406	r =	1·401	1·402	mn	103·2
h =	1·405	1·412	s =	1·407	1·407	nr	114·1
i =	1·361	1·368	de	120·1°		cd	112·3
j =	1·396	1·401	ef	118·4		og	116·8
k =	1·369	1·371	fr	122·4			

(10)

3-Methyl-3-pyrazolin-5-one, $C_4H_6N_2O$
W. H. de Camp and J. M. Stewart, *Acta Cryst. B*, **27**, 1227 (1971).
$\sigma \sim 0{\cdot}004$, $R = 0{\cdot}071$, planar, hydrogen bonds N—H···O′ = 2·63 and 2·71 Å

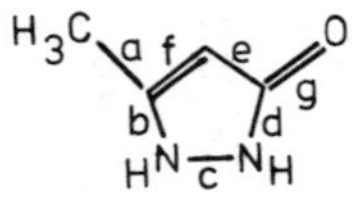

a = 1·491	ab	119·9°
b = 1·333	af	131·2
c = 1·365	bc	108·0
d = 1·337	cd	109·7
e = 1·410	de	106·1
f = 1·357	ef	107·2
g = 1·285	dg	122·0

(11)

5-Amino-2-mercapto-1,3,4-thiadiazole, $C_2H_3N_3S_2$
T. C. Downie, W. Harrison and E. S. Raper, *Acta Cryst. B*, **28**, 1584 (1972).
$\sigma \sim 0{\cdot}007$, $R = 0{\cdot}106$, ~planar (±0·02 Å), thione form

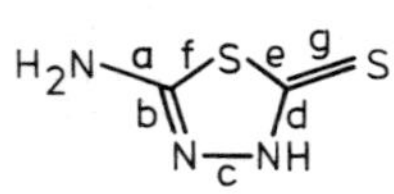

a = 1·339	ab	123·1°
b = 1·305	bf	113·7
c = 1·382	bc	110·2
d = 1·334	cd	117·8
e = 1·738	de	108·4
f = 1·746	ef	89·6
g = 1·678	dg	128·1

(12)

1,2,4-Triazole, $C_2H_3N_3$
P. Goldstein, J. Ladell and G. Abowitz, *Acta Cryst. B*, **25**, 135 (1969).
$\sigma \sim 0{\cdot}003$, $R = 0{\cdot}051$ at −160°C, planar

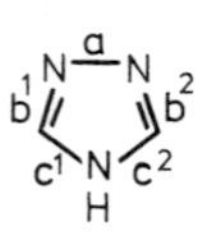

	Uncorr.	Corr.		
a	= 1·354	1·359	ab^1	102·1°
b^1	= 1·319	1·323	ab^2	110·2
b^2	= 1·326	1·331	b^1c^1	114·6
c^1	= 1·352	1·359	b^2c^2	110·1
c^2	= 1·320	1·324	c^1c^2	103·0

(13)

Recently, the structure of triphenylverdazyl (**14**), a stable free radical, was solved. As expected, the bonds N(sp^2)—C(sp^3) = c = 1·443 Å are longer than N(sp^2)—C(sp^2) = e = 1·414 Å. Likewise, the N(sp^2)—C(sp^3) bonds in cyclo-trimethylene-trinitramine (**15**) amount to about 1·455 Å.

2,4,6-Triphenylverdazyl, $C_{20}N_4H_{17}$
D. E. Williams, *Acta Cryst. B*, **29**, 96 (1973).
$\sigma \sim 0{\cdot}007$, $R = 0{\cdot}076$, stable radical, CH_2-group 0·59 Å out of plane

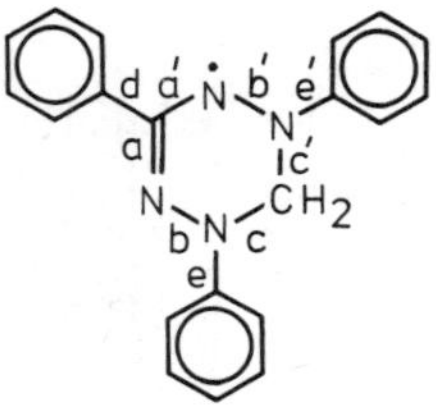

a = 1·338	aa′	126·8°
a′ = 1·337	ab	114·2
b = 1·353	a′b′	114·5
b′ = 1·348	bc	117·4
c = 1·440	b′c′	118·2
c′ = 1·445	cc′	106·1
d = 1·485	be	117·3
e = 1·417	b′e′	117·5
e′ = 1·411	ad	115·5

Twisting angles: e 13·1°, e′ 23·0°, d 1·6°

(14)

cyclo-Trimethylene-trinitramine, $C_3H_6N_6O_6$
C. S. Choi and E. Prince, *Acta Cryst. B*, **28**, 2857 (1972).
$\sigma \sim 0{\cdot}004$, $R = 0{\cdot}021$ (neutron), puckered ring

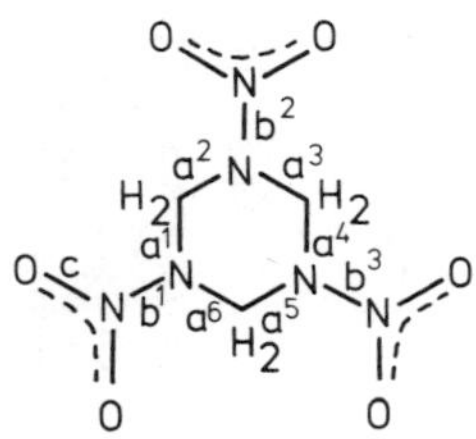

Average

a = 1·454	N—C—N	109·3°
b = 1·380	C—N—C	114·8
c = 1·210	ab	117·9
	bc	117·2
	O—N—O	125·4

(15)

The structures of a few triazenes (**16**, **17** and **18**) and tetrazenes (**19**, see also **44**) are determined and they mostly show nearly ideal N—N single and double bonds. About equal N—N bond lengths were only found for *p*-bromodiazoaniline (**18**), indicating an equilibrium of

$$\underset{\text{H}}{\text{—N}}\text{—N=N—} \leftrightarrow \text{—N=N—}\underset{\text{H}}{\text{N}}\text{—}$$

The same was found for *p*-dibromodiazoaminobenzene[26], but in this case the effect can be explained by disorder. For the strongly related 2,4-dibromo compound (**16**) the N—N bonds are quite different (1·25 and 1·45 Å)

2,4-Dibromodiazoaminobenzene, $C_{12}H_9Br_2N_3$
Y. A. Omel'chenko and Y. D. Kondrashev, *Sov. Phys. Crystallogr.*, **10**, 690 (1966).
$\sigma \sim 0{\cdot}01$, $R_{hk0} = 0{\cdot}109$, $R_{hk3} = 0{\cdot}136$, planar

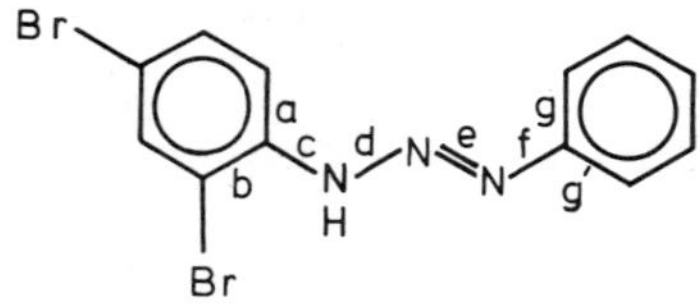

a	= 1·38	ac	122·1°
b	= 1·38	bc	117·1
c	= 1·43	cd	114·9
d	= 1·45	de	107·7
e	= 1·25	ef	110·0
f	= 1·42	fg	125·2
g	= 1·38	fg′	113·0
g′	= 1·38		

(16)

1-*p*-Tolyl-3-(α-cyano)benzylidenetriazene
J. W. Schilling and C. E. Nordman, *Acta Cryst. B*, **28**, 2177 (1972).
$\sigma \sim 0{\cdot}02$, $R = 0{\cdot}077$, planar

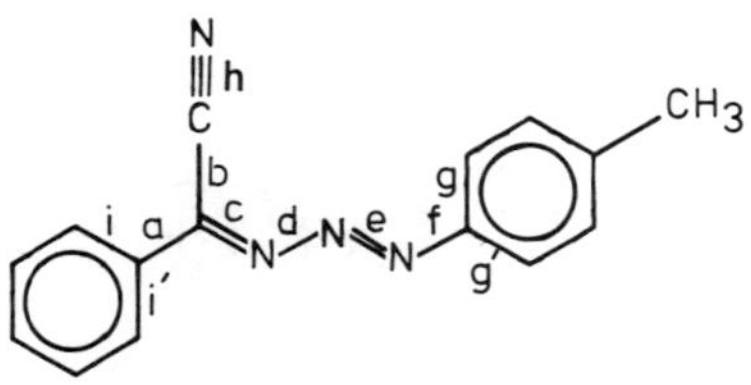

a	= 1·469	ab	120·5°
b	= 1·373	ac	119·7
c	= 1·315	bc	119·6
d	= 1·403	cd	109·7
e	= 1·239	de	109·5
f	= 1·414	ef	114·8
⟨g⟩	= 1·393	fg	127·6
h	= 1·112	fg′	116·4
⟨i⟩	= 1·382	bh	173·5
		ai	119·3
		ai′	118·0

(17)

p-Bromodiazoaniline, $C_{12}H_{10}BrN_3$
Y. A. Omel'chenko and Y. D. Kondrashev, *Sov. Phys. Crystallogr.*, **12**, 359 (1967).
$\sigma \sim 0{\cdot}02$, $R_{h0l} = 0{\cdot}114$, $R_{h3l} = 0{\cdot}16$, planar

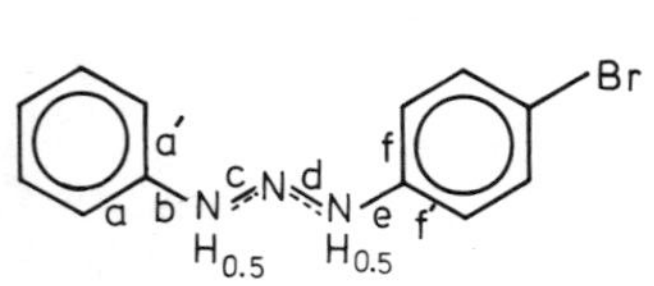

a	= 1·38	ab	121·5°
a′	= 1·37	a′b	118·5
b	= 1·45	bc	120
c	= 1·27	cd	115
d	= 1·25	de	115·5
e	= 1·47	ef	125
f	= 1·36	ef′	114·5
f′	= 1·37		

(18)

1,4-Bis-(*N*-ethyl-1,2-dihydrobenzthiazol-2-ylidine)tetrazene, $C_{18}H_{18}N_6S_2$
R. Allmann, *Acta Cryst.*, **22**, 246 (1967).
$\sigma \sim 0{\cdot}004$, $R = 0{\cdot}084$, centrosymmetric, ~planar (±0·1 Å except ethyl-groups)

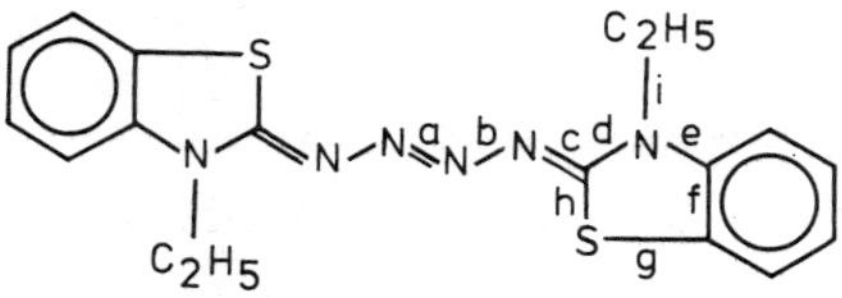

a	= 1·257	ab	110·6°
b	= 1·400	bc	109·1
c	= 1·302	cd	122·8
d	= 1·371	ch	126·2
e	= 1·390	de	115·1
f	= 1·405	ef	112·2
g	= 1·759	fg	111·2
h	= 1·759	gh	90·6
i	= 1·472	hd	110·9
		di	122·3

(19)

As an example for a heterocyclic compound with four connected nitrogen atoms the structure of azidopurine (**20**) is given. By resonance, the N═N

7*H*-Tetrazolo[5,1-i]-purine, $C_5H_3N_7$
J. P. Glusker, D. van der Helm, W. E. Love, J. A. Minkin and A. L. Patterson, *Acta Cryst. B*, **24**, 359 (1968).
$\sigma \sim 0{\cdot}009$, $R = 0{\cdot}096$, planar

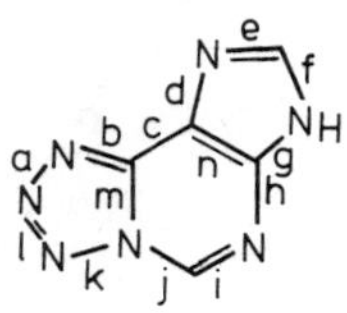

a	= 1·37	h	= 1·36	ab	104°	hi	116
b	= 1·33	i	= 1·29	bm	109	ij	121
c	= 1·41	j	= 1·39	mk	108	jm	125
d	= 1·37	k	= 1·38	kl	105	de	103
e	= 1·33	l	= 1·28	la	114	ef	114
f	= 1·33	m	= 1·36	mc	115	fg	107
g	= 1·36	n	= 1·40	cn	115	gn	105
				nh	128	nd	111

(20)

'Tetrazene' = 1-amino-1-[(1H-tetrazol-5-yl)azo] guanidine hydrate (and hydrobromide), $C_2H_6N_{10}$—H_2O (and $C_2H_6N_{10}$—HBr)
J. R. C. Duke, *Chem. Comm.*, 2 (1971).
Hydrate: two modifications: $R = 0{\cdot}026$ and $0{\cdot}046$, $\sigma = 0{\cdot}002$–$0{\cdot}007$, ~planar
Hydrobromide: $\sigma = 0{\cdot}005$–$0{\cdot}009$, $R = 0{\cdot}034$, *: site of H in hydrobromide

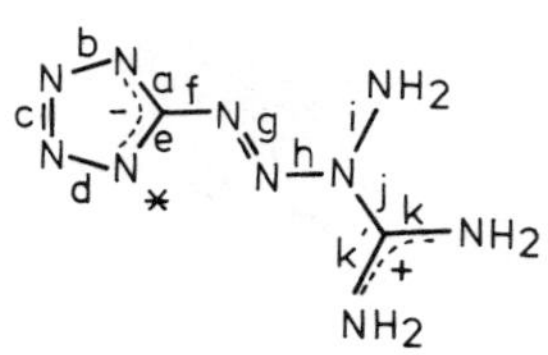

	·H₂O Average	·HBr		·H₂O Average	·HBr
a	= 1·334	1·311	g	= 1·259	1·254
b	= 1·338	1·355	h	= 1·358	1·340
c	= 1·315	1·298	i	= 1·390	1·389
d	= 1·349	1·333	j	= 1·384	1·377
e	= 1·322	1·335	k	= 1·307	1·299
f	= 1·412	1·396	k′	= 1·314	1·307

(21)

bond is increased to 1·28 Å and the N—N bonds are decreased to 1·375 Å. The same is found for 'tetrazene' (**21**), in which molecule a tetrazol ring and an open tetrazene chain occur simultaneously. Within the ring, the lengths of N═N and N—N are nearly compensated (1·315 and 1·344 Å), but in the open chain they are well distinguished (1·259 and 1·374 Å), although affected by resonance.

III. STRUCTURAL EXAMPLES

A. *Hydrazine Derivatives and Other Compounds with* N—N *Single Bonds*

No structure determinations of aromatic hydrazo compounds could be found in literature, but as hydrazobenzene itself is *N*,*N'*-diphenylhydrazine, the structures of some hydrazine derivatives shall be reported here. In the parent compound, hydrazine (**22**), N is sp^3-hybridized and the hydrogen atoms are arranged in a *gauche*-form (the azimuthal angle around the N—N bond is 90–95° relative to the *cis*-configuration[6], N—N = 1·45 Å). About the same holds for tetrafluorohydrazine (N—N = 1·47, N—F = 1·37 Å, ∡NNF = 104°, ∡FNF = 108°, azimuthal angle 65°[27]; a more recent paper[28] reports longer bonds: N—N = 1·52(2), N—F = 1·393(8) Å with two different angles NNF of 99·0° and 103·5°, ∡FNF = 103·7° and azimuthal angle = 69·3°). But if the molecule becomes crowded as in (**23**), or if resonance is possible, as in diformylhydrazine (**24**), the nitrogen atoms become sp^2-hybridized and the N—N bond shrinks to about 1·40 Å.

Hydrazine[6, 7], H_4N_2

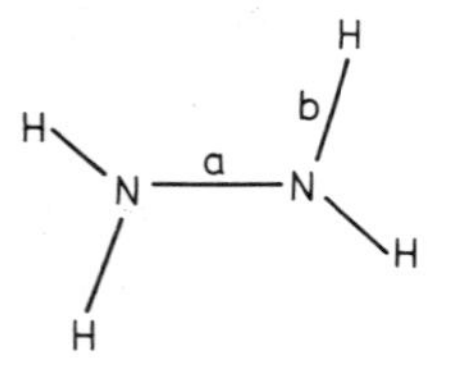

	r_0^6	r_g^7
a =	1·453(5)	1·449(4)
b =	1·020 −1·025	1·022(6)
ab	112·0 −112·5°	112·0(15)

(22)

Tetrakis(trifluoromethyl)hydrazine[19], $C_4F_{12}N_2$

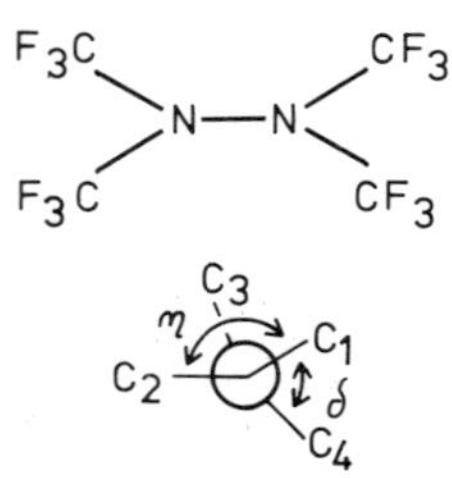

N—N	1·402(20)
C—N	1·433(10)
C—F	1·325(5)
CNC	121·2° ± 1·5°
NNC	119·0° ± 1·5°
FCF	108·2° ± 0·5°
$\eta = 2\delta$	= 170° ± 5°

(23)

Diformylhydrazine[20], $C_2H_4N_2O_2$
$R_{hk0} = 0{\cdot}068$, $R_{h0l} = 0{\cdot}107$, planar, centrosymmetric

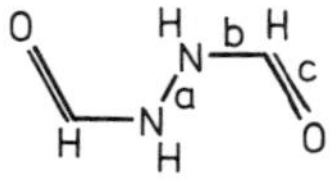

a = 1·392(7)		ab	121·3°
b = 1·325(4)		bc	124·9
c = 1·214(5)			

(24)

In hydrazinium salts, e.g. hydrazinium sulphate (**25**), the cation $N_2H_6^{2+}$ has a staggered conformation like ethane and the N—N bond may be shorter than in hydrazine itself. For hydrazinium acetate, CH_3COO—N_2H_5[29], the cation $N_2H_5^+$ is reported to be in an almost eclipsed form with N—N = 1·462(12) Å. Hydrazine is able to react

Hydrazinium sulphate, $[H_6N_2]^{2+}.[SO_4]^{2-}$
P. G. Jönsson and W. C. Hamilton, *Acta Cryst. B*, **26**, 536 (1970).
$R = 0{\cdot}032$ (X-ray) and 0·040 (neutron), staggered form

$H_3N{-}NH_3 \cdot SO_4$

	Average Neutron	X-ray
N—N	1·426(3)	1·418(2)
S—O	1·475(5)	1·477(1)
NNH	109·8°	108·7
HNH	110·2	110·0

(25)

with carboxylic acids to form hydrazides as ammonia forms amides. The nitrogen atom bound to the carboxyl group becomes sp^2-hybridized because of its interaction with the C═O double bond, i.e. most hydrazide moieties are planar and the C—N bonds are quite short (1·32–1·36 Å). Most important are the hydrazides of carbonic acid and the derived thiocompounds. The mono-hydrazide of carbonic acid is capable of forming salts in which the hydrazine carboxylate anion acts as a bidentate ligand. In the Ca salt (**26**), the N—N distance is about an sp^2-sp^3 single one with 1·425 Å. In carbohydrazide (**27**, X = O), the di-hydrazide of carbonic acid, one hydrazide group is found to be *trans* and the other *cis* to the C═O group. The same is true for thiocarbhydrazide (**27**, X = S). Both compounds behave as bidentate ligands, but even more important in metal complex chemistry is thiosemicarbazide (**28**), the mixed hydrazide-amide of thiocarbonic acid. N—N is found to be *trans* to C═S.

Semicarbazide and thiosemicarbazide are used for the identification of aldehydes and ketones because they condense with the carbonyl group to form sparingly soluble semicarbazones (**29**) and thiosemicarbazones (**30**).

Calcium hydrazine carboxylate monohydrate, $C_2H_6CaN_4O_4$—H_2O
A. Braibanti, A. M. Manotti Lanfredi, M. A. Pellinghelli and A. Tiripicchio *Acta Cryst. B*, **27**, 2261 (1971).
$\sigma \sim 0{\cdot}01$, $R = 0{\cdot}103$, ~planar hydrazine carboxylate moieties

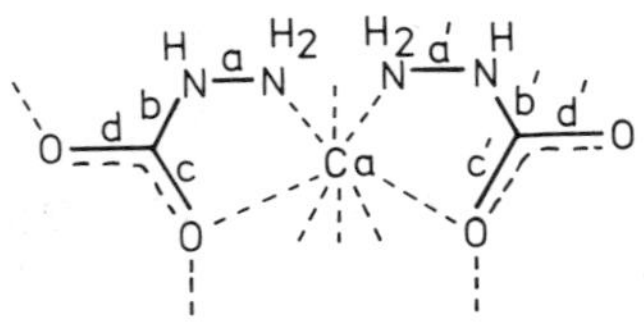

a	= 1·430	ab	121·1°
a′	= 1·417	a′b′	118·5
b	= 1·352	bc	118·3
b′	= 1·361	b′c′	118·2
c	= 1·266	bd	115·9
c′	= 1·258	b′d′	116·6
d	= 1·269	cd	125·7
d′	= 1·272	c′d′	125·2
⟨Ca—O⟩	= 2·44		
⟨Ca—N⟩	= 2·59		

(26)

Carbohydrazide and thiocarbohydrazide, CH_6N_4O and CH_6N_4S
P. Domiani, M. A. Pellinghelli and A. Tiripicchio, *Acta Cryst. B*, **28**, 2495 (1972), and A. Braibanti, A. Tiripicchio and M. Tiripicchio Cammelli, *Acta Cryst. B*, **25**, 2286 (1969).
X = O: $\sigma = 0{\cdot}002$–$0{\cdot}003$, $R = 0{\cdot}040$ ~planar
X = S: $\sigma \sim 0{\cdot}011$, $R = 0{\cdot}079$

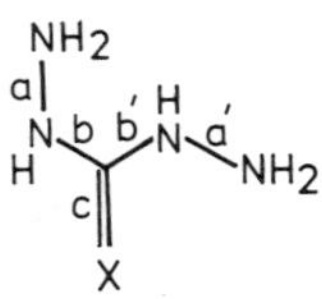

	X = O	X = S		X = O	X = S
a	= 1·417	1·407	ab	121·6°	118·8
a′	= 1·416	1·404	a′b′	121·2	122·4
b	= 1·349	1·335	bc	121·1	118·3
b′	= 1·350	1·315	b′c	122·4	124·2
c	= 1·242	1·724	bb′	116·4	117·5

(27)

Thiosemicarbazide, CH_5N_3S
I: F. Hansen and R. Grønbæk Hazell, *Acta Chem. Scand.*, **23**, 1359 (1969).
II: G. D. Andreetti, P. Domiano, G. F. Gaspari, M. Nardelli and P. Sgarabotto, *Acta Cryst. B*, **26**, 1005 (1970).
I: $\sigma = 0{\cdot}005$–$0{\cdot}006$, $R = 0{\cdot}06$, planar (±0·03 Å), II: $\sigma \sim 0{\cdot}002$, $R = 0{\cdot}024$

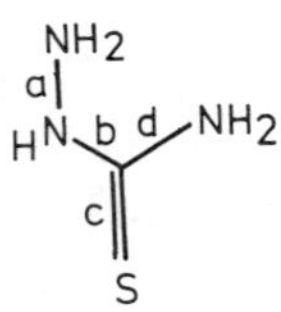

	I		II		I	II
	Uncorr.	Corr.				
a	= 1·412	1·415	1·411	ab	120·8	121·1
b	= 1·325	1·334	1·326	bc	118·8	119·2
c	= 1·701	1·713	1·708	bd	119·0	118·8
d	= 1·320	1·331	1·316	cd	122·1	122·0

(28)

In these, both nitrogens are sp^2-hybridized and by resonance the N—N bond length is decreased to about 1·38 Å, as in the two given examples.

Dimethylglyoxal bisguanylhydrazone dihydrochloride dihydrate, $C_6H_{14}N_8 \cdot 2\,HCl \cdot 2\,H_2O$
J. W. Edmonds and W. C. Hamilton, *Acta Cryst. B*, **28**, 1362 (1972).
$\sigma \sim 0{\cdot}002$, $R = 0{\cdot}048$, centrosymmetric

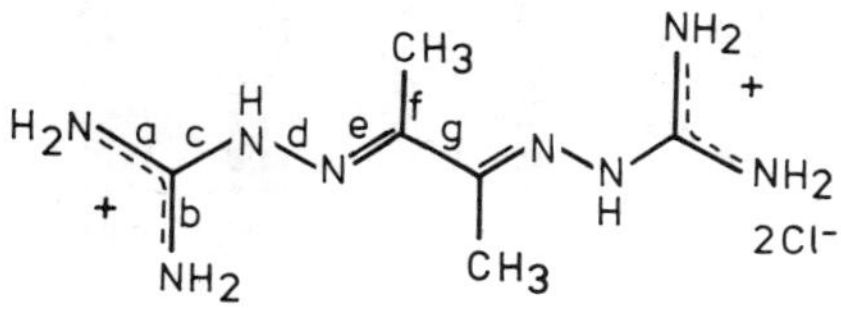

a = 1·323		ab	121·5°
b = 1·324		ac	118·7
c = 1·356		bc	119·8
d = 1·381		cd	117·3
e = 1·300		de	117·1
f = 1·503		ef	124·2
g = 1·478		eg	115·1
		fg	120·7

(29)

2-Formylthiophene thiosemicarbazone, $C_6H_7N_3S_2$
M. Mathew and G. J. Palenik, *Acta Cryst. B*, **27**, 59 (1971).
$\sigma \sim 0{\cdot}004$, $R = 0{\cdot}052$, ~planar

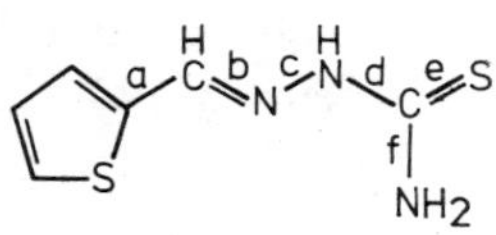

a = 1·436	ab	120·8°
b = 1·286	bc	115·2
c = 1·374	cd	119·4
d = 1·345	de	119·0
e = 1·695	df	118·0
f = 1·315	ef	123·0

(30)

As an example for almost single N—N bond (1·41 Å) in a heterocyclic ring, two 2-pyrazolin-5-one derivatives of high accuracy are reported here (**31**). These structures are a good example for the consistency of lengths for chemically equivalent bonds in different structures. Note also the short C═N bonds of about 1·29 Å.

4-(4-*N*,*N*-Diethylaminophenylimino)-3-methyl-1-phenyl-2-pyrazolin-5-one (UPAD), $C_{20}H_{22}N_4O$ (R = H)
4-(2,6-Dimethyl-4-*N*,*N*-diethylaminophenylimino)-3-methyl-1-phenyl-2-pyrazolin-5-one (HPAD), $C_{22}H_{26}N_4O$ (R = CH_3)
D. L. Smith and E. K. Barret, *Acta Cryst. B*, **27**, 2043 (1971).

UPAD: $\sigma \sim 0{\cdot}003$, $R = 0{\cdot}054$, HPAD: $\sigma \sim 0{\cdot}002$, $R = 0{\cdot}042$

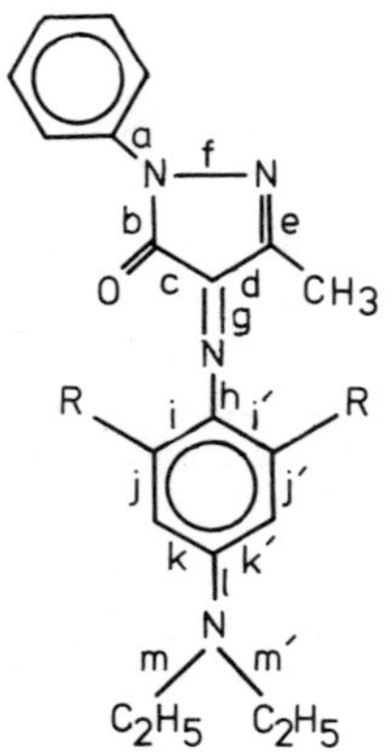

	UPAD	HPAD		UPAD	HPAD
a	= 1·414	1·414	ab	128·3	129·1°
b	= 1·369	1·377	af	118·7	118·4
c	= 1·499	1·494	bf	112·9	112·4
d	= 1·447	1·451	bc	103·8	103·7
e	= 1·290	1·292	cd	103·6	104·4
f	= 1·407	1·413	de	112·8	112·0
g	= 1·297	1·292	ef	106·9	107·5
h	= 1·374	1·399	cg	135·0	131·4
⟨i⟩	= 1·403	1·412	dg	121·3	123·4
⟨j⟩	= 1·364	1·378	gh	131·0	124·4
⟨k⟩	= 1·414	1·408	hi	128·0	125·0
l	= 1·355	1·370	hi′	115·4	115·7
⟨m⟩	= 1·488	1·469	⟨kl⟩	121·8	121·7
			⟨lm⟩	122·2	121·2
			mm′	115·7	115·6

(31)

In azines, the central N—N bond of the —C(H)=N—N=C(H)— moiety is rather long (1·40 Å) and the C=N bonds are accordingly short (1·27 Å) i.e. there is only little resonance. As examples anisaldehyde azine (**32**) and salicylaldehyde azine (**33**) are given here. The influence of the intramolecular O—H···N hydrogen bonds in (**33**) is best seen for bonds *c* and *e*. Benzalazine[30] itself was determined with low accuracy only: N—N = 1·380, N=C = 1·264 and C—C = 1·465 Å, $\sigma \sim 0{\cdot}017$ Å.

Anisaldehyde azine, $C_{16}H_{16}N_2O_2$

J. L. Galigné and J. Falgueirettes, *Acta Cryst. B*, **24**, 1523 (1968).

$\sigma \sim 0{\cdot}008$, $R = 0{\cdot}095$

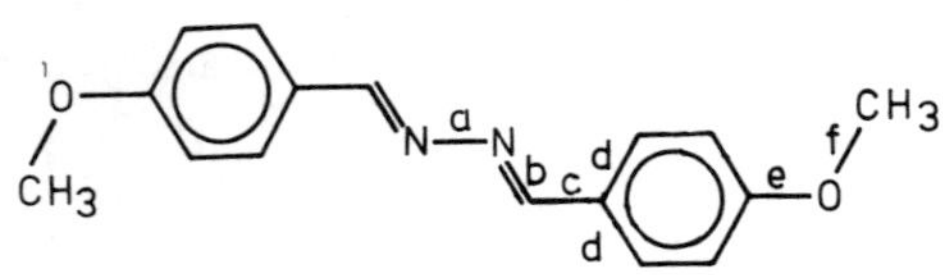

Average			
a = 1·410		ab	111·3°
b = 1·283		bc	120·8
c = 1·495		cd	123·0
d = 1·387		cd′	118·0
d′ = 1·412		dd′	119·0
e = 1·392		ef	118·4
f = 1·445			

(32)

Salicylaldehyde azine, $C_{14}H_{12}N_2O_2$

G. Arcovito, M. Bonamico, A. Domenicano and A. Vaciago, *J. Chem. Soc.* (*B*), 733 (1969).

$\sigma = 0{\cdot}004$–$0{\cdot}005$, $R = 0{\cdot}087$, centrosymmetric, ~planar, strong intramolecular hydrogen bond (k)

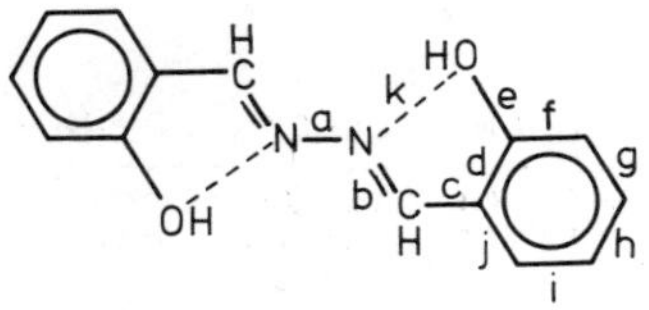

a = 1·407		ab	113·5°
b = 1·274		bc	121·6
c = 1·448		cd	122·6
d = 1·429		cj	120·1
e = 1·341		dj	117·3
f = 1·393		de	121·2
g = 1·362		ef	118·3
h = 1·411		df	120·5
i = 1·377		ek	86·3
j = 1·397		bk	88·4
k = 2·645			

(33)

B. *Azo Compounds*

Azo compounds are more stable than hydrazo compounds and therefore a good number of different structures of azo compounds have been determined. Most molecules are planar with a preference for the *trans*-form —N⁄N—. A compilation of some simple diazenes[23], such as difluorodiazene F—N=N—F (**34**), reports N=N distances of 1·21–1·29 Å, most distances being in the range of 1·22–1·24 Å. Similarly, 1·22–1·25 Å is also found for aliphatic azo compounds (**35, 36, 37**), whereas simple aromatic azo compounds or compounds with C=O groups bound to N=N have slightly longer N=N bonds of about 1·24–1·26 Å[31, 32]. This indicates some resonance in aromatic azo compounds (bond order $n \sim 1{\cdot}9$ for N=N and ~1·1 for N—C).

Diimide[I] and difluorodiazene (*trans* and *cis*)[II], H_2N_2 and F_2N_2
I: A. Trombetti, *Can. J. Phys.*, **46**, 1005 (1968).
II: R. K. Bohn and S. H. Bauer, *Inorg. Chem.*, **6**, 309 (1967).

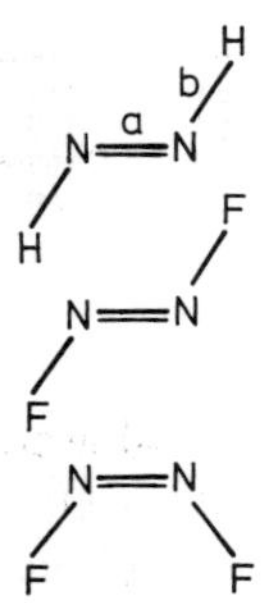

	H_2N_2	F_2N_2 *trans*	F_2N_2 *cis*
a =	1·238(7)	1·230(3)	1·214(3)
b =	—	1·396(2)	1·410(2)
ab	109	105·5	114·4°
		all planar	

(34)

Azomethane and fluoro derivatives[23], $C_2H_6N_2$, $C_2H_3F_3N_2$ and $C_2F_6N_2$
The *cis*-form of $C_2F_6N_2$ is questionable

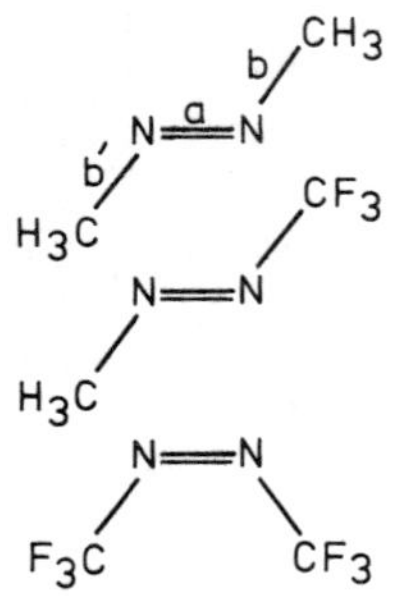

	$C_2H_6N_2$	$C_2H_3F_3N_2$	$C_2F_6N_2$
a =	1·254(3)	1·219(8)	1·236(5)
b =	1·474(3)	1·476(12)	1·490(6)
b′ =	—	1·440(12)	—
ab	111·9	126·2	133·0°
ab′	—	110·5	—

(35)

Azobisisobutyronitrile, $C_8H_{12}N_4$ and $C_8D_{12}N_4$ ($R = CH_3$) and azobis-3-cyano-3-pentane, $C_{12}H_{20}N_4$ ($R = C_2H_5$)
A. B. Jaffé, D. S. Malament, E. P. Slisz and J. M. McBride, *J. Amer. Chem. Soc.*, **94**, 8515 (1972).
(See also: G. Argay and K. Sasvári, *Acta Cryst. B*, **27**, 1851 (1971).)
$R = CH_3$: $\sigma \sim 0{\cdot}003$, two modifications, R = 0·092 and 0·061; $R = CD_3$: R = 0·070
$R = C_2H_5$: $\sigma \sim 0{\cdot}004$, R = 0·076, disordered; planar except for C_2H_5 (and CH_3)

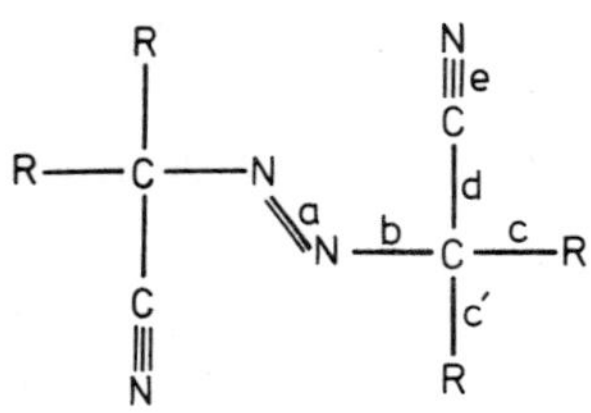

		Average	
	$R = CH_3$	$R = CD_3$	$R = C_2H_5$
a =	1·220	1·218	1·218
b =	1·497	1·488	1·496
c =	1·519	1·523	1·538
d =	1·474	1·483	1·477
e =	1·135	1·130	1·122
ab	114·5	114·8	115·2°
bd	112·0	112·0	111·9
de	176·4	176·5	177·2
bc	106·8	106·9	107·2
cd	111·0	110·1	109·7

(36)

1,2,8,9-Tetraazacyclotetradeca-1,8-dien-3,10-ylen-diacetate, $C_{14}H_{24}N_4O_4$
R. Allmann and I. Kawada, *Angew. Chemie*, **80**, 1001 (1968); *Int. Ed.*, **7**, 944 (1968), (refined to $R = 0{\cdot}046$ in *Cryst. Struct. Comm.*, **3**, 83 (1974)).
$\sigma \leqslant 0{\cdot}01$, R = 0·149, centrosymmetric

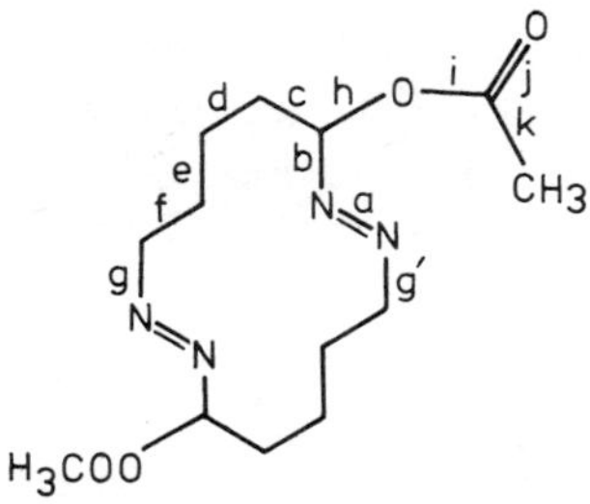

a = 1·22		ag′	112°
b = 1·46		ab	116
c = 1·53		bc	109
d = 1·49		cd	113
e = 1·50		de	115
f = 1·54		ef	112
g = 1·47		fg	110
h = 1·43		bh	114
i = 1·34		ch	108
j = 1·20		hi	117
k = 1·48		ij	123
		jk	125

(37)

For molecules with several azo groups the N═N bonds do increase further (1·32 Å in (**42**)), but without the expected decrease for the C—N bond lengths. The N═N—C angles for aromatic compounds are about 113°, for aliphatic ones about 115°. The N—C—C angles to the adjacent phenyl rings deviate appreciably from 120°: they are about 125° *cis* to the azo group and about 115° for the *trans* side, i.e. the average dimensions for *trans*-azo compounds are:

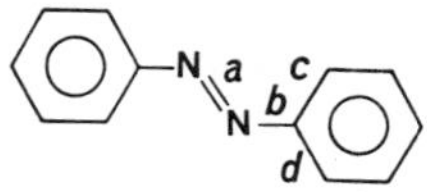

a = 1·25 *b* = 1·42 Å ∢*ab* = 113°,
∢*bc* = 125°
∢*bd* = 115°

cis-Azobenzene (**38**) is overcrowded and, to get rid of a too short C—H···H—C contact, the phenyl groups are twisted out of plane by 53·3° and the N═N—C angle is opened to 122°. By these and some other minor deformations, the C···C distance is increased to 3·35 Å as compared with about 1·5 Å in the impossible planar *cis*-form.

cis-Azobenzene[31], $C_{12}H_{10}N_2$
$\sigma = 0{\cdot}003$–$0{\cdot}004$, R = 0·039, phenyl rings twisted by 53·3° about the C—N bonds, twofold symmetry-axis

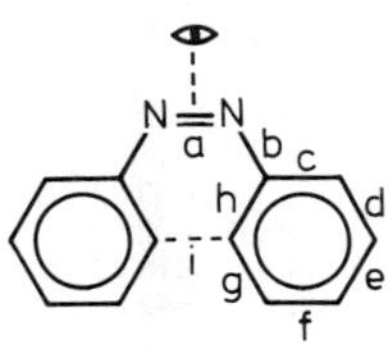

	Uncorr.	Corr.		
a =	1·251	1·253	ab	121·9°
b =	1·443	1·449	bc	117·3
c =	1·396	1·410	cd	118·7
d =	1·370	1·378	de	121·7
e =	1·364	1·374	ef	119·0
f =	1·376	1·389	fg	120·8
g =	1·369	1·377	gh	120·0
h =	1·374	1·385	hc	119·8
i =	3·348		hb	122·5

(38)

All other reported azo compounds are of the *trans*-form (see examples **39–41**). For *p*-azotoluene[33] and one part of the molecules of *trans*-azobenzene (**39**), disordered structures were found because the molecules may pack in either of two possibilities.

trans-Azobenzene, $C_{12}H_{10}N_2$
C. J. Brown, *Acta Cryst.*, **21**, 146 (1966).
$\sigma \sim 0{\cdot}003$, $R = 0{\cdot}069$; two different centrosymmetric molecules, one of which is disordered and shows unreliable distances

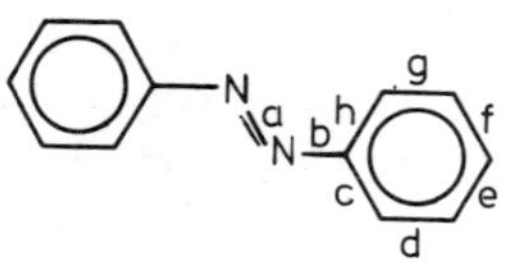

	Ordered		Disordered	Ordered	
	Uncorr.	Corr.		Molecule only	
a	= 1·243	1·247	1·172	ab	113·6°
b	= 1·433	1·434	1·472	bc	115·5
c	= 1·384	1·388	1·373	bh	124·1
d	= 1·390	1·391	1·373	ch	120·3
e	= 1·368	1·372	1·365	cd	119·4
f	= 1·389	1·393	1·385	de	120·4
g	= 1·379	1·380	1·375	ef	119·9
h	= 1·385	1·390	1·383	fg	120·3
				gh	119·6

(39)

4,4′-Dichloroazobenzene[32], $C_{12}H_8Cl_2N_2$
$\sigma = 0{\cdot}004$–$0{\cdot}005$, $R = 0{\cdot}049$, centrosymmetric, nearly planar

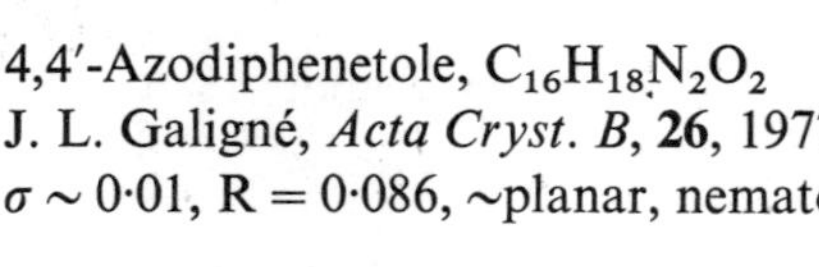

a	= 1·252	ab	112·6°
b	= 1·443	bc	114·2
c	= 1·378	bh	125·9
d	= 1·385	cd	120·8
e	= 1·382	de	118·7
f	= 1·381	ef	121·7
g	= 1·387	fg	119·3
h	= 1·399	gh	119·6
i	= 1·737	ei	119·4

(40)

4,4′-Azodiphenetole, $C_{16}H_{18}N_2O_2$
J. L. Galigné, *Acta Cryst. B*, **26**, 1977 (1970).
$\sigma \sim 0{\cdot}01$, $R = 0{\cdot}086$, ~planar, nematogenic

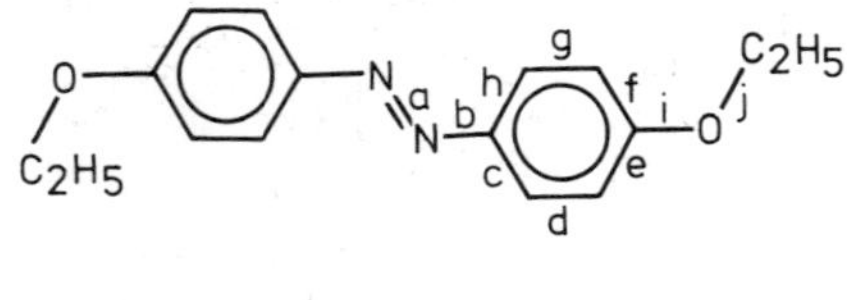

Average of two halves			
a	= 1·255	ab	114·3°
b	= 1·426	bc	115·2
c	= 1·393	bh	125·2
d	= 1·383	ei	115·5
e	= 1·396	fi	124·5
f	= 1·418	ij	117·3
g	= 1·392	i	= 1·361
h	= 1·388	j	= 1·453

(41)

For three other structures with lower accuracy only the N=N, N—C and ∡N—N—C values shall be given here: *p*-azotoluene[33]: 1·244, 1·433 Å, 113·8°; *p,p'*-dibromoazobenzene[34]: 1·276, 1·428 Å, 112·2°, $\sigma \sim 0{\cdot}02$ Å; and azobenzene-2-sulphenylcyanide[35]: 1·22, 1·44 and 1·42 Å, 116° and 107°, $\sigma \sim 0{\cdot}03$ Å. In 4-phenylazo-azobenzene (**42**) two azo groups are present with rather long N=N and C—N bonds. For the dimethylaniline

4-Phenylazoazobenzene, $C_{18}H_{14}N_4$
R. D. Gilardi and I. L. Karle, *Acta Cryst. B*, **28**, 1635 (1972).
$\sigma = 0{\cdot}0035$–$0{\cdot}005$, R = 0·078, ~planar (±0·05 Å)

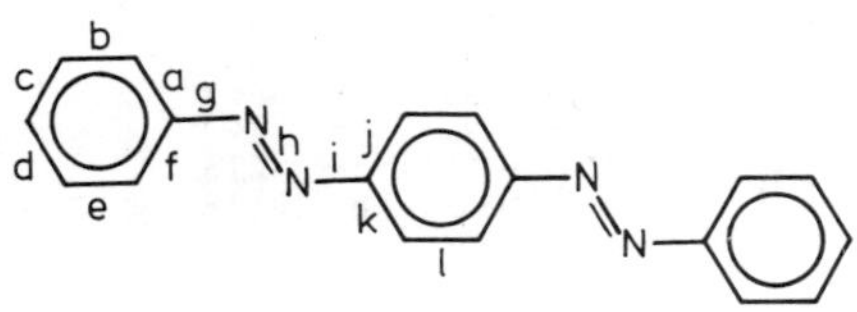

a	= 1·382	j	= 1·393
b	= 1·379	k	= 1·391
c	= 1·382	l	= 1·395
d	= 1·385	ga	113·4°
e	= 1·389	gf	125·8
f	= 1·392	gh	115·5
g	= 1·461	hi	111·7
h	= 1·232	ij	126·8
i	= 1·449	ik	113·0

(42)

ring in methyl orange (**43**), a simple azo dye, the quinonoide form is of some importance, apparent by the short bonds *g*, *i* and *k* and the increase of *h* and *j*.

Methyl orange—H_2O—C_2H_5OH, $C_{14}H_{14}NaN_3O_3S$—H_2O—C_2H_6O
A. W. Hanson, *Acta Cryst. B*, **29**, 454 (1973).
$\sigma = 0{\cdot}006$–$0{\cdot}009$, R = 0·049, ~planar (±0·27 Å)

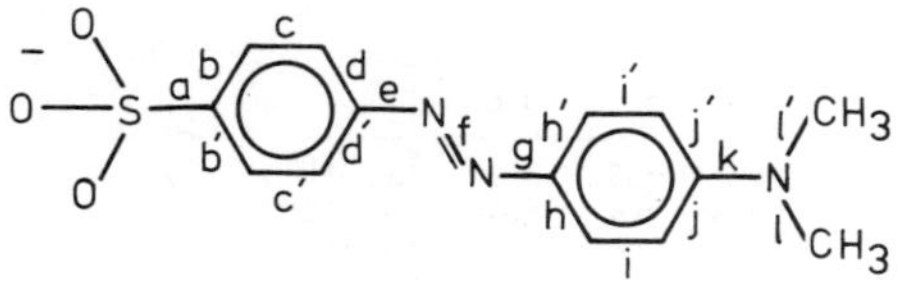

a	= 1·780	⟨S—O⟩	= 1·451
⟨b⟩	= 1·380	⟨Na—O⟩	= 2·41
⟨c⟩	= 1·388	de	116°
⟨d⟩	= 1·376	d'e	125
e	= 1·440	ef	112·6
f	= 1·244	fg	115·0
g	= 1·407	gh	115
⟨h⟩	= 1·392	gh'	127
⟨i⟩	= 1·372	⟨jk⟩	121·3
⟨j⟩	= 1·406	⟨kl⟩	121·1
k	= 1·365	ll'	117·8
⟨l⟩	= 1·446		

(43)

An azo group bound to heterocyclic moieties is given in the next example (**44**). The N—N bond (1·39 Å) therein has about the same bond order

($n \sim 1{\cdot}1$) as the C—N bonds have in the foregoing structures. Somewhat

1,1′-Azo-2-phenylimidazo[1,2-a]-pyridinium dibromide, $[C_{26}H_{20}N_6]^{2+}\cdot$ $2\,Br^-$
D. J. Pointer and B. J. Wilford, *J. Chem. Soc. Perkin II*, 2259 (1972).
$\sigma \leqslant 0{\cdot}02$, R = 0·112, centrosymmetric, phenyl rings twisted by 47·5°

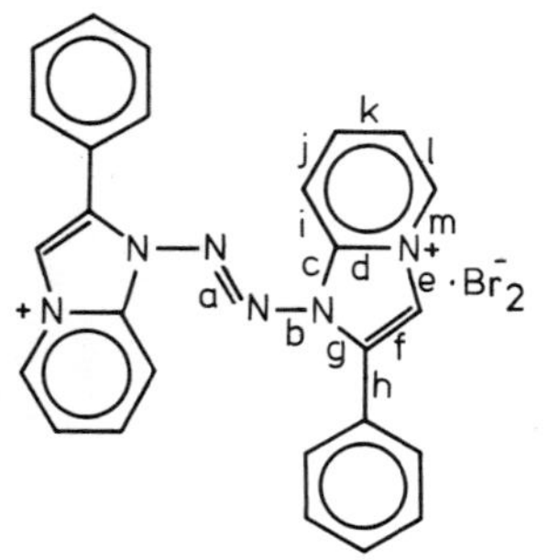

a = 1·249		ab	111·1°
b = 1·388		bc	130·3
c = 1·352		bg	118·8
d = 1·375		cg	110·4
e = 1·410		cd	106·0
f = 1·338		de	109·0
g = 1·416		ef	108·1
h = 1·467		fg	106·4
i = 1·390		fh	129·4
j = 1·395		gh	124·2
k = 1·413		dm	123·5
l = 1·363		di	119·5
m = 1·357			

(44)

longer N═N bonds (~1·27 Å) are reported for the two asymmetric compounds *o*-bromobenzene-*anti*-diazocyanide (**45**) and *p*-chlorobenzene-*trans*-diazoimido-glyoxynitrile (**46**), but the overall dimensions are very similar to other azo compounds.

o-Bromobenzene-*anti*-diazocyanide, $C_7H_4BrN_3$
I. Bø, B. Klewe and C. Rømming, *Acta Chem. Scand.*, **25**, 3261 (1971).
$\sigma = 0{\cdot}008$–$0{\cdot}011$, R = 0·059, planar

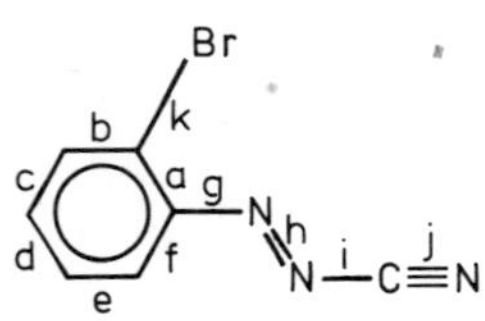

	Uncorr.	Corr.		Uncorr.	Corr.
a	= 1·388	1·369	ag	116·7°	
b	= 1·384	1·388	fg	125·7	
c	= 1·366	1·371	ak	121·4	
d	= 1·383	1·389	bk	117·8	
e	= 1·352	1·355	gh	113·6	
f	= 1·380	1·383	hi	113·0	
g	= 1·400	1·403	ij	172·5	
h	= 1·269	1·276	j	= 1·140	1·142
i	= 1·362	1·364	k	= 1·906	1·914

(45)

p-Chlorobenzene-*trans*-diazoimidoglyoxynitrile, $C_8H_5ClN_4$
H. H. Erichsen and C. Rømming, *Acta Chem. Scand.*, **22**, 1430 (1968).
$\sigma \sim 0{\cdot}01$, $R = 0{\cdot}09$, nearly planar

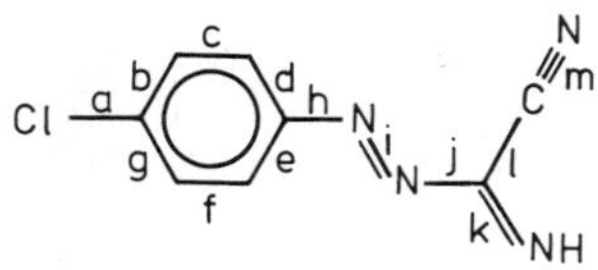

a = 1·735		dh	113·9°
b = 1·374		eh	124·4
c = 1·395		hi	112·9
d = 1·400		ij	111·2
e = 1·381		jk	114·5
f = 1·368		jl	119·1
g = 1·391		kl	126·4
h = 1·447		lm	177·2
i = 1·264			
j = 1·441		l	= 1·432
k = 1·273		m	= 1·136

(46)

Azodicarbonamide, $NH_2OC—N_2—CONH_2$[36], was determined with lower accuracy ($\sigma \sim 0{\cdot}02$ Å): N=N = 1·24, N—C = 1·48 Å, ∡N—N—C = 109·5°, and the reported N—C bond length seems to be too long.

In potassium *syn*-methyl-diazotate (**47**), some resonance exists between

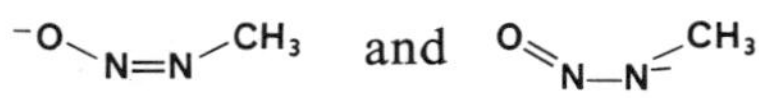

(47)

of which the first form is preferred.

Potassium-*syn*-methyl-diazotate, $CH_3N_2O—K$
R. Huber, R. Langer and W. Hoppe, *Acta Cryst.*, **18**, 467 (1965).
$\sigma \sim 0{\cdot}008$, $R = 0{\cdot}084$

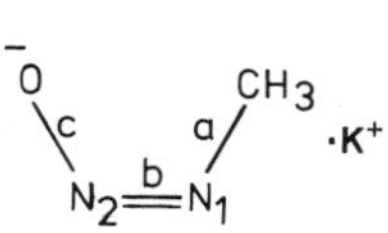

a = 1·477		K—O	= 2·650
b = 1·246			2·728
c = 1·306			2·785
		KN_1	= 2·914
ab	116·2°		2·957
bc	119·8	KN_2	= 2·900
			2·901

(47)

When the azo group is bound to transition metal ions or atoms, several kinds of bonding are observed. If only the lone-pair electrons form dative σ-bonds to the metal atom, the dimensions of the complexed azo compound are about the same as for the free one. Thus, for the cuprous chloride azomethane complex $Cu_2Cl_2—C_2H_6N_2$[37], the following dimensions are reported: N=N = 1·26(3), N—C = 1·46(3) Å, ∡N—N—C = 118°, Cu—N = 1·99(2) Å. Infinite Cu—Cl double chains (Cu—Cl = 2·368 and 2·319(6) Å in a single zig-zag chain, 2·547(6) Å between chains) are linked through *trans*-azomethane molecules by dative σ-bonds (compare azomethane, **35**).

In bis-(*t*-butylisocyanide)(azobenzene)nickel(0) (**48**), olefins like π-bonds to Ni occur, in which the π^*-orbitals of the azo group are used. Azobenzene is in the *trans*-form in this complex and the N═N bond is weakened to nearly a single one (1·385 Å; compare *trans*-azobenzene, **39**).

Bis(*t*-butylisocyanide)(azobenzene)nickel(0), $C_{12}H_{10}N_2$—$(C_5H_9N)_2$—Ni
R. S. Dickson, J. A. Ibers, S. Otsuka and Y. Tatsuno, *J. Amer. Chem. Soc.*, **93**, 4636 (1971).
R = 0·07, dihedral angle NiNN to NiCC = 1·2° only (i.e. ~planar surrounding of Ni, which is about perpendicular to the plane of *trans*-azobenzene), phenyl rings twisted by ~14°, dihedral angle of the two N—N—C(phenyl) planes = 26·8°

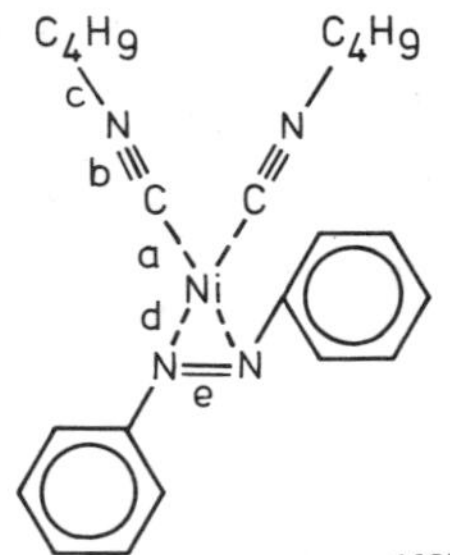

Average	
a =	1·840(5)
b =	1·154(6)
c =	1·441(6)
d =	1·897(4)
e =	1·385(5)

(48)

In μ-methazo-bis(tricarbonyl iron), $C_2H_6N_2$—2 $Fe(CO)_3$ (**49**), the two N and two Fe atoms approximately form a tetrahedron possessing a long N═N bond (1·366 Å). The same structural feature was found for benzo[c]-cinnoline bis(tricarbonyl iron), $C_{12}H_8N_2$—2 $Fe(CO)_3$[38], in which the N—N bond in the phenanthrene moiety measures 1·399(8) Å, as compared with 1·292 Å in phenanthrene (**6**) itself.

μ-Methazo-bis(tricarbonyliron), $C_2H_6N_2 \cdot 2\ Fe(CO)_3$
R. J. Doedens and J. A. Ibers, *Inorg. Chem.*, **8**, 2709 (1969).
R = 0·028, about C_{2v}—2mm symmetry, N—N perpendicular to Fe—Fe

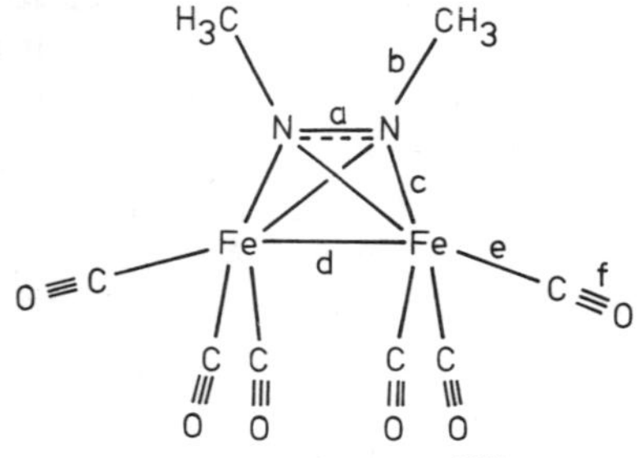

a	= 1·366(8)
⟨b⟩	= 1·476(6)
⟨c⟩	= 1·878(3)
d	= 2·496(3)
⟨e⟩	= 1·775(4)
⟨f⟩	= 1·143(4)
ab	123·0°

(49)

C. Azoxy Compounds and Nitroso Dimers

The lone-pair electrons of the azo group 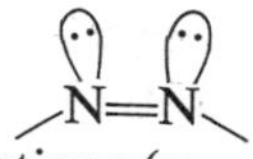 are able to form dative bonds not only to metal atoms and cations (see above) but also to oxygen atoms. With one such N → O bond the azoxy group 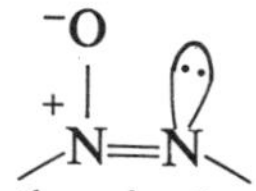 is obtained, in which the N═N bond seems to be even stronger than in the azo group. This strengthening may be caused by the different electronegativities of the two nitrogen atoms, leading to a somewhat ionic N═N bond. Of the only two structure determinations of azoxy compounds which have been reported up to now, the structure of ethyl-*p*-azoxybenzoate (**51**) exhibits unreliable N═N and N—C distances of 1·16 and 1·56 Å; only the N—O bond of 1·29 Å seems to be acceptable. Hence, the data from the structure of *p*-azoxyanisol (**50**) must be taken for standard values until better structure determinations are available. The N═N bond amounts to 1·22 Å, indicating an isolated double bond. The N—C bonds (1·50 Å) are quite long as compared with about 1·42 Å in aromatic azo compounds or 1·45 Å in azobenzene dioxide (**52**). As one nitrogen atom is tetravalent, some increase of one C—N bond is to be expected, but this may not explain why the other C—N bond is also increased to about the same length. The N—O bond (1·28 Å) is in accordance with that in nitroso dimers (1·26 Å, see **52**–**55**) and may be compared with 1·243 Å in NO_3^- and 1·405 and 2 × 1·206 Å in HNO_3[21].

p-Azoxyanisole, $C_{14}H_{14}N_2O_3$
W. R. Krigbaum, Y. Chatani and P. G. Barber, *Acta Cryst. B*, **26**, 97 (1970).
$\sigma \sim 0{\cdot}005$, R = 0·091, dihedral angle between benzene rings: 22·6°

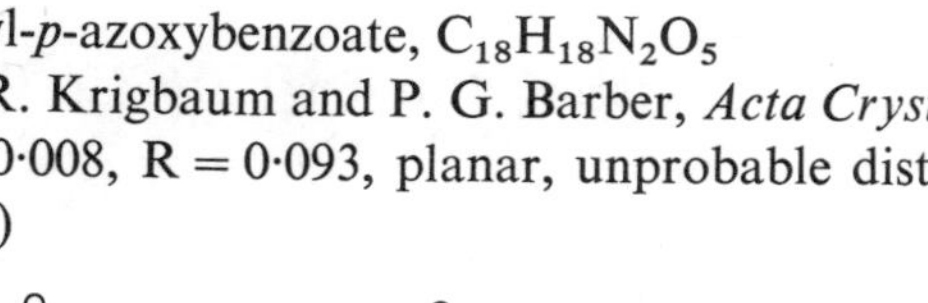

a = 1·372		ac	116·1°
b = 1·386		bc	122·8
c = 1·496		cd	117·9
d = 1·279		de	130·3
e = 1·218		ce	111·8
f = 1·496		ef	114·7
g = 1·373		fg	130·6
h = 1·398		fh	109·9

(50)

Ethyl-*p*-azoxybenzoate, $C_{18}H_{18}N_2O_5$
W. R. Krigbaum and P. G. Barber, *Acta Cryst. B*, **27**, 1884 (1971).
$\sigma \sim 0{\cdot}008$, R = 0·093, planar, unprobable distances (*e* too short; *c*, *f* too long)

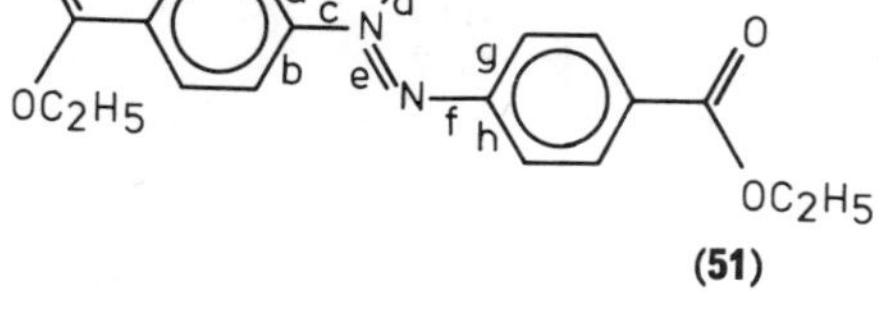

a = 1·358		ac	112·8°
b = 1·381		bc	121·8
c = 1·507		cd	116·2
d = 1·291		de	134·7
e = 1·155		ce	109·1
f = 1·559		ef	113·5
g = 1·393		fg	129·0
h = 1·353		fh	110·0

(51)

If both nitrogen atoms of an azo group form N → O bonds, repulsion of the two positively charged N atoms is to be expected. So in fact the N=N bonds in the nitroso dimers $\left(\text{—N=O} + \text{O=N—} \rightarrow \text{—}\overset{\overset{\text{}^{-}\text{O}}{\uparrow}}{\underset{+}{\text{N}}}\text{=}\overset{\overset{\text{O}^{-}}{\uparrow}}{\underset{+}{\text{N}}}\text{—}\right)$ are found to be about 1·30–1·32 Å, which, however, is still quite a strong bond, not to be compared with the very weak N—N bond of 1·750 Å in the dimer of nitrogen dioxide, N_2O_4[39]. The N → O bonds are about 1·26 Å and the N—C bonds 1·44–1·45 Å, which is even somewhat longer than for a pure single N(sp^2)—C(sp^2) bord (in (**53**) and (**54**) C is sp^3-hybridized). This indicates that there is only very little resonance, if any at all, between the dinitroso moiety and the rest of the molecules.

cis-Azobenzene dioxide (**52**) has greater twisting angles of the phenyl groups (67°) than *cis*-azobenzene (**38**) has (53°). Dinitrosomethane exists both in the *cis*-form (**53**) and in the *trans*-form[40], but the structure determinations are not very accurate (in the *trans*-form: N=N = 1·22(4), N → O = 1·25(2), N—C = 1·57(8) Å, ∡N—N—O = 126°, ∡N—N—C = 109°, ∡O—N—C = 125°). As an example of a *trans*-dimer with fairly representative dimensions that of 2-nitronitrosoethane (**54**) is given here.

cis-Azobenzene dioxide = nitrosobenzene dimer, $C_{12}H_{10}N_2O_2$
D. A. Dietrich, I. C. Paul and D. Y. Curtin, *Chem. Comm.*, 1710 (1970).
$\sigma \sim 0{\cdot}005$, R = 0·057, nonplanar: twisting around b and b′ = 64·8 and 111·7°, angles around N are within 1·2° = 120°, torsion angle around a = 17·9°

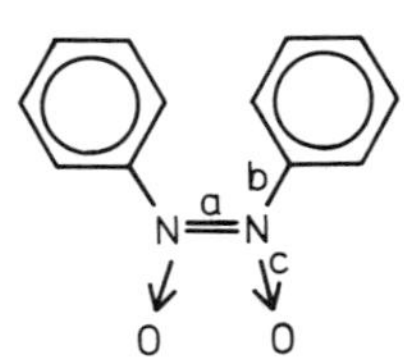

a = 1·323
⟨b⟩ = 1·454
⟨c⟩ = 1·264

(52)

cis-Dinitrosomethane, $C_2H_6N_2O_2$=$(CH_3NO)_2$
G. Germain, P. Piret and M. van Meersche, *Acta Cryst.*, **16**, 109 (1963).
$\sigma = 0{\cdot}02–0{\cdot}03$, R = 0·156

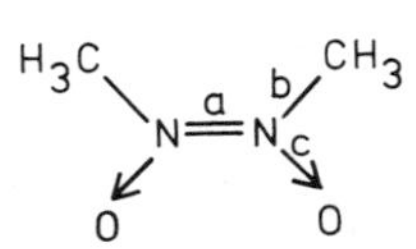

a = 1·31
⟨b⟩ = 1·47
⟨c⟩ = 1·31
ab 121·5°
ac 119·0
bc 120·0

(53)

2-Nitronitrosoethane *trans*-dimer, $(C_2H_4N_2O_2)_2$
F. P. Boer and J. W. Turley, *J. Amer. Chem. Soc.*, **91**, 1371 (1969).
$\sigma = 0{\cdot}004$–$0{\cdot}006$, R = 0·108, centrosymmetric, the carbon atoms are staggered, (survey on 5 other nitroso dimers)

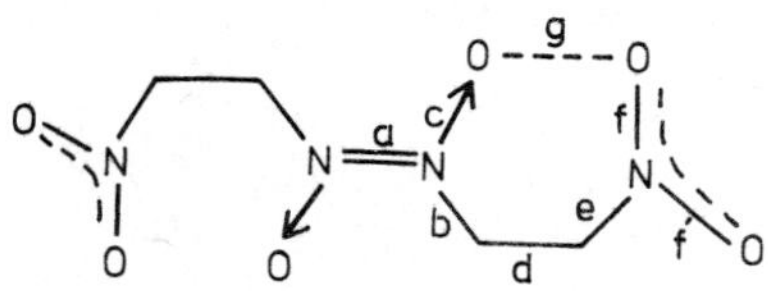

a	= 1·304	ab	117·4°
b	= 1·470	ac	121·1
c	= 1·262	bc	121·4
d	= 1·504	bd	111·3
e	= 1·485	de	112·3
⟨f⟩	= 1·202	ef	119·8
g	= 3·077	ef′	116·9

(54)

In 1,8-dinitrosonaphthalene (**55**), the N═N bond is quite long, but this may be partly to reduce the strain in the 5-membered ring. The N → O bonds are similar to those in *trans*-4,4′-azopyridine-*N*-oxide[41] (N → O = 1·283(11) Å, and normal dimensions for the azo group: N═N = 1·228, C—N = 1·419 Å, ∡N—N—C = 113·3°).

1,8-Dinitrosonaphthalene, $C_{10}H_6N_2O_2$
C. K. Prout, T. S. Cameron, R. M. A. Dunn, O. J. R. Hodder and D. Viterbo, *Acta Cryst. B*, **27**, 1310 (1971).
$\sigma \sim 0{\cdot}008$, R = 0·078, ~planar

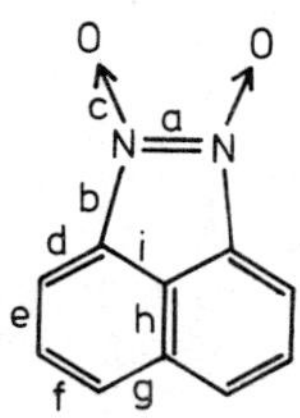

Average			
a = 1·376		ab	108·7°
b = 1·435		ac	122·3
c = 1·265		bc	129·0
d = 1·381		bi	106·2
e = 1·429		di	121·1
f = 1·394		de	115·6
g = 1·435		ef	123·7
h = 1·397		fg	120·3
i = 1·400		gh	114·7
		hi	124·8

(55)

IV. REFERENCES

1. B. P. Stoicheff, *Can. J. Phys.*, **35**, 730 (1957).
2. B. P. Stoicheff, *Can. J. Phys.*, **32**, 630 (1954).
3. C. C. Costain, *J. Chem. Phys.*, **29**, 864 (1958).
4. L. S. Bartell, *J. Chem. Phys.*, **23**, 1219 (1955); **35**, 1945 (1961).
5. L. S. Bartell, *J. Chem. Phys.*, **38**, 1827 (1963); **23**, 1854 (1955).
6. A. Yamaguchi, I. Ichishima, T. Shimanouchi and S.-I. Mizushima, *J. Chem. Phys.*, **31**, 843 (1959).
7. Y. Marino, I. Iljima and Y. Murata, *Bull. Chem. Soc. Japan*, **33**, 46 (1960).

8. R. W. James, *Phys. Z.*, **33,** 737 (1932).
9. D. W. J. Cruickshank, *Acta Cryst.*, **9,** 757 (1956); **14,** 896 (1961).
10. V. Shomaker and K. N. Trueblood, *Acta Cryst.*, B **24**, 63 (1968).
11. R. Allmann, *Z. Kristallogr.*, **132,** 129 (1970).
12. W. R. Busing and H. A. Levy, *Acta Cryst.*, **17,** 142 (1964).
13. R. F. Stewart, E. R. Davidson and W. T. Simpson, *J. Chem. Phys.*, **42,** 3175 (1963).
14. P. Coppens, *Science*, **158,** 1579 (1967).
15. P. Coppens (Ed.), 'Experimental and theoretical studies of electron densities', *Transactions Am. Cryst. Assoc.*, **8** (1972).
16. L. Pauling, *The Nature of the Chemical Bond*, Cornell University Press, 1960.
17. M. S. J. Dewar and H. N. Schmeising, *Tetrahedron*, **11,** 96 (1960).
18. T. Nishikawa, *J. Phys. Soc. Japan*, **12,** 668 (1957).
19. L. S. Bartell and H. K. Higginbotham, *Inorg. Chem.*, **4,** 1346 (1965).
20. Y. Tomiie, C. H. Koo and I. Nitta, *Acta Cryst.*, **11,** 774; 875 (1958).
21. W. J. Orville-Thomas, *The structure of small molecules*, Elsevier, 1966.
22. D. W. J. Cruikshank and R. A. Sparks, *Proc. Roy. Soc.* (*London*) *A*, **258,** 270 (1960).
23. C. H. Chang, R. F. Porter and S. H. Bauer, *J. Amer. Chem. Soc.*, **92,** 5313 (1970).
24. A. Langseth and B. P. Stoicheff, *Can. J. Phys.*, **34,** 350 (1956).
25. E. G. Cox, D. W. J. Cruikshank and J. A. S. Smith, *Proc. Roy. Soc.* (*London*) *A*, **247,** 1 (1958).
26. Y. D. Kondrashev, *Sov. Phys. Crystallogr.*, **6,** 413 (1961).
27. D. R. Lide Jr. and D. E. Mann, *J. Chem. Phys.*, **31,** 1129 (1959).
28. R. K. Bohn and S. H. Bauer, *Inorg. Chem.*, **6,** 304 (1967).
29. S. A. Hady, I. Nahringbauer and I. Olovson, *Acta Chem. Scand.*, **23,** 2764 (1969).
30. U. C. Sinha, *Acta Cryst.*, *B* **26,** 889 (1970).
31. A. Mostad and C. Rømming, *Acta Chem. Scand.*, **25,** 3561 (1971).
32. H. Hope and D. Victor, *Acta Cryst. B*, **25,** 1849 (1969).
33. C. J. Brown, *Acta Cryst.*, **21,** 153 (1966).
34. A. G. Amit and H. Hope, *Acta Chem. Scand.*, **20,** 835 (1966).
35. K. K. Kakati and B. Chaudhuri, *Acta Cryst. B*, **24,** 1645 (1968).
36. J. H. Bryden, *Acta Cryst.*, **14,** 61 (1961).
37. I. D. Brown and J. D. Dunitz, *Acta Cryst.*, **13,** 28 (1960).
38. R. J. Doedens, *Inorg. Chem.*, **9,** 429 (1970).
39. D. W. Smith and K. Hedberg, *J. Chem. Phys.*, **25,** 1282 (1956).
40. M. van Meersche and G. Germain, *Acta Cryst.*, **12,** 818 (1959).
41. E. L. Eichhorn, *Acta Cryst.*, **12,** 746 (1959).

CHAPTER 3

Thermochemistry of hydrazo, azo and azoxy groups

ROBERT SHAW

Physical Sciences Division, Stanford Research Institute, Menlo Park, California, U.S.A.

I. INTRODUCTION

The title of this chapter is something of a misnomer because I have included some data on hydrazine derivatives that, strictly speaking, are not hydrazo compounds and because I have been unable to find any information on the thermochemistry of azoxy compounds. A more accurate title would therefore be *The Thermochemistry of Hydrazines, Azo Compounds and Related Free Radicals*.

The thermochemical quantities under consideration are heat of formation $\Delta H^0_{f_{298}}$, entropy S^0_{298} and heat capacity C^0_p, from 298 to 1500 K for the ideal gas state. Although the joule is now the recommended unit of energy, English-speaking chemists and English-language chemical journals still predominantly use the kilocalorie. The units used throughout this review will therefore be kcal/mole for ΔH^0_f and gibbs/mole for S^0 and C^0_p.

At first sight, N—N thermochemistry is a neglected field of research. For example, Stull, Westrum and Sinke's excellent monograph, *The Chemical Thermodynamics of Organic Compounds*[1], has no text or tables on hydrazines or azo compounds other than hydrazine itself. Very little additional data are given in JANAF[2] or in Cox and Pilcher's monograph[3]. One of the primary sources for what is happening in the thermochemist's world is the *Bulletin of Thermochemistry and Thermodynamics*, published annually by IUPAC. The Bulletins for the last three years contain no details of work in progress or published in N—N thermochemistry. However, there is some activity, notably at Rice University in Texas. There, Engel and Wood[4] have some interesting results on azo compounds (discussed later), and I understand from them that Dr. F. D. Rossini, one of the founders of modern American thermochemistry, plans to burn azomethane at Rice later this year.

II. ESTIMATION BY GROUP ADDITIVITY

The limited quantities of experimental data place an added premium on those obtained by estimation. Group additivity is simple, fast and accurate, and is quickly gaining wide acceptance as the best method for estimation of thermochemical data. It is used to estimate thermochemical data that are lacking, to check measured data for consistency with those for chemically related compounds, and to point out where further experiments are needed.

Group additivity postulates that the chemical thermodynamic properties of molecules consist of contributions from the individual groups that make up the molecule. It is therefore an extension of the series: atom additivity, bond additivity, ..., and turns out to be an excellent compromise between simplicity and accuracy. For a detailed treatment of the additivity principle as applied to thermochemistry, see an early paper by Benson and Buss[5] and a more recent *Chemical Review* article[6]. The latter contains all the group values that could be derived from gas-phase thermochemical data up to 1969. The group values permit the estimation of ΔH_f^0 to an accuracy of better than ± 1 kcal/mole, S^0 to an accuracy of ± 1 gibbs/mole, and C_p^0 to an accuracy of $\pm 0{\cdot}5$ gibbs/mole.

As an example of the use of group additivity, suppose it is necessary to estimate $\Delta H_{f_{298}}^0$ of methylhydrazine. Methylhydrazine is composed of the groups [C—(N)(H)$_3$], [N—(N)(C)(H)] and [N—(N)(H)$_2$]. The group values (see Table 1) are: $-10{\cdot}1$, $20{\cdot}9$ and $11{\cdot}4$, giving a total of $22{\cdot}2$ compared with the experimental value of $21{\cdot}5$ kcal/mole (see Table 2).

The Benson and Buss paper[5] contains bond contributions that permit estimation for N—N compounds of ΔH_f^0 to ± 6 kcal/mole, S^0 and C_p^0 at 25°C to ± 3 gibbs/mole. The accuracy of bond additivity is, therefore, not as good as that of group additivity, but it is often good enough for a first-order approximation. Even atom additivity can be used for S^0 and C_p^0 at

25°C and has about the same accuracy as bond additivity. The Benson and Buss paper also contains the atom contributions.

The next step in developing group additivity is to extend it to the liquid phase. As yet there are no groups available for heats of formation of liquids. However, group additivity has been used[7] for the estimation of the heat capacities of liquids at 25°C with an improvement in precision from about ±4 to better than 1·5 gibbs/mole.

If the required groups are not available, then C_p^0 of the liquid can be estimated to within a few cal/(mole-K) from ΔC_p^0 (liquid minus gas) = 12 cal/(mole-K) if the C_p^0 of the ideal gas is known. This rule breaks down for long, straight-chain molecules. The longer the chain, the less the rule is obeyed.

III. KINETICS AND THERMOCHEMISTRY

One of the main uses of thermochemistry, a well-developed science, is to help understand kinetics, which is still something of an art. The relationship between thermochemistry and kinetics has been treated in detail in a particularly useful monograph[8]. Therefore, only the main aspects will be discussed here, followed by specific cases of importance in N—N chemistry.

From Transition State Theory, the rate constant of an elementary chemical reaction is given by

$$k = \frac{\mathscr{K}T}{h}\exp(-\Delta G_T^{0\ddagger}/RT) \tag{1}$$

where k is the rate constant, $\mathscr{K}$ is Boltzmann's constant, h is Planck's constant, T is absolute temperature, R is the gas constant, and $\Delta G_T^{0\ddagger}$ is the standard free energy change at temperature T in going from the initial state to the transition state. From thermodynamics

$$\Delta G_T^{0\ddagger} = \Delta H_T^{0\ddagger} - T\Delta S_T^{0\ddagger} \tag{2}$$

where $\Delta H_T^{0\ddagger}$ and $\Delta S_T^{0\ddagger}$ are the enthalpy and entropy changes on going from the reactants to the transition state. The entropy and heat changes at T are related to those at 298 K by

$$\Delta H_T^{0\ddagger} = \Delta H_{298}^{0\ddagger} + \int_{298}^{T} \Delta C_p^{0\ddagger}\,\mathrm{d}T \tag{3}$$

$$\Delta S_T^{0\ddagger} = \Delta S_{298}^{0\ddagger} + \int_{298}^{T} \frac{\Delta C_p^{0\ddagger}}{T}\,\mathrm{d}T \tag{4}$$

where $\Delta C_p^{0\ddagger}$ is the standard state reaction heat capacity change, that is,

$$\Delta G_T^{0\ddagger} = \Delta H_{298}^{0\ddagger} + \int_{298}^{T} \Delta C_P^{0\ddagger}\, \mathrm{d}T - T\Delta S_{298}^{0} - T \int_{298}^{T} \Delta C_P^{0\ddagger}\, \mathrm{d}\ln T \qquad (5)$$

Although $\Delta C_p^{0\ddagger}$ is generally a function of temperature, for most purposes it may be taken as constant even over a wide temperature range. For example, Benson and co-workers[9] recently showed that for the reaction $A + BC \rightarrow AB + C$, where A, B and C are atoms, $\Delta C_P^{0\ddagger}$ is -3 ± 1 gibbs/mole from 200 to 4000 K. If $\Delta C_p^{0\ddagger}$ is constant then equation (5) can be integrated directly to give

$$\Delta G_T^{0\ddagger} = \Delta H_{298}^{0\ddagger} - T\Delta S_{298}^{0\ddagger} + \Delta C_p^{0\ddagger}[(T - 298) - T\ln(T/298)] \qquad (6)$$

and the art of estimating the rates of elementary chemical reactions becomes the art of measuring or estimating $\Delta H_{298}^{0\ddagger}$, $\Delta S_{298}^{0\ddagger}$ and $\Delta C_p^{0\ddagger}$. Since many elementary chemical reactions obey the Arrhenius law over a limited temperature range, their rate constants can be broken down into pre-exponential Arrhenius A-factors and activation energies:

$$k = A\exp(-E/RT) \qquad (7)$$

There is a useful relation between Arrhenius parameters and thermochemistry. For example, in the reaction pair,

$$\mathrm{R + H_2NNH_2 \longrightarrow RH + HNNH_2} \qquad (8)$$

it can be shown that the overall entropy change for the reaction ΔS_8^0 is related to the A-factor of the forward reaction A_8 and to the A-factor of the reverse reaction A_{-8} by $\exp(\Delta S_8^0/R) = A_8/A_{-8}$. Similarly, the heat of reaction ΔH_8^0 is related to the activation energy of the forward reaction E_8 and to the activation energy of the back reaction E_{-8} by $\Delta H_8^0 = E_8 - E_{-8}$. Therefore, if the Arrhenius parameters of the forward reaction are known, the Arrhenius parameters for the back reaction can be calculated exactly from the known or estimated thermochemical properties of the reactants and products with no assumption necessary for transition state properties.

An important example of the interaction between thermochemistry and kinetics is the kinetic method for determining bond dissociation energies. This subject has been reviewed in detail by Kerr[10]. To take a specific case, the N—N bond dissociation energy in Me_2NNMe_2 is the enthalpy, ΔH_9^0, of the reaction

$$\mathrm{Me_2NNMe_2 \longrightarrow Me_2N + NMe_2} \qquad (9)$$

TABLE 1. Group values[a] for calculating thermochemical properties of hydrazine and substituted hydrazines

Group	$\Delta H^0_{f\,298}$ (kcal/mole)	S^0_{298} (gibbs/mole)	C^0_p/(gibbs/mole)					
			300 K	400 K	500 K	600 K	800 K	1000 K	1500 K
N—(N)(H)$_2$	11·4	29·1	6·1	7·4	8·4	9·3	10·5	11·5	13·2
N—(N)(C)(H)	20·9	9·6	4·8	5·8	6·5	7·0	7·8	8·3	9·0
N—(N)(C)$_2$	29·2	−13·8							
N—(N)(C$_B$)[b](H)	22·1	11·4							
C—(N)(H)$_3$	−10·1	30·4	6·2	7·8	9·4	10·8	13·0	14·8	17·6
C—(N)(C)(H)$_2$	−6·6	9·8	5·2	6·9	8·3	9·4	11·1	12·3	
C—(N)(C)$_2$(H)	−5·2	−11·7	4·7	6·3	7·6	8·4	9·6	10·2	
C—(N)(C$_3$)	−3·2	−34·1	4·4	6·2	7·3	7·9	8·5	8·5	
C$_B$—N	−0·5	−9·7	4·0	5·2	5·9	6·3	6·5	6·6	

[a] Data from Reference 6.
[b] C$_B$ is a carbon atom in benzene.

TABLE 2. Thermochemical properties of hydrazine and substituted hydrazines[a]

Compound	$\Delta H^0_{f_{298}}$ (kcal/mole)	S^0_{298} (gibbs/mole)	C^0_p/(gibbs/mole)						
			300 K	400 K	500 K	600 K	800 K	1000 K	1500 K
NH_2NH_2	22·8	54·4	12·6	15·1	16·9	18·3	20·6	22·3	26·0
CH_3NHNH_2	21·5	66·5	17·1	21·0	24·3	27·1	31·3	34·6	39·8
$CH_3NHNHCH_3$	21·8	74·4	22·0	27·2	31·8	35·6	41·6	46·2	53·2
$(CH_3)NNH_2$	20·0	71·8	23·5				47·8		
$PhNHNH_2$	48·7	87·0	30·9				63·1		

[a] This table was compiled from data in References 2 to 6.

That is, $D(Me_2N—NMe_2) = \Delta H_9^0 \approx E_9 - E_{-9} = E_9$ with the reasonable assumption that E_{-9} is negligible. Thus the N—N bond dissociation energy in Me_2NNMe_2 can be obtained by measuring the activation energy for reaction (9).

IV. HYDRAZINE AND SUBSTITUTED HYDRAZINES

The thermochemical properties of hydrazines and substituted hydrazines are listed in Table 2 and the group values for estimating them are in Table 1. The emphasis of this section will be on the kinetics based on that thermochemistry. Golden and co-workers[11] have concluded that monomethylhydrazine can decompose in three ways, each unimolecular:

N—N *Homolysis*

$$CH_3NHNH_2 \longrightarrow CH_3NH + NH_2 \qquad (10)$$

Four-centre concerted loss of ammonia

$$CH_3NHNH_2 \longrightarrow CH_2{=}NH + NH_3 \qquad (11)$$

Four-centre concerted loss of molecular hydrogen

$$CH_3NHNH_2 \longrightarrow CH_3N{=}NH + H_2 \qquad (12)$$

Let us consider each type of reaction in turn and then come back to the individual hydrazines.

A. N—N *Homolysis*

The key to N—N homolysis is the thermochemistry of the radical fragments produced by breaking the N—N bond. The thermochemical properties of the amino and substituted amino radicals are given in Table 3. The relevant group values for calculating them are given in Table 4.

The bond dissociation energy (also called the bond strength) of $R^1R^2NNR^3R^4$ is given by the heat of reaction, which is equal to the sum of the heats of formation of the product radicals minus the heat of formation of the reactant hydrazine; that is $D(R^1R^2N—NR^3R^4) = \Delta H^0 = \Delta H_f^0(R^1R^2N) + \Delta H_f^0(R^3R^4N) - \Delta H_f^0(R^1R^2NNR^2H_4)$. The values of the N—N bond strengths calculated from heats of reaction for a number of substituted hydrazines are given in Table 5. The results in Table 5 show the progressive weakening in the N—N bond as the nitrogens are increasingly substituted. The Arrhenius *A*-factors for N—N fission in the aliphatic hydrazines are probably[12] of the order of $10^{16.5}$ s^{-1}. In the phenyl hydra-

TABLE 3. Thermochemical properties of the amino and substituted amino radicals[a]

Compound	$\Delta H^0_{f_{298}}$ (kcal/mole)	S^0_{298} (gibbs/mole)	C^0_p/(gibbs/mole)						
			300 K	400 K	500 K	600 K	800 K	1000 K	1500 K
$\dot{N}H_2$	46	46·5	8·0	8·2	8·5	8·8	9·5	10·2	11·6
$CH_3\dot{N}H$	45	59·0	11·3	13·3	15·1	16·8	19·5	21·7	25·3
$(CH_3)_2N^{\cdot}$	38	66·2	15·7	19·2	22·7	25·7	30·5	34·3	39·9
$Ph\dot{N}H$	55	75·3	24·8	32·8	39·2	44·2	51·4	56·3	
$Ph\dot{N}CH_3$	53·5	83·3	30·2	39·3	47·0	53·3	62·6	69·0	

[a] This table was compiled from data in References 11 and 12.

TABLE 4. Group values for calculating thermochemical properties of amino and substituted amino radicals[a]

Group	ΔH_f^0 (kcal/mole)	S_{298}^0 (gibbs/mole)	C_p^0/(gibbs/mole)						
			300 K	400 K	500 K	600 K	800 K	1000 K	1500 K
N·—(H)(C)	55·3	30·8	5·1	5·5	5·7	6·0	6·5	6·9	7·7
C—(N·)(H)$_3$	−10·1	30·4	6·2	7·8	9·4	10·8	13·0	14·8	17·6
C—(N·)(C)(H)$_2$	−6·6	9·8	5·3	6·9	8·3	9·4	11·1	12·3	
C—(N·)(C)$_2$H	−5·2	−11·7	4·7	6·3	7·6	8·4	9·6	10·2	
C—(N·)(C)$_3$	−3·2	−34·1	4·4	6·2	7·3	7·9	8·5	8·5	
N·—(C)$_2$	56·2	9·8	3·3	3·6	3·9	4·1	4·5	4·7	4·7
C_B[b]—N·	−0·5	−9·7	4·0	5·2	5·9	6·3	6·5	6·6	
N·—(C_B)(H)	39·0	27·4	4·6	5·4	6·0	6·4	7·2	7·7	8·6
N·—(C_B)(C)	45·6	7·2	3·9	4·2	4·7	5·0	5·6	5·8	5·9

[a] This table was compiled from data in References 11 and 12.
[b] C_B is a carbon atom in benzene.

TABLE 5. N—N Bond strengths (in kcal/mole) in hydrazine and substituted hydrazines, $R^1R^2N—NR^3R^4$

R^3R^4N	R^1R^2N			
	NH_2	CH_3NH	$(CH_3)_2N$	PhNH
NH_2	70	66	61	51
CH_3NH	66	64	59	49
$(CH_3)_2N$	61	59	54	44
PhNH	51	49	44	34

zines, stiffening in the transition state must reduce the entropy increase from the reactant hydrazine to the transition state, resulting in an *A*-factor of around $10^{15.5}\ s^{-1}$.

B. Four-centre Concerted Loss of Ammonia

Golden and co-workers[11] have observed deamination of monomethylhydrazine in their very low pressure pyrolysis experiments.

$$\underset{21.5}{CH_3NHNH_2} \longrightarrow \underset{26}{CH_2{=}NH} + \underset{-11}{NH_3}$$

The heat of formation of methyleneimine has not been measured, but it can be estimated to be 26 kcal/mole from $\Delta H_f^\circ(C_2H_5CH{=}NC_2H_5) \sim 0$ kcal/mole. Substituting values for the heats of formation of monomethylhydrazine and ammonia, the heat of reaction is –6·5 kcal/mole; that is, the deamination is 6·5 kcal/mole exothermic.

Unsymmetrical dimethylhydrazine may also undergo deamination. The heat of formation of the resultant imine was estimated by group additivity, giving a heat of reaction of –13 kcal/mole (i.e., exothermic)

$$\underset{20}{(CH_3)_2NNH_2} \longrightarrow \underset{18}{CH_2{=}NCH_3} + \underset{-11}{NH_3}$$

Symmetrical dimethylhydrazine can eliminate methylamine in a reaction that is almost thermoneutral

$$\underset{21.8}{CH_3NHNHCH_3} \longrightarrow \underset{26}{CH_2{=}NH} + \underset{-5.5}{CH_3NH_2}$$

Trimethylhydrazine can undergo two possible eliminations

$$\underset{19\cdot8}{(CH_3)_2NNHCH_3} \longrightarrow \underset{18}{CH_2{=}NCH_3} + \underset{-5\cdot5}{CH_3NH_2}$$

$$\underset{19\cdot8}{(CH_3)_2NNHCH_3} \longrightarrow \underset{26}{CH_2{=}NH} + \underset{-4\cdot5}{(CH_3)_2NH}$$

Loss of methylamine is 7·3 kcal/mole exothermic and is therefore favoured thermochemically over loss of dimethylamine, which is 1·7 kcal/mole endothermic.

The last member of the series, tetramethylhydrazine, has only one possible route for amine elimination. It is 4·5 kcal/mole exothermic.

$$\underset{18}{(CH_3)_2NN(CH_3)_2} \longrightarrow \underset{18}{CH_2{=}NCH_3} + \underset{-4\cdot5}{(CH_3)_2NH}$$

The heats of reaction unfortunately do not tell the whole story. Concerted four-centre eliminations usually have significant activation energy barriers in addition to any endothermicity to be overcome. Benson and Haugen[13] have developed a method for calculating the activation energies for the reverse of four-centre eliminations, that is, the addition of HX to olefins. They considered the cases where X = F, Cl, Br or I. Similar principles have since been applied to the addition of HONO to nitro-olefins to give activation energies for the reverse reaction, the decomposition of nitro-alkanes. The basis of the calculational method is semi-ion pairs. Although these principles can no doubt be applied to the deamination of hydrazines, there are some complicating features, and the calculations are outside the scope of this review. However, in the absence of further results it is fairly safe to generalize the activation energies for deamination of the alkyl hydrazines will be in the range 54 ± 10 kcal/mole with *A*-factors in the region of 10^{13} s^{-1}.

C. Four-centre Concerted Loss of Molecular Hydrogen

Golden and co-workers[11] showed that monomethylhydrazine can eliminate molecular hydrogen by a four-centre mechanism:

$$\underset{21\cdot5}{CH_3NHNH_2} \longrightarrow \underset{42\cdot1}{CH_3N{=}NH} + \underset{0}{H_2}$$

The heat of formation of the azo compound was calculated from group additivity (see later section on azo compounds), making the reaction

20·6 kcal/mole endothermic. The analogous reaction for hydrazine is

$$\underset{22\cdot8}{NH_2NH_2} \longrightarrow \underset{50\cdot2}{HN{=}NH + H_2}$$

and is 27·4 kcal/mole endothermic.

V. AZO COMPOUNDS

There are at least four reasons for studying the thermochemistry of azo compounds. First, there is the intrinsic interest in the data itself and its 'fit' by group additivity. Second, the unimolecular decomposition of azoalkanes has long been an active subject for theoreticians and experimentalists in the field of gas-phase kinetics. Thirdly, azo compounds are a convenient source of alkyl radicals:

$$RN_2R \longrightarrow R + RN_2 \tag{13}$$

$$RN_2 \longrightarrow R + N_2 \tag{14}$$

RN_2 may be unstable with respect to R and N_2; that is, the second step may have negative heat of reaction, the excess energy going into vibrational energy of R. It is clearly important from a kinetics point of view to know

TABLE 6. Heats of formation (in kcal/mole) of azo compounds measured by Engel and Wood[4] and estimated by group additivity

Compounds		ΔH_f liquid	ΔH_{vap}	ΔH_f^0 gas obs	ΔH_f^0 gas est	Difference obs − est
$[CH_3CH_2CH_2N]_2$	*trans*	3·2	9·6	12·8	12·0	0·8
$[(CH_3)_2CHN]_2$	*trans*	0·4	8·5	8·9	8·4	0·5
	cis[a]	0·4	10·5	10·9	9·4	1·5
$[(CH_3)_3CN]_2$	*trans*	−18·4	8·7	−9·7	−9·6	−0·1
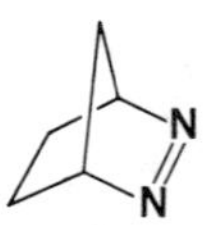		36·8	10·5	47·3	48·6	−1·3
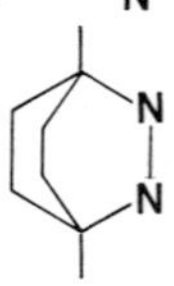		5·6	16·7	22·2	21·7	0·5

[a] Dr. Engel has pointed out that the data for this compound are preliminary.

exactly how much energy is likely to go into R. Fourthly, there is the question of the energy difference between the *cis* and *trans* isomers.

First the data: as Benson and Walsh have commented[6] '(Thermochemical) information on azo compounds is particularly lacking.' The only azo compounds for which experimentally determined heats of formation have been published are difluorodiazine FNNF, diimide HNNH and azoisopropane $(CH_3)_2CHNNCH(CH_3)_2$. However, Engel and Wood[4] recently measured the heats of combustion and vaporization of a large group of azo compounds. Their results are summarized in Table 6.

Engel and Wood's value for the heat of formation of azoisopropane is 11 kcal/mole lower than Coates and Sutton's 1948 value[14] of 19·4 kcal/mole. A consequence of this change is that ΔH_f^0 of azomethane calculated from graph values (Table 7) becomes 34·0 kcal/mole instead of the previous value[3] of 44·7 kcal/mole. A further consequence is that the heat of hydrogenation of azomethane then becomes 12·4 kcal/mole, which is very different from the heat of hydrogenation of diimide (27·4 kcal/mole).

TABLE 7. Group values for calculating heats of formation of trans azoalkanes and azo radicals[a]

Group	ΔH^0_{f298} (kcal/mole)
N_A[b]—$(N_A\cdot)$(C)	56·4
N_A—(C)	27·0
C—$(N_A)(H)_3$	−10·1
C—$(N_A)(C)(H)_2$	−6·0
C—$(N_A)(C)_2(H)$	−3·4
C—$(N_A)(C)_3$	−3·0

[a] This table was compiled from data in References 4 and 12, after considerable discussion with Benson. There are some typographical errors on p. 32 of O'Neal and Benson's monograph[12]. The groups [C—(N)$(H)_2$], [C—(N)-(H)(C)] and [C—(N)$(C)_2$] in the third column of Table V-8 are all missing a carbon atom. They should read [C—(C)(N)$(H)_2$], [C—(N)(H)$(C)_2$] and [C—(N)$(C)_3$]. Also in the text $DH_1^0 - DH_2^0$ should read $DH_1^0 + DH_2^0$. On p. 31, the sign in front of [C—(N_A)-$(H)_x(C)_{3-x}$] should be minus.

[b] N_A is an azo nitrogen atom.

TABLE 8. Thermochemical properties of aliphatic azo compounds[a]

Compound		$\Delta H^0_{f_{298}}$ (kcal/mole)	S^0_{298} (gibbs/mole)	C^0_p/(gibbs/mole)						
				300 K	400 K	500 K	600 K	800 K	1000 K	1500 K
FNNF	*trans*	19·4								
	cis	16·4								
HNNH	*cis*	50·2	52·2	8·8	9·8	10·9	11·9	13·5	14·8	16·9
CH_3NNCH_3	*trans*	34·0	71·0	20·1				35·9		
[*n*-PrN]$_2$	*trans*	12·0								
[*i*-PrN]$_2$	*trans*	8·4	103·3	41·9						
	cis	10·4								
[*t*-BuN]$_2$	*trans*	–9·6	109·3	53·9						
	cis	10·0								

[a] This table is compiled from data in References 2, 4, 15 and 16.

However, Engel and Wood's data are very self-consistent as shown by the reasonable agreement between their experimental results and those calculated by group additivity for the cyclic compounds (Table 7). Therefore the more recent work by Engel and Wood is favoured. Rossini's planned

TABLE 9. Maximum energy of second alkyl radical in decomposition of azoalkanes

Azoalkane	Minimum bond strength (kcal/mole) D(RN_2—R)	Heat of reaction (kcal/mole) for $RN_2R \rightarrow 2R + N$	Maximum bond strength (kcal/mole) D(R—N_2)	Maximum energy (kcal/mole) of R
$[CH_3N]_2$	52·5	34·0	−18·5	18·5
$[EtN]_2$	50·0	30·0	−20·0	20·0
$[i\text{-}PrN]_2$	47·5	26·6	−20·9	20·9
$[t\text{-}BuN]_2$	43·5	24·0	−19·5	19·5
$[PhCH_2N]_2$	37·4	4·0	−33·0	33·0

measurement of the heat of combustion of azomethane should prove conclusive.

The thermochemical properties of the aliphatic azo compounds are summarized in Table 8. The 'new' value of the heat of formation of azomethane, 34 kcal/mole, compared with the 'old' value of 44 kcal/mole means that the decomposition of the CH_3N_2 radical has an activation energy of −18 kcal/mole rather than the previous *E* of −28·3 kcal/mole. The maximum excess energy in the alkyl radical R from decomposition of $RN_2\cdot$ is given in Table 9.

It should be emphasized that all the above thermochemical kinetic considerations apply to gas-phase reactions only. In the liquid phase, a very wide range in rates of decomposition has been observed[17]. This range is so wide that there may be concerted two-bond rupture in these systems rather than the one-bond rupture in simple azo compounds.

The question of the stability of *cis* and *trans* isomers is an interesting one. In the case of difluorodiazine FN=NF, ΔH^0_{f298} of the *trans* isomer was found[15] to be +19·4 kcal/mole, that is 3·0 kcal/mole less stable than an 'active' isomer, probably the *cis* compound. On the other hand, Mill and Stringham[16] have concluded that *cis* azoisobutane is about 20 kcal/mole less stable than the *trans* isomer in the liquid phase. With azoisopropane, in the liquid state, Engel and Wood[4] have found that the *cis* and *trans* forms have the same ΔH^0_{f298}, namely, 0·45 kcal/mole. The same authors have measured heats of vaporization of 10 kcal/mole for the *cis* and 7·9 kcal/mole for the *trans*. In the gas phase, the *cis* form is therefore about 2 kcal/mole less stable than the *trans*.

VI. ACKNOWLEDGMENTS

I am greatly indebted to P. S. Engel and J. L. Wood for communicating their results prior to publication and for their helpful comments, and to A. S. Rodgers for drawing my attention to Wood and Engel's work. It is also a pleasure to acknowledge the work done by S. W. Benson in reviewing the manuscript and analysing the data on azo compounds. I thank H. E. O'Neal and T. Mill for helpful comments, Kitta Reeds for editing, and Joyce Hayes for typing the manuscript.

VII. REFERENCES

1. D. R. Stull, E. F. Westrum and G. C. Sinke, *The Chemical Thermodynamics of Organic Compounds*, John Wiley, New York, 1969.
2. *JANAF Thermochemical Tables* (Ed. D. R. Stull), Dow Chemical Company, Midland, Michigan, U.S.A.
3. J. D. Cox and G. Pilcher, *Thermochemistry of Organic and Organometallic Compounds*, Academic Press, New York, 1970.
4. P. S. Engel and J. L. Wood, unpublished work, 1973.
5. S. W. Benson and J. H. Buss, *J. Chem. Phys.*, **29**, 546 (1958).
6. S. W. Benson, F. R. Cruickshank, D. M. Golden, G. R. Haugen, H. E. O'Neal, A. S. Rodgers, R. Shaw and R. Walsh, *Chem. Rev.*, **69**, 279 (1969).
7. R. Shaw, *J. Chem. Eng. Data*, **14**, 461 (1964).
8. S. W. Benson, *Thermochemical Kinetics*: *Methods for the Estimation of Thermochemical Data and Rate Parameters*, John Wiley, New York, 1968.
9. S. W. Benson, D. M. Golden and R. Shaw, *Quarterly Progress Report on EPA Grant No. R-800798* October 1972.
10. J. A. Kerr, *Chem. Rev.*, **66**, 465 (1966).
11. D. M. Golden, R. K. Solly, N. A. Gac and S. W. Benson, *Int. J. Chem. Kinetics*, **4**, 433 (1972).
12. S. W. Benson and H. E. O'Neal, *Kinetic Data on Gas-phase Unimolecular Reactions*, NSRDS-NBS 21, U.S. Government Printing Office, Washington, D.C., 1970.
13. S. W. Benson and G. R. Haugen, *J. Amer. Chem. Soc.*, **87**, 4036 (1965).
14. G. E. Coates and L. E. Sutton, *J. Chem. Soc.*, 1187 (1948).
15. G. T. Armstrong and S. Marantz, *J. Chem. Phys.*, **38**, 169 (1963).
16. T. Mill and R. S. Stringham, *Tetrahedron Letters*, **23**, 1853 (1969).
17. J. W. Timberlake and M. L. Hodges, American Chemical Society Meeting, Boston, U.S.A., 1972.

CHAPTER **4**

Preparative procedures

J. W. TIMBERLAKE and J. C. STOWELL

Louisiana State University in New Orleans,
New Orleans, Louisiana, U.S.A.

I. PREPARATION OF HYDRAZINES

A detailed review of the literature to 1967 is available[1], as well as several other general reviews[2,3,4]. Most of the recent developments are in the area of alkyl substituted hydrazines where many special methods have been devised to overcome the lack of specificity in simple alkylation. The aryl and acyl hydrazine chemistry has been well developed before 1967 and is covered in the reviews[1,2].

A. *Alkylations*

I. Alkylation of hydrazine and alkyl hydrazines

Alkyl hydrazines may be prepared by direct alkylation of hydrazine with halides, sulphates, epoxides, aziridines, and active olefins. In many cases the method is limited due to the inability to control the number and location of alkyl groups introduced, giving mixtures of mono and polysubstituted hydrazines[5]. The use of excess hydrazine affords moderate yields of mono-substitution products; for example, one mole of *n*-butyl bromide with 10 moles of hydrazine hydrate gave *n*-butylhydrazine in 71% yield[6]. *N,N*-Dialkylhydrazines may be prepared in fair yield using a smaller excess of hydrazine, e.g. 1 mole of benzyl chloride and 2 moles of hydrazine gave *N,N*-dibenzylhydrazine (35%)[7]. Tetrabenzylhydrazine was prepared in 14% yield using equimolar amounts of chloride and hydrazine[7].

Steric hindrance may overcome the preference for *N,N*-alkylation and lead to *N,N'*-dialkyl products, e.g. treatment of 1-phenyl-2-bromopropane with an equimolar amount of isopropylhydrazine gave mostly *N*-(1-phenyl-2-propyl)-*N'*-isopropylhydrazine in ~29% yield[8].

Several other recent examples are illustrated in equations (1–4).

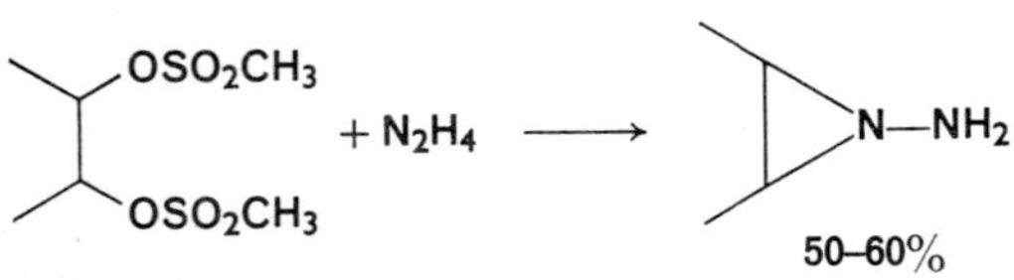

(1)[9]

$$(CH_3)_2NCH_2CH_2CH_2Cl + \underset{\text{excess}}{N_2H_4} \xrightarrow[K_2CO_3]{100°C} \underset{58\%}{(CH_3)_2NCH_2CH_2CH_2NHNH_2} \quad (2)^{10}$$

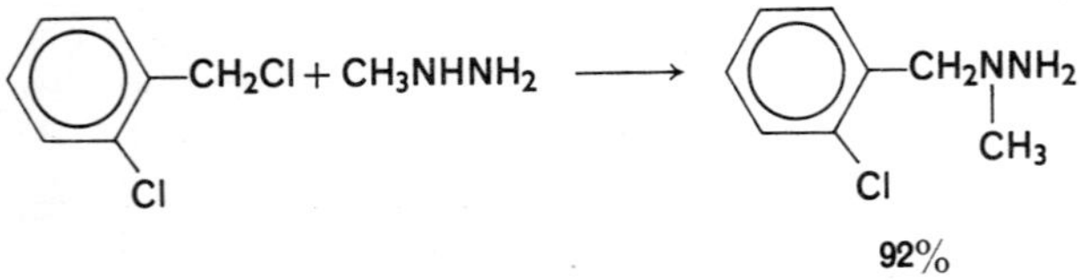

(3)[11]

$$(CH_3)_2NNH_2 + CH_2{=}CHCN \longrightarrow \underset{53–56\%}{(CH_3)_2NNHCH_2CH_2CN} \quad (4)^{10}$$

2. Alkylation of *N,N′*–diacylhydrazines[1]

N,N′-Disubstituted hydrazines may be prepared by alkylating a blocked hydrazine such as *N,N′*-dibenzoylhydrazine[12], *N,N′*-diformylhydrazine[13], or *N*-methyl-*N,N′*-bis(benzyloxycarbonyl)hydrazine[14]. The dialkylated

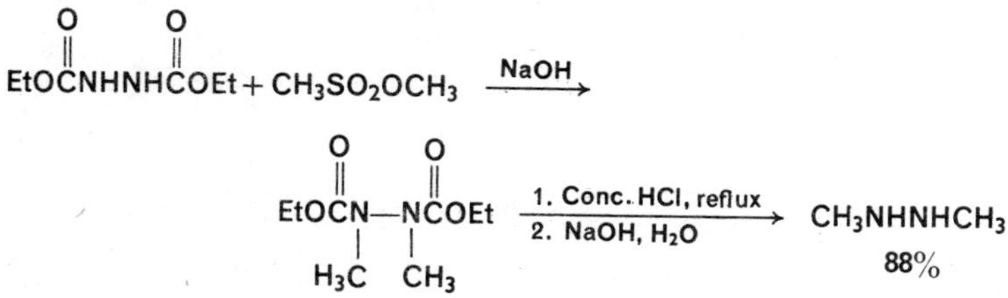

(5)[15]

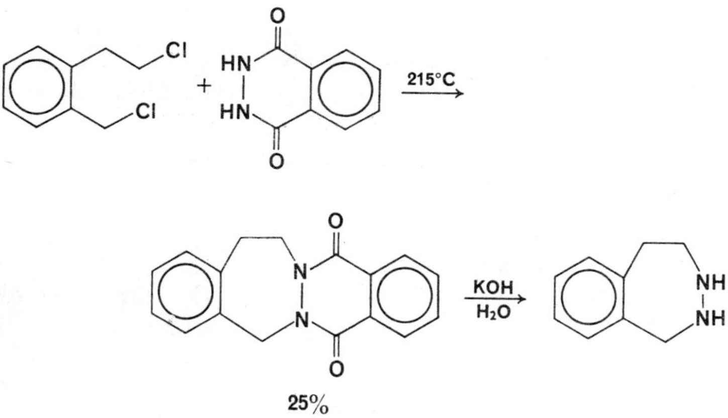

(6)[16]

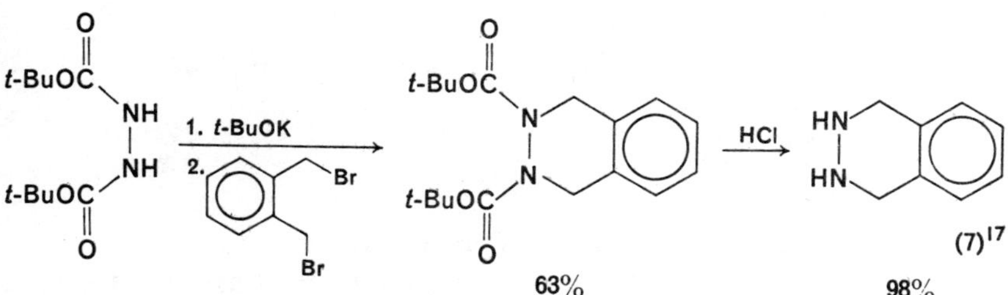

(7)[17]

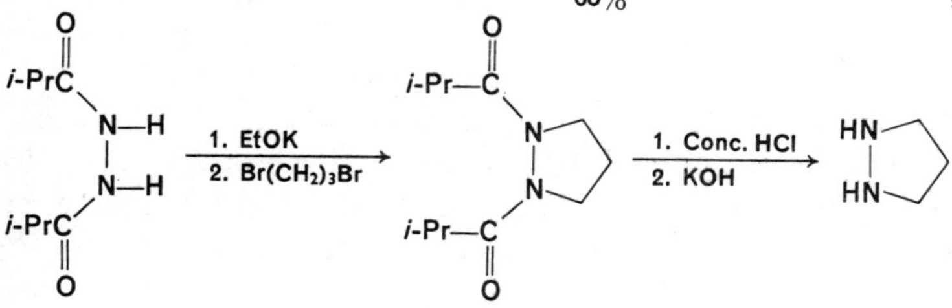

(8)[18]

product may then be hydrolysed to the *N,N'*-dialkylhydrazine as in equations (5–8).

A related method is the alkylation of hydrazide dianions produced from azo compounds. Treatment of diethyl azodicarboxylate with two equivaents of potassium followed by benzyl chloride gave diethyl *N,N'*-dibenzylbicarbamate in 60% yield[19].

Diethyl azodicarboxylate may also be dialkylated with di-*t*-butylmercury[20].

3. Alkylation of acylhydrazines

N-Methyl-*N*-(3,3-diphenylallyl)hydrazine has been prepared by successive alkylations of *t*-butyl carbazate[21].

B. *Lithium Reagent Addition to Azo Compounds*

Addition of *t*-butyllithium to 2,3-diazabicyclo[2.2.1]hept-2-ene followed by quenching with ammonium chloride affords 2-*t*-butyl-2,3-diazabicyclo-[2.2.1]heptane in over 90% yield[22]. Similar additions are known with azobenzene[23] and *t*-butylazidoformate[24].

C. *Grignard Reagent Addition to Hydrazones*

Heating a solution of a formaldehyde hydrazone in THF with a Grignard reagent followed by hydrolysis affords substituted hydrazines[25].

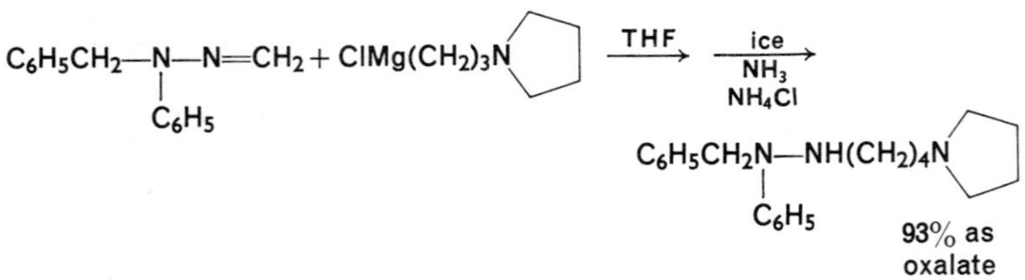

(9)

D. *Cycloaddition Reactions of Azo Compounds*

Diels–Alder reactions between diethyl azodicarboxylate and dienes to give hydrazines have been known since 1925[1,26]. More recently it has been shown that *cis* azo compounds are much more dienophilic[27]. Cyclic examples include indazolone[28]; 3,6-pyridazinedione; 1,4-phthalazinedione[29,30]; and 4,5-dihydro-3,6-pyridazinedione[30] which are thermally unstable well below room temperature. They react rapidly with dienes at low temperature and so are prepared and used *in situ* to form hydrazines[28–32] see equation 10.

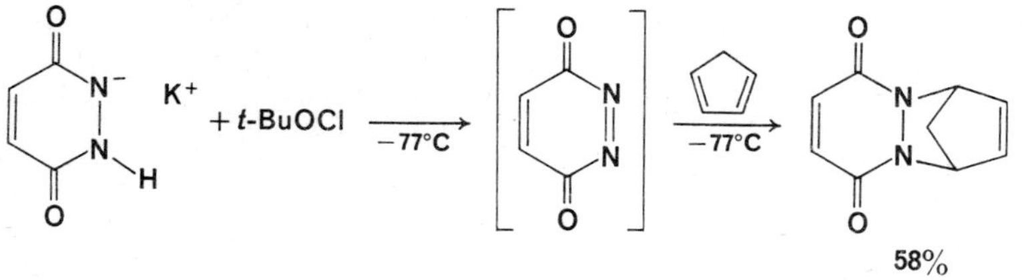

(10)

The more stable isolable dienophiles, *N*-phenyltriazoline-3,5-dione[33,34] and *N*-methyltriazoline-3,5-dione react rapidly at low temperatures with many olefins such as cyclopentadiene, butadiene, cycloheptatriene (11),

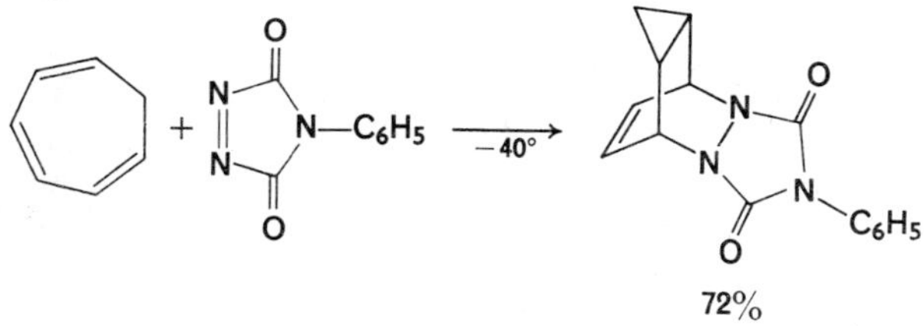

(11)

anthracene, styrene and cyclooctatetraene[35–39]. A homo Diels–Alder reaction occurs with bicycloheptadiene (12)[36]. 2 + 2 Cycloadditions occur with

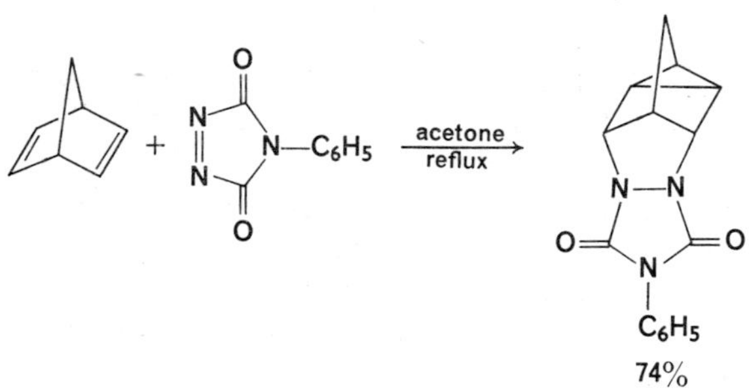

(12)

some olefins to afford diazetidines[40–42], while others give more complex products[42,43].

The cyclic adducts from these azo compounds may in many cases be hydrolysed to give novel cyclic hydrazine compounds (13, 14)[44–49]. In some cases, base hydrolysis was not successful. An alternative method involves bis(2,2,2-trichloroethyl)azodicarboxylate as dienophile, and finally reductive cleavage by zinc–copper couple in methanol with *in situ* mercuric oxide oxidation to the azo compound[49a]. A further alternative is to use

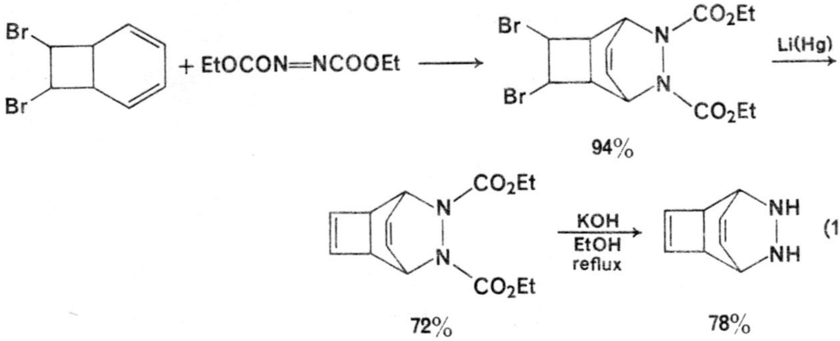

(13)[44]

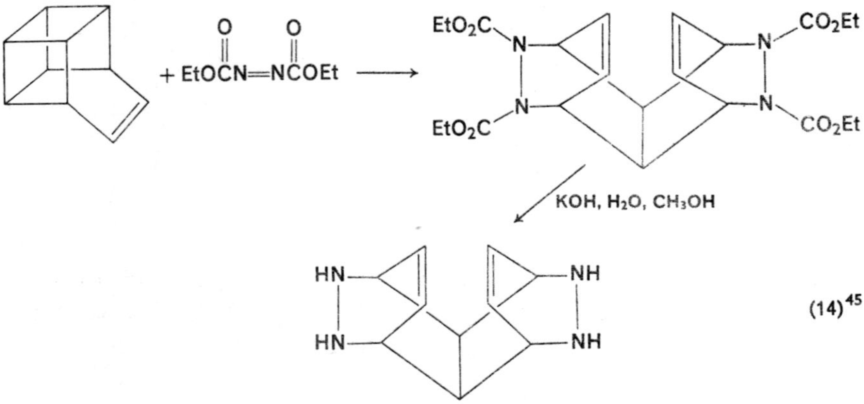

(14)[45]

1,3,4-thiadiazoline-2,5-dione as the dienophile. The resulting adducts may be hydrolysed at 25°C with dilute lithium hydroxide[49b].

E. Reductions

1. Reduction of *N*-nitroso compounds

Nitrosation of secondary amines followed by reduction is a common route to *unsym*-dialkyl-, alkylaryl-, and diarylhydrazines. A great variety of reducing agents has been used including zinc in acetic acid, sodium in ethanol, aluminium, lithium aluminium hydride, and hydrogenation catalysed by Pd, Pt, Rh, and Raney Ni[1]. A simple electrolytic process for preparative scale reduction of *N*-nitrosamines has been demonstrated on 20 examples (15)[50].

$$i\text{-}Bu_2NNO \xrightarrow[\text{HCl—H}_2\text{O—EtOH}]{\text{electrolysis}} i\text{-}Bu_2NNH_2 \qquad (15)$$

78% isolated
95% by analysis

Lithium aluminium hydride reductions of nitrosamines often show a long induction period and then a sudden vigorous reaction[51,52]. In some cases this may be avoided by slow addition to the nitrosamine to a refluxing suspension of LiAlH4 in ether while monitoring gas evolution (equation 16)[11].

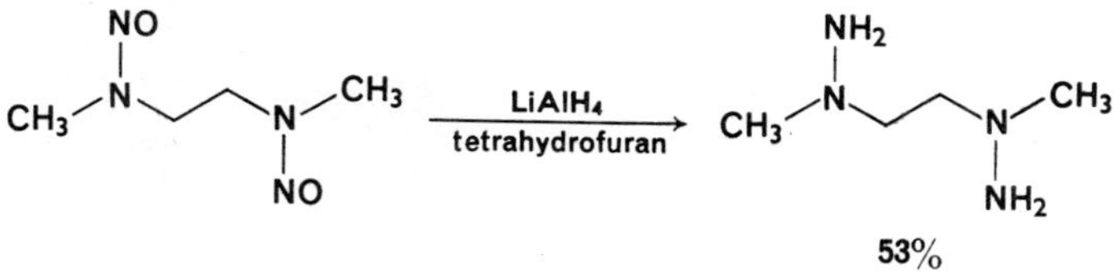

(16)[53]

In cases where one alkyl group is *t*-butyl, LiAlH$_4$ fails but Zn in acetic acid[54] or aluminium amalgam in wet ether[55] give reduction to the hydrazines. Attempts to reduce 4-[4-methoxybenzyl(nitroso)amino]benzoic acid by the usual methods failed, but this and several other examples were readily reduced in ethanol at 0°C with zinc dust, ammonium carbonate and aqueous ammonia[56].

2. Reduction of azines, hydrazones, and azo compounds

Reduction of C=N bonds has been carried out using catalytic hydrogenation[57], lithium in liquid ammonia[58], LiAlH$_4$[10,55,59], NaBH$_4$[60], Na, Na amalgam and Raney Ni-hydrazine[1]. Recent applications of these methods are illustrated in 17–20.

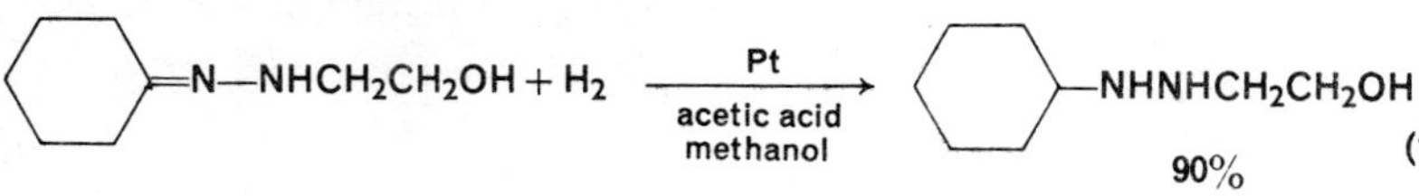

(17)[57]

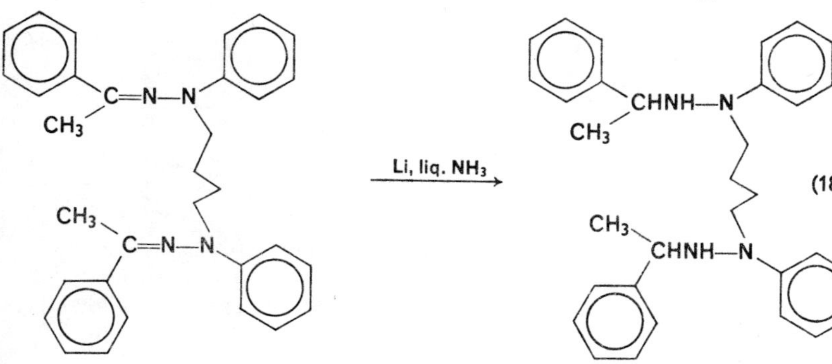

(18)[58]

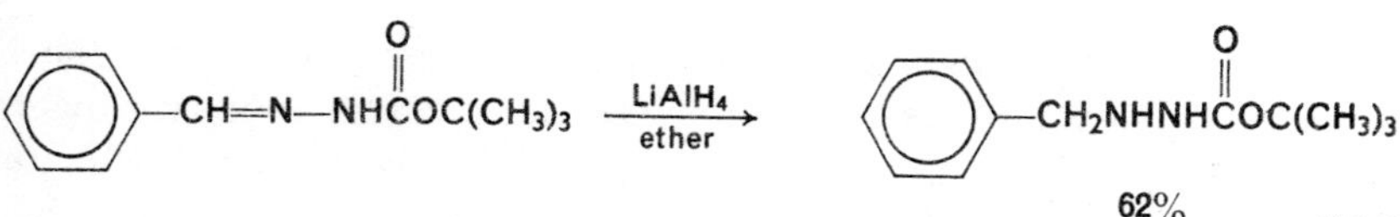

(19)[55]

Seven examples of di and trialkylhydrazines were prepared via the hydrazones from 1-acetyl-1-methylhydrazine. The hydrazone formation was quantitative and the remaining steps were carried out in 38 to 82% overall yield[60] (equation 20). Similar compounds were prepared directly

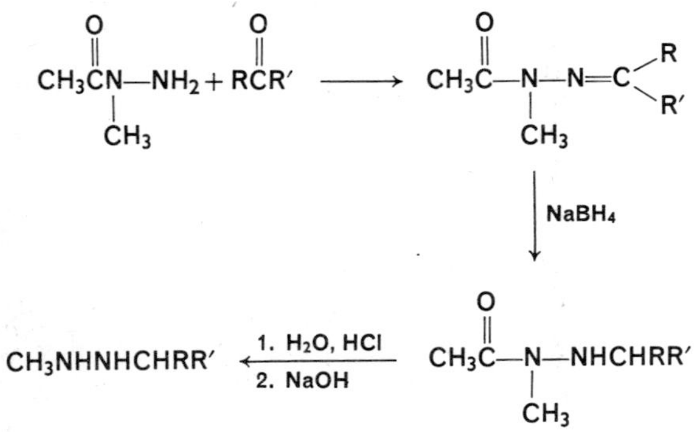

(20)

from the hydrazines in a 'one flask' procedure using diborane (equation

$$CH_3NHNH_2 + C_6H_5CHO \xrightarrow[\text{2. } B_2H_6\text{, ambient temp.}\ \text{3. HCl gas}]{\text{1. Acetic acid}\ CH_3OCH_2CH_2OCH_2CH_2OCH_3} CH_3NHNHCH_2C_6H_5 \cdot n\ HCl \tag{21}$$

21)[61]. Using an excess of the carbonyl compound and sodium cyanoborohydride as the reducing agent, tetraalkylhydrazines may be prepared (equations 22, 23)[62].

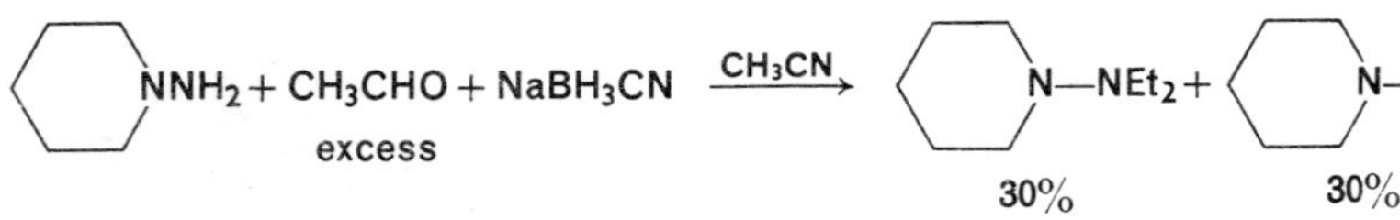

(22)

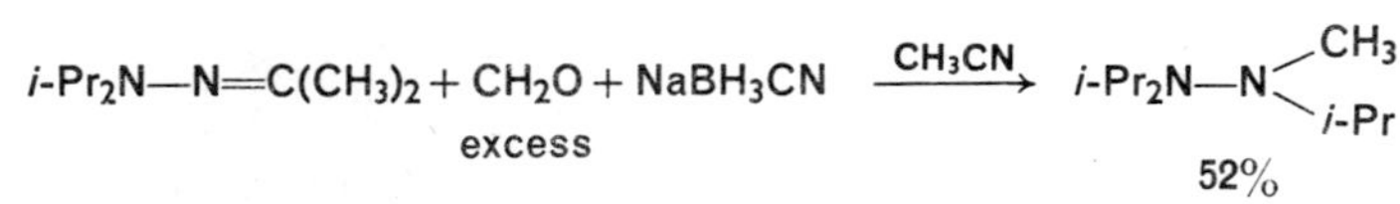

(23)

Numerous azines were hydrogenated to hydrazines and then oxidized to azo compounds without characterization of the intermediates. These are discussed under azo synthesis, Section II, A.

Since some azo compounds are available from precursors other than hydrazo compounds or from oxidation reactions where the intermediate hydrazo compound is not readily isolated (Section II, E) it is useful to reduce azo compounds. Azomethane was reduced using zinc in aqueous sodium hydroxide in 90% yield[63]. Symmetrical azo compounds where the alkyl groups are less bulky than *sec*-butyl may be reduced in good yield using excess hydrazine hydrate and Raney-Ni catalyst in ethanol at 60°C, e.g. *N,N'*-di-*n*-butylhydrazine was prepared in 84% yield[64]. The very bulky *N,N'*-di-*t*-butylhydrazine is available in 92% yield from the Pd catalysed hydrogenation of the azo compound[65].

3. Reduction of acylhydrazines

Acylhydrazines may be reduced to alkylhydrazines using $LiAlH_4$ or diborane (equations 24–26). Some acylhydrazines of the type

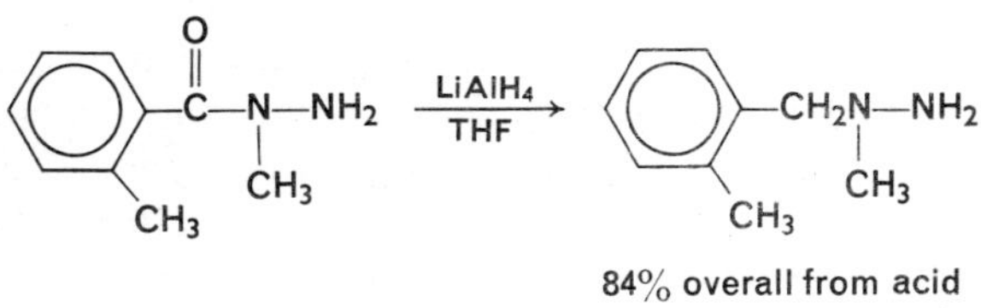

(24)[11]

$R^1CONHNR^2R^3$ give low yields or no alkylhydrazines when $LiAlH_4$ in tetrahydrofuran is used[67], but diborane at higher temperatures is satisfactory. For example, *N,N'*-dipropionylhydrazine was treated with diborane

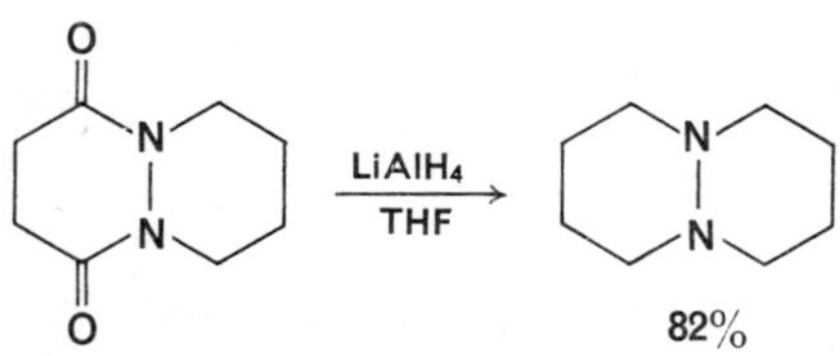

(25)[18]

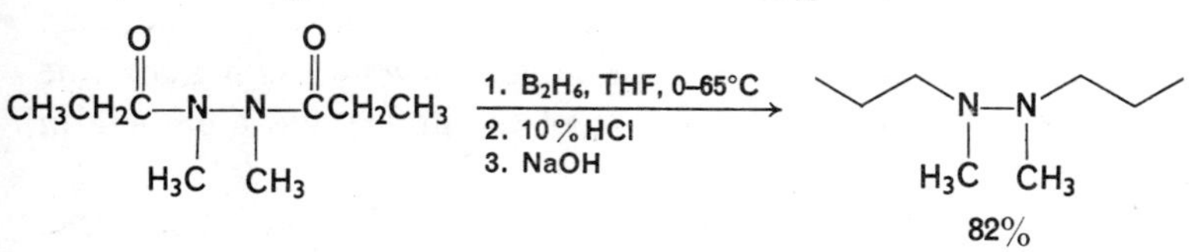

(26)[66]

at 0–5°C, heated to 134°C and then hydrolysed with dil. HCl to give *N,N'*-di-*n*-propylhydrazine (65%)[66]. When methylal is used as solvent, $LiAlH_4$ gives good results, e.g. *N*-acetyl-*N'*-phenylhydrazine was converted to *N*-ethyl-*N'*-phenylhydrazine (89%)[7,59]. Formylhydrazines and ethoxycarbonylhydrazines are reduced easily even in ether[68]. An *N*-aminonaphthalimide was reduced to an *N*-aminodihydroisoquinoline in 50% yield using $NaBH_4$[55].

Spialter and co-workers reported seven examples of 1,2-disubstituted hydrazines prepared by $LiAlH_4$ reduction of both acyl and hydrazone functions in one step[69]. This method is, of course, limited to *N*-primary alkyl-*N'*-primary or secondary alkyl hydrazines, but is particularly useful for preparation of *N,N'*-dialkylhydrazines with two different substituents.

$$CH_3\overset{\overset{O}{\|}}{C}NHNH_2 + CH_3COCH_3 \longrightarrow CH_3\overset{\overset{O}{\|}}{C}NHN{=}C(CH_3)_2 \xrightarrow[\text{ether}]{LiAlH_4} \underset{61\%}{CH_3CH_2NHNHCH(CH_3)_2} \quad (27)$$

F. N-Amination

Primary, secondary and tertiary amines are capable of displacing good leaving groups from ammonia derivatives. Primary and secondary amines in 4- to 8-fold excess react with chloramine in aqueous solution to give hydrazines in 40–75% yield[70]. Similar results were obtained when gaseous chloramine was passed into an excess of anhydrous primary or secondary amines[70]. Recently five examples of secondary amines were converted to hydrazines in 91 to 99% yield by slow addition of excess gaseous chloramine containing ammonia and nitrogen to a solution of the amine in methanol[71].

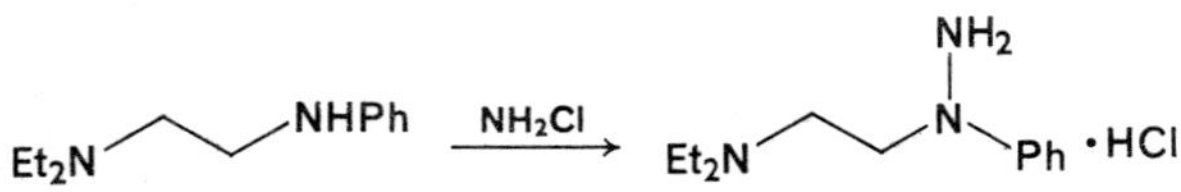

95%

(28)

Chloramine and excess tertiary amines give *N,N,N*-trialkylhydrazinium salts in high yield calculated on chloramine consumption[70, 72]. A simplified procedure allows generation of the chloramine *in situ* by passing chlorine gas diluted with nitrogen into a solution of the amine and ammonia in isopropyl alcohol[73].

Anionic nitrogen nucleophiles also may give good yields with chloramine.

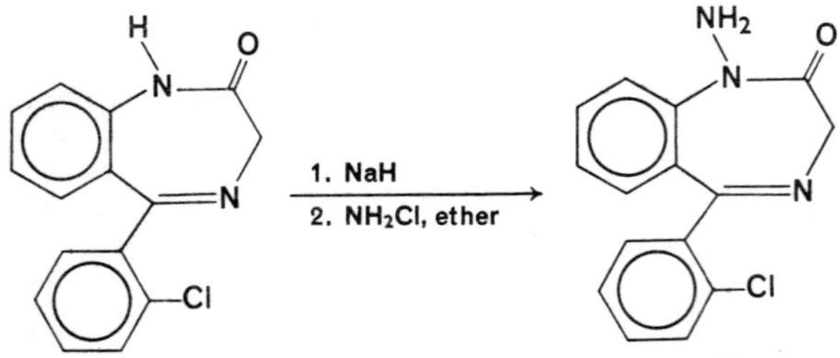

78%

(29)[74]

Hydroxylamine-*O*-sulphonic acid with potassium carbonate or KOH is used similarly[1,75], for example butylamine was converted to butylhydrazinium sulphate in 49–56% yield[75a].

Other derivatives which offer more organic solubility than hydroxylamine-*O*-sulphonic acid and more stability than chloramine are *O*-mesitoylhydroxylamine[76] and *O*-2,4-dinitrophenylhydroxylamine[77]. These give hydrazines in high yield from amines or amide anions, and allow isolation of products (equation 31), which are unstable toward the basic conditions used with hydroxylamine-*O*-sulphonic acid. *O*-Mesitylenesulfonylhydroxylamine has been used similarly, but caution must be observed since an explosion was reported[78].

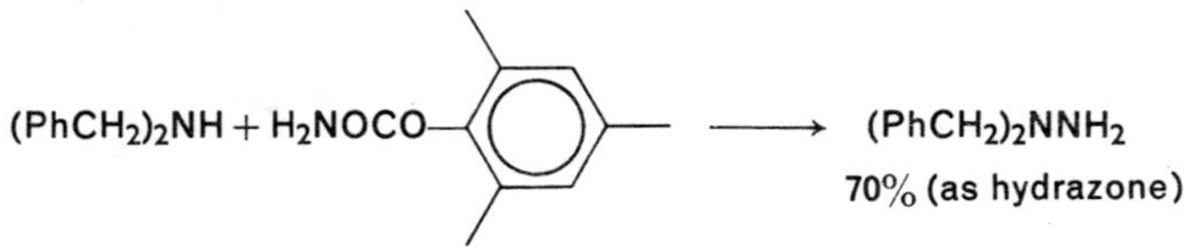

(30)[76]

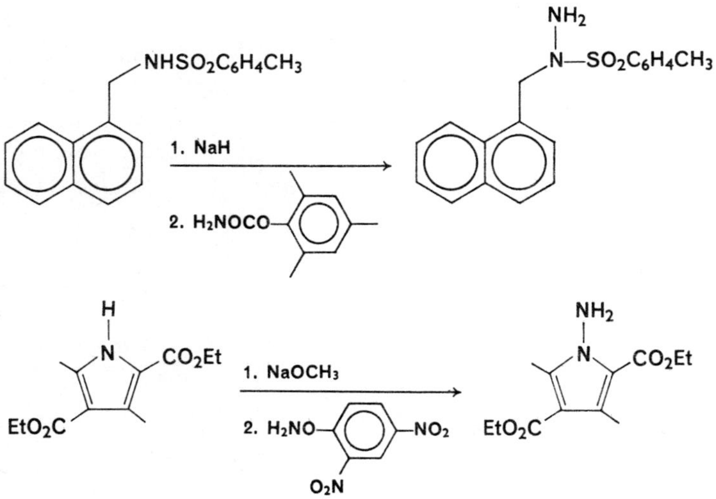

(31)[76]

(32)[77]

t-Butylamine was converted to *t*-butylhydrazine in 51% yield using the aminating reagent **1**[78a].

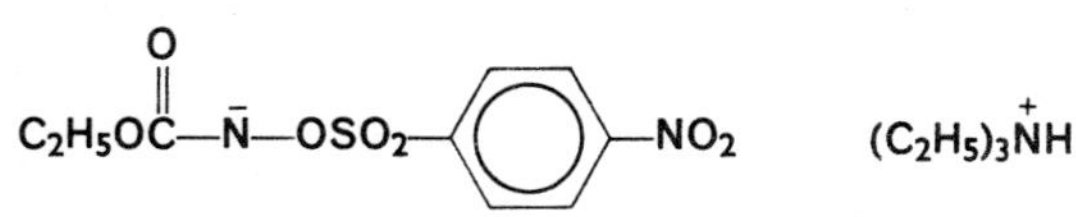

(1)

The yields are much lower when *N*-amination is used in the preparation of *N*,*N'*-dialkylhydrazines, thus $MeNHOSO_3H$ and isopropylamine gave *N*-methyl-*N'*-isopropylhydrazine in 8% yield[79]. The use of *N*-chloro primary amines in place of chloramine usually gives elimination to aldimines rather than hydrazines. This may be successful when there is no β-hydrogen, for example *N*,3,5-tri-*t*-butylaniline was treated with butyllithium and *N*-chloro-*N*,3,5-tri-*t*-butylaniline to give the tetrasubstituted hydrazine[79a]. Intramolecular reactions are somewhat more favourable such as the cyclization of *N*-chloro-1,4-diaminobutane which gives pyrazolidine in 33% yield[79b]. Likewise, 2,4,6-trimethyltriazine with *t*-butyl hypochlorite and sodium carbonate gives 2,4,6-trimethyl-1,3,5-triazabicyclo[3.1.0]-hexane in 20–40% yield[80]. To improve the coupling of primary amines, a modification was devised in which they are held together initially as ureas or sulphamides. This method is the subject of Section I, H.

G. Hydrazines from Diaziridines

The reaction of chloramine with amines usually requires a 4- to 8-fold excess of amine over aminating agent to limit the reaction of aminating agent with the desired hydrazine product. This problem may be avoided with primary amines by first converting the amine to an imine, second treating with chloramine, and finally hydrolysing the resulting diaziridine[81,82], e.g. see equation (33). The intermediate diaziridine is insensitive toward excess aminating agent.

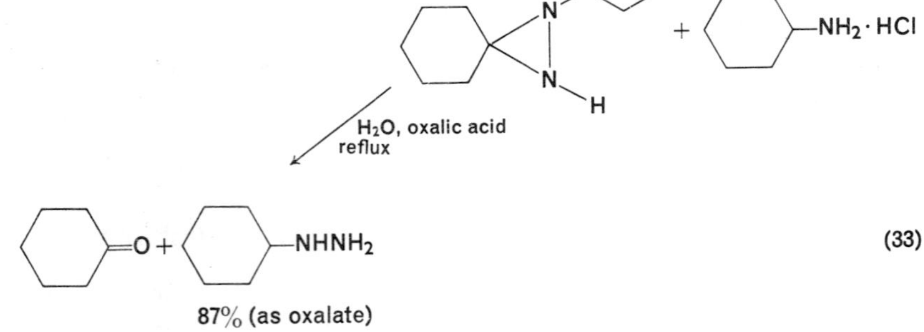

(33)

An alternative procedure using cyclohexanone, hydroxylamine-*O*-sulphonic acid, and excess alkylamine affords the alkylhydrazines without isolation of any intermediates[83].

Another advantage of this method is the applicability of *N*-chloro primary amines to the preparation of *N,N'*-dialkylhydrazines[84].

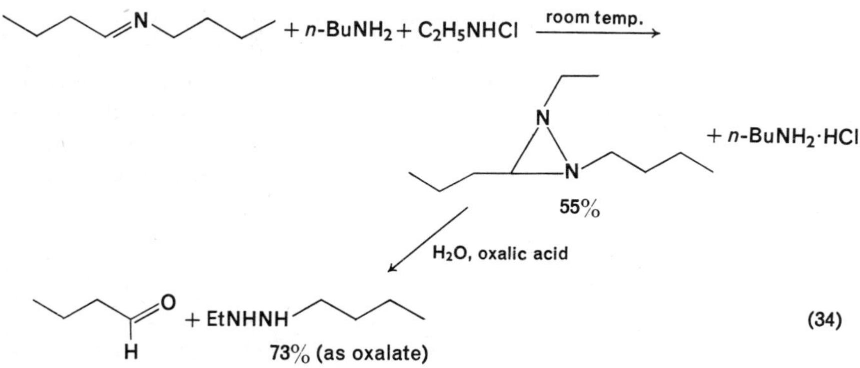

(34)

H. Hydrazines from Ureas and Sulphamides

Early workers reported that treatment of phenylurea with hypochlorous acid followed by sodium hydroxide gave a small amount of *p*-chlorophenylhydrazine[85]. More recently methylhydrazine was obtained in 70% yield by treatment of methylurea with NaOH and NaOCl at 80°C[86]. When *N*-mono-substituted acyl- or alkylureas are first treated with chlorine gas in water, the *N*-chloroureas are obtained in good yield. Treatment of these with aqueous sodium hydroxide at 5°C gives the mono-substituted hydrazines in excellent yield. By this process, *t*-butyl-, *t*-amyl, and 1-methylcyclohexylhydrazine were prepared in 92, 76 and 52% yields respectively. When methanolic sodium methoxide is used as the base, the corresponding methyl carbazates are obtained[86a]. Unsubstituted urea was shown to be proceeding via two mechanisms, Hofmann rearrangement and a nitrogen analogue of the Favorskii reaction[87]. In the case of *N,N'*-di-*t*-alkylureas where only the Favorskii mechanism is possible, the very stable intermediate diaziridinones have been isolated. These intermediates are readily hydrolysed to afford the hydrazines in good yield[88]. Although the method is successful only for *N,N'*-di-*t*-alkylhydrazines, e.g. *t*-butyl or *t*-amyl[89], it is complementary to the facile reduction of azines.

The ready availability of ureas from isocyanates and amines suggests that this method will be useful for preparation of hydrazines with two

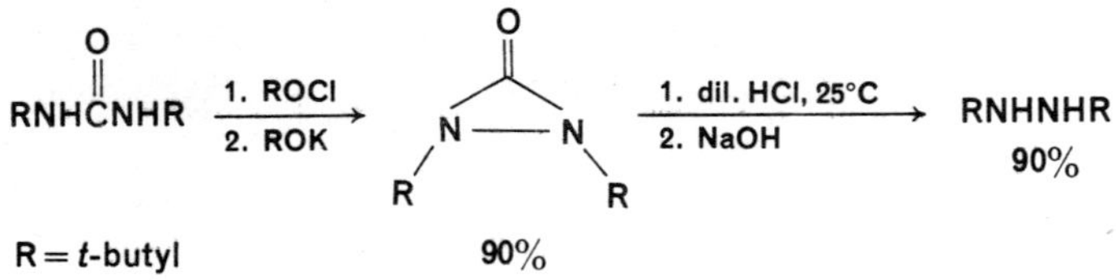
(35)

different substituents. The intermediate diaziridinones are also available from the reaction of nitroso compounds with isocyanides[90].

In a related method, the leaving group is the oxygen of an oxaziridine. Treatment of pentamethyleneoxaziridine with an isocyanate followed by sodium hydroxide leads to a hydrazine, possibly via a transient diaziridinone (equation 36)[91,92]. Treatment with other bases leads to carbazates

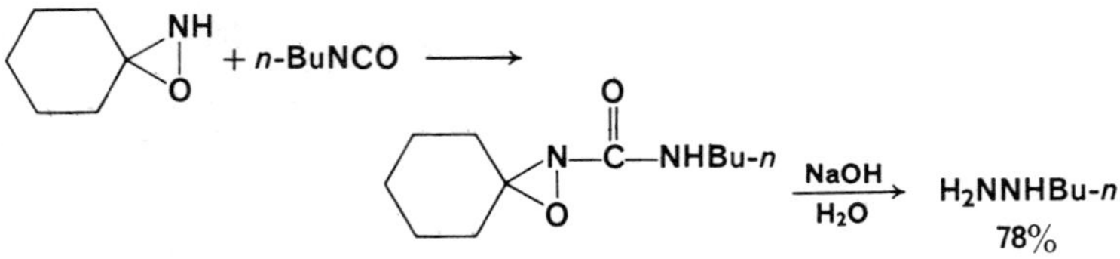
(36)

and semicarbazides. The leaving group may also be a sulphonate, e.g. *N*-methyl-*N'*-hydroxyurea was treated with chlorosulphonic acid followed by aqueous NaOH to give methylhydrazine (71 %). Similarly *N*-methyl-*N*-hydroxyurea was treated with tosyl chloride and then aqueous NaOH to give methylhydrazine (80 %), in this case necessarily via a diaziridinone[87].

Treatment of *N*,*N*-dimethylurea with aqueous NaOH and NaOCl affords *N*,*N*-dimethylhydrazine (82 %)[87].

Reaction of sulphamides with NaOH and NaOCl followed by acid hydrolysis affords hydrazines. There is no limitation to *t*-alkyl substituents, and examples of monosubstituted, *N*,*N*-disubstituted, and *N*,*N'*-disubstituted hydrazines have been prepared (equations 37–38)[93,94]. The

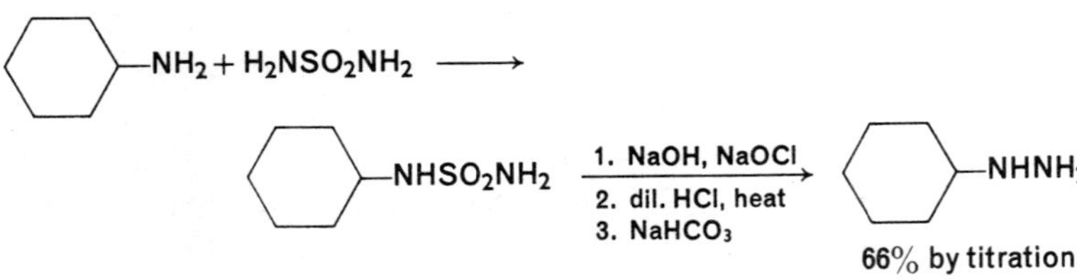
(37)

$$C_2H_5NHSO_2NHC_2H_5 \xrightarrow[\text{2. dil. HCl, heat; 3. NaHCO}_3]{\text{1. NaOH, NaOCl}} \underset{96\%\ \text{by titration}}{C_2H_5NHNHC_2H_5} \qquad (38)$$

lower yields of hydrazo compounds in some cases, e.g. *N,N'*-diisopropylhydrazine, 57%, may be due to further oxidation of some of the hydrazine by part of the first mole of NaOCl used[95]. In the case of the *N,N'*-di-*t*-butyl compound, a good yield was obtained by using 2 moles of NaOCl and then reducing the resulting azo compound[65].

N,N-Dialkylsulphamides behave similarly:

$$Et_2NSO_2NH_2 \xrightarrow[\text{2. HCl; 3. NaHCO}_3]{\text{1. NaOH, NaOCl}} \underset{\text{79\% by titration}}{Et_2NNH_2} \qquad (39)$$

More examples of these reactions and a discussion of the intermediates are given under azo compounds in Section II, E.

II. PREPARATION OF ACYCLIC AZOALKANES

Extensive reviews are available covering the methods of preparation of aliphatic and aromatic azo compounds before 1967[2,96–99]. For the most part this review concentrates on recent methods with reference to older ones that appear to be of general utility.

The preparation of cyclic azo compounds is covered in another chapter and will not be reviewed here.

A. *Oxidation of 1,2-Dialkylhydrazines (Preparation from Aldehydes and Ketones)*

It should be noted that any synthesis of a 1,2-dialkylhydrazine potentially provides an azoalkane by oxidation (see Section I).

Perhaps the most general method for preparation of azoalkanes with primary and/or secondary alkyl groups is the scheme outlined in equation 40[100–102]. A variety of symmetrical azoalkanes (**1**, R and R′ = alkyl or

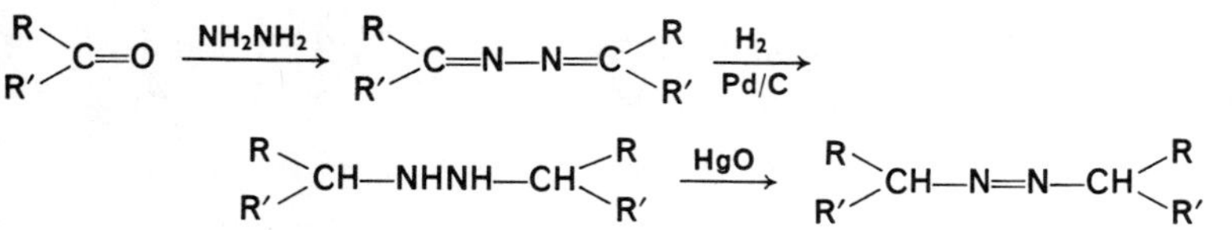

(40)

(**1**)

aryl; R = H, and R′ = alkyl or aryl) have been prepared using this basic condensation–reduction–oxidation approach, although, the overall yields are generally less than 50%. For example, even using what is reported to be an improved oxidation procedure employing freshly prepared yellow mercuric oxide, yields for 1,1′-diarylazoethanes (based on hydrazines) ranged from 10% (**1**, R = CH_3; R′ = *m*-$CF_3C_6H_4$) to 65% (**1**, R = CH_3; R′ = *p*-$CH_3C_6H_4$)[101]. The low yields, however, could be the result of formation of diastereomers of different solubility[103]. A series of *p*-substituted phenylazoamethanes (**1**, R = *p*-X-C_6H_4; R′ = H), where no diastereomers are possible, have been prepared in overall yields of ~40% from the benzaldehydes. The oxidation step using commercially available yellow mercuric oxide proceeds in 60%[104] yield.

A similar approach is useful for making mixed or unsymmetrically substituted azoalkanes. In cases where the hydrazines are readily available (equation 41[105] and 42[106]) the synthesis is straightforward, although, low yields are again apparently unavoidable.

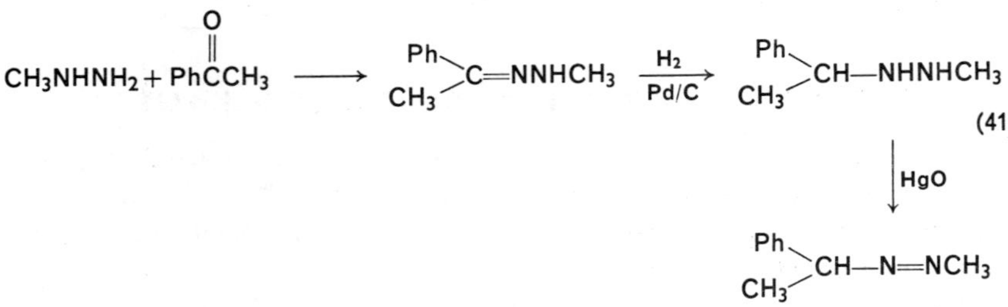

(41)

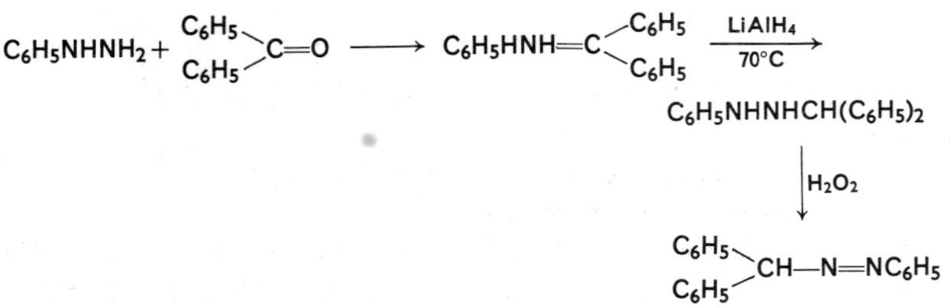

(42)

For less accessible hydrazines a more elaborate scheme was developed by Overberger and DiGiulio[107] (equations 43 and 44) to prepare a number of examples of substituted azoalkanes **2** and **3** where R, R′, R″ and R‴ are combinations of H and alkyl groups. More recently, modifications of this

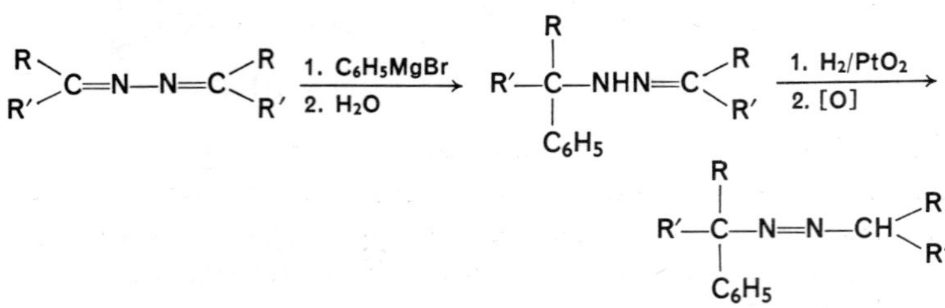

(43)

(2)

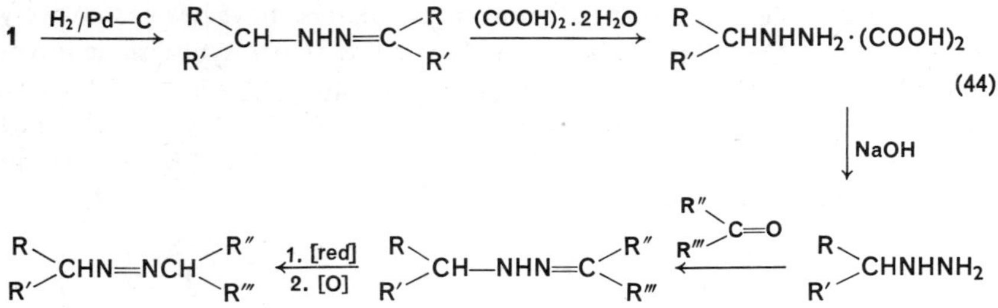

(44)

(4) (3)

procedure have been used to prepare **5**[108], **6**[109], **7**[110], and **8**[111]. Both **7** and **8** were obtained optically pure by prior resolution of the intermediate hydrazine (**3**, $R = C_6H_5$; $R' = CH_3$).

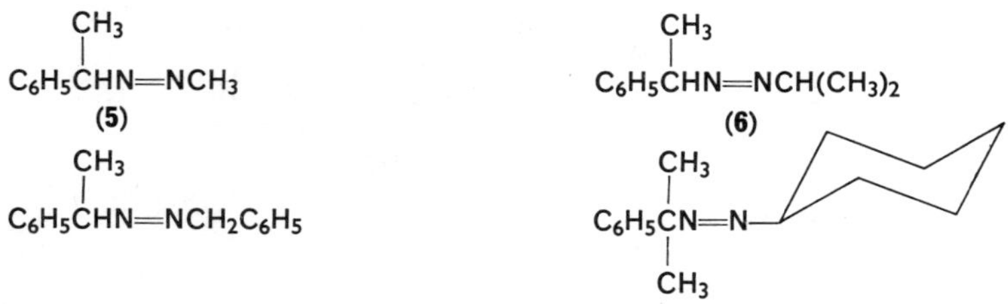

(5) (6) (7) (8)

A highly successful procedure has been used to make a series of five unsymmetrical azoalkanes (equation 45); the 1-acyl-2-alkylidenehydrazines (**9**) were formed in 75–95%, the reduction to alkylhydrazines (**10**) in 61–74% and the oxidation to azoalkanes (**11**) in 23–59% yield[69].

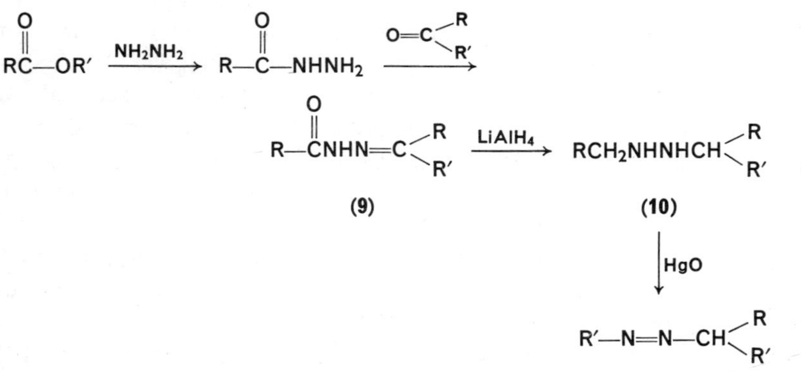

(45)

(9) (10) (11)

Azo-bis-nitriles, a derivatized form of azoalkanes, have been extensively researched by Overberger and co-workers[97]. The basic synthesis involves the addition of HCN to the azine (**12**, equation 46) followed by oxidation. The cyano group can be transformed to an ester[112–114] (**13**, 74%)[114] and reduced to the alcohol (**14**, 67%)[114] and re-esterified (**15**, 77%)[114] without disrupting the azo-linkage.

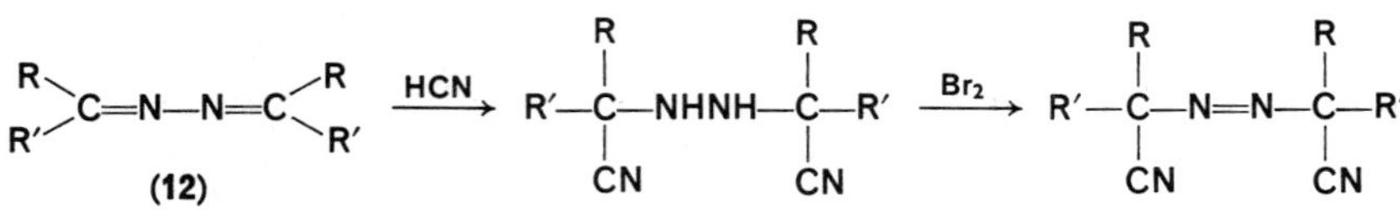

(12)

(46)

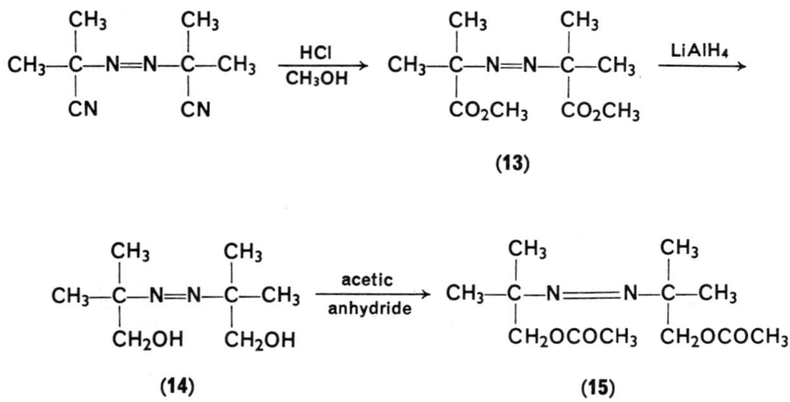

(13)

(14) (15)

(47)

Other interesting α-substituted azoalkanes are known, most of them prepared by displacement reactions on chloroazoalkanes (**17**)[115–116]. For example, acetone azine (**16**) undergoes 1,4-addition with chlorine at −60°C to give 2,2′-dichloro-2,2′-azopropane (**17**) in 85% yield[116]. Treatment of **17** with the appropriate nucleophile yields the functionalized derivatives **18**a–j[114–120]. It should be noted, however, that not all nucleophiles yield substituted derivatives. When **17** is treated with phenyl magnesium bromide[115], sodium iodide[115] or sodium nitrite[121], acetone azine (**16**) is the only isolable product.

In view of the ease with which chloroazoalkanes are prepared, the modification discovered by Rüchardt (equation 49) illustrates a useful method of preparing tertiary azoalkanes[122].

Another α-substituted azoalkane of surprising stability has been reported by Schmitz and Ohme (equation 50)[123].

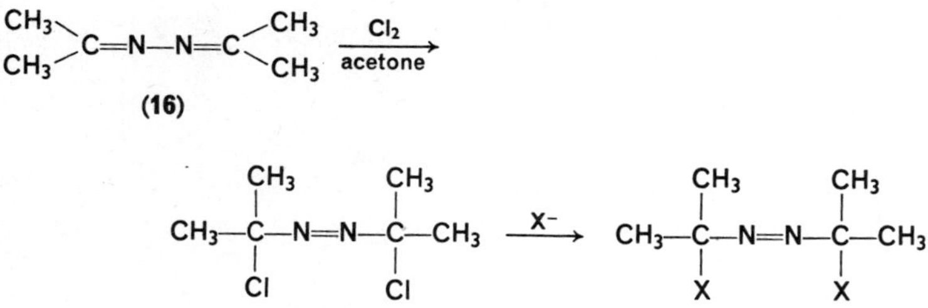

(16) (17) (18a–j) (48)

a, X = SCN[116]
b, X = CH_3CO_2[116]
c, X = CH_3COS[116]
d, X = C_6H_5S[116]
e, X = CN[116]
f, X = N_3[116]
g, X = C_6H_5O[117]
h, X = CH_3[118]
i, X = CH_3O[119]
j, X = CH_3S[120]

(49)

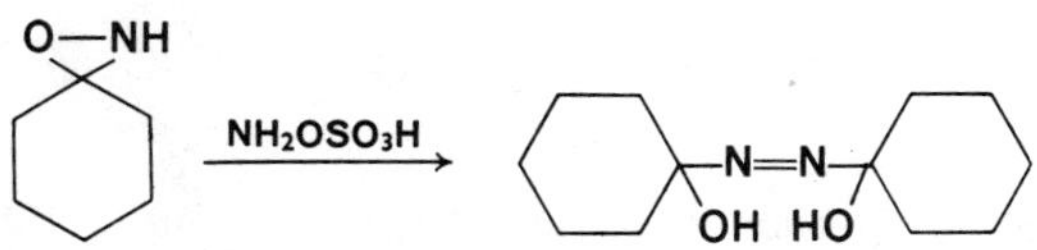

(50)

Diazenes of the type RN═NH have frequently been cited as reaction intermediates but have only recently been isolated and characterized[124]. Although there are several methods of synthesis, one of the most convenient is outlined in equation 51. Condensation of carbonyl compounds with methyl carbazate (**19**) gives carbomethoxyhydrazones (**20**) which can be reduced, oxidized and decarboxylated (equation 51)[125]. No yields for this method were reported.

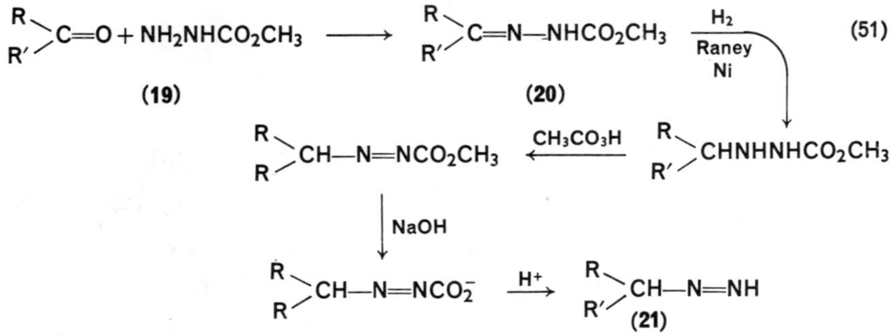

(51)

a, $R = R' = CH_3$
b, $R = H, R' = CH_3$
c, R and $R' = —(CH_2)_5—$
d, $R = (CH_3)_3C, R' = CH_3$

B. Oxidation of Amines

One of the most interesting recent methods for synthesis of symmetrical tertiary substituted azoalkanes was introduced by Stevens in 1961[126]. The reaction provided the first useful preparation of 2,2′-azoisobutane (**22**) in 48% yield via the oxidative coupling of *t*-butylamine with iodine pentafluoride (equation 52). This compound had been prepared previously in 13%

$$(CH_3)_3CNH_2 \xrightarrow[\text{CH}_2\text{Cl}_2,\ 0°\text{C}]{\text{IF}_5} (CH_3)_3CN{=}NC(CH_3)_3 \quad (52)$$

Pyridine $\qquad$ **(22)**

yield (equation (53)[127] by a method reported to be 'laborious and inconsistent'[114].

$$(CH_3)_3CNH_2 \xrightarrow{\text{KOBr}} (CH_3)_3NHBr \xrightarrow{\text{Ag}_2\text{O}} \mathbf{22} \quad (53)$$

The IF_5 method has since been used to prepare many different alkyl substituted azoalkanes **23**a–l. Although the reaction involves only one step, the workup procedure is somewhat involved (owing to the formation of both iodine and HF) and probably accounts for the highly variable yields, 4% for **23**l and 60% for **23**h.

This reagent can only be used on tertiary alkyl primary amines. With primary alkyl amines the powerful dehydrogenating ability of IF_5 leads to nitriles (equation 54)[126].

$$RCH_2NH_2 \xrightarrow[\text{pyridine}]{IF_5,\ CH_2Cl_2} RC{\equiv}N\ (35\text{–}90\%) \qquad (54)$$

There are other reports of successful oxidative couplings of amines. The reaction in equation 55 goes in 86% but probably is a special case[134].

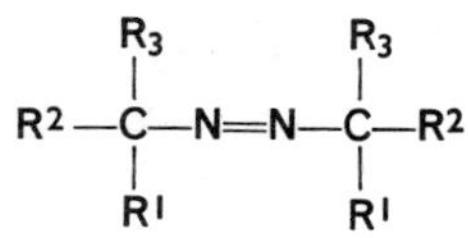

(23)

a, $R^1 = R^2 = CH_3$; $R_3 = C_6H_5$[128]

b, $R^1 = R^2 = CH_3$; $R_3 = X{-}C_6H_4$, $X = CH_3$, C_2H_5, $(CH_3)_2CH$, $(CH_3)_3C$, Br[129]

c, $R^1 = CH_3$, $R^2 = C_2H_5$, $R_3 = C_6H_5$[130–131]

d, $R^1 = CH_3$, $R^2 = (CH_3)_2CH$, $R^3 = C_6H_5$[130–131]

e, $R^1 = CH_3$, $R^2 = (CH_3)_3C$, $R^3 = C_6H_5$[130–131]

f, $R^1 = CH_3$, $R^2 = C_2H_5$, R^3 = Cyclohexyl[131]

g, $R^1 = R^2 = CF_3$, $R^3 = C_6H_5$[132]

h, $R^1 = R^2 = CH_3$, R^3 = Cyclo-C_3H_5[118]

i, $R^1 = CH_3$ $R^2 = R^3$ = cyclo-C_3H_5[118]

j, $R^1 = R^2 = R^3$ = cyclo-C_3H_5[118]

k, $R^1 = R^2 = CH_3$, $R^3 = CH_3CH_2CH_2$[133]

l, $R^1R^2R^3C$ = 1-adamantyl[133]

A similar reaction with *t*-butylamine gives no 2,2′-azoisobutane[121]. However, the conversion of adamantylamine to the *N*-bromo derivative (81%) followed by treatment with potassium *t*-butoxide gives 1,1′-azoadamantane (**23**l in 81%, equation 56)[135].

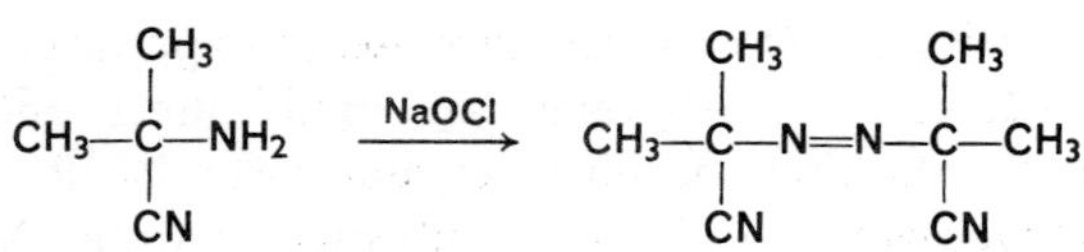

(55)

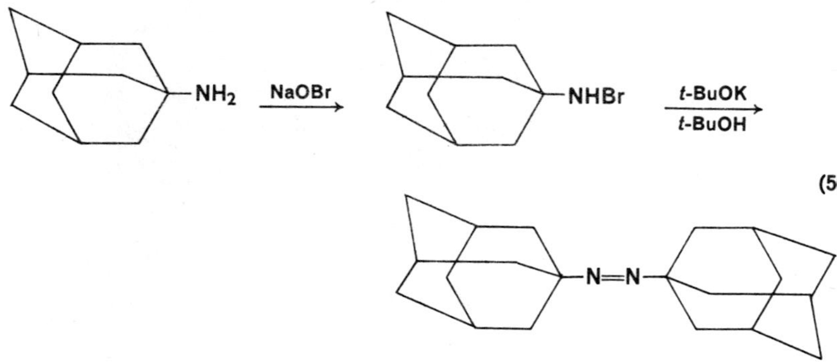

(56)

(231)

C. Oxidation of Ureas

It has recently been shown that ureas can be converted to diaziridinones by treatment with *t*-butylhypochlorite and base. These cyclic ureas can be readily hydrolysed to hydrazines and oxidized to azoalkanes (equation 57)[88].

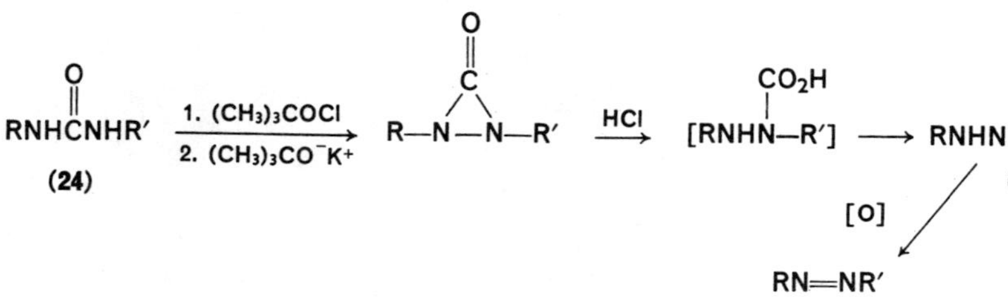

(57)

The procedure, as outlined, allows preparation of symmetrical or unsymmetrical azoalkanes; the critical factor being the availability of substituted urea (24)[88]. These unsymmetrical ureas can be obtained from isocyanates and amines (equation 58).

$$RN{=}C{=}O \xrightarrow{R'NH_2} RNH\overset{O}{\overset{\|}{C}}NHR' \qquad (58)$$

Rüchardt, Hinz and Oberlinner have used this approach (equation 58) to synthesize **25**a–e[136, 137]. Although the initial ureas and diaziridinone intermediates can be obtained in reasonable yields (**24**, R = $C_6H_5C(CH_3)_2$, R′ = 1-methylcyclopentyl 85%), the subsequent hydrolysis and oxidation is sometimes, although not always[88], a low yield reaction. For example, **25**d is prepared from **24**d in 4% yield.

R—N=N—R′

(25) a, R = $(CH_3)_3C$, R′ = 1-adamantyl
b, R = $(CH_3)_3C$, R′ = 1-bicyclo [2.2.2]octyl
c, R = $(CH_3)_3C$, R′ = 1-norbornyl
d, R = $C_6H_5C(CH_3)_2$, R′ = 1-methylcyclopentyl
e, R = $C_6H_5C(CH_3)_2$, R′ = 1-methylcyclohexyl

A similar method (equation 59) has been applied by Fowler to the preparation of **26**a, b[138]. The diaziridinones, if intermediates, were not isolated and the addition of first base and then *t*-butyl hypochlorite to the ureas produced both azo compounds **26**a and b in approximately 20%. Compound **26**b was optically active having been prepared from previously resolved 2-bromo-9-methyl-9-aminofluorene.

$$RNHC(=O)NHC_6H_5 \xrightarrow[2.\ (CH_3)_3COCl]{1.\ (CH_3)_3CO^-K^+} R{-}N{=}N{-}C_6H_5 \qquad (59)$$

(26)

a, R = $C_6H_5{-}C(CH_3)(C_2H_5){-}$

b, R =

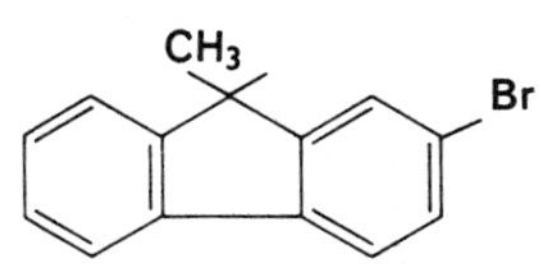

D. *From Hydrazodicarboxylates*

The preparation of cyclic azoalkanes via cycloadditicn reactions with dialkylazodicarboxylates or 4-phenyl-1,2,4-triazoline-3,5-dione[139–142] are quite well known. Discussions in this section are limited to acyclic azoalkanes and will therefore not deal with these reactions. (See Chapter 11.)

Azoalkanes with primary alkyl groups are most easily produced by the reduction–oxidation procedure of the hydrazine aldehyde adduct (see earlier discussion). This technique is limited to compounds with functional groups which are not susceptible to oxidation and reduction, e.g. R = alkyl. Crawford and co-workers[143–144] have designed a synthesis for several allylic derivatives and, while this multi-step procedure gives only moderate yields, there appears to be no better alternative (equation 60). Commercially available hydrazodicarboxylate esters can be mono or dialkylated as shown in equation 60 and 61 and upon hydrolysis and oxidation afford 3,3′-azo-

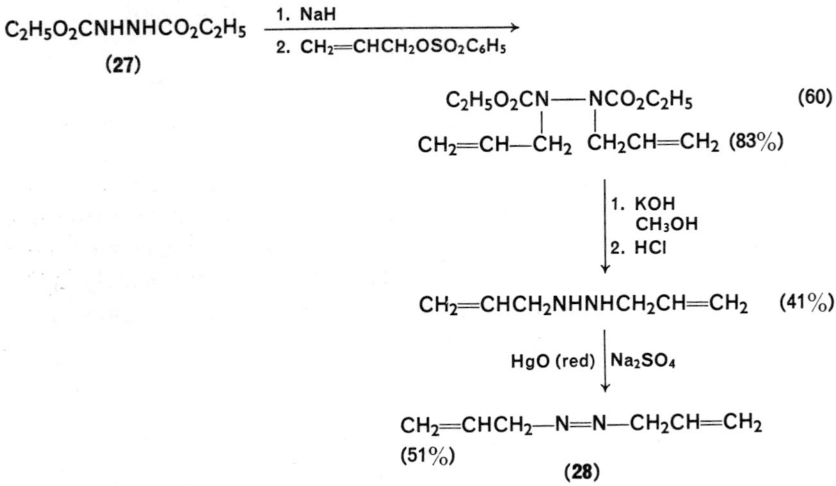

1-propene (**28**) in overall 21% yield. Several examples of mixed azoalkanes

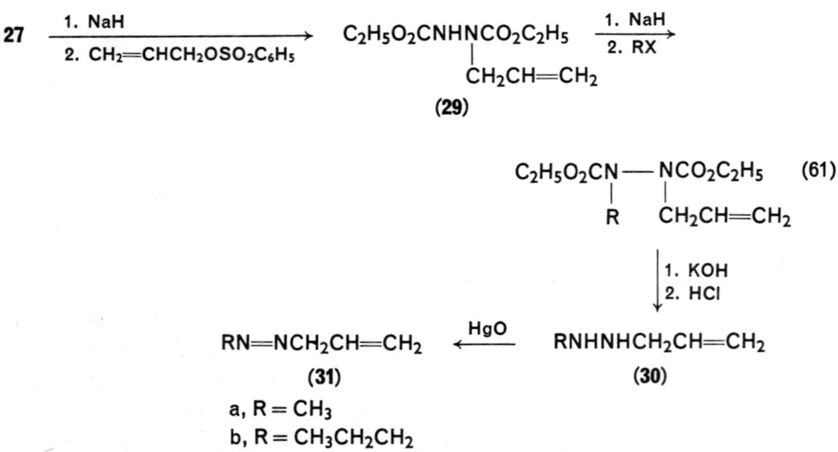

31a, b and **33** were prepared analogously[143, 144]. Compounds **30**a and b were obtained in overall yields of 15% and 9% (based on **29** prepared in 60% equation 61), and **32** was prepared from *t*-butylhydrazine in 10% (equation 62). It is possible that yields could be improved by using dibenzyl hydrazodicarboxylate **34** (also commercially available) which can be more

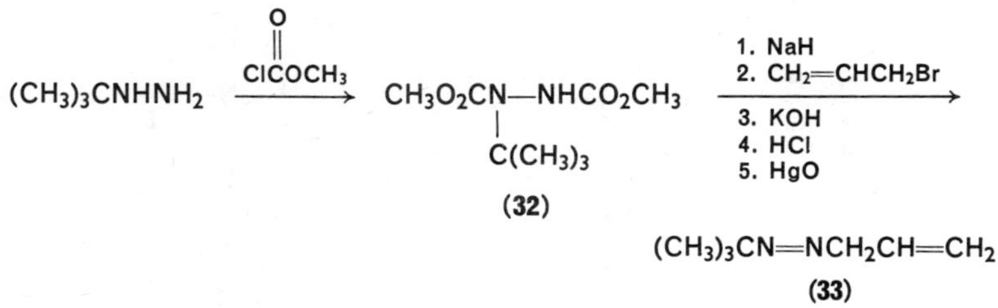

(62)

readily hydrolysed. A possible extension of the Crawford method is outlined in equation 63. However, it would only be useful for primary azoalkanes and, except for mixed systems, would probably not offer any significant improvement over the aldehyde hydrazine-reduction–oxidation method, equation 40 (see references 142 and 145 for examples of dibenzylazodicarboxylate in cycloadditions).

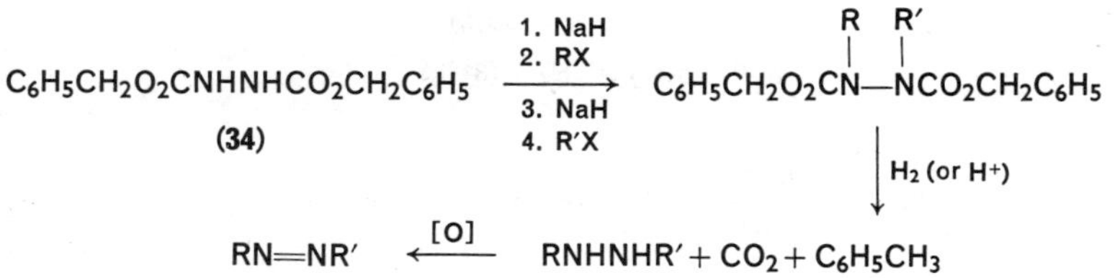

(63)

E. *Oxidation of* N,N′*-Dialkylsulphamides*

Probably the most significant recent contribution to the synthesis of azoalkanes has been pioneered by Ohme, Schmitz and Preuschhof[92–94]. The method can be modified for making symmetrical or unsymmetrical alkyl or aryl azo compounds. Furthermore, it is not limited to tertiary alkyl groups. This approach, outlined in equation 64, is postulated to involve chlorination of nitrogen, formation of the anion (**36**) and ring closure to the thiadiaziridine-1,1-dioxide (**37**)[92, 94]. While the dioxide intermediate (**37**) is not isolated under these conditions, the hydrolysis product 1,2-dialkyl hydrazine (**38**) can be by carefully controlling the amount of sodium hypochlorite present (see Section I.-H). Yields for **39**a–e (equation 64) range from 30–80%[93].

Precursor dialkyl sulphamides are prepared in good yields (60–80%) directly from amines and sulphuryl chloride (equation 65) or by generation of a pyridine-sulphuryl chloride adduct followed by treatment with amine (equation 66)[92–94]. The latter method can be used when the amine is scarce. An improved synthesis of *N,N*-dialkylsulphamides has also been reported[146].

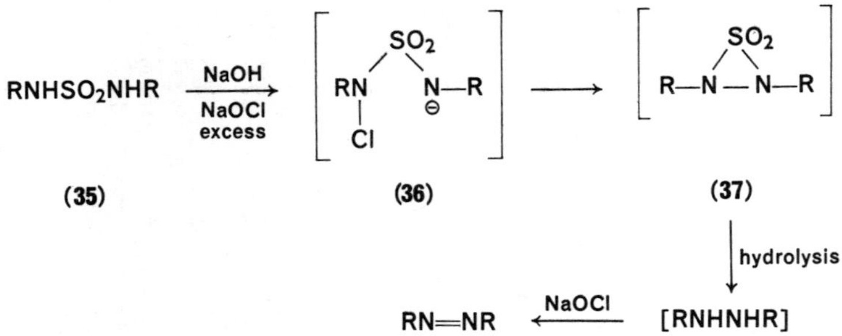

(35) (36) (37) (38) (39) (64)

a–e

a, R = CH_3CH_2— (54%)

b, R = $CH_3(CH_2)_2$— (54%)

c, R = $CH_3(CH_2)_3$— (54%)

d, R = cyclohexyl- (80%)

e, R = p-$O_2NC_6H_4$— (31%)

$$4\ RNH_2 \xrightarrow[\text{pentane}]{SO_2Cl_2} RNHSO_2NHR + 2\ R\overset{+}{N}H_3Cl^- \tag{65}$$

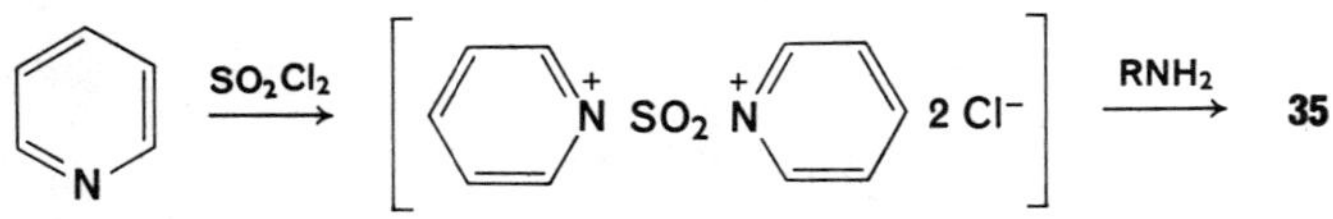

(66)

Several modifications of the Ohme–Schmitz–Preuschhof method have been proposed[65, 147]. One of these[65] employs a mixed solvent system and has been used to prepare **39**f–l[65, 148–150]. The synthesis of **40**j and **40**k is

RN=NR

(39)

f, R = t-butyl

g, R = (S,S)—$C_6H_5CH(CH_3)$—

h, R = 1-adamantyl

i, R = s-butyl

j, R = CH_2=CH—$C(CH_3)_2$—

k, R = CH≡C—$C(CH_3)_2$

l, R = n-butyl (59%)

important in that it shows that unsaturated azoalkanes can be obtained which are not available from hydrogenation and oxidation of azines[149].

If the chlorinating agent and the base are changed to *t*-butylhypochlorite and sodium hydride, good yields (~80%) of thiadiaziridine 1,1-dioxides (**37**) can be achieved[147, 151]. These compounds are rapidly converted into the

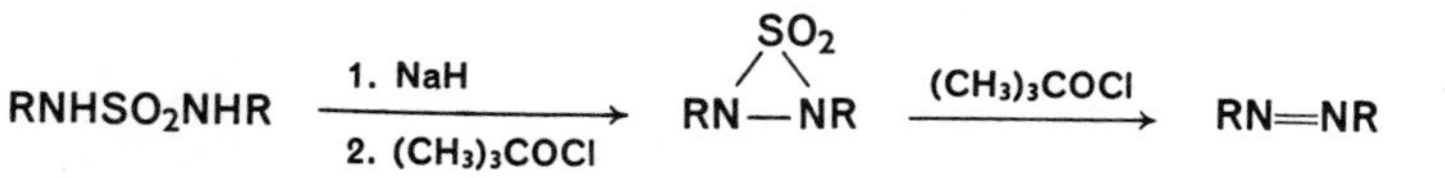

(67)

(**35**) (**37**) (**39**)

m, R = $(CH_3)_3CC(CH_3)_2$—

n, R = $(CH_3)_3CCH_2C(CH_3)_2$—

azoalkane with excess hypochlorite. This method has been used to prepare **39**l (R = *n*-butyl, 37%), **39**i (R = *sec*-butyl, 64%), **39**f (R = *t*-butyl, 85%), **39**m (R = *t*-heptyl, 36%), **39**n (R = *t*-octyl, 78%)[147].

Porter and co-workers[152–153] have prepared several unsymmetrical azo-alkanes by trapping sulphurylaniline (**41**, Lossen-type rearrangement of **40**)[154] with the appropriate amine (equation 68). The sulphamide (**42**) is then converted into azo compound via the Ohme–Schmitz–Preuschhof

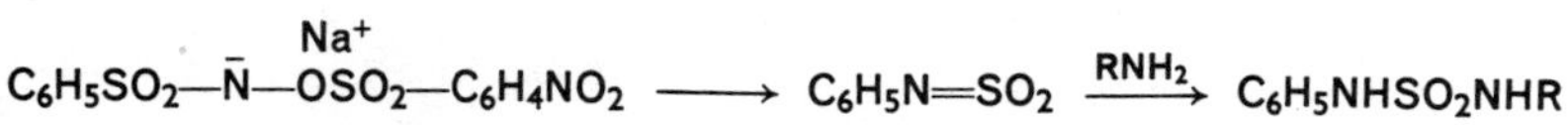

(68)

(**40**) (**41**) (**42**)

a, R = $C_6H_5C(CH_3)_2$—

b, R = $C_6H_5C(C_2H_5)(CH_3)$—

c, R = $CH_2{=}CH{-}C(CH_3)_2$—

method (equation 64). Interesting work on the photo-isomerization to the much less stable *cis*- forms of these azoalkanes has recently been observed by a number of workers[155]. In connection with this work it has been found that diisopropylsulphamide gives both *cis*- and *trans*-azoisopropane when treated with base and sodium hypochlorite[156]. It is not known how general this approach is since many *cis*-azo compounds isomerize rapidly to the *trans*-isomer.

The sulphamide method appears also to have potential for preparation of unsymmetrical alkyl azo compounds. Treatment of tertiary alcohols with chlorosulphonyl isocyanate gives monoalkyl sulphamyl chlorides[157]. Using this method to synthesize a mixed sulphamide has in one case resulted in a mixed azoalkane (equation 69)[157a].

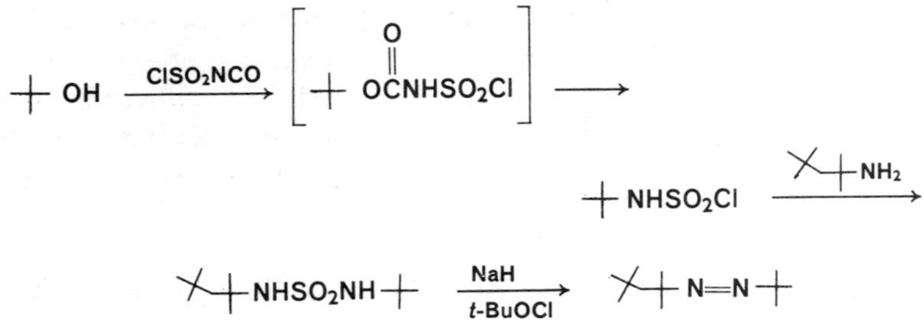

(69)

III. PREPARATION OF AZOXY COMPOUNDS

Extensive reviews are available on the synthesis of aliphatic and aromatic azoxy compounds[1,2,96,99]. This section reviews the recent methods for the aliphatic cases.

A. *Oxidation of Azo Compounds*

Oxidation of azo compounds with peracetic acid, *m*-chloroperbenzoic acid, or hydrogen peroxide gives azoxy compounds in good yield[1,96] (equations 70–72). Seven azo compounds, prepared from triazolinedione

$$trans\text{-}RN{=}NR + R'CO_3H \longrightarrow trans\text{-}R\overset{\overset{O}{\uparrow}}{N}{=}NR \qquad (70)$$

R	*Yield, %*	*Reference*
t-Bu	—	158
Benzyl	67	158
1-Adamantyl	95	135
n-Bu	35	150
Cyclododecyl	93	159

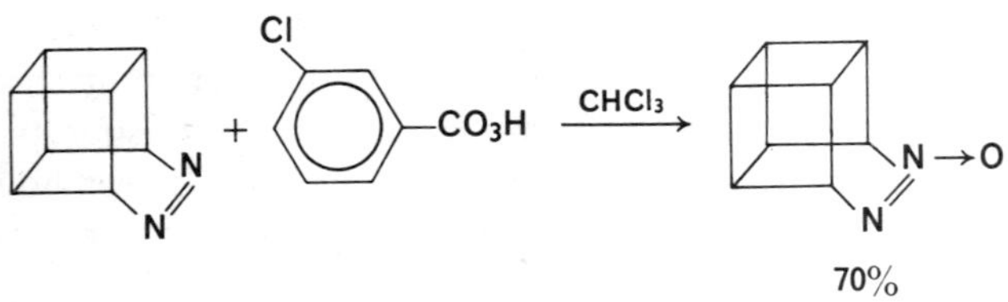

(71)[160]

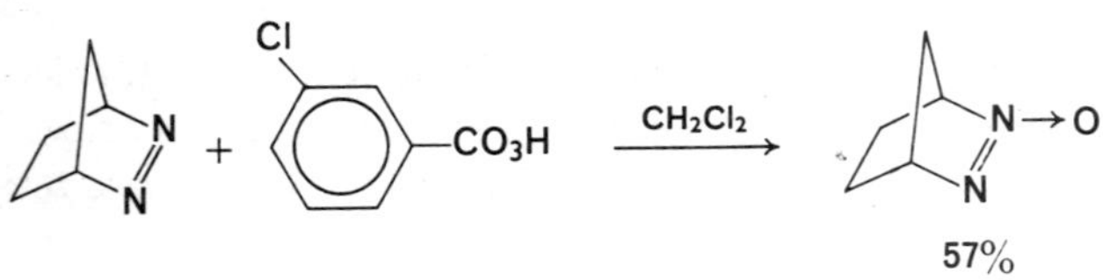

(72)[150]

Diels-Alder adducts, were oxidized in 8 to 98% yields without isolation of the azo compounds (equation 73)[161]. In one case the intermediate azo

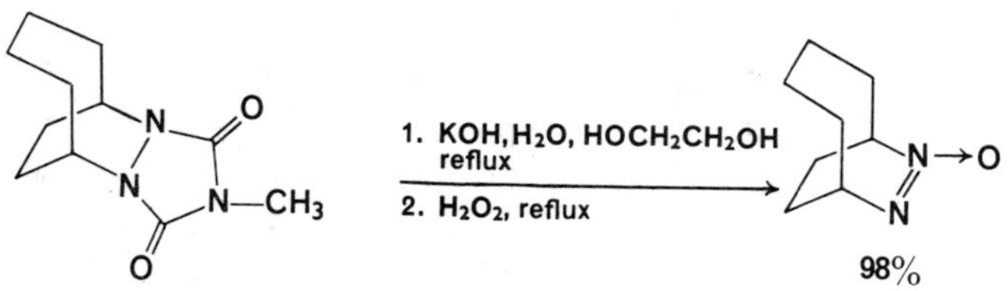

(73)

compound was very unstable even at –78°C, but the stable azoxy compound could be prepared at 120°C by oxidative hydrolysis (equation 74)[162].

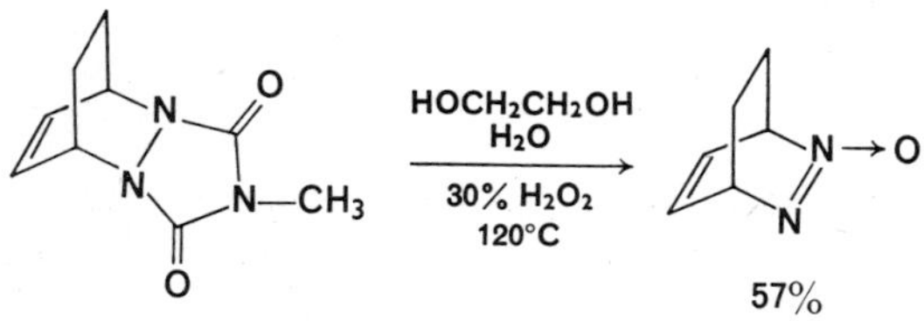

(74)

An acyclic *cis*-isomer was prepared by stereospecific peracid oxidation: *cis*-azoisopropane gave *cis*-azoxyisopropane in 72% yield[163]. Treatment of the unsymmetrical phenylazomethane with perbenzoic acid gave predominantly oxidation on the nitrogen bearing the methyl group and only a trace of the other isomer[158]. Oxidation of α,β-unsaturated azo compounds gave azoxy compounds, with the location of the oxygen dependent on the alkyl substitution[158a].

B. *Reduction of Azodioxy Compounds*

Azodioxy compounds (nitroso dimers) may be reduced to azoxy compounds using stannous chloride or catalytic hydrogenation[1,96].

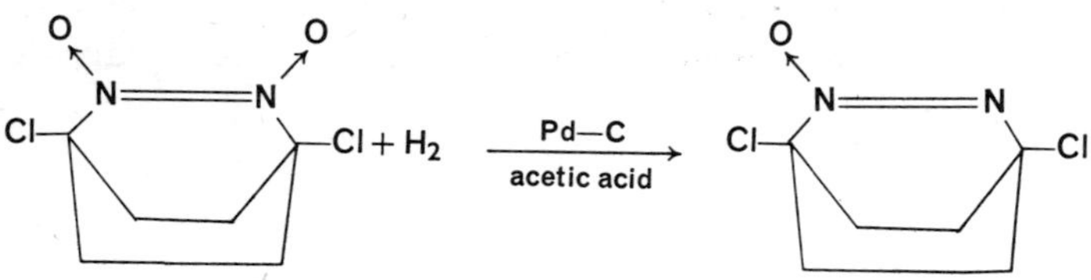

(75)[164]

C. *Condensation of Nitroso Compounds with Hydroxylamines*

Heating equimolar amounts of nitroso compounds with hydroxylamines gives azoxy compounds (equation 77)[1,96]. The lower yields with secondary alkyl nitroso compounds are the result of competing oxime tautomerization.

$$R\overset{\overset{O}{\uparrow}}{N}{=}\underset{\underset{O}{\downarrow}}{N}R + H_2 \xrightarrow{Pd} R\overset{\overset{O}{\uparrow}}{N}{=}NR \qquad (76)^{159,164}$$

R = cyclohexyl 66–72%

R = cyclododecyl 79%

$$RNO\ dimer + RNHOH \longrightarrow R\overset{\overset{O}{\uparrow}}{N}{=}NR \qquad (77)$$

R	*Yield*, %	*Reference*
1,1,5-Trimethylhexyl	75	165
Cyclohexyl	47	159
Cyclododecyl	21	159
1-Adamantyl	70–75	135

In the adamantyl case, it was found that the reaction is faster in the presence of acetic acid and much faster with KOH. A few examples with two different substituents were prepared from methylhydroxylamine and nitroso compounds; and in each case the oxygen appeared on the nitrogen originating from the nitroso compound (equation 78)[158]. The generality of this was not checked by, e.g. treating nitrosomethane with *N*-*t*-butylhydroxylamine to obtain the opposite isomer. Trifluoronitrosomethane condensed rapidly

$$RNO + CH_3NHOH \longrightarrow CH_3N{=}\overset{\overset{O}{\uparrow}}{N}R + \text{trace } CH_3\overset{\overset{O}{\uparrow}}{N}{=}NR \qquad (78)$$

R = *t*-Bu or Ph

with hydroxylamines at –70°C to give azoxy compounds[166].

This same condensation may be taking place in the partial oxidation of hydroxylamines to azoxy compounds using air with a cobalt catalyst or using lead tetraacetate[159].

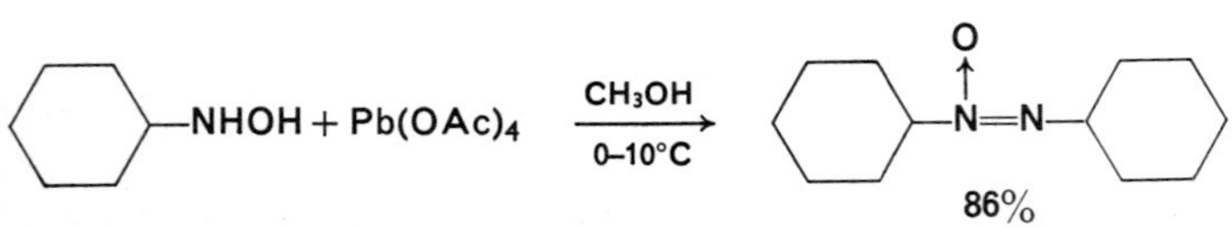

(79)

A similar condensation occurs when dichloroamines and nitroso compounds are treated with potassium hydroxide[166a].

D. Alkylation of Diazotates

An unambiguous synthesis of azoxyalkanes containing two different primary or secondary alkyl groups was developed recently by Moss[167]. Conversion of amines to diazotates[168] followed by alkylation on nitrogen gave *trans*-azoxyalkanes with the oxygen specifically on the nitrogen bearing the group from the alkylating agent. Meerwein's reagent or alkyl iodides serve as suitable alkylating agents (equation 80). The competing O-alkyla-

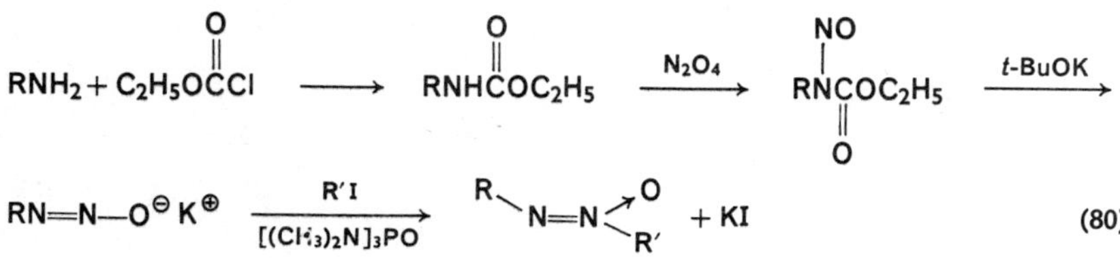

(80)

R	R′	Yield, % [a]
n-C_8H_{17}	C_2H_5	44
n-C_8H_{17}	*sec*-C_4H_9	32
sec-C_8H_{17}	C_2H_5	64[b]
$PhCHCH_3$	n-C_4H_9	33

[a] Yields are calculated by g.l.c. analysis, based on nitrosourethane.
[b] Alkylation with Meerwein's reagent.

tion leads to ethers and olefins. This method is suitable for preparation of optically pure azoxy compounds with a chiral centre α to the nitrogen on either side (equations 81, 82)[169]. In this way L-dihydroelaiomycin (**44**)

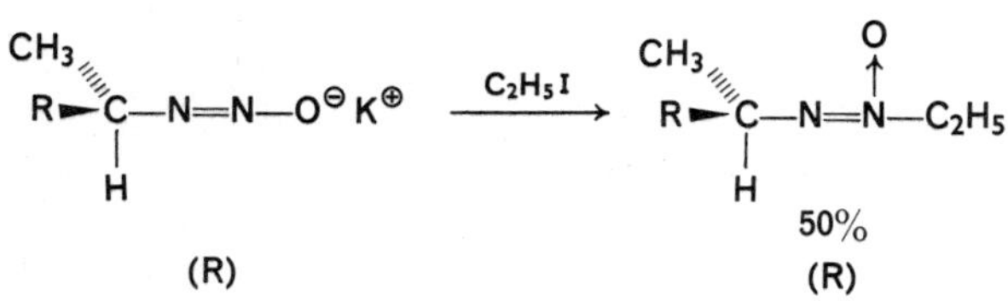

(81)

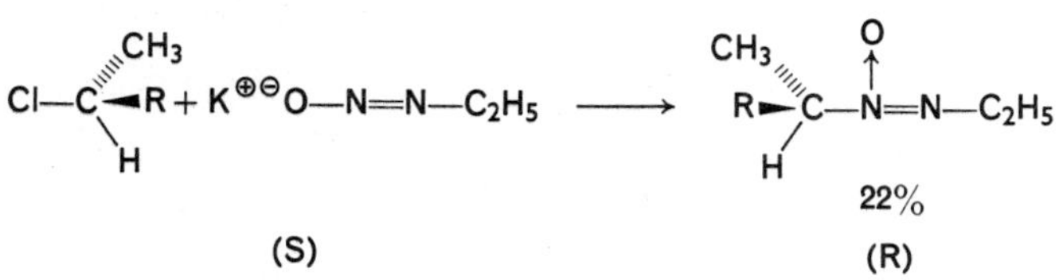

(82)

R = $n \cdot C_6H_{13}$

was prepared in 12 steps from L-threonine in 8% overall yield[170]. The final steps were alkylation of diazotate **43** with *n*-octyl iodide followed by cleavage of the tetrahydropyranyl protecting group.

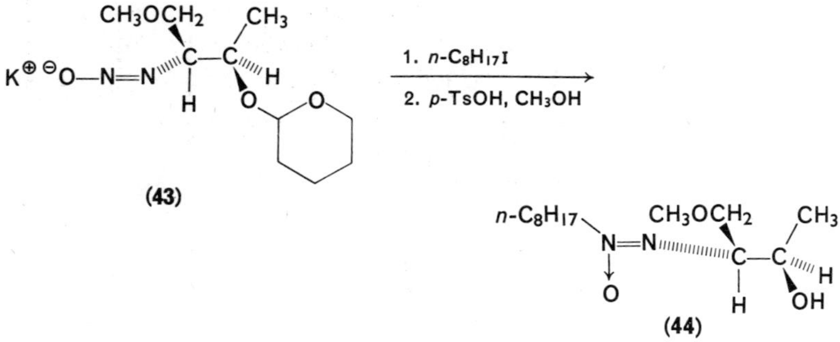

(83)

The unsymmetrical azoxy compounds may be alkylated α to the nitrogen bearing the oxygen. In addition, the carbanion may be condensed with an aldehyde, and the resulting alcohol dehydrated to give α,β-unsaturated azoxy compounds regiospecifically.[170a].

E. Grignard Reagent Displacements on Diimide N-Oxides

Fluoride or tosylate leaving groups may be replaced by alkyl groups affording aryl alkyl azoxy compounds with the oxygen on the aryl side (equations 84, 85)[171,172]. When the aryl group was replaced by an alkyl,

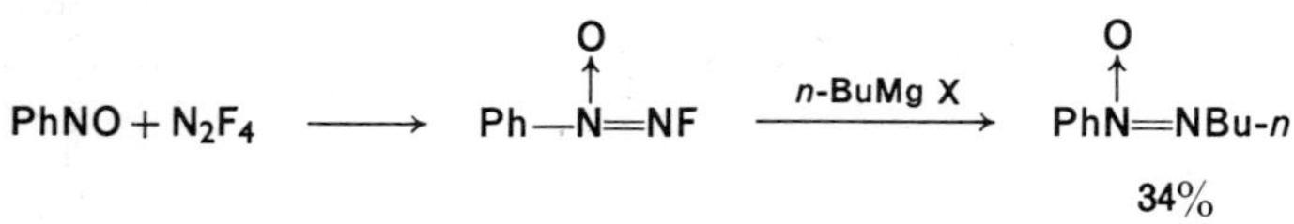

(84)

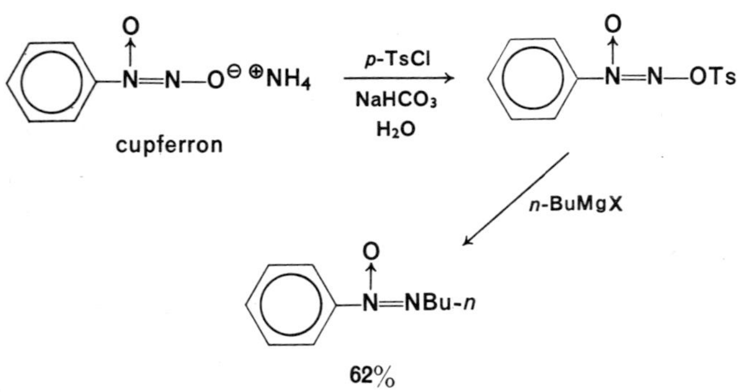

(85)

no azoxy compounds could be found.

F. Photochemical Syntheses

Ultraviolet irradiation of an acyclic *trans* alkyl azoxy compound gives some of the *cis*-isomer and some ring closed product, the oxadiaziridine[150,163,173,174]. The oxadiaziridine thermally reverts to some of each of the *cis*- and *trans*-isomers of the azoxy compound from which pure *cis* may be isolated in preparatively useful quantities. For example, *cis*-azoxyisopropane was prepared in up to 32% yield from the *trans*-isomer[163]. When there are two different substituents, more products appear via oxygen migration. Irradiation of *trans*-**45** gave some of all four possible compounds[174].

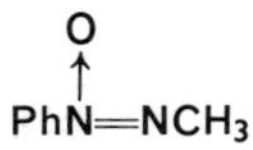

(45)

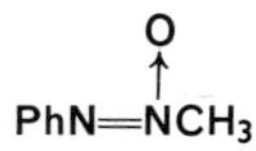

(46)

$$\textit{trans}\text{-}\mathbf{45} \xrightarrow{h\nu} \underset{14{\cdot}2\%}{\textit{trans}\text{-}\mathbf{45}} + \underset{20{\cdot}5\%}{\textit{cis}\text{-}\mathbf{45}} + \underset{9{\cdot}6\%}{\textit{trans}\text{-}\mathbf{46}} + \underset{3{\cdot}2\%}{\textit{cis}\text{-}\mathbf{46}} \tag{86}$$

Some cyclic azoxy compounds were prepared by ultraviolet irradiation (equations 87–89).

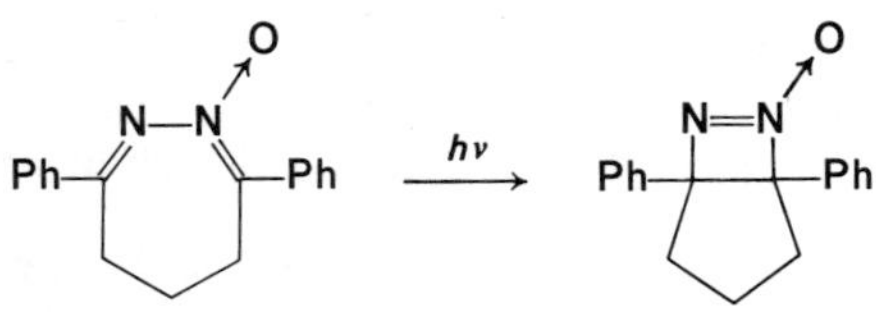

28%

(87)[175]

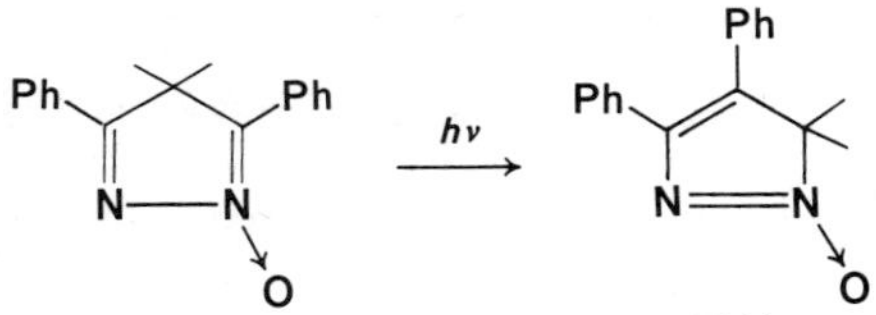

70%

(88)[175]

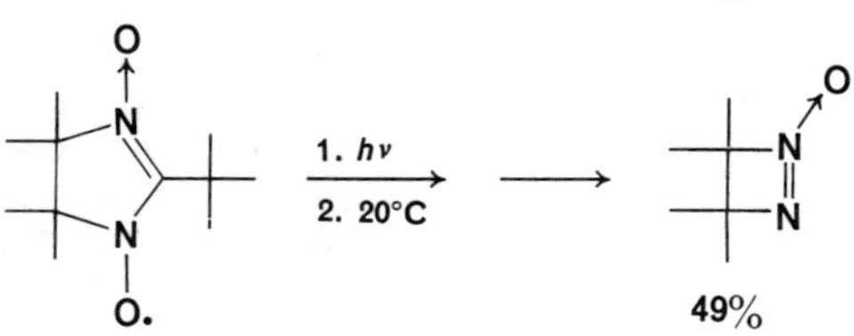

49%

(89)[176]

IV. REFERENCES

1. E. Müller, *Methoden der Organischen Chemie* (Houben-Weyl, Ed.), vol. 10/2, Georg Thieme Verlag, Stuttgart, 1967.
2. P. A. S. Smith, *The Chemistry of Open Chain Nitrogen Compounds*, Vol. 2, W. A. Benjamin, Inc., New York, 1966.
3. R. Ohme and A. Zubek, *Z. Chem.*, **8,** 41 (1968).
4. S. R. Sandler and W. Karo, *Organic Functional Group Preparations*, Vol. 1, Academic Press, New York, 1968, pp. 363–379.
5. F. Klages, G. Nober, F. Kirchner and M. Bock, *Justus Liebigs Ann. Chem.*, **547,** 1 (1941).
6. H.-H. Stroh and H.-G. Scharnow, *Chem. Ber.*, **98,** 1588 (1965).
7. M. J. S. Dewar and W. B. Jennings, *J. Amer. Chem. Soc.*, **95,** 1562 (1973).
8. J. H. Biel, A. E. Drukker, T. F. Mitchell, E. P. Sprengler, P. A. Nuhfer, A. C. Conway and A. Horita, *J. Amer. Chem. Soc.*, **81,** 2805 (1959).
9. R. K. Müller, D. Felix, J. Schreiber and A. Eschenmoser, *Helv. Chim. Acta*, **53,** 1479 (1970).
10. E. F. Elslager, E. A. Weinstein and D. F. Worth, *J. Med. Chem.*, **7,** 493 (1964).
11. D. E. Butler, S. M. Alexander, J. W. McLean and L. B. Strand, *J. Med. Chem.*, **14,** 1052 (1971).
12. H. Hatt, *Org. Syn., Coll.* **2,** 208 (1943).
13. F. Hrabak, *Coll. Czech. Chem. Commun.*, **34,** 4010 (1969).
14. P. Zeller, H. Gutmann, B. Hegedus, A. Kaiser, A. Langemann, and M. Müller, *Experientia*, **19,** 129 (1963).
15. C. D. Hurd and F. F. Cesark, *J. Amer. Chem. Soc.*, **89,** 1417 (1967).
16. E. Schmitz and R. Ohme, *Chem. Ber.*, **95,** 2012 (1962).
17. L. A. Carpino, *J. Amer. Chem. Soc.*, **85,** 2144 (1963).
18. H. Stetter and H. Spangenberger, *Chem. Ber.*, **91,** 1982 (1958).
19. A. Zweig and A. K. Hoffmann, *J. Amer. Chem. Soc.*, **85,** 2736 (1963).
20. W. P. Neumann and U. Blaukat, *Angew. Chem., Int. Ed., Engl.*, **8,** 611 (1969.)
21. W. D. Ollis, I. O. Sutherland and Y. Thebtaranonth, *Chem. Commun.*, 1199 (1970).
22. S. F. Nelsen and R. T. Landis, *J. Amer. Chem. Soc.*, **95,** 2719 (1973).
23. E. M. Kaiser and G. J. Bartling, *Tetrahedron Lett.*, 4357 (1969).
24. L. A. Carpino, P. H. Terry and P. J. Crowley, *J. Org. Chem.*, **26,** 4336 (1961).
25. D. Bar, F. Erb-Debruyne, P. Marcincal, A. Marcincal-Lefebvre and O. Labiau, *Bull. Soc. Chim. Fr.*, 1783 (1971).
26. O. Diels, J. H. Blom and W. Kohl, *Justus Liebigs Ann. Chem.*, **443,** 243 (1925).
27. E. Fahr and H. Lind, *Angew. Chem., Int. Ed., Engl.*, **5,** 372 (1966).
28. E. F. Ullman and E. A. Bartkus, *Chem. and Ind.*, 93 (1962).
29. T. J. Kealy, *J. Amer. Chem. Soc.*, **84,** 966 (1962).
30. R. A. Clement, *J. Org. Chem.*, **27,** 1115 (1962).
31. D. L. Forster, T. L. Gilchrist, C. W. Rees and E. Stanton, *Chem. Commun.*, 695 (1971).
32. J. F. W. McOmie and D. H. Perry, *Chem. Commun.*, **248** (1973).
33. J. C. Stickler and W. H. Pirkle, *J. Org. Chem.*, **31,** 3444 (1966).
34. R. C. Cookson, S. S. Gupte, I. D. R. Stevens and C. T. Watts, *Org. Syn.*, **51,** 121 (1971).

35. D. G. Farnum and J. P. Snyder, *Tetrahedron Lett.*, 3861 (1965).
36. R. C. Cookson, S. S. H. Gilani and I. D. R. Stevens, *J. Chem. Soc.*, **C**, 1905 (1967).
37. J. Sauer and B. Schröder, *Chem. Ber.*, **100**, 678 (1967).
38. J. A. Berson and R. F. Davis, *J. Amer. Chem. Soc.*, **94**, 3658 (1972).
39. S. S. H. Gilani and D. J. Triggle, *J. Org. Chem.*, **31**, 2397 (1966).
40. J. Firl and S. Sommer, *Tetrahedron Lett.*, 4713 (1972) and references cited therein.
41. D. J. Pasto and A. F. T. Chen, *Tetrahedron Lett.*, 713 (1973).
42. D. J. Pasto, A. F. T. Chen and G. Binsch, *J. Amer. Chem. Soc.*, **95**, 1553 (1973).
43. T. J. Katz and N. Acton, *J. Amer. Chem. Soc.*, **95**, 2738 (1973).
44. R. Askani, *Chem. Ber.*, **102**, 3304 (1969).
45. K. Shen, *Chem. Commun.*, 391 (1971).
46. R. Askani, *Tetrahedron Lett.*, 3349 (1970).
47. L. A. Paquette, *J. Amer. Chem. Soc.*, **92**, 5765 (1970).
48. L. A. Paquette, *J. Amer. Chem. Soc.*, **93**, 7110 (1971).
49. E. L. Allred and K. J. Voorhees, *J. Amer. Chem. Soc.*, **95**, 620 (1973).
49a. M. F. Semmelhack, J. S. Foos and S. Katz, *J. Amer. Chem. Soc.*, **95**, 7325 (1973).
49b. E. J. Corey and B. B. Snider, *J. Org. Chem.*, **38**, 3632 (1973).
50. P. E. Iversen, *Acta Chem. Scand.*, **25**, 2337 (1971).
51. H. Zimmer, L. F. Audrieth, M. Zimmer and R. A. Rowe, *J. Amer. Chem. Soc.*, **77**, 790 (1955).
52. C. G. Overberger, L. C. Palmer, B. S. Marks and N. R. Byrd, *J. Amer. Chem. Soc.*, **77**, 4100 (1955).
53. S. F. Nelsen and R. Fibiger, *J. Amer. Chem. Soc.*, **94**, 8497 (1972).
54. D. C. Iffland and E. Cerda, *J. Org. Chem.*, **28**, 2769 (1963).
55. L. A. Carpino, A. A. Santilli and R. W. Murray, *J. Amer. Chem. Soc.*, **82**, 2728 (1960).
56. B. T. Hayes and T. S. Stevens, *J. Chem. Soc.*, **C**, 1088 (1970).
57. H. Dorn and A. Zubek, *Z. Chem.*, **12**, 129 (1972).
58. F. E. Henoch and C. R. Hauser, *Can. J. Chem.*, **47**, 157 (1969).
59. K. Kratzl and K. P. Berger, *Monatsch. Chem.*, **89**, 83 (1958).
60. F. E. Condon, *J. Org. Chem.*, **37**, 3615 (1972).
61. J. A. Blair and R. J. Gardner, *J. Chem. Soc.*, **C**, 1714 (1970).
62. S. F. Nelsen and G. R. Weisman, *Tetrahedron Lett.*, 2321 (1973).
63. J. Thiele, *Chem. Ber.*, **42**, 2575 (1909).
64. B. V. Ioffe, Z. I. Sergeeva and Y. Y. Dumpis, *Zh. Org. Khim.*, **5**, 1735 (1969); Eng. Edit. 1683 (1969).
65. J. C. Stowell, *J. Org. Chem.*, **32**, 2360 (1967).
66. H. Feuer and F. Brown, Jr., *J. Org. Chem.*, **35**, 1468 (1970).
67. R. L. Hinman, *J. Amer. Chem. Soc.*, **78**, 1645, 2463 (1956).
68. R. T. Beltrami and E. R. Bissell, *J. Amer. Chem. Soc.*, **78**, 2467 (1956); K.-H. Linke, R. Turley and E. Flaskamp, *Chem. Ber.*, **106**, 1052 (1973).
69. L. Spialter, D. H. O'Brien, G. L. Untereiner and W. A. Reusch, *J. Org. Chem.*, **30**, 3278 (1965).
70. W. Theilacker and E. Wegner, *Newer Methods of Preparative Organic Chemistry*, Vol. 3 (Ed. W. Foerst), Academic Press, New York, 1964, p. 303.

71. H. Schubert, F. Fohringen and H. Noack, *Z. Chem.*, **10,** 68 (1970).
72. G. M. Omietanski and H. H. Sisler, *J. Amer. Chem. Soc.*, **78,** 1211 (1956).
73. R. Appel and D. Hänssgen, *Chem. Ber.*, **103,** 3733 (1970).
74. W. Metlesics, R. F. Tavares and L. H. Sternbach, *J. Org. Chem.*, **30,** 1311 (1965).
75. M. Somei and M. Natsumi, *Tetrahedron Lett.*, 461 (1974).
75a. R. Gösl and A. Meuween, *Org. Syn.*, **43,** 1 (1963); *Chem. Ber.*, **92,** 2521 (1959).
76. L. A. Carpino, *J. Amer. Chem. Soc.*, **82,** 3133 (1960); *J. Org. Chem.*, **30,** 321 (1965).
77. T. Sheradsky, *Tetrahedron Lett.*, 1909 (1968); T. Sheradsky and G. Salemnick, *Tetrahedron*, **28,** 3833 (1972); G.-F. Huang, T. Okamoto, M. Maeda, and Y. Kawazoe, *Tetrahedron Lett.*, 4541 (1973).
78. Y. Tamura, J. Minamikawa, Y. Kita, J. H. Kim and M. Ikeda, *Tetrahedron*, **29,** 1063 (1973); Y. Tamura, J. Minamikawa, Y. Miki, S. Matsugashita and M. Ikeda, *Tetrahedron Lett.*, 4133 (1972); R. Y. Ning, *Chem. and Engineering News*, Dec. 17, 1973, p. 36.
78a. W. Lwowski, New Mexico State University, private communication.
79. N. V. Khromov-Borisov and T. N. Kononova, *Probl. Poluch. Poluprod. Prom. Org. Sin., Akad. Nauk SSSR, Otd. Obsch. Tekh. Khim.*, 10 (1967); *Chem. Abstr.*, **68,** 48947v (1968).
79a. S. F. Nelsen and R. T. Landis, II, *J. Amer. Chem. Soc.*, **95,** 8707 (1973).
79b. A. Luttringhaus, J. Jander and R. Schneider, *Chem. Ber.*, **92,** 1756 (1959).
80. A. T. Nielsen, R. L. Atkins, D. W. Moore, D. Mallory and J. M. LaBerge, *Tetrahedron Lett.*, 1167 (1973).
81. E. Schmitz and D. Habisch, *Chem. Ber.*, **95,** 680 (1962).
82. E. Schmitz, *Dreiringe mit Zwei Heteroatomen*, Springer-Verlag, Berlin, 1967, p. 95.
83. R. Ohme, E. Schmitz and L. Sterk, *J. Prakt. Chem.*, **37,** 257 (1968).
84. E. Schmitz and K. Shinkowski, *Chem. Ber.*, **97,** 49 (1964).
85. G. R. Elliot, *J. Chem. Soc.*, **123,** 804 (1923).
86. D. W. Lum and I. L. Mador, U.S. patent 2,917,545; *Chem. Abstr.*, **54,** 6549 (1960).
86a. C. S. Sheppard and L. E. Korczykowski, *U.S. Patent* 3,657,324 (1972).
87. R. Ohme and H. Preuschhof, *J. Prakt. Chem.*, **312,** 349 (1970).
88. F. D. Greene, J. C. Stowell and W. R. Bergmark, *J. Org. Chem.*, **34,** 2254 (1969).
89. F. D. Greene, W. R. Bergmark and J. G. Pacifi, *J. Org. Chem.*, **34,** 2263 (1969).
90. F. D. Greene and J. F. Pazos, *J. Org. Chem.*, **34,** 2269 (1969).
91. E. Schmitz, R. Ohme and S. Schramm, *Tetrahedron Lett.*, 1857 (1965).
92. Reference 82, p. 55.
93. R. Ohme and H. Preuschhof, *Justus Liebigs Ann. Chem.*, **713,** 74 (1968).
94. R. Ohme, H. Preuschhof and H-U. Heyne, *Org. Syn.*, **52,** 11 (1972).
95. R. Ohme and E. Schmitz, *Angew. Chem., Int.. Ed.*, Engl., **4,** 433 (1965).
96. E. Müller, *Methoden der Organischen Chemie* (Houben-Weyl), Vol. 10/3, Georg. Thieme Verlag, Stuttgart, 1965.
97. C. G. Overberger, J. P. Anselme and J. G. Lombardino, *Organic Compounds with Nitrogen–Nitrogen Bonds*, Ronald Press, New York, N.Y., 1966.

98. H. Zollinger, *Azo and Diazo Chemistry*, Interscience Publishers Inc., New York, N.Y., 1961.
99. S. R. Sandler and W. Karo, *Organic Functional Group Preparations*, Vol. 2, Academic Press, New York, 1971.
100. S. G. Cohen, S. J. Groszos, and D. B. Sparrow, *J. Amer. Chem. Soc.*, **72,** 3947 (1950).
101. J. R. Shelton and C. K. Liang, *Synthesis*, 204 (1971).
102. S. E. Scheppele and S. Seltzer, *J. Amer. Chem. Soc.*, **90,** 358 (1968).
103. F. D. Greene, M. A. Berwick and J. C. Stowell, *J. Amer. Chem. Soc.*, **92,** 867 (1970).
104. B. K. Bandlish, Ph.D. Thesis, Louisiana State University in New Orleans, 1973.
105. S. Seltzer and F. T. Dunne, *J. Amer. Chem. Soc.*, **87,** 2628 (1965).
106. S. G. Cohen and C. H. Wang, *J. Amer. Chem. Soc.*, **77,** 3628 (1955).
107. C. G. Overberger and A. V. DiGiulio, *J. Amer. Chem. Soc.*, **80,** 6562 (1958).
108. A. Tsolis, S. G. Mylonakis, M. T. Nieh and S. Seltzer, *J. Amer. Chem. Soc.*, **94,** 829 (1972).
109. S. Seltzer, *J. Amer. Chem. Soc.*, **85,** 14 (1963).
110. K. R. Kopecky and T. Gillan, *Can. J. Chem.*, **47,** 2371 (1969).
111. R. C. Neuman and E. S. Alhadeff, *J. Org. Chem.*, **35,** 3401 (1970).
112. J. Thiele and K. Heuser, *Justus Liebigs Ann. Chem.*, **290,** 1 (1896).
113. A. F. Bickel and W. A. Waters, *Rec. Trav. Chim.*, **69,** 312 (1950).
114. G. A. Mortimer, *J. Org. Chem.*, **30,** 1632 (1965).
115. S. Goldschmidt and B. Acksteiner, *Justus Liebigs Ann. Chem.*, **618,** 173 (1958).
116. E. Benzing, *Justus Liebigs Ann. Chem.*, **631,** 1 (1960).
117. A. Ohno and Y. Ohnishi, *Tetrahedron Lett.*, 4405 (1969).
118. J. W. Timberlake and J. C. Martin, *J. Org. Chem.*, **33,** 4054 (1968).
119. J. W. Timberlake and M. L. Hodges, *Tetrahedron Lett.*, 4147 (1970).
120. J. W. Timberlake, A. W. Garner and M. L. Hodges, *Tetrahedron Lett.*, 309 (1973).
121. J. W. Timberlake, unpublished results.
122. W. Duismann, H. D. Beckhaus, and C. Rüchardt, *Tetrahedron Lett.*, 265 (1974).
123. E. Schmitz, R. Ohme and S. Schramm, *Angew. Chem., Int. Ed. Engl.*, **2,** 157 (1963).
124. E. M. Kosower, *Accounts of Chem. Research*, **4,** 193 (1971).
125. T. Tsuji and E. M. Kosower, *J. Amer. Chem. Soc.*, **93,** 1992 (1971).
126. T. E. Stevens, *J. Org. Chem.*, **26,** 2531 (1961).
127. E. Farenhorst and E. C. Kooyman, *Rec. Trav. Chim.*, **72,** 993 (1953).
128. S. F. Nelsen and P. D. Bartlett, *J. Amer. Chem. Soc.*, **88,** 137 (1966).
129. J. R. Shelton, J. F. Gormish, C. K. Liang, P. L. Samuel and P. Kovacic, *Can. J. Chem.*, **46,** 1149 (1968).
130. P. D. Bartlett and J. M. McBride, *Pure Appl. Chem.*, **15,** 89 (1967).
131. D. J. Severn and E. M. Kosower, *J. Amer. Chem. Soc.*, **91,** 1710 (1969).
132. J. B. Levy and E. J. Lehmann, *J. Amer. Chem. Soc.*, **93,** 5790 (1971).
133. M. Prochazka, O. Ryba and D. Lim, *Coll. Czech. Chem. Commun.*, **33,** 3387 (1968).
134. Reference 97, p. 772.

135. H. Stetter and E. Smulders, *Chem. Ber.*, **104,** 917 (1971).
136. J. Hinz and C. Rüchardt, *Justus Liebigs Ann. Chem.*, **765,** 94 (1972).
137. A. Oberlinner and C. Rüchardt, *Tetrahedron Lett.*, 4685 (1969).
138. J. S. Fowler, *J. Org. Chem.*, **37,** 510 (1972).
139. M. Martin and W. R. Roth, *Chem. Ber.*, **102,** 811 (1969).
140. B. T. Gillis and J. D. Hagarty, *J. Org. Chem.*, **32,** 330 (1967).
141. M. Heyman, U. T. Bandurco and J. P. Snyder, *Chem. Commun.*, 297 (1971).
142. M. L. Heyman and J. P. Snyder, *Tetrahedron Lett.*, 2859 (1973) and references cited therein.
143. B. H. Al-Sader and R. J. Crawford, *Can. J. Chem.*, **48,** 2745 (1970).
144. R. J. Crawford and K. Takagi, *J. Amer. Chem. Soc.*, **94,** 7406 (1972).
145. B. K. Bandlish, J. N. Brown, J. W. Timberlake and L. M. Trefonas, *J. Org. Chem.*, **38,** 1102 (1973).
146. M. Bermann and J. R. VanWazer, *Synthesis*, 576 (1972).
147. J. W. Timberlake, M. L. Hodges and K. Betterton, *Synthesis*, 632 (1972).
148. C. C. Wamser and P. L. Chang, *J. Amer. Chem. Soc.*, **95,** 2044 (1973).
149. P. S. Engel and D. J. Bishop, *J. Amer. Chem. Soc.*, **94,** 2148 (1972).
150. F. D. Greene and S. S. Hecht, *J. Org. Chem.*, **35,** 2482 (1970).
151. J. W. Timberlake and M. L. Hodges, *J. Amer. Chem. Soc.*, **95,** 634 (1973).
152. N. A. Porter and P. M. Iloff, *Chem. Commun.*, 1575 (1971).
153. N. A. Porter and L. J. Marnett, *J. Amer. Chem. Soc.*, **95,** 4361 (1973).
154. W. Lwowski and E. Scheiffele, *J. Amer. Chem. Soc.*, **87,** 4359 (1965).
155. See reference 154 for a partial review of *cis*-azoalkane chemistry.
156. P. S. Engel, L. D. Fogel, and C. Steel, *J. Amer. Chem. Soc.*, **96,** 327 (1974).
157. J. B. Hendrickson and J. Joffee, *J. Amer. Chem. Soc.*, **95,** 4084 (1973).
157a. J. W. Timberlake, M. L. Hodges and A. W. Garner, *Tetrahedron Lett.*, 3843 (1973).
158. J. P. Freeman, *J. Org. Chem.*, **28,** 2508 (1963).
158a. B. T. Gillis and J. D. Hagarty, *J. Org. Chem.*, **32,** 95 (1967).
159. H. Meister, *Justus Liebigs Ann. Chem.*, **679,** 83 (1964).
160. J. P. Snyder, L. Lee and D. G. Farnum, *J. Amer. Chem. Soc.*, **93,** 3816 (1971).
161. V. T. Bandurco and J. P. Snyder, *Tetrahedron Lett.*, 4643 (1969).
162. J. P. Snyder, L. Lee, V. T. Bandurco, C. Y. Yu and R. J. Boyd, *J. Amer. Chem. Soc.*, **94,** 3260 (1972).
163. J. Swigert and K. G. Taylor, *J. Amer. Chem. Soc.*, **93,** 7337 (1971).
164. W. Lüttke and V. Schabacker, *Justus Liebigs Ann. Chem.*, **687,** 236 (1965).
165. J. G. Aston, D. F. Menard and M. G. Mayberry, *J. Amer. Chem. Soc.*, **54,** 1535 (1932).
166. V. A. Ginsburg, L. L. Martynova, N. F. Privezentseva and Z. A. Buchek, *J. Gen. Chem. USSR*, **38,** 2422 (1968).
166a. F. Sullivan, E. Luck and P. Kovacic, *Chem. Commun.*, 217 (1974).
167. R. A. Moss, M. J. Landon, K. M. Luchter and A. Mamantov, *J. Amer. Chem. Soc.*, **94,** 4392 (1972).
168. R. A. Moss, *J. Org. Chem.*, **31,** 1082 (1966).
169. R. A. Moss and G. M. Love, *J. Amer. Chem. Soc.*, **95,** 3070 (1973).
170. R. A. Moss, private communication.
170a. R. A. Moss and G. M. Love, *Tetrahedron Lett.*, 4701 (1973).
171. T. E. Stevens and J. P. Freeman, *J. Org. Chem.*, **29,** 2279 (1964).
172. T. E. Stevens, *J. Org. Chem.*, **29,** 311 (1964).

173. S. S. Hecht and F. D. Greene, *J. Amer. Chem. Soc.*, **89**, 6761 (1967).
174. K. G. Taylor and T. Riehl, *J. Amer. Chem. Soc.*, **94**, 250 (1972).
175. W. M. Williams and W. R. Dolbier, Jr., *J. Amer. Chem. Soc.*, **94**, 3955 (1972).
176. E. F. Ullman, L. Call and S. S. Tseng, *J. Amer. Chem. Soc.*, **95**, 1677 (1973).

CHAPTER 5

Mass spectra of hydrazo, azo and azoxy compounds

S. W. TAM

Department of Chemistry, Chung Chi College, The Chinese University of Hong Kong, Shatin, N.T., Hong Kong.

I. INTRODUCTION

Although the functional groups of hydrazo, azo and azoxy compounds all contain the chain of C—N—N—C atoms, the modes of fragmentation upon electron impact depend chiefly on the nature of the bonds between these atoms as well as on the nature of the terminal carbons. Other types of compounds, whose functional groups also possess the C—N—N—C sequence of atoms, include those of the hydrazones and azines. In order to illustrate fully the effects of the bond nature on the fragmentation modes upon electron impact, the mass spectra of some hydrazones and azines are also presented.

Of the hydrazo, azo and azoxy compounds whose mass spectra have been studied, the majority were aromatic. For example, azobenzene and substituted azobenzenes have been studied much more widely than any other azo compounds.

Most of the spectra of these compounds registered peaks of rearrangement ions. In fact, skeletal rearrangement and hydrogen and carbon scrambling are common processes in the fragmentation of many of these compounds. In the following discussion, the fragmentation modes which involved or were directed by the functional groups are divided into two sections which are concerned (*i*) with ions derived by simple fragmentation, and (*ii*) with ions formed in conjunction with skeletal and/or other rearrangements. Other fragmentations observed in these spectra which were not affected by the functional groups are, in the main, not discussed.

As stressed by some authors of the original papers[1], the structures of the ions are generally not known, even though their composition may have been substantiated by exact mass measurements. The empirical structural representations are employed in order that fragmentation processes may be related to, or rationalized with, the original molecules prior to ionization[2].

II. SIMPLE FRAGMENTATIONS

This section deals with direct fragmentations involving the functional groups and taking place without any apparent rearrangement. The ions observed in these fragmentations should bear a direct relationship with the functional groups. Their usefulness from the diagnostic point of view depends very much on (*a*) the relative intensities of these ions in their mass spectra, and (*b*) the consistency with which these ions appear in the spectra of compounds containing the same functional group. The discussion in this section is divided according to the functional groups, and special attention is focused on fragmentations which are characteristic of the presence of the functional groups.

A. Azo Compounds

As mentioned earlier, azobenzene[3,4] and substituted azobenzenes[1a,5–8] have been studied more extensively than any other azo compounds. The mass spectra of other aromatic azo compounds that have been reported included those of 1- or 2-naphthyl or hydroxynaphthyl derivatives[7]. Of the aliphatic azo compounds, azomethane has been studied[9] in detail. Reports of the mass spectra of polyfluorinated azomethanes[10, 11] have also been recorded.

I. Azobenzenes

The mass spectra of all azobenzenes and other aromatic azo compounds showed prominent molecular peaks. In general, azobenzenes fragment upon electron impact in a well-defined and characteristic manner. The mass spectrum of azobenzene (**1**) shows characteristic fragmentation processes which are summarized below. The presence of a metastable ion for a process is indicated by an asterisk (*). The cleavage of the C—N bond

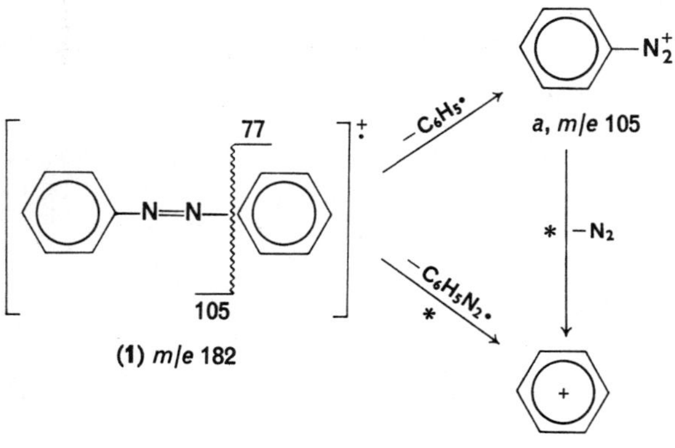

(**1**) *m/e* 182

a, *m/e* 105

b, *m/e* 77

generating two ions, *a* and *b*, is the most significant process among the simple fragmentations. Ion *a* further fragments by loss of a neutral nitrogen molecule to furnish the phenyl ion *b*. These processes are typical of symmetrical azobenzene derivatives.

Monosubstituted azobenzenes behave similarly but, due to the substitution of one of the phenyl rings, two different ions are generated as exemplified by *p*-methylazobenzene (**2**)[1a]. In general, one could expect that

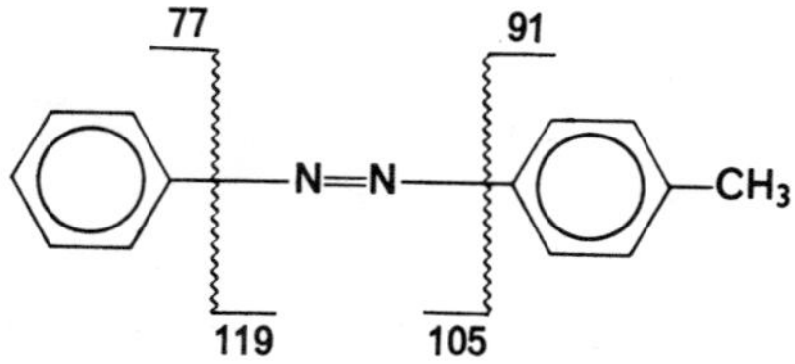

(**2**) *m/e* 196

for fragmentation processes the abundance of $A > B$ and $C > D$. This is

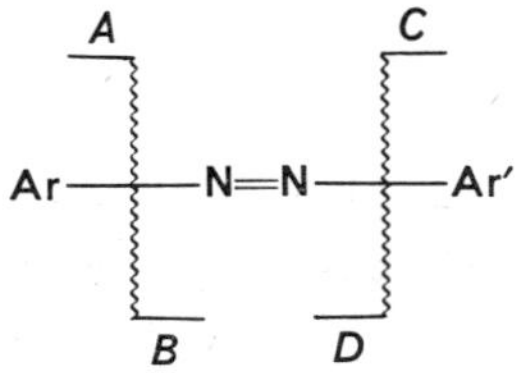

true for all *para*- and *meta*- substituted derivatives and, in general, no distinction could be observed in the mass spectra of isomers where the substituents on the benzene ring are found in different positions; *ortho* substituents exert to certain degrees the so-called '*ortho*-effect' and this becomes much more significant in the cases of *o*-OMe and *o*-CO_2Et substituted azobenzenes[1a]. A detailed discussion on the effects of these two substituents in the *ortho* position is given in the next section.

The mass spectrum of decafluoroazobenzene[8] also closely resembles that of azobenzene. There are very few nitrogen-containing fragment ions and the majority of the ions are fragments of fluorocarbons.

2. Azomethanes

The mass spectrum of azomethane (**3**) has been studied in detail by Prášil and Frost[9]. The major fragmentation is again similar to azobenzene where the weakest bond of the molecule, i.e. the C—N bond, is cleaved generating the ion *c* and the methyl cation. Based on the ionization potentials of N_2

$$\underset{(\mathbf{3})\ m/e\ 58}{[CH_3{-}N{=}N{-}CH_3]^{+\cdot}} \xrightarrow{-CH_3^{\bullet}} \underset{c,\ m/e\ 43}{CH_3N_2^+} \xrightarrow{-N_2} \underset{m/e\ 15}{CH_3^+}$$

(15·58 eV) and of CH_3 (9·83 eV), competitive cleavage of the fragment ion *c* via the following process is most unlikely and this has been substantiated by the fact that only very little N_2^+ at *m*/*e* 28 was observed.

$$\underset{c,\ m/e\ 43}{CH_3N_2^+} \longrightarrow CH_3 + N_2^+$$

Loss of a hydrogen atom from the molecular ion gives *d* which further fragments according to the process shown below.

$$\underset{(\mathbf{3})\ m/e\ 58}{[CH_3{-}N{=}N{-}CH_3]^{+\cdot}} \xrightarrow{-H^{\bullet}} \underset{d,\ m/e\ 57}{CH_3N_2CH_2^+} \xrightarrow{-CH_3^{\bullet}} \underset{m/e\ 42}{CH_2N_2^{+\cdot}}$$

The mass spectra of some polyfluorinated azomethanes have been recorded[10, 11]. Hexafluoroazomethane produces $CF_3^{+\cdot}$ as the most prominent ion. The intensity of the $CF_3^{+\cdot}$ ion from CF_3—N═N—CF_2H and CF_3—N═N—CH_3 is high but it is less abundant than CF_2H^+ and $CH_3^{+\cdot}$, respectively. In contrast to azomethane itself, all these fluorinated analogues yield a prominent $N_2^{+\cdot}$ ion[10].

B. Azoxy Compounds

A study[12] of the skeletal rearrangement processes of some aromatic azoxy compounds upon electron impact has been reviewed[13]. As in the azo compounds, the majority of the azoxy compounds whose mass spectra have been studied are aromatic substances. The mass spectra of fluorinated azoxymethanes have also been reported[10, 11].

1. Aromatic azoxy compounds

In general, the position of the *N*-oxide in the azoxy moiety can be determined by examining the mass spectra of these compounds. The most prominent fragmentation process occurs by cleaving the C—N bond α to

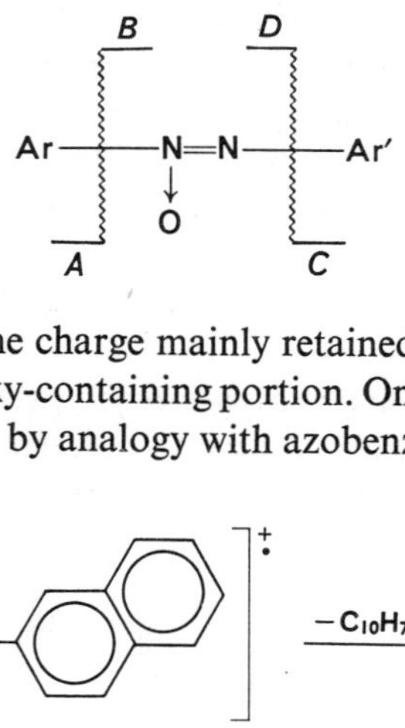

the *N*-oxide group with the charge mainly retained in the aromatic system which is free from the azoxy-containing portion. One can predict with a high degree of confidence that, by analogy with azobenzenes, for fragmentation

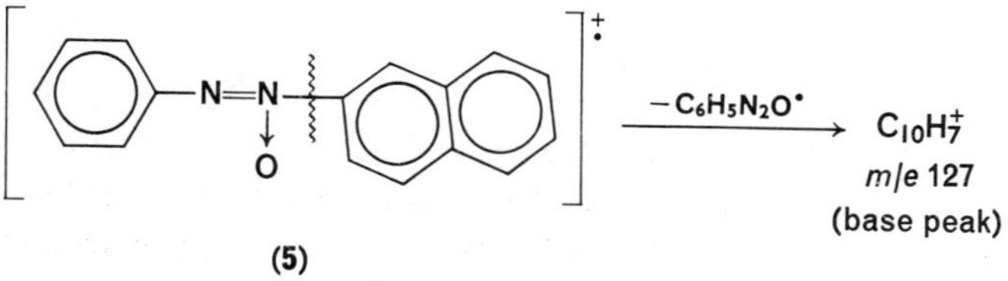

(4)

(5)

processes the relative intensities of $A > B$ and $C > D$, and A is usually the base peak of the spectrum. As an illustration[14], the base peak of the spectrum of compound **4** is the phenyl ion at *m/e* 77 and that of compound **5** is the naphthyl ion at *m/e* 127.

Studies of the metastable ions revealed that the formation of these most prominent ions did not occur in single steps as shown above, but rather via two steps as generalized below. Nevertheless, its usefulness in ascertaining the position of the *N*-oxide group is noteworthy.

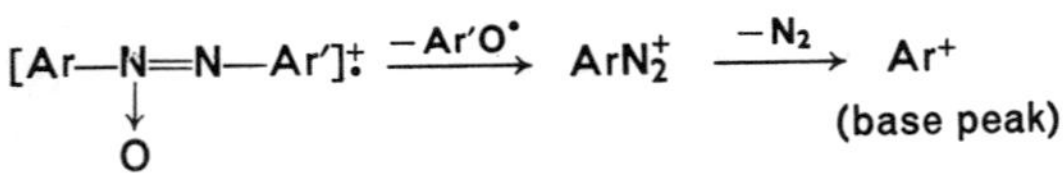

(6)

Other fragmentations which occur by direct cleavages include the loss of an oxygen atom from the molecular ion. The relative intensities of the $[M\text{-}16]^{+\cdot}$ ions vary in different compounds, but this ion is particularly prominent (the second largest peak) in the spectrum of 3,3′,4,4′,5,5′-hexamethoxyazoxybenzene[15].

One of the main features of aromatic azoxy compounds is the occurrence of several pronounced skeletal rearrangements upon electron impact. In order to avoid these rearrangement ions, negative-ion mass spectra of some aromatic azoxy compounds have been recorded[16]. The mass spectra are very simple, but quite different from the fragmentation modes in the positive-ion spectra. Loss of an oxygen atom from the molecular ion also occurs in the negative-ion spectra. The molecular ion is the base peak in all the spectra recorded[16] and fragment ions are generally of low relative intensities.

Since rearrangement processes are absent, all fragment ions registered in negative-ion spectra can be directly correlated with the structures. In general, azoxy compounds (**6**) fragment by three distinct processes, which are summarized as follows:

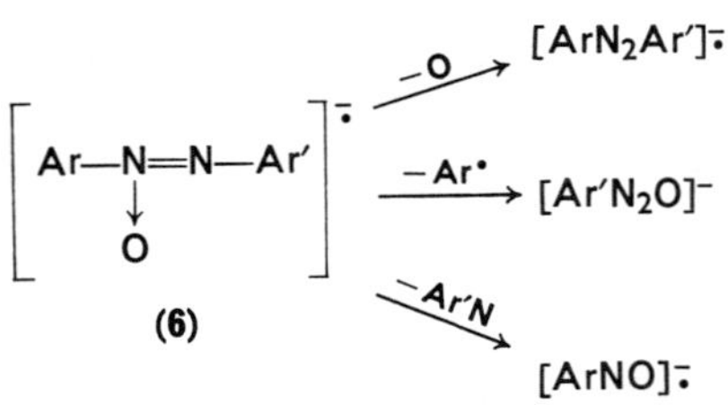

The process where the cleavage of the C—N bond α to the N→O moiety occurs is also common to all negative-ion spectra and may be used for diagnostic purposes to determine the N—O position in the azoxy group.

2. Polyfluorinated azoxymethanes

By analogy with hexafluoroazomethane, hexafluoroazoxymethane[10] produces CF_3^+ as the most abundant ion. It also yields NO^+ ion and other less abundant oxygenated ions. A pronounced ion corresponding to N_2^+ is also observed. The mass spectrum of $CF_3—N{=}N(\rightarrow O)—CH_3$ has also been recorded[10, 11].

C. Hydrazo Compounds

As far as the author is aware, no mass spectrum of a simple hydrazo compound has yet been reported. Related compounds which have been studied in mass spectrometry are mainly substituted hydrazine derivatives, e.g. *N,N*-disubstituted hydrazines[17], phenylhydrazine[17] and nitrophenylhydrazines[18], *N,N*-dimethyl-*N′,N′*-dibenzoylhydrazine (**7**)[19], and other *N,N′*-dibenzoylhydrazines (**8**)[20].

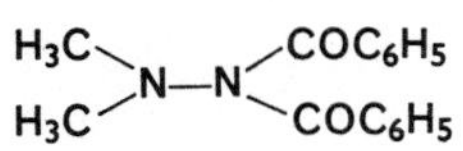

(**7**)

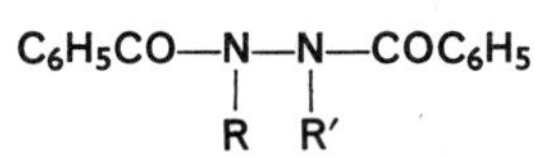

(**8**)

1. Substituted hydrazines

The mass spectra of these hydrazine derivatives reveal that simple fragmentations which do not involve some forms of skeletal rearrangement or hydrogen transfer are rare. *N,N*-Dimethylhydrazine shows a substantial $[M\text{-}15]^+$ peak in its spectrum[17] but whether this is a direct loss of a methyl radical from the molecular ion, or an expulsion of NH with the transfer of a hydrogen atom, has yet to be established. *N*-Benzoylhydrazines (**9**) fragment very readily and, in most cases, the benzoyl ion (*m/e* 105) is the base peak of the spectrum[19, 20]. The charge is retained to a much lesser extent in the complementary fragment ion of the nitrogen-containing species.

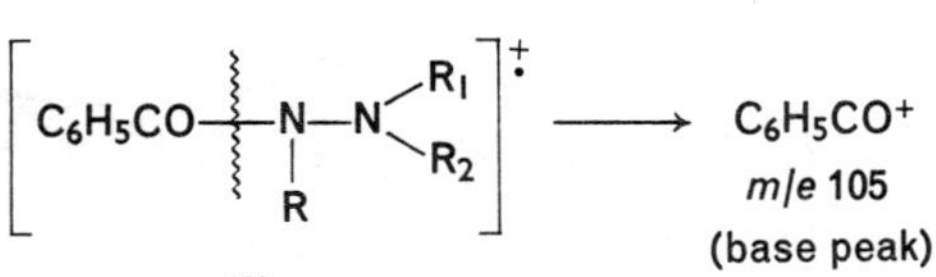

(**9**)

It is interesting to note that the fission of the N—N bond does not occur without involving the transfer of hydrogen atoms. This is also true in the case of substituted aroylhydrazones[21].

2. Aliphatic hydrazones

With the introduction of a C=N double bond to the C—N—N—C sequence which is the basic skeleton of the functional group of the three types of compounds being discussed, significant changes in the fragmentation modes are expected. Indeed, the most prominent fragmentation process in simple aliphatic dimethylhydrazones, as exemplified by butyraldehyde and valeraldehyde *N,N*-dimethylhydrazones (**10**)[22], is β cleavage of the molecular ion which leads to the formation of a stable ion, *e*, present as the base peak in these spectra. This type of allylic cleavage of the molecular

(**10**) R = H or CH_3 *e*, *m/e* 85

ion is not, however, a universally significant feature for other hydrazones; fragmentations involving transfer of one or two hydrogen atoms are much more common.

3. Aromatic azines

Further introduction of a second C=N double bond to the C—N—N—C skeleton gives azines where, in contrast to the other classes of compounds, direct fission of the N—N bond is probable. The mass spectra of a series of aromatic aldazines and ketazines[23] have been studied and they all show fragment ions of moderate intensities derived by the simple cleavage of the central N—N bond. As an illustration, the molecular ion of acetophenone azine (**11**) (*m/e* 236) fragments to yield *f* (*m/e* 118) whose composition is half that of the parent compound[24].

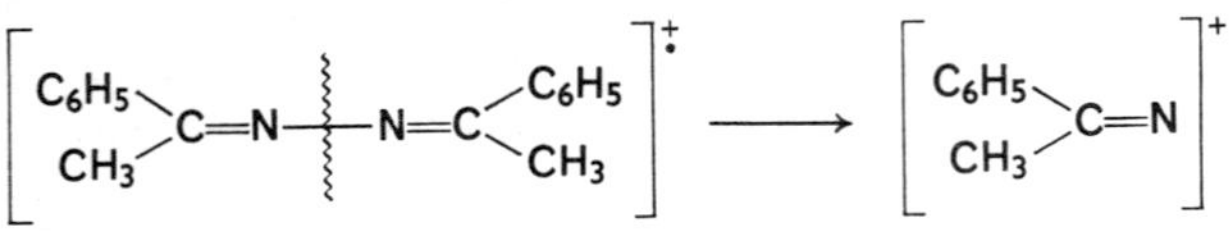

(**11**) *m/e* 236 *f*, *m/e* 118

III. FRAGMENTATIONS INVOLVING REARRANGEMENTS

Skeletal rearrangements or hydrogen migrations in mass spectrometry have been subjects of intensive study[25] in recent years. Many of the fragmentations which cannot occur by simple cleavage of the molecules can be visualized as taking place through rearrangement processes.

It is a characteristic feature of the mass spectra of the types of compounds under review that skeletal rearrangements commonly occur. Some of the rearrangement ions are purely due to the functional groups, e.g. as in the case of azoxy compounds[12], but other rearrangement ions are derived from the presence of particular substituents, e.g. *o*-OMe or *o*-CO_2Et in azo compounds[1a]. This section is concerned with fragmentations accompanied by rearrangements (either skeletal or hydrogen migration or both) which involve or are affected by the functional groups. Most processes which do not fall within this scope are omitted.

A. Azo Compounds

One of the well-established processes of skeletal rearrangements in mass spectrometry is the expulsion of a bridging group in the form of a small neutral molecule as represented by the general equation

$$ABC^{+\cdot} \longrightarrow AC^{+\cdot} + B$$

This type of fragmentation is common when *A* and *C* are aromatic systems. Aromatic azo compounds fall into this category and fragmentation according to the equation

$$[\mathrm{Ar{-}N_2{-}Ar'}]^{+\cdot} \longrightarrow [\mathrm{Ar{-}Ar'}]^{+\cdot} + \mathrm{N_2}$$

is expected. Substituents in the aromatic systems *meta* or *para* to the C—N bonds normally do not affect this fragmentation process but, in some cases, the substituents or part of the substituents may be cleaved from the molecular ion prior to the loss of the nitrogen molecule. '*ortho*-Effects' may become significant on certain occasions when the cleavage according to the above equation is replaced by other fragmentation processes.

1. Azobenzenes

Azobenzenes have been studied extensively[1a, 3–8], and they all show skeletal rearrangement ions in their spectra. The fragmentation processes summarized below for azobenzene are characteristic and common to all azobenzenes except for the *o*-OMe and *o*-CO_2Et analogues[1a]. Expulsion of a neutral nitrogen molecule gives ion *g* which further fragments with loss of two hydrogen atoms to yield an ion which may be the biphenylene radical ion *h*[1a, 3, 5]. Study of the mass spectra of azobenzene-2,2′,4,4′,6,6′-d_6 (**12**) revealed that loss of the two hydrogen atoms from *g* occurs after the

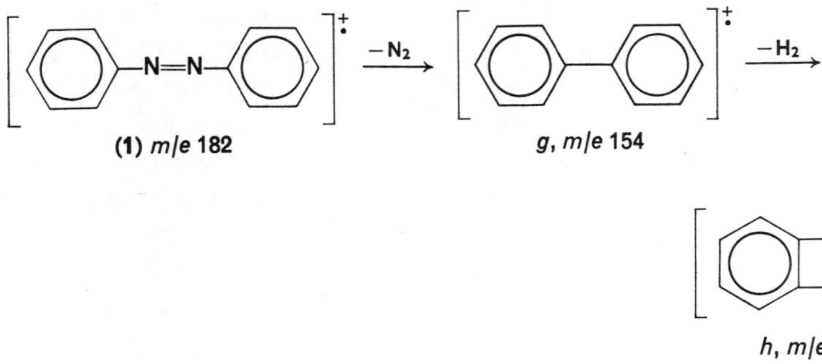

scrambling of the hydrogen and deuterium atoms[4]. The spectrum of 4,4′-dimethylazobenzene-2,2′,6,6′-d_4 (**13**) further indicates that complete randomization between the ring and methyl hydrogens has taken place[4] prior to the elimination of the two hydrogen atoms.

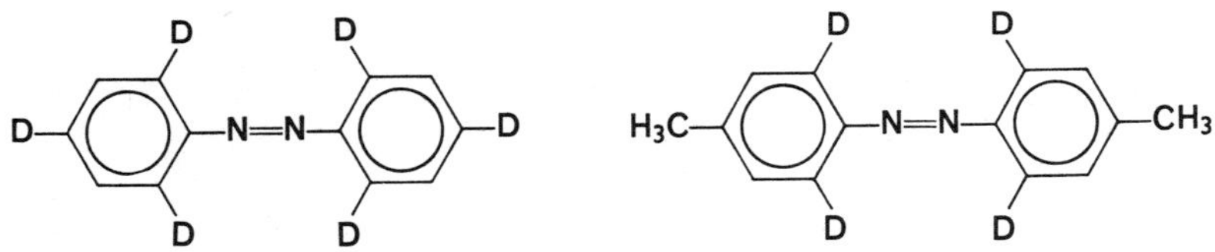

(12) (13)

All monosubstituted azobenzenes fragment by elimination of a nitrogen molecule but other modes of fragmentation depend on the nature of the substituents[1a]. The biphenylene radical ion *h* (*m*/*e* 152) is observed in all spectra and is formed by eliminating N_2, $R^{\cdot}$ and $H^{\cdot}$ from the molecular ion. However, the order of eliminations varies depending upon the nature of the substituents[1a].

The ion corresponding to *g* is absent from the mass spectrum of decafluoroazobenzene, however. The rearrangement with the expulsion of a nitrogen molecule is not an appreciable process in this polyfluorinated compound[8].

The presence of the *o*-OMe substituent exerts a strong '*ortho*-effect' and an additional peak corresponding to an amine ion (*i*) is observed in the spectra of *o*-methoxyazobenzenes. This additional peak is absent from the spectra of the *m*- and *p*-methoxy analogues. The genesis of the ion *i* can be rationalized by N=N cleavage with the migration of two hydrogen atoms from the OMe group to the β nitrogen as illustrated below[1a].

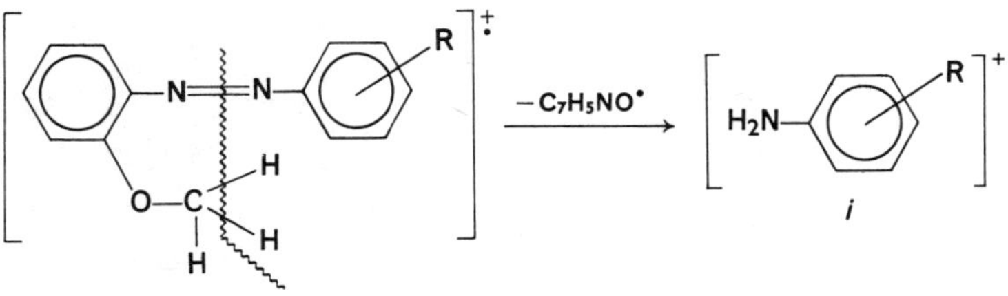

This rearrangement process has been substantiated by deuterium labelling[1a], and the ions *i* are observed without exception in all *o*-methoxy-azobenzenes.

2. Azomethanes

The rearrangement process, $ABC^{+\cdot} \rightarrow AC^{+\cdot} + B$ is not limited to aromatic systems, but is also observed in aliphatic systems. The ethane ion (*m/e* 30) has been recorded in the mass spectrum of azomethane[9]. The formation of the ethane ion must undoubtedly occur via the rearrangement process

$$\underset{(3)}{[CH_3{-}N_2{-}CH_3]^{+\cdot}} \longrightarrow \underset{m/e\ 30}{[CH_3{-}CH_3]^{+\cdot}} + N_2$$

3. Aromatic azines

The main difference between the basic structures of azo compounds (C—N=N—C) and of azines (C=N—N=C) is that (*i*) there is an extra degree of unsaturation in azines and (*ii*) the double bonds are located at different positions in the C—N—N—C skeleton as shown. Upon electron impact, azine molecules also fragment with loss of a bridging neutral molecule according to the rearrangement process $ABC^{+\cdot} \rightarrow AC^{+\cdot} + B$. However, instead of giving off a nitrogen molecule, loss of HCN or an alkyl nitrile molecule from the azine molecular ions is commonly observed. The fragmentation of benzaldazine[23,26,27] as described below is characteristic and common to most aromatic aldazines except where strong '*ortho*-effects' are operative.

Expulsion of HCN from the molecular ion of benzaldazine (**14**) gives the radical ion of an anil (**15**) from which a hydrogen atom is cleaved generating the ion *j* (*m/e* 180). The ion *j* is also derived from the azine molecular ion via the sequential loss of $H^{\cdot}$ and HCN in two steps. That both routes are operative is supported by the observation of appropriate metastable ions in the spectrum. Further loss of [HCN + H˙] from *j* yields the biphenylene radical ion *h* (*m/e* 152) which is also a prominent ion in the mass spectra of azobenzenes.

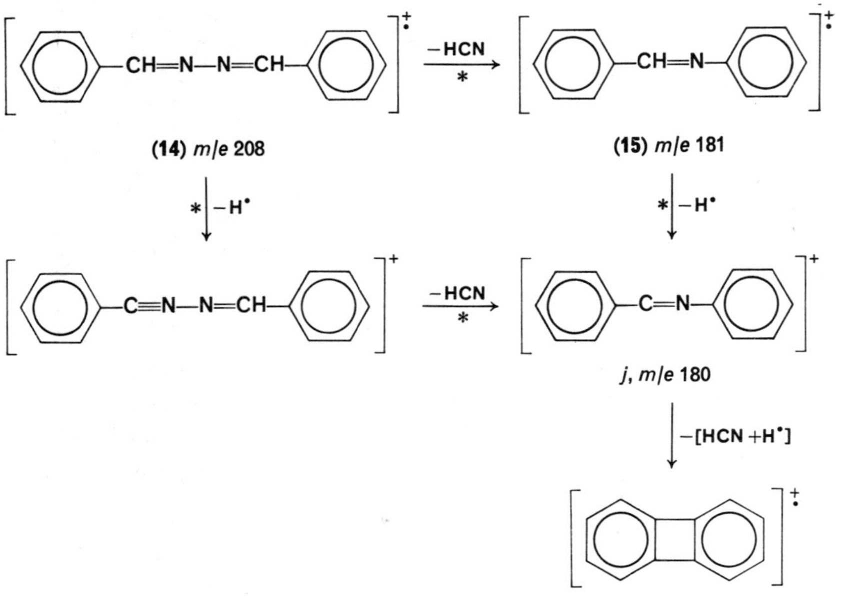

The spectrum of benzaldazine below *m/e* 181 is in fact similar to that[28] of the anil (**15**) which indicates that the process as shown above for **14** → *h* is operative.

Studies of the deuterium-labelled compounds, **16**[23,26] and **17**[26,27], revealed that loss of the methine hydrogen atom from the molecular ion occurs after the scrambling of the aromatic hydrogen and the methine hydrogen. This is in contrast to the d_1-derivative of the anil (**15**)[28] where randomization of the aromatic hydrogen and the methine deuterium atom does not occur in forming the $[M\text{-}1]^+$ ion.

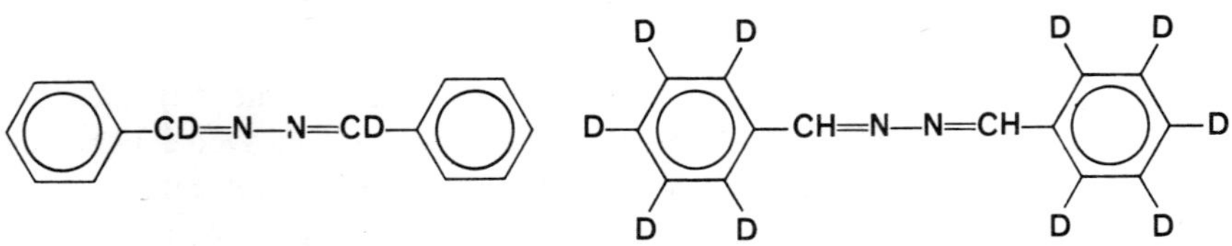

(**16**) (**17**)

In the case of acetophenone azine (**18**)[23,24,29,30] where the two methine hydrogens are substituted by two methyl groups, both fragmentation processes, $[M^{+\cdot} \overset{*}{-} \mathrm{MeCN} \overset{*}{-} \mathrm{Me}]$ and $[M^{+\cdot} \overset{*}{-} \mathrm{Me} \overset{*}{-} \mathrm{MeCN}]$, occur. These fragmentation modes are similar to those observed in **14**.

In contrast to benzaldazine (**14**), the relatively minor $[M\text{-}1]^+$ ion from acetophenone azine (**18**) is specifically derived by expelling an aromatic hydrogen from one of the phenyl rings; this postulation[23] has been confirmed by the study of the spectra of deuterated derivatives **19** and **20**[29]. In the same study it has been shown that the elimination of the methyl radical, generating the $[M\text{-}15]^+$ ion from **18** occurs without much loss of the structural identity of the original azine molecule.

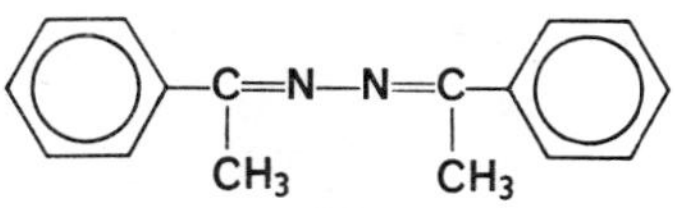

(**18**)

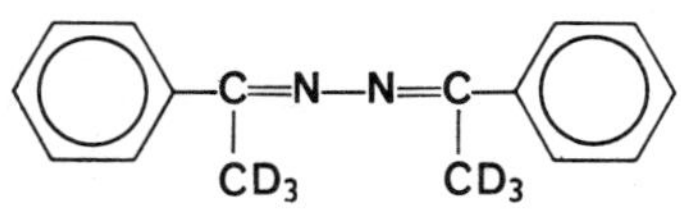

(**19**)

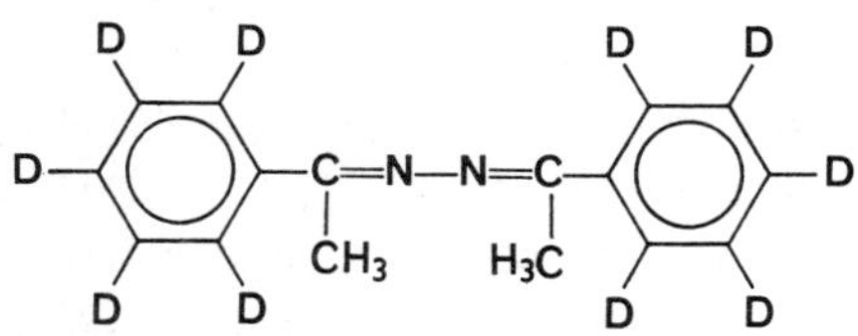

(**20**)

B. Azoxy Compounds

Aromatic azoxy compounds are known to exhibit numerous rearrangement ions in their mass spectra[12–14]. Bowie and co-workers[12,14] generalized these fragmentation–rearrangement modes into four processes.

(1) $[M^{+\cdot}\text{-}N_2O]$. This rearrangement process belongs to the general type of $ABC^{+\cdot} \rightarrow AC^{+\cdot} + B$, similar to that observed in azo compounds and azines, furnishing biaryl radical ions. However, such ions, when observed, are generally of low intensity in the spectra of azoxy compounds.

(2) $[M^{+\cdot}\text{-}CO - N_2 - H^{\cdot}]$. The process is particularly diagnostic and occurs in all spectra of azoxy compounds. The loss of carbon monoxide followed by the nitrogen molecule may take place in successive steps, $[M^{+\cdot}\text{-}CO - N_2 - H^{\cdot}]$, or in a concerted manner, $[M^{+\cdot}\text{-}(CO + N_2) - H^{\cdot}]$, depending on the nature of the substituents. The occurrence of metastable ions, substantiating these fragmentation processes, is commonly observed.

Expulsion of carbon monoxide from azoxybenzene (**21**) upon electron impact involves rearrangement of the molecular ion. The possibility of forming *o*-hydroxyazobenzene (**22**) as an intermediate in the rearrangement process is excluded by the fact that the mass spectrum[1a] of **22** exhibits the loss of a nitrogen molecule prior to the loss of carbon monoxide, $[M^{+}-N_2 - CO - H^{\cdot}]$[12]. Consequently, the ion *k* has been proposed as an alternative intermediate to account for this rearrangement process.

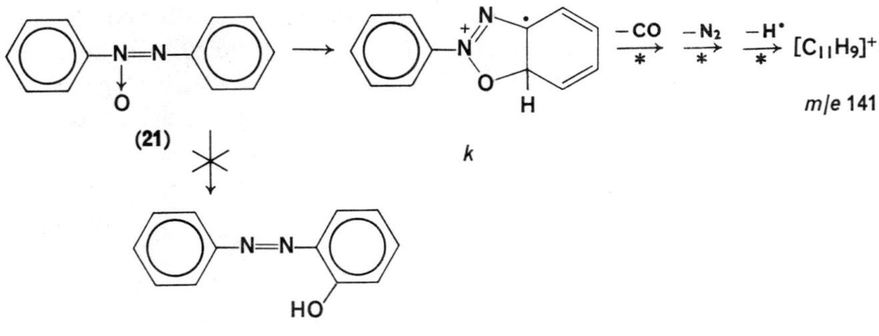

(21) *k* (22)

The mass spectra of *o*-substituted azoxy compounds may be entirely different from those of the *m*- and *p*-analogues. *p,p′*-Dimethylazoxybenzene and *m,m′*-dimethylazoxybenzene fragment in a similar manner to azoxybenzene itself. However, *o,o′*-dimethylazoxybenzene (**23**) exhibits a prominent $[M\text{-}15]^{+}$ ion, *l* (*m/e* 211).

In fact, the rearrangement involving loss of CO and N_2 only occurs from the ion *l*, resulting in the overall process $[M^{+\cdot}\text{-}CH_3^{\cdot} - CO - N_2]$[14]. A pronounced $[M\text{-}OH]^{+}$ ion, *m* (*m/e* 209) is also observed which is absent from the spectra of the other isomers[14].

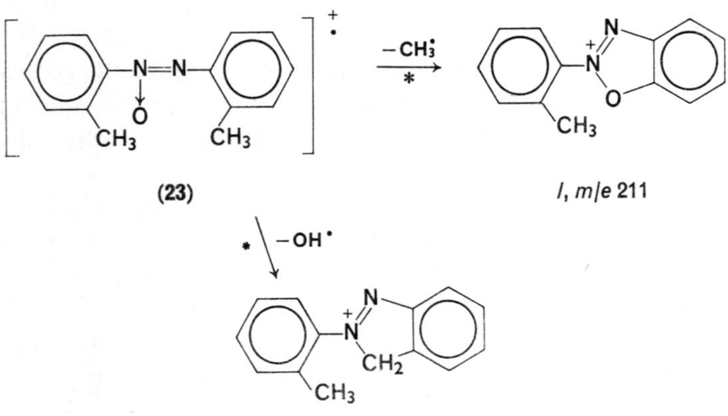

(23)

l, *m/e* 211

m, *m/e* 209

Compound **24**, where the **A** ring is unsubstituted and the two *ortho* positions of the **B** ring are occupied by methyl substituents, shows even more significant peaks for $[M\text{-}CH_3]^+$ and $[M\text{-}OH]^+$ ions in its spectrum. These observations substantiate the above fragmentation modes as suggested and indicate that interaction occurs between the oxygen and the *o*-methyl substituents on the **B** ring[14].

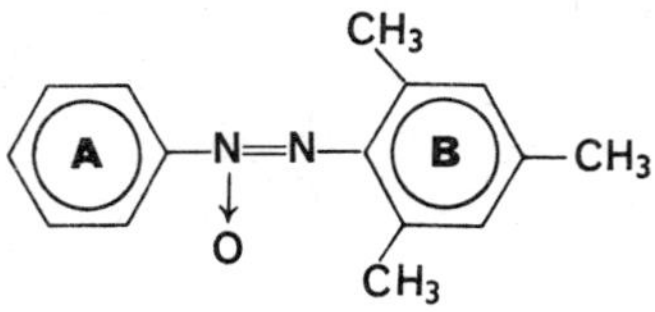

(24)

(3) $[M^{+\cdot}\text{-}(\text{Ring } \mathbf{B}+\text{O})^{\cdot}]$. The process involves the migration of the oxygen atom to ring **B**, probably via intermediate ions, *k* and *n*; cleavage of the N—O bond in ion *n* generates the rearrangement ion $[M^{+\cdot}\text{-}(\text{ring } \mathbf{B}+\text{O})^{\cdot}]$[14].

(4) $[M^{+\cdot}\text{-}(\text{Ring } \mathbf{B}+\text{NO})^{\cdot}]$. Cleavage of the N=N bond in ion *n* yields another rearrangement ion $[M^{+\cdot}\text{-}(\text{ring } \mathbf{B}+\text{NO})^{\cdot}]$[14].

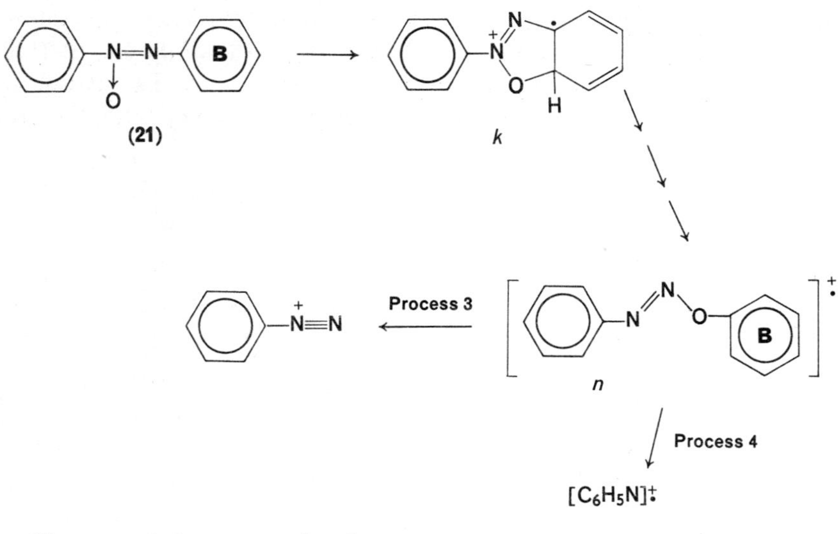

However, it is noteworthy that processes 3 and 4 are observed in only some of the spectra[14].

C. Substituted Hydrazines

In the cases of *N*-substituted and *N,N*-disubstituted hydrazines where one of the amino groups remains free, losses of N, NH, NH_2 and NH_3 have been observed[17]. Apart from the loss of NH_2 which may be realized by the direct fission of the N—N bond, elimination of the other groups must involve some form of hydrogen migration.

Loss of NH is significant in this class of compounds and pronounced $[M\text{-}15]^+$ peaks are usually observed in spectra of *N*-amino compounds[17]. Phenylhydrazine, *N*-methyl-*N*-phenylhydrazine, and *N*,*N*-dimethylhydrazine all show the $[M\text{-}NH]^+$ ions with moderate intensities. Expulsion of the NH group from the molecular ion of *N*,*N*-diphenylhydrazine is so prominent that this ion forms the base peak of its spectrum[17].

N-Methyl-*N*-phenylhydrazine has the base peak at $[M\text{-}17]^{+\cdot}$ which could be represented by ion *o*[17].

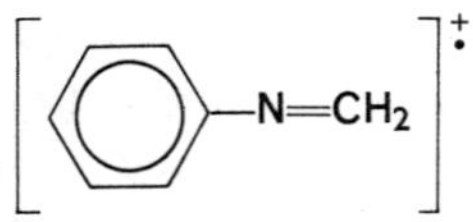

o, *m*/*e* 105

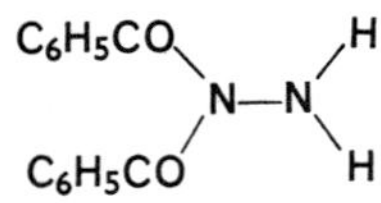

(25)

N,*N*-Dibenzoylhydrazine (**25**) or *N*,*N'*-dibenzoylhydrazine (**8**, R = R' = H) fragments in a very simple manner[20]. Substitution of alkyl or aryl groups on one or both nitrogen atoms increases the complexity of the spectra and fragmentations involving rearrangements become important. Two hydrogen atoms are transferred from the same methyl group in compound **7** to the

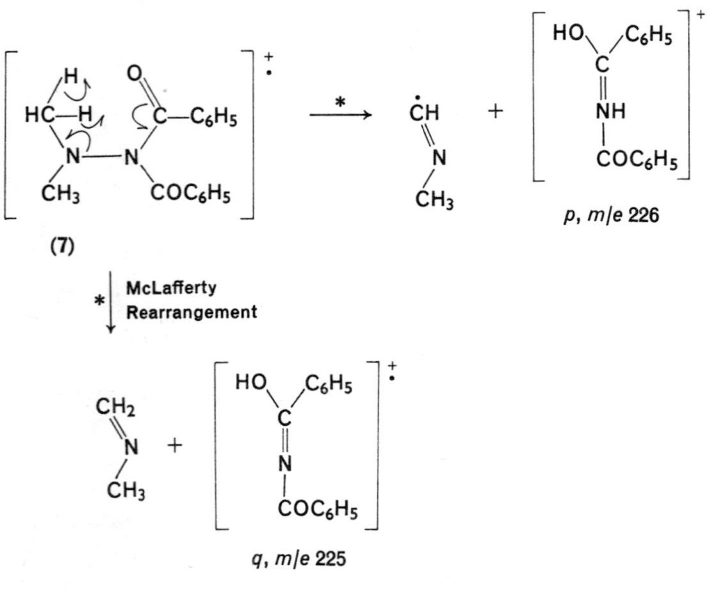

benzoyl moiety with fission of the N—N bond to yield the ion *p* (*m/e* 226). The concerted nature of this process is supported by the appearance of a metastable ion[19].

The proposed mode of fragmentation for the double hydrogen migration is substantiated by deuterium labelling[19] in an experiment where one of the methyl groups was substituted by a CD_3 group. Transfer of a single hydrogen also occurs but it is likely to take place via a normal McLafferty rearrangement[19], as illustrated above.

The analogue of compound (**7**), *N,N'*-dimethyl-*N,N'*-dibenzoylhydrazine (**26**), fragments in a different manner. $[M\text{-}CH_2N]^+$ is a characteristic peak in the spectra of *N,N'*-dialkylated-*N,N'*-dibenzoylhydrazines and its formation can be rationalized as the result of a 1,2-benzoyl migration, followed by a McLafferty-type rearrangement[20].

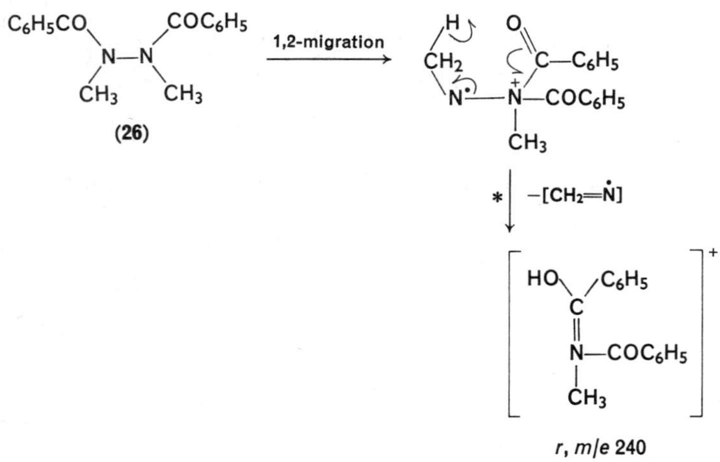

Other fragmentation modes which involve N—N bond fission are always accompanied by single or double hydrogen migration. This is also true for other *N*-alkyl or *N*-aryl-*N,N'*-dibenzoylhydrazines[20].

IV. CONCLUSIONS

The types of azo and azoxy compounds which have been studied in mass spectrometry are mainly substituted derivatives of simple aromatic systems. Nevertheless, many of the fragmentation modes, with or without rearrangements, are characteristic and typical for these compounds. This is chiefly due to the fact that the adjacent nitrogen atoms together with double bonds in the C—N—N—C skeleton provide appropriate sites where charges are localized after the ionization of the molecules upon electron impact[22]. Skeletal rearrangements and hydrogen migrations are common processes, the majority of which are well-studied and documented.

V. REFERENCES

1a. J. H. Bowie, G. E. Lewis and R. G. Cooks, *J. Chem. Soc.* (*B*), 621 (1967).
1b. R. J. Liedtke, A. M. Duffield and C. Djerassi, *Org. Mass Spectrom.*, **3,** 1089 (1970).
2. For detailed discussions, see H. Budzikiewicz, C. Djerassi and D. H. Williams, *Mass Spectrometry of Organic Compounds*, Holden-Day, San Francisco, 1967, pp. 6–8.
3. J. H. D. Eland and C. J. Danby, *J. Chem. Soc.*, 5935 (1965).
4. P. C. Wszolek, F. W. McLafferty and J. H. Brewster, *Org. Mass Spectrom.*, **1,** 127 (1968).
5. J. A. Völlmin, P. Pachlatko and W. Simon, *Helv. Chem. Acta*, **52,** 737 (1969).
6. N. S. Vul'fson, V. A. Puchkov and Yu. S. Nekrasov, *Izvest. Akad. Nauk S.S.S.R., Ser. Khim*, 1881 (1967).
7. Yu. S. Nekrasov, V. A. Puchkov and N. S. Vul'fson, *Zh. Obshch. Khim.*, **38,** 1385 (1968).
8. G. F. Lanthier and J. M. Miller, *Org. Mass Spectrom.*, **6,** 89 (1972).
9. Z. Prášil and W. Frost, *J. Amer. Chem. Soc.*, **90,** 3344 (1968).
10. S. S. Dubov, A. M. Khokhlova and N. P. Rodionova, *Zh. Vses. Khim. Obshchestva im D. I. Mendeleeva*, **7,** 692 (1962).
11. S. S. Dubov and A. M. Khokhlova, *Zh. Obshch. Khim.*, **34,** 586 (1964).
12. J. H. Bowie, G. E. Lewis and R. G. Cooks, *Chem. Comm.*, 284 (1967).
13. Reference 2, pp. 524–525.
14. J. H. Bowie, R. G. Cooks and G. E. Lewis, *Austral. J. Chem.*, **20,** 1601 (1967).
15. H. I. Bolker and F. L. Kung, *J. Chem. Soc.* (*C*), 2298 (1969).
16. J. H. Bowie, S.-O. Lawesson, B. S. Larsen, G. E. Lewis and G. Schroll, *Austral. J. Chem.*, **21,** 2031 (1968).
17. R. G. Gills, *Org. Mass Spectrom.*, **5,** 1107 (1971).
18. F. Benoit and J. L. Holmes, *Can. J. Chem.*, **47,** 3611 (1969).
19. Q. N. Porter and A. E. Seif, *Org. Mass Spectrom.*, **4,** 361 (1970).
20. Q. N. Porter and A. E. Seif, *Austral. J. Chem.*, **25,** 523 (1972).
21. D. G. I. Kingston, H. P. Tannenbaum, G. B. Baker, J. R. Dimmock and W. G. Taylor, *J. Chem. Soc.*, (*C*), 2574 (1970).
22. D. Goldsmith and C. Djerassi, *J. Org. Chem.*, **31,** 3661 (1966).
23. R. G. Cooks and S. W. Tam., *Org. Mass Spectrom.*, **1,** 583 (1968).
24. S. E. Scheppele, R. D. Grigsby, E. D. Mitchell, D. W. Miller and G. R. Waller, *J. Amer. Chem. Soc.*, **90,** 3521 (1968).
25. A number of good reviews on this subject are available, e.g. (*a*) P. Brown and C. Djerassi, *Angew. Chem. Inter. Ed., Engl.*, **6,** 477 (1967); (*b*) R. G. Cooks, *Org. Mass Spectrom.*, **2,** 481 (1969); (*c*) J. H. Bowie, B. K. Simons and S.-O. Lawesson, *Rev. Pure Appl. Chem.* (*Australia*), **19,** 61 (1969).

26. S. E. Scheppele, R. D. Grigsby, K. F. Kinneberg, E. D. Mitchell and C. A. Mannan, *Org. Mass Spectrom.*, **3,** 557 (1970).
27. N. P. Buu-Hoi and G. Saint-Ruf, *Bull. Soc. Chim. France*, 661 (1968).
28. J. H. Bowie, R. G. Cooks, J. W. Fisher and T. McL. Spotswood, *Austral. J. Chem.*, **21,** 2021 (1968).
29. S. E. Scheppele, R. D. Grisby, D. W. Whitaker, S. D. Hinds, K. F. Kinnerberg and R. D. Mitchum, *Org. Mass Spectrom.*, **3,** 571 (1970).
30. N. P. Buu-Hoi and G. Saint-Ruf, *Bull. Soc. Chim. France*, 1861 (1970).

CHAPTER 6

Chiroptical properties of the azo and the azoxy chromophores

G. SNATZKE and F. SNATZKE

Lehrstuhl für Strukturchemie, Ruhr-Universität, Bochum, Germany

I. THE AZO CHROMOPHORE

A. MO-Treatment and Electron Absorption Spectra

The geometry and electron spectra of azo compounds can easily be understood on the basis of a simple MO treatment[1] which is only quantitatively but not qualitatively changed by refined and more sophisticated calculations[2–6]. The two nitrogens are assumed to be sp^2-hybridized and the remaining p-electrons are used to build up a π-orbital between the two N-atoms. This leads to a coplanar but non-linear arrangement for azo compounds and the barrier of rotation is high enough to allow isolation of *cis*- and *trans*-isomers in some cases. In general the *trans*-diastereomer is energetically favoured over the *cis* one, though exceptions are possible. The bond angle Θ (N—N—C) is 111·9° ± 0·5° for azomethane (**1**) as found from

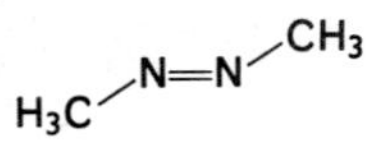

(**1**)

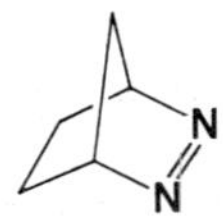

(**2**)

electron diffraction measurements[7] which is in best agreement with such an sp^2-hybridization.

The two lone pairs on the nitrogen atoms interact with each other giving rise to two linear combinations corresponding to two irreducible representations of the respective point groups (*trans*: C_{2h}; *cis*: C_{2v}). As can be seen from Figure 1 overlap for the symmetrical combination $(n_1 + n_2)$ is

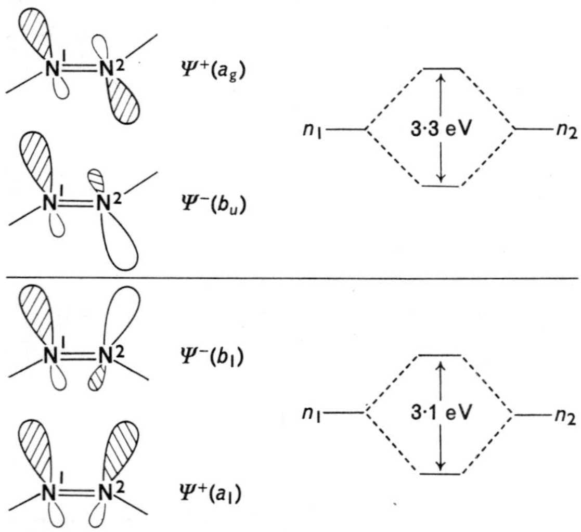

FIGURE 1. Simple MO-picture[1] of the interacting non-bonding orbitals at N^1 and N^2 of an azo compound and energy splitting as determined from photoelectron spectroscopy[8,9]. *Top*: *trans*-azo compounds; *Bottom*: *cis*-azo compounds

negative for the *trans*-, and positive for the *cis*-azo alkanes, so for the *trans*-isomer

$$\psi^+(a_g) = \mathcal{N}^+ . (n_1 + n_2)$$

is the higher lying,

$$\psi^-(b_u) = \mathcal{N}^- . (n_1 - n_2)$$

the lower lying n-orbital combination, whereas for the *cis*-isomers this is reversed, $\psi^+(a_1)$ being the lower, $\psi^-(b_1)$ the higher MO. The group theory notations follow from the symmetry properties of these orbitals (cf. Tables 1 and 2). Most of the calculations derived at the same result as this simple picture, but there was considerable discrepancy on the magnitude of this splitting (cf. the comparison in ref. 5).

Photoelectron spectroscopy is a means to measure directly orbital energies by application of Koopmans' theorem and gives an energy difference of 3·3 eV for the *trans*-case [azomethane (**1**)[8]] and 3·1 eV for a *cis*-compound [diazabicycloheptene (**2**)[9]]. Of course this method can only give the energy splitting for these two n-orbitals but not their assignments, which are however not changed by calculations like MINDO/1, MINDO/2, CNDO or Extended Hückel[5]. Furthermore it is clearly shown by photoelectron spectroscopy that the π-orbital energy is in between the two n-orbital energies[8, 9]. In the naive one-electron picture one would then assume that the bands corresponding to the respective transitions into the π^*-orbital shall be found in the sequence $n \rightarrow \pi^*$-, $\pi \rightarrow \pi^*$-, $n \rightarrow \pi^*$-. Calculations taking into account configurational interaction predict however, that for *trans*-azo alkanes like **1** the $\pi \rightarrow \pi^*$-band shows up at shortest wavelengths (cf. 4). Also for *cis*-azo alkanes the $\pi \rightarrow \pi^*$-band is calculated to appear at shorter wavelengths than the two $n \rightarrow \pi^*$-bands, with some $n \rightarrow \sigma^*$-transitions lying also at relative low energy[4].

From symmetry considerations (cf. Table 1) follows that for *trans*-azo compounds the $\psi^+ \rightarrow \pi^*$-transition corresponding to the band at longest wavelengths is forbidden whereas the $\psi^- \rightarrow \pi^*$-transition is symmetry allowed and out-of-plane polarized ($^1A_1 \rightarrow {}^1A_u$). For the long-wavelengths band at 340–380 nm ε is indeed very small (e.g.[10] **1**: $\varepsilon = 5$; *trans*-t-butyl analogue: $\varepsilon = 13{\cdot}5$; 1-adamantanyl analogue: $\varepsilon = 15{\cdot}8$). For the *cis*-compounds the situation is reversed (cf. Table 2), the long-wavelengths band belonging to an allowed $\psi^- \rightarrow \pi^*$-transition, which is also out-of-plane polarized ($^1A_1 \rightarrow {}^1B_2$), whereas the short-wavelengths band corresponds to the forbidden $\psi^+ \rightarrow \pi^*$-transition. In agreement with this prediction the

TABLE 1. Character table for point group C_{2h}

C_{2h}	E	C_2^z	i	σ	Transitions of *trans*-azo compounds	Components of $\vec{\mu}$	Components of $\vec{m}$
A_g	+1	+1	+1	+1		—	z
B_g	+1	−1	+1	−1	$\psi^+ \rightarrow \pi^*$	—	x, y
A_u	+1	+1	−1	−1	$\psi^- \rightarrow \pi^*$	z	—
B_u	+1	−1	−1	+1	$\pi \rightarrow \pi^*$	x, y	—

TABLE 2. Character table for point group C_{2v}

C_{2v}	E	C_2^z	$\sigma_v(xz)$	$\sigma_v(yz)$	Transitions of *cis*-azo-compounds	Components of $\vec{\mu}$	Components of $\vec{m}$
A_1	+1	+1	+1	+1		z	—
A_2	+1	+1	−1	−1	$\psi^+ \rightarrow \pi^*$	—	z
B_1	+1	−1	+1	−1	$\pi \rightarrow \pi^*$	x	y
B_2	+1	−1	−1	+1	$\psi^- \rightarrow \pi^*$	y	x

long-wavelengths band of 3-methyl-1-pyrazoline (**3**) is more than one order of magnitude stronger ($\varepsilon = 225$)[4] than that of **1**. For the spiro-pyrazoline derivatives **5** and **7** we[11] measured $\varepsilon = 340$ and 320, respectively, for the same band at 329 nm (in dioxan).

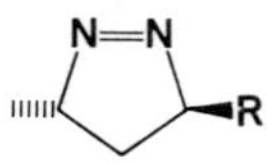

(3) R = H

(4) R = CH_3 (or enantiomer)

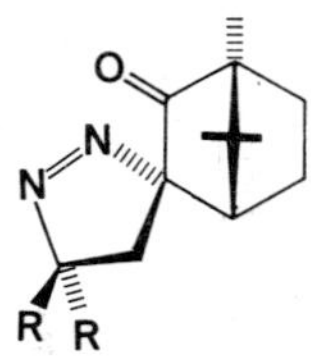

(5) R = H

(6) R = CH_3

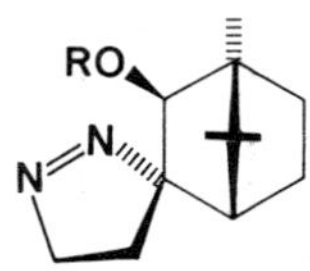

(7) R = H

(8) R = Ac

Conjugation with one or two benzene rings introduces a strong bathochromic shift for the longest-wavelengths band, e.g. to 449 nm for *trans*-azobenzene (**9**)[13] and 440 nm for *cis*-azobenzene (**10**)[14]. Intensity borrowing

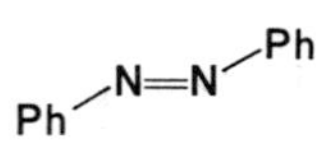

(9)

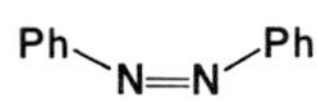

(10)

leads to a pronounced hyperchromicity even for the *trans*-compound ($\varepsilon = 405$)[13], but the corresponding band of the *cis*-azobenzene (**10**) still has ε three times bigger (1250)[14]. There is no complete agreement about the assignment of the other bands[2,15].

B. Chiroptical Properties

1. General

Optical activity[16]—and especially a Cotton effect (Circular Dichroism: CD; anomalous Optical Rotatory Dispersion: ORD)—can occur only if light excites the electron on a helical 'path', i.e. by generating an electric ($\vec{\mu}$) as well a magnetic transition moment ($\vec{m}$), which must not be perpendicular one to each other. The rotational strength R which is in principle the area under a particular CD-band is given by the equation $R = \mu m \cos(\vec{\mu},\vec{m})$.

Rotational strength is acquired if e.g. (*a*) the chromophoric system itself is chiral ('chirality of the first sphere'[18], both $\vec{\mu} \neq 0$ and $\vec{m} \neq 0$, $\Theta(\vec{\mu},\vec{m})$ near 0° or 180°, 'inherently chiral chromophore'[19]), (*b*) the ring into which the azo chromophore is incorporated is helical or chiral ('chirality of the second sphere'), or (*c*) if further rings or substituents are chirally arranged around the chromophore ('chirality of the third, fourth, ..., sphere'). The first sphere can become chiral either by twisting the azo grouping or by chiral interaction with another group containing π- or n-orbitals which strongly interact with the chromophoric system; only examples of the last type are known hitherto.

A second chiral sphere can occur only if the corresponding ring is at least six-membered as long the —N=N— grouping is planar; pyrazolines have at least a local C_s-symmetry, their ring is, therefore, achiral.

That the planar azo chromophore is inherently achiral follows quite clearly from Tables 1 and 2. In the *trans*-compound of C_{2h}-symmetry the $\psi^+ \rightarrow \pi^*$-transition is electrically forbidden ($\vec{\mu} \equiv 0$), but magnetically allowed (degenerate, $m_x \neq 0$, $m_y \neq 0$), whereas the $\psi^- \rightarrow \pi^*$-transition is electrically allowed ($\vec{\mu} \neq 0$), but acquires no magnetic moment ($\vec{m} \equiv 0$). For the *cis*-compounds the $\psi^- \rightarrow \pi^*$-transition is magnetically ($m_x \neq 0$) and electrically ($\mu_y \neq 0$) allowed, the two moment vectors are, however, perpendicular to one another, so again R will be zero. The $\psi^+ \rightarrow \pi^*$-transition is magnetically allowed ($m_z \neq 0$), but has no electric transition moment ($\vec{\mu} \equiv 0$).

Twisting of the —N=N— chromophore reduces the symmetry to C_2 and both the $^1A \rightarrow {}^1A$- and $^1A \rightarrow {}^1B$-transition have parallel electric and magnetic transition moment vectors, a prerequisite for 'inherent chirality'[19].

2. Azoalkanes

Kosower and Severn[10, 20] were the first who have thoroughly investigated the chiroptical properties (ORD and CD) of azoalkanes, viz. those of **11** and **12**. The CD of **11** (or enantiomer, absolute configuration unknown) is +0·85 at 383 nm (anomalous ORD found at 387 nm) with an anisotropy factor[16] $g' = \Delta\varepsilon/\varepsilon = 0{\cdot}025$. A second Cotton effect with very small negative rotational strength was observed at approx. 320–330 nm. It could arise by equilibrium with a solvated species[21], by the presence of two different vibronic band systems[22], or could be interpreted as an indication of the splitting between the two n-orbitals. As a change of the solvent has no great influence upon

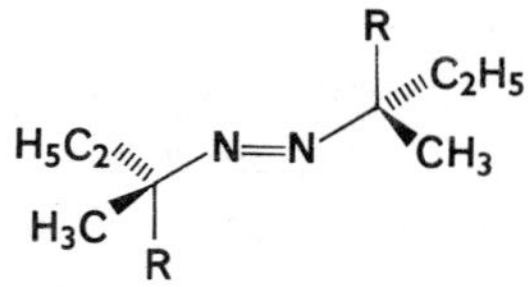

(11) R = Ph

(12) $R = C_6H_{11}$

the CD and most calculations give a larger splitting, furthermore the hydrogenated compound **12** shows only one such CD-band ($\Delta\varepsilon_{max} = +1{\cdot}35$ at 379 nm; $g' = 0{\cdot}061$), the vibronic origin of the second band is favoured by the authors[10, 20]. For the homologue compound **13** a positive Cotton effect is described[23] (ORD-curve) to appear at somewhat longer wavelengths (393 nm) than that of **11** or **12**. On the other hand the CD (dioxan solution) of the dicarboxylic acid **14** and its dimethyl ester **15** appears at shorter

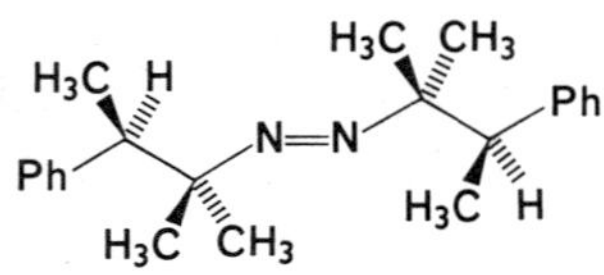

(13)

wavelengths[24] (**14**: $\Delta\varepsilon_{max} = 0{\cdot}73$ at 350 nm; **15**: $\Delta\varepsilon_{max} = -0{\cdot}87$ at 347 nm). Neither in dioxan nor in methanol solution could a second band at approx. 330 nm be detected, but ORD indicated the presence of another Cotton effect at still shorter wavelengths. It was interpreted as originating from the COOR-chromophore, which can, however, not be the reason in case of **7**, where (in ethanol) we[11, 12] also could measure two Cotton effects, one at

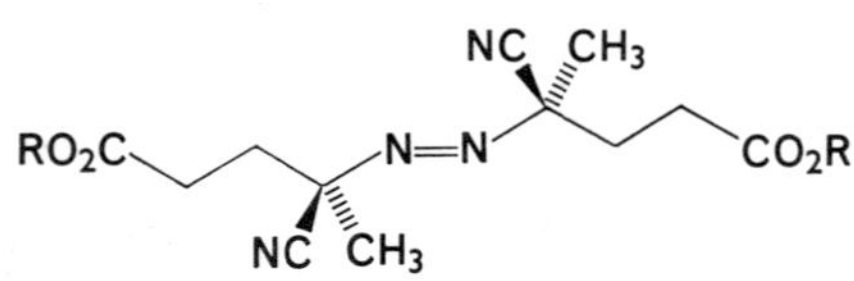

(14) R = H

(15) $R = CH_3$

(or enantiomers)

321 nm ($\Delta\varepsilon_{max} = -1{\cdot}26$), and another one at 222 nm ($\Delta\varepsilon_{max} = +1{\cdot}92$). In the u.v.-spectrum no corresponding band can be detected but at still shorter wavelengths there is a very strong absorption band. The 220 nm band of **7**

can thus be assigned the $\psi^- \rightarrow \pi^*$-($^1A_1 \rightarrow {}^1A_2$-transition, which by all calculations is placed at somewhat longer wavelengths than the very intense $\pi \rightarrow \pi^*$-transition. Interestingly enough for the acetate **8** of **7** the $^1B_2$-band at 322 nm shows a positive CD[12] ($\Delta\varepsilon_{max} = 1{\cdot}51$), whereas that at short wavelengths (239 nm) is negative (–0·84, stronger negative at still shorter wavelengths). For **8** the COOR-chromophore could indeed be responsible for the 239 nm CD-band, whereas the Cotton effect at short wavelengths (indicated by the ORD-spectrum) of **11** could not be characterized exactly[10] and may come either from the azo chromophore or the benzene ring. The simple *trans*-dimethylpyrazoline **4** of unknown absolute configuration shows a positive Cotton effect at 340 nm[25], and for the *gem*-diphenyl substituted pyrazoline **16** a surprisingly large negative Cotton effect ($\Delta\varepsilon_{max} = -3{\cdot}32$) at 339 nm (dioxan solution) is recorded[26]. Perhaps one of the phenyl rings interacts here in a similar way as a C═O-group with the azo chromophore (cf. Section B, 4).

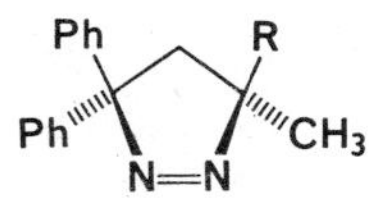

(16) R = CH_2OH

(17) R = CO_2

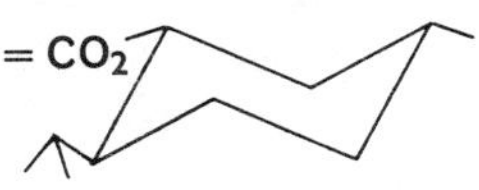

3. Phenylazo compounds

As in the u.v.-spectrum also in the CD-spectrum conjugation with the phenyl ring leads to a bathochromic shift of the longest wavelengths $n \rightarrow \pi^*$-band. In case of **18** prepared from the same enantiomer of 2-phenyl-2-methyl propylamine as **11** a positive CD (+0·51, calculated for 100%

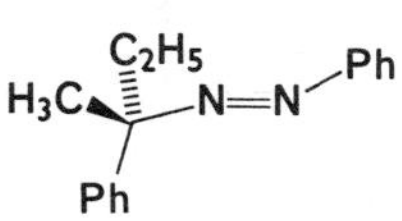

(18)

optical purity from the data given by the authors[27]) is observed at 416 nm[27]. Buckingham and Guthrie[28] were able to investigate the influence of *cis*/*trans* isomerism upon chiroptical properties in case of the two 3β-phenylazo cholestanes **19** and **20**. The *trans*-compound **19** showed in the u.v.-spectrum

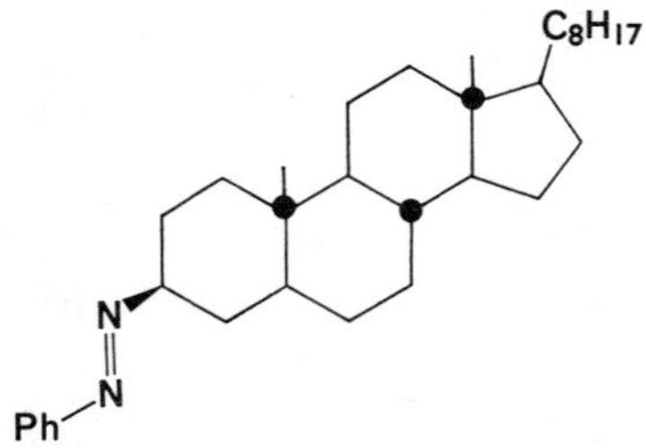

(19)

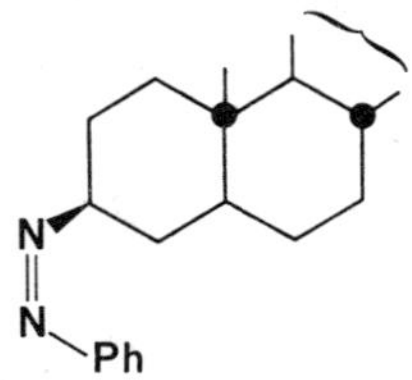

(20)

the longest-wavelengths $n \rightarrow \pi^*$-band at 404 nm ($\varepsilon = 180$) and the conjugation band at 269 nm ($\varepsilon = 10300$), whereas for the *cis*-isomer **20** the corresponding values are 394 nm ($\varepsilon = 310$) and 244·5 ($\varepsilon = 5650$), which data are in agreement with the assignment of the stereochemistry of **19** and **20** from their preparation. The Cotton effects of **19** and **20** have opposite signs as judged from their ORD-curves, and the amplitude *a* for the *cis*-compound **20** is seven times that of the *trans*-compound **19**. Unfortunately neither signs nor magnitudes of the rotations are cited in the paper[28]. The larger rotational strength of **20** within the $n \rightarrow \pi^*$-band as compared with that of **19** was believed[28] to be due to more severe steric hindrance in the *cis*-compound **20**. Since we are however dealing with transitions of different symmetry properties restricted rotation must not be the only explanation for this difference in magnitudes of the Cotton effects.

The 17-phenylazo steroids **21–25** all show[29] Cotton effects (ORD) around 400 nm, the configuration is chemically proved only for **21**, that of the

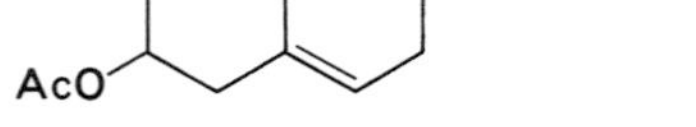

(21) R = H

(22) R = OAc

(23) R = OCH_3

(24) R = OC_2H_5

(25) R = O_2CPh

(26) R = OOH

others is assumed to be the same from reaction mechanism and comparison of ORD-data. This Cotton effect is not solvent dependent for the steroids **21–25**, however in the case of **26** it is negative in carbon tetrachloride and positive in more polar solvents. In mixtures of CCl_4 and dioxan the anomalous ORD-curve can gradually be changed from a positive to a negative Cotton effect, and all curves meet in one point for which the name 'isorotatic point' was proposed[29]. With some reservation this behaviour of **26** was explained[29] by assuming an equilibrium between a solvated and an unsolvated form, which latter is internally hydrogen bonded (cf. Figure 2).

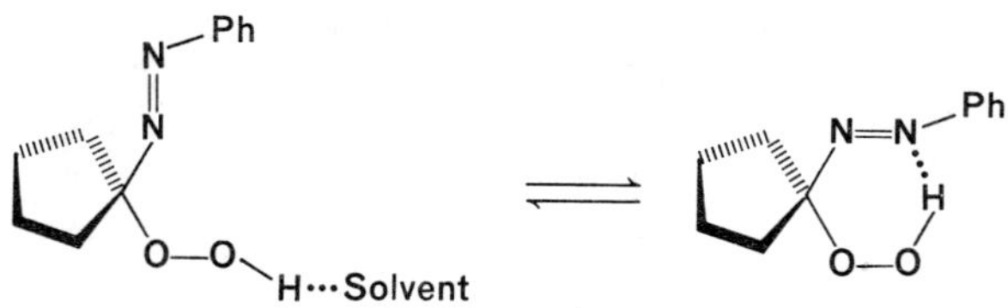

FIGURE 2. Projection of ring D and the 17-substituents of **26** for solvated and nonsolvated species according to ref. 29.

The 3-phenylazo derivatives **27–33** of 3-deoxymethyl glucoside have been

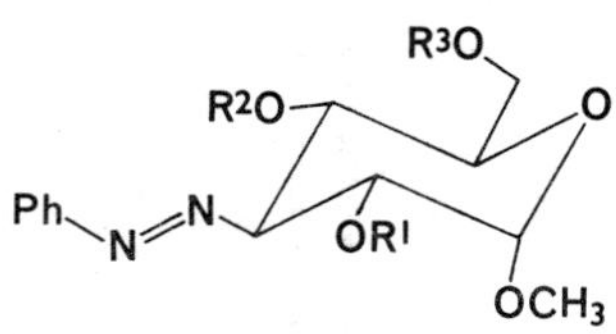

(27) $R^1 = R^2 = R^3 = H$

(28) $R^1 = CH_3$, $R^2 = R^3 = H$

(29) $R^1 = Ac$, $R^2 = R^3 = H$

(30) $R^1 = CH_3$, $R^2 = R^3 = Ac$

(31) $R^1 = R^2 = R^3 = Ac$

(32) $R^1 = H$, $\left.\begin{matrix}R^2\\R^3\end{matrix}\right\} = Ph—CH$

(33) $R^1 = Ac$, $\left.\begin{matrix}R^2\\R^3\end{matrix}\right\} = Ph—CH$

the first azo compounds whose chiroptical properties in the region of the longest-wavelengths $n \rightarrow \pi^*$-band have been described[30]. The Cotton effect (ORD) of the benzylidene derivatives **32** and **33** is positive and solvent effects have been observed in the CD-spectra. The others show no pronounced influence of the solvent, and the Cotton effect is positive for **27–29**, but negative for **30** and **31**. It is interesting to note that acetylation of the OH at C-2 of the glucose has no influence upon the sign of the Cotton effect whereas such an acetylation at the OH at C-4 inverts the sign. Such a phenomenon must not be explained solely by a change of conformation as can be inferred from a comparison with the CD of **7** and **8**.

4. Azo compounds with chiral first sphere

The pyrazoline derivative **34** of a 16-ene-20-keto steroid shows a very

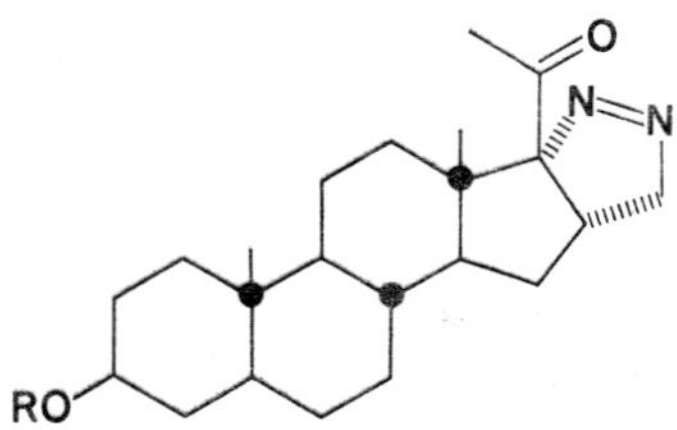

(34)

strong CD-band at 338 nm[11,31] ($\Delta\varepsilon_{max} = +11{\cdot}00$). This is of the same magnitude as that of some homoconjugated ketones which not only show very high $\Delta\varepsilon$-values but also intense $n \rightarrow \pi^*$-absorptions due to intensity borrowing[32]. The Cotton effect of the carbonyl group can be seen independently and appears at 302 nm. The isomeric pyrazoline ketone **35** can of

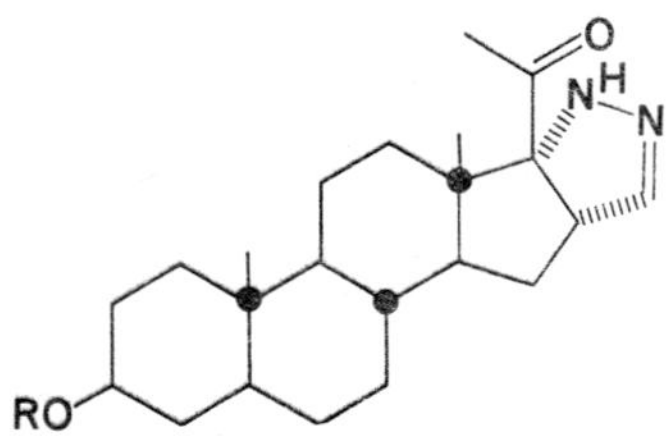

(35)

course not show any Cotton effect at 330 nm, the CD of its ketone group is strongly but not excessively positive[11,31]. In order to study this unusual behaviour in more detail some other pyrazolines substituted in α-position by a carbonyl group have been prepared[11,12]. In these (spiro compounds **5**, **6**, **36**, **37** and derivatives thereof) the conformation is more rigid than for **34**.

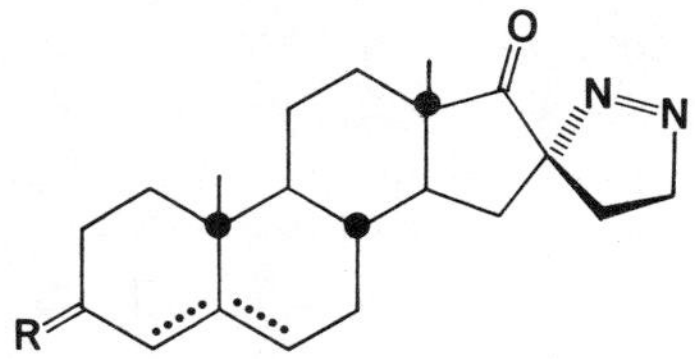

(36) R = 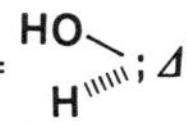 ; Δ^5

(37) R = O=; Δ^4

All these show the same spectral characteristics and the relationship between the chirality of the α-C-acyl substituted pyrazoline ring and the sign of this CD-band at 330 nm is given by Figure 3[33]. Figure 4 shows some

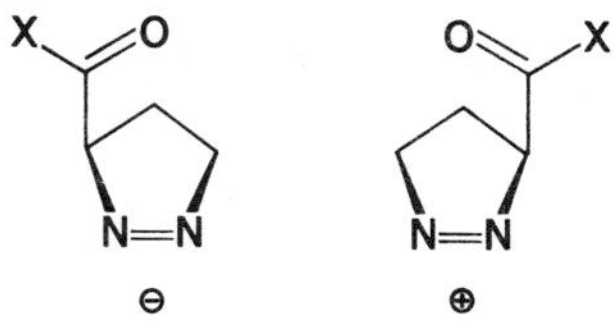

FIGURE 3. Enantiomeric arrangements of pyrazolines substituted by —COX in α-position to the —N=N— chromophore leading to negative and positive Cotton effects above 220 nm, respectively.

actual CD-curves. Interestingly enough ε of **5** and **7** is practically the same, which means that the magnitude of the electric transition moment vector for the longest-wavelengths $n \rightarrow \pi^*$-band is not changed by the presence of the C=O group. Furthermore a second CD is found around 230 nm, but its rotational strength is about one order of magnitude smaller than for the 330 nm band. For ketones like **5** or **6** it has opposite sign to that of the CD at 330 nm, for lactones like **40** or **54** the sign is the same. As **5** in this region of the spectrum also does not show an intense u.v.-absorption band we are inclined to assign this CD-band to the second $n \rightarrow \pi^*$-transition, in full agreement with the theoretical calculations. The magnitude of the CD of the carbonyl group remains practically unchanged by the introduction of the pyrazoline and so we have to explain why only one of these three mentioned Cotton effects becomes stronger.

As mentioned above only the ${}^1A_1 \rightarrow {}^1B_2$-transition ($C_{2v}$-symmetry) shows both an electric and a magnetic transition moment, but they are perpendicular to each other (cf. Table 2). We can now assume that owing to the influence of the orbitals on the carbonyl group in the neighbourhood of the —N=N— chromophore the direction of one (or of both) of the two moments is changed somewhat, so that they are not orthogonal any more. As the u.v. spectrum shows, the magnitude of the electric transition moment vector of the —N=N— chromophore is not influenced at all.

As can be seen from the equation given for the rotational strength R any change of the angle between $\vec{\mu}$ and $\vec{m}$ (originally 90°) will lead to a strong Cotton effect ('inherent chirality'[18, 19]). The short-wavelengths $^1A_1 \rightarrow {}^1A_2$-transition is electrically forbidden and can, therefore, not acquire larger rotational strength by such a mechanism, and since the same is true for the carbonyl $n \rightarrow \pi^*$-transition, the intensity of both corresponding CD-bands remains unchanged.

Enhancement of rotational strength is not confined to the presence of a C=O group from a ketone, but can also be caused by a lactone or ester group in this same position (Figure 3: X = OR), as is shown, e.g. by the strong CD of the correct sign ($\Delta\varepsilon_{max} = +19{\cdot}5$) at 332 nm for the lactone **38**[34]. To the second CD at 235 nm (+6·35) the lactone $n \rightarrow \pi^*$-transition

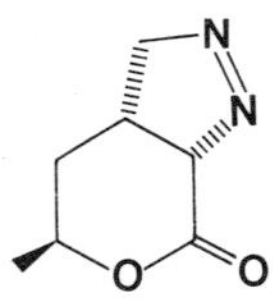

(38)

may also contribute. The CD of **38** supports therefore the view that we are dealing indeed with a chiral first sphere for such compounds.

A rule like that of Figure 3 for a somewhat 'exotic' chromophore may seem to be without any practical application. This is, however, not the case because many naturally occurring terpenoids contain a methylene–γ-lactone grouping to which even in small-scale operations CH_2N_2 can be added easily to form spiro pyrazoline lactones. The absolute configuration at the spiro atom can be inferred unequivocally from the rule of Figure 3, and from this we are able to determine the stereochemistry in the vicinity of the lactone grouping. Indeed very often use has been made of this method and only a few examples will be cited here explicitely.

O-Acetyl asperilin[35] **39** is a sesquiterpenoid with *eudesmane* skeleton. The two carboxylic rings are *trans*-fused, whereas the lactone ring is attached in *cis* to ring B. The β-side (front side) of the methylenelactone double bond is therefore heavily shielded, so attack of diazomethane is only possible from the α-side (rear-attack) leading to the spiro compound **40**[33, 36], whose CD is strongly positive in accord with Figure 3. The second CD-band at approx. 230 nm is also positive but of smaller magnitude than the first one.

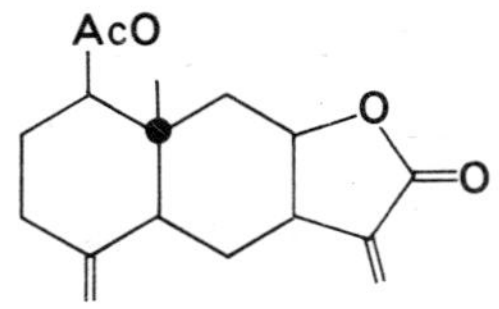

(39)

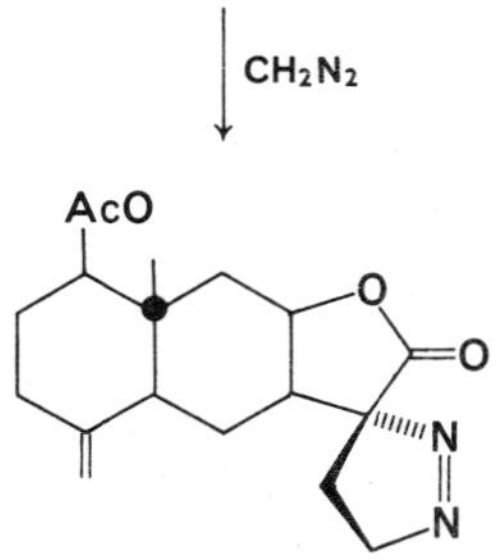

(40)

Another example with this skeleton is farinosin, which was originally believed to have structure **41**[37]. The CD of the corresponding spiro pyrazoline is strongly negative[33], and thus in disagreement with the fact that attack by diazomethane has to proceed from the less hindered side.

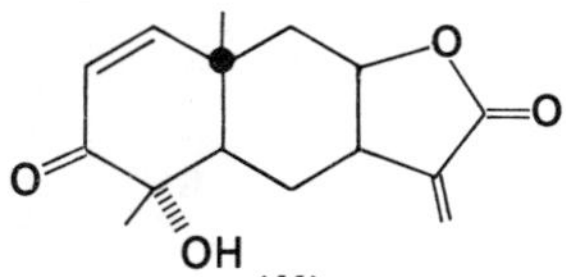

(41)

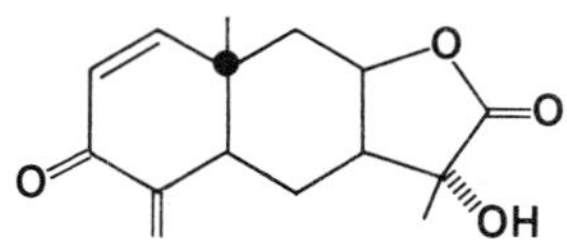

(42)

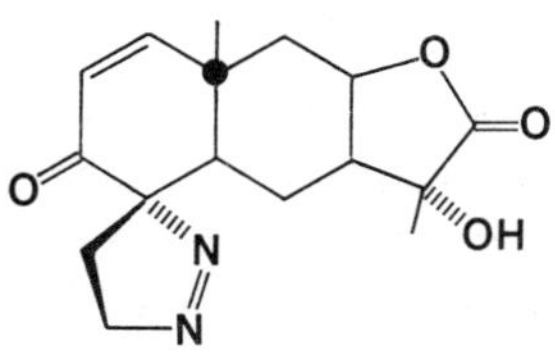

(43)

Though CD alone cannot determine the correct structure at least it clearly ruled out **41**. Indeed in a later publication[38] the revised structure **42** was proved for farinosin and following the rule, rear attack onto the activated methylene group leads to **43**.

In case of *guaiane* derivatives with a *cis*-fused methylene lactone ring like damsin (**44**)[39] the addition of diazomethane takes also place mainly from the rear side to form **45**[33, 36, 40], though careful chromatography led to the

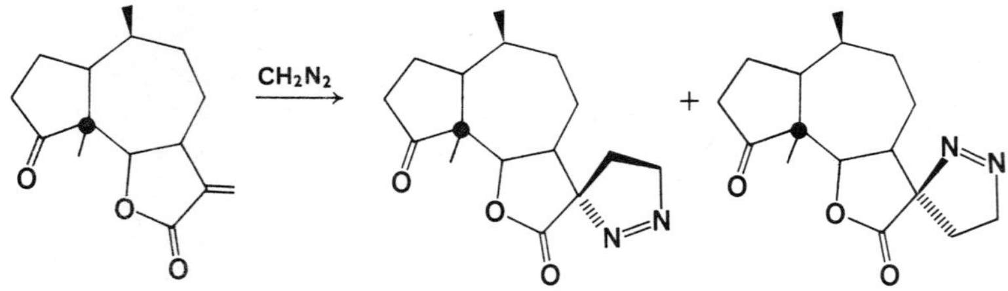

(44) **(45)** **(46)**

isolation of approx. 5% of the other stereoisomer **46**[40]. The CD of both compounds follows again the rule of Figure 3.

Stöcklin, Waddell and Geissman[41] have proposed a rule which allows a correlation of the sign of the $n \rightarrow \pi^*$-Cotton effect of such methylenelactones with their stereochemistry. This rule could not be applied, however, to such compounds where this CD-band cannot be detected well in the spectrum. In all such cases it was possible, however, to draw conclusions about the lactone stereochemistry from the CD of the corresponding spiro pyrazoline lactones.

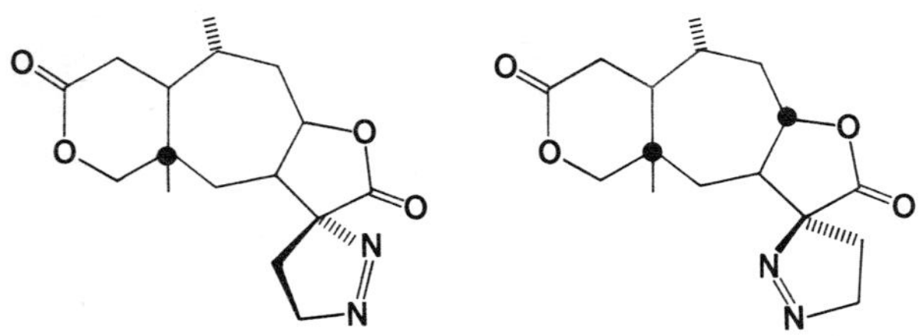

(47) **(48)**

Interesting is the comparison of the CD of the pyrazoline derivatives **47** of floribundin[42] and **48** of veermerin[42], which differ only in the stereochemistry of the lactone ring: floribundin has a *cis*- and veermerin a *trans*-fusion. In case of the first, diazomethane must attack from the rear and this is proved by the strong positive CD at 319 nm for **47**. No prediction can be made for the *trans*-lactone veermerin, but the CD of **48** shows that the addition of CH_2N_2 occurred here from the front side (negative CD at 324·5 nm).

Methylene lactones of the *germacranolide* series also form spiro pyrazoline lactones and the strong positive CD of **49**[33,36], a derivative of costunolide, indicates front attack. Laurenobiolide[43] where the lactone ring is directed to another position of the ten-membered ring is, however, attacked from the rear to form **50**, and the same is true for one of its epoxides[43].

Compounds where the COOR grouping is not fixed in its conformation with respect to the pyrazoline ring also show this characteristic CD, its

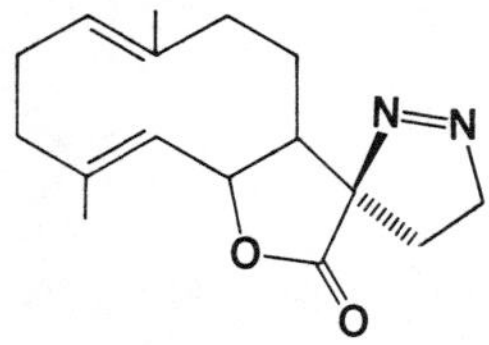

(49)

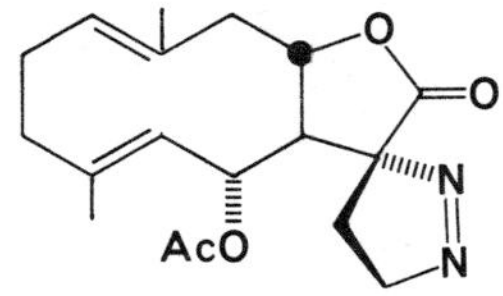

(50)

magnitude is however not any more so big. Thus for the pyrazoline derivative **51** of methyl iliciate the CD is only +1·25 at 326 nm[33, 36], but as in case of the ester **17**[26] it still follows the rule of Figure 3.

A special case is the phenylazo compound **52** which shows[12] several CD-bands in the range between 450 and 200 nm (cf. Figure 4). The strong CD-band at 327 nm (+16·72) corresponds to the conjugation band and judged from its position by comparison with the u.v.-spectra of **19** and **20**

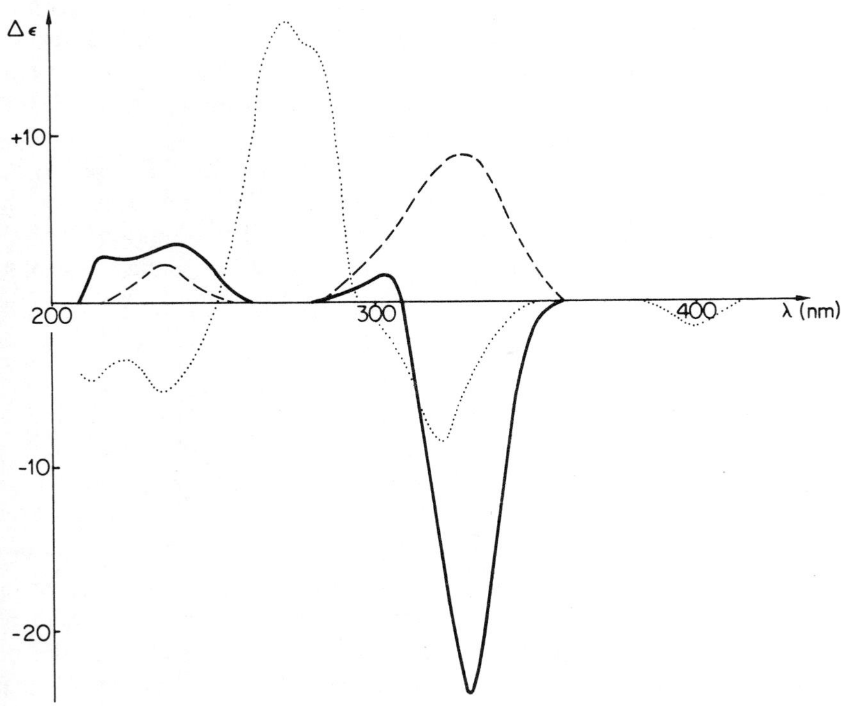

FIGURE 4. CD-spectra of **5** (——), **40** (----), and **52** (. . . .).

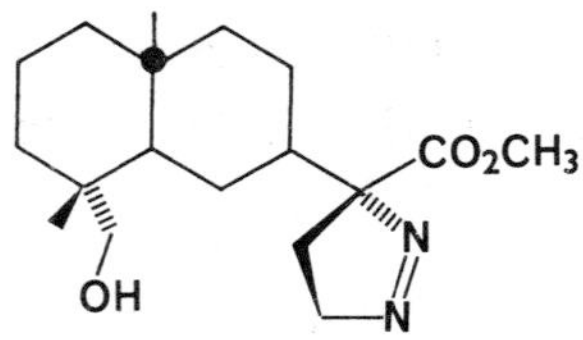

(51)

trans-configuration of the —N═N— group is proved for **52**. The $n \rightarrow \pi^*$-band at longest wavelengths (399 nm, $\Delta\varepsilon_{max} = -1{\cdot}55$) is only moderately

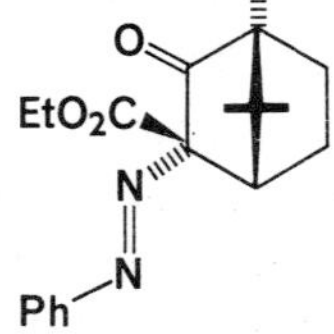

(52)

strong. One reason may be that two C═O groups are present in the position necessary for such an enhancement of the CD, and their influences will compensate to a great extent. Furthermore in a *trans*-azo-compound we are dealing with a $^1A_g \rightarrow {}^1B_g$-transition which cannot acquire enhancement of the Cotton effect by the mechanism discussed before.

Another nice example of the application of the rule of Figure 3 is the determination of the absolute configuration at the chiral centre of the lactone **53**, a hydrolysis product of the tuliposides B[44]. Addition of diazomethane to 50 mg of **53** resulted in a mixture of the two spiro pyrazoline

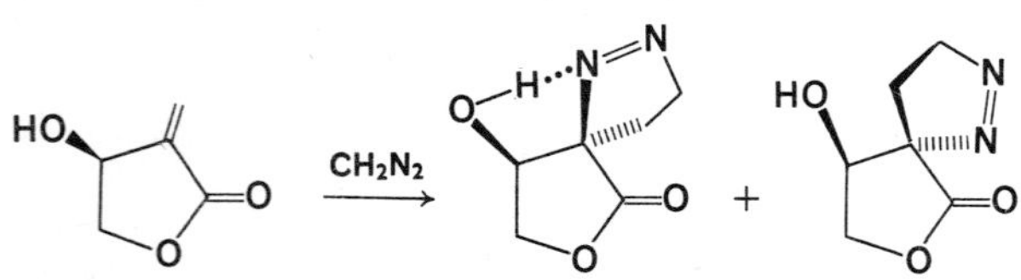

(53) **(54)** **(55)**

lactones **54** and **55** which could be separated by column chromatography. One of these, viz. **55** which was the main product, showed in the i.r.-spectrum only the bands of a free OH-group (3618 cm^{-1}) whereas the other isomer **54** contains an internal hydrogen bridge ($\nu_{OH} = 3571$ cm^{-1}). In **54** the OH and the nitrogen connected to the lactone ring must be in *cis*-position, whereas in **55** these two groups are *trans*-positioned. Infrared spectroscopy gives thus the relative configuration, the absolute at the 'auxiliary' centre of chirality is found from the CD to be R for **54** and S for **55**. This settles then the absolute configuration of **53** as S.

Chiroptical data of such pyrazoline derivatives of methylenelactones have also been determined of, e.g. jurineolide[33,36], dihydrocostuslactone[36], pinnatifidin[33,36], uvedalin[45], axivalin[46], pyrethresin[43], parthemollin[47], methyl damsinate[40], ambrosiol[41], ambrosin[41], psylostachyin[41], psylostachyin C[41], and psylotropin[41]. The CD-spectra of many derivatives of **5** which do, however, not contain any more the —N=N— chromophore are found in reference 12.

5. Biopolymer derivatives

Kagan and Vallee[48] used the arsanilazo grouping coupled to a tyrosine moiety as a probe to investigate the conformation of carboxypeptidase A. Cotton effects were measured in the range of the absorption bands of this azo chromophore and they are sensitive to the removal of the catalytically essential Zn(II) atom to the binding of different inhibitors or substrates as well as to modifications of the enzyme.

6. Induced Cotton effects

a. Magnetic optical rotatory dispersion (MORD) Compounds whose molecules are either achiral or which are present as racemates can show rotation of the plane of linearly polarized light (MORD), or a magnetic circular dichroism (MCD) by applying a longitudinal strong magnetic field ('Faraday-effect')[16]. Shashoua[49] has measured the MORD-spectra of several compounds but was not able to detect any anomalous dispersion in the MORD-curve of azobenzene around 440 nm. This failure of observing such an anomalous dispersion in the range of absorption may in his case be due to the equipment used (too small magnetic field).

b. Complexes with biopolymers. It is well known[50] that complex formation between achiral molecules of dyes and biomolecules like proteins or nucleic acids gives rise to induced Cotton-effects in the range of the absorption bands of the dye. Some aromatic azo compounds have been used as a 'probe' for the determination of conformation of proteins, e.g. carbonic anhydrase is known to complex with sulphamides. Using prontosil (**56**), lutezol (**57**) and especially neoprontosil (**58**) Coleman[51,52] was able to show that induced Cotton-effects appear within the absorption bands of the azo compounds, but only if the Zn(II)- or Co(II)-complexes were used. With other metals like Mn(II), Ni(II), Cu(II), Cd(II), or Hg(II) which also are known to bind to the active site of the enzyme no such induction was found. Investigation of the complexes of different isozymes and species variants of carbonic anhydrase revealed that both sign and magnitude of

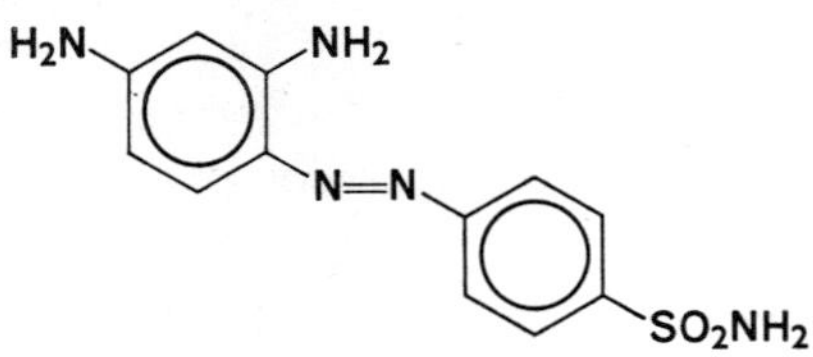

(56)

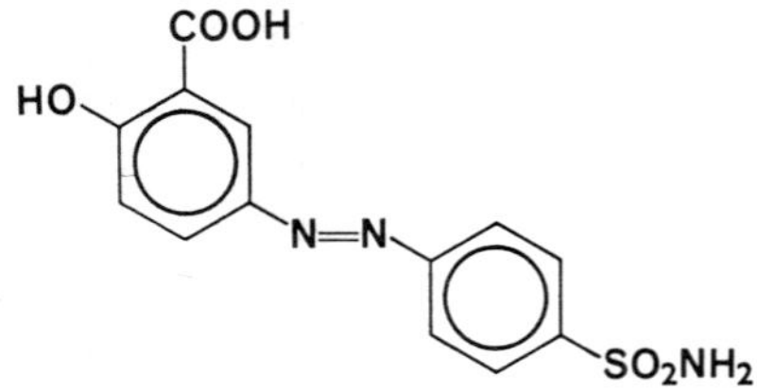

(57)

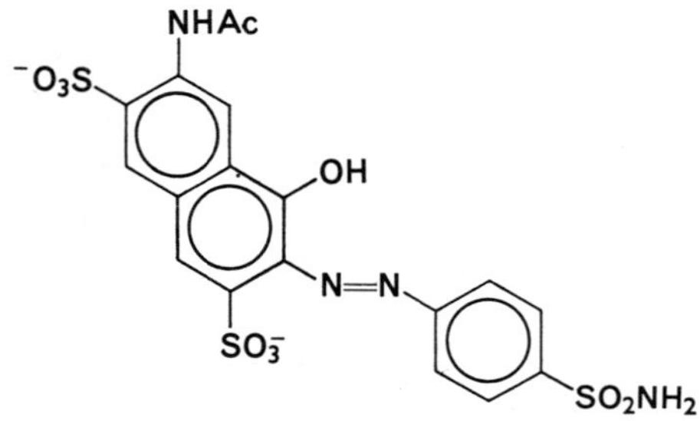

(58)

the strong CD in the visible are unique for each isozyme. There are however also marked differences in the appearance of the CD-bands of complexes which differ only in the presence of either Zn(II) or Co(II).

Einarsson and Zeppezauer[53] by a similar method investigated the induced CD spectrum of 5-(4′-sulphamylphenylazo)-8-hydroxy-quinoline (**59**)

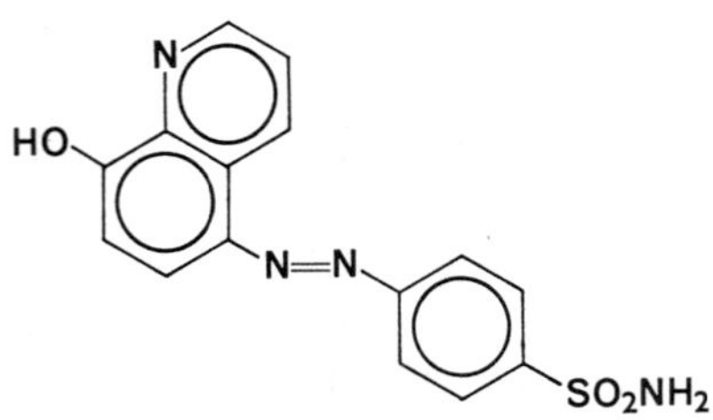

(59)

bound to human carbonylic anhydrases B and C as well as to alcohol dehydrogenase. Binding to the latter enzyme does not change the u.v.-spectrum of the dye, whereas such a change is noticed for the carbonic anhydrase complexes. From their measurements the authors[53] concluded that the hydroxyquinoline dye **59** is probably not able to give structure informations about the catalytical active state of liver alcohol dehydrogenase.

Complexes between phenylazo compounds and cyclodextrins also give Cotton-effects within the absorption bands of the azo dyes[59].

II. THE AZOXY CHROMOPHORE

A. MO-Treatment

In contrast to their aromatic counterparts aliphatic azoxy compounds have been made easily available only quite recently[54–56], stimulated by the fact that some antibiotics with this chromophore have been isolated, such as LL-BH 872 α (**60**) or elaiomycin (**61**)[57, 58]. A short theoretical treatment

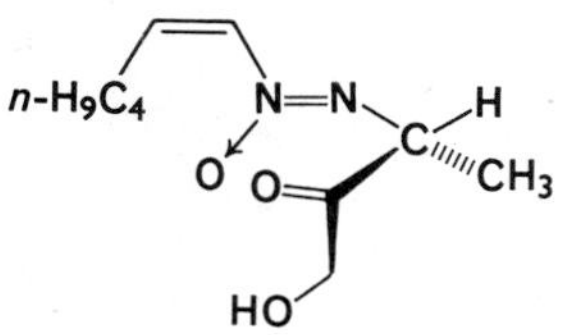

(60)

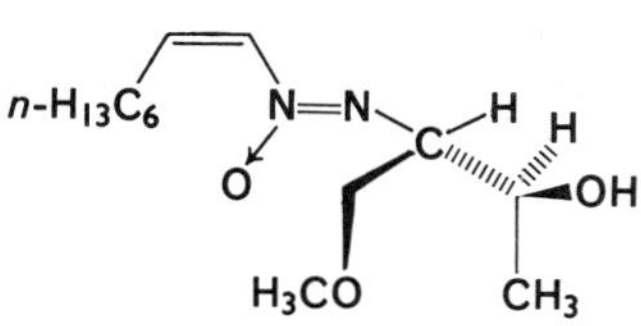

(61)

of the isolated azoxy chromophore has been given by Wagnière and co-workers[6]. According to this the highest three occupied orbitals are (in sequence of increasing energy) a non-bonding orbital localized mainly on the oxygen (n_O), a non-bonding orbital localized mainly on that nitrogen not bearing the oxygen atom (n_N), and a π-orbital. One can therefore assume that the three transitions at lowest energy are of $\pi \rightarrow \pi^*$, $n_N \rightarrow \pi^*$, and $n_O \rightarrow \pi^*$-type (not necessarily in this order, cf. the discussion of the transitions of the —N=N— chromophore) and indeed three distinct bands can be identified in the CD- and u.v.-spectra of simple alkyl derivatives.

B. Chiroptical Properties

Compound **62** shows in the CD-spectrum a shoulder at approx. 280 nm, another band of the same sign at 255 nm, and a stronger CD around 230 nm.

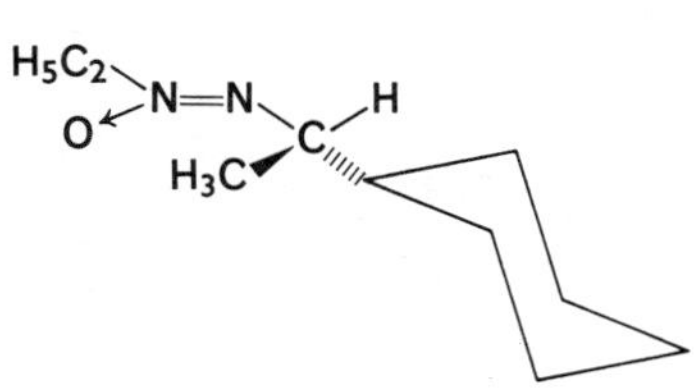

(62)

At shorter wavelengths below 190 nm the onset of a fourth Cotton effect is indicated[58]. In the u.v.-spectrum[54, 55] one sees only one strong band around 217–232 nm (ε approx. 8000, at short wavelengths for *cis*, at long wavelengths for *trans*-azoxy compounds; '*cis*' and '*trans*' refer here to the alkyl groups), and a shoulder around 280 nm.

TABLE 3. Character table of point group C_s

C_s	E	σ	Transitions of the azoxy chromophore	Components of $\vec{\mu}$	Components of $\vec{m}$
A'	+1	+1	$n_N \to \pi^*$, $n_O \to \pi^*$	x, y	z
A''	+1	−1	$\pi \to \pi^*$	z	x, y

Table 3 shows a character table of point group C_s. Both $n \to \pi^*$-transitions are similarly electrically allowed (x- and y-polarized), the $\pi \to \pi^*$-transition is also allowed but out-of-plane polarized (z). Though McGahren and Kunstmann[58] have observed a blue shift on going from an unpolar to a polar solvent for the 220 nm CD-band of **62** they have assigned this band to a $\pi \to \pi^*$-transition, because in case of very polar ground and excited states the usual rules for the solvent shifts may indeed break down. According to Table 3 it would, however, not be unreasonable to assign this transition to an $n_O \to \pi^*$ transition. Whatever type of transition is chosen for these three bands, however, in any case in point group C_s the electric and magnetic transition moments corresponding to a particular transition are perpendicular to each other; we can thus expect a tremendous enhancement of the rotational strengths for this chromophore also, if a π-system such as that of a carbonyl group is present as a substituent at the α-atom, and this is indeed the case.

Elaiomycin (**61**) gives a CD-spectrum[57] whose rotational strength is similar to that of the model compound **62**[58] (or its enantiomer **E-62**[58]), all

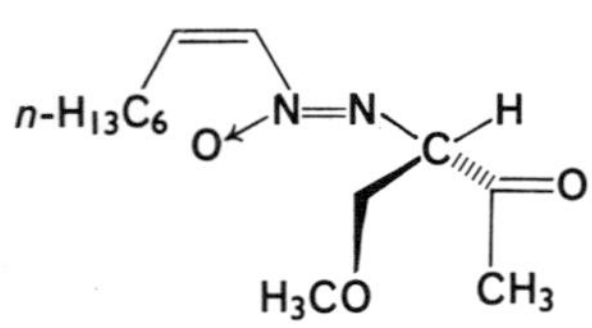

(63)

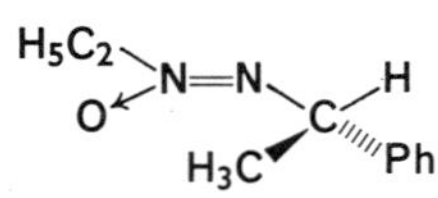

(64)

bands are, however, redshifted due to the conjugation with one more C=C double bond. The presence of the carbonyl group as in LL-BH 872 α (**60**)[57] and in oxidized elaiomycin **63**[57], or that of a phenyl ring as in **64**[58] (and its

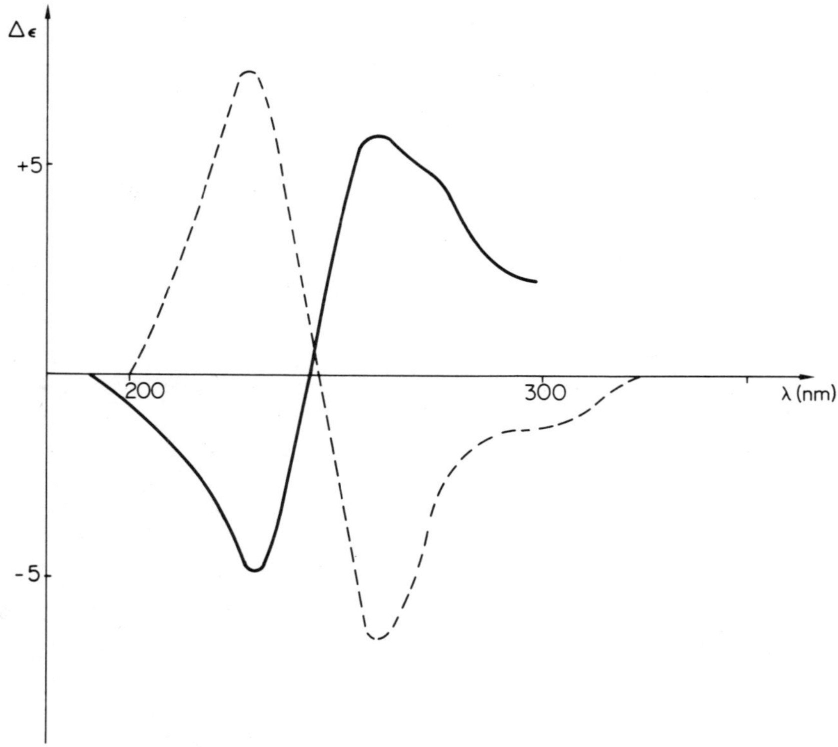

FIGURE 5. CD-spectra of **60**-acetate (——) and **63** (----).

enantiomer **E-64**[58]) as a substituent at the α-atom next to the azoxy chromophore enhances the rotational strengths approximately by two orders of magnitude. We propose therefore that also for this chromophore a chirality of the first sphere is introduced by the carbonyl group or the benzene ring by changing the angle between the electric and the magnetic transition moment vectors, and Figure 5 clearly shows the enantiomorphous behaviour of the CD-curves of LL-BH 872 α-acetate (**60**-acetate) and oxidized elaiomycin (**63**) (both in trifluorethanol solution) which represents the enantiomeric stereochemistry at the chiral centres. If we assume, for example, a conformation in which the two alkyl groups are staggered with respect to the NO-bond and the hydrogen is *anti*-periplanar to this same group then the phenyl ring of **64** is in the same position relative to the azoxy chromophore as the carbonyl of **63**, but enantiomeric to that of the carbonyl group of **60** or its dihydro derivative **65**. In accordance with this the first

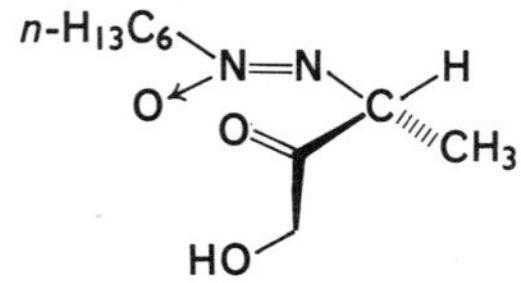

(65)

CD-band at approximately 280 nm as well as that at 240–245 nm of **64** are of opposite signs to those of **65**, the next band at approx. 215–225 nm has, however, the same sign for both compounds; most probably the $^1B_{1u}$-transition of the benzene ring makes already a strong contribution to this band, so the sign of this CD-band is not any more solely determined by the azoxy chromophore transitions.

These azoxy compounds thus follow a similar type of rule as pyrazolines with a chiral first sphere (cf. Figure 3) and demonstrate well the general applicability of the principles[18] used to explain the correlation between CD and stereochemistry.

III. ACKNOWLEDGMENT

We thank Prof. Dr. W. Herz (Tallahassee), Prof. Dr. T. A. Geissman (Los Angeles) and Prof. Dr. H. M. Walborsky (Tallahassee) very much for providing us with samples and information.

IV. REFERENCES

1. See, for example, H. Suzuki, *Electron Absorption Spectra and Geometry of Organic Molecules*, Academic Press, New York, 1967, pp. 499–503.
2. M. B. Robin and W. T. Simpson, *J. Chem. Phys.*, **36,** 580 (1962).
3. G. Kortüm and H. Rau, *Ber. Bunsenges.*, **68,** 973 (1964).
4. M. B. Robin, R. R. Hart and N. A. Kuebler, *J. Am. Chem. Soc.*, **89,** 1564 (1967).
5. E. Haselbach and A. Schmelzer, *Helv. Chim. Acta*, **54,** 1575 (1971).
6. J. Kuhn, W. Hug, R. Geiger and G. Wagnière, *Helv. Chim. Acta*, **54,** 2260 (1971).
7. C. H. Chang, R. F. Porter and S. H. Bauer, *J. Am. Chem. Soc.*, **92,** 5313 (1970).
8. E. Haselbach and E. Heilbronner, *Helv. Chim. Acta*, **53,** 684 (1970).
9. F. Brogli, W. Eberbach, E. Haselbach, E. Heilbronner, V. Hornung and D. M. Lemal, *Helv. Chim. Acta*, **56,** 1933 (1973).
10. D. J. Severn and E. M. Kosower, *J. Am. Chem. Soc.*, **91,** 1710 (1969).

11. G. Snatzke and J. Himmelreich, *Tetrahedron*, **23**, 4337 (1967).
12. G. Snatzke, H. Langen and J. Himmelreich, *Ann. Chem.*, **744**, 142 (1971).
13. H. H. Perkampus, I. Sandeman and C. J. Timmons (Eds.), *DMS-UV-Atlas of Organic Compounds*, Butterworth/Verlag Chemie, London/Weinheim, 1966–71, spectrum C 7/5.
14. As ref. 13, spectrum C 7/17.
15. H. H. Jaffé, S.-J. Yeh and R. W. Gardner, *J. Mol. Spectr.*, **2**, 120 (1958).
16. For a recent survey of the principles see the review by F. Ciardelli and P. Salvadori in ref. 17, p. 3.
17. F. Ciardelli and P. Salvadori (Eds.), *Fundamental Aspects and Recent Developments in Optical Rotatory Dispersion and Circular Dichroism*, G. Heyden, London, 1973.
18. G. Snatzke and F. Snatzke, in ref. 17, p. 173.
19. A. Moscowitz, K. Mislow, M. A. W. Glass and C. Djerassi, *J. Am. Chem. Soc.*, **84**, 1945 (1962).
20. E. M. Kosower and D. J. Severn, *Tetrahedron Lett.*, 3125 (1966).
21. M. Legrand, in ref. 17, p. 285.
22. O. E. Weigang, Jr., in ref. 17, p. 50.
23. M. J. Tremelling and J. M. McBride, *J. Org. Chem.*, **37**, 1073 (1972).
24. C. G. Overberger and D. A. Labianca, *J. Org. Chem.*, **35**, 1762 (1970).
25. R. J. Crawford, A. Mishra and R. J. Dummel, *J. Am. Chem. Soc.*, **88**, 3959 (1966).
26. H. M. Walborsky, personal communication.
27. N. A. Porter, M. E. Landis and L. J. Marnett, *J. Am. Chem. Soc.*, **93**, 795 (1971).
28. J. Buckingham and R. D. Guthrie, *Chem. Comm.*, 1241 (1967).
29. J. Buckingham and R. D. Guthrie, *Chem. Comm.*, 570 (1967).
30. E. O. Bishop, G. J. F. Chittenden, R. D. Guthrie, A. F. Johnson and J. F. McCarthy, *Chem. Comm.*, 93 (1965).
31. L. Velluz, M. Legrand and M. Grosjean, *Optical Circular Dichroism*, Verlag Chemie, Weinheim, 1965, p. 216.
32. See G. Snatzke and F. Snatzke, in ref. 17, p. 109.
33. G. Snatzke, *Riechstoffe, Aromen, Körperpflegemittel*, **19**, 98 (1969).
34. G. Snatzke and E. Otto, *Tetrahedron*, **25**, 2041 (1969).
35. W. Herz and N. Vismanathan, *J. Org. Chem.*, **29**, 1022 (1964).
36. M. Suchý, L. Dolejš, V. Herout, F. Šorm, G. Snatzke and J. Himmelreich, *Coll. Czechoslov. Chem. Comm.*, **34**, 229 (1969).
37. T. A. Geissman and R. Mukherjee, *J. Org. Chem.*, **33**, 656 (1968).
38. W. Herz, P. S. Subramaniam and T. A. Geissman, *J. Org. Chem.*, **33**, 3743 (1968).
39. M. Suchý, V. Herout and F. Šorm, *Coll. Czechoslov. Chem. Comm.*, **28**, 2257 (1963).
40. R. W. Doskotch and C. D. Hufford, *J. Org. Chem.*, **35**, 486 (1970).
41. W. Stöcklin, T. G. Waddell and T. A. Geissman, *Tetrahedron*, **26**, 2397 (1970).
42. W. Herz, K. Aota, M. Holub and Z. Samek, *J. Org. Chem.*, **35**, 2611 (1970).
43. H. Tada and K. Takeda, *Chem. Comm.*, 1391 (1971).
44. R. Tschesche, F.-J. Kämmerer and G. Wulff, *Chem. Ber.*, **102**, 2057 (1969).
45. W. Herz and S. V. Bhat, *J. Org. Chem.*, **35**, 2605 (1970).
46. G. D. Anderson, R. S. McEwen and W. Herz, *Tetrahedron Lett.*, 4423 (1972).

47. W. Herz, S. V. Bhat and A. L. Hall, *J. Org. Chem.*, **35,** 1110 (1970).
48. H. M. Kagan and B. L. Vallee, *Biochemistry*, **8,** 4223 (1969).
49. V. E. Shashoua, *J. Am. Chem. Soc.*, **86,** 2109 (1964).
50. E. R. Blout, in ref. 17, p. 352.
51. J. E. Coleman, *J. Am. Chem. Soc.*, **89,** 6757 (1967).
52. J. E. Coleman, *J. Biol. Chem.*, **243,** 4574 (1968).
53. R. Einarsson and M. Zeppezauer, *Acta Chem. Scand.*, **24,** 1098 (1970).
54. R. A. Moss, M. J. Landon, K. M. Luchter and A. Mamantov, *J. Am. Chem. Soc.*, **94,** 4392 (1972).
55. K. G. Taylor and R. Riehl, *J. Am. Chem. Soc.*, **94,** 250 (1972).
56. J. Swigert and K. G. Taylor, *J. Am. Chem. Soc.*, **93,** 7337 (1971).
57. W. J. McGahren and M. P. Kunstmann, *J. Am. Chem. Soc.*, **92,** 1587 (1970).
58. W. J. McGahren and M. P. Kunstmann, *J. Org. Chem.*, **37,** 902 (1972).
59. F. Cramer and H. Hettler, *Naturwiss.*, **54**, 625 (1967).

CHAPTER 7

Basicity, hydrogen bonding and complex formation involving hydrazo, azo and azoxy groups

P. J. KRUEGER

University of Calgary, Calgary, Alberta, Canada, T2N 1N4

I. INTRODUCTION

Hydrazo compounds (RNHNHR′) can be considered to be derivatives of hydrazine, with a lone pair of electrons on each nitrogen atom which can potentially act as bases, accepting protons from Lowry–Brönsted acids. Strong proton donors in ionizing solvents would therefore be expected to lead to the ions $(RNHNH_2R')^+$ and $(RNH_2NH_2R')^{2+}$ in a stepwise fashion, with the ammonium-like ions probably hydrogen bonded to the anions, whereas weaker proton donors might only be expected to lead to a hydrogen-bonded complex without complete proton transfer, as in equation (1).

$$AH + RNHNHR' \rightleftharpoons AH\cdots NHRNHR' \qquad (1)$$

While an intermolecular reaction is shown in equation (1), it can also be intramolecular.

Just as the strengths of ordinary amines as bases vary considerably, depending on the nature of the solvent, the temperature, and the inductive (polar), resonance (mesomeric) and steric effects of the group attached to the nitrogen atom, so the base strengths of the substituted hydrazines vary. Basicity of amines has been fully discussed in a previous volume[1], and a detailed discussion of the factors involved will therefore not be given here. The only additional feature in hydrazines arises out of the two adjacent lone pairs and the N—N bond. The hydrazides (—CONHNH_2) and semicarbazides $\left(\text{>N—NHCONH}_2\right)$ can also be considered to be substituted hydrazines.

A number of important classes of compounds involve the azomethine linkage $\left(\text{>C=N—}\right)$. These include the oximes $\left(\text{>C=N—OH}\right)$, the hydrazones $\left(\text{>C=NNH}_2\right)$, the phenylhydrazones $\left(\text{>C=NNHC}_6\text{H}_5\right)$ and the semicarbazones $\left(\text{>C=NNHCONH}_2\right)$. The last class, as well as the semicarbazides, would be expected to exhibit some hydrogen bonding and complex forming properties that could be related to the amide group; that is not covered in this review.

The material on azo (—N=N—) and azoxy (—N(O)=N—) compounds relates largely to aromatic compounds, as these have received most attention to date. As far as the azo dyes are concerned, only material relating fairly directly to the —N=N— bond has been included. Diazonium salts and diazoalkanes are both outside the scope of this review.

While unsubstituted triazene (HN=NH—NH_2) does not yet appear to have been isolated, organic derivatives of triazene and tetrazene -1 and -2 are well known. This review includes material on the organic triazenes and triazene oxide. Organic azides are not included.

Every effort has been made to include references to material published up to the end of 1972, with some 1973 publications being reviewed as well. However, the bibliography is not an exhaustive survey of the literature and emphasis was placed on more recent publications. A number of general reviews dealing with hydrogen bonding are available[2,3].

Brief comments on the basic structural features of these compounds are provided in the next section, since they have a direct bearing on many of the complexes that can form.

II. STRUCTURAL CONSIDERATIONS

A. Hydrazines and Related Compounds

The conformation of hydrazine is known to be *gauche* from experimental measurements[4] and theoretical calculations show that the *trans* form is unfavourable[5]. The analysis of i.r. and Raman spectra of alkylhydrazines indicates that here too only conformations of the *gauche* type need to be considered in all phases[6]. Only when hydrazine is fully substituted with very electronegative groups is the *trans* form important[6].

B. Azomethine Compounds

Geometrical isomerism about the $>C{=}N-$ linkage is similar to that about the $>C{=}C<$ bond, with the terms *cis* and *trans* of alkene chemistry being replaced by *syn* and *anti*, respectively. In the case of aldehyde derivatives the isomer with the substituent on N on the same side of the double bond as the aldehydic H atom is labelled *syn*; for ketone derivatives the group which is *syn* to the N substituent must be specified.

Since the azomethine linkage is highly polarized

$$\left(>C{=}N- \leftrightarrow >C^{+}{-}N^{-}-\right)$$

interconversion of stereoisomers as a result of rotation about the C—N bond occurs more readily than rotation about a C=C double bond. While the isolation of *syn* and *anti* forms of imines themselves is rare, if electronegative groups are attached to N the individual geometric isomers are considerably more stable and thus more readily isolated and characterized. Compounds containing an azomethine linkage are therefore quite unlike compounds with saturated N atoms, where the occurrence of stable stereoisomers is rare. Oximes, hydrazones and semicarbazones can be considered as substituted azomethines as far as geometrical isomerism about the C=N double bond is concerned. Proton magnetic resonance techniques are very useful in the identification of *syn* (**1a**) and *anti* (**1b**) isomers of oximes[7] and hydrazones (**2a** and **2b**)[8].

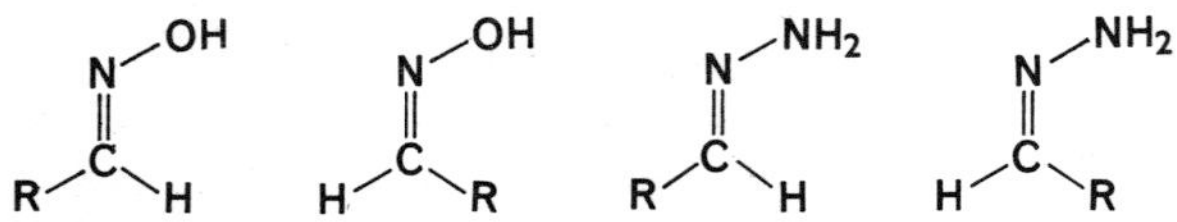

(**1a**) *syn* (**1b**) *anti* (**2a**) *syn* (**2b**) *anti*

In monoarylazomethines, benzalanilines and their conjugate acids (Schiff bases) the conformation of the phenyl ring attached to nitrogen is periperpendicular to the rest of the molecule, as illustrated in (**3**)[9]. From

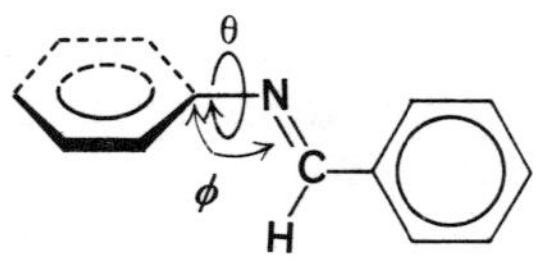

(**3**)

n.m.r. measurements on substituted azomethines and MO-LCAO calculations taking into account the energy of π-electron delocalization, the repulsion energy of protons α to the C=N bond and *ortho* on the phenyl ring, the minimum energy conformation has $\theta = 42$–$50°$ and $\phi = 120°$[10].

C. Azo Compounds

Azo compounds involve a double bond and therefore exist as *cis*- and

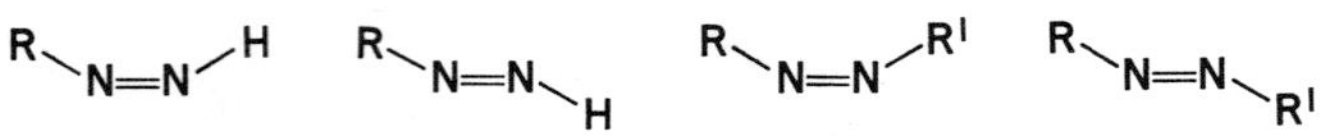

(**4a**) (**4b**) (**5a**) (**5b**)

trans-isomers[11], illustrated by (**4a, 5a**) and (**4b, 5b**). Aliphatic azo compounds of the type RN=NH have not been studied extensively in terms of hydrogen bonding and complex formation. This may be related to the fact that they decompose readily into nitrogen and a hydrocarbon.

Aromatic azo compounds are resonance stabilized (equation 2); evidence

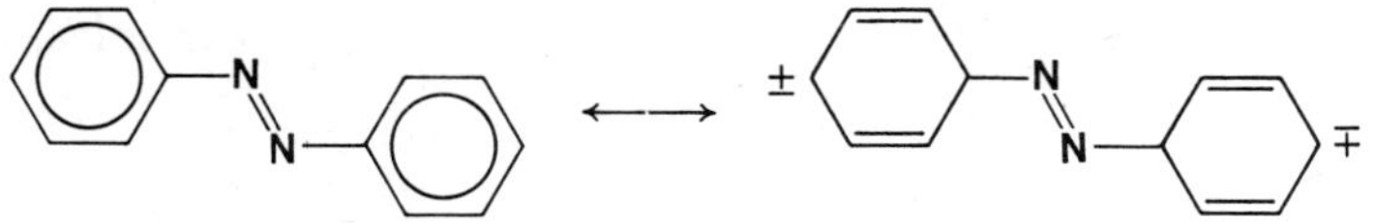

(2)

for this comes from the shortening of the carbon–nitrogen bond in *trans*-azobenzene from its usual value of 1·47 Å to 1·41 Å as a result of conjugation[12]. In addition to the *trans* isomer shown in equation (2), the geometrical *cis* isomer (**6**) exists, and is normally stable and readily isolated. For

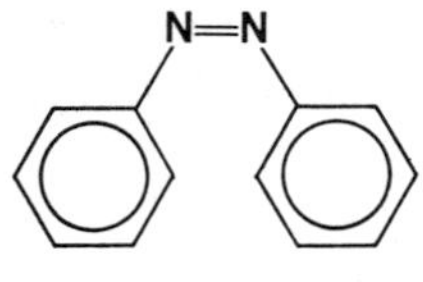

(**6**)

azobenzene itself the *trans* form is known to be considerably more stable than the *cis* form[11], which is directly related to the resonance stabilization in the almost coplanar structure of the former, while in the *cis* isomer the benzene rings are rotated out of the plane containing the nitrogen atoms by about 50°, for steric reasons[13].

Tautomerism involving the azo group is well known, and must be borne in mind in the interpretation of experimental data in terms of structural information. Thus, isomerization of phenylhydrazones involves hydrazone

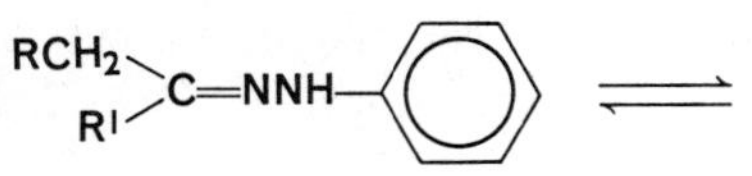

(**7a**)

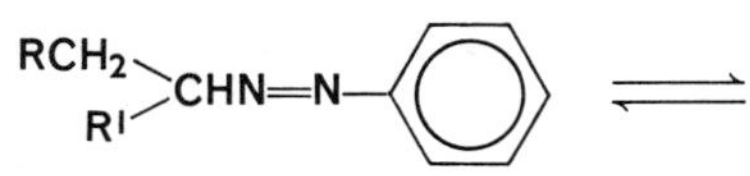

(**7b**)

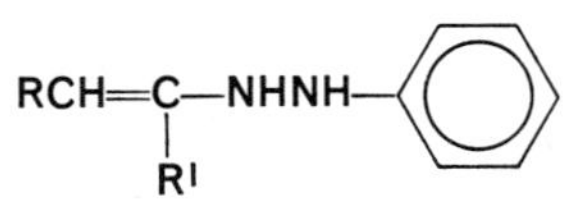

(**7c**)

(**7a**), azo (**7b**) and ene-hydrazine (**7c**) forms[14]. I.r. and n.m.r. measurements show that phenylhydrazones exist in the hydrazone form in non-polar solvents or as pure liquids[14], but shift to the ene-hydrazine form in polar solvents, as based on the detection of carbon–carbon double bonds in the i.r. spectra of methanol solutions[15]. All three tautomeric forms have been noted in polarographic studies in aqueous methanols[16].

Hydrazone-azo tautomerism can be strongly influenced by synergetic tautomerism in another portion of the molecule, e.g. keto-enol, as in equation (3). The equilibrium moves toward (**8b**) in polar solvents, in

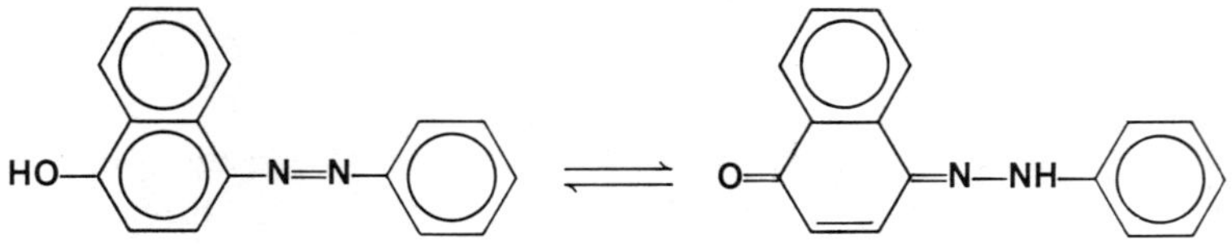

(3)

(**8a**) (**8b**)

parallel with observations on simple keto-enol equilibria[17]. On the other hand, *o*- and *p*-hydroxyazobenzene exist largely in the hydroxyazo form in solution and in the solid state, while hydroxyazoanthracenes appear to be in the hydrazone form[18].

D. Azoxy Compounds

Azoxy compounds exist as *cis* and *trans* isomers and under appropriate conditions they can be prepared, in general, with the oxygen attached to either nitrogen atom, leading to the isomers (**9a–9d**)[19a]. In *cis*-azoxybenzene,

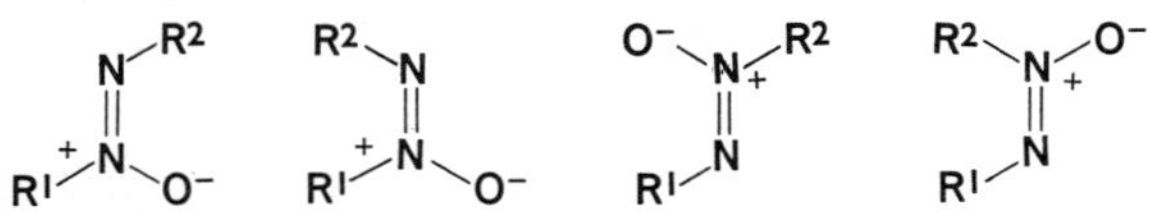

(**9a**) (**9b**) (**9c**) (**9d**)

the phenyl rings are rotated 56° out of the plane of the N=N double bond[20].

E. Triazenes

Diazoaminobenzenes (1,3-diphenyltriazenes) are probably the most common triazenes that have been studied. Tautomerism in unsymmetrically substituted diazoaminobenzenes, as illustrated by equation (4), has been

$$XC_6H_4\text{—NH—N=N—}C_6H_5 \rightleftharpoons XC_6H_4\text{—N=N—NH—}C_6H_5 \qquad (4)$$

(**10a**) (**10b**)

investigated in dilute CCl_4 solution by means of the i.r. NH absorption bands, using an elegant ^{15}N substitution technique[21,22]. Tautomer (**10a**) predominates when electron-withdrawing substituents like NO_2 are in the *para* position, while tautomer (**10b**) predominates when electron-donating groups like CH_3 or Cl are in the *para* position. For *meta* derivatives tautomer (**10b**) is always favoured. Since most *meta* substituents withdraw electrons inductively, the shift toward **10b** is readily understood in terms of the higher basicity of the imino N in **10b** relative to **10a**.

The kinetics of protolysis of some *p*-substituted 1-phenyl-3-methyl triazenes, and the reaction products, have been interpreted in terms of a mechanism in which tautomerism plays a key role[23].

X-ray diffraction techniques are obviously extremely valuable to establish tautomeric structures in the solid state, particularly if hydrogen atoms can be located. In 2,4-dibromodiazoaminobenzene crystals only a single tautomer analogous to **10a** has been found[24]. Here the NH group does not engage in hydrogen bonds and the formation of centrosymmetric dimers is ascribed to charge-transfer interaction between the Br atoms and the triazo group.

The geometrical configuration of triazenes commonly appears to be *trans* in the crystalline state (structure **11**). In solution dipole moment measure-

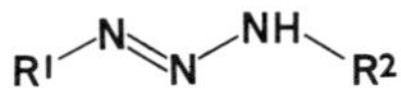

(11)

ments have not settled this point conclusively[25]. Other structural aspects of triazene chemistry have been summarized recently[19b].

F. Triazene Oxides

Many substances for which a hydroxytriazene structure is arbitrarily assumed actually exist in the triazene *N*-oxide structure as a result of

$$R^1—N{=}N—N(OH)R^2 \rightleftharpoons R^1—NH—N{=}\overset{+}{N}(\overset{-}{O})R^2 \qquad (5)$$

tautomerism (equation 5)[19c]. Spectroscopic evidence involving the so-called ^{15}N-labelled 1,3-diphenyl-3-hydroxytriazene, which has the *N*-oxide structure and a chelate hydrogen bond, is presented in Section IV, F.

III. BASICITIES

A. Hydrazines and Related Compounds

The basicity of hydrazine itself ($pK_a = 7{\cdot}95$)[26] is intermediate between that of ammonia and hydroxylamine[27a]. Correlations of base strengths with the substituent constant σ^* which have been established[28] show that hydrazine fits well in the relationship for the class of primary amines as a whole[29]. On the basis of these correlations it was concluded that solvation of the ammonium group through N^+H bonds played an important role in determining amine base strengths[29].

The electron-withdrawing properties of an amino group are such that its substitution for a hydrogen atom in ammonia should decrease the basicity of the molecule considerably. A fall of only one pK_a unit suggests a compensating base-strengthening factor related to repulsion between the two filled lone-pair orbitals. Attachment of a phenyl or other group which withdraws electrons conjugatively leads to a large drop in basicity[30].

While a number of pK_a values of hydrazines have been compiled and evaluated previously in a IUPAC report[31], Table 1 compares pK_a values of selected substituted hydrazines with a few reference compounds. One of the striking features is that substitution of an alkyl group for hydrogen in the hydrazine molecule lowers the base strength, and the introduction of further alkyl groups decreases the basicity even further[32]. Hydrazines can also pick up a second proton, forming the ion $H_3N^+—N^+H_3$ or its analogue, which have been well-characterized by X-ray diffraction. The basicity of the intermediate ion $N_2H_5^+$ is low, apparently close to that of water[27a].

Correlations of the basicity of alkyl hydrazines with substituent constants[28, 33] have been extended as more data became available. Thus for *N,N*-dialkylhydrazines the following correlation holds (equation 6),

$$pK_a = (-2{\cdot}92 \pm 0{\cdot}22) \sum \sigma^* + (7{\cdot}02 \pm 0{\cdot}09) \quad (6)$$

where the sum of the substituent constants for all the substituents is used[34]. This study also showed that alkylhydrazines are protonated at the most alkylated nitrogen atom. For negatively substituted mono-alkylhydrazines, including fluoroalkylhydrazines, the linear free energy relationship shown in equation (7) is obtained (correlation coefficient = −0·986; standard deviation = 0·132)[35]. On the basis of the slope of this correlation, which is

$$pK_a = -2{\cdot}36 \sum \sigma^* + 7{\cdot}45 \quad (7)$$

unlike that found for secondary amines[32] or for all classes of amines in another correlation[46], it was concluded that negatively substituted mono-alkylhydrazines protonate at the β-nitrogen atom, the α-NH unit introducing a reduction factor of 1·37 as far as the transmission of electronic effects is concerned[35].

Phenylhydrazine is a considerably weaker base than either hydrazine or ammonia because of delocalization of the N lone-pair electrons into the phenyl ring. The pK_a values of ring-substituted phenylhydrazines (Table 2) can be correlated approximately with the Hammett σ values of the sub-

TABLE 1. pK_a Values of hydrazines and selected reference compounds in aqueous solution

Compound		pK_a/Temperature (°C)
Hydrazine		7·95/25[26], 8·07/30[32]
Acetyl-		3·24/24[40]
1-Acetyl-2-isopropylidene-1[*N*-methylpiperidyl-(4′)]-		8·60/20[26]
1-Acetyl-1-[*N*-methylpiperidyl-(4′)]-		8·70/20[36]
1-Acetyl-2-[*N*-methylpiperidyl-(4′)]-		8·70/20[36]
1-Acetyl-2-[*N*-methylpiperidylene-(4′)]-		7·45/20[36]
1-Acetyl-2-phenyl-		1·3[37]
1-Allyl-2-benzyl-		5·98/25[34]
1-Allyl-2-cyclohexyl-		6·95/25[34]
n-Amyl-		7·75/25[38]
Benzyl-		6·35/25[38], 6·83/25[32]
n-Butyl-		7·82/25[38]
Carbethoxymethyl-		5·97/25[39]
2-Cyanoethyl-		5·91/25[39]
1,1-Diallyl-		6·32/25[34]
1,1-Dibenzyl-		5·67/25[a,34]
1,1-Diethyl-		7·71/30[32], 7·60/25[34]
1,2-Diethyl-		7·78/30[32]
1,1-Di-*i*-propyl-		8·38/25[34]
1,1-Dimethyl-		7·21/30[32], 7·09/25[34]
1,2-Dimethyl-		7·52/30[32]
1,1-Di-*n*-propyl-		7·00/25[38], 7·41/25[34]
Ethyl-		7·99/30[32], 7·95/25[38]
1-Ethyl-1-*i*-propyl-		7·98/25[34]
Hydroxyethyl-		7·12/25[39]
1-*i*-Propylidene-2-[*N*-methylpiperidyl-(4′)]-		8·90/20[36]
1-*i*-Propylidene-2-[*N*-*i*-propylpiperidyl-(4′)]-	(pK_1)	9·55/20[36]
	(pK_2)	6·30/20[36]
1-*i*-Propyl-2-[*N*-methylpiperidyl-(4′)]-	(pK_1)	9·07/20[36]
	(pK_2)	6·32/20[36]
N-*i*-Propylpiperidyl-(4′)-	(pK_1)	9·65/20[36]
	(pK_2)	5·75/20[36]
1-[*N*-*i*-Propylpiperidyl-(4′)]-2-methyl-	(pK_1)	9·1/20[36]
	(pK_2)	6·20/20[36]
Methyl-		7·87/30[32]
1-Methyl-1-allyl-		6·68/25[34]
1-Methyl-1-benzyl-		6·44/25[34]
1-Methyl-1-cyclohexyl-		7·77/25[34]
1-Methyl-1-ethyl-		7·30/25[38], 7·32/25[34]
1-Methyl-1-*i*-propyl-		7·52/25[34]
1-Methyl-2-[*N*-methylpiperidyl-(4′)]-	(pK_1)	8·70/20[36]
	(pK_2)	5·95/20[36]

TABLE 1 (*cont.*)

Compound		pK_a/Temperature (°C)
N-Methylpiperidyl-(4′)-	(pK_1)	8·85/20[36]
	(pK_2)	5·70/20[36]
1-[*N*-Methylpiperidyl-(4′)-2-phenyl]-		8·70/20[36]
1-(2,2,3,3,4,4,5,5-Octafluoropentane)-		5·34/25[39]
2-Phenoxyethyl-		6·80/25[39]
2-Phenylethyl-		6·75/25[38]
3-Phenylpropyl-		6·80/25[38]
n-Propyl-		7·92/25[38]
1-Propyl-1-*i*-propyl-		7·65/25[34]
2-(3,3,4,4-Tetrafluorobutane)-		5·59/25[39]
Tetramethyl-		6·30[41]
2,2,2-Trifluoroethyl-		5·38/25[39]
2-(1,1,1-Trifluoro-2-phenylethane)-		4·88/25[39]
Trimethyl-		6·56/30[32]
Vinyl-		5·46/25[39]
Semicarbazide		3·65/25[42]
Reference compounds		
Ammonia		9·2445/25[43]
Methylamine		10·6532/25[44]
Ethylenediamine		9·927/25[45]

[a] In 60% methanol.

stituents[47], but a much better fit is obtained for *meta*-substituents only, using Taft's σ^0 values (inductive substituent constants)[50]. The deviations observed for *para*-substituents are interpreted in terms of protonation occurring at the terminal amino group[51], but with a considerable degree of conjugative electron transfer to the amino group attached to the ring, which is then relayed to the protonation site by an inductive mechanism. The pK_a values of ring-substituted α-methyl phenylhydrazines are also correlated by the Hammett σ values of ring substituents[48]. The second protonation stage has also been investigated for some ring-substituted phenylhydrazines; here pK_as range from $-5{\cdot}2$ in phenylhydrazine itself to $-9{\cdot}2$ in *p*-nitrophenylhydrazine[52].

The pK_a values of some hydrazides are found in Table 3. These are much weaker bases because of the conjugative electron-withdrawing group attached to the hydrazino system.

TABLE 2. pK_a Values of substituted phenylhydrazines in aqueous solution at 25°C

Compound	pK_a
Phenylhydrazine	5·27[47], 5·20[30]
3,4-Benzo-	5·06[30]
2-Bromo-	5·54[47]
3-Bromo-	5·84[47]
4-Bromo-	5·05[47]
2-Bromo-α-methyl-	4·68[48]
3-Bromo-α-methyl-	4·30[48]
4-Bromo-α-methyl-	4·62[48]
2-Bromo-4-methyl-	4·32[a,48]
3-Bromo-4-methyl-	5·14[48]
2-Bromo-4-methyl-α-methyl-	4·95[a,48]
3-Bromo-4-methyl-α-methyl-	4·62[48]
2-Carbethoxy-	4·66[47]
3-Carbethoxy-	4·81[47]
4-Carbethoxy-	4·65[47]
2-Carbethoxy-α-methyl-	4·91[48]
3-Carbethoxy-α-methyl-	4·45[48]
4-Carbethoxy-α-methyl-	4·15[48]
2-Carboxy-	3·45[47]
3-Carboxy-	3·54[47]
4-Carboxy-	4·13[47]
2-Carboxy-α-methyl-	2·63[48]
3-Carboxy-α-methyl-	2·90[48]
4-Carboxy-α-methyl-	3·41[48]
2-Chloro-	4·65[47]
3-Chloro-	4·92[47], 4·78[30]
4-Chloro-	5·10[47], 4·96[30]
2-Chloro-α-methyl-	4·78[48]
3-Chloro-α-methyl-	4·43[48]
4-Chloro-α-methyl-	4·64[48]
3-Cyano-	4·47[30]
4-Cyano-	4·25[30]
2,4-Dibromo-	4·27[a,48]
2,4-Dibromo-α-methyl-	4·65[48]
2,3-Dichloro-	4·10[a,48]
2,4-Dichloro-	4·30[48]
2,5-Dichloro-	4·44[48]
2,6-Dichloro-	4·79[48]
3,4-Dichloro-	4·80[48]
2,4-Dimethyl-	5·72[48]
2,4-Dimethyl-α-methyl-	6·03[48]
2,4-Dinitro-	2·68[a,48]
2-Ethoxy-	5·36[47]

TABLE 2 (*cont.*)

Compound	pK_a
3-Ethoxy-	5·14[47]
4-Ethoxy-	5·59[47]
2-Ethoxy-α-methyl-	5·25[48]
3-Ethoxy-α-methyl-	4·72[48]
4-Ethoxy-α-methyl-	5·53[48]
α-Ethyl-	5·16[48]
α-Methyl-	4·98[48]
2-Methyl-	5·32[47]
3-Methyl-	5·43[47], 5·26[30]
4-Methyl-	5·49[47], 5·32[30]
2-Methyl-α-methyl-	5·29[48]
3-Methyl-α-methyl-	5·09[48]
4-Methyl-α-methyl	5·30[48]
3-Methyl-α-(3-methylphenyl)-	4·48[a,48]
4-Methyl-α-(4-methylphenyl)-	3·90[a,48]
2-Methyl-4-bromo-	4·86[a,48]
3-Methyl-4-bromo-	5·20[48]
2-Methyl-4-bromo-α-methyl-	4·21[a,48]
3-Methyl-4-bromo-α-methyl-	4·73[a,48]
2-Methoxy-	5·53[47]
3-Methoxy-	5·30[47]
4-Methoxy-	5·71[47]
2-Methoxy-α-methyl-	5·42[48]
3-Methoxy-α-methyl-	4·90[48]
4-Methoxy-α-methyl-	5·60[48]
2-Nitro-	3·10[47]
3-Nitro-	4·39[47], 4·36[30]
4-Nitro-	3·94[47], 3·70[30]
2-Nitro-α-methyl-	4·32[48]
4-Nitro-α-methyl-	2·65[a,48]
α-Phenyl-	3·80[a,48]
2-Sulpho-	2·33[47]
3-Sulpho-	2·43[47]
4-Sulpho-	2·61[47]
4-Sulpho-α-methyl-	2·03[48]
Reference compound	
Aniline	4·603[49]

[a] In water/dioxan.

B. Azomethines and Related Compounds

The basicities of a large number of aromatic azomethines have been reported recently[55], but they are not repeated here to conserve space.

TABLE 3. pK_a Values of carboxylic acid hydrazides ($RCONHNH_2$) in aqueous solution at 25°C

Substituent		pK_a
(1) R = XC_6H_4		
X = *p*-NMe_2		13·03 ± 0·04[53]
o-NH_2	(pK_1)	1·85[54]
	(pK_2)	3·47[54]
	(pK_3)	12·80[54]
m-NH_2	(pK_1)	2·57[54]
	(pK_2)	3·75[54]
	(pK_3)	12·45[54]
p-OMe		12·83 ± 0·03[53]
p-Me		12·75 ± 0·03[53]
H	(pK_1)	3·03/20°[54]
		12·52 ± 0·03[53], 12·45/20°[54]
p-Cl		12·09 ± 0·03[53]
m-Cl		11·95 ± 0·03[53]
p-COOEt		12·38 ± 0·04[53]
m-NO_2		11·36 ± 0·03[53]
p-NO_2		11·26 ± 0·03[53]
o-NO_2		11·34 ± 0·05[53]
(2) R = alkyl		
Me		13·04 ± 0·04[53]
CH_2Ph		13·17 ± 0·03[53]
$CHPh_2$		13·00 ± 0·03[53]
CH_2OPh		12·23 ± 0·04[53]
CH_2OMe		12·14 ± 0·04[53]
CH_2OPh(*p*-Cl)		12·19 ± 0·03[53]
CH_2CN		11·04 ± 0·05[53]
(3) R = other		
3-Pyridyl		11·49 ± 0·03[53]

There are very few accurate measurements available on the basicity of imines, but they seem to be weaker bases than the corresponding amines. For example, *p*-chloro-benzaldehyde anil has pK_a = 2·8, relative to pK_a = 4·6 for aniline[27b]. The lower pK_a of imines is ascribed to two opposing effects, the change from sp^3 to sp^2 hybridization for the nitrogen atom (decreasing basicity by about 5 pK_a units) and the inability of the unshared pair of electrons on the nitrogen atom to overlap with the aromatic π-electrons (increasing the basicity by about 3 pK_a units, relative to aniline)[56]. Semicarbazones are weaker bases than semicarbazide itself, by about 5 pK_a

units, again because of the change in hybridization of the nitrogen atom[27b, 56].

The pK_a values of some hydrazones, which are also relatively weak bases, are listed in Table 4. The pK_as of aromatic aldehydes and ketones can be correlated with the Hammett σ values of substituents, and the generalized behaviour interpreted in terms of the following resonance structures

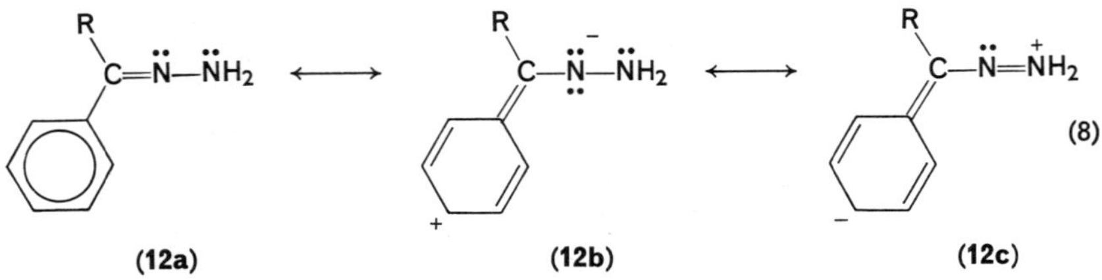

(equation 8)[57]. Electron-releasing substituents favour **12b**, with the accumulation of negative charge on N increasing the basicity, while electron-withdrawing substituents lower the basicity below that of the parent hydrazone[57]. The sp^3 nitrogen atom of hydrazones must therefore interact substantially with the groups attached to the carbon atom[19d].

2,4-Dinitrophenylhydrazones of benzaldehydes and acetophenones are acidic, with pK_a values in the range 10·4–11·6[58].

Oximes are weak acids. The pK_a values of simple oximes range from 10 to 12, while α-keto groups tend to increase their acid strength (pK_a 7–10). α-Dioximes are also stronger acids than monoximes[19e]. Table 5 lists the pK_as for a variety of oximes. Because of the double bond oximes also have a weak basicity, the conjugate acid of acetone oxime having a pK_a = 1·75[37].

TABLE 4. pK_a Values of hydrazones in methanol at 22°C[57]

Parent ketone	pK_a
Acetophenone	4·70 ± 0·04
Benzophenone	3·85 ± 0·03
4,4′-Dimethoxybenzophenone	4·38 ± 0·05
4,4′-Dichlorobenzophenone	3·13 ± 0·03
4-Methoxyacetophenone	4·94 ± 0·05
Phenyl 2-thienyl ketone	3·80 ± 0·04

TABLE 5. pK_a Values of oximes in aqueous solution

Compound	pK_a/Temperature (°C)
Acetophenone oxime	11.48[59]
Acetoxime	12·42/24·9[60]
syn-Benzaldoxime	10·68[61]
Diacetyl monoxime	9·30/25[62]
Diethyl α-hydroxyiminomalonate	5·48/20·6[63]
Diethyl ketoxime	12·59/24·9[60]
Dimethyl glyoxime	10·72[64]
syn-2,4-Dinitrobenzaldoxime	9·43[61]
Ethyl α-hydroxyiminocyanoacetate	4·66/19·4[63]
syn-Furfuraldoxime	10·82[59]
anti-Furfuraldoxime	10·85[59]
Glyoxime	9·11[64]
4-Hydroxybenzaldoxime	8·93/25[62]
Hydroxyiminoacetone	8·30/25[62]
Hydroxyiminoacetophenone	8·25/25[62]
Hydroxyiminoacetylacetone	7·38/25[62]
3-Hydroxyimino-4-methylpentane-2-one	9·50/25[62]
3-Hydroxyiminopentane-2-one	9·38/25[62]
2-Hydroxyiminomethylpyridine	10·10/25[62]
2-Hydroxyiminomethylpyridine methiodide	7·82/25[62]
3-Hydroxyiminomethylpyridine methiodide	9·10/25[62]
4-Hydroxyiminomethylpyridine methiodide	8·23/25[62]
Methyl ethyl ketoxime	12·45/24·9[60]
syn-3,4-Methylenedioxybenzaldoxime	10·85[61]
syn-2-Methoxybenzaldoxime	10·89[61]
syn-3-Methoxybenzaldoxime	10·59[61]
syn-4-Methoxybenzaldoxime	10·92[61]
Methylglyoxime	9·69[64]
syn-2-Nitrobenzaldoxime	10·06[61]
syn-3-Nitrobenzaldoxime	10·15[61]
anti-3-Nitrobenzaldoxime	10·74[59]
syn-4-Nitrobenzophenone oxime	10·85[59]
anti-4-Nitrobenzophenone oxime	10·47[59]
2-Oxo-*n*-butyraldoxime	8·37/25[62]
Salicylaldoxime	9·17/25[62]
Salicylhydroxamic acid	7·43/25[62]

C. Azo Compounds

Base strengths of aliphatic azo compounds have not been measured since they isomerize rapidly to hydrazones under strongly acidic conditions. The pK_as for azobenzenes are well known; Table 6 lists some of them in aqueous solution and Table 7 gives comparable data in 20% ethanol/80%

TABLE 6. pK_a Values of azobenzenes in aqueous solution

Compound		pK_a/Temperature (°C)
Azobenzene		−2·48/25[a,65]
4-Amino-		2·82/25[68], 2·76/22[67]
4-Anilino-		0·99/22[69]
2-Carboxy-4′-hydroxy-	(pK_1)	−1·26/25[65]
	(pK_3)	8·2/25[65]
4-Carboxy-4′-hydroxy-	(pK_1)	−1·64/25[65]
	(pK_3)	7·9/25[65]
4-Diethylamino-4′-hydroxy-		−1·89/20[70]
4-Diethylamino-4′-methoxycarbonyl		−2·70/20[70]
4-Dimethylamino-	(pK_1)	3·299/18[71], 3·226/25[71] 3·122/35[71], 3·5/25[65]
	(pK_2)	−4·37/25[65]
4-Dimethylamino-4′-hydroxy-	(pK_1)	−1·81/25[65]
	(pK_2)	3·4/25[65]
	(pK_3)	8·6/25[65]
4-Dimethylamino-4′-methoxycarbonyl	(pK_1)	2·60/20[70]
	(pK_2)	−3·10/20[70]
4-Hydroxy-	(pK_1)	−0·93/25[65]
	(pK_2)	8·2/25[65]
4-Hydroxy-2,2′,4′,6,6′-pentamethyl-	(pK_1)	−1·30/25[65]
	(pK_2)	9·1/25[65]
4-Nitro-		−3·47/25[72]

[a] The pK_a of *trans*-azobenzene is −2·95 and that of *cis*-azobenzene −2·25[66]; those for some isomers of other derivatives have also been reported[67].

H_2SO_4 solution. It is noteworthy that the *trans* and *cis* isomers of azobenzene itself have different basicities (pK_as of −2·95 and −2·25, respectively) and that they are weaker bases than water[67].

The site of protonation has led to much controversy and only the more recent evidence will be summarized here. Earlier proposals that the proton is bound symmetrically to both N atoms[77], that both N atoms together act as a single basic site[67] or that the proton is bound to the —N=N— group in the form of a π-complex[78] cannot be reconciled with the n.m.r. spectra and pK_as of substituted azobenzenes, particularly those with *ortho*-

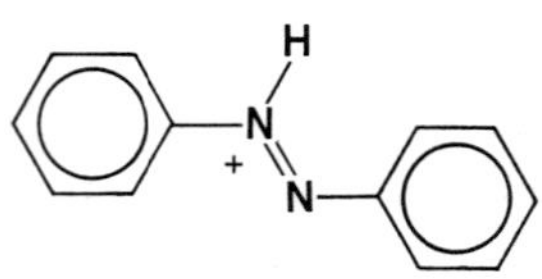

(13)

TABLE 7. pK_a Values of azobenzenes in 20% ethanol/80% H_2SO_4 at 25°C[a,73]

Compound	$-pK_a$
Azobenzene	2·90 ± 0·02, 2·93
4-Acetyl-	3·98 ± 0·04
4-Acetyl-4′-methoxy-	2·32 ± 0·02
3-Acetyl-3′-nitro-	5·69 ± 0·02
4-Acetyl-3′-nitro-	5·96 ± 0·02
3-Bromo-	3·83 ± 0·03
4-Bromo-	3·47 ± 0·02
3-Bromo-4′-methoxy-	2·06 ± 0·02
4-Bromo-4′-methoxy-	1·74 ± 0·03
3-Bromo-3′-nitro-	5·52 ± 0·02
4-Bromo-3′-nitro-	5·04 ± 0·01
3-Bromo-4′-hydroxy-	1·67 ± 0·01
4-Bromo-4′-hydroxy-	1·42 ± 0·01
4-Cyano-	4·52 ± 0·02
4-Cyano-3′-nitro-	6·50 ± 0·04
2,2′-Dimethoxy-	0·43
2,4-Dimethoxy-	−0·31
2,4′-Dimethoxy-	0·91
2,6-Dimethoxy-	−0·02
4,4′-Dimethoxy-	0·75 ± 0·03, 0·73
2,4-Dimethoxy-2′-methyl-	−0·34
2,4-Dimethoxy-4′-methyl-	−0·41
2,6-Dimethoxy-2′-methyl-	−0·32
2,6-Dimethoxy-4′-methyl-	−0·46
4,4′-Dimethoxy-2-methyl-	0·45
2,2′-Dimethyl-	2·97
2,4-Dimethyl-	1·75
2,4′-Dimethyl-	2·09
2,6-Dimethyl-	3·64
4,4′-Dimethyl-	1·85
2,2′-Dimethyl-4,4′-dimethoxy-	0·73
2,2′-Dimethyl-4-methoxy-	1·43
2,2′-Dimethyl-4-methoxy-4′-nitro-	2·56
3,3′-Dinitro-	6·57 ± 0·07
4,4′-Dinitro	4·02
4-Ethoxy-	1·28 ± 0·02
4-Ethoxy-3′-nitro-	2·48 ± 0·03
2,4,6,2′,4′,6′-Hexamethyl-	2·38
2-Hydroxy-	2·04
4-Hydroxy-	1·02 ± 0·01
2-Hydroxy-2′-methyl-	2·04
2-Hydroxy-6-methyl-	1·31
2-Methoxy-	1·37

TABLE 7 (*cont.*)

Compound	$-pK_a$
4-Methoxy-	1·36 ± 0·03, 1·34
4-Methoxy-4′-hydroxy-	0·56 ± 0·03
2-Methoxy-2-methyl-	1·20
2-Methoxy-6-methyl-	−0·03
2-Methoxy-4′-methyl-	1·16
4-Methoxy-2′-methyl-	1·66
4-Methoxy-3′-nitro-	2·54 ± 0·03
4-Methoxy-4′-nitro	2·52
2-Methyl-	2·59
3-Methyl-	2·70 ± 0·03
4-Methyl-	2·35 ± 0·02, 2·32
4-Methyl-4′-hydroxy-	0·84 ± 0·03
2-Methyl-4-methoxy-	0·59
4-Methyl-4′-methoxy-	1·03 ± 0·02, 1·10
2-Methyl-4-methoxy-4′-nitro-	1·57
2-Methyl-4′-methoxy-4-nitro	3·47
2-Methyl-4-nitro-	4·94
2-Methyl-4′-nitro-	4·18
3-Methyl-3′-nitro-	4·32 ± 0·02
4-Methyl-3′-nitro-	3·83 ± 0·02
3-Nitro-	4·63 ± 0·02
4-Nitro-	4·70 ± 0·02, 4·02
2,4,2′,4′-Tetramethoxy-	−1·42
2,4,2′,4′-Tetramethyl-	1·90
2,4,6,4′-Tetramethyl-	2·53
2,4,6,2′-Tetramethyl-	2·87
2,6,2′,6′-Tetramethyl-	3·44
2,4,2′-Trimethyl-	2·97
2,4,4′-Trimethyl-	1·54
2,4,6-Trimethyl-	2·83
2,6,2′-Trimethyl-	3·48
2,6,4′-Trimethyl-	3·15

[a] Values without uncertainty limits from reference 74; uncertainty ±0·07 pK_a units. Temperature not stated. Additional pK_a values for azobenzenes are found in references 46, 75 and 76.

substituents, but they provide strong support for the very classical view that protonation takes place on a distinct N atom with the formation of an N—H σ-bond (**13**)[74, 79]. Substituent effects on pK_as over a very large range, and including *ortho*-substituents, have been paramatrized using a modified Hammett equation[74].

TABLE 8. pK_a Values of aminoazobenzenes in 50% aqueous ethanol at 26°C[50, 80, 81, a]

Compound	pK_a
(1) Aminoazobenzenes (AB)	
2-AB	1·8
3-AB	3·0
4-AB	2·35
2,3′-Dimethyl-4-AB	2·92
2,4′-Dimethyl-4-AB	2·92
3,4′-Dimethyl-4-AB	2·39
4′,5-Dimethyl-2-AB	2·5
2′,3-Dimethyl-4-AB	2·29
(2) 4-Dimethylaminoazobenzenes (DAB)	
DAB	2·28
3′-Acetyl-DAB	2·03
4′-Acetyl-DAB	2·16
3′-Acetylamino-DAB	2·27
4′-Acetylamino-DAB	2·25
3′-Acetyl-2-methyl-DAB	2·73
4′-Acetyl-2-methyl-DAB	2·88
4′-Acetyl-3-methyl-DAB	3·27
2′-Chloro-DAB	1·74
3′-Chloro-DAB	2·01
4′-Chloro-DAB	2·00
3′-Chloro-2-methyl-DAB	2·67
2,2′-Dimethyl-DAB	2·64
2′,5′-Dimethyl-DAB	2·0
2′-Ethyl-DAB	1·85
4′-Ethyl-DAB	2·30
4′-Fluoro-DAB	2·00
4′-Isopropyl-DAB	2·31
2′-Methoxy-DAB	2·20
4′-Methoxy-DAB	2·40
2-Methyl-DAB	3·08
2′-Methyl-DAB	2·04
3-Methyl-DAB	3·48
3′-Methyl-DAB	2·33
4′-Methyl-DAB	2·36
2′-Nitro-DAB	1·5
3′-Nitro-DAB	1·67
4′-Nitro-DAB	1·81
2′-Nitro-2-methyl-DAB	2·12
3′-Nitro-2-methyl-DAB	2·32
3′-Nitro-3-methyl-DAB	3·18
2′,4′,6′-Tribromo-DAB	1·0
3′-Trifluoromethyl-DAB	1·84

TABLE 8 (*cont.*)

Compound	pK_a
(3) 4-Methylethylaminoazobenzenes (MEAB)	
MEAB	2·58
3′-Acetyl-MEAB	2·28
4′-Acetyl-MEAB	2·35
3′-Acetylamino-MEAB	2·47
2′-Chloro-MEAB	2·14
4′-Ethyl-MEAB	2·72
4′-Fluoro-MEAB	2·40
2′-Nitro-MEAB	1·75
3′-Nitro-MEAB	2·00
3′-Trifluoromethyl-MEAB	2·15
(4) 4-Diethylaminoazobenzenes (DEAB)	
DEAB	3·08
3′-Nitro-DEAB	2·39
(5) Others	
4-Benzylmethylaminoazobenzene	1·6
4-Ethylaminoazobenzene	2·58
4-Methylaminoazobenzene	2·37
3′-Methyl-4-methylaminoazobenzene	2·44

[a] Additional pK_a values are reported in reference 82.

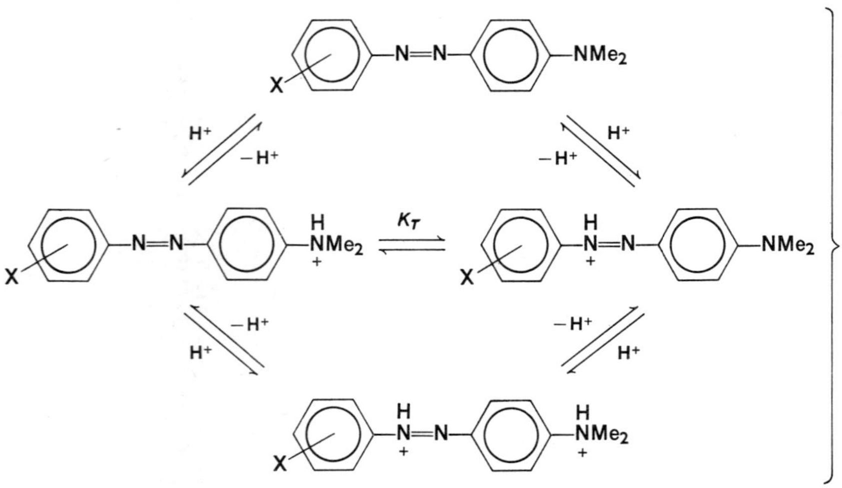

(9)

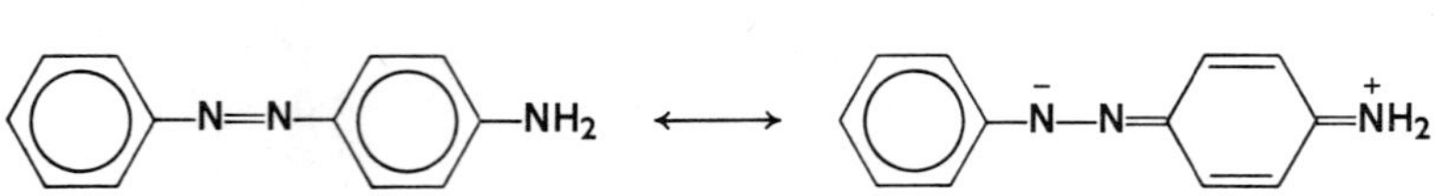

(10)

p-Aminoazobenzenes, whose pK_as are given in Table 8, form a particularly interesting class of compounds, and their protonation behaviour is summarized below (equation 9)[83]. The ratio of azonium to ammonium ions can be determined from u.v. spectra[81,84,85]. Values for K_T range from 0·39 in the *p*-NO_2 and *p*-CN compounds to 2·70 for *p*-OH[83], and rise to above 7 in *p*-Me-*p'*-NEt_2 and *p*-Cl-*p'*-NEt_2[82]. In *N*-pyrrolidino azobenzenes K_T drops to near zero, and the order of K_T is explained on the basis of cumulative resonance and steric effects[82]. An earlier publication reports a number of pK_1 and pK_2 values for mono- and di-protonated azobenzenes[86]. The base strength of the β nitrogen atom in the azo link is probably enhanced by the resonance effect (equation 10)[80,86].

D. Azoxy Compounds

Azoxy compounds are even weaker bases than azo compounds. Protonation is believed to occur on the oxygen atom in azoxybenzenes, and substituent effects are correlated by the Hammett equation[87]. Table 9 lists

TABLE 9. pK_a Values of azoxybenzenes in 20% ethanolic H_2SO_4 at 25°C[87]

Compound[a]	$-pK_a$
Azoxybenzene	6·45 ± 0·03
4-Bromo-	6·94 ± 0·03
4'-Bromo-	7·01 ± 0·04
4-Bromo-4'-methyl-	6·90 ± 0·03
4-Chloro-	6·96 ± 0·05
4,4'-Dibromo-	7·77 ± 0·03
4,4'-Dichloro-	7·69 ± 0·03
4,4'-Dimethoxy-	5·23 ± 0·03
4,4'-Dimethyl-	5·47 ± 0·05
4'-Ethoxy-	6·04 ± 0·03
4-Methoxy-	6·15 ± 0·04
4'-Methoxy-	6·10 ± 0·04
4-Methyl-	6·16 ± 0·04
4'-Methyl-	6·04 ± 0·03
4-Methyl-4'-bromo-	6·95 ± 0·04
4'-Nitro-	9·83 ± 0·07

[a] The numbering scheme is as follows:

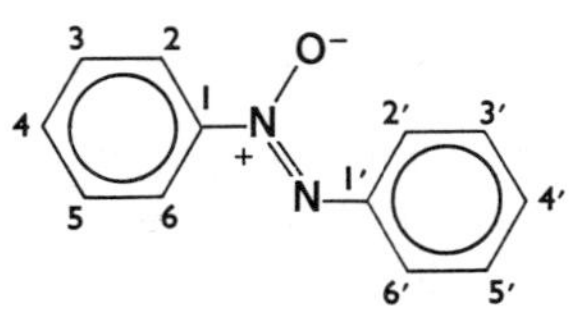

TABLE 10. pK_a Values for monoxides and dioxides of *trans-p*-dimethylaminoazobenzene in 20% ethanolic H_2SO_4 at 25°C[88]

Compound	pK_{a2}	pK_{a1}
p-Dimethylaminoazobenzene	2·96 ± 0·03	−5·34 ± 0·02
(14) *p*-Oxidodimethylaminoazobenzene	4·11 ± 0·02	−4·65 ± 0·04
(15) α-*p*-Dimethylaminoazoxybenzene	1·93 ± 0·05	−8·51[a]
(16) β-*p*-Dimethylaminoazoxybenzene	2·62 ± 0·04	−8·02 ± 0·03
(17) β-*p*-Oxidodimethylaminoazoxybenzene	4·03 ± 0·03	−8·00 ± 0·05
(18) α-*p*-Oxidodimethylaminoazoxybenzene	3·71 ± 0·04	−8·41 ± 0·03

[a] Extrapolated value.

the available pK_a data for substituted azoxybenzenes while Table 10 contains corresponding data for the monoxides and dioxides of *trans*-dimethylaminoazobenzene (**14–18**). In each compound the dimethylamino or the oxidodimethylamino group is protonated first in dilute acid and further protonation of the azo or azoxy group occurs only at high acid concentrations[87].

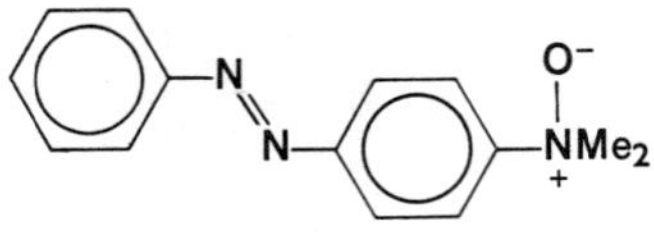

(14)

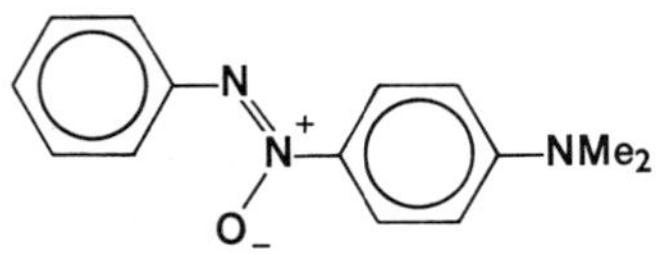

(15)

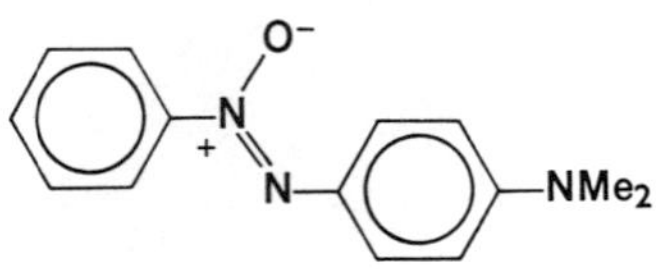

(16)

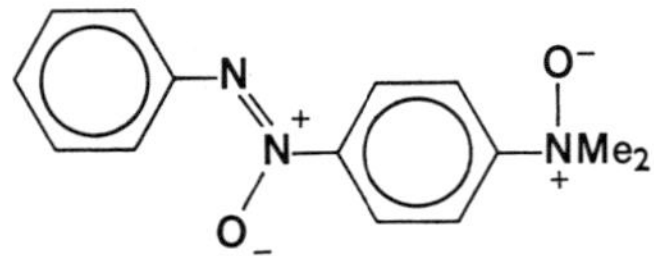

(17)

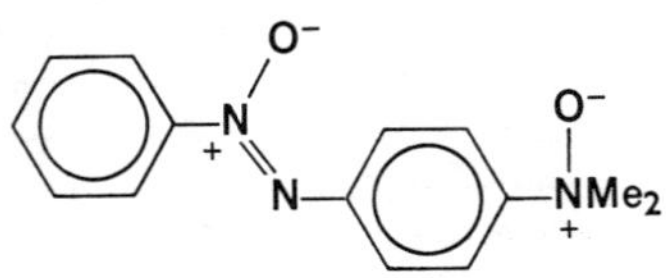

(18)

The pK_a values for phenyl–naphthyl and naphthyl–naphthyl azoxy compounds have been interpreted in terms of the extent of delocalization of the unshared electrons on the O atom through the aryl rings[89].

E. *Triazenes and Triazene Oxides*

Disubstituted triazenes are distinctly acidic (form isolable cuprous salts) but also possess some basic properties (are decomposed by mineral acids)[19b]. Little quantitative information appears to be available. The pK_a values of some 3-hydroxy-1-alkyl-3-phenyltriazenes are given in Table 11, and those for a series of 3-hydroxy-1,3-diphenyltriazenes in Table 12 (but see also Section IV, F regarding their structure).

TABLE 11. pK_a Values of 3-hydroxy-1-alkyl-3-phenyltriazenes[a] in dioxan solution at 25°C[90]

Substituents[b]			pK_a
R^1	R^2	R^3	
Me	H	H	12·11
Me	Me	H	12·52
Me	H	Me	12·30
Me	OMe	H	12·69
Me	Cl	H	11·52
Me	H	Cl	11·31
Et	H	H	12·28
Et	Me	H	12·67
Et	OMe	H	12·85
Et	Cl	H	11·65
Et	H	Cl	11·44
n-Pr	H	H	12·37
n-Pr	Me	H	12·79
n-Pr	Cl	H	11·74

[a] Tautomerism to the corresponding triazene-*N*-oxide is not precluded.
[b] See Section IV, F, structure (**87**) for positions of R groups.

IV. HYDROGEN BOND FORMATION

A. *Hydrazines and Related Compounds*

Much early evidence with regard to the association of hydrazine and substituted hydrazines was obtained from freezing-point diagrams (thermal analysis). Table 13 lists some representative systems, and the mole ratios of the components established in the complex. Clearly intermolecular hydrogen bond formation between hydroxylic or phenolic proton donors and the basic hydrazines plays an important role in stabilizing these

TABLE 12. pK_a Values of 3-hydroxy-1,3-diphenyltriazenes[a] in dioxan solution at 25°C[91]

Substituents[b]			pK_a
R^1	R^2	R^3	
H	H	H	11·41
Cl	H	H	10·72
H	Cl	H	10·52
H	H	Cl	10·65
Br	H	H	10·86
Me	H	H	11·78
H	Me	H	11·55
H	H	Me	11·86
Me	H	Me	12·18
NHAc	H	H	11·66
CO_2	H	H	10·97
Ac	H	H	10·96
OMe	H	H	10·95
SO_3Na	H	H	9·99

[a] Tautomerism to the corresponding triazene *N*-oxide is not precluded.

[b] See Section IV, F, structure (**88**) for positions of R groups.

complexes, as has subsequently been established by other techniques. The role of hydrogen bonding in complex formation was also noted in the systematic study of refractive index[103], viscosity, density, vapour pressure and conductivity[94] of binary liquid mixtures.

The self-association of hydrazine has been studied by Raman spectroscopy, the temperature dependence of the 3189 cm^{-1} band leading to a ΔH^0 value of $-1·450$ kcal/mole for hydrogen bond formation[104]. Competitive hydrogen bonding of hydrazine to dimethylsulphoxide was also employed in this investigation to establish the identity of the 3189 cm^{-1} band[104]. Similarly, $\Delta H^0 = -0·418$ kcal/mole for intermolecular hydrogen bond formation in methylhydrazine[105].

Proton magnetic resonance studies on *N,N*-dimethyl-, *N,N'*-dimethyl and *N,N,N'*-trimethylhydrazine in cyclohexane support a monomer–dimer equilibrium with association constants of 1·8, 1·0 and 0·8 (mole fraction)$^{-1}$, respectively, for these compounds at 28·5°C[106]. Evidence was also obtained for a 1:1 hydrogen bonded complex of *N,N,N'*-trimethylhydrazine with carbon tetrachloride, with a dimerization constant of 0·4 (mole frac-

TABLE 13. Binary compounds with hydrazines, from freezing point diagrams (thermal analysis)

Components		Mole ratio (A:B)
A	B	
Water	Hydrazine	1:1[92]
Ethanol	Hydrazine	2:1[96]
Phenol	Hydrazine	1:1[93], 2:1[94], 4:1[95]
Phenol	Phenylhydrazine	1:1[98]
p-Chlorophenol	Phenylhydrazine	1:2, 1:1, 3:1[99]
m-Nitrophenol	Phenylhydrazine	1:1[97]
p-Nitrophenol	Phenylhydrazine	1:1[97]
β-Naphthol	Phenylhydrazine	1:1[98]
o-Cresol	Phenylhydrazine	2:1[98]
m-Cresol	Phenylhydrazine	2:1[98]
p-Cresol	Phenylhydrazine	2:1[98]
Acetic acid	Phenylhydrazine	1:1[100]
o-Cresol	Hydrazobenzene	?[101]
m-Cresol	Hydrazobenzene	?[101]
p-Cresol	Hydrazobenzene	?[101]
2,4-Dinitrophenol	Hydrazobenzene	1:1, 2:1[102]
Pyridine	Hydrazobenzene	1:1, 2:1[102]

tion)$^{-1}$[106]. Subsequently it has been claimed, on the basis of further p.m.r. studies in different solvents, that the cyclic dimer scheme is incorrect for *N,N*-dimethyl- and *N,N′*-dimethylhydrazine, and that the associated species is a cyclic trimer for the former, while both cyclic trimers and tetramers occur for the latter[107]. These studies were interpreted in terms of a hydrogen bond strength of 2 kcal/mole for each NH···N bond in *N,N′*-dimethylhydrazine, and 2·4 kcal/mole for each NH···N bond in *N,N*-dimethylhydrazine[107].

Nuclear quadrupole resonance spectra of hydrazines have also been linked with hydrogen bonded configurations in the solid state. Thus atomic orbital occupation numbers lead to the following order of the hydrogen bonding ability of the amino N atom: hydrazine > alkylhydrazines > *N*-aminoheterocycles > arylhydrazines[108].

Hydrogen bond energies in *N,N′*-diformyl- and *N,N′*-diacetylhydrazine have been obtained from heats of sublimation based on the temperature dependence of the vapour pressure of the solid; subtraction of the contributions from dispersion and electrostatic interaction energies leads to hydrogen-bond energies of 5·0–6·1 and 3·03 kcal/mole, respectively[109, 110].

The i.r. spectrum of *N,N'*-diformylhydrazine confirms the presence of a strong hydrogen bond, since the NH stretching frequency lies at 3100 cm^{-1}, whereas in ordinary amides like *N*-methylacetamide it appears at 3250 cm^{-1} [111]. The lower hydrogen bond strength in *N,N'*-diacetylhydrazine is reflected in a smaller NH stretching shift in the i.r., to 3235 cm^{-1} [112].

TABLE 14. Hydrogen bond lengths in hydrazine, substituted hydrazines and their salts in the crystalline state

Compound	Bond type[a]	R(AH···B) (Å)[b]
Cyclopropanecarbohydrazide[113]	NH···O	2·94
	NH···N	3·16, 3·26
Diacetylhydrazine[114]	NH···O	2·877
Diformylhydrazine[115]	NH···O	2·799
Dihydrazinium sulphate[116]	NH···N	3·03, 3·14
	NH···O	2·76–3·14
n-Dodecanoic acid hydrazide[117]	NH···O	3·00
	NH···N	3·17
Hydrazine[118]	NH···N	3·19, 3·30 (within chains)
	NH···N	3·25–3·67 (between chains)
Hydrazinium bis(dihydrogen phosphate)[119]	NH···O	2·634–2·825
Hydrazine bis(ethanol)[96]	NH···O	3·041, 3·060
	OH···N	2·730
Hydrazine bis(methanol)[120]	OH···O	2·931–3·104
	OH···N	2·747, 2·724
Hydrazine dihydrochloride[121]	NH···Cl	3·10
Hydrazine dihydrofluoride[122]	NH···F	2·62
Hydrazine hydrate[123]	NH···O	3·11, 3·15
	OH···N	2·79
Hydrazine hydrochloride[124]	NH···N	2·95
	NH···Cl	3·12–3·62
Hydrazine tetramethanol[124a]	NH···O	2·959, 3·025
	OH···N	2·682
Hydrazinium acetate[125]	NH···N	2·923
	NH···O	2·716–3·081
Hydrazinium dihydrogen phosphate[126]	NH···O	2·812–3·252
	NH···N	2·907
Hydrazinium hydrazine dithiocarboxylate[127]	NH···S	3·38 (in layers)
	NH···N	3·078 (between layers)
	NH···N	2·968 (between cations)
Hydrazinium hydrogen oxalate[128]	NH···N	2·858
	NH···O	2·872–3·424
Hydrazinium perchlorate hemihydrate[129]	OH···N	2·837, 2·897

TABLE 14 (*cont.*)

Compound	Bond type[a]	R(AH···B) Å[b]
Hydrazinium sulphate[130]	N[1]H···O	2·689–2·782
	N[2]H···O	2·830–2·991
Isonicotinic acid hydrazide[131]	NH···N	3·04, 2·97
n-Nonanoic acid hydrazide[132]	NH···O	2·896
	NH···N	3·188
Phenylhydrazine[133]	NH···N	3·23
Phenylhydrazine hydrochloride[134]	NH···Cl	3·134, 3·162
	NH···N	3·007, 3·420
Thiosemicarbazide[135]	NH···S	3·359–3·816

[a] NH···N includes N^+H···N, etc.
[b] Where more than two different A···B distances exist, only the shortest and the longest are given here.

The importance of hydrogen bonds in crystals has long been recognized[2]. Thus X-ray diffraction, neutron diffraction and n.m.r. studies of crystals have also provided much information on hydrazino compounds, the last two techniques being particularly powerful because of the ability to locate hydrogen atoms. Table 14 lists some typical hydrogen bond lengths observed in crystals of hydrazine, its salts, and in some substituted hydrazines. While a number of other geometric factors are important in determining hydrogen bond strength, these have been discussed fully[2] and only the AH···B distance is given in Table 14.

Hydrazine complexes and salts pack in the crystalline state in such a manner that a large number of hydrogen bonding possibilities frequently exist. Thus, in hydrazinium perchlorate hemihydrate each $N_2H_5^+$ ion is surrounded by at least eleven atoms at possible hydrogen bonding distances[129] while at the same time no hydrogen bonds between either water molecules and the ClO_4^- ion, or between $N_2H_5^+$ ions are observed. In $N_2H_6(H_2PO_4)_2$ the phosphate layers are held together by the $N_2H_6^{2+}$ ions, each one forming six N^+—H···O bonds[119]. Liminga's review of these crystal structures[136] emphasizes that 'abnormal' hydrogen bond configurations occur frequently, i.e., cases where the coordination number of the hydrogen atom in the hydrogen bond is greater than two (bi- or tri-furcated bonds) are common. This has been confirmed by neutron diffraction, where the hydrogen atoms can be located precisely. It has also been demonstrated that strong hydrogen bonding between hydrazine and kaolin exists; hydrazine enters the kaolin lattice reversibly from aqueous solution, greatly expanding the interlayer distances. Some four to five hydrazine molecules per $Al_4(OH)_8Si_4O_{10}$ unit enter the lattice[136a].

I.r. and Raman spectra of crystalline hydrazo compounds confirm the dominance of hydrogen bonding. Thus, in the i.r. spectra of hydrazinium fluoride and the fully deuterated analogue, the lower NH stretching modes and the increased frequency of deformation and bending modes in the $N_2H_5^+$ and $N_2D_5^+$ ions when compared with corresponding chloride salts are in accord with strong N^+—H···F hydrogen bonds[137]. In hydrazinium sulphate ($N_2H_6SO_4$) and dihydrazinium sulphate [$(N_2H_5)_2SO_4$] i.r. spectra of the powdered solids are in accord with correlation curves for hydrogen-bonded NH_3^+ groups[138]. Both the Raman and i.r. spectra of hydrazinium hydrogen oxalate and dihydrazinium oxalate can be explained on the basis of NH···O, NH···N, N^+—H···O and N^+—H···N hydrogen bonds involving the $N_2H_5^+$ ion[139]. The spectrum of the fully deuterated analogue of the former has also been investigated[140].

A strong intramolecular OH···N hydrogen bond has been found in **19**,

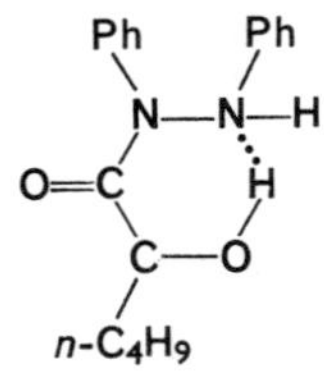

(19)

as based on vicinal p.m.r. coupling constants[141]. In dilute CCl_4 solution **19** exhibits OH absorption at 3495 cm^{-1}. In another study, i.r., u.v. and n.m.r. spectra have shown that the diazonium coupling products of aroylacet-anilides, $RC_6H_4COC(:NNHPh)CONHPh$, with R = *p*-OMe, *p*-Me, *m*-Me, H, *p*-Cl, *p*-Br and *m*-Cl, exist exclusively in the hydrazo form with hydrogen bonds[142].

In the p.m.r. spectrum of polycrystalline hydrazinium hydrogen oxalate observed over the range 190–340 K, the second moment of the resonance line width decreases as the temperature is raised from 210 to 280 K. This is ascribed to the onset of reorientation of the NH_3^+ group with a rate fast in comparison with the line width[143]. Similar deuteron magnetic resonance studies of deuterated hydrazinium sulphate show that the ND_3^+ group reorients itself rapidly about the N—N bond at room temperature[144].

The electrical conductivity of $LiN_2H_5SO_4$ is anisotropic, with the largest value along the *c* axis of the crystal, and this has been explained in terms of proton transfer along ···NH···NH···NH··· chains, which are parallel to that axis[145].

Provided that the imino group adjacent to the acyl group remains unsubstituted, molecular weight determinations show that molecular association of hydrazides in solution is general[146]. This has been attributed to hydrogen-bond formation between —CONH— groups similar to that in amides[146]. Hydrogen bonding in hydrazides has also been confirmed from their i.r. spectra, through NH frequency shifts[147]. In dilute solution dipole moment measurements have been interpreted in terms of bifurcated intramolecular hydrogen bonding between the primary amino group and the carbonyl oxygen atom in aliphatic and aromatic acid hydrazides, and in semicarbazide, but not in thiosemicarbazide[148].

B. Azomethines and Related Compounds

Changes in the electron density at N in substituted benzalanilines (**20**)

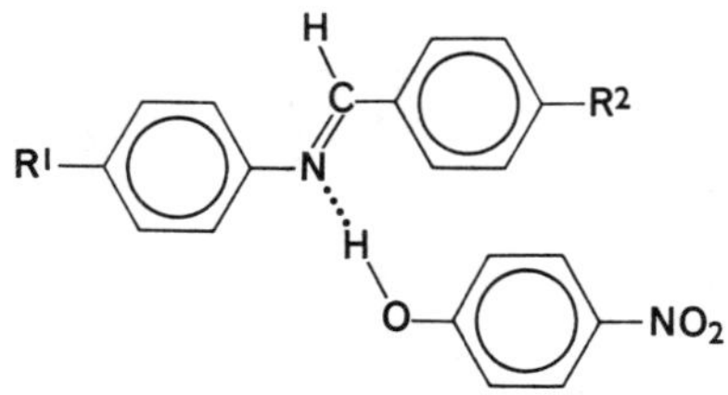

(20)

can be monitored through intermolecular OH···N association with a common proton donor like *p*-nitrophenol under standardized conditions[149]. The association constants (Table 15) reflect the inductive and resonance polar effects of R^1 and R^2. For variations in R^1 with $R^2 = H$, the $\log K_{assoc}$ values are linearly related to the Hammett σ constants of R^1, but for variations in R^2 with $R^1 = H$ the Taft σ^+ substituent constants must be

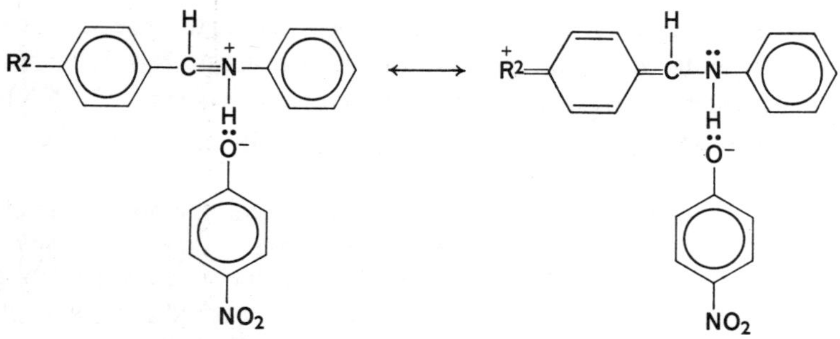

(11)

(21a) **(21b)**

TABLE 15. Association constants of *p*-nitrophenol with substituted benzalanilines (**20**) in dilute CCl_4 solution at 27·5°C[a], [149]

Substituents R^1	R^2	$K_{assoc} \times 10^{-1}$ (l/mole)
H	NMe_2	28
H	OMe	8·8
H	Me	6·1
H	H	3·9
H	Cl	2·9
H	Br	2·5
H	NO_2	1·2
NMe_2	H	14
OMe	H	6·3
Me	H	6·1
Cl	H	2·5
Br	H	2·1
NO_2	H	1·3
NMe_2	NMe_2	80
OMe	OMe	12
Me	OMe	12
NMe_2	NO_2	3·6
Cl	Cl	1·5

[a] Enthalpies of hydrogen bond formation ranged from −5·5 to −7·2 kcal/mole.

employed for a linear relationship with $\log K_{assoc}$. This has been attributed to stabilization of the hydrogen bond by resonance interaction with R^2, as follows (equation 11).

Polarographic studies indicate that the *N*-benzylidene–phenol complex similar to **20** is reduced at a potential more positive than the free Schiff base, reflecting the polarization implied by resonance structure **21b**[150].

Solvents which are capable of hydrogen bonding with the >C=NNHR group in hydrazones stabilize the *syn* isomer with respect to the *anti*-isomer[151].

Salicylidene anilines (**22**) have no 'free' OH stretching vibration; instead a broad weak band with fine structure is found from 2700 to 3100 cm^{-1} [152], in which a weak band near 2730 cm^{-1} has been assigned to a very strongly intramolecularly hydrogen bonded OH in a six-membered chelate ring[153]. This band is unperturbed by the presence of *p*-nitrophenol and persists in

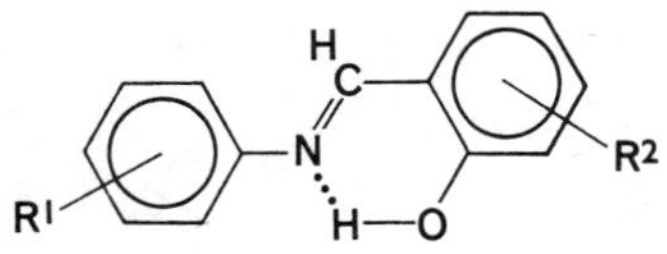

(22)

copper chelates[152]. Polarographic studies show that intramolecular hydrogen bonds exist in **23**, with hydrogen bond strengths estimated to be 6–10 kcal/mole[150].

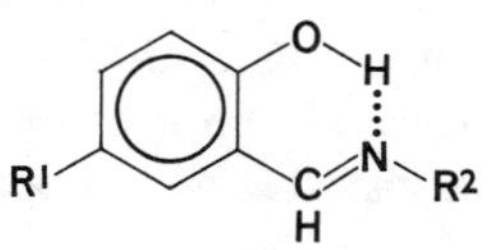

(23)

(a) $R^1 = H$, $R^2 = Me$
(b) $R^1 = H$, $R^2 = CH_2CH_2N{=}CHC_6H_4$(*o*-OH)
(c) $R^1 = H$ or other group, $R^2 = Ph$
(d) $R^1 = H$, $R^2 =$ *o*-OHPh

If the OH group is on the phenyl ring that carries the N atom (**24**) only a

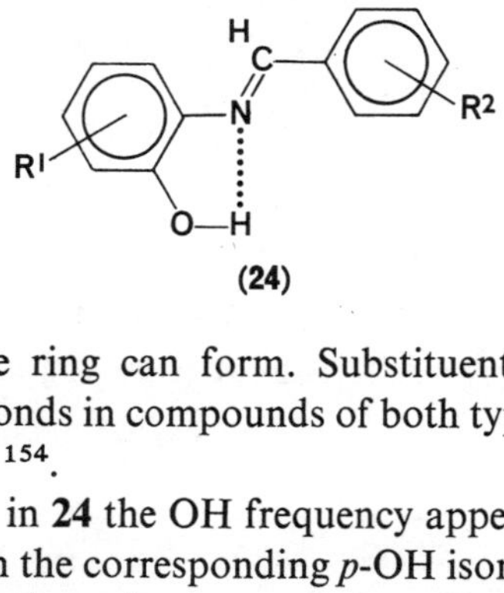

(24)

five-membered chelate ring can form. Substituent effects on the intramolecular hydrogen bonds in compounds of both type **23** and **24** have been studied extensively[153, 154].

When $R^1 = R^2 = H$ in **24** the OH frequency appears at 3443 cm^{-1}, fully 175 cm^{-1} lower than in the corresponding *p*-OH isomer[155]. Enlargement of the five-membered chelate ring to a six-membered ring in *o*-hydroxybenzalaniline (**22**, $R^1 = R^2 = H$) strengthens the hydrogen bond so that ν_{OH} falls to 3180 cm^{-1}, somewhat lower even than in *N*-2-hydroxybenzylaniline ($\nu_{OH} = 3250$ cm^{-1}) where the N atom is in an sp^3 hybridization state but has some of its electron density delocalized into the phenyl ring. Salicylidene-*o*-hydroxybenzylamine (**25**) and salicylidene-*o*-aminophenol (**26**) each exhibit two OH stretching bands. In **25** one of the OH groups is essentially 'free' while the other exhibits the normal chelate frequency, while in **26** the chelate ν_{OH} is about 130 cm^{-1} higher because the second OH group can form an intramolecular hydrogen bond to the C=N bond, thereby withdrawing some charge density from the N atom[155].

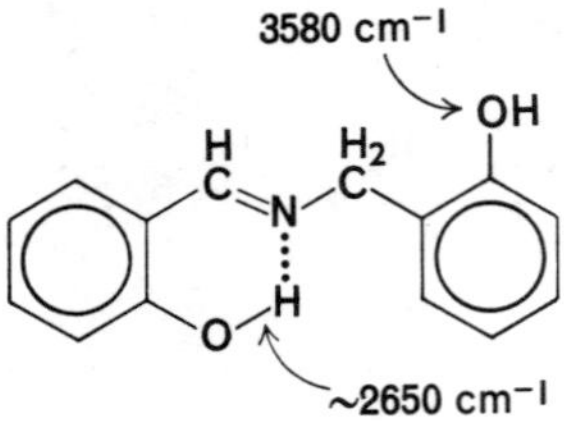

(25)

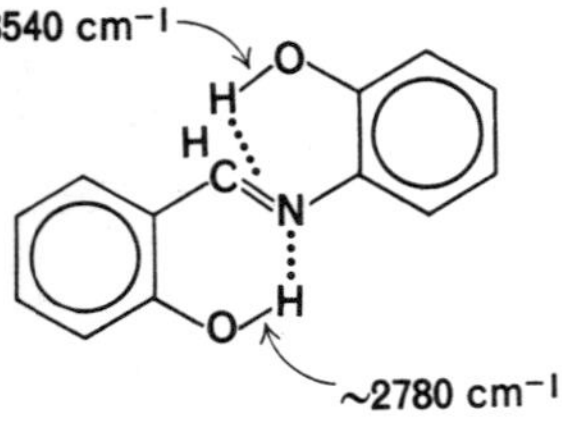

(26)

Chemical shifts of OH and methine protons in n.m.r. spectra are correlated with the Hammett σ constants of substituents in both five- and six-membered chelate rings in *o*-hydroxy derivatives of Schiff bases, and the weaker nature of the five-membered ring is demonstrated by its apparent rupture in dimethylsulphoxide solution[156].

For compounds of the general structure **27** the following series yield

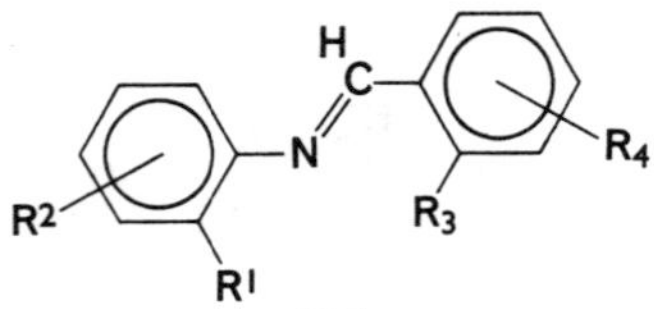

(27)

	R^1	R^2	R^3	R^4
(a)	OH	5-Me	H	H, 2′-, 3′-, 4′-(OMe, Cl, NO_2), 4′-NMe_2
(b)	OH	H	H	H, 3′-, 4′-(OMe, Cl), 4′-(NMe_2, NO_2)
(c)	OH	5-NO_2	H	H, 4′-(NMe_2, OMe, Cl, NO_2)

linear plots of ν_{OH} (range ~3380–3470) versus the Hammett σ values of R^4, with identical positive slopes. The relative displacement of these lines reflects the variation in the acidity of the OH group, which increases in the order $R^2 = 5\text{-}CH_3 < 5\text{-H} < 5\text{-}NO_2$. Changes in R^4 (and to a much lesser extent in R^2 in the 5-position) control the electron density on the nitrogen atom and hence its basicity, the second factor determining the strength of the intramolecular hydrogen bond and the shift of ν_{OH}. No 'free' OH band is ever observed. The only significant deviation from these linear relationships occurs when $R^4 = 4'\text{-}NO_2$, and the lower ν_{OH} is explained by direct resonance interaction of the type **28**[153, 154].

No similar correlation with substituent effects for **27** when $R^1 = H$ and $R^3 = OH$ has been established because of the broad and diffuse OH···N

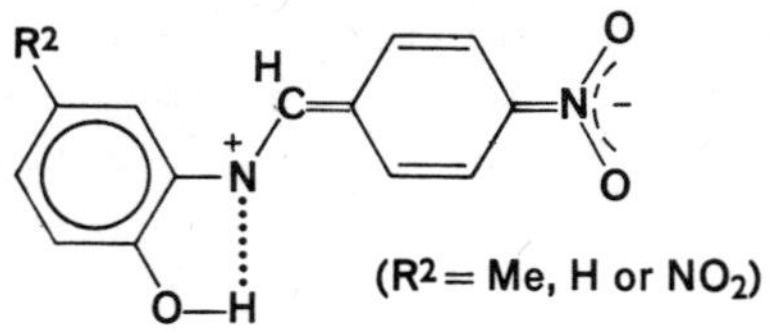

(28)

band in the six-membered chelate ring, which is overlapped by CH and other vibrations[154]. The breadth of this band and its structure also reflects benzenoid–quinonoid tautomerism which leads to some intramolecular NH···O= hydrogen bonding (*vide supra*). The effect of chelation in five- and six-membered rings in these compounds on the in-plane δ(OH) and δ(OD) bending modes, and on the out-of-plane γ(OD) bending vibration has also been noted. The latter shifts to higher frequency as the hydrogen bond becomes stronger, and these frequencies can be correlated with the Hammett σ values of ring substituents, in different structural series[154]. These data clearly confirm that the hydrogen bond in a six-membered chelate ring is stronger than in a five-membered ring.

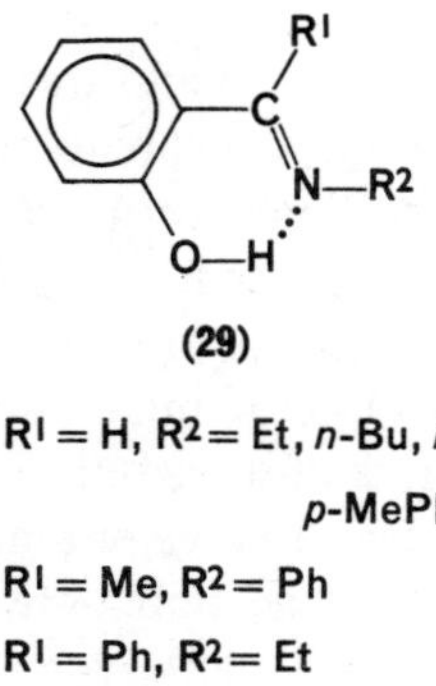

(29)

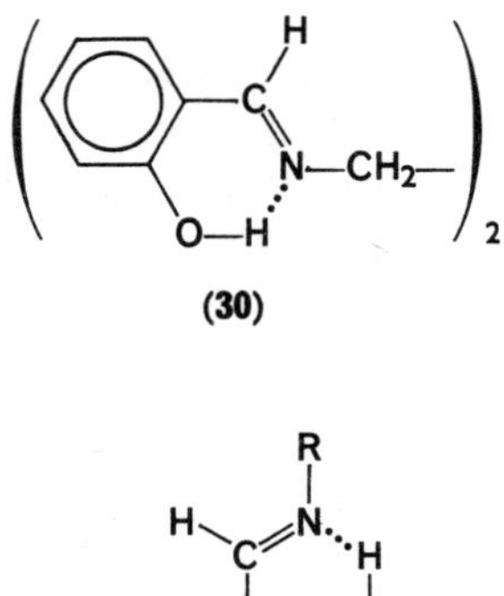

(30)

(a) R^1 = H, R^2 = Et, *n*-Bu, *i*-Bu, *p*-MePh, *p*-NO_2Ph
(b) R^1 = Me, R^2 = Ph
(c) R^1 = Ph, R^2 = Et

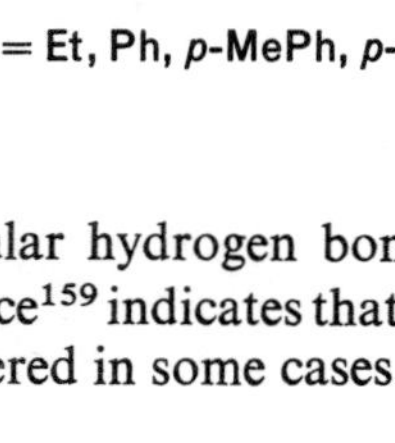

(31) R = Et, Ph, *p*-MePh, *p*-NO_2Ph

I.r. and u.v. spectra confirm intramolecular hydrogen bonding in the following Schiff bases[157, 158], but other evidence[159] indicates that benzenoid–quinonoid tautomerism must also be considered in some cases.

Electron and n.m.r. spectroscopy have established that benzenoid–quinonoid tautomerism exists in Schiff bases derived from *o*-hydroxy-

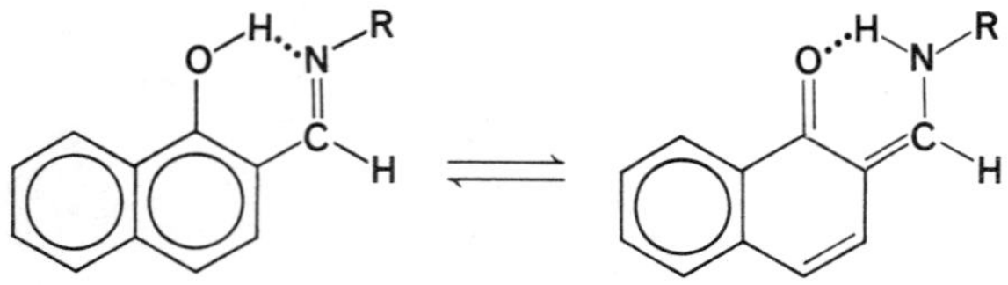

(32a) (32b) (12)

naphthaldehydes (equation 12)[160–162]. The i.r. spectrum in the 2400–3300 cm^{-1} region cannot provide a solution in this case as to the location of the H atom, due to its broad and overlapped nature. Furthermore, while a band at 1620–1640 cm^{-1} could be attributed to the C═N stretching vibration in **32a**[163], this is difficult to distinguish unequivocally from the C═O stretching vibration in **32b**[164]. Appreciable spectral changes occur in the region 1500–1650 cm^{-1} when the solvent is changed from CCl_4 to dioxan to $CHCl_3$ for those azomethines which undergo tautomerism (e.g., the anils of 2-hydroxy-1- and 1-hydroxy-2-naphthaldehyde)[150], while those which exist solely or predominantly in only one form, either benzenoid (e.g., derivatives of 2-methoxy-1-naphthaldehyde, 1-methoxy-2-naphthaldehyde, 3-hydroxy-2-naphthaldehyde)[161] or quinonoid (e.g., 2-hydroxy-1-anthraldehyde)[165] have identical spectra in all three solvents.

When R^2 in **23** is a phenyl group, the phenol-imine form predominates[166]. 2,3-Naphthalene Schiff bases also have the phenol-imine form, presumably because the keto-amine form does not provide sufficient hydrogen bond stabilization to compensate for the resonance energy lost in its formation[162, 167]. However, a very short-lived (1 msec) keto-amine tautomer of (**22**, $R^1 = R^2 = H$), but not for the *para* isomer, is observed during photochemical isomerizations in acidic solution[167a]. Schiff bases of amino acids with salicylaldehyde, 3-hydroxypyridine-2-aldehyde and 3-hydroxy-4-aldehyde exist in the form of two tautomeric species, although largely in the keto-amine form[168]. Various investigations attest to the strong hydrogen bonding in compounds like **29**[169].

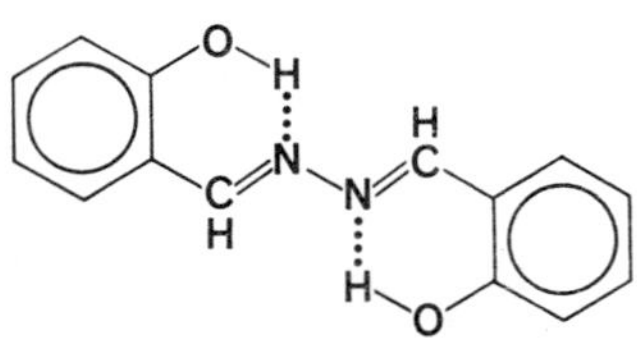

(33)

X-ray diffraction studies of salicylaldehyde azine (**33**) show a centrosymmetric structure for the molecule, with two intramolecular OH···N bonds with length 2·645 Å[170]. The data indicate significant contribution from the quinonoid resonance form. 4-Formyl-pyridine thiosemicarbazone exists in the form of dimers linked by hydrogen bonded thioamide groups in the crystalline state, but these dimers are then held together by NH···N=C hydrogen bonds ($R_{NH\cdots N} = 2{\cdot}961$ Å)[171].

A series of anils from aniline-^{15}N identify two bands at 1630 and 1584 cm^{-1} (in benzylideneaniline-^{15}N) as symmetric and asymmetric stretching vibrations of the conjugated fragment C_{ar}=C_{ar}—C=N, with the C=N group contributing more to the higher frequency band[159]. In *o*-hydroxybenzylideneaniline these bands are displaced to lower frequency, for comparable isotopic substitution, reflecting an intramolecular OH···N hydrogen bond. The intensity of the 1630 cm^{-1} band increases markedly as the solvent polarity increases.

Photoexcitation at 77 K induces the transfer of protons from the phenol-imine form (**32a**) of anils of *o*-hydroxybenzaldehyde and *o*-hydroxynaphthaldehyde to the corresponding keto-amine form (**32b**)[172] and intramolecular hydrogen bond energies of 12–16 kcal/mole are estimated. The strength of the hydrogen bond in *o*-hydroxyazomethines where six-membered chelate rings can form has also been correlated with increasing C=C double bond order in the bond adjacent to C=N[172]. The above hydrogen bond energy may be compared with $-5{\cdot}80 \pm 0{\cdot}17$ kcal/mole obtained for the formation of an intermolecular OH···N bond between *β*-naphthol and pyridine[173]. Proton transfer in this quasi-aromatic ring is reversible, as demonstrated by luminescence studies[174]. The rate of proton transfer is not affected by the substitution of *p*-Me or *p*-OMe groups in the aniline ring of salicylidene aniline, but the substitution of *p*-NO_2 or a second *o*-OH group lowers it, the former by reducing the electron density on the azomethine nitrogen atom and thus making the keto-amine form less stable, and the latter presumably by introducing competitive although less stable five-membered ring formation by hydrogen bonding. The rate of proton transfer is higher in solution than in the powdered state or the melt. Proton-donor and proton-acceptor solvents lower the rate of proton transfer because of intermolecular interactions which interfere with formation of the six-membered chelate ring. Proton transfer back to the phenol-imine form is accompanied by the release of energy and the partial conversion of electronic energy to vibrational energy[174]. Another study of the luminescence of frozen solutions of azomethines with intramolecular hydrogen bonds has shown that, in addition to the electronic effect of substituents, steric factors which lead to non-coplanar ring systems decrease the fluorescence, supporting the hypothesis that it is connected with the phototransfer of a proton[175].

The six-membered chelate ring in Schiff bases can be broken by competitive intramolecular proton acceptor groups, as in equation (13), where

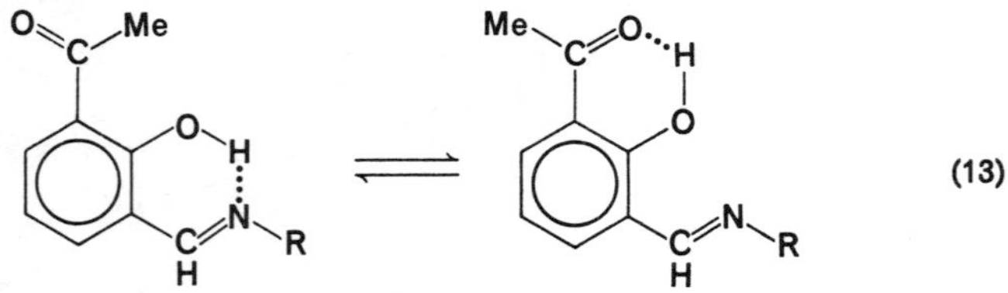

(34a) (34b)

even at very low concentrations in CCl_4 the OH stretching vibration is doubled[176] in spite of the complete absence of intermolecular association. Identification of the components follows from deuteration (OD) and from comparison with the spectra of *o*-hydroxyacetophenone. Similar competition is provided by replacing the COMe group by an NO_2 group[176].

Schiff bases derived from aliphatic β-diketones and aliphatic amines can potentially have the structures (**35a**), (**35b**) or (**35c**) (equation 14). N.m.r.

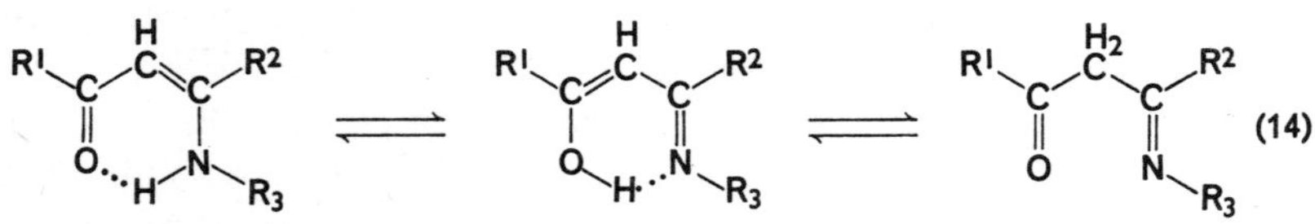

(35a) keto-enamine (35b) enol-imine (35c) keto-imine

spectra show that they exist predominantly as **35a** rather than **35b** or **35c**, indicating that NH···O hydrogen bonding in **35a** is appreciably stronger than OH···N hydrogen bonding in **35b**, particularly since the latter would have significant extra resonance stabilization[177]. The equilibrium position in equation (14) is unaffected by variation in the solvent or in R^3[178]. Salicylaldehyde anils exist in enol-imine form analogous to **35b**[172,177].

Carbonyl and NH frequency shifts in the i.r. spectra of α-phenylhydrazono ketones $R^1COC(:NNHPh)R^2$ show that six-membered chelate rings exist

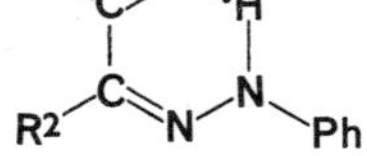

(36)

(36)[179]. Substituent and solvent effects on the u.v. spectra of nitrophenylhydrazones of substituted benzaldehydes indicate that intramolecular NH···O hydrogen bonds to *o*-NO_2 groups can play a significant role in modifying the properties of some of these compounds[180].

Diphenylketimine (**37a**) exhibits a monomeric NH stretching vibration at 3266 cm^{-1} in CCl_4 solution which shifts to 3234 cm^{-1} on association with dimethylsulphoxide[181]. Self-association in CCl_4 shifts ν_{NH} to 3223 cm^{-1}. The association constants of diphenylketimine with 2,2,2-trifluoroethanol, *p*-chlorophenol, dimethylsulphoxide and hexamethylphosphotriamide have been found to be 31·5 ± 3, 80 ± 4, 0·75 and 2·5 l/mole, respectively, in the ternary solvent CCl_4 at 25°C. Self-association has been interpreted in terms of the following species:

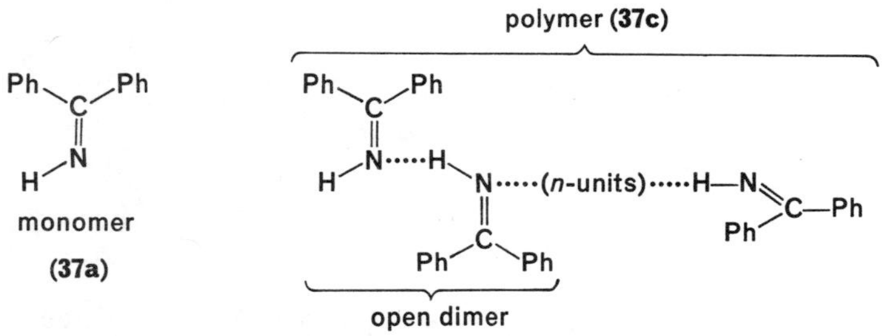

(37a)

(37b)

with the monomer **37a** being the dominant species at concentrations below 0·1M, the open dimer **37b** and the monomer being in equilibrium betweeo 0·1M and 0·2M, and the monomer–polymer **37c** equilibrium dominating in the range 0·2–2·5M[181]. Complexation of diphenylketimine with the proton-acceptor solvents acetonitrile, dioxan, pyridine and triethylamine, and with the proton donor solvents chloroform and *o*-cresol, has also been explored[182]. *o*-Cresol is more complexed with the imine than with *N*-methylaniline because of the localization of an electron pair on N in the former compound[182].

A comparative i.r. study of self-association of ethyl propyl ketimine, dibutyl ketimine, propyl phenyl ketimine and diphenyl ketimine in dilute

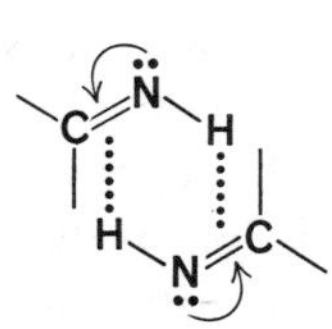

(38)

CCl_4 solution led to dimerization constants of 5·00, 1·30, 3·65 and 0·136 l/mole, respectively, at 25°C, with enthalpies of association for all four compounds in the range −2·0 to −2·3 kcal/mole[183]. While the possibility of some cyclic association like **38** could not be excluded, an open dimer is again considered to be more likely, particularly since graphical separation of the overlapped band envelopes suggests that the terminal NH of the open dimer absorbs around 3250 cm^{-1} [183]. The association of ketimines in general is therefore slightly stronger than that of comparable aliphatic amines. While their proton-donor strength is comparable to that of amines, their proton-acceptor ability is far more pronounced because of the trigonal hybridization of the lone pair[183].

In *p*-chlorobenzene-*anti*-diazoimidoglyoxynitrile (**39**) the molecule is

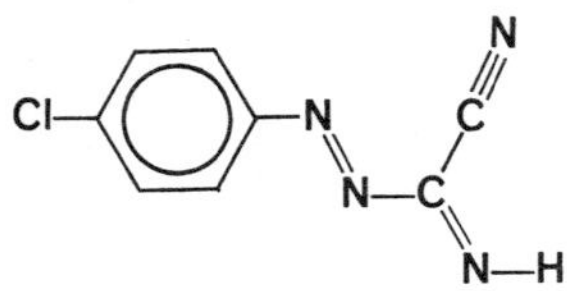

(39)

nearly planar and the imide NH acts as a proton donor and the imide N atom as an acceptor, the azo group not being involved[184]. Chains of hydrogen bonded molecules have an NH···N distance of 3·261 Å[184].

The acidity of the NH in *O*-ethylbenzimidates (**40**) increases in the order

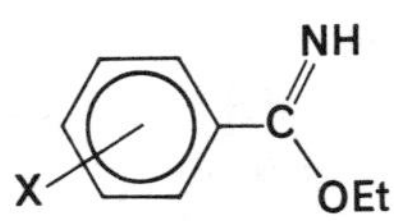

(40)

X = *m*-Me < *p*-Me < H < *m*-Cl < *p*-NO_2, which has been established on the basis of NH frequency shifts in a series of solvents, relative to its position in hexane[185]. These shifts are satisfactorily correlated by the Hammett σ values of X. The NH frequency in CCl_4 shifts little from X = *p*-NO_2 (3346 cm^{-1}) to X = *m*-CH_3 (3342 cm^{-1}), which is ascribed to the high degree of conjugation that exists between the N lone pair electrons and the aromatic ring throughout this series[185]. Doubling of the NH stretching frequency in *O*-ethylphenylacetamidates (**41**) in dilute CCl_4 solution is attributed to two conformations, one of which has an intramolecular NH···π hydrogen bond (equation 15)[186].

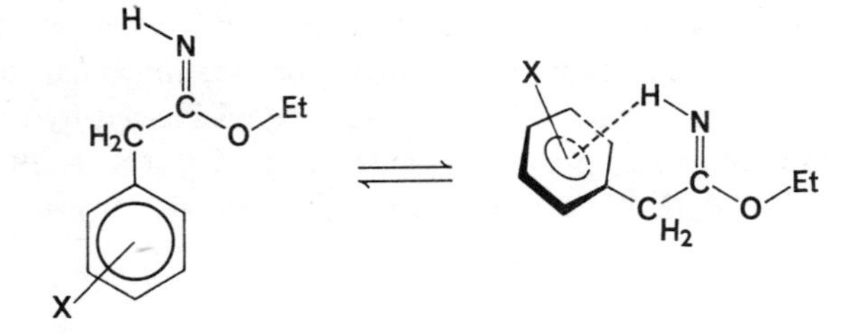

(15)

(41a) $\nu_{NH} = 3335–3336\ cm^{-1}$ $2\nu_{NH} = 6518–6520\ cm^{-1}$

(41b) $\nu_{NH \cdots \pi} = 3322–3326\ cm^{-1}$ $2\nu_{NH \cdots \pi} = 6490–6500\ cm^{-1}$

The equivalence of the 3- and 5-positions in pyrazoles has been related to tautomeric proton transfer in a hydrogen-bonded cyclic dimer (equation 16)[187]. Cryoscopic molecular weight determinations show that association

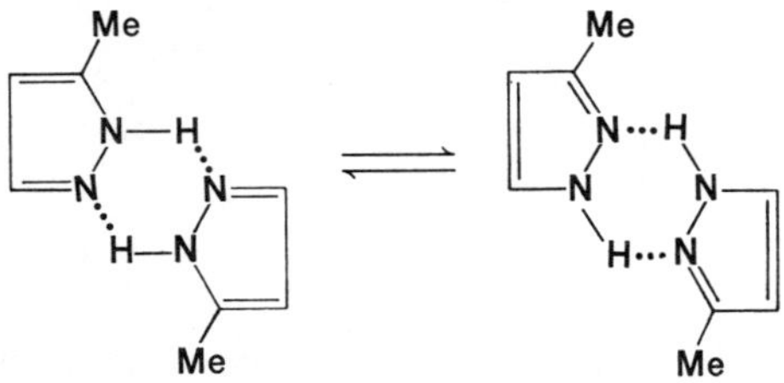

(16)

(42a) **(42b)**

also occurs in the amidines (**43**) and the benzimidazoles (**44**)[188, 189].

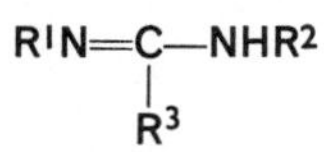

(43)

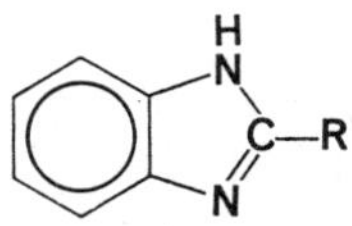

(44)

Indazole (**45**) also behaves like **42**, but in pyrazoline (**46**) the tautomeric possibility is removed and the compound is much more volatile than

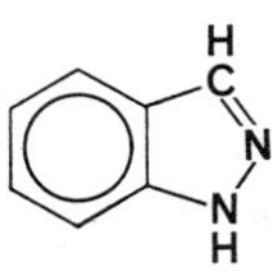

(45)

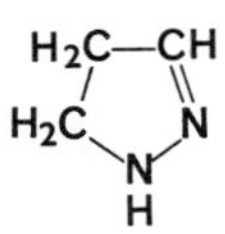

(46)

pyrazole (**42**), showing that it is less associated[189]. Where intramolecular hydrogen bonding can occur this either suppresses molecular association, as in 2-benzoylbenzimidazole (**47**) and its phenyl methyl hydrazone (**48**), or eliminates it almost completely as in the phenylhydrazone of 2-benzoylbenzimidazole (**49**), 2-*o*-hydroxyphenylbenzimidazole (**50**), or 2-*o*-aminophenylbenzimidazole (**51**)[189].

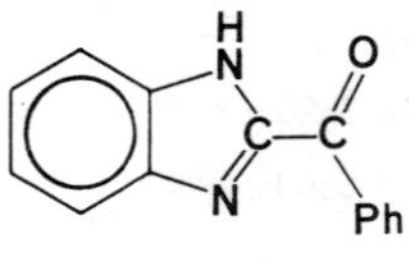

(47)

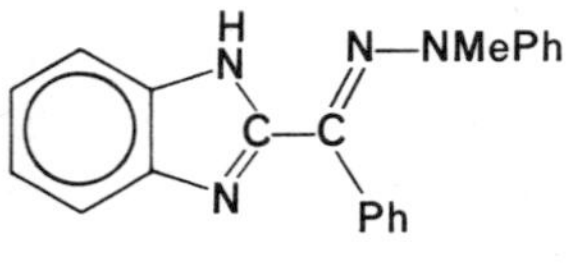

(48)

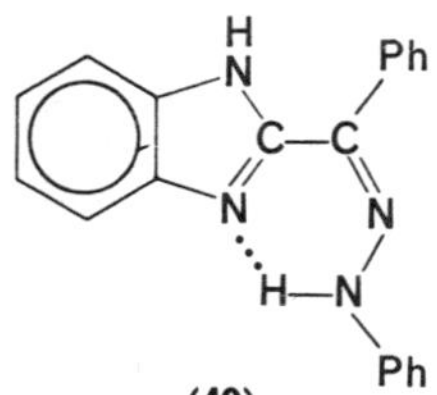

(49)

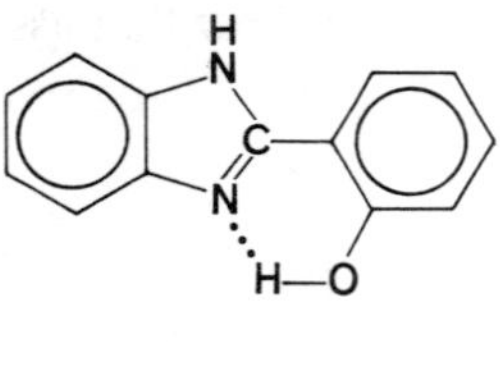

(50)

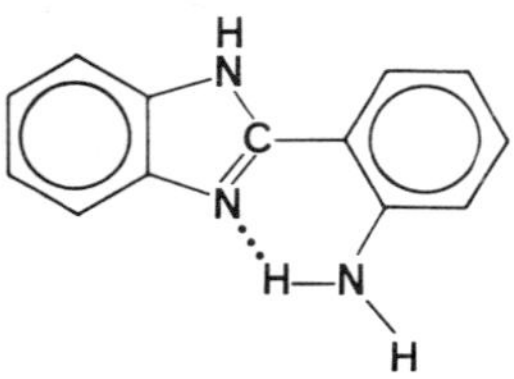

(51)

Tautomerism is considered to be most improbable in 2-methylbenzimidazole (**44**, R = Me) and the broad strong i.r. absorption band between 2600–3050 cm^{-1} in the solid, reflecting unusually strong NH···N bonding, is ascribed to a resonance stabilized linear polymer[190]. In chloroform solution these hydrogen bonds are weaker, the NH band having shifted to 3180 cm^{-1}. In solid indazole (**45**) the NH···N band appears at 3150 cm^{-1}, while in chloroform some absorption at 3450 cm^{-1} indicates that not all the NH bonds are involved in association[190].

N-Unsubstituted 1,2,4-triazoles (**52**) have a broad symmetrical infrared band in the range 1900–2500 cm^{-1} which is assigned to very strong hydrogen

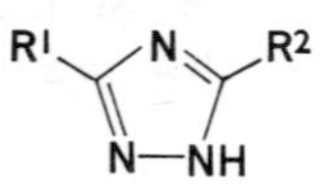

(52)

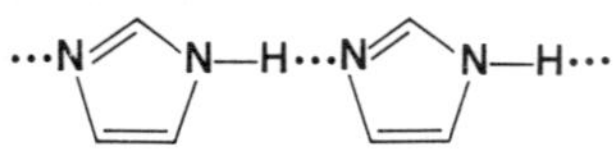

(53)

bonding and proton tunnelling[191]. A symmetrical double minimum potential, with rapid proton interchange between the two potential wells, has also been well documented for imidazole (**53**)[192].

The reaction of amidines with an available hydrogen atom on nitrogen (**54**) with isocyanates (**55**) gives dipolar molecules (**56**) which lead to carbamoyl-amidines (**57**) by proton transfer (equation 17)[193]. There is

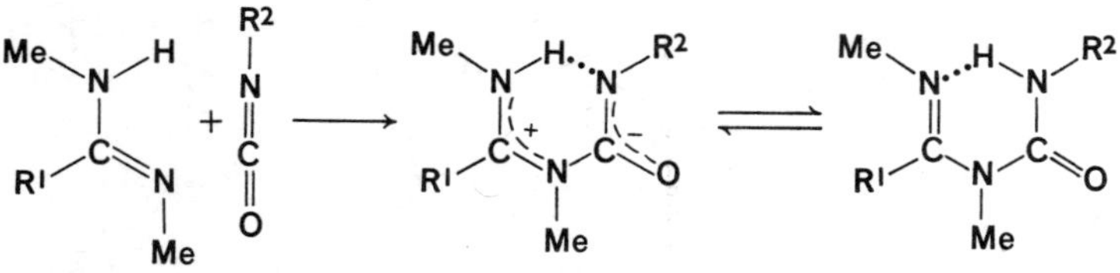

(17)

(**54**) (**55**) (**56**) (**57**)

strong intramolecular hydrogen bonding in **57**, as ν_{NH} is shifted to 3000 cm^{-1} or below. The donor and acceptor N atoms are forced close together by the five sp^2-hybridized atoms in the quasi ring system. As R^1 is changed from H to Me to Et the NH frequency is displaced downward, signifying a strengthening of the hydrogen bond as the bridge is compressed due to steric interference of peripheral substituents. When $R^1 = i$-Pr or t-Bu there is no indication of a hydrogen bond, substituents of this size having forced **57** into another conformation unfavourable for intramolecular NH···N bonding. Increasingly strongly electron-withdrawing substituents at R^2 (Me, Ph, p-ClC_6H_4, p-$O_2NC_6H_4$) increase the NH bond moment and strengthen the NH···N bond, again observed by a decrease in $\nu_{NH\cdots N}$. While the reaction also proceeds with isothiocyanates (O replaced by S in equation 17) there is no evidence for intramolecular hydrogen bonding in the product, the size of the sulphur atom having caused such great steric hindrance that another conformation is thermodynamically more favourable[193].

In azo-*bis*-*N*-chloroformamidine (**58**) crystals NH···N hydrogen bonds

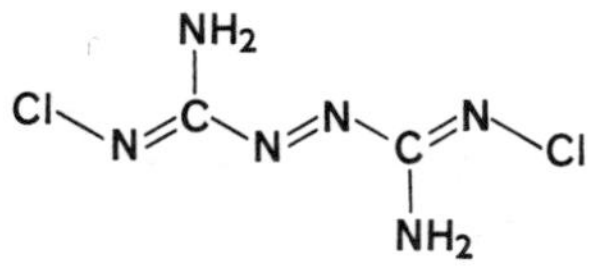

(**58**)

are found[194]. These involve the primary amino group as a proton donor and the terminal azomethine N atoms as proton acceptors. The azo group is not involved.

Hydrogen bonding involving the N atom of the azomethine linkage in oximes is well established from X-ray diffraction studies. Thus OH···N bond lengths of 2·78, 2·82, 2·75 and 2·83 Å have been found in crystals of acetoxime[195], *p*-chlorobenzaldoxime[196], cyclohexanoxime[197] and dimethylglyoxime[198], respectively. The high vapour pressure of acetoxime crystals is attributed to the formation of planar trimers with strong intramolecular OH···N bonds, with only van der Waals forces acting between trimers[195]. While *syn*-*p*-chlorobenzaldoxime exists in the form of dimers in crystals[196], the *anti* form forms polymers in the solid state[199]. Cyclohexanone oxime is also trimeric in the solid state[197], each trimer unit being stabilized by three OH···N hydrogen bonds.

Hydrogen bonds in formamidoxime include weak NH···N hydrogen bonds ($R_{NH\cdots N} = 3{\cdot}12$ Å) involving the primary amino group as a proton donor and the azomethine N atom as a proton acceptor, as well as stronger OH···N bonds[200].

For arylazo oximes structure **59** was originally assumed[201], although

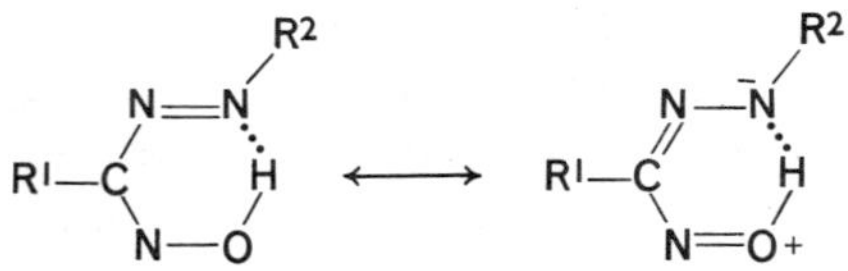

(59a) **(59b)**

recent studies, while they did not exclude this, found no specific evidence in favour of such a chelate[202]. However, the N—O stretching frequency is unusually high due to the contribution from resonance forms like **59b**, which should enhance chelation in the appropriate isomer. U.v. spectra confirm the azo structure for **59** and cryoscopic measurements in $CHCl_3$ show that extensive molecular association exists[202].

Electronic effects have a substantial influence on the tendency of oximes to associate in solution[203]. The association factor $f = M/M_0$, where M is the apparent molecular weight in solution and M_0 is the molecular weight of the monomer, increases linearly with the dipole moment of monomeric oximes in the series $CH_3(R)C{=}NOH$, where R = Me, Et, *i*-Pr, *t*-Bu, cyclohexyl and Ph, thus showing that the same factors govern both changes[203]. For *p*-substituted acetophenone oximes f decreases linearly with the Hammett σ values of substituents, showing that the strength of the hydrogen bond is determined mainly by the charge density on the electron-donor atom[203].

While the phenomenon of association of oximes in solution through hydrogen bond formation was first noted a long time ago from i.r. spectroscopic studies[204], the nature of the associated species remained in doubt for some time because of the large number of possible cyclic structures

(60a–d) that may exist, in addition to linear polymers. Based on the 'free' OH stretching frequency at 3610 cm^{-1} in CCl_4 solution, it has been concluded that in the concentration range 0·01–0·3 moles/l the Mecke–Kempter relationship[205] is not obeyed; thus linear chain polymers are ruled out for

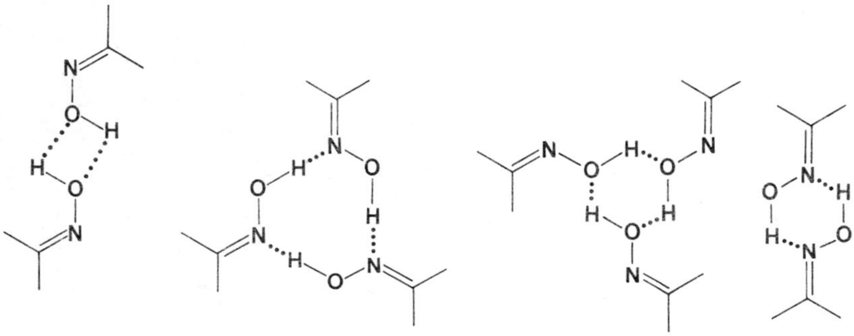

(60a) **(60b)** **(60c)** **(60d)**

the four isomeric *n*-octanone oximines, in spite of previous claims that this relationship was obeyed for other oximes[206]. While shifts in the C=N and N—O stretching frequencies were carefully evaluated, no definite choicę among the cyclic associates could be made, although **60c**, with OH···O bonds, was favoured for the trimer and **60d** with OH···N bonds, for the dimer. The *n*-octanone oxime i.r. data are accommodated by an interpretation in which cyclic dimers and cyclic trimers compete in the association mechanism. Both of the association constants K_{12} and K_{23}, were found to decrease in the order caprylaldoxime, octanone-2-oxime, octanone-3-oxime and octanone-4-oxime, with dimers always more abundant than trimers, and the dimerization tendency increasing for the oximes in the order given[205]. The association energy for dimer formation increased from 7·0 to 12·8 kcal/mole on going from caprylaldoxime to octanone-4-oxime, while that for trimer formation fell from 6·4 to 5·2 kcal/mole for the compounds in the same order[205].

I.r. spectra of $(CD_3)_2C{=}NOH$ in rigid glasses at temperatures down to −180°C show that, in addition to the two broad bands observed in the 3100–3200 cm^{-1} region in the polycrystalline solid[206a], a number of other temperature-dependent bands with frequencies in the normal CH stretching range, and down to 2760 cm^{-1}, appear[206b]. These are held to indicate species more highly associated than cyclic dimers and trimers, possibly hydrogen bonded chain-like structures[206b].

Other data for self-association of oximes in dilute CCl_4 solution, based on i.r. measurements, are found in Table 16, which also includes informa-

TABLE 16. Association characteristics of oximes in dilute CCl_4 solution at 25°C[207]

Compound	$\nu_{OH}(cm^{-1})$			Dimerization constant[b]	Association constant[b] (with THF)
	Monomer	Cyclic dimer	RNOH···THF[a]		
Acetaldoxime	3600	3296	3336	150	46
Acetoxime	3606	3275	3344	330	43
Butanone-2-oxime	3604	3279	3342	535	50
Cyclohexanone oxime	3603	3278	3340	310	30
Benzaldehyde oxime	3594	—	3312	—	57
Acetophenone oxime	3595	3246, 3289	3300	267	52
Benzophenone oxime	3585	3283	3300	245	55

[a] THF = tetrahydrofuran.
[b] Concentrations in mole fraction units.

tion on the tendency of these same oximes to associate with a common base tetrahydrofuran, in the solvent CCl_4[207]. The data were interpreted on the basis of a preponderance of cyclic dimers. Vapour density measurements on acetaldoxime gave an enthalpy of association of $-10{\cdot}1 \pm 0{\cdot}5$ kcal/mole[208]. Cyclic dimerization involving two OH···N bonds per dimer unit was assumed in this study[208].

No difference has been found between the i.r. ν_{OH} values for *syn* and *anti* benz- and acetaldoxime monomers in dilute CCl_4 solution[209], whereas pronounced differences exist in the crystalline state, where the associated *anti* species have much lower OH frequencies[209, 210], which is ascribed to their stronger hydrogen bonds. From X-ray diffraction work it appears that *syn* aldoximes form dimers whereas the *anti* isomers associate in long chains[196, 211] in the crystalline state. Pyridine-2-aldoxime and pyridine-4-aldoxime were found to have unusually low hydrogen-bonded OH frequencies in the solid state from which was inferred that here the OH is bonded to the pyridine N atom[209]. Deuteration and association sensitive bands in the 1250–1450 cm^{-1} region of oximes (coupled OH and CH deformation vibrations) have also been related to conformations, the degree of coupling being more pronounced for the *syn* isomers of monomeric aldoximes[209].

A recent i.r. spectroscopic study has established that *syn* benzaldoxime is more acidic than *anti* benzaldoxime, from their relative tendencies to associate with proton-accepting solvents[212]. This is supported by charge-density calculations. For association with dioxan, ΔH^0 for hydrogen-bond formation with the *syn* isomer is $-1{\cdot}4 \pm 0{\cdot}4$ kcal/mole, and with the *anti* isomer $-1{\cdot}1 \pm 0{\cdot}3$ kcal/mole. Proton-donor solvents therefore favour the *anti* form and proton-acceptor solvents the *syn* form. For dimerization in CCl_4 the following thermodynamic parameters were obtained[212]:

	syn	*anti*
ΔG^0 (kcal/mole)	$-3{\cdot}4 \pm 0{\cdot}1$	$-3{\cdot}6 \pm 0{\cdot}1$
ΔH^0 (kcal/mole)	$-6{\cdot}8 \pm 0{\cdot}5$	-10 ± 2
ΔS^0 (cal/deg mole)	-11 ± 2	-21 ± 7

The role of proton-donor and proton-acceptor solvents in alteration of the *syn–anti* interconversion rate was also investigated[212].

Intramolecular OH···O bonding has been found in diethyl α-hydroxy-iminomalonate (**61**) in the pure liquid state and in non-polar solvents,

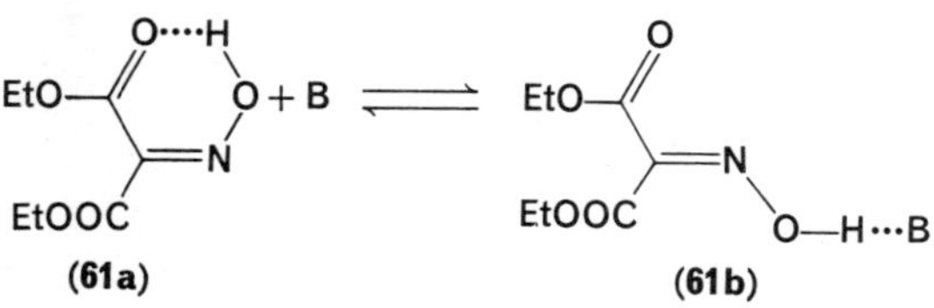

(18)

while in polar proton-accepting solvents like dimethylsulphoxide-d_6 the intramolecular hydrogen bond is broken and complex **61b** is favoured (equation 18)[63]. Oximes with electron-withdrawing substituents in conjugation with the oximino group do not form OH···N=C hydrogen bonds[63], since the extent of such association depends on the basicity of the oxime, as has been shown by the correlation of the association constants with MO calculations of the π-charge at the N atom[213]. The effect of such electronic charge withdrawal can also be seen in the low values of the C=N stretching frequency found in compounds like diacetyl monoxime, relative to values found in unconjugated oximes[214].

A large difference in acidity exists between *amphi*-chloroglyoxime ($pK_a = 8{\cdot}13$) and *anti*-chloroglyoxime ($pK_a = 3{\cdot}92$), the latter of which is a fairly strong acid due to the electron-withdrawing effect of the chlorine atom[64]. The dramatically weaker acidity of the *amphi* isomer is ascribed to an intramolecular OH···N hydrogen bond (**62**) which makes it more difficult for the acidic proton to leave[64].

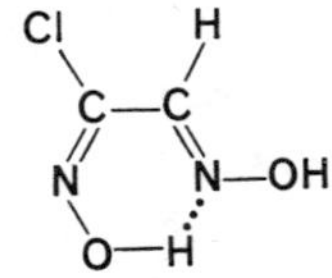

(62)

C. Azo Compounds

The azo group can act as a proton acceptor in hydrogen bonds, and a notable example in the crystalline state occurs in α,4-dibromo-α(4-methyl-2-nitrophenylazo)acetanilide (**63**) where an intramolecular NH···N bond

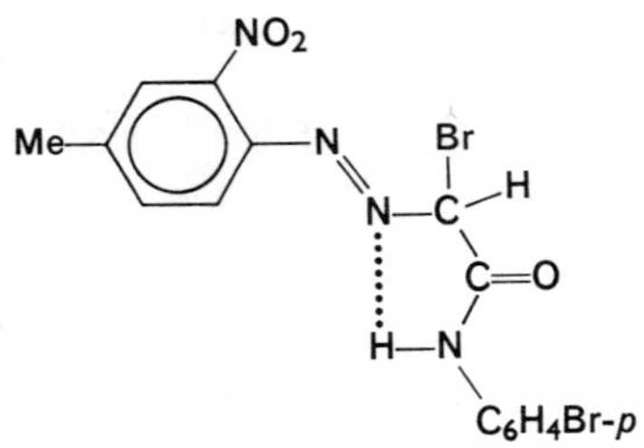

(63a)

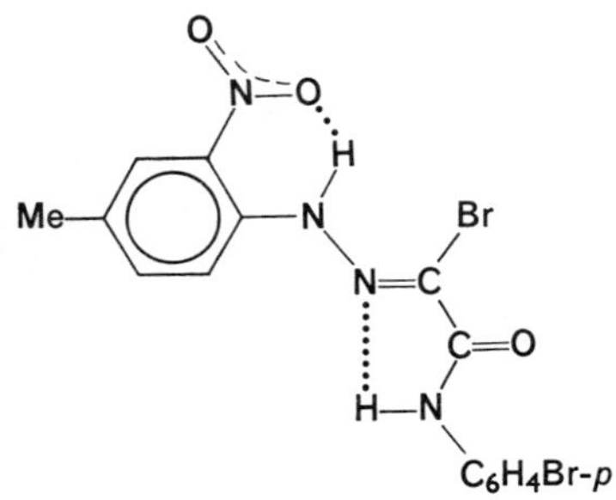

(63b)

with an N to N distance of 2·75 Å is inferred from X-ray studies[215]. While hydrogen atoms were not detected directly, tautomerism to **63b** could not be excluded. Earlier, dielectric measurements had shown that azobenzene pairs with one and with two molecules of *o*-nitrophenol or 2,4-dinitrophenol[216] and thermal analysis showed that azobenzene interacts with water[217].

The role of hydrogen bonding in dye aggregation has been accepted for some time. More recent studies indicate that monomer–dimer equilibria exist for ionic 1-phenylazo-2-naphthols in the 10^{-4}–10^{-6} molar concentra-

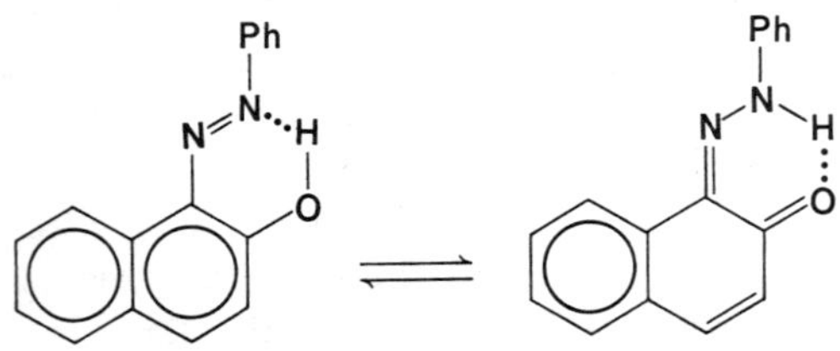

(64a) (64b)

(19)

tion range in aqueous solutions and in methanol[218]. Furthermore, self-association in these compounds under the conditions stated appears to be exclusively in the polar hydrazone tautomer, which has an intramolecular hydrogen bond in the monomeric state (**64b**) (equation 19)[219]. Hueckel molecular orbital calculations support this, by showing that while hydrogen bonding solvents stabilize both tautomers, the effect on **64b** is three times as great as that on **64a**[220]. Other conclusions from these HMO calculations are: (*i*) that intramolecular hydrogen bonding also stabilizes both tautomers, but again **64b** is stabilized more than **64a**, by a factor of two, (*ii*) that as substituents on the phenyl ring not bearing the O atom change from electron-releasing to electron-withdrawing the stability of **64b** increases, (*iii*) the larger the ring system bearing the O atom, the more stable the tautomer of form **64b**, and (*iv*) hydrazo–hydrazo dimers are three times as stable as azo–azo dimers[220]. While azo and hydrazo tautomers of compounds like **64** dominate the electronic spectra of neutral solutions, in the crystalline state *o*-hydroxyazoaromatics show a pronounced red-shift, which is held to have its origin in intermolecularly hydrogen-bonded hydrazone-like aggregates[221]. Even the amorphous to crystalline transformation in azonaphthols with hydrogen-bonding substituents is accompanied by a dramatic colour change, which is not observed for azo compounds without such substituents[221].

The role of hydroxylic solvents[222] and proton-accepting solvents[223] in shifting the electronic spectrum and the apparent dipole moment[224] of *p*-hydroxyazobenzenes clearly involves hydrogen bonding; even 4-methylamino-2′-carboxyazobenzene, with an intramolecular hydrogen bond, is affected[222]. Solvent blue-shifts of the $n \rightarrow \pi^*$ transition of the azo group have been used to evaluate solute–solvent interactions[225, 226]. I.r. OH shifts demonstrate hydrogen bonding between ethanol and pyridazine[225], which affects the skeletal vibrations of pyridazine in a manner which suggests that the interaction is at the nitrogen site[227].

The absorption curves of methyl orange and similar *p*-aminoazobenzenes in organic and aqueous solvents have been resolved into two overlapping bands, and the blue-shift of the absorption maximum when organic solvents are added to an aqueous solution of methyl orange, or when it is bound to bovine serum albumin or a surfactant micelle, arises from a change in the relative intensities in the component bands[228]. The low-frequency component is assigned to a $\pi \rightarrow \pi^*$ transition of a solvate in which the solvent hydrogen bonds specifically to the azo nitrogens, while the high-frequency component is assigned to a $\pi \rightarrow \pi^*$ transition of a solvate where such interaction is absent. The low-frequency component is favoured by aqueous solvent composition, and by a decrease in temperature[228]. Thus a shift in the absorption maximum accompanying the binding of these compounds to a protein or micelle should be interpreted as a shift in an equilibrium instead of a shift in transition energy.

The effect of salts on the absorption spectra of some *p*-aminoazobenzenes has been ascribed to their effect on hydrogen bonds of the solvent water to the dye molecules[229].

The specific association of various sterically hindered hydroxylic compounds with azobenzene in dilute CCl_4 solution has been examined by means of i.r. OH shifts, some of which are tabulated in Table 17, together with the relevant association constants[230]. The enthalpy of hydrogen bond formation between di-*t*-butyl carbinol and azobenzene was established to be –2·6 kcal/mole, and the lower association constants for these proton donors are due to the entropy factor. The OH frequency shifts were not found to be linearly related to the enthalpies of hydrogen-bond formation.

Cooling of 4-phenylazo-1-naphthol in methylcyclohexane was found to shift the tautomeric equilibrium toward the α-naphthoquinone form, while cooling it in alcohol shifted the equilibrium toward the azo form, this being ascribed to the change in dielectric constant and to hydrogen bonding[231].

Electron-withdrawing X groups in **65** should (*a*) increase the OH···N bond strength if they were acting on the OH bond through the ring system only, and (*b*) decrease the OH···N bond strength as the electron density

TABLE 17. Proton donor/acceptor hydrogen bonding of sterically hindered hydroxylic compounds to azobenzene in CCl_4 solution at 25°C[230]

Compounds			ν_{OH} (cm^{-1})	K_{assoc} (l/mole)	$\Delta\nu/\nu \times 10^3$ [a]
R^1	R^2	R^3			
(1) *Alcohols*: $R^1R^2R^3COH$					
H	*t*-Bu	*t*-Bu	3650	0·6	35·6
Et	*i*-Pr	*i*-Pr	3625	1·0	4·1
i-Pr	*t*-Bu	*t*-Bu	3635	0·6	5·9
(2) *Phenols*: $R^1R^2C_6H_3OH$					
H	H		3610	1·0	65·1
H	2-*i*-Pr		3612	0·5	69·2
H	2-*t*-Bu		3610	1·0	65·1
2-Me	6-*t*-Bu		3615	0·5	47·0
2-*i*-Pr	6-*i*-Pr		3620	0·4	38·7
2-*t*-Bu	6-*t*-Bu		3645	<0·1	—

[a] Fractional downward frequency displacement on hydrogen-bond formation, from monomeric ν_{OH} value given in column 4.

on the proton-accepting N atom is depleted. An analysis of the pK_a values of *o*-hydroxyazobenzenes (**65**) and corresponding *p*-hydroxy reference

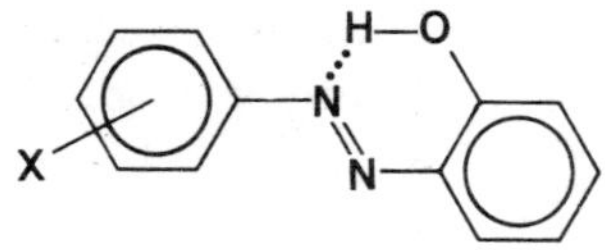

(65)

compounds shows that about one half of the total electronic effect of X is due to the second factor[232]. The greatly increased pK_a of 2,4-dihydroxy-azobenzene in comparison with reference compounds is ascribed to an internal hydrogen bond[233]. Since an *ortho*-Cl substituent in *o*-chloro-*o'*-hydroxy azo compounds like **66** increases the pK_a (weakens the acidity)

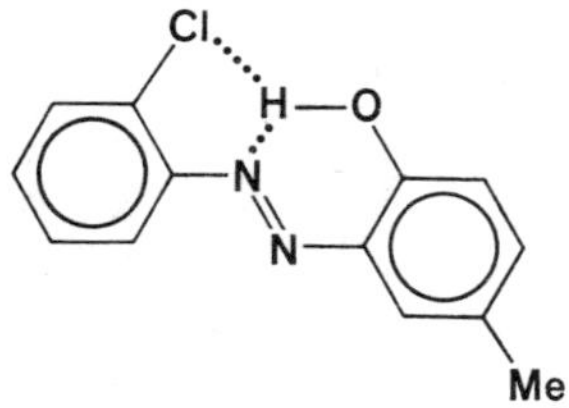

(66)

relative to the corresponding *para*-Cl compound, a bifurcated intra-molecular hydrogen bond (strong OH···N interaction and weak OH···Cl interaction) is suggested[234]. Ionization of the OH would then require the expenditure of additional energy to rupture the OH···Cl bond. Acid–base equilibria involving 3,4-dihydroxyazobenzene have been discussed in terms of hydrogen-bonded forms (**67a** and **67b**) in equation (20)[235]. In

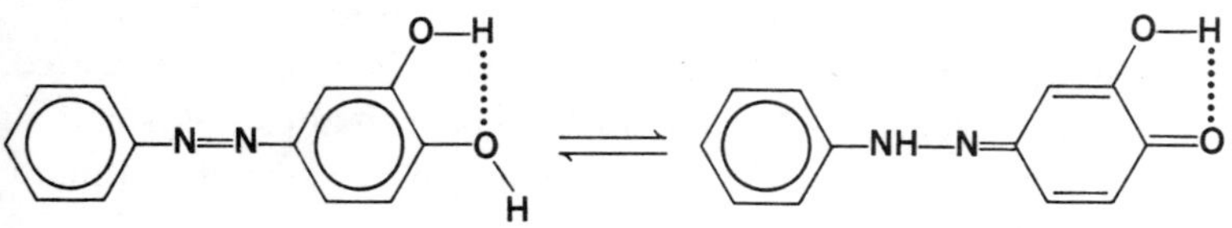

(20)

(67a) **(67b)**

strongly alkaline solutions the azo form is held to exist, although it is less stable than **67b**[235]. The dependence of hydrogen-bond strength in a wide range of azo dyes on pK_a values of the OH groups has also been investi-gated[236].

Strong chelate hydrogen bonding in *o*-hydroxyazo and *o*-aminoazo compounds shows up prominently in i.r. spectra[76, 237], following the pattern already discussed for *o*-hydroxyazomethines. The vibrational spectra of *o,o'*-dihydroxyazonaphthols have been discussed in terms of the following hydrogen-bonded tautomers (**68a, 68b** and **68c**)[237a].

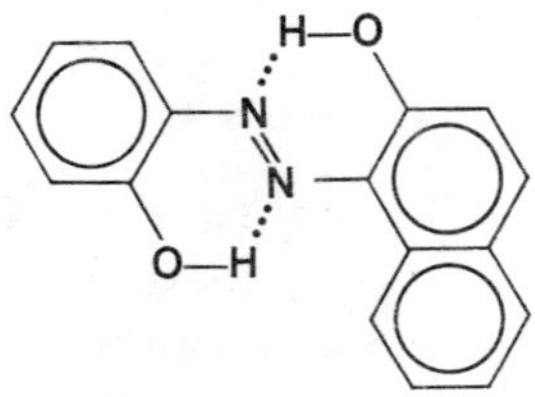

(**68a**)

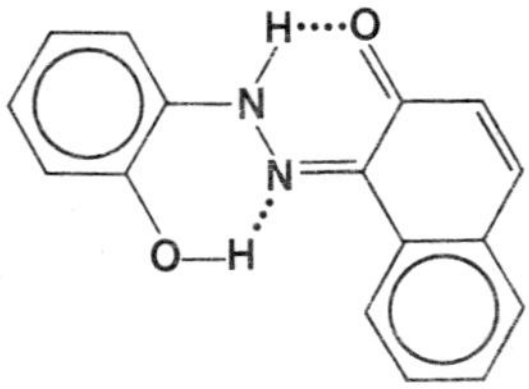

(**68b**)

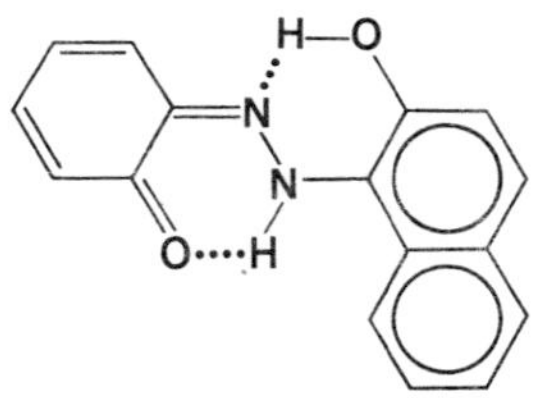

(**68c**)

From p.m.r. shielding parameters for dilute CCl_4 solutions of 2-hydroxyazobenzene, 1-phenylazo-2-naphthol and selected derivatives the intramolecular hydrogen bond in the 2-naphthol is determined to be considerably stronger than the hydrogen bond in 2-hydroxyazobenzene[194]. This can only be due to the attachment of the OH group to a larger resonance system, thereby increasing its acidity[238]. Decreased proton shielding on going from 2-hydroxyazobenzene to salicylidene aniline (**22**, $R^1 = R^2 = H$) shows that the nitrogen atom of an azomethine bond is more basic in a six-membered chelate ring than a nitrogen atom in an azo bond in an analogous system[238]. A number of other proton chemical shift studies of hydrogen bonding in hydroxyazo compounds have been reported[239].

Specific increases in the hydrazone form for dyes containing *ortho*-carbonyl groups have been noted and explained in terms of hydrogen-bond formation. Thus 1-phenylazo-2'-carboxy-2-naphthol (**69a**) exhibits a relatively strong bathochromic u.v. shift relative to the *p*-isomer, and this is ascribed to greater stability of **69b** with a bifurcated hydrogen bond to two carbonyl groups in two six-membered chelate rings[240]. Since 2,3-

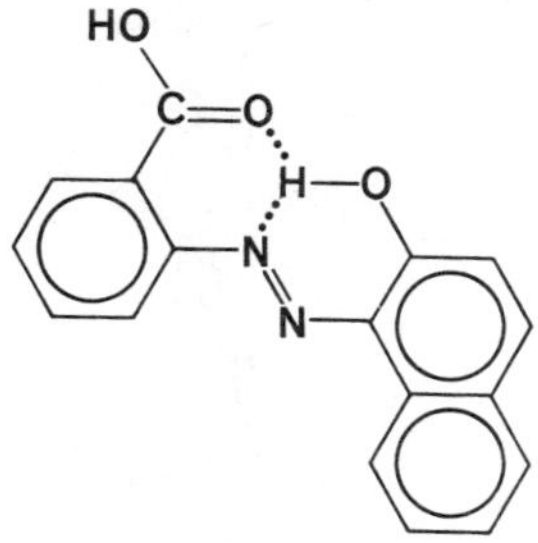

(69a)

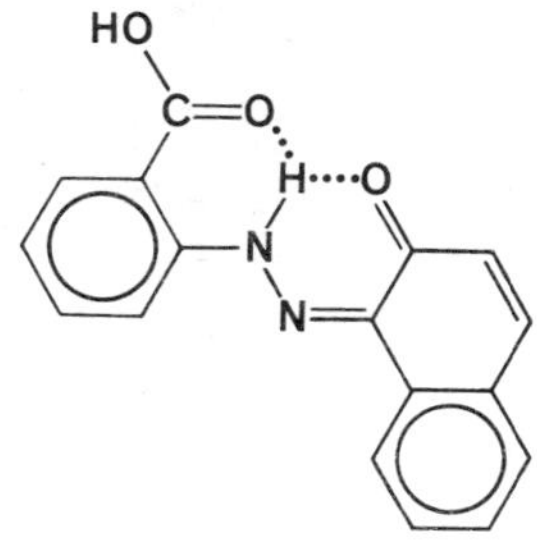

(69b)

naphthoquinone is unstable and unknown, the hydrazone tautomer is improbable for 2-phenylazo-3-naphthol. The *cis-O*-methyl ether of this compound behaves like the parent unsubstituted azo compound when subjected to thermal *cis* → *trans* isomerization (activation energy = 23 kcal/mole), while for the hydroxy compound the activation energy is only 12–14 kcal/mole[241]. Quantum yields for the *cis* → *trans* conversion were the same for both compounds, but for *trans* → *cis* they were one to two orders of magnitude smaller for the *o*-hydroxy compound. Hydrogen-bond stabilization of both the ground and excited transition state has been indicated as a likely explanation of these observations[241]. Another study shows that thermal *cis* → *trans* isomerization proceeds through the hydrazone in two consecutive reactions[242], implicating hydrogen bonding in this tautomer as a relevant factor.

Temperature jump relaxation methods indicate that for 'normal' acid–base reactions the rate is diffusion controlled, but not where an internal hydrogen bond is involved[243]. Here the rate becomes dependent on the ionization constant of the weak acid, as demonstrated for a series of *ortho*-hydroxy azo compounds subjected to deprotonation by OH^- [243].

In acetone solution 1-phenylazo-2-naphthol (**70**) is predominantly in the hydrazone form, while the corresponding 4-phenylazo compound (**71**) is mainly in the azo tautomer, based on n.m.r. spectra (equations 21 and 22)[244]. The smaller ΔH^0 for **70** is attributed to the stabilization of tautomer **70a** relative to **70b** by intramolecular hydrogen bonding. The enthalpy difference favours the hydrazone form in both cases, and the entropy term favours the less polar azo form in both cases. In **71** the entropy term dominates, while in **70** the enthalpy term is the most important[244].

Chelate rings formed by NH···N bonds are less stable than corresponding rings formed by OH···N bonds, in general. In 2-*N*-acetylamino-azobenzene (**72**) u.v. spectra show that the chelate ring is stable in dioxan, but perturbed in methanol, whereas the chelate ring in 2-hydroxy-4-methylazobenzene is

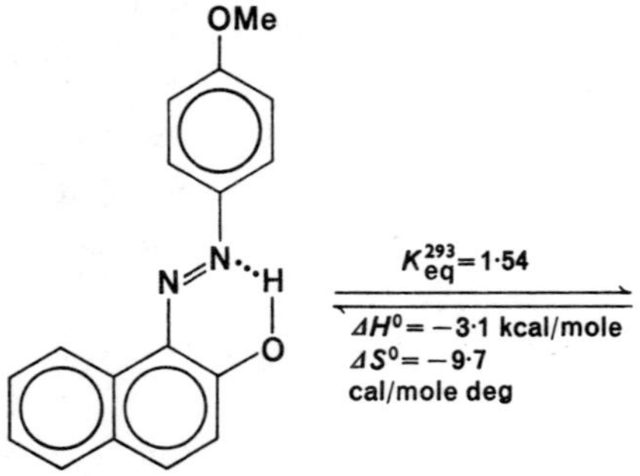

(70a) (70b) (21)

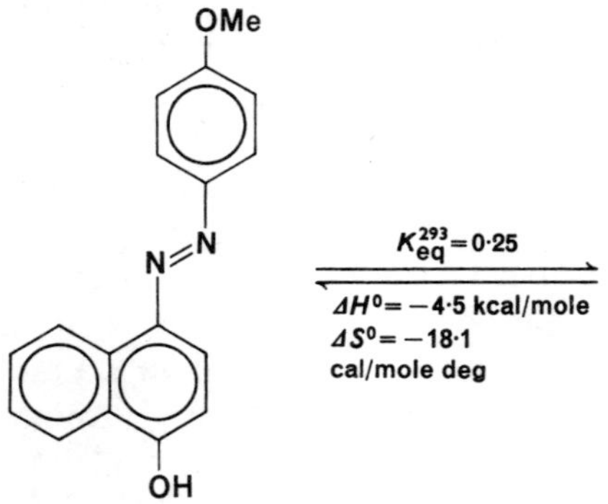

(71a) (71b) (22)

stable to dioxan and hydroxylic solvents[245]. Chelate rings involving NH···N=N bonds in azo compounds have also been treated theoretically[246] and have been studied in the i.r.[50, 81, 247]. On the basis of u.v. spectra of 2-amino-4′-dimethylaminoazobenzene (**73**) both five- and six-membered chelate rings with NH···N bonds have been suggested (equation

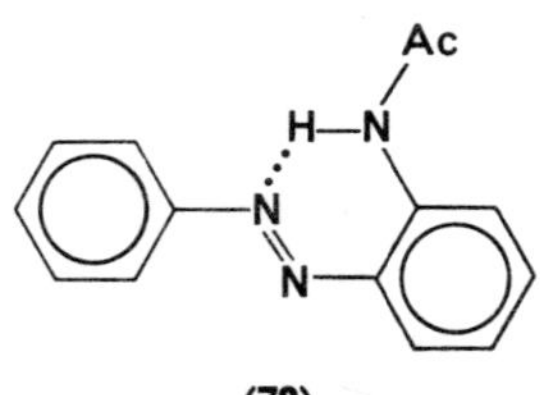

(72)

23)[84]. While a five-membered ring is more strained and usually weaker than a six-membered ring, structure **73a** would be stabilized by the greater

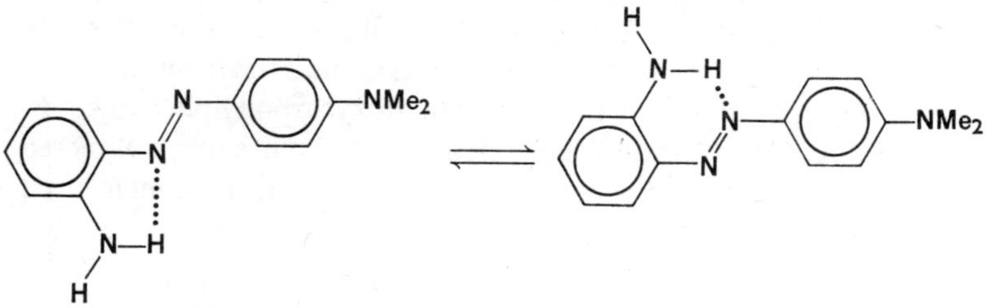

(73a) (73b) (23)

electron density on the β nitrogen atom in the azo link[84]. 1-Arylazo-2-amino-8-naphthols **(74)** are particularly colour insensitive to acids and bases, a fact which has been ascribed to two internal hydrogen bonds in the molecule, resulting in great coplanarity[248]. The absorption maximum

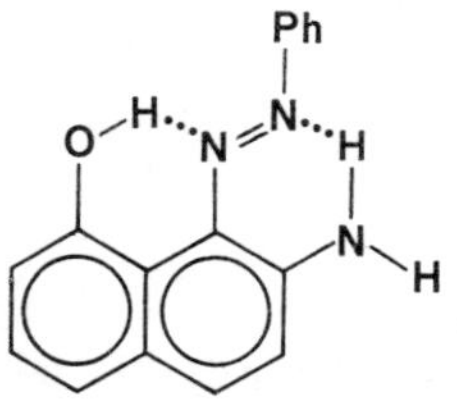

(74)

lies at 58 nm longer wavelength than in the corresponding 1-arylazo-2-naphthylamine. If the transfer of a hydrogen atom from O to N in any resonance structure were implied, the wavelength shift corresponds to a net decrease of ~7 kcal/mole in the transition energy, a value consistent with hydrogen bond strengths[248]. I.r. spectra of phenolazopyrazolone dyes also indicate the presence of several types of chelate hydrogen bonds[249].

Quinolinol arylazo derivatives **(75)** have a diffuse room temperature reflectance spectrum which is dominated by a hydrogen bonded azo

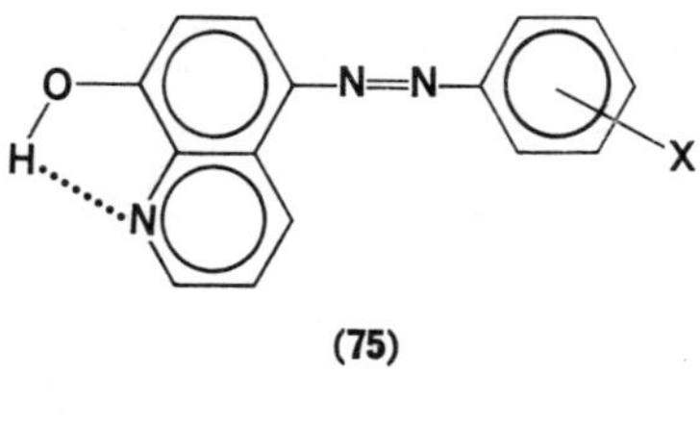

(75)

tautomer[250]. These compounds do, however, exhibit thermochromism in NaCl solution[250]. Spectra of 1-aroylazo-2-naphthols mixed with NaCl show that the solid is an azo–hydrazone mixture[251], and a similar situation exists when these compounds are supported on fibres[252]. In the latter case the nature of the fibres is a major factor in determining the position of the tautomeric equilibrium[252]. In 5-(2′-methoxycarbonylphenylazo)-8-hydroxyquinoline (**76**) an intramolecular hydrogen bond in a five-membered ring (**76a**) similar to that in **75** is balanced against a stronger six-membered ring in the hydrazo tautomer **76b** (equation 24)[253].

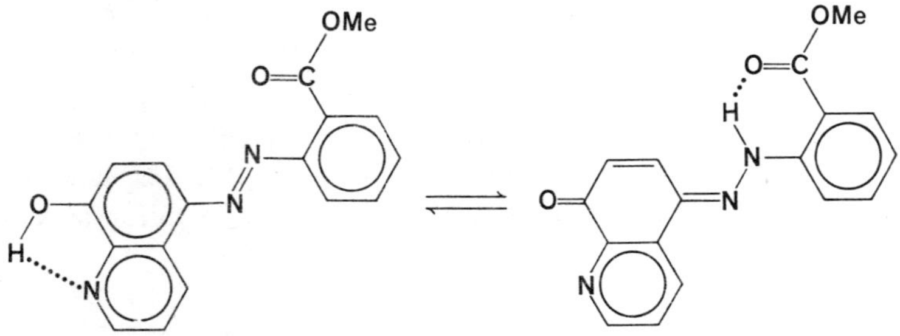

(24)

(76a) **(76b)**

Intramolecular hydrogen bonds involving *o*-hydroxy or *o*-carboxy groups and the N=N bond increase the stability of these compounds to reductive cleavage by live yeast, provided that the N=N bond is included in the chelate ring[254]. The estrogenic activity of hydroxyazobenzenes has also been discussed in terms of molecular structures imposed by intramolecular hydrogen bonds[255], as has the carcinogenic nature of some azo compounds[256].

The fluorescence of *o*-hydroxyazo compounds at low temperatures has been traced to the presence of hydrogen bonds (intra- or intermolecular)[257], and the presence of intramolecular hydrogen bonds in azo dyes has been shown to lead to an increased mobility on paper electrophoresis[258]. X-ray crystallographic studies show that intramolecular hydrogen bonding with the formation of chelate rings can become very extensive as in α-(2-nitro-4-chlorophenylazo)-acetoacetanilide (**77**), where three chelate rings are

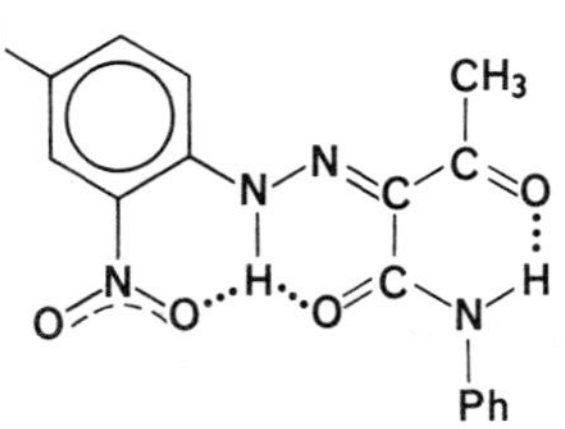

(77)

formed[259]. Since hydrogen atom positions were not determined directly, the possibility of tautomerism to the azo form was not precluded, but the central chelate would still remain, with bifurcated enol OH···N=N and OH···ONO bonds. The dyestuff properties of **77** are attributed to the formation of a large planar resonance system[259].

D. *Azoxy Compounds*

Relatively little information is available about the hydrogen bonding characteristics of azoxy compounds. Enthalpies of hydrogen bond formation between phenol and some diazine *N*-oxides in dichloromethane solution are as follows: pyridine *N*-oxide ($-5{\cdot}9 \pm 0{\cdot}2$); pyridazine *N*-oxide ($-5{\cdot}0 \pm 0{\cdot}3$); pyrimidine *N*-oxide ($-4{\cdot}9 \pm 0{\cdot}3$); pyrazine *N*-oxide ($-4{\cdot}2 \pm 0{\cdot}3$ kcal/mole)[260]. The same procedure gave $\Delta H^0 = -5{\cdot}9 \pm 0{\cdot}2$ kcal/mole for hydrogen-bond formation of phenol with pyridine. These data were interpreted on the basis of the oxygen atom being the electron-donor site.

E. *Triazenes*

In the crystalline state, X-ray diffraction shows that diazoaminobenzene (α form) exists as infinite helical chains of molecules which involve intermolecular NH···N hydrogen bonds ($R_{NH\cdots N} = 3{\cdot}25$ Å)[261]. This appears to be characteristic of triazenes, since *p*-bromo- (β form)[262], *p,p'*-dibromo-[263] and *p,p'*-dimethyldiazoaminobenzene[264] exhibit the same crystal structure, all with NH···N bonds. The NH···N bond lengths in the first two compounds named are 3·20 and 3·26 Å, respectively. The α-form of *p*-bromodiazoaminobenzene crystallizes in the form of dimers held together by NH···N bonds, with $R_{NH\cdots N} = 3{\cdot}24$ and 3·27 Å[265]. Hydrogen bonding in these dimers and in the spiral chains takes form **78** where resonance

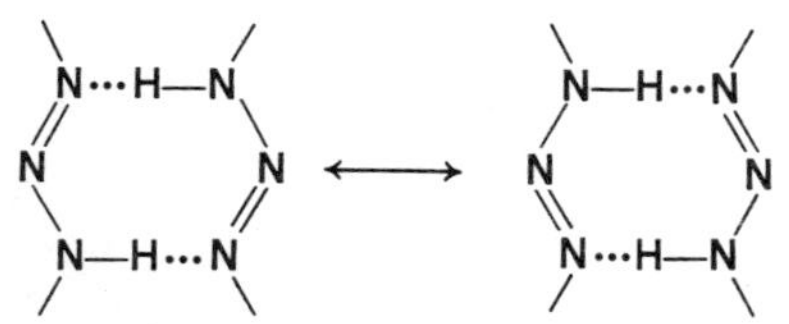

(78a) (78b) (25)

stabilization may play a role (equation 25). The labile character of the triazene system is illustrated by the equal N—N bond lengths that have been found in *p,p'*-dibromodiazoaminobenzene, suggesting central positions for the hydrogen atoms in the hydrogen bonds[263]. Individual molecules of this compound are also nearly planar in the crystal state[263].

In solution the absence of colour in diazoaminobenzenes is related to deviations from planarity[266]. Thus 2,6,2′,6′-tetrasubstituted derivatives, except for the tetramethyl, are colourless, but when two or more of these positions are unsubstituted the compound is yellow. Bulky substituents *ortho* to the triazo linkage shift the absorption maximum to longer wavelengths.

Cryoscopic measurements in 0·005–0·1 molal benzene and naphthalene solutions are in agreement with cyclic dimer formation in the form **78** for diazoaminobenzene[188, 267]. The tendency to associate has been related to tautomeric capability, since *N*-alkylation leads to normal molecular weights from cryoscopic measurements[268]. In tetrachloroethylene the dimerization constant for diazoaminobenzene at room temperature is $1{\cdot}2 \pm 0{\cdot}2$ l/mole, based on i.r. NH bands at 3322 cm^{-1} (monomer) and 3180 cm^{-1} (dimer)[269]. It is suggested that the hydrogen exchange required in tautomerism takes place in the cyclic hydrogen-bonded complex. I.r. spectra confirm that the 1-phenyl-benzo-1,2,3-triazoles possess a cyclic diazoimino structure (**79**) with two possible tautomers and not structure **80**[270]. These spectra also favour ordinary NH···N linear association like **81** in the solid state[270]. A high degree of linear polymerization of type

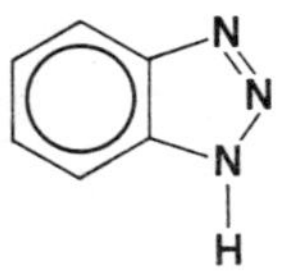

(79)

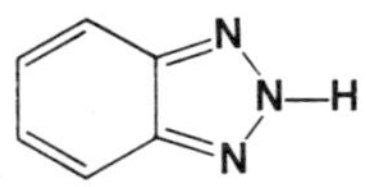

(80)

81 for **79** in naphthalene was inferred more than 30 years ago from cryoscopic measurements[267].

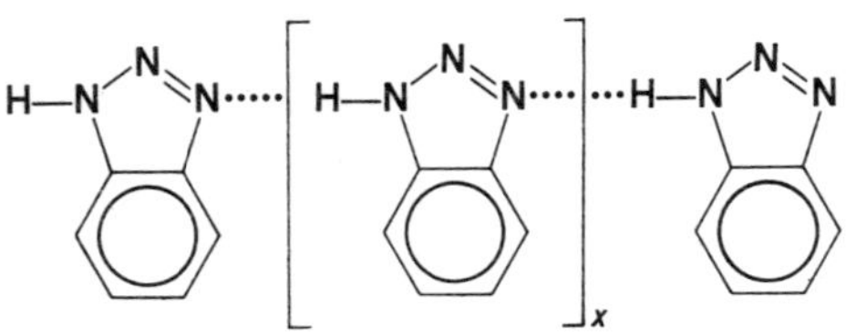

(81)

Intramolecular hydrogen bonding can affect the tautomerism in diazoaminobenzenes (**82**) significantly, while at the same time impeding intermolecular association at higher concentrations. In the tautomeric equilibrium (equation 26) the hydrogen atom will prefer to reside on the N atom

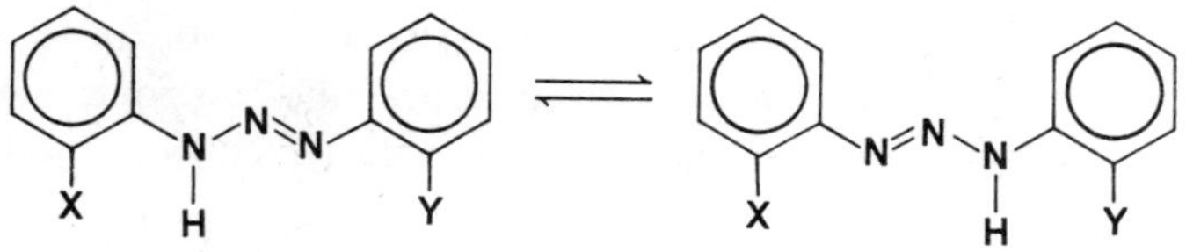

(82a) (82b)

(26)

which will permit intramolecular hydrogen-bond formation with either X (**82a**) or Y (**82b**). In dilute CCl_4 and $CHCl_3$ solution such chelation is marked by an NH stretching frequency shift of less than 10 cm^{-1} if $X = NO_2$, Cl, Br or OMe, and Y = H, but the band undergoes significant intensification, thus confirming structure **82a** for these compounds[271]. When X = COMe, Y = H or Cl and X = COOEt, Y = H these compounds are exclusively in form **82a**, the intramolecularly hydrogen-bonded NH being displaced in these cases from the normal value of about 3330 to 3257–3220 cm^{-1}, and the carbonyl band being lowered 20–30 cm^{-1}. When X = COOEt, Y = Cl less than 5% of tautomer **82b** is suggested, which increases to about 22% when X = COOEt, $Y = NO_2$ and to 25–33% when X = COMe, $Y = NO_2$. In the solid state only tautomer **82a** is found when X = COOEt, $Y = NO_2$, and in solid $X = COCH_3$ derivatives the ratio **82a**:**82b** increases relative to what it is in solution.

In the solid state, when X = Y = H and *m*- or *p*-halogen substituents are present, the NH absorption falls to 3190 cm^{-1} due to intermolecular association, but in the *m*- or *p*-NO_2 compound the sharper and more intense band at 3280 cm^{-1} indicates intermolecular bonding to the nitro group[271].

A comparative study of the isomeric nitrobenzotriazoles shows that the vicinal 4(7)-nitrobenzotriazole (**83**) is more volatile, more rapidly reduced polarographically, a weaker acid and less associated in solution than the other isomers[272]. This is attributed to chelation (equation 27) which restricts tautomerism in favour of **83b**, thus impeding withdrawal or other involvement of the hydrogen atom.

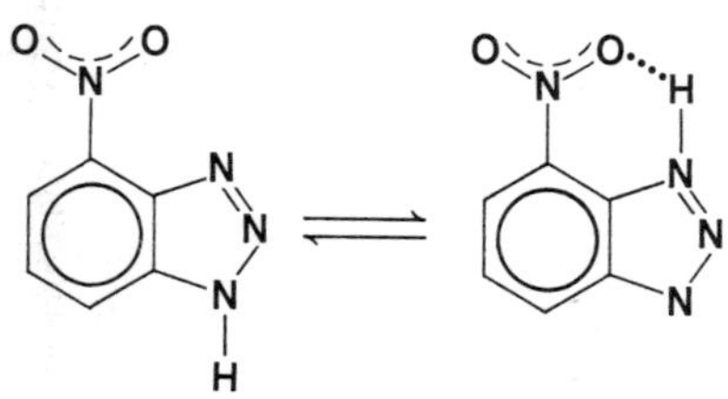

(83a) (83b)

(27)

I.r. spectra of the solid state and p.m.r. spectra of dialkyltriazanium chlorides in dimethylsulphoxide-d_6 and water show that they have the structure [($H_2N—N^+R_2—NH_2$)(Cl^-)] where $R_2 = Me_2$, Et_2, or $(CH_2)_4$[273]. Chemical shifts as a function of concentration indicate strong intermolecular and intramolecular hydrogen bonds in both solvents.

F. Triazene Oxides

Early cryoscopic data suggested that 1-methyl-3-phenyltriazene-1-oxide exists primarily as a monomer in ~0·1M benzene solution, but that intermolecular association occurred at higher concentrations[274].

The compound formally named 1,3-diphenyl-3-hydroxytriazene is a misnomer, at least in CCl_4 solution, as demonstrated by ^{15}N shifts in i.r. spectra[275]. No 'normal' NH or OH stretching frequencies are observed in dilute CCl_4 solution, but a band is found at 3250·6 cm^{-1}; this is consistent with strongly intramolecularly bonded OH or NH groups, and structures

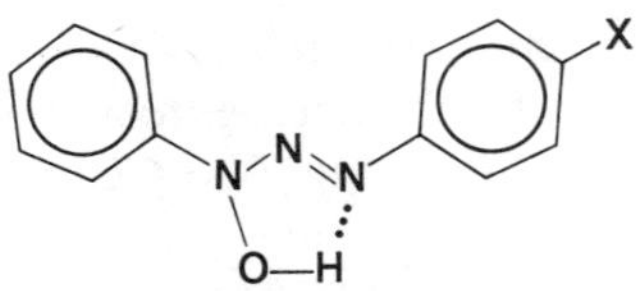

(**84a**) X = H
(**85a**) X = OMe

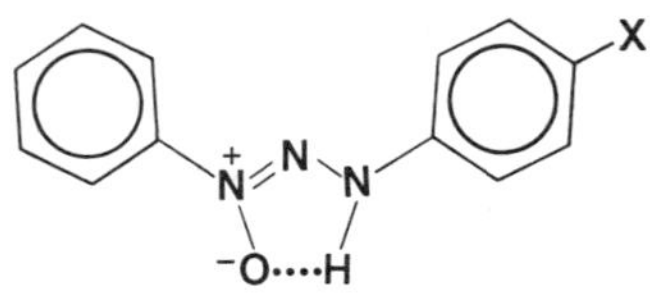

(**84b**) X = H
(**85b**) X = OMe

84a or **84b** are possible. Coupling of *N*-phenylhydroxylamine with aniline-^{15}N gave a labelled product with a band at 3243·7 cm^{-1} with the same width and integrated intensity as in the unlabelled **84**. This shift of 6·9 cm^{-1} agrees well with the calculated 7·3 cm^{-1} ^{15}N isotopic shift on the NH stretching frequency in **84b** and excludes **84a**. Further, diazotized *p*-methoxyaniline-^{15}N with *N*-phenylhydroxylamine gave labelled **85** with a band 3237·6 cm^{-1}, in contrast to one at 3244·8 cm^{-1} in unlabelled **85**, thus showing that the structure is exclusively **85b**, the intramolecularly hydrogen-bonded *N*-oxide. It is noteworthy that the electron-releasing *p*-methoxy group would be expected to favour the form with the azo linkage next to the substituted phenyl ring. In labelled *p*-methoxydiazoaminobenzene structure **86a** is favoured over structure **86b**[275]:

	ν_{NH}	$\Delta\nu_{1/2}$
(**86a**) *p*-$CH_3OC_6H_4{}^{15}N{=}NNHC_6H_5$	3329·0 cm^{-1}	11·8 cm^{-1}
(**86b**) *p*-$CH_3OC_6H_4{}^{15}NHN{=}NC_6H_5$	3316·1	11·8
(**86c**) Unlabelled parent compound	3328·6	

The *N*-oxide structure (**84b**) has been verified by other i.r. study[276, 277].

I.r. spectroscopic studies of 1,3-diphenyltriazene-*N*-oxide and ring-substituted 1-alkyl-3-phenyltriazene-1-oxides over a concentration range in CCl_4 show that, in addition to the monomer band due to form **84b**, another broad band due to intermolecular association appears at somewhat lower frequency (~3200 cm^{-1}) as the concentration is increased[278]. However, *o*-substituents on the aromatic ring appear to prevent such association, presumably due to steric hindrance[278].

The weak acidity of hydroxytriazenes (**87**) in dioxan has been attributed to an intramolecularly hydrogen bonded structure (**87a**)[90]. Spectrophotometric procedures have been used to study the equilibrium and to obtain pK_a values (equation 28 and Table 11). Any substituent which weakens the intramolecular hydrogen bond increases the acidic character of these compounds. This may occur by a decrease in the polarity of the OH bond,

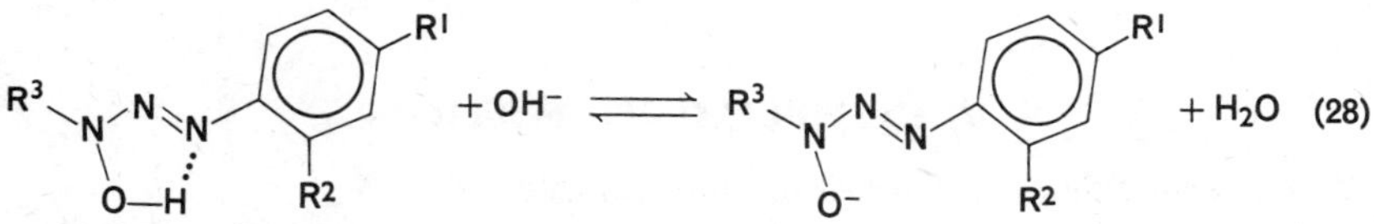

(**87a**) 305 mμ
strongly acidic solution

(**87b**) 335 mμ
strongly basic solution

by a decrease in the basicity of the azo N atom, or by steric interference from an *ortho* substituent (or by a combination of these factors). Thus when $R^1 = R^2 = H$, the pK_a values increase in the order $R^3 = Me < Et < n\text{-}Pr$, reflecting the increasing inductive electron-donating effect. When $R^1 = R^2 = H$ and R^3 changes from Me to Ph the pK_a drops from 12·11 to 11·41 as the basicity of the proton acceptor N atom falls, making the hydroxylic proton more available. Electron-donor groups like Me or OMe at R^1 increase the pK_a as the intramolecular hydrogen bond becomes stronger, while if $R^1 = Cl$ the pK_a decreases. When $R^2 = Cl$ the pK_a decreases because of apparent steric interference with the intramolecular hydrogen bond[90]. This method has been extended to substituted 3-hydroxy-1,3-diphenyltriazenes (**88**), where the dominating influence of the intramolecular hydrogen bond on pK_as is supported (Table 12)[91].

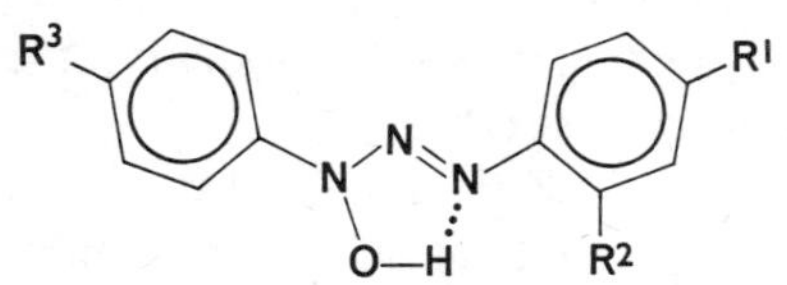

(**88**)

The ^{15}N n.m.r. spectra of *N*-nitrosohydrazines ($R^1N(NO)NR^2R^3$) show that only the *anti* isomer (**89**) exists when either or both R^2 and R^3 are

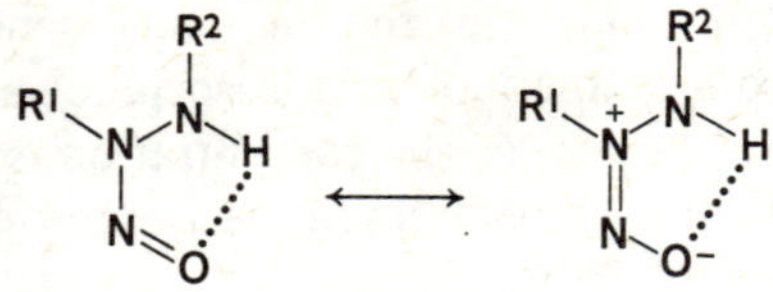

(89)

hydrogen, while both isomers are present when $R^2 = R^3 = Me$[279]. Stabilization of **89** by intramolecular hydrogen bonding is proposed as an explanation for this observation, as well as for the fact that the *syn* isomer of *N*-nitroso-*N*-benzylhydroxylamine is not observed, even at low temperature[279].

V. COMPLEX FORMATION

A. General Comments on Metal–Ion Complexes

A very comprehensive collection of stability constants for metal–ion complexes with both organic and inorganic ligands is available, covering the literature up to the end of 1968[280]. In addition, a review of hydrazine complexes of transition metals has appeared in 1970[281] and metal complexes of Schiff bases have been covered extensively[282]. The role of organonitrogen compounds in the formation of metal carbonyl and related complexes has also been summarized recently[283]. Other specialist reviews of metal–ion complexation with macrocyclic nitrogenous ligands[284] and with α-diimines, hydrazones and oximes[285] are available. This topic is therefore not dealt with further here.

Hydrogen bonding plays an important role in many complexes. Thus, in nickel dimethylglyoxime the Ni^{2+} ion is tetracoordinated to the azomethine N atoms in a square planar complex and extremely short intramolecular OH···O bonds with an O···O length of 2·44 Å are observed in the crystal[286]. The OH stretching frequency of 1800 cm^{-1} has been interpreted as arising from a symmetrical OH···O bond[287]. Intramolecular hydrogen bonding also occurs in the 1:1 complex of Cu^{2+} with *o,o′*-dihydroxyazobenzene and in the solid adducts of ethylenediamine and ethanolamine with this complex[288].

From a study of the Cu^{2+}, Zn^{2+} and Cd^{2+} complexes of *o*-hydroxy -and *o,o′*-dihydroxyazomethines a relation between intramolecular hydrogen bond formation, chelate ring size and the stability of the metal–ion chelates has been derived[289]. Again, six-membered chelate rings are more stable than five-membered rings, the latter forming only when stabilized by a six-membered ring in the same chelate. From C=N stretching vibrations in the infrared spectra it follows that in five-membered chelate rings the π-electrons of the C=N double bond also take part in the chelation, whereas in six-membered rings this effect is minimal[289].

B. Addition Products with Lewis Acids

Amino groups not only accept protons from Lowry–Brönsted acids, but they form adducts with compounds that can accept an electron pair, of which $H_3N \rightarrow BF_3$ is the classical example. Hydrazines and azo compounds follow the same pattern.

Borane complexes that have been prepared include $BH_3 \cdot N_2H_4$[290], $2\,BH_3 \cdot N_2H_4$[291] and $2\,BH_3 \cdot NH_2NMe_2$[292]. X-ray diffraction studies on hydrazine monoborane have been published, and the observed dipole moment is consistent with a large proportion of the *trans* form of the molecule[290]. In dioxan it associates with increasing concentration[290]. Both phenylhydrazine and 1,1-diphenylhydrazine form 1:1 adducts with borane[293].

Trimethylborane forms a 2:1 adduct with hydrazine, $2\,BMe_3 \cdot N_2H_4$, which is stable at or below –78°C and which evolves trimethylborane at higher temperatures to form the 1:1 adduct $BMe_3 \cdot N_2H_4$[294]. Direct reactions between BX_3 (X = Me, F) and the methylhydrazines $MeNHNH_2$, Me_2NNH_2, Me_2NNHMe and Me_2NNMe_2 give stable 1:1 adducts (except for BMe_3 with Me_2NNMe_2, which does not react), while the more unstable 1:1 adducts $Me_2NNH_2 \cdot BCl_3$ and $Me_2NNHMe \cdot BCl_3$ can be prepared in solution[295]. N.m.r. spectra confirm that adduct formation occurs at the more highly methylated nitrogen atom in 1,1-dimethylhydrazine and in trimethylhydrazine, except that in the $MeNHNH_2 \cdot BF_3$ case two isomeric adducts are found, corresponding to a 5:1 molar ratio of complexation at the —NHMe group and the —NH_2 group, respectively[295]. Structural studies on some trifluoromethylphosphino hydrazines have also been reported[295]. The adduct $BF_3 \cdot N_2H_4$ has been prepared[296].

Theoretical calculations (CNDO/2) confirm that the BF_3 is attached to the methylated nitrogen atom in $Me_2NNH_2 \cdot BF_3$, and indicate that at ambient temperature the *gauche*/*trans* ratio in the complex is 35:65, as opposed to 75:25 in the free molecule[297].

Spectrophotometric measurement of the dissociation constants for various complexes of $SbCl_5$ with electron donors in 1,2-dichloroethane lead to the following decreasing order of electron donor strength (values of $K_{D \cdot SbCl_5}$ in parentheses)[298]: *trans*-azobenzene (2×10^{-8}) > tetrahydrofuran ($1{\cdot}2 \times 10^{-7}$) > *trans-p*-nitroazobenzene (6×10^{-6}) > acetone ($1{\cdot}4 \times 10^{-5}$) > *trans*-azoxybenzene (4×10^{-5}) > acetonitrile ($1{\cdot}4 \times 10^{-3}$). The generally weaker donor characteristics of the azoxy group relative to the azo group

TABLE 18. Dissociation constants of azomethine and azo complexes with $SbCl_3$ in 1,2-dichloroethane[300]

Ligand (D)	($D/SbCl_3$) ratio in complex	$K_{D \cdot SbCl_3}$
Benzylidene aniline	1:1	$1{\cdot}7 \times 10^{-2}$
Benzylidene-4-dimethylaminoaniline	1:1	$1{\cdot}5 \times 10^{-4}$
	1:2	$5{\cdot}8 \times 10^{-2}$
4-Dimethylaminobenzylidene aniline	1:1	$4{\cdot}8 \times 10^{-5}$
Azobenzene	1:2	4·3
2,2′-Dimethylazobenzene	1:2	1·5
4-Methylazobenzene	1:2	$8{\cdot}5 \times 10^{-1}$
4,4′-Dimethylazobenzene	1:2	$5{\cdot}0 \times 10^{-2}$
4-Ethoxyazobenzene	1:2	$2{\cdot}2 \times 10^{-2}$
4-Hydroxyazobenzene	1:2	$2{\cdot}2 \times 10^{-2}$
4-Aminoazobenzene	1:1	$2{\cdot}1 \times 10^{-2}$
4-Dimethylaminoazobenzene	1:1	$2{\cdot}0 \times 10^{-3}$

are also reflected in the fact that $PdCl_2$ and $PdBr_2$ form 1:2 complexes with *trans*-azomethane and *trans*-azobenzene, whereas similar complexes could not be prepared with azoxybenzene[299]. Further evidence for co-ordination to a single nitrogen lone pair was found in these $PdCl_2$ and $PdBr_2$ complexes[299].

TABLE 19. Dissociation constants for azo donor–halogenide acceptor complexes in acetonitrile at 20°C[301]

Acceptors	Donors		
	$C_6H_5N{=}NC_6H_5$	p-$MeOC_6H_4N{=}NC_6H_5$	p-$Me_2NC_6H_4N{=}NC_6H_5$
$SbCl_5$	3×10^{-5}	$1{\cdot}5 \times 10^{-6}$	1×10^{-8}
BF_3	$7{\cdot}5 \times 10^{-4}$	$4{\cdot}0 \times 10^{-5}$	1×10^{-7}
$GaBr_3$	$1{\cdot}8 \times 10^{-3}$	$8{\cdot}0 \times 10^{-5}$	5×10^{-7}
$GaCl_3$	$3{\cdot}0 \times 10^{-3}$	$1{\cdot}0 \times 10^{-4}$	9×10^{-7}
$SnCl_4$	$3{\cdot}5 \times 10^{-3}$	$2{\cdot}2 \times 10^{-4}$	9×10^{-7}
BBr_4	9×10^{-3}	$(0{\cdot}5–5) \times 10^{-4}$	6×10^{-7}
$TiCl_4$	—	$6{\cdot}0 \times 10^{-4}$	$1{\cdot}3 \times 10^{-6}$
BCl_3	$1{\cdot}8 \times 10^{-2}$	1×10^{-3}	$1{\cdot}5 \times 10^{-6}$
$AlBr_3$	(1×10^{-1})	$(10^{-3}–10^{-4})$	—
$AlCl_3$	(2×10^{-1})	$(10^{-3}–10^{-4})$	4×10^{-6}
$InCl_3$	—	10^{-2}	$1{\cdot}5 \times 10^{-5}$

Table 18 compares the dissociation constants of some azomethine and azo complexes with $SbCl_3$ in 1,2-dichloroethane[300]. For each series electron-donating ring substituents increase the stability of the complex; for a given substituent the azomethine complex appears to be more stable than the azo complex. Similar data for the electron donors azobenzene, *p*-methoxyazobenzene and *p*-dimethylaminoazobenzene with eleven different halogenides are found in Table 19. For a given acceptor, the spectrophotometric dissociation constants decrease in the ligand order: azobenzene > *p*-methoxyazobenzene > *p*-dimethylaminobenzene. It should be noted, however, that for the substituted donors complexation occurs first at the —OMe and —NMe_2 groups, and further complexation at the azo group is only possible with the strongest acceptors, $SbCl_5$, and then only with an excess of reagent[301]. In general, acceptor strengths decrease in the order[301]: $SbCl_5 > BF_3 > GaBr_3 > GaCl_3 \sim SnCl_4 > BBr_3 > TiCl_4 > BCl_3 > AlBr_3 > AlCl_3 > InCl_3$. The N═N stretching vibration at 1442 cm^{-1} in the Raman spectrum, which is forbidden in the i.r. for *trans*-azobenzene, becomes intense in the i.r. near 1390 cm^{-1} in the $SbCl_5$ complex because of the unsymmetrical nature of the linkage in the complex[301]. The u.v. spectrum of this complex is also similar to that of protonated azobenzene.

Other complexes that have been isolated and the respective donor/acceptor molar ratios are:

(*a*) Of hydrazine with: $SnBr_4$(6/1)[302]; SnI_4(8/1)[302]; TiF_4(1/1)[303].

(*b*) Of phenylhydrazine with: TiF_4(1/1)[303], (6/1)[304]; $TiCl_3$(4/1), (1/1)[304]; $TiCl_4$(2/1), (3/1), (4/1), (8/1)[304]; $AlCl_3$(2/1)[305]; $SnBr_4$(4/1)[302]; SnI_4(8/1)[302].

(*c*) Of *p*-tolylhydrazine and 1,2-diphenylhydrazine with: TiF_4(1/1)[303].

VI. REFERENCES

1. J. W. Smith, in *The Chemistry of the Amino Group* (Ed. S. Patai), Chap. 4, Interscience, London, 1968, pp. 161–204.
2. G. C. Pimentel and A. L. McClellan, *The Hydrogen Bond*, W. H. Freeman and Co., San Francisco, 1960.
3. A. S. N. Murthy and C. N. R. Rao, *Appl. Spectrosc. Rev.*, **2**, 69 (1968); G. C. Pimentel and A. L. McClellan, *Ann. Rev. Phys. Chem.*, **22**, 347 (1971).
4. D. W. Scott, G. D. Oliver, M. E. Gross, W. N. Hubbard and H. M. Huffman, *J. Amer. Chem. Soc.*, **71**, 2293 (1949) and references cited therein.
5. W. G. Penny and G. B. B. M. Sutherland, *Trans. Faraday Soc.*, **30**, 902 (1934).
6. J. R. Durig and W. C. Harris, *J. Chem. Phys.*, **55**, 1735 (1971); J. R. Durig, R. W. MacNamee, L. B. Knight and W. C. Harris, *Inorg. Chem.*, **12**, 804 (1973).
7. G. J. Karabatsos and R. A. Taller, *Tetrahedron*, **24**, 3347 (1968).
8. G. J. Karabatsos and C. E. Osborne, *Tetrahedron*, **24**, 3361 (1968).

9. E. Haselbach and E. Heilbronner, *Helv. Chim. Acta*, **51,** 16 (1968).
10. R. S. Tsekhanskii, L. I. Vinogradov, S. N. Nikolaev and I. I. Saidashev, *Vop. Stereokhim.*, 86–8 (1971); *Chem. Abstr.*, **77,** 139053c (1972).
11. J. P. Freeman, *J. Org. Chem.*, **28,** 2508 (1963); G. S. Hartley, *Nature*, **140,** 281 (1937); G. S. Hartley, *J. Chem. Soc.*, 633 (1938).
12. J. J. de Lange, J. M. Robertson and I. Woodward, *Proc. Roy. Soc.* (*London*), **A171,** 398 (1939); C. C. Hampson and J. M. Robertson, *J. Chem. Soc.*, 409 (1941).
13. J. H. Collins and H. H. Jaffe, *J. Amer. Chem. Soc.*, **84,** 4708 (1962).
14. A. J. Bellamy and R. D. Guthrie, *J. Chem. Soc.*, 2788 (1965); G. J. Karabatsos and R. A. Taller, *J. Amer. Chem. Soc.*, **85,** 3625 (1963).
15. R. R. Shagidullin, F. K. Sattarova, T. V. Troepol'skaya and Y. P. Kitaev, *Izv. Akad. Nauk SSSR, Otd. Khim. Nauk*, 385 (1963); *Chem. Abstr.*, **58,** 13761 (1963).
16. Y. P. Kitaev and T. V. Troepol'skaya, *Izv. Akad. Nauk SSSR, Otd. Khim. Nauk*, 454 (1963); *Chem. Abstr.*, **59,** 7347 (1963).
17. E. Fischer and Y. F. Frei, *J. Chem. Soc.*, 3159 (1959); Y. Yagi, *Bull. Chem. Soc. Japan*, **37,** 1875 (1964).
18. N. V. Sidgwick, *The Organic Chemistry of Nitrogen*, 3rd ed. (revised by I. T. Miller and H. D. Springall), Clarendon Press, Oxford, 1966, pp. 583–585 and references cited therein.
19. P. A. S. Smith, *The Chemistry of Open-Chain Nitrogen Compounds*, Vol. 2, W. A. Benjamin, New York, 1966, (a) pp. 321–323, (b) pp. 336–340, (c) pp. 340–343, (d) pp. 150–151, (e) pp. 30–31.
20. D. L. Webb and H. H. Jaffe, *J. Amer. Chem. Soc.*, **86,** 2419 (1964).
21. S. Weckherlin and W. Lüttke, *Tetrahedron Letters*, 1711 (1964).
22. T. Mitsuhashi and O. Simanura, *Chem. Ind.* (*London*), 578 (1964).
23. V. Zverina, M. Remes, J. Divis, J. Marhold and M. Matrka, *Coll. Czech. Chem. Comm.*, **38,** 251 (1973).
24. Yu. A. Omel'chenko and Yu. A. Kondrashev, *Soviet Physics-Cryst.*, **10,** 690 (1966).
25. R. J. W. Le Fevre and T. H. Liddicoet, *J. Chem. Soc.*, 2743 (1951).
26. N. Yni, *Bull. Inst. Phys. Chem. Res.* (*Tokyo*), **20,** 256 (1941); *J. Org. Chem.*, **23,** 1587 (1958).
27. P. A. S. Smith, *The Chemistry of Open-Chain Nitrogen Compounds*, Vol. 1, W. A. Benjamin, New York, 1965, (a) p. 9, (b) p. 294.
28. For a full discussion of all substituent constants referred to in this review see: H. H. Jaffe, *Chem. Revs.*, **53,** 191 (1953); C. D. Ritchie and W. F. Sager, in *Progress in Physical Organic Chemistry*, Vol. 2 (Ed. S. G. Cohen, A. Streitwieser and R. W. Taft), Interscience, 1964, pp. 323–400 and in particular, tables of various substituent constants on pp. 334–337.
29. H. K. Hall, *J. Amer. Chem. Soc.*, **79,** 5441 (1957).
30. A. Fischer, D. A. Happer and J. Vaughan, *J. Chem. Soc.*, 4060 (1964).
31. D. D. Perrin, *Dissociation Constants of Organic Bases in Aqueous Solution*, Butterworths, London, 1965.
32. R. L. Hinman, *J. Org. Chem.*, **23,** 1587 (1958).
33. F. E. Condon, *J. Amer. Chem. Soc.*, **87,** 4481 (1965).
34. W. Broser and V. Bollert, *Z. Naturforsch.*, **B26,** 618 (1971).
35. R. Pollet and H. Vanden Eynde, *Bull. Soc. Chim. Belges*, **77,** 341 (1968).

36. A. Ebnöther, E. Jucker, A. Lindenmann, E. Rissi, R. Steiner, R. Süess and A. Vogel, *Helv. Chim. Acta*, **42,** 533 (1959).
37. N. F. Hall, *J. Amer. Chem. Soc.*, **52,** 5115 (1930).
38. H. H. Stroh and H. G. Scharnow, *Chem. Ber.*, **98,** 1588 (1965).
39. R. Pollet and H. Vanden Eynde, *Bull. Soc. Chim. Belges*, **77,** 341 (1968).
40. C. R. Lindegren and C. Niemann, *J. Amer. Chem. Soc.*, **71,** 1504 (1949).
41. J. B. Class, J. G. Aston and T. S. Oakwood, *J. Amer. Chem. Soc.*, **75,** 2937 (1953).
42. Unpublished measurements by M. R. Sprinkle quoted in Reference 37.
43. R. G. Bates and G. D. Pinching, *J. Res. Nat. Bur. Stand.*, **42,** 419 (1949).
44. D. Northcott, data given in W. Van der Linde, D. Northcott, W. Redmond and R. E. Robertson, *Can. J. Chem.*, **47,** 279 (1969).
45. D. H. Everett and B. R. W. Pinsent, *Proc. Roy. Soc. (London)*, **A215,** 416 (1952).
46. E. Folkers and O. Runquist, *J. Org. Chem.*, **29,** 830 (1964).
47. H.-H. Stroh and G. Westphal, *Chem. Ber.*, **96,** 184 (1963).
48. H.-H. Stroh and G. Westphal, *Chem. Ber.*, **97,** 83 (1964).
49. A. I. Biggs and R. A. Robinson, *J. Chem. Soc.*, 388 (1961).
50. E. Sawicki and D. Gerber, *J. Org. Chem.*, **21,** 410 (1956).
51. R. F. Evans and W. Kynaston, *J. Chem. Soc.*, 3151 (1963).
52. P. J. Brignell, C. D. Johnson and A. R. Katritzky, *J. Chem. Soc.*, (*B*), 1233 (1967).
53. E. V. Titov, N. G. Korzhenevskaya and V. I. Rybachenko, *Ukr. Khim. Zh.*, **34,** 1253 (1968); *Chem. Abstr.*, **70,** 109624e (1969).
54. V. D. Canic, *Naturwiss.* **46,** 575 (1959).
55. V. I. Minkin and V. A. Bren, *Reakts. Sposobnost Org. Soedin.*, **4,** 112 (1967); *Chem. Abstr.*, **69,** 100231u (1968).
56. E. H. Cordes and W. P. Jencks, *J. Amer. Chem. Soc.*, **84,** 832 (1962).
57. H. F. Harnsberger, E. L. Cochran and H. H. Szmant, *J. Amer. Chem. Soc.*, **77,** 5048 (1955).
58. L. A. Jones and N. L. Nueller, *J. Org. Chem.*, **27,** 2356 (1962).
59. O. L. Brady and M. M. Chokshi, *J. Chem. Soc.*, 2271 (1929).
60. C. V. King and A. P. Marion, *J. Amer. Chem. Soc.*, **66,** 977 (1944).
61. O. L. Brady and R. F. Goldstein, *J. Chem. Soc.*, 2403 (1926).
62. A. L. Green and B. Saville, *J. Chem. Soc.*, 3887 (1956).
63. D. S. Jackson, *J. Chem. Soc.*, (*B*), 785 (1969).
64. H. E. Ungnade, G. Fritz and L. W. Kissinger, *Tetrahedron*, **19,** Suppl. 1, 235 (1963).
65. I. M. Klotz, H. A. Fiess, J. Y. Chen Ho and M. Mellody, *J. Amer. Chem. Soc.*, **76,** 5136 (1954).
66. F. Gerson, E. Heilbronner, A. van Veen and B. M. Wepster, *Helv. Chim. Acta*, **43,** 1889 (1960); M. Isaks and H. H. Jaffe, *J. Amer. Chem. Soc.*, **86,** 2209 (1964).
67. J. H. Collins and H. H. Jaffe, *J. Amer. Chem. Soc.*, **84,** 4708 (1962); M. Isaks and H. H. Jaffe, *J. Amer. Chem. Soc.*, **86,** 2209 (1964).
68. K. N. Bascombe and R. P. Bell, *J. Chem. Soc.*, 1096 (1956).
69. E. Högfeldt and J. Bigeleisen, *J. Amer. Chem. Soc.*, **82,** 15 (1960).
70. A. Meretoja and R. Linko, *Suomen Kemistilehti*, **23B,** 26 (1950).
71. M. Kilpatrick and C. A. Arenberg, *J. Amer. Chem. Soc.*, **75,** 3812 (1953).

72. L. P. Hammett and M. A. Paul, *J. Amer. Chem. Soc.*, **56,** 827 (1934); L. P. Hammett and A. J. Deyrup, *J. Amer. Chem. Soc.*, **54,** 2721 (1932).
73. S. J. Yeh and H. H. Jaffe, *J. Amer. Chem. Soc.*, **81,** 3274 (1959).
74. E. Haselbach, *Helv. Chim. Acta*, **53,** 1526 (1970).
75. C. S. Frankovskii, N. V. Melamed and L. D. Koval'chuk, *J. Org. Chem. U.S.S.R.*, **3,** 1795 (1967).
76. A. G. Catchpole and H. M. Catchpole, *Spectrochim. Acta*, **18,** 1353 (1962).
77. H. H. Jaffe and R. W. Gardner, *J. Amer. Chem. Soc.*, **80,** 319 (1958); S.-J. Yeh and H. H. Jaffe, *J. Amer. Chem. Soc.*, **81,** 3279, 3283 (1959).
78. G. Cilento, *J. Org. Chem.*, **24,** 2015 (1959).
79. F. Gerson and E. Heilbronner, *Helv. Chim. Acta*, **45,** 51 (1962).
80. E. Sawicki and F. E. Ray, *J. Org. Chem.*, **19,** 1686 (1954).
81. E. Sawicki, *J. Org. Chem.*, **21,** 605 (1956).
82. S. Yamamoto, N. Nishimura and S. Hasegawa, *Bull. Chem. Soc. Japan*, **46,** 194 (1973).
83. F. Gerson and E. Heilbronner, *Helv. Chim. Acta*, **45,** 42 (1962).
84. E. Sawicki, *J. Org. Chem.*, **22,** 915 (1957).
85. E. Sawicki, *J. Org. Chem.*, **22,** 621 (1957).
86. W. T. Rogers, T. W. Campbell and R. W. Maatman, *J. Amer. Chem. Soc.*, **73,** 5122 (1951).
87. C.-S. Hahn and H. H. Jaffe, *J. Amer. Chem. Soc.*, **84,** 949 (1962).
88. R. E. Ellerhorst and H. H. Jaffe, *J. Org. Chem.*, **33,** 4115 (1968).
89. A. Dolenko, K. Mahendran and E. Buncel, *Can. J. Chem.*, **48,** 1736 (1970).
90. S. M. Dugar and N. C. Sogani, *Zeit. Naturforsch*, **B21,** 657 (1966).
91. D. N. Purohit and N. C. Sogani, *J. Chem. Soc.*, 2820 (1964).
92. V. I. Semishin, *J. Gen. Chem. U.S.S.R.*, **8,** 654 (1938).
93. A. Curtius and B. Thun, *J. Prakt. Chem.*, **4,** 190 (1891).
94. I. M. Bokhovkin, *Zhur. Obshchei Khim.*, **29,** 3531 (1959); *Chem. Abstr.*, **54,** 17024h (1960); V. I. Semishin, *J. Gen. Chem. U.S.S.R.*, **9,** 788 (1939).
95. R. Cazaneuve and S. Moreau, *Compt. Rend.*, **129,** 1255 (1899).
96. R. Liminga, *Acta Chem. Scand.*, **21,** 1206 (1967).
97. D. E. Dionis'ev and F. T. Pozharskii, *Sbornick Statei Obshchei Khim., Akad. Nauk SSSR*, **1,** 219 (1953); *Chem. Abstr.*, **48,** 12534d (1954).
98. N. A. Pushin and L. Maric, *Bull. Soc. Chim. Belgrade*, **16,** 5 (1951).
99. N. A. Pushin and G. M. Dimitrijević, *Z. Physik. Chem.*, **A184,** 231 (1939).
100. N. A. Pushin and I. I. Rikovski, *Z. Physik. Chem.*, **A161,** 336 (1932).
101. K. Hrynakowski and F. Adamanis, *Roczniki Chem.*, **18,** 538 (1938); *Chem. Abstr.*, **33,** 7286[4] (1939).
102. P. Laurant, *Ann. Chim.*, **10,** 397 (1938); *Chem. Abstr.*, **33,** 1578[1] (1939).
103. N. A. Pushin, P. Matavulj and I. I. Rikovski, *Bull. Soc. Chim. Belgrade*, **14,** 41 (1949).
104. J. R. Durig, S. F. Bush and E. E. Mercer, *J. Chem. Physics*, **44,** 4238 (1966).
105. J. R. Durig, W. C. Harris and D. W. Wertz, *J. Chem. Physics*, **50,** 1449 (1969).
106. J. R. Cook and K. Schug, *J. Amer. Chem. Soc.*, **86,** 4271 (1964).
107. N. G. Yunda, G. V. Lagodzinskaya and G. B. Manelis, *Izv. Akad. Nauk SSSR, Ser. Khim.* (6), 1284 (1972); *Chem. Abstr.*, **77,** 93594z (1972).
108. E. G. Sauer and P. J. Bray, *J. Chem. Physics*, **56,** 820 (1972).
109. K. Suzuki, S. Onishi, T. Koide and S. Seki, *Bull. Chem. Soc. Japan*, **29,** 127 (1956).

110. S. Takagi, R. Shintani, H. Chihara and S. Seki, *Bull. Chem. Soc. Japan*, **32,** 137 (1959).
111. T. Miyazawa, *J. Chem. Soc. Japan*, **22,** 1629 (1954).
112. A. Yamaguchi, *J. Chem. Soc. Japan*, **79,** 880 (1958).
113. D. B. Chestnut and R. E. Marsh, *Acta Cryst.*, **11,** 413 (1958).
114. R. Shintani, *Acta Cryst.*, **13,** 609 (1960).
115. Y. Tomiie, C. H. Koo and I. Nitta, *Acta Cryst.*, **11,** 158 (1958).
116. R. Liminga and J.-O. Lundgren, *Acta Chem. Scand.*, **19,** 1612 (1965).
117. L. H. Jensen, *J. Amer. Chem. Soc.*, **78,** 3993 (1956).
118. R. L. Collin and W. N. Lipscomb, *Acta Cryst.*, **4,** 10 (1951).
119. R. Liminga, *Acta Chem. Scand.*, **20,** 2483 (1966); A. Kvick, P. G. Jonsson and R. Liminga, *Acta Chem. Scand.*, **26,** 1087 (1972).
120. R. Liminga, *Arkiv. Kemi.*, **28,** 471 (1968).
121. J. Donohue and W. N. Lipscomb, *J. Chem. Physics*, **15,** 115 (1947).
122. M. Kronberg and D. Harker, *J. Chem. Physics*, **10,** 309 (1942).
123. R. Liminga, *Arkiv. Kemi.*, **28,** 483 (1968); R. Liminga and I. Olovsson, *Acta Cryst.*, **17,** 1523 (1964).
124. K. Sakurai and Y. Tomiie, *Acta Cryst.*, **5,** 293 (1952).
124a. R. Liminga and A. M. Sorensen, *Acta Chem. Scand.*, **21,** 2669 (1967).
125. S. A. Hady, I. Nahringbauer and I. Olovsson, *Acta Chem. Scand.*, **23,** 2764 (1969).
126. P. G. Jonsson and R. Liminga, *Acta Chem. Scand.*, **25,** 1729 (1971).
127. A. Braibanti, A. M. M. Lanfredi, A. Tiripicchio and F. Logiudice, *Acta Cryst.*, **B25,** 93 (1969).
128. N. A. K. Ahmed, R. Liminga and I. Olovsson, *Acta Chem. Scand.*, **22,** 88 (1968); A. Nilsson, R. Liminga and I. Olovsson, *Acta Chem. Scand.*, **22,** 719 (1968).
129. R. Liminga, *Acta Chem. Scand.*, **21,** 1217 (1967).
130. P. Jönsson and W. Hamilton, *Acta Cryst.*, **B26,** 536 (1970).
131. S. Du Pre and H. Mendel, *Acta Cryst.*, **8,** 311 (1955).
132. L. H. Jensen and E. C. Lingafelter, *Acta Cryst.*, **14,** 507 (1961).
133. S. Srinivasan and S. Swaminathan, *Zeit. fuer Krist.*, **127,** 442 (1968).
134. C. H. Koo, *Bull. Chem. Soc. Japan*, **38,** 286 (1965).
135. G. D. Andreeti, P. Domiano, G. F. Gaspari, M. Nardelli and P. Sgarabotto, *Acta Cryst.*, **B26,** 1005 (1970).
136. R. Liminga, *Arkiv. Kemi.*, **28,** 483 (1968).
136a. A. Weiss, W. Thielepape, W. Ritter, H. Schaefer and G. Goering, *Z. Anorg. Allgem. Chemie*, **320,** 183 (1963).
137. P. Glavic and D. Hadzi, *Spectrochim. Acta*, **A28,** 1963 (1972).
138. A. V. R. Warrier and P. S. Narayanan, *Ind. J. Pure and Appl. Physics*, **5,** 216 (1967).
139. A. Selvarajan, *Ind. J. Pure and Appl. Phys.*, **8,** 342 (1970).
140. J. Lindgren, J. De Villepin and A. Novak, *Chem. Phys. Letters*, **3,** 84 (1969).
141. G. A. Neville and D. V. C. Awang, *Org. Mag. Res.*, **2,** 341 (1970).
142. A. S. A. S. Shawali, A. M. Dewidar and M. M. Naoum, *Ind. J. Chem.*, **10,** 464 (1972).
143. J. Tegenfeld and I. Olovsson, *Acta Chem. Scand.*, **25,** 101 (1971).
144. T. P. Morton and F. L. Howell, *Mol. Physics*, **20,** 1147 (1971).

145. J. Vanderkooy, J. D. Cuthbert and H. E. Petch, *Can. J. Phys.*, **42,** 1871 (1964).
146. H. T. Hayes and L. Hunter, *J. Chem. Soc.*, 326 (1940).
147. J. B. Jensen, *Acta Chem. Scand.*, **10,** 667 (1956); S. Laptii, T. B. Lutsenko, V. N. Vatulev, A. P. Grekov and V. Y. Veselov, *Zh. Prikl. Spektrosk.*, **16,** 367 (1972); *Chem. Abstr.*, **73,** 133769y (1972); C. G. Macarovici, *Rev. Roum. Chim.*, **12,** 151 (1967); *Chem. Abstr.*, **68,** 73594k (1968).
148. H. Lumbroso and J. Barassin, *Bull. Chim. Soc. France*, 3190 (1964).
149. J. Weinstein and E. McIninch, *J. Amer. Chem. Soc.*, **82,** 6064 (1960).
150. M. Uehara and J. Nakaya, *Bull. Chem. Soc. Japan*, **43,** 3136 (1970).
151. G. J. Karabatsos, B. L. Shapiro, F. M. Vane, J. S. Fleming and J. S. Ratka, *J. Amer. Chem. Soc.*, **85,** 2784 (1963).
152. J. E. Kovacic, *Spectrochim. Acta*, **A23,** 183 (1967).
153. A. W. Baker and A. T. Shulgin, *J. Amer. Chem. Soc.*, **81,** 1523 (1959).
154. J. Klicnar, F. Kristek, V. Bekarek and M. Vecera, *Coll. Czech. Chem. Commun.*, **34,** 553 (1969).
155. H. H. Freedman, *J. Amer. Chem. Soc.*, **83,** 2900 (1961).
156. V. Bekarek, J. Klicnar, F. Kristek and M. Vecera, *Coll. Czech. Chem. Commun.*, **33,** 994 (1968).
157. L. A. Kazitsyna and V. V. Mischenko, *Dokl. Akad. Nauk SSSR*, **150,** 555 (1963); *Chem. Abstr.*, **59,** 8270g (1963).
158. M. St. C. Flett, *Spectrochim. Acta*, **10,** 21 (1957).
159. V. I. Minkin, O. A. Osipov and V. N. Sheinker, *Russ. J. Phys. Chem.*, **44,** 12 (1970).
160. E. Lippert, J. Muszik and W. Voss, *J. Chim. Phys.*, **61,** 228 (1964).
161. M. D. Cohen, Y. Hirsberg and G. M. J. Schmidt, *J. Chem. Soc.*, 2060 (1964).
162. G. O. Dudek, *J. Amer. Chem. Soc.*, **85,** 694 (1963).
163. L. Clougerty, *J. Org. Chem.*, **22,** 462 (1957); I. Farbian, *Bull. Soc. Chim. France*, **41,** 1499 (1956).
164. E. Lippert, D. Samuel and E. Fischer, *Ber. Bunsenges. Phys. Chem.*, **69,** 155 (1965).
165. O. A. Osipov, V. I. Minkin and V. N. Sheinker, *Zhur. Obshch. Khim.*, **38,** 1903 (1968).
166. G. O. Dudek and E. P. Dudek, *J. Amer. Chem. Soc.*, **88,** 2407 (1966).
167. O. A. Osipov, V. I. Minkin and V. A. Kogan, *Zh. Fiz. Khim.*, **37,** 1492 (1963); *Chem. Abstr.*, **59,** 11218 (1963).
167a. D. G. Anderson and G. Wettermark, *J. Amer. Chem. Soc.*, **87,** 1433 (1965).
168. D. Heinert and A. E. Martell, *J. Amer. Chem. Soc.*, **84,** 3257 (1962); **85,** 183 (1963); **85,** 188 (1963).
169. E. R. Blout and R. M. Gofstein, *J. Amer. Chem. Soc.*, **67,** 13 (1945); F. C. McIntire, *J. Amer. Chem. Soc.*, **69,** 1377 (1947); L. N. Ferguson and I. Kelly, *J. Amer. Chem. Soc.*, **73,** 3707 (1951); O. Gerngross and A. Olcay, *Chem. Ber.*, **96,** 2550 (1963); F. E. Prichard, *Spectrochim. Acta*, **20,** 925 (1964).
170. G. Arcovito, M. Bonamico and A. Domenicano, *J. Chem. Soc.*, (*B*), 733 (1969).
171. R. Restivo and G. J. Palenik, *Acta Cryst.*, **B26,** 1397 (1970).
172. D. N. Shigorin, I. Y. Pavlenishvili, G. V. Panova, B. N. Bolotin and N. N. Shapet'ko, *Russ. J. Phys. Chem.*, **40,** 822 (1966).
173. B. B. Bhowmik, *J. Phys. Chem.*, **74,** 4442 (1970).

174. Y. I. Kozlov, D. N. Shigorin, R. N. Nurmukhametov and V. A. Puchkov, *Russ. J. Phys. Chem.*, **37,** 1315 (1963).
175. B. M. Bolotin and B. M. Krasovitskii, *Prom. Khim. Reaktivov Osobo Chist. Veshchestv.*, No. 8, 121 (1967); *Chem. Abstr.*, **69,** 72561f (1968).
176. H. Hoyer, *Z. Physik. Chem.*, **36,** 320 (1963).
177. N. M. D. Brown and D. C. Nonhebel, *Tetrahedron*, **24,** 5655 (1968).
178. G. O. Dudek and R. H. Holm, *J. Amer. Chem. Soc.*, **83,** 2099 (1961).
179. F. A. Isherwood and R. L. Jones, *Nature*, **175,** 419 (1955); D. A. Shapiro, R. A. Abramovitch and S. Pinchas, *J. Amer. Chem. Soc.*, **78,** 2144 (1956); E. M. Tanner, *Spectrochim. Acta*, **15,** 20 (1959).
180. Z. Rappoport and T. Sheradsky, *J. Chem. Soc.*, (*B*), 277 (1968).
181. E. Gentric, A. Le Narvor, J. Lauransan and P. Saumagne, *J. Chim. Phys.*, **69,** 87 (1972).
182. A. Perrier-Datin and M. L. Josien, *Compt. Rend.*, **C266,** 1475 (1968).
183. G. Geiseler and A. Janossy, *J. Mol. Struct.*, **11,** 393 (1972).
184. H. H. Ericksen and C. Romming, *Acta Chem. Scand.*, **22,** 1430 (1968).
185. R. Pujol and F. Mathis, *Compt. Rend.*, **B269,** 1049 (1969).
186. R. Mathis, B. Baccar, J. Barrans and F. Mathis, *J. Mol. Struct.*, **7,** 355 (1971).
187. H. T. Hayes and L. Hunter, *J. Chem. Soc.*, 1, 1941.
188. N. E. White and M. Kilpatrick, *J. Phys. Chem.*, **59,** 1044 (1955).
189. L. Hunter and J. A. Marriott, *J. Chem. Soc.*, 777 (1941).
190. D. G. O'Sullivan, *J. Chem. Soc.*, 3278 (1960).
191. E. J. Browne and J. B. Polya, *J. Chem. Soc.*, (*C*), 1056 (1969).
192. J. Brickmann and H. Zimmermann, *Ber. Bunsenges. Physik. Chem.*, **70,** 521 (1966).
193. G. Schwenker, *Angew. Chem.* (*Int. Ed.*), **10,** 426 (1971).
194. J. H. Bryden, *Acta Cryst.*, **11,** 158 (1958).
195. T. K. Bierlein and E. C. Lingafelter, *Acta Cryst.*, **4,** 450 (1951).
196. B. Jerslev, *Nature*, **166,** 741 (1950).
197. Y. Okaya, A. Shimada and I. Nitta, *Bull. Chem. Soc. Japan*, **29,** 210 (1956).
198. L. L. Merritt and E. Lanterman, *Acta Cryst.*, **5,** 811 (1952).
199. B. Jerslev, quoted in reference 94.
200. D. Hall and F. J. Llewellyn, *Acta Cryst.*, **9,** 108 (1956).
201. L. Hunter and C. B. Roberts, *J. Chem. Soc.*, 823 (1941).
202. K. C. Kalia and A. Chakravorty, *J. Org. Chem.*, **35,** 2231 (1970).
203. A. Reiser, in *Hydrogen Bonding* (Ljubljana Symposium, 1957) (Ed. D. Hadzi and H. W. Thompson), Pergamon Press, London, 1959, pp. 443–447.
204. A. M. Buswell, W. H. Rodebush and M. F. Roy, *J. Amer. Chem. Soc.*, **60,** 2444 (1938).
205. G. Geiseler and J. Fruwert, *Zeit. Physik Chem.* (*Frankfurt*), **26,** 111 (1960).
206. S. Califano and W. Lüttke, *Zeit. Physik. Chem.* (*Frankfurt*), **5,** 240 (1955).
206a. W. C. Harris and S. F. Bush, *J. Chem. Phys.*, **56,** 6147 (1972).
206b. T. Di Paolo and C. Sandorfy, *Can. J. Chem.*, **51,** 1441 (1973).
207. A. Ossart, H. Sauvaitre and P. Pineau, *J. Chim. Phys.*, **64,** 953 (1967).
208. A. Reiser, A. Kimla and J. Hajek, *Coll. Czech. Chem. Comm.*, **19,** 4 (1954).
209. D. Hadzi and L. Premru, *Spectrochim. Acta*, **A23,** 35 (1967).
210. A. Palm and H. Werbin, *Can. J. Chem.*, **31,** 1004 (1953); **32,** 858 (1954).
211. B. Jerslev, *Nature*, **180,** 1410 (1957); K. Folting, W. N. Lipscomb and B. Jerslev, *Acta Cryst.*, **17,** 1263 (1964).

212. H. Sauvaitre, *J. Mol. Struct.*, **11,** 403 (1972).
213. H. Sauvaitre and J. Deschamp, *Tetrahedron Letters*, 1761 (1967).
214. M. Kimura, Y. Kuroda, H. Takagi and M. Kubo, *Bull. Chem. Soc. Japan*, **33,** 1086 (1960); Y. Kuroda and M. Kimura, *Bull. Chem. Soc. Japan*, **36,** 464 (1963).
215. C. J. Brown, *J. Chem. Soc.*, (*A*), 405 (1967).
216. P. Laurent, *Compt. Rend.*, **199,** 582 (1934).
217. A. Kofler, *Zeit. Physik. Chem.* (*Leipzig*), **190,** 287 (1942).
218. A. R. Monahan and D. F. Blossey, *J. Phys. Chem.*, **74,** 4014 (1970); A. R. Monahan, N. J. Germano and D. F. Blossey, *J. Phys. Chem.*, **75,** 1227 (1971).
219. A. R. Monahan, A. F. De Luca and A. T. Ward, *J. Org. Chem.*, **36,** 3838 (1971).
220. J. E. Kuder, *Tetrahedron*, **28,** 1973 (1972).
221. A. R. Monahan and J. B. Flannery, *Chem. Phys. Letters*, **17,** 510 (1972).
222. E. P. J. Elenius, *Acta Chem. Scand.*, **16,** 256 (1962).
223. H. Baba, *Bull. Chem. Soc. Japan*, **31,** 169 (1958).
224. D. A. Ibbitson, D. C. Colinese and C. W. Stone, *J. Chem. Soc.*, (*B*), 570 (1971).
225. G. J. Brealey and M. Kasha, *J. Amer. Chem. Soc.*, **77,** 4462 (1955).
226. A. Balasubramanian and C. N. R. Rao, *Spectrochim. Acta*, **18,** 1337 (1963).
227. H. Takahashi, K. Mamola and E. K. Plyler, *J. Mol. Spectroscopy*, **21,** 217 (1966).
228. R. L. Reeves, R. S. Kaiser, M. S. Maggio, E. A. Sylvestre and W. H. Lawton, *Can. J. Chem.*, **51,** 628 (1973).
229. V. V. Pal'chevskii, L. A. Polyanskaya, E. M. Semenova and E. E. Yudovich, *Opt. and Spectr.*, **26,** 558 (1969).
230. S. Singh and C. N. R. Rao, *J. Amer. Chem. Soc.*, **88,** 2142 (1966).
231. E. Fischer and Y. F. Frei, *J. Chem. Soc.*, 3159 (1959).
232. B. I. Stepanov and B. A. Korolev, *J. Gen. Chem. U.S.S.R.*, **38,** 1317 (1968).
233. M. V. Melamed, O. B. Shevchuk and C. S. Frankovskii, *J. Org. Chem. U.S.S.R.*, **6,** 1256 (1970).
234. B. A. Korolev, S. L. Polyakova and B. I. Stepanov, *J. Gen. Chem. U.S.S.R.*, **38,** 947 (1968).
235. L. I. Gen, M. Z. Yampol'skii and E. R. Oskotskaya, *J. Anal. Chem. U.S.S.R.*, **26,** 1875 (1971).
236. M. Z. Yampol'skii, V. D. Salikhov and T. M. Devyatova, *Tr. Kom. Anal. Khim., Akad. Nauk SSSR*, **17,** 107 (1969); *Chem. Abstr.*, **72,** 22568e (1970).
237. D. N. Shirogin, I. Y. Pavlenishvili and V. I. Kokachasvili, *Russ. J. Phys. Chem.*, **40,** 1613 (1966); G. D. Bagratishvilli and D. N. Shigorin, *Fiz. Sbornik L'vov. Univ.*, No. 3, 190 (1957); *Chem. Abstr.*, **55,** 16148h (1961).
237a. Y. M. Dedkov and A. V. Kotov, *Zh. Prikl. Spektrosk.*, **14,** 431 (1971).
238. L. W. Reeves, *Can. J. Chem.*, **38,** 748 (1960).
239. H. Kamada, S. Toda and C. Nagata, *Nippon Kagaku Zasshi*, **88,** 24 (1967); *Chem. Abstr.*, **66,** 120585d (1967); N. V. Startseva, C. Frankovskii, B. I. Ionin and N. G. Chebotova, *Zh. Org. Khim.*, **7,** 337 (1971).
240. J. N. Ospenson, *Acta Chem. Scand.*, **4,** 1351 (1951).
241. G. Gabor and E. Fischer, *J. Phys. Chem.*, **66,** 2478 (1962).
242. E. Fischer and Y. F. Frei, *J. Chem. Soc.*, 3159 (1959).
243. J. Stuehr and M. C. Rose, *J. Amer. Chem. Soc.*, **93,** 4350 (1971); W. H.

Inskeep, D. L. Jones, W. F. Silvfast and E. M. Eying, *Proc. Nat. Acad. Sci. U.S.*, **59,** 1027 (1968).
244. F. D. Saeva, *J. Org. Chem.*, **36,** 3842 (1971).
245. L. Skulski and W. Waclawek, *Bull. Acad. Pol. Sci., Ser. Chim.*, **19,** 277 (1971).
246. J. Ladik, A. Messmer and J. Redly, *Acta Chim. Acad. Sci. Hung.*, **38,** 393 (1963); *Chem. Abstr.*, **59,** 15146b (1963); B. Geza, L. Janos and A. Messmer, *Magy. Kem. Folyoirat*, **71,** 388 (1965); *Chem. Abstr.*, **64,** 569d (1966).
247. K. Antos and S. Kovac, *Chem. Zvesti.*, **16,** 187 (1962); *Chem. Abstr.*, **58,** 4050c (1963).
248. D. L. Ross and E. Reissner, *J. Org. Chem.*, **31,** 2571 (1966).
249. S. Toda and M. Koizumi, *Nippon Kagaku Zasshi*, **81,** 1292 (1960); *Chem. Abstr.*, **56,** 567a (1962); **88,** 18 (1967); *Chem. Abstr.*, **66,** 120417a (1967).
250. Y. Matsunaga, *Bull. Chem. Soc. Japan*, **44,** 878 (1971).
251. C. Dehari, Y. Matsunaga and K. Tani, *Bull. Chem. Soc. Japan*, **43,** 3404 (1970).
252. Y. Matsunaga and N. Miyajima, *Bull. Chem. Soc. Japan*, **45,** 3345 (1972).
253. E. Sawicki, *J. Org. Chem.*, **22,** 743 (1957).
254. R. Mecke and D. Schmähl, *Naturwissenschaften*, **42,** 153 (1955).
255. T. Takahashi, *J. Chem. Soc. Japan, Pure Chem. Sec.*, **74,** 673 (1953); *Chem. Abstr.*, **48,** 3570c (1953).
256. H. Druckrey, D. Schmähl and P. Danneberg, *Naturwissenschaften*, **39,** 393 (1952).
257. R. N. Nurmukhametov, D. N. Shigorin, Y. I. Kozlov and V. A. Puchkov, *Opt. and Spectr.*, **11,** 327 (1961).
258. J. Franc and M. Wurst, *Coll. Czech. Chem. Comm.*, **25,** 657 (1960).
259. H. C. Mez, *Ber. Bunsenges. Physik. Chem.*, **72,** 389 (1968).
260. N. Kulevsky and L. Lewis, *J. Phys. Chem.*, **23,** 3502 (1972).
261. V. F. Gladkova and Yu. D. Kondrashev, *Soviet Physics Cryst.*, **13,** 936 (1969).
262. Yu. A. Omel'chenko and Yu. D. Kondrashev, *Soviet Physics-Cryst.*, **12,** 359 (1967).
263. Yu. D. Kondrashev, *Soviet Physics-Cryst.*, **6,** 413 (1962).
264. Yu. D. Kondrashev, *Soviet Physics-Cryst.*, **9,** 323 (1964).
265. Yu. A. Omel'chenko and Yu. D. Kondrashev, *Soviet Physics-Cryst.*, **17,** 837 (1973).
266. V. O. Lukashevich and E. S. Lisitsyna, *Dokl. Akad. Nauk SSSR*, **160,** 129 (1965); *Chem. Abstr.*, **62,** 11711e (1965).
267. T. G. Heafield and L. Hunter, *J. Chem. Soc.*, 420 (1942).
268. L. Hunter, *J. Chem. Soc.*, 320 (1937).
269. L. K. Dyall, *Australian J. Chem.*, **13,** 230 (1960).
270. D. G. O'Sullivan, *J. Chem. Soc.*, 3653 (1960).
271. V. C. Farmer, R. L. Hardie and R. H. Thompson, in *Hydrogen Bonding* (Ljubljana Symposium, 1957) (Ed. D. Hadzi and H. W. Thompson), Pergamon Press, London, 1959, pp. 475–482.
272. N. L. Miller and E. C. Wagner, *J. Amer. Chem. Soc.*, **76,** 1847 (1954).
273. K. Utvary, *Monatsh. Chem.*, **99,** 2324 (1968).
274. M. Elkins and L. Hunter, *J. Chem. Soc.*, 1346 (1938).

275. T. Mitsuhashi, Y. Osamuru and O. Simamura, *Tetrahedron Letters*, 2593 (1965).
276. L. G. Abakumov, G. A. Abakumov and A. Domrachev, *Doklady Akad. Nauk SSSR*, **197**, 1315 (1971).
277. A. Chakravorty, B. Behera and P. S. Zacharias, *Inorg. Chim. Acta*, **2**, 85 (1968).
278. B. Behera and P. S. Zacharias, *Spectrochim. Acta*, **A27**, 2273 (1971).
279. T. Axenrod, P. S. Pregosin and G. W. A. Milne, *Tetrahedron Letters*, 5293 (1968).
280. L. G. Sillen and A. E. Martell, *Stability Constants of Metal–Ion Complexes*, Special Publ. No. 17, The Chemical Society, London, 1964; Supplement No. 1, Special Publ. No. 25, The Chemical Society, London, 1971.
281. F. Bottomley, *Quart. Rev.*, **24**, 617 (1970).
282. S. Yamada, *Coord. Chem. Rev.*, **1**, 415 (1966); E. Sinn and C. M. Harris, *Coord. Chem. Rev.*, **4**, 391 (1969); R. M. Holm, G. W. Everett and A. Chakravorty, *Progr. Inorg. Chem.*, **7**, 83 (1966).
283. M. Kilner, in *Advances in Organometallic Chemistry*, Vol. 10 (Ed. F. G. A. Stone and R. West), Academic Press, N.Y., 1972, pp. 115–198.
284. N. F. Curtis, *Coord. Chem. Rev.*, **3**, 3 (1968).
285. L. F. Lindoy and S. E. Livingstone, *Coord. Chem. Rev.*, **2**, 173 (1967).
286. L. E. Godycki and R. E. Rundle, *Acta Cryst.*, **6**, 487 (1953); D. E. Williams, G. Wohlauer and R. E. Rundle, *J. Amer. Chem. Soc.*, **81**, 755 (1959).
287. R. E. Rundle and M. Parasol, *J. Chem. Phys.*, **20**, 1487 (1952).
288. E. J. Gonzales and H. B. Jonassen, *J. Inorg. Nucl. Chem.*, **24**, 1595 (1962).
289. F. Umland, B. K. Podder and H. Stegemeyer, *Z. Anal. Chem.*, **216**, 125 (1966).
290. J. Goubeau and E. Ricker, *Z. Anorg. Allgem. Chem.*, **310**, 123 (1961).
291. S. Hermanek, J. Plesek and V. Gregor, *Coll. Czech. Chem. Comm.*, **31**, 1281 (1966).
292. C. Belinski, G. Francais, C. Horney and F. X. Lalau-Keraly, *Compt. Rend.*, **259**, 3737 (1964).
293. R. J. Baumgarten and M. C. Henry, *J. Org. Chem.*, **29**, 3400 (1964).
294. W. G. Paterson and M. Onyszchuk, *Can. J. Chem.*, **29**, 2324 (1961).
295. L. K. Peterson and G. L. Wilson, *Can. J. Chem.*, **49**, 3171 (1971).
296. R. C. Aggarwal and M. Onyszchuk, *Can. J. Chem.*, **41**, 876 (1963).
297. M. Graffenil, J.-F. Labarre, C. Leibovici and E. Taillandier, *J. Mol. Structure*, **15**, 367 (1973).
298. V. Gutmann and U. Mayer, *Monatsh. Chem.*, **98**, 294 (1967).
299. A. L. Balch and D. Petridis, *Inorg. Chem.*, **8**, 2247 (1969).
300. N. B. Kupletskaya, A. A. Nilson, L. A. Kazitsyna and O. A. Reutov, *Vestn. Mosk. Univ. Khim.*, **24**, 69 (1969); *Chem. Abstr.*, **71**, 25164k (1969).
301. V. Gutmann and A. Steininger, *Monatsh. Chem.*, **96**, 1173 (1965).
302. R. C. Aggarwal and R. C. Makhija, *Z. Anorg. Allg. Chem.*, **349**, 101 (1966).
303. R. C. Aggarwal and S. K. Singh, *Ind. J. Chem.*, **7**, 924 (1969).
304. R. C. Aggarwal and D. S. Bhusri, *Ind. J. Chem.*, **4**, 19 (1966).
305. S. Prasad and L. P. Pandey, *J. Proc. Inst. Chem.*, (India), **36**, 293 (1964); *Chem. Abstr.*, **63**, 5545g (1965).

CHAPTER **8**

Directing and activating effects

D. A. R. Happer and J. Vaughan

University of Canterbury, Christchurch, New Zealand

I. INTRODUCTION

The three functional groups covered in this volume, despite their common features of a pair of linked nitrogen atoms and a relative ease of interconversion, do not easily lend themselves to a profitable comparative account of electronic effects. We have preferred to discuss each group in a separate section.

All three groups are best described as bridging groups and, for any one of them, the rest of the molecule containing it can be thought of as two moieties, one on either side of the bridge. The moieties flanking the bridging group are most commonly alkyl or aryl groups. The simplest structures, in which one of the moieties is hydrogen, are not considered in this review because in the case of the azo- and azoxy groups they are too unstable to have been the subject of relevant study, and the hydrazino group is better regarded as chemically separate from the hydrazo grouping.

None of the functional groups covered in this chapter has been favoured by extensive studies of its activating and directing effects. A consequent problem of the reviewer, not met so frequently when dealing with more common functional groups, is that of the interesting result for which subsequent confirmation is absent. Furthermore, there are early isolated results in conflict with later results which might be equally isolated. The difficulty in each type of problem lies, of course, not in reporting the published results but in the confidence one has in them.

Where suitable quantitative data are available the electronic effects of the functional groups are discussed in terms of Hammett substituent constants. Of the various scales available[1], those used in this chapter are the σ^0, σ^- and σ^+. The σ^0 scale has been preferred to the more commonly used σ scale (based on benzoic acid ionizations) because the latter contains a significant resonance component and all groups under discussion are, potentially at least, conjugatively-releasing (+R) groups.

II. DIRECTING AND ACTIVATING EFFECTS OF THE AZO GROUP

A. General Considerations

The most commonly encountered azo compounds are the azobenzenes, and these have been the most widely studied. Consequently we are in a much better position to comment on the directing and activating effects of the phenylazo- and other arylazo-substituents than on aliphatic azo groups. The bulk of the quantitative information available on substituent effects of azo groups is, in fact, derived from measurements of the effect of *meta*- and *para*-phenylazo- or arylazo groups on reactions of aromatic side-chains. For the present survey, the results of such studies have been interpreted, where possible, in terms of Hammett substituent constants.

There are some indications (*vide infra*) that azo-containing groups may behave as +R as well as −R groups, depending on the electronic requirements of the reaction. Unfortunately, lack of information on substituent effect of alkylazo groups prevents us from deciding to what extent this flexibility is due to (*i*) the azo group itself and (*ii*) interaction between two aromatic systems, with the azo group operating merely as an electron-conducting bridge.

Azo compounds can exist as two distinct geometrical isomers, *cis*- and *trans*-. The latter are much more stable and, in subsequent discussion, an azo compound for which the isomeric nature is unspecified should be taken as the *trans*-isomer.

B. Reactions of Aromatic Side-chains

In this section we consider the effects of *meta*- and *para*-arylazo substituents on side-chain reactivities. The arylazo group is sufficiently polar for proximity effects to become significant in the *ortho*-position, where its inherent behaviour may be masked by field effects and by even more direct interaction with the side-chain.

The amount of available information is not large but is adequate as a qualitative index to behaviour. It is noteworthy, however, that many of the reactions studied are those of amines, phenols or their derivatives with the arylazo in the *para*-position; this is presumably because such compounds are readily accessible through diazonium coupling reactions. Data are conventionally discussed in terms of substituent constants although, as will be seen, the use of discrete σ^0, σ^+ and σ^- values may not be well suited to these groups.

Furthermore, for reasons which will emerge later, the effects associated with substituted phenylazo groups are discussed in terms of transmission coefficients of the azo linkage rather than substituent constants of the individual groups. The phenylazo group is therefore considered first, and separate from other arylazo groups.

1. The phenylazo substituent

Until relatively recently the only value listed for the substituent constant of this group was that of +0·64 given by Hammett for the *para*-phenylazo group[2]. It was based on the rate of reaction of phenolate ions with ethylene oxide and propylene oxide[3], and was presumably a σ^- value. During the last ten years or so, however, a number of other values have been calculated. These appear to form a body of reliable data and, while we can find no values based on series that would give true σ^0 values, there have been three different studies of benzoic acid ionizations. The results are in reasonable agreement. Syz and Zollinger[4] and Exner and Lakomy[5], have calculated σ values based on the ionization of benzoic acids in 80% methyl cellosolve. The values for σ_m are +0·299 and +0·29, and for σ_p the figures obtained are +0·349 and +0·33 respectively. In addition, Ryan and Humffray[6] report a σ_p of +0·31, based on the same equilibrium in 50% ethanol. The values for σ_m should be very close to σ_m^0 and they agree well with other reported σ_m values: +0·248 (ionization of phenols in 50% ethanol)[4], +0·281 (basicities of anilines in 50% ethanol)[4], +0·28 (rates of diazonium coupling in water)[7] and +0·32 (quarter-wave potentials of phenylferrocenes in acetonitrile)[8].

The values for σ_p may not represent true σ^0 values because, as has been said, there is some indication that the phenylazo group can behave as a +R group. Variation is therefore to be expected* and the reported values are lower limits for σ_p^0; the magnitude of the true value will be greater, by some indeterminate amount. Ryan and Humffray report σ_p values of +0·39 and +0·37, derived from alkaline hydrolysis (in 60% aqueous acetone) of aryl acetates[6] and benzoates[9] respectively. They state that the results are best correlated by a Yukawa–Tsuno relationship, and if we take their r values together with a reasonable figure for σ_p^- then their data point to a value for σ_p^0 of around +0·30. However, Fischer and co-workers[10] have found that, for the alkaline hydrolysis of aryl benzoates in 60% ethanol (a larger number of substituents but not including the azo group), a much smaller r value is needed and on this basis a better figure for σ_p is around +0·35, with an upper limit of +0·37 to +0·39.

Finally, Lifschits and co-workers[11] calculated a value for σ_p of +0·44 after measurements on the basicities of some polymethine dyes. This may have some σ^- component but, on the basis of values obtained for other −R substituents, it should not be large. In view of the other values, however, this σ_p^0 value seems to be a little high. The weight of evidence would imply that σ_p^0 might not be very different from σ_p^0 based on benzoic acid ionization, i.e. around +0·34.

While it is common for the *para*-phenylazo group to give values intermediate between σ^0 and σ^-, it is not easy to decide whether the variations are to be attributed to solvent effects, to reaction type, or to normal scatter (in all the reactions studied, ρ values have been high). The highest reported

* Yukawa and co-workers[12] report an r value of 0·27 for benzoic acid ionizations.

values have been close to +0·7, the individual figures being +0·695 from ionization of phenols in 50% ethanol[4], +0·67 to +0·68 from ^{19}F chemical shifts for *p*-fluorobenzenes[13], and 0·672 from reaction of 1-chloro-2-nitro-4-*x*-benzenes with sodium methoxide/methanol[14]. Lower values include +0·613 (ionization of anilines in 50% ethanol)[4], +0·56 (chemical shifts for —OH proton of phenols in DMSO)[15], and +0·55 (reaction of 1-chloro-2-nitro-4-*x*-benzenes with piperidine in DMSO)[16]. These results are just representative; many others are available. Some, however, are merely duplicates of those mentioned (e.g., phenol ionization) and others (most) are based on data or correlations considered to be too unreliable for our present purpose. Overall, the evidence suggests a σ_p^- value of around +0·68 for normal reactions in protic solvents, but occasionally lower values may be needed, particularly in non-protic solvents.

The phenylazo substituent has the potential to behave as a +R group, and qualitative and semi-quantitative data on electrophilic substitution in azobenzene indicate a reactivity towards electrophiles which is comparable to that of benzene, i.e. $\sigma_p^+ \approx 0$. However, the lowest σ_p value so far reported is for the diazonium coupling reaction with R-acid[7] ($\sigma_p = +0{\cdot}190$) and this value is based on very few Hammett points. Furthermore, if we accept the suggested *r* value of Yukawa[12] for the ionization of benzoic acids (0·27) and an upper limit for σ^0 of +0·37–0·39, then this would place a lower limit to σ_p^+ of around +0·2, a figure considerably different from that predicted on the basis of substitution rates. We favour a figure for σ_p^+ of greater than 0·2 in the absence of further evidence to the contrary.

Alternative interpretations of the substitution results are discussed in a later section; relatively few substituents are capable of behaving as both +R and −R groups and, of them, the phenylazo group is the only one that has a positive value for σ_R^0 (the reported value for σ_I is +0·25)[17].

2. The arylazo substituent

Electronic effects for arylazo groups may be examined in two ways. The first, and more obvious method, is to consider each arylazo group as a separate substituent, to evaluate its substituent constant by normal methods, and to discuss this value in terms of structure. The second, and more generally useful approach, is to attempt to interpret the observed electronic effect of an arylazo group in terms of a combination of the electronic effect of the phenylazo group as a reference point, and the introduced electronic effect of the further substituent in the phenylazo group. The advantage of the latter approach is that it permits the estimation of substituent constants for arylazo groups for which experimental data are not available.

Consider the general aromatic system (**1**) which represents a typical

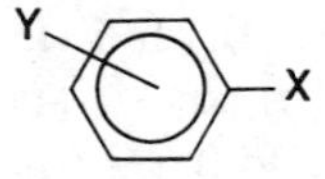

(1)

reaction series used for the study of linear free-energy relationships. Here X is presumed to be the reacting side-chain and the variable substituent is Y (more strictly, Y—C_6H_4—). For reactions of azobenzene derivatives, Y will be the Ar—N=N— group and an examination of electronic effects in terms of the first approach would lead to individual substituent constants for each arylazo group.

The second approach to the azobenzene series assumes that the azo linkage is not part of the substituent, so that the —N=N—C_6H_4— group is regarded as the central system which transfers the effect of a variable aryl group to the reaction site in X. We adopt this treatment, which is the more frequently met in the literature. The efficiency of the transfer process, involving —N=N—C_6H_4— is expressed as a transmission coefficient π' which is defined as the ratio of the Hammett reaction constants (ρ) for the 'Y = arylazo' series and the corresponding 'Y = aryl' series.

In theory we have to consider transmission through three distinct systems, although in practice almost all studies to date have been on systems of type **4**.

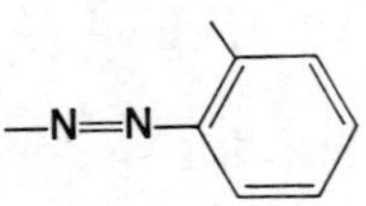

(2)

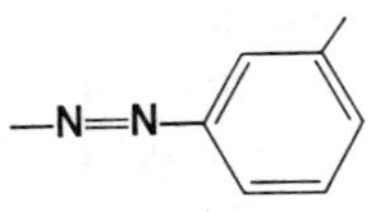

(3)

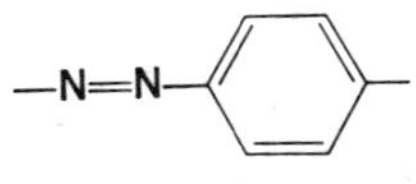

(4)

In the *ortho*-azophenylene system (**2**), complications will arise if the reaction site can interact more directly with the phenylazo group. The only reactions studied so far are the ionizations of 2-arylazophenols. Korolev and Stepanov[18, 19] have proposed, on the basis of pK_a values of 2-arylazo-4-methylphenols (in 50% ethanol) that the electronic effect of the aryl group is transmitted through two competing pathways: the normal one, via the azo group and the phenol ring, and also a direct one from an azo nitrogen to the phenolic hydrogen.

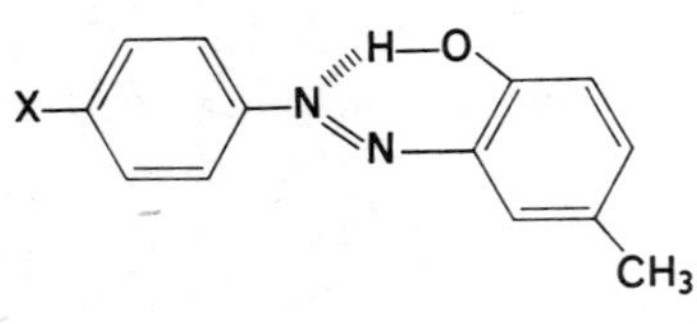

They have attempted to resolve the combined ρ values of +1·163 and have arrived at +0·396 for the normal transmission and +0·757 for the more direct interaction. The figure of +0·396 is less than the reaction constant determined by them for the 4-arylazophenols ($\rho = +0{\cdot}62$) and, since the distance between the substituent and the reaction site is actually less for the 2-arylazo compounds, it is perhaps surprising to find that ρ is smaller than for the 4-azo series. We have re-examined the data of Korolev and Stepanov and have placed a different interpretation on it. Their analysis was based on the observation that for 2-arylazophenols, full σ^- values were not needed for satisfactory correlation of strong *para*-R substituents, whereas such values gave good correlations in the case of 4-arylazophenols. The observation of intermediate substituent constants for strong *para*-R groups was attributed by these workers to two competing transmission pathways, one following a σ^0 correlation and the other σ^-. If, however, one considers, not differences in values, but $\Delta\Delta G$ values for these substituents, then it becomes clear that the extent of resonance interaction, calculated on this basis, is about the same for the two systems. Accordingly, in the absence of better evidence than that provided by Korolev and Stepanov, we are reluctant to accept any significant, direct electron transfer between azo and hydroxy groups. Socha and co-workers measured the pK_as of a series of 5-methyl and 5-chloro-2-hydroxyazobenzenes in the same solvent and obtained fairly similar ρ values: +1·223 for the 5-methyl- and +1·316 for the 5-chloro-compounds respectively[20, 21].

A ρ value of around +1·2 for the 2-arylazophenol system coupled with a figure of +2·67 for phenols[22] gives a value for the transmission coefficient π' of about 0·45. This figure is high, and is comparable with that for the ethylene bridge. Cohen and Jones[22] have pointed out, however, that reaction constants for *ortho*-substituted compounds often show differences from those of unsubstituted compounds caused by factors such as differing solvation or intermolecular hydrogen bonding. Since the latter phenomenon is likely to occur in the 2-arylazo-phenol system, the reference ρ value of +2·67, which is based on *meta*- and *para*-substituted phenols, may not be valid. From the reported pK_a values for matched pairs of 5-chloro- and 5-methyl-2-arylazophenols, it appears that the contribution to acidity from hydrogen bonding is about 0·2 pK units and that the ρ value used for the calculation of π' could be about 10% low. This would cut the magnitude of π' to about 0·4, a figure that is still appreciably greater than that for the 4-arylazophenols*.

* Socha, Horska and Vecera[23] have since shown that, for the hydrolysis of the arylazo acetates, $\rho_o \approx \rho_p$.

From a slightly different viewpoint, the abnormally high value might be the result of a tautomeric equilibrium.

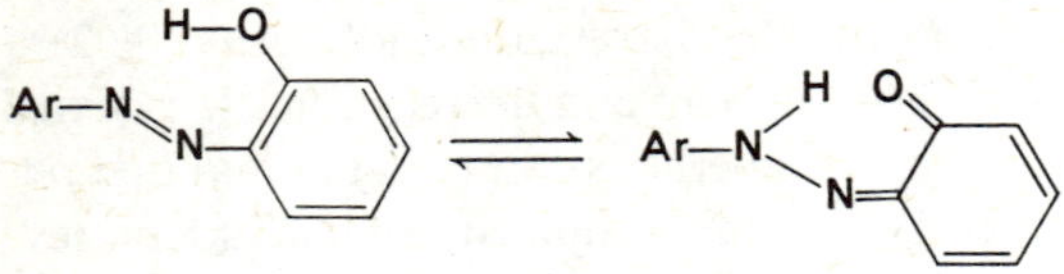

The dienone form is here stabilized much more than in the *para*-arylazophenols because of hydrogen bonding:

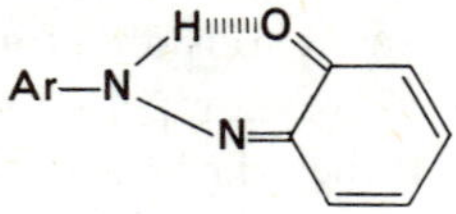

Mechanistically, however, this suggestion and that of Korolev and Stepanov are equivalent for intramolecular proton transfer.

In contrast to the sparse data available on the *ortho*-azophenyl group, there is much more information on transmission across the *para*-azophenyl group. There is, unfortunately, some scatter in the recorded results and this might be due either to variations in π' from reaction to reaction, or to insufficiently reliable data. The major problem in interpretation arises from a range of substituents which is not good for our purposes. Most reaction series are based on 4,4′-disubstituted azobenzenes and most of the reactions are those of phenols and amines; there is therefore uncertainty about the correct substituent constant (σ^0, σ^- or something intermediate) for *para*-R groups and reaction constants may be unreliable.

Jaffe[24] estimated a value for π' of 0·14, based on molecular orbital calculations, which was in good agreement with the limited experimental data available to him. Reaction series reported to have transmission coefficients of this order include the basicities of 4-arylazo-*N*,*N*-dimethylanilines[25] (0·13), ionization of 4-arylazophenols in 20% ethanol[26] (0·16) and the quarter-wave potentials of 4-arylazophenylferrocenes[8] (0·13). None of these series, however, covered a satisfactory range of substituents, bearing in mind the more recent tendency to base Hammett correlations on 'well-behaved' substituents. From this point of view, more reliable data come from the work of Socha and his colleagues[20] on the ionization of 4-arylazophenols in 50% ethanol, and on the chemical shifts of the hydroxylic hydrogen of the same series in DMSO[15]. They report considerably higher values for the transmission coefficients: 0·275 and 0·26 respectively. Frankovskii and Melamed[27] report a π' figure of 0·333 for the phenylazo-phenol series in 60% dioxan–water and a closely similar value of 0·346 for the *p*-(*o*-chlorophenylazo) phenol series under the same conditions. In a subsequent paper Frankovskii reports a value for π' of 0·290 for n.m.r. chemical shifts in DMSO[28], a figure which agrees well with that of Socha.

The difference, for phenol ionization, between the two studies may be associated with the substituents (and substituent constants) chosen by the investigators. Frankovskii's correlation uses Hammett σ values and includes *para*- −R substituents which may require exalted values for satisfactory correlation. If, however, we base our regression line on non-interacting *meta*-substituents, the results are much more in keeping with the values of Socha's group.

The effect of *para*-arylazo groups on the i.r. stretching frequencies of phenols in carbon tetrachloride has also been investigated. The results of Catchpole, Foster and Holden[29] suggest a transmission coefficient of ca. 0·6, but the choice of substituents is not good, and the precision of the measurements is low in terms of the overall magnitude of the observed effects. The same authors also looked at stretching frequencies, under the same conditions of a few 3,3′5,5′-tetramethylazophenols. For these, the transmission coefficient is approximately halved and this effect is attributed to partial steric inhibition of resonance.

There is a recent report by Colinese, Ibbitson and Stone[30] of ν_{O-H} values for a series of 4-X-4′-hydroxyazobenzenes and transmission coefficients around 0·3 may be derived from the regression lines. However, the available points form a set that is far from ideal and it is to be noted that Catchpole attributed a σ^- constant to the 4-nitro substituent. Our calculation of π' for these data is based on the use of σ^0 for the 4-nitro substituent and, while the use of this value for the later work would give more compatible values, in both cases the derived transmission coefficients must be regarded as approximate only.

Catchpole also reports pK_a data on the phenols[29]. Insufficient information is given for the calculation of transmission coefficients but the presence of the methyl groups in the tetramethyl compound certainly appears to reduce the magnitude of π' to about one-third.

In addition to equilibrium studies, a few kinetic investigations have been carried out. Hashida and co-workers[7] have measured the rates of coupling with R-acid of a number of *p*-substituted *p*-arylazobenzene diazonium ions; the value of π' is 0·198, but the absence of *meta*-substituents reduces the reliability of this figure. Socha, Horska and Vecera have measured the rates of hydrolysis of 4-arylazophenyl acetates and of 2-arylazo-5-methylphenyl acetates[23]. Unfortunately, these measurements were carried out in 50% ethanol and the only data available on arylacetates are from hydrolysis in

60% acetone. The dielectric constants of the two solvents are not too different, and the reaction constant for 50% ethanol should be not far from that in 60% acetone. Using Ryan and Humffray's ρ value of +1·69[6], this gives $\pi' = 0{\cdot}22$ for both the *ortho-* and *para*-arylazo systems, which is appreciably lower than that for phenol ionizations. On the other hand, measurements by Socha on rates of halide displacement by piperidine, in benzene[31] and in DMSO[16], gave values for π' of 0·45 and 0·30 respectively.

Even the most reliable data, therefore, do not lead to acceptably constant values for π', but the results do fit a pattern. The reactions that force the highest negative charge on the aromatic system (nucleophilic aromatic substitution) have the highest values for π', while reactions that involve very little charge, such as the saponification of arylacetates*, have low values. Phenol ionizations, placing an intermediate charge on the ring by comparison with the others, have an intermediate π'. High values for the transmission coefficient are therefore consistent, in the present context, with situations in which the azo group is required to carry high proportions of charge.

Of other bridging groups of similar type to the —N=N—C_6H_4— group the closest are the —C_6H_4—, —CH=CH—C_6H_4—, —N=CH—C_6H_4— and —CH=N—C_6H_4— linkages. For the —C_6H_4— group (the least relevant of these) Jaffe[24] reports a calculated value for π' of 0·177, a figure lower than his experimental figure of 0·303. In the light of our discussion on the azophenylene group, this apparent discrepancy may not be important and the 0·3 figure has received experimental confirmation from other workers[32].

Transmission effects in the stilbene series have been investigated by Veschambre and Kergomard in ionization studies of amines and phenols in 50% ethanol[33]. Their data yield π' values of 0·14 and 0·15 respectively (together with a π' value of 0·12 for the —C≡C—C_6H_4— group). They further estimate on the basis of data of Syz and Zollinger[34], that π' is around 0·16 for the dimethylaniline series. These figures are reasonably consistent with the data for the phenylene bridge if we assume an attenuation of ca. 0·5[24] associated with the —CH=CH— linkage between the two rings.

Minkin and co-workers have calculated π' values, for transmission through the azomethine bridge, from ionization measurements, in 98% methanol, on phenols and carboxylic acids[35]. They report values of 0·184 and −0·256 for the *p*-N=CH—C_6H_4— group (phenols and carboxylic acids) and 0·148 and 0·185 for the *p*-CH=N—C_6H_4— bridge. The change in sign for the anil bridge with carboxylic acid ionization is noteworthy but

* Socha's data require a Yukawa–Tsuno *r* factor of less than 0·1.

is not observed in 80% methanol. Bekarek and co-workers have reported values based on n.m.r. chemical shifts of the —OH of *p*-substituted phenols in DMSO[36]; Their data give $\pi' = 0{\cdot}180$ (—N═CH—C_6H_4—) and 0·133 (—CH═N—C_6H_4—). While these reverse Minkin's findings, the actual differences between the two sets of results are not great.

In the interpretation of these results, it is significant that aromatic Schiff's bases are not planar[37], the angle between the aromatic nuclei being about 55°. In contrast with azo systems, therefore, the possibility of direct resonance interaction between the two aromatic nuclei may be ignored.

The π' value for the —CH═N—C_6H_4— system (0·148 in phenol ionization) might appear to be low, on the basis of similarity between this group and the azo group. However, Ryan and Humffray[6] report that σ_p^- for the C_6H_5—CH═N— substituent is only 0·22, so that, even though it is a –R group ($\sigma_p = 0{\cdot}0$ for benzoic acid ionization), the nitrogen would not be expected to carry much negative charge. On the other hand, for the C_6H_5—N═CH— group these authors report $\sigma_p^- = +0{\cdot}54$ and $\sigma_p = +0{\cdot}42$; indeed, the —N═CH—C_6H_4— linkage does have a transmission coefficient (phenol ionization) which is intermediate between those for —N═N—C_6H_4— and —CH═N—C_6H_4—. This suggests that, as predicted it is the nitrogen more distant from the ring that carries the bulk of the negative charge.

The transmission coefficients of the —CH═N—C_6H_4— and —CH═CH—C_6H_4— groups appear to be comparable in magnitude, suggesting that what the azomethine group gains from the higher electronegativity of the imino-nitrogen, it loses as a result of a non-planar system. Coplanarity of the system might therefore be a contributing factor to the apparently anomalously high transmission coefficient of the azophenyl linkage. Confirmatory evidence for this suggestion is presented in the following section.

3. Transmission of resonance effects through the azo group

Structural studies using X-ray diffraction have shown that, although *trans*-azobenzene is approximately planar in the solid state, the C—N and N═N bonds are about the same length as those in azomethane[38]. We can assume that limiting structures such as **5** and **6** are relatively unimportant.

While bond length determination is not a particularly sensitive probe for resonance interactions, the conclusion is in keeping with the expectation that, in such a case, dipolar canonical forms would make only minor contributions. That some interaction does exist is indicated by the coplanarity of the two benzene rings and the azo group; in contrast, benzylidine

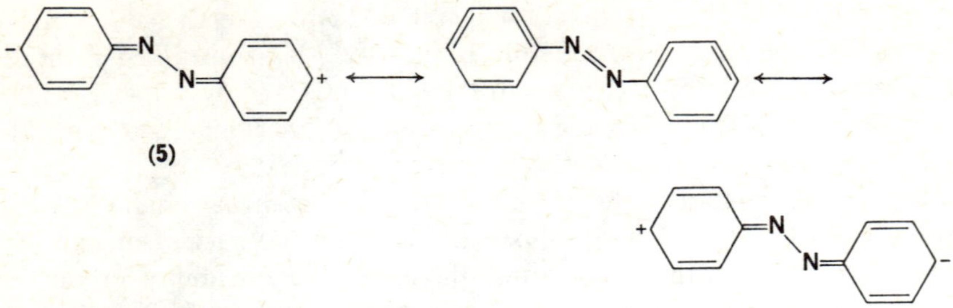

(5)

(6)

aniline, a compound that should have similar packing requirements, is not coplanar, the plane of the anilino ring lying at an angle of 55° to that of the other[37]. Presumably the anilino ring interacts with the nitrogen lone pair rather than the π electron system of the double bond. Such is apparently not the case with azobenzene.

A relatively sensitive method of detection of resonance interaction across the N=N bond makes use of the Hammett equation. A lack of correlation often indicates direct resonance interaction between the substituent group and the reaction site. On a quantitative basis Wepster and co-workers[39] have proposed that comparisons of such interaction are best based, not on the magnitude of $(\bar{\sigma} - \sigma)$, but on that of $\rho(\bar{\sigma} - \sigma)$ because this term is proportional to the free energy of interaction.

In the case of azobenzenes, it is possible to draw resonance forms of a double quinonoid type when we have +R and −R substituents in the 4- and 4′ positions, e.g.;

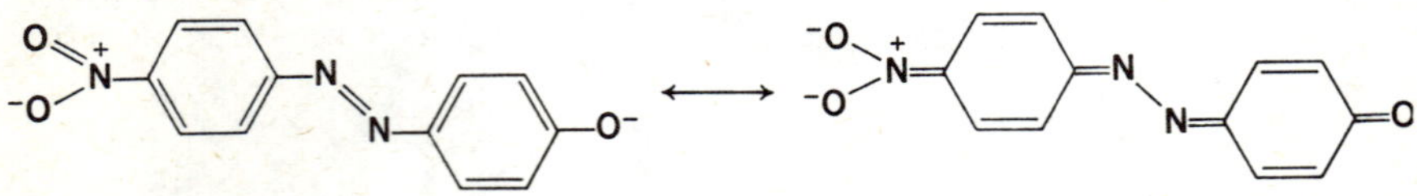

Unfortunately, because of a restricted range of substituents, it is often difficult to extract from the literature a ρ value if all *para*-R substituents are omitted, but these must necessarily be discarded when analysing amine and phenol reactions, from which most of the data are drawn.

The most precise data available are on the correlation of chemical shifts of phenols in DMSO[15,28] and on the ionization of phenols both in 50% ethanol and in 60% dioxan[18–21]. For the first group of results, the derived σ values for the substituents *p*-CN and *p*-Ac are indistinguishable from their σ^0 values. This is not the case for phenol ionization and *para*- −R substituents are more electron-withdrawing than would be predicted on the basis of σ^0 values, although they do not display full σ^- behaviour. Socha, Horska and Vecera[20] have calculated that for the 4-arylazophenols $r = 0{\cdot}487$ and for the 2-arylazo-4-cresols $r = 0{\cdot}286$. Direct interaction between the substituent in one ring and the reaction site in the other is therefore only about 10–15% that in *para*-substituted benzenes.

Veschambre and co-workers[33] have looked closely at stilbene derivatives in both ionization of phenols and protonation of anilines. Their data show that normal substituent constants give satisfactory correlation for the 4-amino-stilbenes but, for the phenols, better correlation is obtained with σ^- values. Syz and Zollinger have measured the pK_as of the dimethylamino-stilbenes[34] and for *para*- –R substituents their data are consistent with values intermediate between σ^0 and σ^-: a satisfying result. Furthermore, if we make allowance for the lower transmission coefficients prevailing in the stilbene series, by comparison with phenol ionization, the resonance interactions between substituent and reaction site are about the same as for azobenzenes.

In conclusion, therefore, resonance interaction between the two aromatic nuclei does exist but is rather weak. This is not unexpected, and calculations made in connection with studies on annulene aromaticity demonstrate that extended conjugation need not be favoured energetically[40]. Comparison of azophenols with stilbenols has shown that where strong +R and –R substituents are in the 4- and 4′-positions, then the azo group is behaving rather like a vinyl group and mainly as a neutral bridge. However, when there is less interaction between substituent and side-chain, the higher π' associated with the azo bridge indicates that the azo nitrogen farther from the reaction site is a more effective charge-carrier than a vinyl carbon.

C. *Electrophilic Aromatic Substitution Reactions in Azo Compounds*

1. General considerations

The accepted mechanism for electrophilic aromatic substitution involves the rate-determining formation of a cyclohexadienonium intermediate (the Wheland intermediate) followed by fast loss of a proton to give the substitution product[41]:

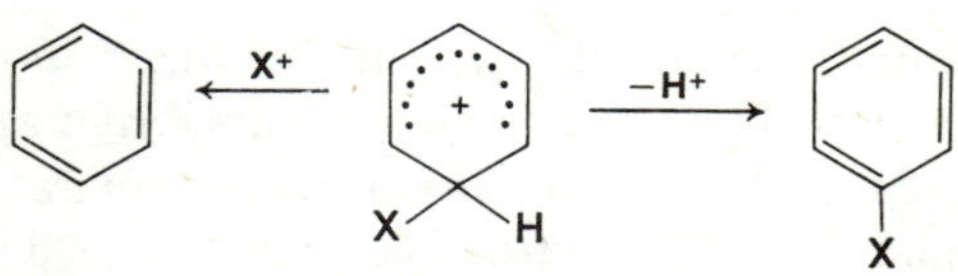

The presence of +R groups *ortho-* or *para-* to the site of attack leads to enhanced reaction rates because of the ability of such substituents to stabilize the transition state for formation of the intermediate. The phenylazo substituent behaves, in most of its reactions, as an electron-withdrawing substituent and has a 'normal' σ_p of about +0·34 based on benzoic acid ionization (see Section II, **B**, 1.). It has been reported, however, that the rate of chlorination of azobenzene in aqueous acetic acid is about 4·6 times that of benzene[42], a figure which implies that, if the mechanism is normal, then $\sigma_p^+ \approx -0{\cdot}1$. The assumption of normality of the substitution mechanism is a doubtful one, as will be seen later.

If we accept, provisionally, that the phenylazo substituent can behave as a +R group, then resonance stabilization of the benzenonium ion intermediate:

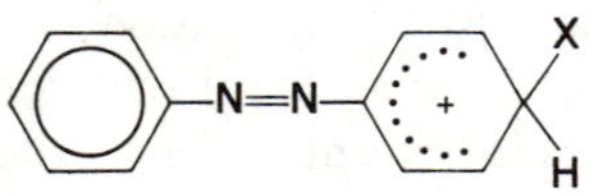

may involve **7** or **8**:

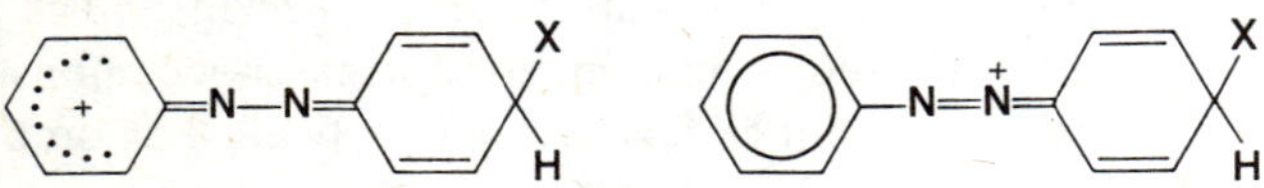

(7) (8)

Geometrical considerations make the two forms mutually exclusive. The involvement of structure **8** implies a lack of coplanarity of the two rings and in azomethines, which are known to be non-planar[37], σ_p for the C_6H_5—CH=N— group is reported by Ryan and Humffray[6] to be about 0·0 (based on benzoic acid ionization). This would imply a σ_p^+ value below zero for this non-planar system.

In seeking a contrast, we consider that the least ambiguous findings are

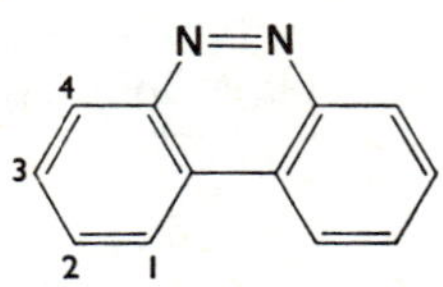

those of Corbett and Holt on benzo[*c*]cinnoline which, although a *cis* azo compound, is roughly planar and should resemble *trans* azobenzene in its behaviour towards electrophiles. Bromination requires a mixture of molecular bromine and silver nitrate in sulphuric acid and the major product is the 4-isomer[43]. In the absence of unambiguous results, it might well be, therefore, that any electron-releasing process involving the azo group will be one which depends on the unshared electrons of the nitrogen and is independent of orbital overlap with the second aromatic nucleus.

2. Reactions

The most commonly encountered azo compounds are those in which both groups are aryl and, as far as electrophilic substitution is concerned, the available information is restricted to scattered data, occasionally of doubtful reliability, on products isolated as a result of substitution reactions carried out on azobenzene or on simple derivatives of it.

The most studied reactions have been nitration and halogenation, especially bromination. In 1928, Burns, McCombie and Scarborough[44] critically reviewed earlier work, and cast doubt on the claims of Janovsky and Erb[45] to have isolated all three mono-bromoazobenzenes following bromination of azobenzene in acetic acid at 50–60°C. In their laboratory they were able to isolate only one monobromo- and one dibromoazobenzene from the reaction and, although they do not give the structure or physical properties of either, these were probably the 4-bromo- and 4,4-dibromo-derivatives, since earlier work by Werigo, using bromine in the absence of a solvent at low temperatures, resulted in the isolation of 4,4-dibromoazobenzene[46].

For nitration also, earlier claims regarding isolation of *ortho*- and *meta*-isomers must be discounted; the only nitro isomers of which the formation by direct nitration can be established with certainty are the 4-nitro- and 4,4′-dinitro-azobenzenes[47]. Nitration may be accompanied by some oxidation to azoxybenzenes, and in cases where substituted azoxybenzenes are formed, some doubt must exist as to which process comes first: nitration or oxidation. Nitration under vigorous conditions is reported to lead to azoxybenzenes with substituents in both rings[47], including trinitro derivatives with a single annular nitro substituent in the *ortho*- or *meta*-position. However, with so much evidence against direct electrophilic attack at such positions, these products could well have arisen from nitration occurring after oxidation of the azo group.

Sulphonation using fuming sulphuric acid at 130°C leads to mixtures of the 4- and 4,4′-sulphonic acids[48]. Sometimes the expected 4,4′-disulphonic acid is accompanied by some of the 3,4′-isomer and this product could be the result of at least partial thermodynamic control of the reaction[49]. This would be in keeping with known sulphonation characteristics and it is the only reported instance of attack on the *meta*-position on an unsubstituted azobenzene ring.

In so far as attack on substituted azobenzenes is concerned, the expected results are found with only few exceptions. A strongly activating substituent such as amino, hydroxy or alkoxy will direct incoming substituents to free positions in the same ring *ortho-* and/or *para-* to itself, although nitration can sometimes lead to the loss of a *p*-phenylazo group as a diazonium ion[50]. Deactivating substituents such as nitro and halogen invariably direct the attacking species to the other nucleus.

An interesting result is observed for the case of the methyl-substituted azobenzenes[44]. The 3-methyl compound is readily attacked by both bromine and nitric acid in acetic acid solution, the incoming group going to the 4-position (for bromination) and the 4- and 4′-positions (in nitration). Nitration of the 2-methyl derivative also led to the expected 4,4′-dinitro compound but bromination gave an unknown compound of formula $C_{13}H_9N_2Br_3$. This compound had none of the properties of an azo- or hydrazo-compound. The reaction conditions would presumably lead to some α-bromination, and a possible structure for the unidentified product is:

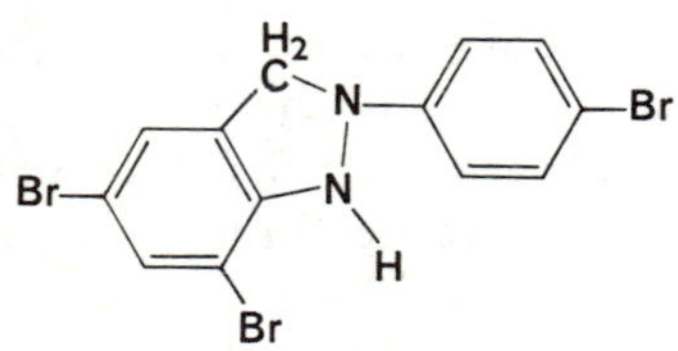

The 4-methylazobenzene proved to be considerably less reactive than either the 2- or 3-methyl compounds. It could be neither brominated nor nitrated in acetic acid solution (although side-chain bromination occurred after prolonged treatment at elevated temperatures). Bromine itself afforded the 4′-bromo derivative while concentrated nitric acid gave the 4′-nitro compound. It would appear, therefore, that the position *para* to the azo group is more activated than an *ortho*, even when the latter is activated by a *meta*-methyl substituent. Furthermore, the activating effect of the methyl group on positions *ortho* to itself does not rival the deactivating effect of the phenylazo group *meta* to these sites. Neither of these conclusions is unexpected. Less easy to explain, however, is the apparent lower reactivity towards electrophiles, of the 4′-position in 4-methylazobenzene by comparison with the 4-position in azobenzene. If this observation is correct*, then it is likely that the mechanism may not be the normal two-step process,

* This observation might well be misleading, because the authors were unaware of the sensitivity of the reaction to hydrogen halides (Robertson, Hitchings and Will)[51].

comprising initial formation of a Wheland intermediate and subsequent loss of a proton to give the product.

This indication that substitution in azobenzenes may be out of pattern has received further support from the study of Robertson, Hitchings and Will[51]. They reported that neither chlorination nor bromination of azobenzene in water-free acetic acid proceeded at a measurable rate in the absence of hydrogen halide. On the other hand, chlorination in 20% aqueous acetic acid was found to occur smoothly at about five times the rate of chlorination of benzene under the same conditions.

These workers explained their bromination results in terms of an initial slow addition of hydrogen bromide to the azobenzene across the azo group. In support, they observed that hydrogen bromide adds very slowly to azobenzene to give a product in which the bromine no longer reacted with silver nitrate. The species actually undergoing substitution therefore, might be this adduct and not azobenzene itself. The observations of Burns and co-workers[44] on 4-methylazobenzene were explained on the basis of the

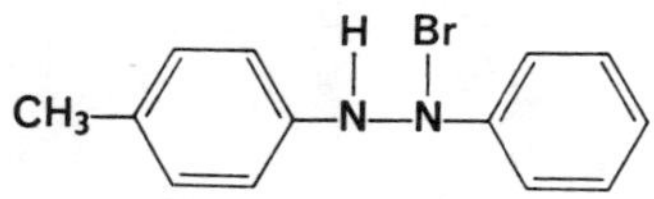

(9)

addition product (**9**) in which the more active *para*-position—that *para* to the NH grouping—is blocked. An alternative halogenation mechanism involving an Orton-type rearrangement of the adduct was proposed, but this was thought to be a minor pathway although likely to be the one leading to the tetrabromobenzidine product isolated by Mills from azobenzene bromination[52].

The addition of hydrogen halide across the azo group does much to account for the increased rate of halogenation, but the explanation may not be entirely satisfactory. In particular, the bromination of 4-methylazobenzene, if proceeding via adduct **9**, should yield 2-bromo-4-methylazobenzene. Furthermore, it has been shown recently that in protonation of substituted azobenzenes there is tautomeric equilibrium involving the two nitrogen atoms[53]; it might be expected that the reaction of a derivative such as 4-methylazobenzene should reveal some evidence of this.

Much of the criticism may be answered on the basis of the superior stability of *para*-quinonoid structures, with the accompanying assumption that in these substrates the advantage is sufficient for 4′-substitution to be predominant.

Although nitration, halogenation and sulphonation of azobenzene can be successful, it does appear that azobenzene is resistant to electrophilic attack on the aromatic ring. Most of the other common electrophiles (with the exception of the hydrogen ion) are weak and the lack of reactivity to Friedel–Crafts reagents, nitrous acid, diazonium salts and mercuric acetate lends support to an assumption that azobenzene behaves as a deactivated substrate towards electrophiles. On the other hand, these reagents might well bind to the azo group, which is by far the most electronegative part of the molecule.

D. Other Aromatic Substitution Reactions

1. Nucleophilic aromatic substitution

Quantitative data on rates of nucleophilic aromatic substitution of 4-X-2-NO_2—1-Cl-benzenes indicate that *p*-phenylazo groups activate the aromatic system towards nucleophilic substitution. Calculated σ_p^- constants vary with the system: +0·672 (methoxide/methanol)[14], +0·66 (piperidine/benzene)[31], +0·55 (piperidine/DMSO)[16]. These differences are probably related to solvent variation, higher σ values often being associated with hydroxylic solvents.

2. Free-radical substitution

The only reaction studied is the phenylation of azobenzene. Miller and co-workers report that the phenylazo group is about equal to the nitro group in activating power, with partial rate factors of 9·7, 0·95 and 8·8 for the *ortho*-, *meta*- and *para*-positions respectively[54]. This high level of activation of the *ortho*- and *para*-positions is attributed to the importance of resonance forms with the unshared electron on nitrogen.

3. Miscellaneous substitutions

Fahey has reported that chlorination or bromination of azobenzene in the presence of palladium(II) salts leads to the formation of *ortho*-substituted

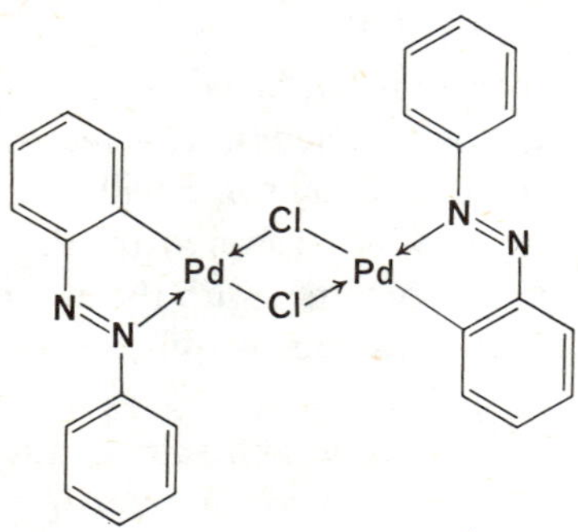

azobenzenes[55]. The specificity arises from interaction of the halogen with a carbon–metal bonded intermediate, di-μ-halobis-2-phenylazophenyl dipalladium(II).

E. Nuclear Magnetic Resonance Spectra of Azo Compounds

The methyl protons of *trans*-azomethane resonate 3·68 p.p.m. downfield relative to TMS[56], and are therefore more shielded than the corresponding atoms in alkenes (1·7 p.p.m.), methyl ketones (2·1 p.p.m.) and acetonitrile (1·97 p.p.m.), the difference being accounted for mainly by the high electronegativity of the nitrogen bonded to the methyl group.

Taft has used ^{19}F n.m.r. shielding of *meta*-substituted fluorobenzene to determine σ_I values for a large number of substituents, including the phenylazo group[17]. From his data he estimated $\sigma_I = +0{\cdot}19$ (CCl_4 or C_6H_{12}) or +0·25 (weakly protic solvents). Shielding of *para*-substituted fluorobenzenes correlates better with σ^- and observations by Suhr on 4-fluoroazobenzene suggested a value of σ^- of +0·67–0·68 for the phenylazo substituent[13].

F. The Protonated Azo (Azonium) Group as a Substituent

The azo group is potentially nucleophilic and is protonated in strongly acidic solutions (*trans*-azobenzene has a pK_a of −2·95 and the *cis*-isomer −2·25 in 20% ethanol)[57]. Although there was originally some debate as to whether one was dealing with σ- or π-protonation it now appears that the former is the case. In azobenzene derivatives there are normally two σ-protonated forms existing in equilibrium and if we have a single substituent such as the phenylazo group we find that, just as in the azoxybenzenes (see following section), we are dealing with two distinct substituents with different structures:

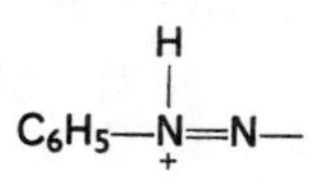

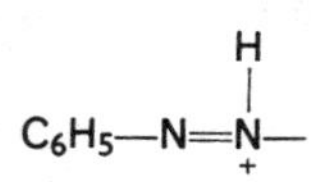

Unlike the azoxybenzene pair, these exist in equilibrium[53].

As a functional group the azonium group bears a certain resemblance to the diazonium cation and it can exert a powerful −R effect. Bunnett and co-workers have reported that it enables the displacement of an alkoxy group by water in nucleophilic aromatic substitution[58].

On the basis of ^{19}F n.m.r. studies on *para*-substituted fluorobenzenes, Suhr has estimated a figure for σ^- of +2·04 in DMSO[13] (under the same conditions, the diazonium group has $\sigma^- = 2{\cdot}63$). Suhr allocates this value to the $C_6H_5{-}N{=}\overset{\overset{H}{|}}{\underset{+}{N}}$ substituent, but he was dealing with an equilibrium mixture of unknown composition.

III. DIRECTING AND ACTIVATING EFFECTS OF THE AZOXY GROUP

A. General Considerations

Like the azo group, the azoxy group is a bridging group which is most commonly met as a link between two aryl groups. For this reason, directing and activating effects relate to an aryl- (usually phenyl-) azoxy unit which, because of the unsymmetrical nature of the azoxy group, will belong to one of two distinct series. However, it seems that the electronic effects of an arylazoxy group depend far more on the site of the azoxy oxygen than on the nature of any annular substituent being carried in the aromatic part of the group.

Of the different systems for naming azoxybenzene derivatives we shall use that of *Chemical Abstracts* for precise description of specific compounds[59]. The numbering is:

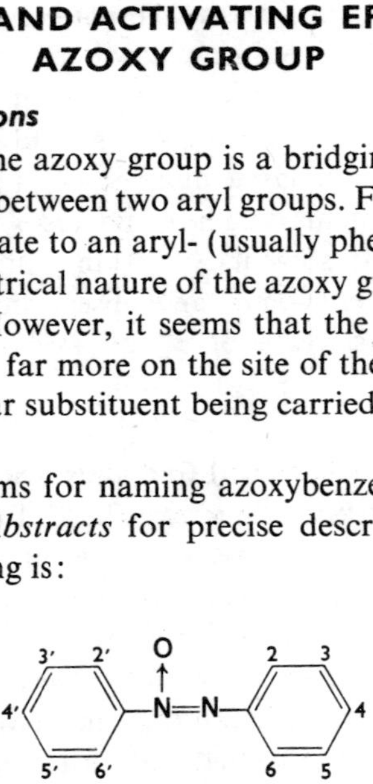

In more general discussion, however, Angeli's terms are sometimes useful to distinguish between isomeric alternatives. Angeli originally referred to the more reactive of an isomeric pair of *para*-substituted azoxybenzenes towards electrophilic reagents as the β-isomer and the less reactive as the α-isomer[60]. In practice, this naming was consistent, the two isomers being:

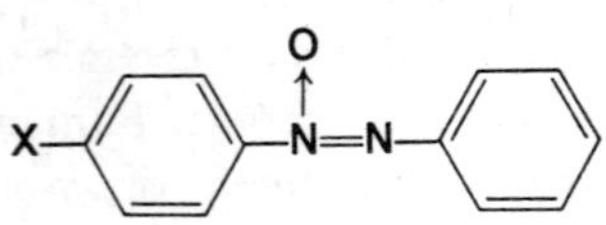

β-isomer

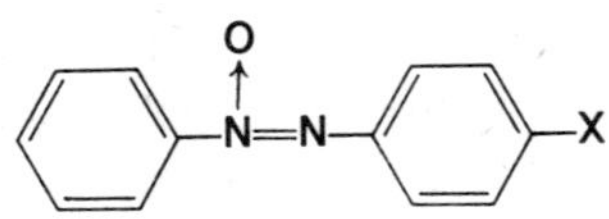

α-isomer

The *Chemical Abstracts* nomenclature is also used to describe the isomeric arylazoxy groups, which are labelled as aryl-NNO-azoxy or aryl-ONN-azoxy, terms which clearly identify the position of the oxygen in the substituent:

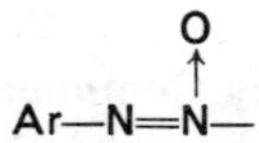

aryl-NNO-azoxy

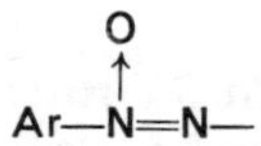

aryl-ONN-azoxy

Isomerism is further complicated by the possibility of *cis–trans* pairs but fortunately, as with azobenzenes, azoxybenzenes with *cis*-aryl groups are much less stable and rarely encountered.

It is to be noted that quantitative data on azoxyarenes are appallingly few, and we have used qualitative results and analogies with other systems in an attempt to derive some measure of substrate reactivity.

B. Reactions of Aromatic Side-chains

The only quantitative information available on side-chain reaction of an azoxy-compound comes from Jaffe and Ellerhorst, who determined the basicities of the two isomeric *para*-dimethylaminoazoxybenzenes in 20% ethanol[61]. They calculated substituent constants (presumably σ^- values) of +0·78 and +0·56 for the phenyl-ONN- and phenyl-NNO-azoxy groups respectively. These results are closely similar to those calculated by Miller[14] in a nucleophilic aromatic substitution reaction if one assumes that one of the two groups had assigned their isomer structures incorrectly. Jaffe and Ellerhorst had, in their structure assignments, already reversed the decision of earlier workers. We therefore arranged for the structure of one isomer (that with m.p. 122°C) to be determined by X-ray crystallography. This proved to be the β-isomer, and not the α-compound as assumed by Jaffe. The reassignment of his structures reconciles his substituent constants with those of Miller.

Since both substituents should be –R groups these figures do not allow the deduction of unperturbed σ^0 values. Jaffe and Ellerhorst did measure the basicities of the two phenyl-azoxy-dimethylaniline *N*-oxides. This is presumably a 'σ^0' reaction, but no reaction constant (ρ) is available. The measured pK_a values for the two isomers differ from that of *p*-oxidodimethylaminoazobenzene by –0·08 (α-isomer) and –0·40 pK_a units (β-isomer) (our assignments). Both arylazoxy substituents therefore have more positive values than that for the *para*-phenylazo substituent, and if the ρ value for ionization of the aryldimethylamino-*N*-oxides were somewhere around unity, then the σ^0 values for the phenyl-ONN- and phenyl-NNO-azoxy groups would be +0·43 and +0·75 respectively. Since the latter figure is higher than the calculated σ^- value, either the experimental data are at fault or the assumed ρ value is too high. Unfortunately, no other basis for estimates seems to be available.

C. Electrophilic Substitution Reactions in Azoxy Compounds

I. General considerations

If we assume that electrophilic aromatic substitution in azoxy compounds occurs through normal rate-determining formation of a Wheland intermediate, then the activating or deactivating effect of an azoxy group will reflect the stabilizing or destabilizing (electron-withdrawing) effect of this substituent on the transition-state leading to intermediate formation.

Both types of azoxy substituent should have moderately strong electron-withdrawing inductive effects (–I) and, in the absence of any strong, stabilizing conjugative (+R) effect, these would lead to *meta*-direction of incoming electrophiles. Both azoxy groups, however, are potential +R groups and therefore have the potential ability to lower the transition state energy for *ortho*- or *para*-attack. For the phenyl-ONN-azoxy group such stabilizing resonance would involve the contributing forms

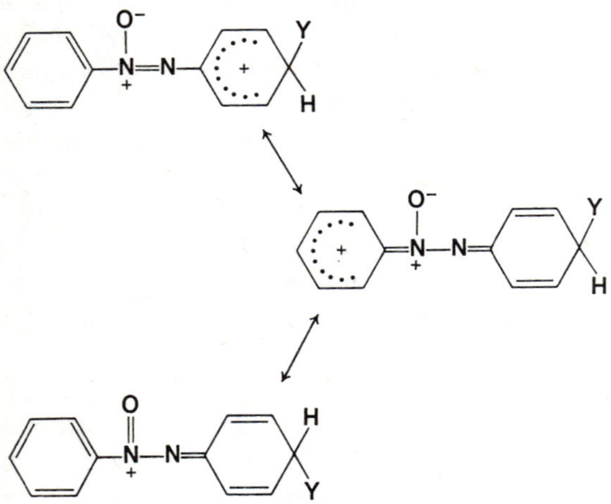

For the phenyl-NNO-azoxy substituent, which has no lone pair adjacent to the aromatic nucleus, the corresponding canonical set would be:

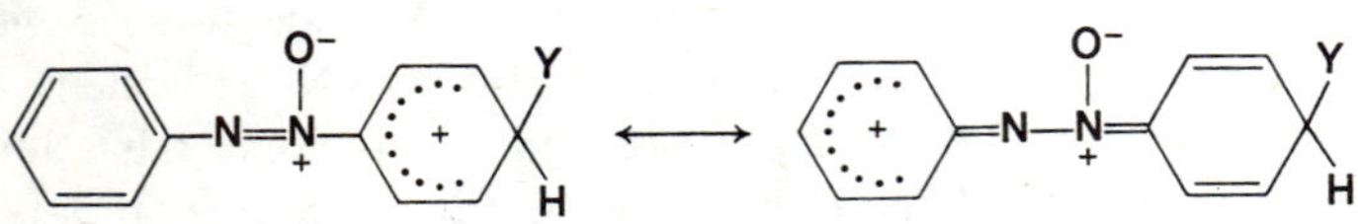

Experimentally, we find that the phenyl-ONN-azoxy group directs attacking electrophiles to the *ortho*- and *para*-positions, while the phenyl-NNO-azoxy group is *meta*-directing and presumably deactivating.

2. Substitution in alkylazoxybenzenes

The only information available on substitution reactions of this type of compound concerns the bromination of the isomeric *p*-chlorobenzylazoxybenzenes. Lythgoe and co-workers have reported that the α-form of *p*-chlorobenzylazoxybenzene, on treatment with equimolar quantity of bromine/silver trifluoracetate in nitrobenzene, gives 3-bromo-4-chlorobenzylazoxybenzene[62]. On the other hand, under the same conditions the *β*-form is brominated *para*- to the azoxy group to give (*p*-bromophenylazoxy)-*p*-chlorotoluene.

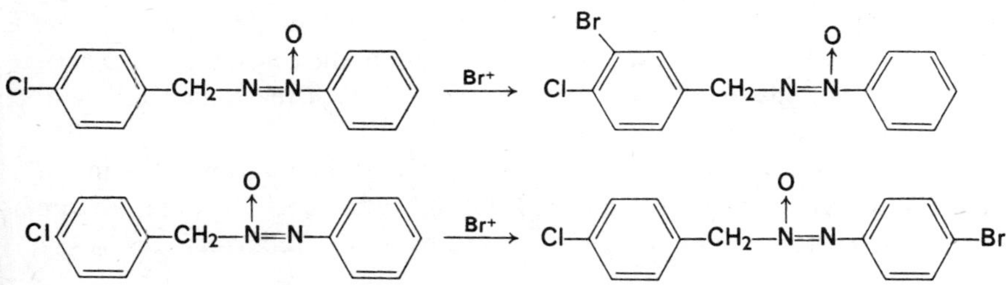

It must be pointed out, however, that the identification of the two isomers was based on these reactions, so that, while it might be true that alkylazoxy groups are similar to arylazoxy groups in their relative effects, some additional evidence is required before this assertion can be made with any justification.

3. Azoxybenzene and substituted azoxybenzenes

The most commonly studied reactions are nitration and halogenation (mainly bromination). Mononitration occurs readily with concentrated nitric acid[47]; use of nitric acid/sulphuric acid leads to polynitration[47] but the situation is complicated by the tendency, under acid conditions, of the azoxy-oxygen to migrate from one nitrogen to the other[63]. Bromination[59] or chlorination[64] in acetic acid introduces a single halogen substituent. The site most susceptible to electrophilic attack is the *para*-position further from the azoxy oxygen (the 4-position). If this is blocked, the most reactive site is the 2-position. When the other ring is attacked, unless activating substituents are present, the reactive positions are those *meta* to the azoxy group. The mercuration of azoxybenzene, using both mercuric acetate and mercuric perchlorate, has been studied[65]. The 2-position appears to be at least twice as reactive as the 4-position, presumably indicating some interaction between the electrophiles and the azoxy group. However, yields are low (< 50%) and the results should perhaps be treated with caution. The only other instance of an *ortho* product being isolated from attack on azoxybenzene occurs in nitration, but the yields of 2-nitro isomer are low: well below those from mercuration[47].

Strongly activating substituents, such as hydroxyl, alkoxyl or amino, control the direction of electrophilic attack in a predictable way. For the methyl group, the position is less clear. With a 4′-methyl substituent, nitration and bromination still give products resulting from attack in the other ring[66]. This is not unexpected. However, 4-methylazoxybenzene resists attack by molecular bromine at 100°C, conditions that might be expected to give at least some 2- or 3-bromo derivative[66].

Sulphonation with fuming sulphuric acid at elevated temperatures is reported to lead to mono-, di-, tri- and tetrasulphonic acids, but the experimental conditions would be favourable for Wallach rearrangements and the products could be sulphonated phenols[67].

Azoxybenzenes react at oxygen with both acid halides and aluminium halides to give azo compounds[68]. Friedel–Crafts reactions are therefore unlikely to be successful and none seem to have been reported.

4. The activating power of the phenylazoxy group

In the absence of quantitative data on electrophilic aromatic substitution in azoxybenzene, known properties lead only to the conclusion that the $C_6H_5{-}\overset{\overset{\displaystyle O}{\uparrow}}{N}{=}N{-}$ group is *ortho–para* directing and more activating than the $C_6H_5{-}N{=}\overset{\overset{\displaystyle O}{\uparrow}}{N}{-}$ group.

However, with some assumptions, approximations and analogies, it is possible to draw further conclusions about the electronic character of the phenylazoxy group. The assumptions made are:

(*a*) When only one substitution product is reported, the identification is accurate and other possible products are formed in much smaller amounts.

(*b*) The mechanism of substitution is normal and invariant.

(*c*) Substituent constants are additive and independent of the reaction, with $\sigma_0^+ \approx \sigma_p^+$.

(*d*) A transmission coefficient (π') of about 0·25 operates across the azoxy group.

The following statements may then be made:

(1) In azoxybenzene the most reactive site towards electrophiles is the 4-position; if this is blocked then substitution occurs at the 2-carbon, in the absence of strong activation by the blocking group.

(2) While nitration of azoxybenzene leads successively to the 4-nitro and the 2,4-dinitro derivatives, the third nitro group attacks the other ring, entering *meta-* to the azoxy group[69].

(3) Nitration of 4-bromo-2-nitroazoxybenzene gives 4-bromo-2,6-dinitroazoxybenzene[69].

From (3), the sum of σ^+ values for the *meta*-bromo, *meta*-nitro, and *ortho*-phenyl-ONN-azoxy groups must be less than σ^+ for the *meta*-4-bromo-2-nitrophenyl-NNO-azoxy group which, if $\pi' = 0{\cdot}25$, should be ca 0·3 units higher than for the unsubstituted phenyl-NNO-azoxy group.

$$\sigma_o C_6H_5\text{—}\overset{\mathrm{O}}{\overset{\uparrow}{\mathrm{N}}}=\mathrm{N}\text{—} + \sigma_m \mathrm{Br} + \sigma_m \mathrm{NO_2} < \sigma_m C_6H_5\text{—}\mathrm{N}=\overset{\mathrm{O}}{\overset{\uparrow}{\mathrm{N}}}\text{—} + 0{\cdot}3$$

$$\sigma_o C_6H_5\text{—}\overset{\mathrm{O}}{\overset{\uparrow}{\mathrm{N}}}=\mathrm{N}\text{—} + 0{\cdot}40 + 0{\cdot}67 < \sigma_m C_6H_5\text{—}\mathrm{N}=\overset{\mathrm{O}}{\overset{\uparrow}{\mathrm{N}}}\text{—} + 0{\cdot}3$$

$$\sigma_o C_6H_5\text{—}\overset{\mathrm{O}}{\overset{\uparrow}{\mathrm{N}}}=\mathrm{N}\text{—} < \sigma_m C_6H_5\text{—}\mathrm{N}=\overset{\mathrm{O}}{\overset{\uparrow}{\mathrm{N}}}\text{—} - 0{\cdot}77$$

From (2) the sum of the substituent constants for two *meta*-nitro groups and one *ortho*-phenyl-ONN-azoxy group must be greater than σ_m for the 2,4-dinitrophenyl-NNO-azoxy substituent which should be about 0·4 units more positive than the phenyl-NNO-azoxy group.

$$\sigma_o C_6H_5\text{—}\overset{\mathrm{O}}{\overset{\uparrow}{\mathrm{N}}}=\mathrm{N}\text{—} + 2\sigma_m \mathrm{NO_2} > \sigma_m C_6H_5\text{—}\mathrm{N}=\overset{\mathrm{O}}{\overset{\uparrow}{\mathrm{N}}}\text{—} + 0{\cdot}4$$

$$\sigma_o C_6H_5\text{—}\overset{\mathrm{O}}{\overset{\uparrow}{\mathrm{N}}}=\mathrm{N}\text{—} > \sigma_m C_6H_5\text{—}\mathrm{N}=\overset{\mathrm{O}}{\overset{\uparrow}{\mathrm{N}}}\text{—} - 0{\cdot}94$$

This suggests that σ^+ for the *ortho*-phenyl-ONN-azoxy group is somewhere between 0·77 and 0·94 units less than σ_m for the phenyl-NNO-azoxy group.

Further information comes from bromination studies on the 4,4′-diethoxy derivative[70]. Both monobromination and mononitration occur at the 3-position, indicating that the $C_6H_5\text{—}\mathrm{N}=\overset{\mathrm{O}}{\overset{\uparrow}{\mathrm{N}}}\text{—}$ group is less electron-withdrawing than the $C_6H_5\text{—N=}\overset{\overset{\text{O}}{\uparrow}}{\text{N}}\text{—}$ group. However, a second bromine will attack the monobromo derivative at the 3′ position. With normal attenuation of the effect of the bromine across the azoxy group, this indicates that the difference in their substituent constants is less than 0·3 units. The $C_6H_5\text{—N=}\overset{\overset{\text{O}}{\uparrow}}{\text{N}}\text{—}$ group is formally similar to the $\text{O=}\overset{\overset{\text{O}}{\uparrow}}{\text{N}}\text{—}$ group but σ_I for the —N=O group is about 0·12 greater than that for the $\text{—N=N—}C_6H_5$ group, and so σ_m for the phenyl-ONN-azoxy substituent should be lower than for the nitro-group by at least this amount. Reactions of pyridines and their *N*-oxides indicate that σ_m for a pyridino nitrogen is appreciably lower than that of its *N*-oxide (+0·6 against + 1·2)[71] and, quantitatively, the phenyl-ONN-azoxy group should be more electron-withdrawing than a phenylazo group. Thus we have a lower limit for σ_m for the phenyl-ONN-azoxy substituent of +0·3 and an upper limit for the phenyl-NNO-azoxy of $\sigma_m = +0{\cdot}6$. The maximum difference between the two is +0·3 units, in agreement with our observations on the bromination of 4,4′-diethoxyazoxybenzene. Other observations indicate that this maximum difference may be close to a real difference. In 4-hydroxyazoxybenzene, the 3-position has sufficient reactivity for preferential attack by benzenediazonium cation. On the other hand, 4′-hydroxyazoxybenzene is not attacked. The behaviour of the former resembles that of 4-chlorophenol and the latter 4-nitrophenol, and the σ_m values are unlikely to be particularly close to each other[72].

To summarize therefore, if we assume a σ_m value of around +0·6 for the phenyl-NNO-azoxy group then we would predict a value for σ_p^+ for the phenyl-ONN-azoxy group of between −0·2 and −0·4. In view of the assumptions made, these limits might easily be too narrow, but the group certainly appears to have a surprisingly strong +R effect.

D. Other Aromatic Substitution Reactions

1. Nucleophilic aromatic substitution

There have been three reports to date of the effect of azoxy groups on nucleophilic aromatic substitutions. Courtney, Geipel and Shriner examined the reaction of 4- and 4′-chloroazoxybenzene with sodium ethoxide in ethanol[64]. From their results, which were based on the amount of chloride ion liberated after a fixed time, they concluded that both azoxy groups are slightly superior to an azo group in their encouragement of halogen displacement, but each is considerably less effective than a nitro group. Of the two azoxy compounds the halogen in the 4-isomer is slightly more labile.

A more quantitative estimate of activating power was made by Miller and co-workers[14] who measured the kinetics of the reaction of 4-substituted 1-chloro-2-nitrobenzenes with sodium methoxide in methanol at 50°C. They determined substituent constants for the two groups and found values of +0·616 (*p*-phenyl-ONN-azoxy) and +0·772 (*p*-phenyl-NNO-azoxy). This reactivity is the reverse of that noted by the earlier group and in Miller's work the corresponding azo compound lay between that of the two azoxy compounds. More recently Hendley and Duffey, using the same semi-quantitative approach as that of Courtney's group, looked at the relative reactivity of 4- and 4′-bromoazoxybenzene with base in ethanol, 90% ethanol and 1-butanol[73]. Their results suggest that the 4′-isomer is the less reactive in ethanol and in butanol. In 90% ethanol the order is reversed but the differences may not be significant.

Miller's results seem to be the most reliable; the reaction products were isolated and identified and his quantitative work led to σ^- values which are in good agreement with those derived by Jaffe and Ellerhorst from the basicities of the dimethylanilines[61].

2. Miscellaneous substitutions

There are no literature reports of direct substitution in either of the aromatic nuclei of azoxybenzenes other than by a simple electrophilic or nucleophilic mechanism.

E. *Nuclear Magnetic Resonance Spectra of Azoxy Compounds*

1. Aliphatic compounds

Freeman has determined the chemical shifts of protons in a number of aliphatic azoxy derivatives[56]. Protons close to the oxygen-bearing nitrogen resonate further downfield than comparable protons adjacent to either the other nitrogen of the azoxy group or a nitrogen of the corresponding azo compound (see Table 1).

The more electron-withdrawing $-\overset{\overset{\displaystyle O}{\uparrow}}{N}{=}N-R$ group is more effective in deshielding adjacent protons than either the $-N{=}\overset{\overset{\displaystyle O}{\uparrow}}{N}-R$ or $-N{=}N-R$ groups. The group nearer the unoxidized nitrogen of the azoxy compound is upfield relative to that of the corresponding azo system; this is attributed

TABLE 1. N.m.r. chemical shifts in aliphatic azoxy and azo compounds

Compound	(p.p.m. from TMS in CCl_4 on A-60)
$CH_3-\overset{\overset{O}{\uparrow}}{N}=N-CH_3$	4·05
$CH_3-\overset{\overset{O}{\uparrow}}{N}=N-CH_3$	3·07
$CH-N=N-CH_3$	3·68
$(H_3C)_3-\overset{\overset{O}{\uparrow}}{N}=N-C(CH_3)_3$	1·48
$(H_3C)_3-\overset{\overset{O}{\uparrow}}{N}=N-C(CH_3)_3$	1·28
$C_6H_5-\overset{\overset{O}{\uparrow}}{N}=N-CH_3$	3·40
$C_6H_5-N=\overset{\overset{O}{\uparrow}}{N}-CH_3$	4·15
$C_6H_5-N=N=CH_3$	3·90

by Freeman to diamagnetic shielding associated with the conical region above the plane of the N—O bond.

With the methyl-NNO-azoxy- and phenyl-NNO-azoxy groups, (chemical shifts of 4·05 and 4·15 respectively) the deshielding is far less than with the —NO_2 group (4·33) and the structural analogy between these groups is therefore limited in its usefulness.

2. Aromatic compounds

Webb and Jaffe[74, 75] have determined the chemical shift of methyl groups in a series of both *cis*- and *trans*- azo and azoxybenzenes. Their results are shown in Table 2.

The upfield shift of around 10 c.p.s. in the *cis*-compounds is attributed to interaction of the π-systems of the rings which is accompanied by an increase in the paramagnetic shielding and a consequent decrease in the chemical shift of the methyl protons. In the unsymmetrical *trans*-methyl-

TABLE 2. N.m.r. chemical shifts in aromatic azoxy compounds

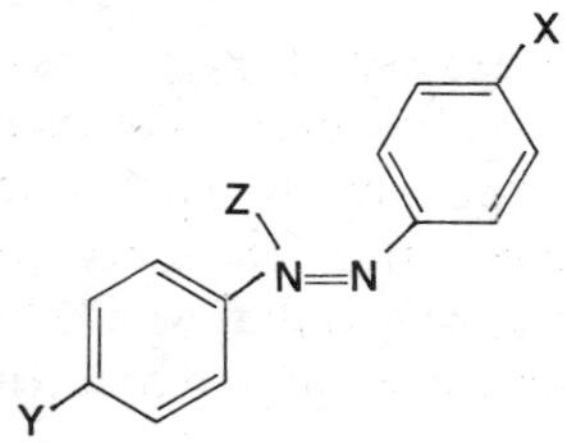

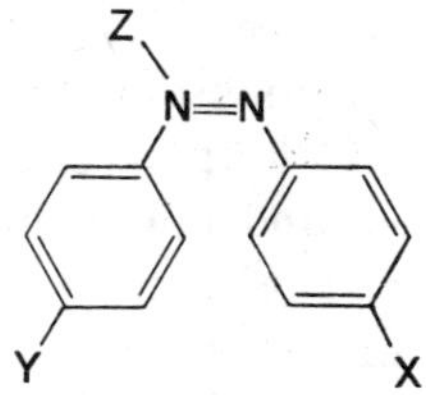

Compound			Chemical shift (c.p.s. relative to TMS)	Compound			Chemical shift (c.p.s. relative to TMS)
X	Y	Z		X	Y	Z	
Me	H	—	124·8	Me	H	—	115·5
Me	Me	—	124·6	Me	Me	—	115·5
Me	H	O	124·6	Me	H	O	112·6
H	Me	O	117·4	H	Me	O	105·4
Me	Me	O	124·5, 118·4	Me	Me	O	112·9, 105·8
MeO	MeO	O	190·6, 195·6	MeO	MeO	O	181·8, 188·4
Me[a]	Me[a]	O	120·7, 128·4	Me[a]	Me[a]	O	98·0, 105·6
NMe_2	H	O	148·5				
H	NMe_2	—	144·0				

[a] *meta*-Substituent.

azoxybenzenes, the methyl group closer to the oxidized nitrogen is shifted upfield by about 7 c.p.s., but the other methyl is about the same as in 4-methylazobenzene. This is the reverse of the results for the aliphatic series (Table 1). A similar reversal of the relative order occurs for the dimethylaminoazoxybenzenes, deshielding being once again greater with the phenyl-ONN-azoxy substituent. However, Rae and Dyall[76] have measured the chemical shifts of the *N*-methyl protons in a series of *para*-substituted *N*,*N*-dimethylanilines (also in benzene solution). They have found that while a Hammett-type correlation is not good, strongly electron-withdrawing substituents give rise to chemical shifts *upfield* by comparison with groups of less electron-withdrawing power. These workers attribute this to specific solvation by benzene (the reversal did not occur in chloroform solvent). Jaffe's observations therefore are in agreement with the phenyl-NNO-azoxy group being the more electron-withdrawing of the two.

IV. DIRECTING AND ACTIVATING EFFECTS OF THE HYDRAZO GROUP

The hydrazo group is grouped with the azo and azoxy groups presumably because of its relatively close relationship to these in terms of interconversion and chemical reactions. In this chapter we are not so much concerned with the reactivity of the group as with its effect on the reactivity of other parts of the molecule, and a more useful relationship is that with hydrazines and amines; for the most part, the behaviour of the hydrazo group is consistent with that of any other amine.

In evaluating the effect of the hydrazo group on reactions at other points in the molecule, one immediately comes up against a problem posed by the high reactivity of the group itself. The amino nitrogens are attractive to electrophiles and the α-diamino arrangement in the hydrazo group is particularly sensitive to both acids and oxidizing agents. Common aromatic electrophilic substitutions such as nitration and halogenation are therefore not observed. Ring-substituted hydrazobenzenes are invariably prepared either through the azo derivatives or by the coupling of suitably substituted nitrogen compounds. Consequently it is of little more than academic interest to obtain, by some indirect method, an estimate of the relative reactivities, towards electrophiles, of the ring carbons in hydrazobenzene. If we take the quantitative measure of the directing activating effect of a substituent to be one of the Hammett substituent constants, we are reduced, in the absence of direct reaction data, to the use of physical measurements or to a comparison with substituents regarded as similar in type. For the hydrazino group, there is indeed a report of σ_I as +0·15, based on aliphatic reactivities, and +0·10 based on ^{19}F n.m.r. studies of *m*-fluorobenzenes[77]. These are 0·05 units higher than figures for the amino group[77], and this is consistent with the view of a hydrazino group as an amino group carrying a –I substituent. The value of σ_I for the R—NHNH— and Ar—NHNH— groups should differ from that of the NH_2NH— group by less than the experimental error in the latter's measurement.

For aromatic systems we must draw an analogy with other groups. There is only one report for σ_m for the NH_2—NH— group (from the ionization of benzenephosphonic acids)[77]. The value of –0·02 is higher than the accepted value for the amino group ($\sigma_m = -0{\cdot}16$) but because it is based on a single measurement too much significance must not be attached to it. Operating from the *para*-position, the +R effect of an amino lone pair becomes important and for the simple NH_2 substituent, a value for σ_R^0 of about –0·42 has been calculated[77]. In the case of the NH_2NH— group, the +R effect of the nitrogen adjacent to the ring is affected by the attached amino group in two ways. The –I effect of the NH_2— will make the lone pair less available for ring activation, but repulsion between the two filled *p*-orbitals on adjacent nitrogens will favour sharing of the lone pair with the aromatic system. The relative importance of these two effects can be assessed only from experimental values of σ_R^0. Katritzky and co-workers have, through i.r. studies, estimated σ_R^0 for the amino group as –0·467 (in good agreement with Taft's value) and for the hydrazino group as –0·487, indicating a balance somewhat in favour of the repulsion effect[78]. On the other hand, they report that for the phenylhydrazo group $\sigma_R^0 = -0{\cdot}44$, consistent with a diminished interaction between the lone pairs in hydrazobenzene. The differences, however, are sufficiently small to allow an assumption that σ_p^0 values for amino and hydrazo groups are about the same, at least within the limits of experimental error. Nevertheless, in cases where σ_p^+ values are expected differences between the groups are likely to be more marked and the alkylhydrazo and hydrazino groups will probably show stronger +R effects than the simple, primary amino substituent.

V. REFERENCES

1. C. D. Ritchie and W. F. Sager, in *Progress in Physical Organic Chemistry*, Vol. II (Ed. S. G. Cohen, A. Streitwieser Jr. and R. W. Taft), Interscience, New York, 1964, pp. 323–400.
2. L. P. Hammett, in *Physical Organic Chemistry*, McGraw-Hill, New York, 1940, p. 188.
3. D. R. Boyd and E. J. Marle, *J. Chem. Soc.*, 2117 (1914).
4. M. Syz and H. Zollinger, *Helv. Chim. Acta*, **48,** 383 (1965).
5. O. Exner and J. Lakomy, *Collection Czech. Chem. Commun.*, **35,** 1371 (1970).
6. J. J. Ryan and A. A. Humffray, *J. Chem. Soc.* (*B*), 842 (1966).
7. Y. Hashida, S. Sekiguchi and K. Matsui, *Kogyo Kagaku Zasshi*, **73,** 184 (1970); *Chem. Abstr.*, **73,** 46613 (1970).
8. W. F. Little, C. N. Reilley, J. D. Johnson, K. N. Lynn and A. P. Sanders, *J. Amer. Chem. Soc.*, **86,** 1376 (1964).
9. J. J. Ryan and A. A. Humffray, *J. Chem. Soc.* (*B*), 468 (1967).
10. Z. S. Chaw, A. Fischer and D. A. R. Happer, *J. Chem. Soc.* (*B*), 1818 (1971).
11. E. B. Lifschits, L. M. Yagupolsky, D. Ya. Naroditskaya and E. S. Kozlova, *Reakts. Sposobnost Org. Soedin., Tartu Gos. Univ.*, **6,** 317 (1969).
12. Y. Yukawa, Y. Tsuno and M. Sawada, *Bull. Chem. Soc. Japan*, **39,** 2274 (1966).
13. H. Suhr, *Ber. Bunsenges. Physik. Chem.*, **68,** 169 (1964).
14. J. Miller in *Aromatic Nucleophilic Substitution*, Elsevier, Amsterdam, 1968, p. 77.
15. J. Socha, V. Bekarek, J. Schreiber and M. Vecera, *Collection Czech. Chem. Commun.*, **35,** 3551 (1970).
16. J. Kavalek, J. Socha, J. Urbanek and M. Vecera, *Collection Czech. Chem. Commun.*, **36,** 209 (1971).

17. R. W. Taft Jr., E. Price, I. R. Fox, I. C. Lewis, K. K. Andersen and G. T. Davis, *J. Amer. Chem. Soc.*, **85,** 709 (1963).
18. B. I. Stepanov, B. A. Korolev and S. L. Polyakova, *J. Gen. Chem.* (*USSR*) (*Engl. Trans.*), **37,** 2028 (1967).
19. B. A. Korolev and B. I. Stepanov, *J. Org. Chem.* (*USSR*) (*Engl. Trans.*), **5,** 1622 (1969).
20. J. Socha, J. Horska and M. Vecera, *Collection Czech. Chem. Commun.*, **34,** 2982 (1969).
21. J. Socha and M. Vecera, *Collection Czech. Chem. Commun.*, **34,** 3740 (1969).
22. L. H. Cohen and M. W. Jones, *J. Amer. Chem. Soc.*, **85,** 3397 (1963).
23. J. Socha, J. Horska and M. Vecera, *Collection Czech. Chem. Commun.*, **34,** 3178 (1969).
24. H. H. Jaffe, *J. Chem. Phys.*, **21,** 415 (1953).
25. M. T. Rogers, T. W. Campbell and R. W. Maatman, *J. Amer. Chem. Soc.*, **73,** 5122 (1951).
26. S-J. Yeh and H. H. Jaffe, *J. Amer. Chem. Soc.*, **81,** 3287 (1959).
27. Ch. S. Frankovskii and N. V. Melamed, *J. Org. Chem.* (*USSR*) (*Engl. Trans.*), **5,** 107 (1969).
28. Ch. S. Frankovskii, N. V. Startseva, B. I. Ionin, E. Sh. El'kis and V. A. Zaplatina, *J. Org. Chem.* (*USSR*) (*Engl. Trans.*), **6,** 1909 (1970).
29. A. G. Catchpole, W. B. Foster and R. S. Holden, *Spectrochim. Acta*, **18,** 1353 (1962).
30. D. C. Colinese, D. A. Ibbotson and C. W. Stone, *J. Chem. Soc.* (*B*), 570 (1971).
31. J. Kavalek, J. Benes, J. Socha, F. Brtnik and M. Vecera, *Collection Czech. Chem. Commun.*, **34,** 2092 (1969).
32. Ch. S. Frankovskii and V. A. Zaplatina, *J. Org. Chem.* (*USSR*) (*Engl. Trans.*), **6,** 1898 (1970).
33. H. Veschambre and A. Kergomard, *Bull Soc. Chim. France*, 336 (1966); H. Veschambre, G. Dauphin and A. Kergomard, *Bull Soc. Chim. France*, 2846 (1967).
34. M. Syz and H. Zollinger, *Helv. Chim. Acta*, **48,** 517 (1965).
35. Yu. A. Zhdanov, V. I. Minkin and E. A. Medyantseva, *J. Gen. Chem.* (*USSR*) (*Engl. Trans.*), **35,** 1286 (1965).
36. V. A. Bekarek, J. Klicnar, F. Kristek and M. Vecera, *Collection Czech. Chem. Commun.*, **33,** 994 (1968).
37. H. B. Bürgi and J. D. Dunitz, *Helv. Chim. Acta*, **53,** 1747 (1970).
38. C. J. Brown, *Acta Cryst.*, **21,** 146 (1966).
39. H. Van Bekkum, P. E. Verkade and B. M. Wepster, *Rec. Trav. Chim.*, **78,** 815 (1959).
40. H. C. Longuet-Higgins and L. Salem, *Proc. Roy. Soc.*, **A251,** 172 (1959).
41. E. Berliner, in *Progress in Physical Organic Chemistry*, Vol. 2, (Ed. S. G. Cohen, A. Streitwieser Jr. and R. W. Taft), Interscience, New York, 1964, pp. 253–321.
42. P. W. Robertson, P. B. D. de la Mare and B. E. Swedlund, *J. Chem. Soc.*, 782 (1953).
43. J. F. Corbett and P. F. Holt, *J. Chem. Soc.*, 5029 (1961); *J. Chem. Soc.* (*C*)., 1638 (1967).
44. J. Burns, H. McCombie and H. A. Scarborough, *J. Chem. Soc.*, 2928 (1928).

45. J. V. Janovsky and L. Erb, *Ber.*, **19,** 2155 (1886); *Monatsh. Chem.*, **8,** 50 (1887); *Ber.*, **20,** 357 (1887).
46. A. A. Werigo, *Ann. Chem.*, **135,** 178 (1865); *Ann. Chem.*, **165,** 199 (1973).
47. A. Werner and E. Stiasny, *Ber.*, **32,** 3256 (1899).
48. P. Griess, *Ann. Chem.*, **154,** 208 (1870).
49. H. Limpricht, *Ber.*, **14,** 1356 (1881); *Ber.*, **15,** 1155 (1882).
50. O. Schmidt, *Ber.*, **38,** 3201 (1905).
51. P. W. Robertson, T. R. Hitchings and G. M. Will, *J. Chem. Soc.*, 808, (1950).
52. C. Mills, *J. Chem. Soc.*, 51 (1894).
53. E. Haselbach, *Helv. Chim. Acta*, **53,** 1526 (1970).
54. J. Miller, D. B. Paul, L. Y. Wong and A. G. Kelso, *J. Chem. Soc. B.*, 62 (1970).
55. D. R. Fahey, *J. Chem. Soc. (D)*, 417 (1970); D. R. Fahey, *J. Organometal. Chem.*, **27,** 283 (1971).
56. J. P. Freeman, *J. Org. Chem.*, **28,** 2508 (1963).
57. F. Gerson, E. Heilbronner, A. van Veen and B. M. Wepster, *Helv. Chim. Acta*, **43,** 1889 (1960).
58. J. F. Bunnett, E. Buncel and K. V. Nambekian, *J. Amer. Chem. Soc.*, **84,** 4136 (1962).
59. L. C. Behr, *J. Amer. Chem. Soc.*, **76,** 3672 (1954).
60. A. Angeli and B. Valori, *Atti Accad. Naz. Lincei, Rend.* [5] **21, I**, 155 (1912).
61. H. H. Jaffe and R. L. Ellerhorst, *J. Org. Chem.*, **33,** 4165 (1968).
62. J. B. Brough, B. Lythgoe and P. Waterhouse, *J. Chem. Soc.*, 4069 (1954).
63. K. H. Schundehutte, in *Houben-Weyl's Methoden der Organischen Chemie* (Ed. E. Muller), Vol. X/3 (Ed. R. Stroh), Georg Thieme, Stuttgart, 1965, p. 774.
64. J. J. Courtney, L. E. Geipel and R. L. Shriner, *Proc. Iowa Acad. Sci.*, **62,** 264 (1955).
65. T. Ukai and Y. Uto, *J. Pharm. Soc. Japan*, **73,** 821 (1953); *Chem. Abstr.*, **48,** 9945 (1954).
66. D. Bigiavi and V. Sabatelli, *Gazz. Chim. Ital.*, **57,** 557 (1927).
67. H. Wilsing, *Ann. Chem.*, **215,** 228 (1882); A. Lachman, *J. Amer. Chem. Soc.*, **24,** 1178 (1902).
68. J. F. Vozza, *J. Org. Chem.*, **34,** 2139 (1969).
69. B. Valori, *Atti Accad. Naz. Lincei, Rend.*, [5] **22 II**, 125 (1913).
70. B. Valori, *Atti Accad. Naz. Lincei, Rend.*, [5] **23 II**, 284 (1914).
71. Reference 14, p. 244.
72. D. Bigiavi and S. Stefanic, *Atti Accad. Naz. Lincei, Rend.*, [6] **9,** 326 (1929).
73. E. C. Hendley and D. Duffey, *J. Org. Chem.*, **35,** 3579 (1970).
74. D. L. Webb and H. H. Jaffe, *Tetrahedron Letters*, 1875 (1964).
75. D. L. Webb and H. H. Jaffe, *J. Amer. Chem. Soc.*, **86,** 2419 (1964).
76. I. D. Rae and L. K. Dyall, *Australian J. Chem.*, in press.
77. Reference 1, p. 335.
78. A. R. Katritzky and R. D. Topsom, *Angew. Chem. Intern. Ed. Engl.*, **9,** 87 (1970).

CHAPTER 9

The transition metal chemistry of azo compounds

M. I. BRUCE* and B. L. GOODALL

Department of Inorganic Chemistry, The University, Bristol, England

* Present address: *Department of Physical and Inorganic Chemistry, University of Adelaide, South Australia 5001.*

Abbreviations

abt	*o*-aminobenzenethiolate
acac	acetylacetonate
azb	2-(phenylazo)phenyl
azbH	azobenzene
bipy	2,2′-bipyridyl
dppe	1,2-bis(diphenylphosphino)ethane
imzH	imidazole
py	pyridine
pzH	pyrazole
THF	tetrahydrofuran

I. INTRODUCTION

The interaction of azo compounds, RN═NR′, with transition metal derivatives has produced complexes of at least four different types:

(*a*) containing the azo compound attached via nitrogen σ-donor bonds;

(*b*) containing the azo compound bonded via π-bonds involving the π electrons of the N═N system;

(*c*) in the case of aromatic derivatives, containing a metallated ligand attached via one nitrogen and a metal–carbon σ-bond to the *ortho* carbon of the ring;

(*d*) containing rearranged nitrogen-donor ligands, e.g., *o*-semidine.

In addition, reaction may occur with other ligands already attached to the transition metal, often giving new organic molecules. With the exception of the first two classes, the reactions are characterized by an interaction of the metal with the *ortho*-position of an aromatic ring.

In general, an important reaction in the chemistry of ligands containing aromatic groups bonded to donor atoms is *ortho*-metallation, whereby organometallic compounds containing aryl carbon–metal σ-bonds are formed, intramolecularly, via displacement of an *ortho* hydrogen of the aromatic ring by a transition metal atom in the complex. The *ortho*-metallation reaction involving azobenzene is one of the most extensively studied. In the resulting 2-(phenylazo)phenyl–metal derivatives the ligand is chelated to the metal (M) via a nitrogen σ-donor bond and a metal–carbon σ-bond forming a five-membered ring (**1**).

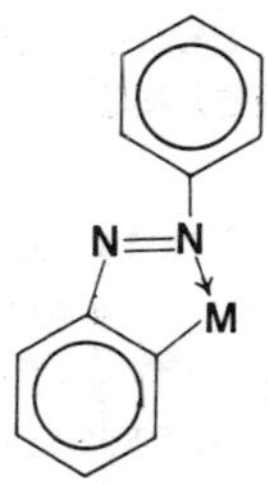

(1)

In this review the relevant chemistry of each transition metal is described and, wherever possible, reaction mechanisms have been proposed together with any supporting evidence. However, in view of the vast number of syntheses reported for complexes incorporating azo compounds as a class, several limitations have had to be imposed, which have resulted in the exclusion of any detailed discussion of complexes of the following types.

(*a*) Coordination complexes of the classical type, containing azo compounds bonded to the early transition metals, and the coinage metals. A variety of derivatives of this type have been obtained from metal halides (of Ti, V, Cu, etc.) and azoalkanes or azobenzene[1].

(*b*) Azo dyes involving metals, for example:

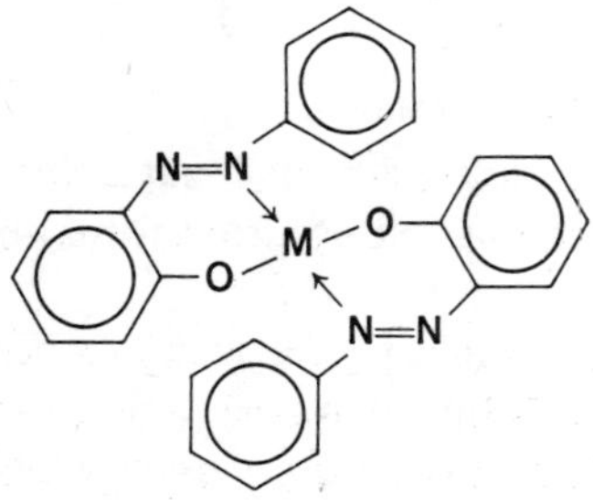

M = Cu(II), Ni(II), Co(II)

The chemistry of these compounds has been extensively studied in connection with their importance as dyestuffs. These complexes are characterized by coordination to the metal via the azo group, and an element other than carbon. Their chemistry has been surveyed recently[2,3].

The following survey, therefore, will discuss in some detail complexes containing azo compounds attached to (usually low-valent) transition metals via the N=N π-system; *ortho*-metallated complexes; alkyl- and aryl-azo metal complexes; derivatives containing rearranged azo compounds as ligands; and the use of transition metals in organic synthetic schemes based on azo compounds. In most cases, the literature has been searched to the end of 1973.

II. COMPLEXES INVOLVING NO METAL–CARBON INTERACTION

A. *Molybdenum*

Azobenzene slowly substitutes the two hydride ligands in dihydridobis-(π-cyclopentadienyl)molybdenum to give air-sensitive deep red crystals[4], [**2**; $M = Mo(\pi\text{-}C_5H_5)_2$]. The tungsten derivative could not be obtained, although similar complexes containing azodibenzoyl have been reported[143].

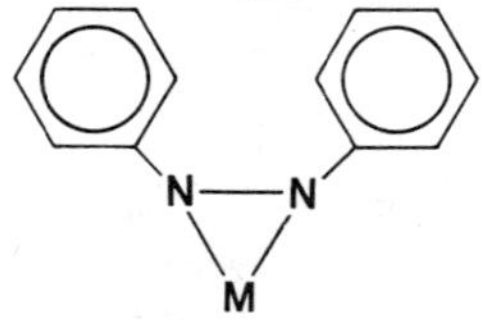

(2)

B. *Iridium*

Diethyl azodicarboxylate reacts with $IrCl(CO)L_2$ ($L = PPh_3$, $PMePh_2$) in benzene solution to form **3**[5] in which the N—N bond is retained,

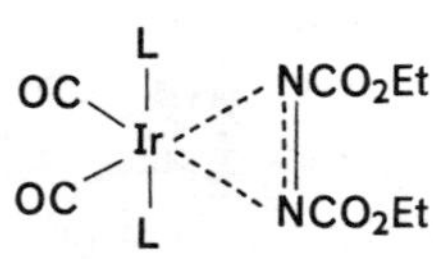

(3)

$L = PPh_3$ or $PMePh_2$

although an earlier report[6] of the complex indicated that it was too unstable to be fully characterized. A similar reaction occurred with 4-phenyl-1,2,4-triazoline-3,5-dione.

The bonding of these azo compounds to the metal is believed to be different from that in a classically π-bonded system, and approximates to a σ-bonded rigid three-membered ring model or to a model which essentially uses sp^2-hybridized nitrogen atoms. The electron-withdrawing substituents lower the energy of the π^*-orbitals to enhance $d\pi$–π^* back-bonding from the metal. The true bonding picture is probably between these two limiting types.

Cleavage of the N—N bond of hexafluoroazomethane with IrCl(CO)-$(PMePh_2)_2$ afforded two isomeric complexes[7], *cis*- and *trans*-IrCl(NCF_3)-(CO)$(PMePh_2)_2$. Initial formation of the adduct IrCl($CF_3N_2CF_3$)(CO)-$(PMePh_2)_2$ was suggested, and subsequent weakening and cleavage of the N—N bond were thought to follow as a result of the back-bonding by iridium *d*-electrons into the azomethane π^*-anti-bonding orbitals.

C. Nickel

Two laboratories simultaneously reported the formation of azobenzene–nickel(0) complexes[8,9] which, on the basis of displacement reactions and spectroscopic properties, appeared to incorporate the azobenzene as a side-bonded, olefin-type ligand. The isocyanide complex was prepared by reacting azobenzene with Ni(t-BuNC)$_2$ at ambient temperatures. Treatment of the appropriate bis(phosphine)nickel dichloride complex with metallic lithium as reducing agent, in the presence of azobenzene, afforded high yields of $(R_3P)_2$Ni(azobenzene) [**2**; M = Ni$(PR_3)_2$]; some reactions of these complexes were also reported[8].

An X-ray structure determination of bis(t-butylisocyanide)-(azobenzene)-nickel(0)[10], [**2**; M = Ni(CNBu-t)$_2$], confirmed the proposed structure (Figure 1). This revealed an N—N bond distance for the coordinated

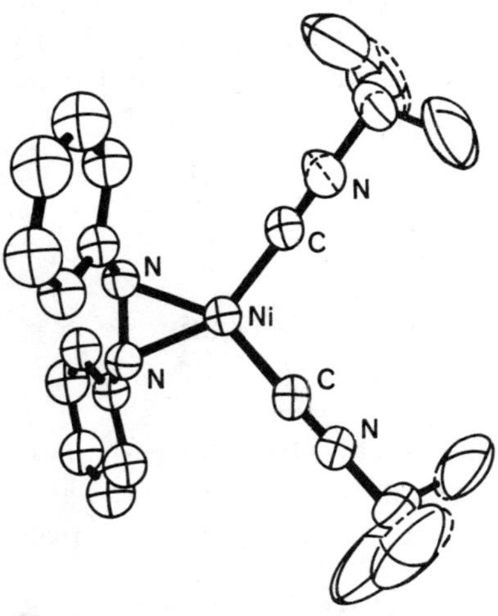

FIGURE 1. Structure of Ni(PhN_2Ph)($CNBu^t$)$_2$. (From ref. 10).

azobenzene of 1·385 Å, which is significantly longer than the corresponding distance in the free ligand, where distances of 1·33 Å and 1·17 Å have been reported[11] for the two different molecules in the asymmetric unit of *trans*-azobenzene. Indeed, the observed N—N distance is very close to the N—N single bond length of about 1·40 Å. Both this bond lengthening and the geometry within the coordinated azobenzene are consistent with the usual description of the π-bonding of electronegatively substituted olefins in transition metal complexes. Furthermore, there is no evidence to suggest that the nitrogen lone pairs affect the azobenzene-to-metal bonding in any way.

On the basis of this bonding model, Ibers and co-workers[10] proposed donation from the azobenzene π-orbitals to vacant orbitals on the metal with concomitant back-donation from filled metal orbitals to the π*-antibonding orbitals of the azo function.

D. Platinum

The complexes originally thought to contain coordinated azobenzene have been reformulated[11a] as the salts $[PhN_2HPh]_2PtCl_6$.

III. IRON CARBONYL DERIVATIVES CONTAINING A CYCLIC Fe_2N_2 SYSTEM

The reaction between organic nitrogen derivatives and iron carbonyls has proved a prolific source of unusual complexes, and the area has been described recently[12, 13]. However, those complexes which contain a cyclic (tetrahedral) Fe_2N_2 system (**4**) are pertinent to this account, and they are described below. Structural data have been summarized by Doedens[13a].

(4)

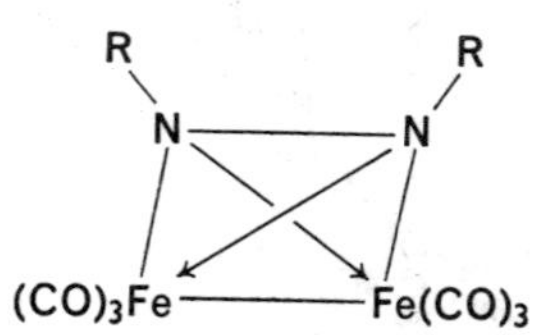

(5)

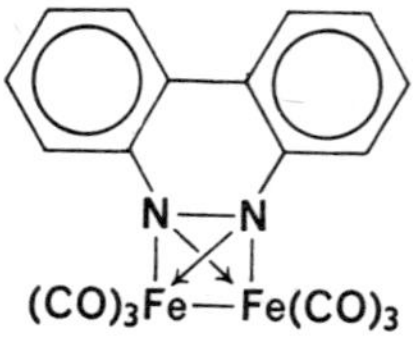

(6)

Although azobenzene reacts with iron carbonyls to give complexes containing *o*-semidine ligands*, azomethane (and also MeNCO) reacts to give complex **5**, R = Me[14]; 2-(methylazo)propene forms **5**, R =

* A more recent report has described the complex **5**, R = Ph, obtained from azobenzene and $Fe(CO)_5$[18].

$CMe{=}CH_2$[15]. Surprisingly, phenylazide reacts to give **5**, R = Ph[16], while 2-azidobiphenyl gave **5**, R = C_6H_4Ph-2, among the products[17]. 2,2′-Diazidobiphenyl afforded complex **6**, also formed from $Fe_2(CO)_9$ and benzo[c]cinnoline[18]. Other compounds of this type have been obtained using substituted cinnolines, 2,3-diazabicyclo[2.2.1]hept-2-ene, and dibenz[1.4.5]oxadiazepine, while 4,5,9,10-tetraazapyrene forms a $Fe_4(CO)_{12}$ complex[18].

Addition of two $Fe(CO)_3$ groups across N═N double bonds in other nitrogen heterocycles has been described. Thus, 3,6-diphenylpyridazine affords **7**[19], and unsymmetrical 1-pyrazolines give complexes of type **8**[20]. An interesting reaction of complex **7** is that with maleic anhydride, which forms the 1:1 adduct **7a** via a new type of cycloaddition reaction[22]. The pyridazine ligand changes from a six-electron donor in **7** to a four-electron

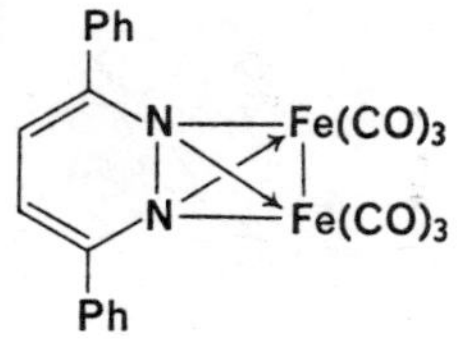

(7)

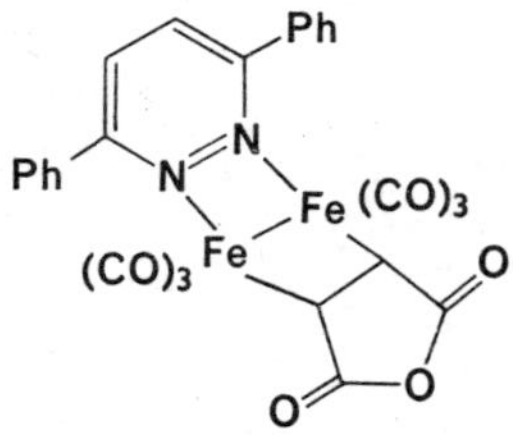

(7a)

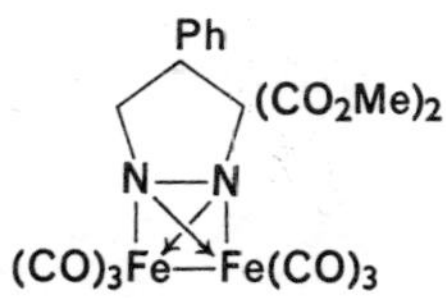

(8)

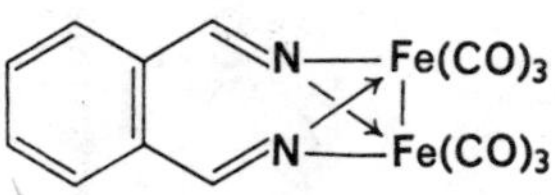

(9a)

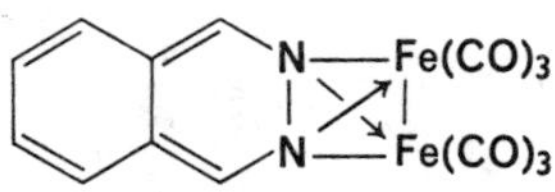

(9b)

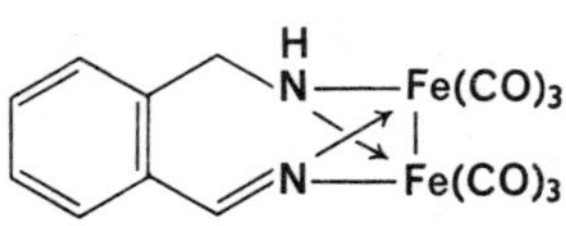

(10)

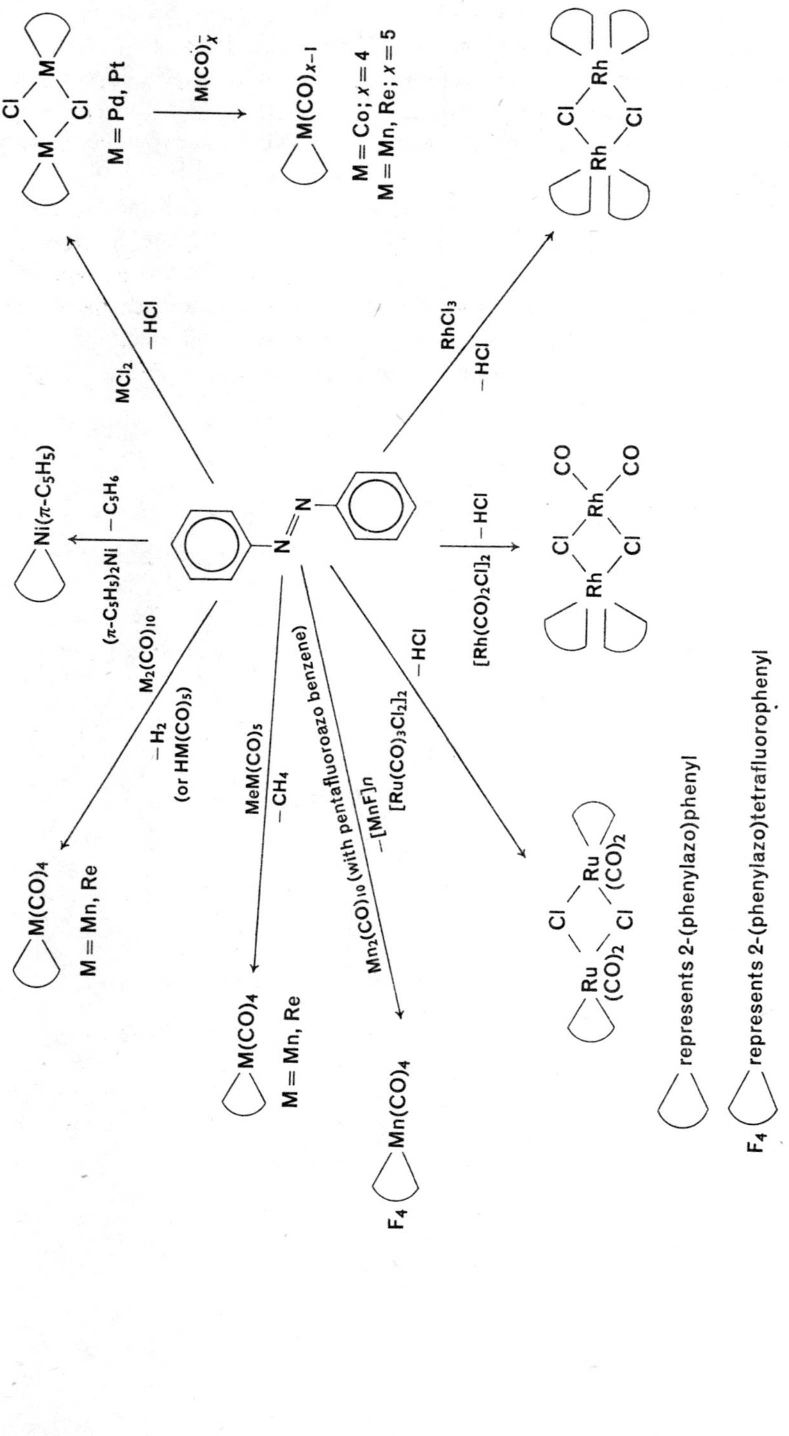

SCHEME 1. Some synthetic routes to 2-(phenylazo)phenylmetal derivatives.

donor in **7a**. Phthalazine reacts with $Fe_3(CO)_{12}$ to give **9**, which may be written in the benzenoid form **9a**, or the *o*-quinonoid form **9b**, the latter preserving an N—N bond. In methanol, the dihydrophthalazine derivative **10**, for which no *o*-quinonoid structure can be written, was formed[23].

IV. COMPLEXES CONTAINING *ORTHO*-METALLATED AZO COMPOUNDS AS LIGANDS

A. General Syntheses

Some preparative methods for these complexes are illustrated in Scheme 1; as shown, the *ortho*-hydrogen which is lost in all of these reactions is eliminated as a variety of small molecules, e.g. H_2*, HCl, C_5H_6 or CH_4. It is not clear how the fluorine is lost in the case of the ring-fluorinated azobenzenes, e.g., pentafluoroazobenzene. It is likely that the fluorine is eliminated either as HF, or as MeF, although an alternative possibility is the formation of a metal fluoride. Although $Mn(CO)_5F$ has not been described, and there is no evidence for its formation in these reactions, the analogous $(\pi\text{-}C_5H_5)Ru(PPh_3)_2F$ is a stable complex[24]. However, it is possible that $Mn_2(CO)_{10}$ and $(\pi\text{-}C_5H_5)Ru(PPh_3)_2Me$ extract a fluorine atom by different mechanisms, since $(\pi\text{-}C_5H_5)Ru(PPh_3)_2Me$ metallates both $C_6F_5N{=}NC_6H_5$ and $C_6F_5N{=}NC_6F_5$, whereas $Mn_2(CO)_{10}$ only reacts with the former. Furthermore, the reaction between $Mn_2(CO)_{10}$ and $C_6F_5N{=}NC_6H_5$ gave two products, one metallated in the C_6F_5-ring and the other in the C_6H_5-ring. It is possible that this reaction proceeds via initial *ortho*-metallation of the C_6H_5-ring, with elimination of the hydrogen atom as $Mn(CO)_5H$ which undergoes reaction with a second mole of pentafluorobenzene with elimination of HF.

B. Molybdenum

The only reported example[6] of a 2-(phenylazo)phenylmolybdenum complex [**1**; $M = Mo(CO)_2(\pi\text{-}C_5H_5)$] was isolated in low yield from the reaction between azobenzene and $[(\pi\text{-}C_5H_5)Mo(CO)_3]_2$ or $(\pi\text{-}C_5H_5)\text{-}Mo(CO)_3Me$. The major product in each of these reactions was one in which the azobenzene nucleus has undergone a rearrangement to yield an *o*-semidine derivative (see Section V, A).

* It is interesting that some hydrides do not react with azobenzene, e.g., $(\pi\text{-}C_5H_5)Ru(PPh_3)_2H$, suggesting that elimination of molecular hydrogen is a relatively difficult process. In reactions with e.g. $Mn_2(CO)_{10}$, where the product $(azb)Mn(CO)_4$ is obtained in only low yield, the *ortho*-hydrogen may be lost as $HMn(CO)_5$.

C. Manganese and Rhenium

The complex $[(azb)PdCl]_2$ readily undergoes ligand-transfer reactions with several metal carbonyl anions[25]. Using $[Mn(CO)_5]^-$, the complex $(azb)Mn(CO)_4$ [**1**; M = $Mn(CO)_4$] was obtained in 24% yield. More recently it has been reported that complexes of this type can be prepared directly from $MeMn(CO)_5$ and the azo compound in almost quantitative yield[26]. Many substituted complexes have been obtained in the course of a study of the mechanism of this reaction[27]. Substitution of the carbonyl groups, e.g., by phosphine ligands, has been reported. Thus triphenylphosphine gives $(azb)Mn(CO)_3(PPh_3)$ virtually quantitatively, also obtained in low yield from azobenzene and $[Mn(CO)_4(PPh_3)]_2$ directly.

Treatment of $(azb)Mn(CO)_4$ with excess $MeMn(CO)_4$ afforded the binuclear complex [**11**; M = $Mn(CO)_4$], in which both nitrogens are used to coordinate two metal atoms which also σ-bond to the two aromatic rings[28].

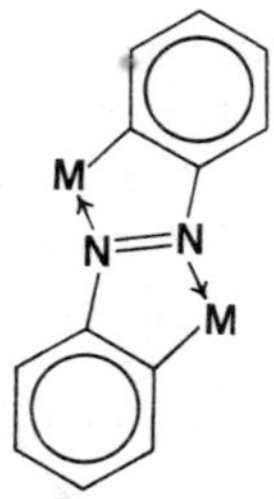

(11)

As expected, the azobenzene chemistry of rhenium generally runs parallel with that of manganese, described above. Tetracarbonyl-2-(phenylazo)phenylrhenium [**1**; M = $Re(CO)_4$] can be prepared either by Heck's ligand-transfer reaction[25], or by the thermal reactions between azobenzene and $Re_2(CO)_{10}$ or $Re(CO)_5Me$[26].

D. Iron

The reactions between iron carbonyls and azobenzene were first reported in 1965[29] but the initial formulation of the products was later shown to be incorrect by X-ray analyses[30]. The products are strongly dependent on both the substitution of the azobenzene and the reaction conditions[31]. While most reactions yielded *o*-semidine derivatives, u.v. irradiation of a mixture of 4,4′-dimethoxyazobenzene and $Fe(CO)_5$ in benzene gave a low yield of the *o*-metallated complex (**12**), as the first example of a complex containing two *o*-metallated azobenzene ligands attached to the same metal atom.

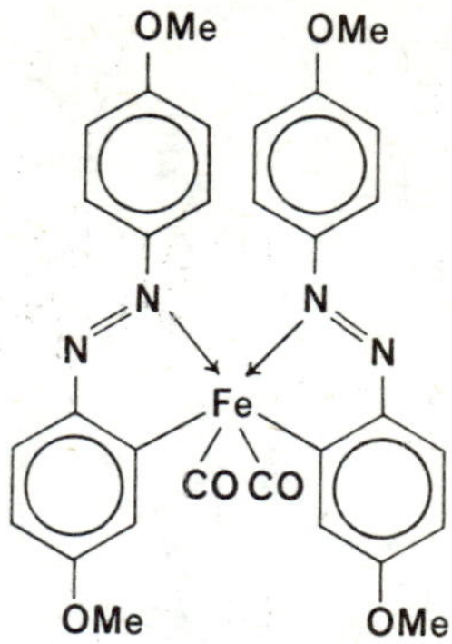

(12)

Azobenzene reacts slowly with both $[(\pi\text{-}C_5H_5)Fe(CO)_2]_2$ and $(\pi\text{-}C_5H_5)$-$Fe(CO)_2Me$[26] to give a light- and air-sensitive green oil [**1**; M = $Fe(CO)(\pi\text{-}C_5H_5)$]; treatment of this with triphenylphosphine afforded deep-purple crystals of (azb)$Fe(PPh_3)(\pi\text{-}C_5H_5)$. It is of interest that these molecules are chiral, and the oily nature of the monocarbonyl has been ascribed to the presence of both optical isomers[26].

E. *Ruthenium*

The reaction of $[Ru(CO)_3Cl_2]_2$ with azobenzene in petroleum at ca. 100°C was found to produce only low yields of **13**[32, 33]; a simple, high-yield route to the same azobenzene derivative was the direct reaction of $[Ru(CO)_3Cl_2]_2$ with molten azobenzene at 100°C.

The ready availability of **13** enabled a study to be made of the effects of other ligands on the bonding of the phenylazophenyl moiety and, in particular, to determine the ease, or otherwise, of synthesis of complexes containing the phenylazophenyl group linked by a metal–carbon σ-bond only. Most of these reactions are summarized in Scheme 2. Monodentate ligands, L, cleaved the halogen bridge to give monomeric complexes (azb)LRu$(CO)_2$Cl. The bidentate ligand, 2,2′-bipyridyl (bipy) gave the ionic complex [(azb)Ru$(CO)_2$(bipy)][(azb)Ru$(CO)_2Cl_2$] (**14**) which contains an unusual example of a cationic Ru—C σ-bond.

The readiness of the phenylazophenyl ligand to unlock its chelate ring was demonstrated in the reaction with cyclopentadienide anion. With sodium cyclopentadienide, complex **13** afforded a red oil; this oil was shown to be identical with the monocarbonyl complex [**1**; M = $Ru(CO)(\pi\text{-}C_5H_5)$] first prepared by irradiating a mixture of azobenzene and $[(\pi\text{-}C_5H_5)Ru(CO)_2]_2$[26]. In contrast, the reaction between **13** and thallium cyclo-

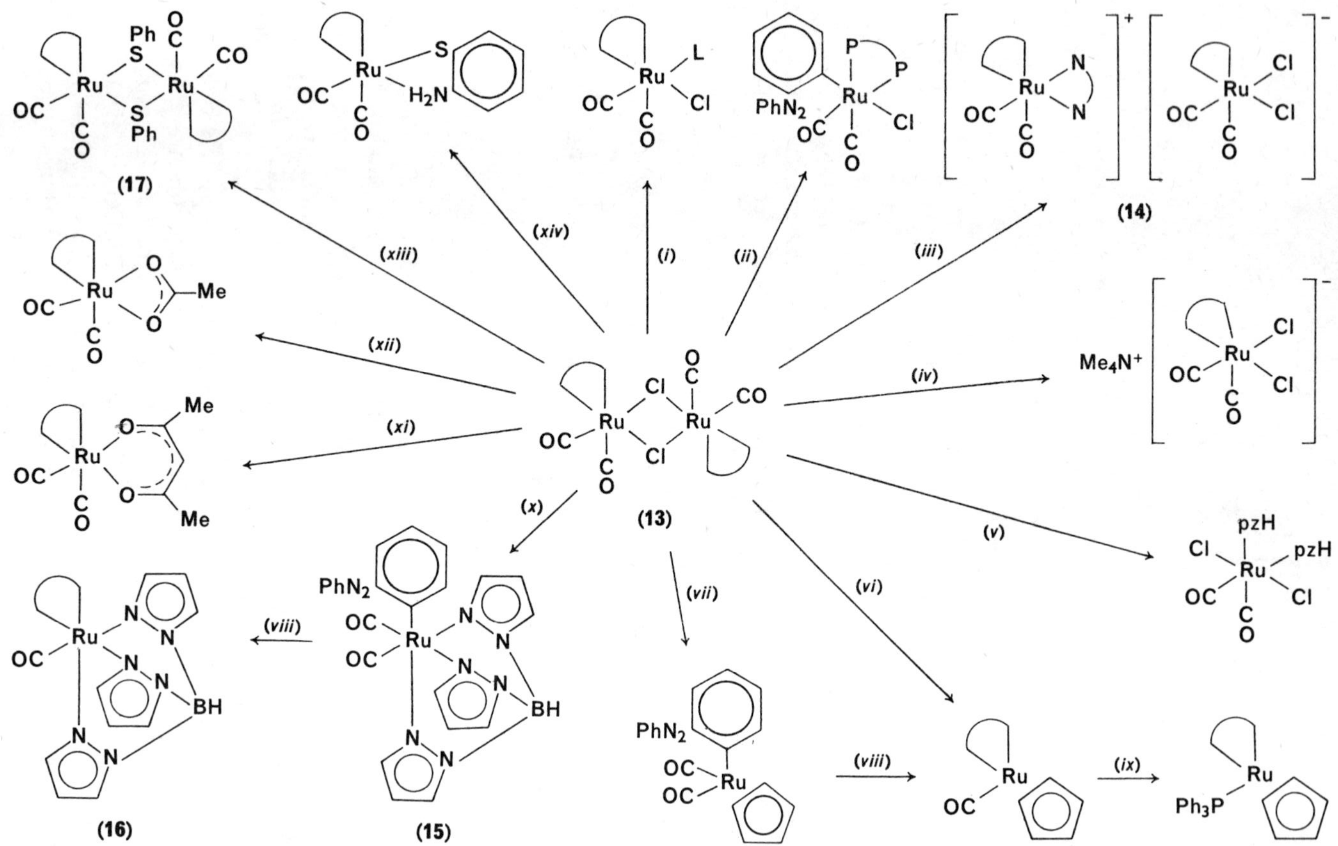

Reagents: (*i*) L=PPh_3, $AsPh_3$, *p*-$MeC_6H_4NH_2$, py, imzH; (*ii*) dppe; (*iii*) bipy; (*iv*) Me_4NCl; (*v*) pyrazole; (*vi*) NaC_5H_5; (*vii*) TlC_5H_5; (*viii*) u.v. light; (*ix*) PPh_3; (*x*) K or Tl [HB(pz)$_3$]; (*xi*) Tlacac; (*xii*) AgOAc; (*xiii*) NaSPh; (*xiv*) Na(abt). ◠ represents 2-(phenylazo) phenyl.

SCHEME 2. Some reactions of $[(azb)Ru(CO)_2Cl]_2$.

pentadienide gave the dicarbonyl, (azb)Ru$(CO)_2(\pi\text{-}C_5H_5)$, which contains the phenylazophenyl-2C ligand bonded by a metal–carbon σ-bond only. U.v. irradiation of solutions of this complex resulted in loss of carbon monoxide and formation of the monocarbonyl complex.

Tris(pyrazolyl)borate anion was found to behave similarly to cyclopentadienide anion, affording the dicarbonyl complex **15**, u.v. irradiation of which gave the monocarbonyl complex **16**. Other anionic chelates, e.g., acetylacetonate or *o*-aminobenzenethiolate, gave monomeric (azb)Ru$(CO)_2$(chelate) complexes; benzenethiolate gave dimeric [(azb)Ru$(CO)_2$(SPh)$]_2$ (**17**), a compound containing SPh bridging groups.

The thermal reaction of azobenzene with dodecacarbonyltriruthenium[33a] afforded three products, the most abundant being the dimeric [(azb)Ru$(CO)_3]_2$ (**18**). Bromination of this affords the bromo analogue of **13**. The nature of the other products awaits the results of X-ray structural determinations.

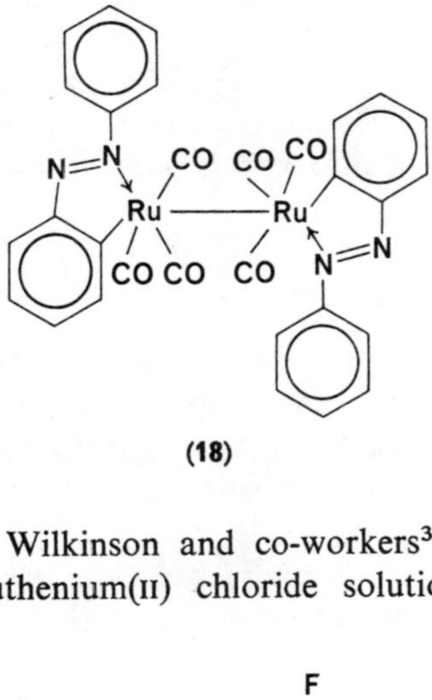

(**18**)

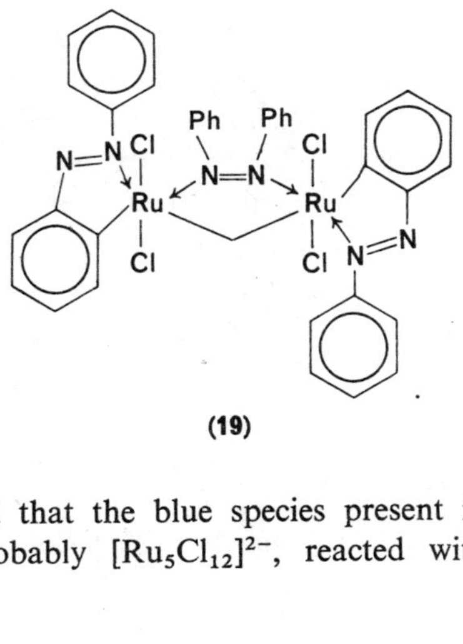

(**19**)

Wilkinson and co-workers[34] found that the blue species present in ruthenium(II) chloride solutions, probably $[Ru_5Cl_{12}]^{2-}$, reacted with

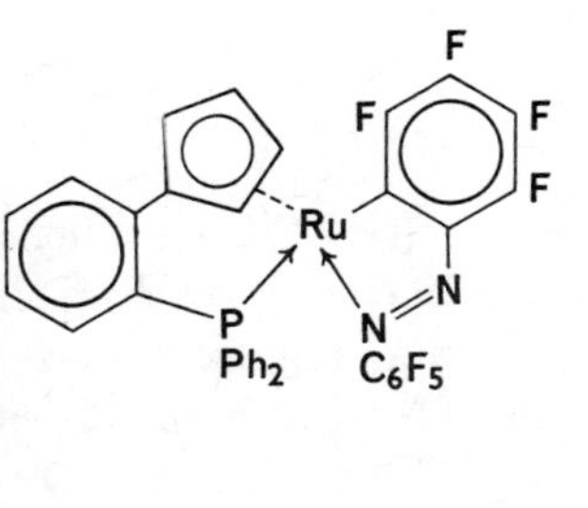

(**20**)

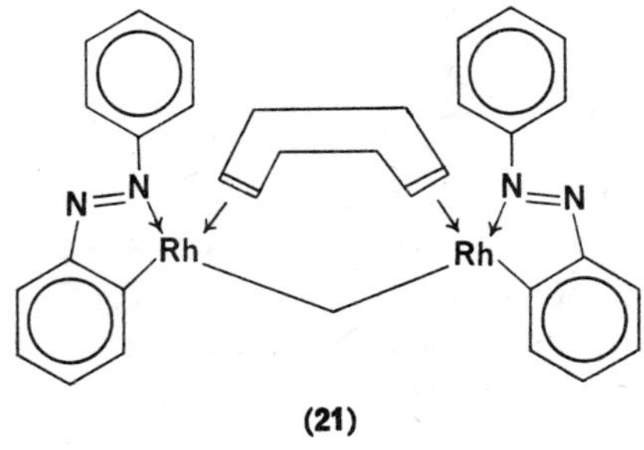

(**21**)

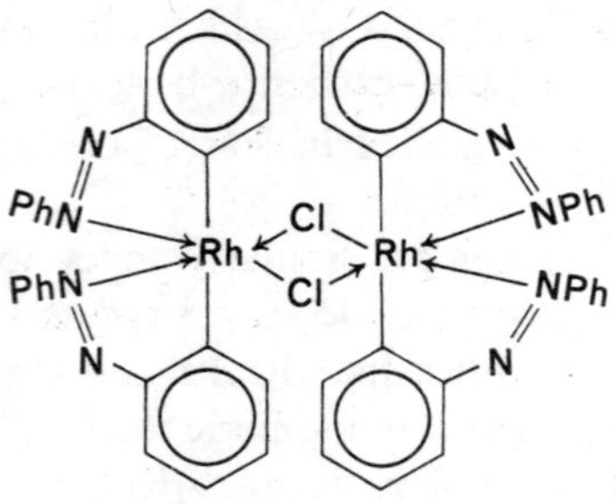

(22)

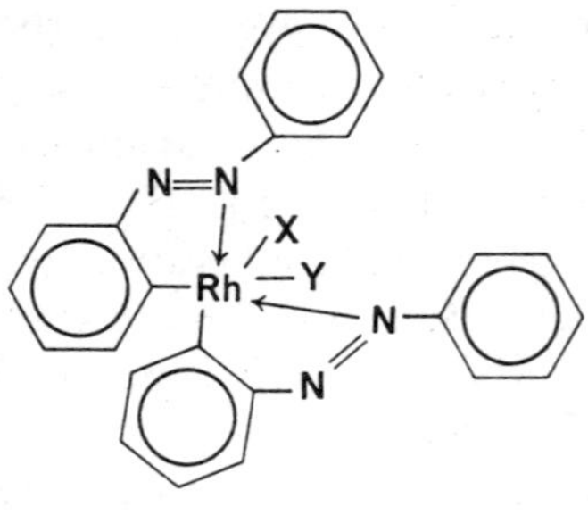

(23)

azobenzene to give a complex thought to have the unusual structure **19**. Treatment of **19** with triphenylphosphine caused bridge cleavage and reduction to [**1**; M = $RuCl(PPh_3)_3$].

Cyclopentadienylruthenium phosphine complexes have proved to be particularly prone to the *o*-metallation reaction[24, 35]. The reactions between azobenzenes and $(\pi\text{-}C_5H_5)Ru(PPh_3)_2Me$ afford complexes of the type $(\pi\text{-}C_5H_5)Ru(PPh_3)$(azb), discussed above. A ready reaction with ring-fluorinated derivatives gave a mixture of the complexes formed by loss of the elements of MeH or MeF. With decafluoroazobenzene, the major product is the unusual complex (**20**)[35a].

F. Cobalt

Tricarbonyl-2-(phenylazo)phenylcobalt [**1**; M = $Co(CO)_3$] can be prepared by the ligand-transfer reaction between the tetracarbonylcobalt anion and $[(azb)PdCl]_2$[8]; the direct reaction between octacarbonyldicobalt and azobenzene affords carbonylated organic derivatives (see Section VI, A).

As with manganese, a carbonyl group in this complex may be easily substituted on reaction with phosphines, giving complexes of the type (azb)$Co(CO)_2$(L) (L = PPh_3, $PMePh_2$, etc.)[28].

G. Rhodium

The thermal reactions between azobenzene and [(1,5-cyclooctadiene)-$RhCl]_2$ and $RhCl_3 \cdot 3\,H_2O$ gave complexes **21** and **22** respectively[36, 37], the latter being one of the few examples of a complex containing two 2-(phenylazo)phenyl moieties bonded to one metal atom. Treatment of **22** with sodium acetate afforded the dark red acetato-complex (**23**; X, Y = OAc) in which the two Rh—C bonds are mutually *cis* and the two Rh—N bonds

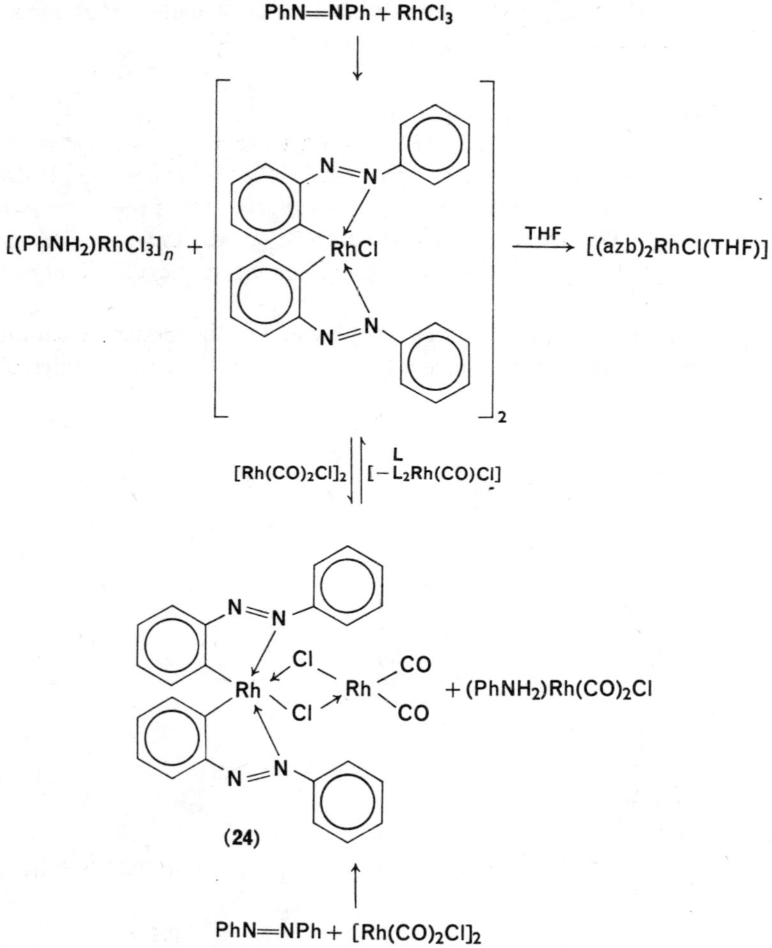

SCHEME 3. L = $AsPh_3$, PPh_3.

trans to one another[38]. It was later found that the reaction between azobenzene and ethanolic rhodium(III) chloride yielded the dimeric complex $[(azb)_2RhCl]_2$ and a second, insoluble complex $[RhCl_3(PhNH_2)_2]_n$ (see Scheme 3)[39]. Recrystallization of **22** from tetrahydrofuran gave the mononuclear adduct [$(azb)_2$RhCl(THF)] (**23**); X = Cl, Y = THF) while treatment of **22** with $[Rh(CO)_2Cl]_2$ afforded $[(azb)Rh(CO)Cl]_2$ (**24**), also obtained from the reaction between azobenzene and $[Rh(CO)_2Cl]_2$[39]. This complex is unusual in containing both Rh(I) and Rh(III) atoms linked by halogen bridges[39a]. From the reaction between azobenzene and $[Rh(CO)_2Cl]_2$ a volatile violet complex was obtained in addition to **24**, identified as a second aniline complex, $RhCl(CO)_2(PhNH_2)$.

H. Iridium

Vaska's complex, $IrCl(CO)(PPh_3)_2$, reacts slowly with azobenzene (refluxing xylene, 3 days, 60%) to give the carbonyl-free complex hydride (**25**; X = H). *o*-Bromoazobenzene reacted similarly to give the corresponding bromide[40]. The former reaction is notable on account of its being an unusual example of oxidative addition of an aromatic C—H bond to iridium (see below).

From a reaction between Vaska's complex and *p*-fluorophenyldiazonium cation, the metallated azo complex **26** was obtained[41], in which migration

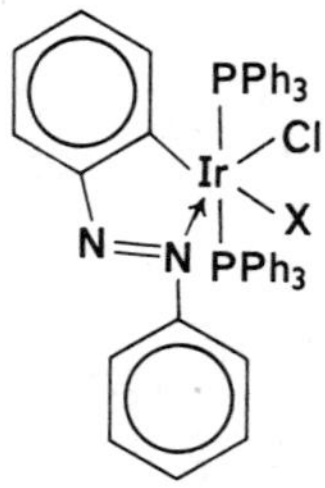

(25)

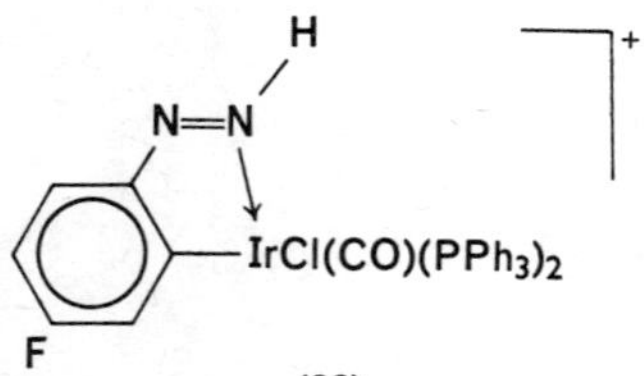

(26)

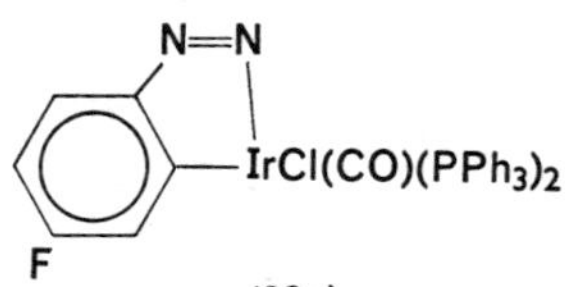

(26a)

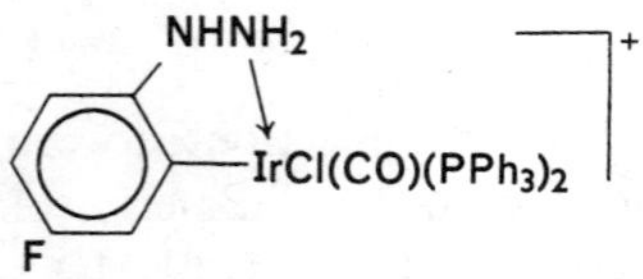

(27)

of an *ortho*-proton to nitrogen has occurred. Two reactions are of interest: treatment with a base results in deprotonation to give **26a**, a neutral complex in which the nitrogen is now σ-bonded to the metal. This reaction is reversed by addition of acid. Secondly, the complex can be hydrogenated under ambient conditions (H_2, 1 atm/25°C, Pd/$BaSO_4$) to the hydrazine complex

27. The relevance of these model reactions to current theories concerning the sequence of reactions occurring during nitrogen fixation has been emphasized[42].

I. Nickel

In 1963 Kleiman and Dubeck[43] reported the first complex containing an *o*-metallated azobenzene ligand prepared from azobenzene and nickelocene, π-cyclopentadienyl-2-(phenylazo)phenylnickel [**1**; M = Ni(π-C_5H_5)]; reaction of the product with lithium tetradeuterioaluminate afforded *o*-deuterio-azobenzene. It was later shown that $[(\pi\text{-}C_5H_5)Ni(CO)]_2$ reacted with azobenzene to give the same product[26]. A structure was put forward in which it was proposed that the azo-group participated in delocalized bonding with the metal, comparable to unsaturated carbon systems. This has long been assumed to be in error, the correct structure finally being proved by X-ray analysis of the related azotoluene complex[44].

Russian workers found that many substituted azobenzenes reacted with nickelocene to afford related products[45,46], often as mixtures of isomers. The reaction of nickelocene with *o*-haloazobenzenes proved to be noteworthy. Firstly, it was found that only the halogen atom was substituted affording [**1**; M = Ni(π-C_5H_5)] alone instead of a possible mixture with two isomers (**28a** and **b**). Further, the rate of formation was shown to

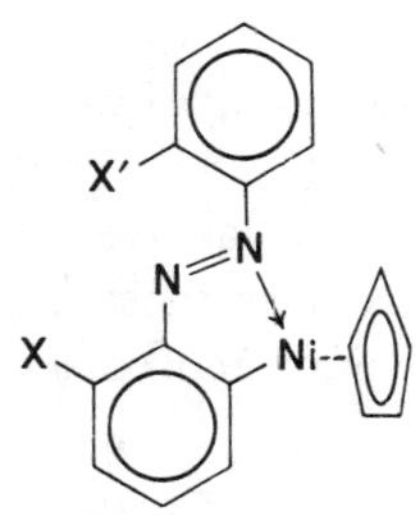

(28) **a**: X = H, X' = Cl
b: X = Cl, X' = H

increase in the series azobenzene, *o*-chloroazobenzene, *o*-bromoazobenzene. Such an increase could be explained either by the fact that the halogen is initially coordinated to the metal atom, whereby the azobenzene would be favourably orientated with respect to the nickel atom, or alternatively by the fact that an electron is transferred from nickelocene to azobenzene whereby an ion pair is formed, followed by halogen elimination and complex formation. In the presence of excess *o*-haloazobenzene, (azb)Ni(π-C_5H_5) was found to undergo further transformations affording a cyclopentacinnoline (see Section VI, A).

2,2′-Dichloroazobenzene readily reacts with nickelocene to give **28a**, which was found[46] to undergo further reaction with excess nickelocene to afford the binuclear complex [**11**; M = Ni(π-C_5H_5)].

J. Palladium and Platinum

Palladium(II) chloride reacts with azobenzene at room temperature, in methanol, to give chloro-2-(phenylazo)phenylpalladium dimer[47] (**29**; M = Pd) which, when treated with four equivalents of triethylphosphine[48] affords the complex PdCl(azb)$(PEt_3)_2$ (**30**). An X-ray structure determina-

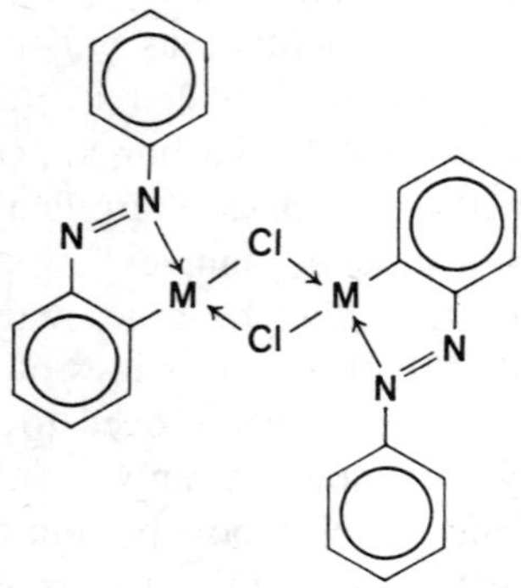

(29)

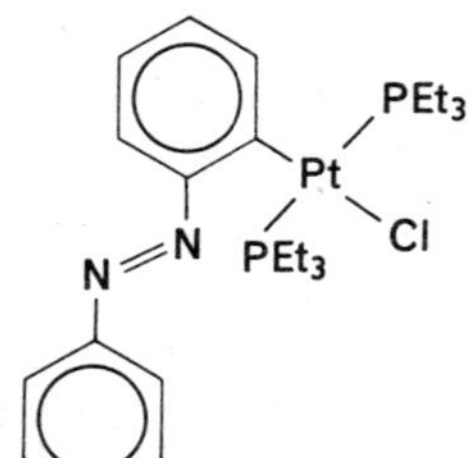

(30)

tion (Figure 2) showed[48] the latter complex to contain the phenylazophenyl-2C ligand, linked to the metal by a metal–carbon σ-bond only; this group assumes a *trans*-configuration and has average bond lengths and angles which do not differ significantly from those of *trans*-azobenzene or *trans*-*p*-azotoluene. Manganese, rhenium and cobalt carbonyl anions undergo

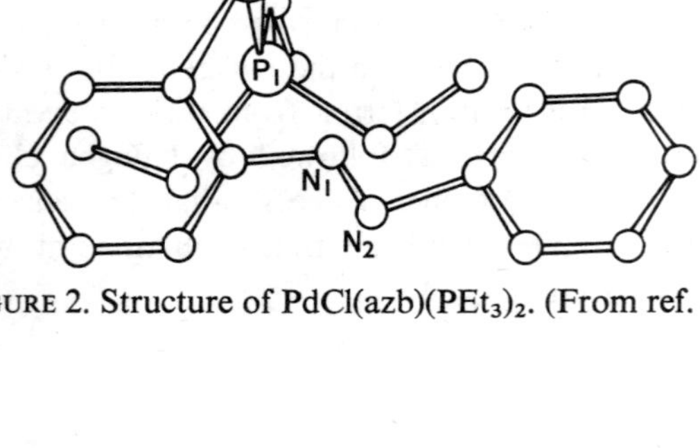

FIGURE 2. Structure of PdCl(azb)$(PEt_3)_2$. (From ref. 48).

ligand-transfer reactions with **29** affording the corresponding 2-(phenylazo)-phenylmetal derivatives[25]. In contrast, the reactions of the azoanisole and azotoluene analogues of **29** with tetracarbonyliron dianion yielded bis(substituted phenylazophenyl)palladium derivatives[31].

Treatment of $[PdCl(azb)]_2$ with sodium or thallium cyclopentadienide causes cleavage of the chlorine bridges and gives π-cyclopentadienyl-2-(phenylazo)phenylpalladium; similarly treatment with sodium methoxide and 2,4-pentanedione affords acetylacetonato-2-(phenylazo)phenyl-palladium[25,36].

The i.r. spectra in the range 600–120 cm^{-1} of palladium(II) chloro- and bromo-bridged complexes analogous to **29** have been reported[49] and the bridging-halogen stretching frequencies were discussed on the basis of a planar asymmetric PdX_2Pd unit. Bridge-splitting reactions with mono- and bi-dentate ligands bearing nitrogen or phosphorus as donor atoms were also reported (see Scheme 4), and the configurations of the resulting products were interpreted on the basis of their far-i.r. spectra.

For the halo-bridged dimers only two ν(PdX) bands in the range 600–120

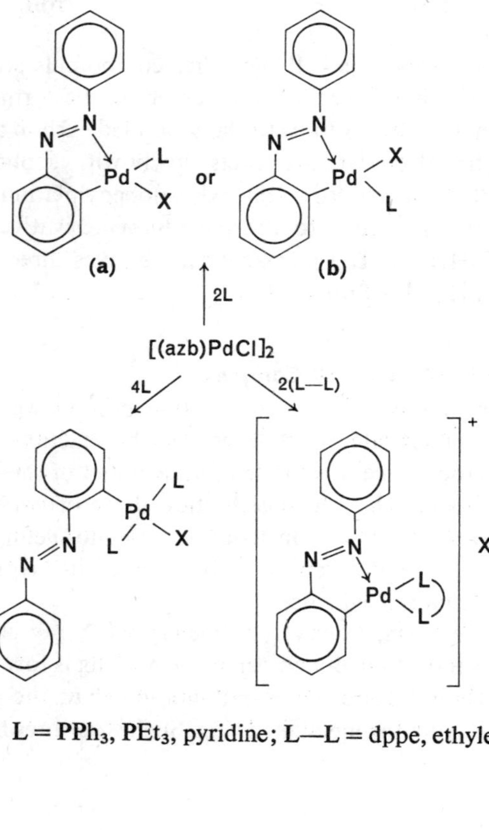

SCHEME 4. L = PPh_3, PEt_3, pyridine; L—L = dppe, ethylenediamine.

cm^{-1} were detected for each compound. On the basis of the higher *trans*-influence of a σ-bonded carbon compared with that of a nitrogen atom, the higher frequency band was attributed to the stretching vibration ν(PdX) *trans* to the nitrogen atom, and the lower one to ν(PdX) *trans* to the carbon atom.

The bridge-splitting reactions can give two possible isomers **a** or **b** (Scheme 4). Study of the i.r. spectra revealed that (azb)Pd(pyridine)Cl was formed as a mixture of both **a** and **b**, while (azb)Pd(pyridine)Br was the only

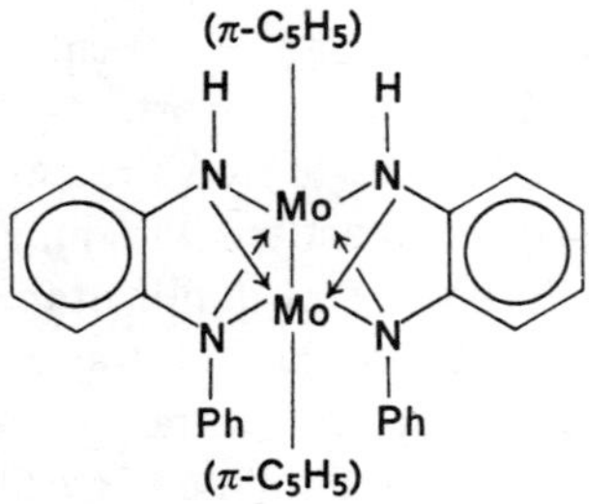

(31a)

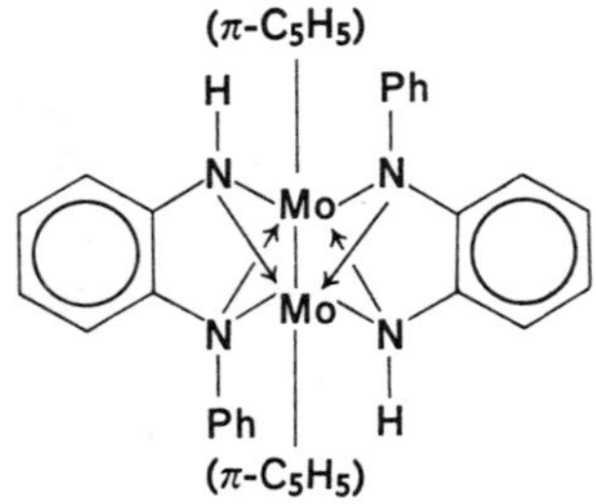

(31b)

complex to exist solely as **a**; all the other compounds studied had configuration **b**. With phosphines a third isomer can be formed **(30)**; in this case the halogen was always *trans* to the σ-bonded carbon atom.

Potassium tetrachloroplatinate reacts slowly with azobenzene at room temperature, affording chloro-2-(phenylazo)phenylplatinum dimer[7] (**29**; M = Pt) which is analogous to the palladium derivative. The reaction between Pt(π-C_3H_5)(π-C_5H_5) and azobenzene gives directly the complex Pt(azb)(π-C_5H_5) [**1**; M = Pt(π-C_5H_5)][28].

K. Structures of o-Metallated Complexes

Although the presence of a metal–carbon σ-bond was easily demonstrated in these *o*-metallated complexes, e.g., by reaction with $LiAlD_4$ to give *o*-deuterioazobenzene, the mode of attachment of the N═N unit was not so clearly shown. The two possibilities, those of two-electron donor bonds formed by the N═N system acting as a pseudo-olefin, or by donation from a nitrogen lone-pair, were initially resolved in favour of the latter largely by circumstantial evidence.

In the last few years, however, single-crystal X-ray studies have unequivocally shown that a five-membered chelate ring is formed, incorporating a metal–carbon σ-bond and a σ-donor bond to the metal from the nitrogen furthest from the metallated ring. Such structures have been found

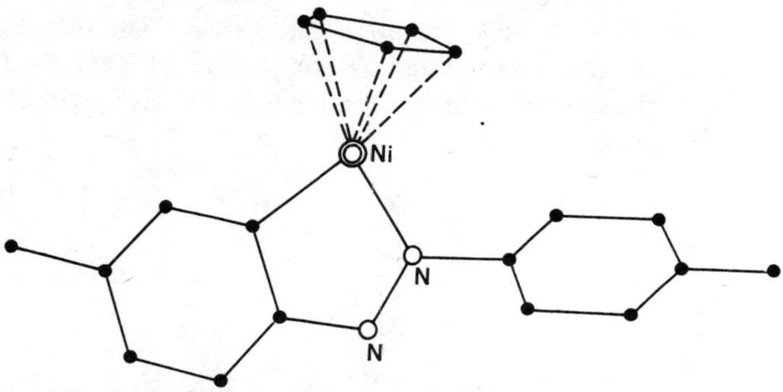

FIGURE 3. Structure of (π-C_5H_5)Ni(C_6H_3MeN=NC_6H_4Me-*p*). (From ref. 44).

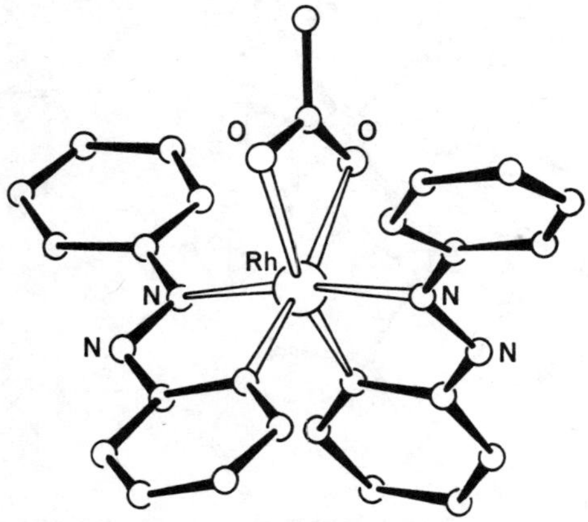

FIGURE 4. Structure of $(azb)_2$Rh(OAc). (From ref. 38).

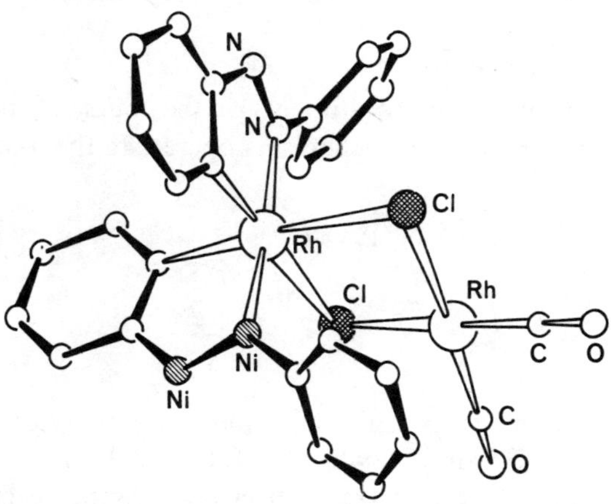

FIGURE 5. Structure of $(azb)_2RhCl_2Rh(CO)_2$. (From ref. 39a).

for (π-C_5H_5)Ni(C_6H_3MeN=NC_6H_4Me) (Figure 3), $(azb)_2$Rh(OAc) (Figure 4), the unusual rhodium(I)-rhodium(III) complex $(azb)_2RhCl_2Rh(CO)_2$ (Figure 5), and the cationic iridium complex, [p-$FC_6H_3N_2$HIrCl(CO)-$(PPh_3)_2$]BF_4 (Figure 6).

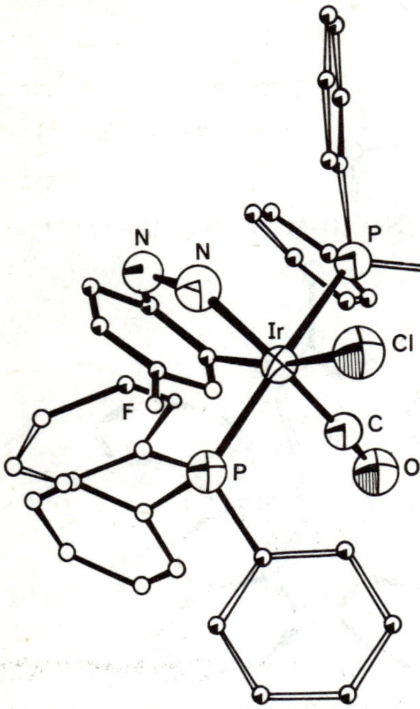

FIGURE 6. Structure of [Ir(HN=NC_6H_3F-*p*)Cl(CO)$(PPh_3)_2$]BF_4. (From ref. 41).

L. Reaction Mechanisms

Studies of the effects of aryl substituents on the course of these reactions have shown that two different mechanisms operate in the *o*-metallation of

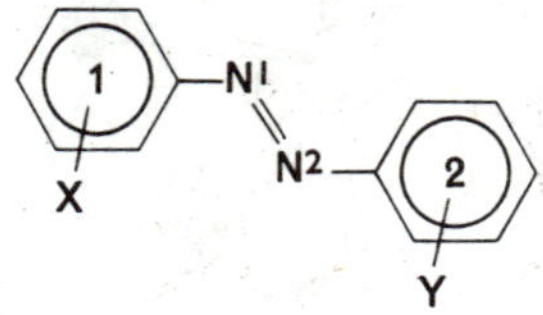

azobenzenes[27]; the first step in this reaction is probably coordination to the metal of either N^1 or N^2, which is followed by an intramolecular metallation in ring (2) or ring (1), respectively. In the reaction between azobenzene and $PdCl_2$, for example, the initial product is the *N*-bonded complex $PdCl_2(azbH)_2$ (also obtained from $PdCl_2(PhCN)_2$ and azobenzene[50]); continued reaction affords the metallated complex $[PdCl(azb)]_2$[51].

Although results of palladation of substituted azobenzenes indicated that metallation of the more electron-rich ring occurred, the examples used were substituted *para* to the azo function[52, 53]. A more recent and detailed study used *meta*-substituted azobenzenes, since the substituent will be *ortho* or *para* to the metal atom if metallation occurs in the substituted ring, and hence any substituent effects may be more directly related to the reaction.

TABLE 1. Isomers formed in metallation reactions of $m\text{-}FC_6H_4N{=}NC_6H_5$

Complex	M	^{19}F Chemical shift[a]	Amount (%)
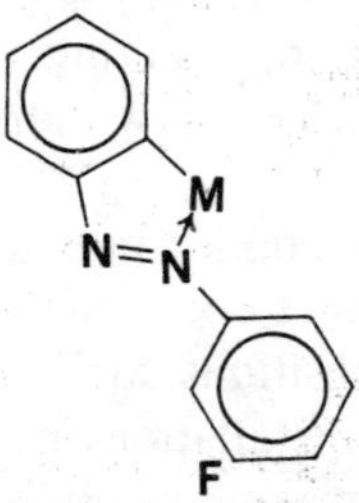	$Pd(\pi\text{-}C_5H_5)$	109·8 p.p.m.	80
	$Mn(CO)_4$	108·7	20
	H	110·1	—
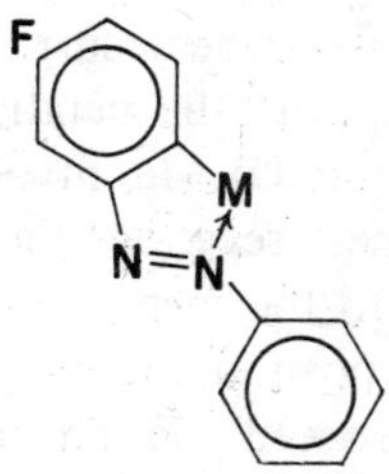	$Pd(\pi\text{-}C_5H_5)$	117·4	20
	$Mn(CO)_4$	—	—
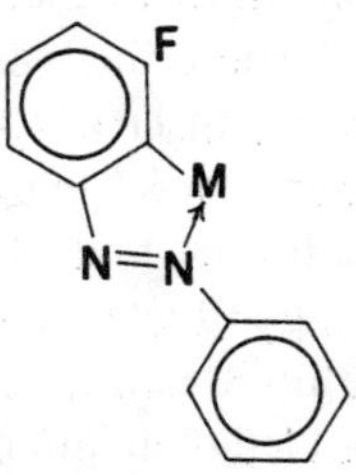	$Pd(\pi\text{-}C_5H_5)$	90·5	<1
	$Mn(CO)_4$	86·2	80

[a] Relative to CCl_3F (0·0 p.p.m.).

The (π-cyclopentadienyl)palladium and tetracarbonylmanganese derivatives of m-$XC_6H_4N{=}NC_6H_5$ (X = F, CF_3, CO_2Et, Me or OMe) have been prepared. There are three possible isomers of each complex, and the configuration, ratio and number of isomers in each case have been determined using ^1H and ^{19}F n.m.r. spectrometry.

The results obtained using m-$FC_6H_4N{=}NC_6H_5$ are illustrative, and are detailed in Table 1, together with their characteristic ^{19}F chemical shifts. These observations are consistent with the accepted *electrophilic* mechanism for palladation[53], in that 80% of the resulting mixture is metallated in the non-fluorinated ring (A). Further, where palladation has occurred in the substituted ring (B), reaction occurs at C(6), i.e., the position most favoured for electrophilic attack. Using $Mn(CO)_5Me$, the major isomer obtained (C) is that substituted on C(2), a result consistent with *nucleophilic* attack on a carbon atom powerfully activated by the inductive effect of fluorine. The formation of the other isomers in these reactions is accounted for by (*i*) coordination to either nitrogen (their donor power being little affected by the *meta* substituent) occurring before metallation, resulting in two competitive reactions, and (*ii*) the presence of the electron-withdrawing arylazo function in each ring.

Results found using substituents other than fluorine, taking into account steric factors, also show that palladation of azobenzenes occurs at the most electron-rich *ortho*-position; when all available positions are strongly deactivated no palladation occurs. On the other hand, metallations involving $Mn(CO)_5Me$ take place at the most electron-deficient *ortho*-position.

These results are consistent with palladation of azobenzene occurring by electrophilic attack on the aromatic ring by the metal; while the metallation reactions of low-valent metal complexes, such as $Mn(CO)_5Me$, involve a nucleophilic mechanism. It is probable that in all these reactions the first step is the formation of a nitrogen σ-donor bond to the metal which is followed by o-metallation in the most favourable position. It has been suggested that reactions involving $Mn_2(CO)_{10}$ proceed via an initial displacement of carbonyl with coordination of the nitrogen to the metal. Lack of back-bonding ability of the ligand atom would increase the nucleophilicity of the metal atom, thereby promoting insertion of the metal into the sterically favoured o-carbon–hydrogen bond. The metal hydride so formed may decompose in the reaction conditions, with elimination of hydrogen. Metallation would be facilitated if a good leaving group, such as methyl, is present, in which case methane is eliminated.

Consideration of the implications of these results has led to an alternative synthesis of metallated derivatives by nucleophilic displacement of a ring fluorine. Fluorocarbons, especially polyfluoroaromatic compounds, are particularly susceptible to nucleophilic attack. The reactions of various transition metal substrates with three fluorinated azobenzenes, $C_6F_5N{=}NC_6H_5$, *o*-$HC_6F_4N{=}NC_6F_5$ and $C_6F_5N{=}NC_6F_5$, were examined. With $Mn_2(CO)_{10}$ and the pentafluoro compound, two products were obtained, containing the azo ligand metallated in the C_6H_5 and C_6F_5 rings, respectively. However, in the reaction with palladium(II) chloride, no evidence was obtained for the formation of any complex other than $[(C_6F_5N{=}NC_6H_4)PdCl]_2$[54]. Related studies with the highly nucleophilic $(\pi\text{-}C_5H_5)Ru(PPh_3)_2Me$ showed that even decafluoroazobenzene entered into the fluorine-abstraction reaction, giving the unusual complex **22**; this azo compound does not react at all with either $Mn_2(CO)_{10}$ or $PdCl_2$[24].

V. COMPLEXES CONTAINING LIGANDS RELATED TO o-SEMIDINE

In the early studies of reactions between azoarenes and iron carbonyls it was found that complexes containing rearranged ligands isomeric with the azoarene were formed. Degradation with $LiAlH_4$ afforded *o*-semidine (*o*-aminodiphenylamine), and X-ray studies revealed the presence of the *o*-$(NH)C_6H_4(NPh)$ ligand, often acting as a six-electron donor. Similar complexes were obtained containing other elements, and these studies are summarized in this section.

A. Molybdenum

The reaction between azobenzene and $[(\pi\text{-}C_5H_5)Mo(CO)_3]_2$ leads to rearrangement of the ligand[26], the resulting complex (**31**) containing the *o*-semidine ligand. N.m.r. data suggested that two isomers, (**a**) and (**b**), were present in equal amounts. A similar product has been obtained from the reaction of azobenzene with $(\pi\text{-}C_5H_5)Mo(CO)_3Me$.

B. Iron and Ruthenium

The reactions between iron carbonyls and various azobenzenes involve extensive rearrangements of the azobenzene nucleus to *o*-semidine; these reactions are strongly dependent on the substituents in the azobenzenes. Thus, pentacarbonyliron (u.v.) or dodecacarbonyltriiron (heat) and azobenzene afforded small amounts of the *o*-semidine derivative **32** (M = Fe, R = Ph); under the same conditions 4,4′-dimethoxyazobenzene gave a similar complex, in much higher yield, while 4,4′-dimethylazobenzene afforded the trinuclear complex **33** (M = Fe, R = *p*-tol) together with

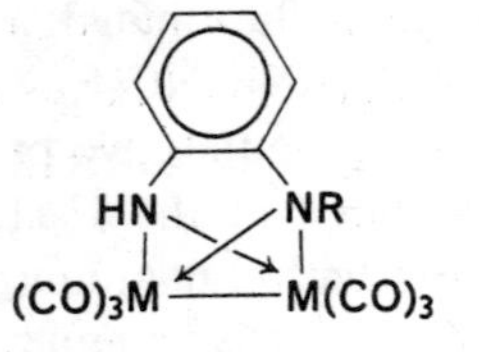

(32)

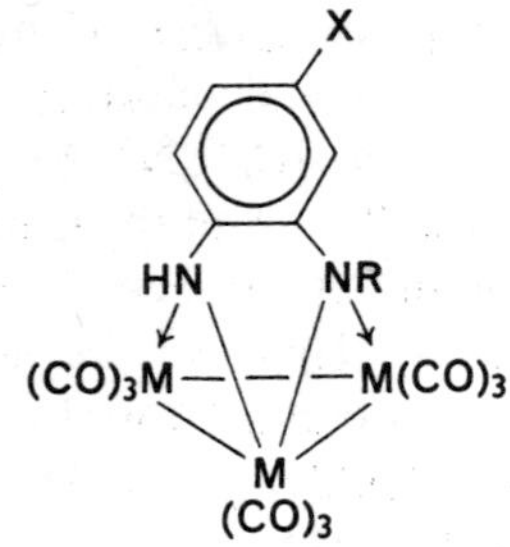

(33)

X = H, R = Ph
X = Me, R = *p*-tol

the *o*-semidine derivative[31]. The latter complex could also be isolated by treating enneacarbonyldiiron with $[(p\text{-MeC}_6\text{H}_4\text{N}{=}\text{NC}_6\text{H}_3\text{Me-}p)\text{PdCl}]_2$.

The presence of *o*-semidine in these complexes was confirmed by oxidative and reductive degradation; also *o*-semidine and *o*-phenylenediamine react wit $Fe_2(CO)_9$ at room temperature affording **32** (M = Fe, R = Ph and H, respectively)[31]. Finally, X-ray evidence[30,55] confirmed this structure in the azotoluene complex, in which the central group of six atoms, $C_2N_2Fe_2$, forms a bicyclo[2.1.1] system. The Fe—Fe bond distance is one of the shortest yet recorded. However, the structures of the trinuclear complexes (**33**) have not yet been confirmed.

The reaction between $Ru_3(CO)_{12}$ and azobenzene affords two complexes **32** (M = Ru, R = H) and **33** (M = Ru, R = Ph) in addition to $[(azb)Ru(CO)_3]_2$ mentioned above. These derivatives are analogous to the iron complexes, but as a result of the well-known stability of the metal–metal bonds in $Ru_3(CO)_{12}$, higher yields of these polynuclear complexes were obtained. The reaction between $Ru_3(CO)_{12}$ and *o*-semidine failed to give the same products, however; instead two new derivatives, **34** and **35**, were

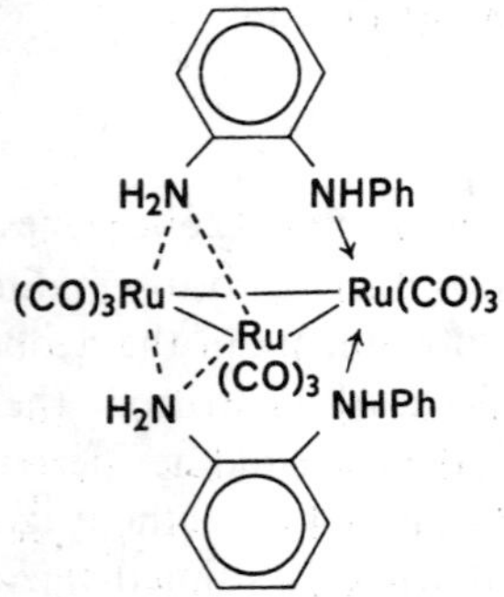

(34)

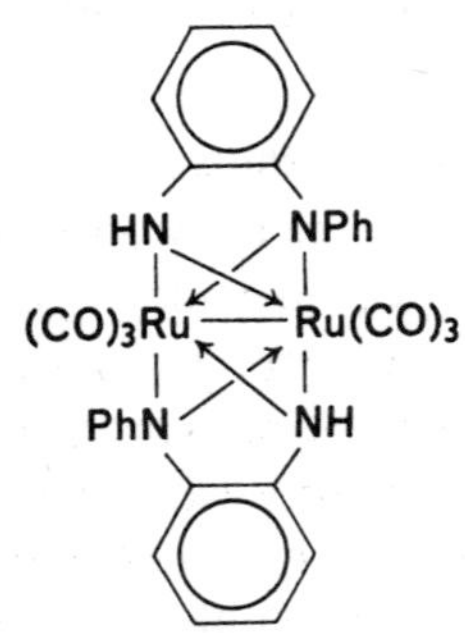

(35)

isolated together with two other products which were not fully characterized.

Trinuclear complexes were also formed with penta- and nona-fluoroazobenzenes, the latter being totally ring-fluorinated. Both complexes were obtained in very low yields compared to the complex derived from azobenzene[54]. A possible explanation for these low yields is the less favourable migration of a C_6F_5N— fragment compared to a C_6H_5N— moiety; this suggestion is consistent with the mechanism proposed for the *o*-semidine rearrangement (see below).

C. Cobalt and Rhodium

o-Semidine derivatives (**36**; M = Co) were isolated from reactions between aromatic azo compounds and π-cyclopentadienylcobalt derivatives[56]. The same complexes could be obtained from the reaction between $(\pi\text{-}C_5H_5)Co(CO)_2$ and *o*-quinonediimine or *o*-phenylenediamine.

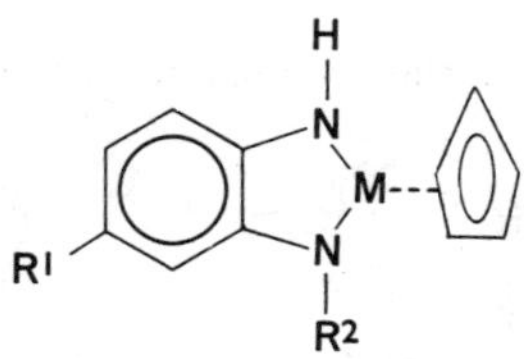

(**36**)

Azobenzene reacts with dicarbonyl-π-cyclopentadienylrhodium to give an *o*-semidine derivative (**36**; M = Rh) analogous to the cobalt complex.

D. The Mechanism of the Metal-Promoted o-Semidine Rearrangement

The mechanism by which azobenzene undergoes its striking rearrangement to the *o*-semidine skeleton under neutral conditions is not yet fully understood. However, it is known that, in acid solution, hydrazobenzene and its derivatives are converted into one or more of five different rearrangement products: benzidine, *o*-benzidine, diphenyline, *p*-semidine and *o*-semidine, depending on substituents. It seems reasonable to suppose that these two reactions occur by similar routes and that in the former case an azoarene–metal complex is formed initially, followed by rearrangement within the coordination sphere of the metal. It is likely that the isolation of the *ortho* isomer reflects the *ortho* activation by the metal, and may be an indication that the reaction proceeds within the coordination sphere.

Several theories relating to the mechanism of the benzidine rearrangement have been advanced, and the relative merits of these are discussed elsewhere in this volume. The latest contribution to this discussion, by Olah[57], suggests an initial diprotonation of the hydrazobenzene. Subsequent rearrangement involves an arenium ion which undergoes intramolecular aromatic alkylation at sites of highest positive charge; the formation of *o*-semidine is depicted in Scheme 5(a).

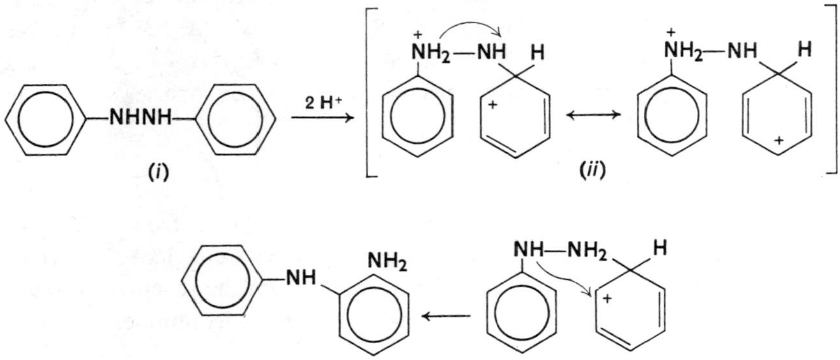

(a) Hydrazobenzene → benzidine rearrangement.

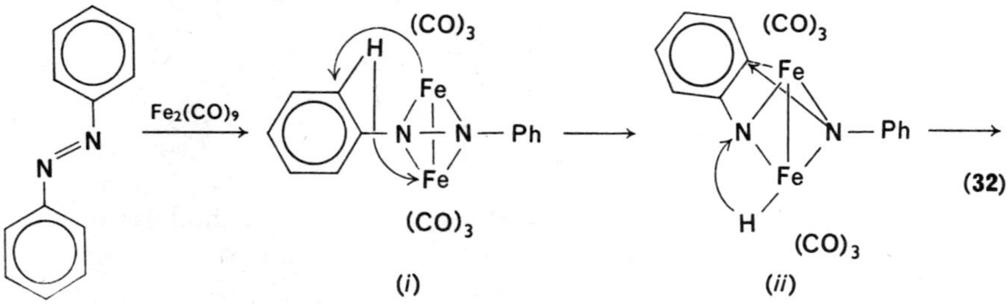

(b) Metal-catalysed azobenzene → *o*-semidine rearrangement.

SCHEME 5. Rearrangements of the azobenzene skeleton.

A possible reaction scheme for the azobenzene → *o*-semidine rearrangement is presented in Scheme 5(b). Complexes similar to the first intermediate, (*i*), are well documented (see Section IV, B) and X-ray studies on such compounds have indicated only a single bond between N^1 and N^2, hence an analogy can be drawn between the complex (*i*) in Scheme (b), and the hydrazobenzene (*i*) in Scheme (a). The resulting metal complex (*i*) then undergoes further transformations, (*i*) → (*ii*) → (**32**) as a result of the interaction of the *ortho* C—H bond with the metal atoms. The resulting N—N bond cleavage may be likened to the bond rupture in the hydrazonium ion (*ii*) in Scheme (a). The phenylnitrene thus generated at N then readily inserts into the newly-formed metal–carbon σ-bond affording the observed product.

VI. ORGANIC SYNTHESES FROM AZOARENES AND METAL COMPLEXES

A. Addition of Unsaturated Molecules to Azoarenes

1. One-carbon units

The octacarbonyldicobalt-catalysed carbonylation of azobenzene was first reported in 1956[58, 59] and was found to afford two products, 2-phenylindazolone (**37**) and 1,3-phenyl-2,4-dioxo-1,2,3,4-tetrahydroquinazoline (**38**), depending on the CO pressure used, together with minor amounts of diphenylurea and the lactone (**39**). At that time the reaction was thought

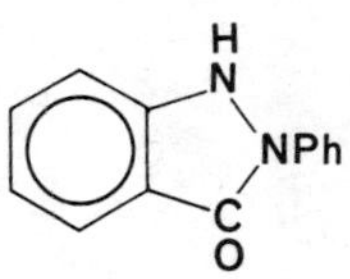

(37)

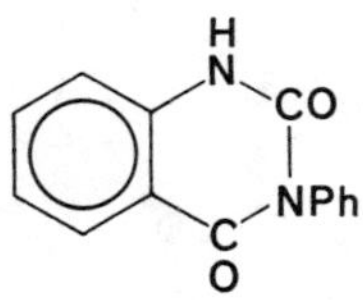

(38)

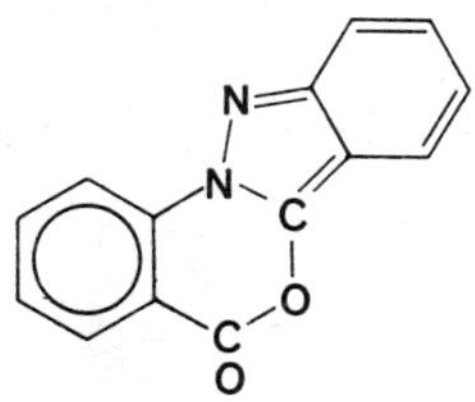

(39)

to proceed via cobalt–nitrogen bonded intermediates which inserted carbon monoxide and then cyclized by reacting with an aromatic hydrogen.

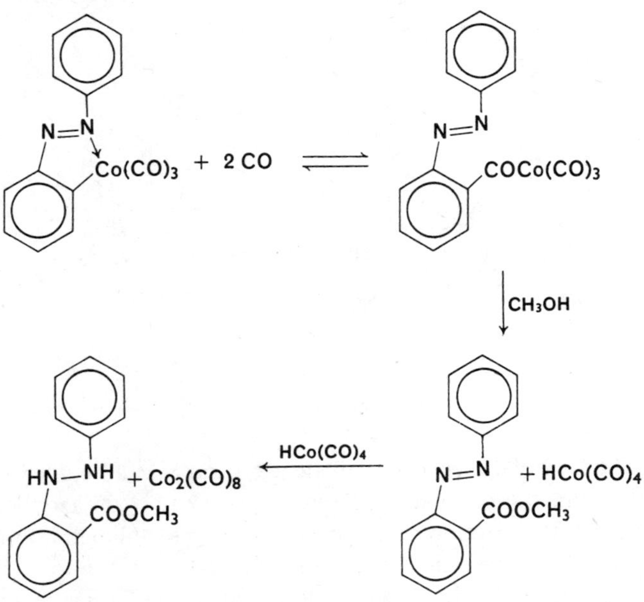

SCHEME 6.

However, Heck[25] showed that carbonylation of tricarbonyl-2-(phenylazo)-phenylcobalt in methanol afforded 2-carbomethoxyhydrazobenzene which he proposed would then form the indazolone (**37**) under the conditions of the original carbonylation reaction (See Scheme 6).

Tsuji and co-workers had shown that various olefinic complexes of palladium chloride reacted with carbon monoxide to give organic carbonyl compounds[60, 61]. For example, carbonylation of π-allylpalladium chloride afforded ethyl 3-butenoate. In these carbonylation reactions it was assumed that a palladium–carbon σ-bond should be formed as a prerequisite of the carbonylation. They later reported[52] that [(azb)PdCl]$_2$, having a palladium–carbon σ-bond, was readily carbonylated affording the same indazolone isolated in the early experiments discussed above. A possible mechanism for this carbonylation was proposed (see Scheme 7).

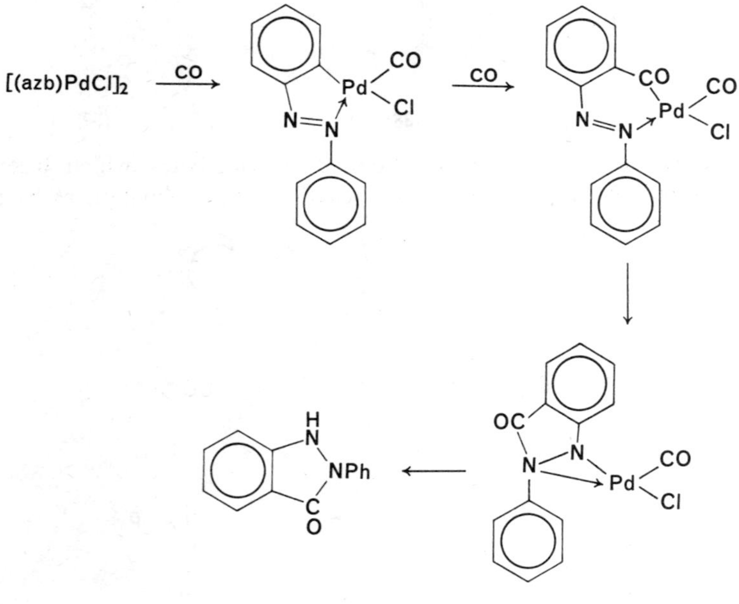

SCHEME 7.

Despite the lack of experimental evidence to support a detailed mechanism for the transition metal-catalysed carbonylation of azobenzene, it seems likely that the reaction does proceed via an *o*-metallated intermediate followed by carbonyl insertion into the metal–carbon σ-bond.

The formation of the dioxoquinazoline (**38**) in the higher temperature $Co_2(CO)_8$-catalysed reaction may be explained by the observation that direct conversion of the indazolone to the dioxoquinazoline occurs under these conditions.

Catalysts other than $Co_2(CO)_8$ and $PdCl_2$ have also been used in carbonylations of azobenzene. Pentacarbonyliron yielded the same product as $Co_2(CO)_8$ but in lower yields, while the nickel carbonyl-catalysed carbonylation is claimed[62] to give primarily **39**, together with small amounts of aniline, 1,3-diphenylurea and the dioxoquinazoline. Two rhodium compounds, $Rh(OH)_3 . H_2O$ and $[(1,5\text{-cyclooctadiene})RhCl]_2$, are also active catalysts for the carbonylation of azobenzene, affording both 1,3-diphenylurea and 2-(3-hydroxyindazol-2-yl)benzoic acid lactone (**39**)[36].

Carbonylation (100 atm/200°C) of *o*-semidine complexes, such as $(\pi\text{-}C_5H_5)Co(sem)$, affords *N*-phenylbenzimidazolone (**40**)[56].

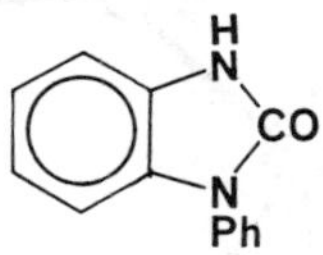

(40)

2. Two-carbon units

o-Chloroazobenzene reacts with $(azb)Ni(\pi\text{-}C_5H_5)$ affording the novel heterocycle 4-phenyl-4*H*-cyclopenta[*c*]cinnoline (**41**) by condensation of one azobenzene residue with a π-cyclopentadienyl group[45]. It was shown that the complexed azobenzene molecule rather than the attacking *o*-chlorinated species is incorporated in the new heterocycle, suggesting

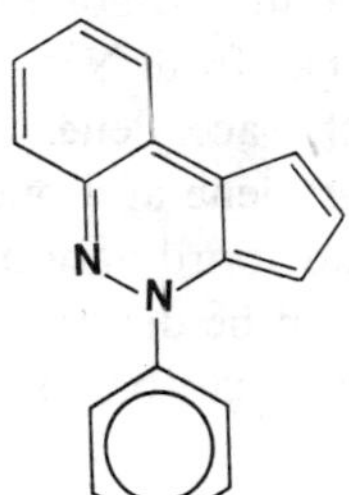

(41)

that the reaction occurs by insertion of a double bond of the cyclopentadienyl group into the nickel–carbon σ-bond.

3. Three-carbon units

In neither of the above cases have any of the proposed organometallic intermediates been isolated. However, the reaction between $(azb)Co(CO)_3$ and hexafluorobut-2-yne caused the insertion of a three-carbon unit into the cobalt–carbon σ-bond. In this work not only was an organic product obtained, but the organometallic intermediate was also isolated; an X-ray crystal structure determination[63] revealed the novel molecular arrangement shown in Figure 7.

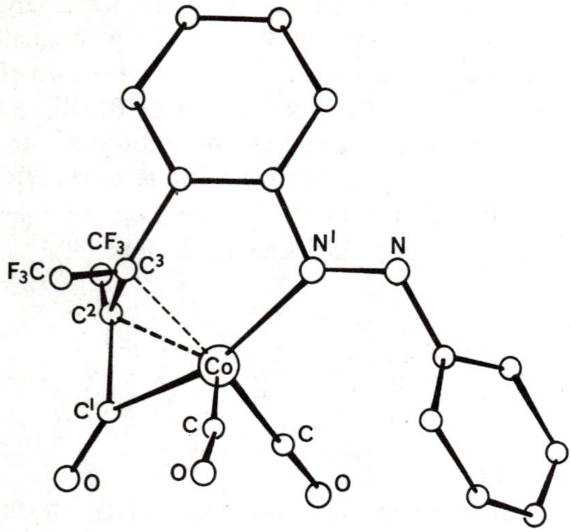

FIGURE 7. Structure of $PhN{=}NC_6H_4C(CF_3)C(CF_3)COCo(CO)_2$. (From ref. 63).

Formally the molecule is given by insertion of one acetylene and one CO molecule into the cobalt–carbon bond of the initial azobenzene complex; concomitant rearrangement results in bonding from the nitrogen adjacent to the metallated ring rather than from N^2. The reaction probably proceeds via initial coordination of the acetylene, displacing N^2, followed by a concerted insertion of the acetylene and one of the carbonyl groups into the cobalt–carbon σ-bond, and coordination of N^1 to the metal.

The resulting system (**42**) can be described as a *trihapto*acryloyl–cobalt system; alternatively, the metal can be considered to be π-bonded to C^2

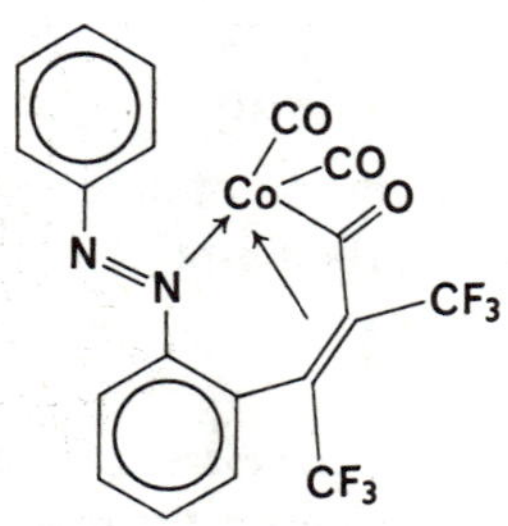

(42)

and C^3 (mean Co—C, 2·0 Å), and σ-bonded to C^1 (Co—C^1, 1·86 Å). There is no significant difference between the C—C bond lengths within the acryloyl system. The Co—N^1 bond length is 2·05 Å, and other dimensions within the azobenzene group are in good agreement with those found in *trans*-azobenzene.

The organic product obtained from this reaction was the *N*-anilinoquinolone (**43**). The acryloyl–cobalt complex could be converted to the *N*-anilinoquinolone either by carbonylation or hydrogenation.

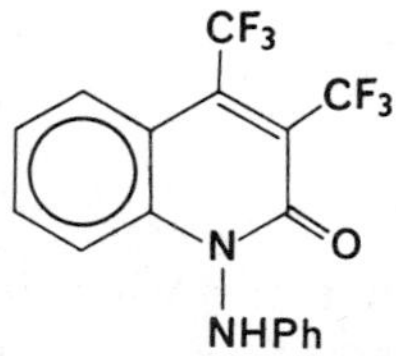

(43)

Thermolysis of **42** in refluxing benzene did not afford the *N*-anilinoquinolone, as was expected, instead a novel intramolecular *o*-metallation reaction was observed, affording **44** (L = CO). The molecule is formed by *o*-metallation of the phenyl ring not previously involved in bonding to the metal, with concomitant elimination of the acyl carbonyl group. The vinylic proton in the resulting complex is probably that lost from the phenyl

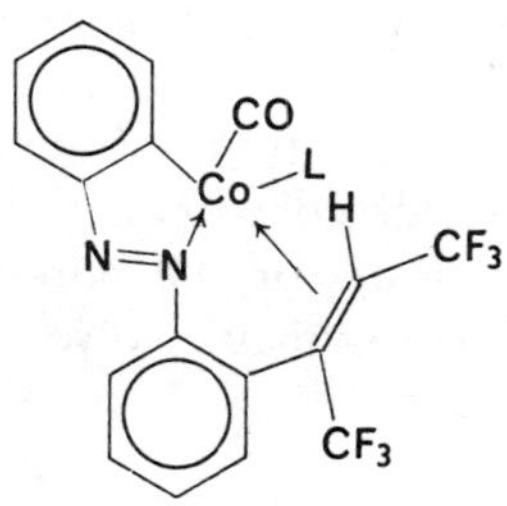

(44)

ring in the metallation reaction, the proton-transfer possibly being assisted by the metal to give the *cis*-isomer, identified by its ^1H n.m.r. spectrum. Treatment of **42** with methyldiphenylphosphine caused a rapid intramolecular *o*-metallation reaction, similar to that described above, affording **44** (L = $PMePh_2$). Azobenzene was also recovered from the reaction mixture.

Bromination of **42** afforded **45**, and traces of a second product which was identified spectroscopically as **46**[28].

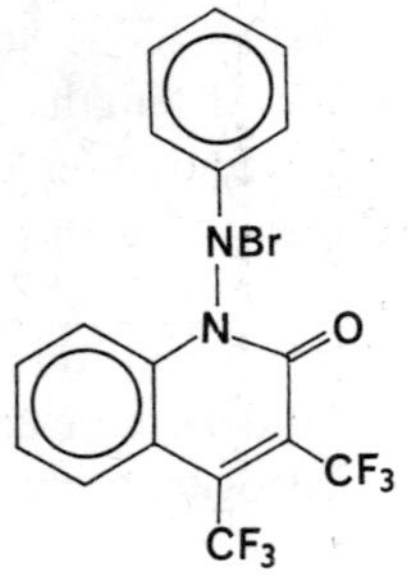

(45)

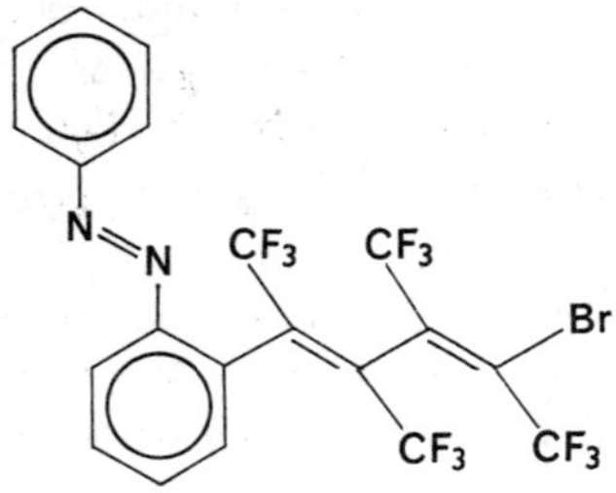

(46)

B. Hydrogenation

Normally azobenzenes do not react with $LiAlH_4$, although reduction to hydrazobenzenes has been noted with certain metal halides; however, degradation of $[(azb)Rh(CO)Cl]_2$ with $LiAlH_4$ afforded aniline[39]. Furthermore, it was found that addition of catalytic amounts of $[Rh(CO)_2Cl]_2$ to a mixture of azobenzene and $LiAlH_4$ rapidly gave a quantitative yield of aniline-free hydrazobenzene. The catalyst was extremely effective with no signs of exhaustion being observed during the reduction of over 60,000 times excess of the azo compound, the reaction being a convenient source of the hydrazo compound free from amines. Substituents were shown to have a profound effect on the reduction; azophenetole afforded 4-ethoxyaniline while azoxyanisole gave hydrazoanisole, both quantitatively.

In the presence of hydrogen and the catalyst system $RhCl_3py_3$—$NaBH_4$, azobenzene was reduced to hydrazobenzene and slowly to aniline[64]. Cobalt complexes also act as catalysts for the borohydride reduction of azo compounds[65], and both reactions may involve *o*-metallated intermediates.

C. Halogenation

Azobenzene is selectively halogenated by chlorine and bromine *ortho* to the azo function, affording mono- and poly-halogenated azobenzenes, when solutions are treated with the respective halogen in the presence of a palladium(II) catalyst[66]. Thus, by bubbling chlorine into such a solution 2-chloroazobenzene (12%), 2,6-dichloroazobenzene (22%), 2,2′-dichloroazobenzene (30%), 2,6,2′-trichloroazobenzene (33%) and 2,6,2′,6′-tetrachloroazobenzene (3%) were obtained. Chlorinated azobenzene–palladium complexes could also be isolated from the reaction mixture, suggesting that the specificity for the *ortho* halogenation is derived from the interaction of the halogen with a metal–carbon σ-bonded intermediate.

D. Other Reactions Involving Metal–Carbon Bond Cleavage

Several reactions resulted in cleavage of the nickel–carbon σ-bond in [**1**; $M = Ni(\pi\text{-}C_5H_5)$][67]; trifluoroacetic acid gave azobenzene, mercuric acetate and mercuric chloride yielded the corresponding *o*-mercurated azobenzenes while peroxybenzoic acid afforded *o*-hydroxyazobenzene together with a small amount of **41**.

Carbonylation of $[(azb)_2RhCl]_2$ does not give an organic carbonyl-containing derivative; instead the two azobenzene moieties combine to give bis-azo-compounds[37].

VII. ARYLAZO AND ARYLDIIMINE COMPLEXES OF TRANSITION METALS

Reactions between diazonium salts and transition metal complexes, usually containing hydride ligands, have afforded complexes containing the ArN=N or ArN=NH groups attached to the metal atom. Some related complexes have been obtained using phenylhydrazines.

A. Molybdenum and Tungsten

The anions $[(\pi\text{-}C_5H_5)Mo(CO)_3]^-$ reacted with $RN_2^+BF_4^-$ to give $(\pi\text{-}C_5H_5)Mo(CO)_2N{=}NAr$ (Ar = Ph, *p*-tol, *p*-$MeOC_6H_4$, *p*-NO_2); the phenyl complexes were also formed from $(\pi\text{-}C_5H_5)M(CO)_3H$ (M = Mo or W) and $PhNHNH_2$, or PhN_2Cl in THF at −40°C[68–70]. Related complexes were obtained using the corresponding $[RB(pz)_3Mo(CO)_3]^-$ anions, and are much more stable chemically than the cyclopentadienyl derivatives. Analogous cationic tris(pyrazolyl)methane derivatives have also been made[71].

Alkylazo derivatives have been prepared by 1,3-addition of metal hydrides to diazoalkanes: thus, the reaction between $(\pi\text{-}C_5H_5)M(CO)_3H$ and (trimethylsilyl)diazomethane afforded $(\pi\text{-}C_5H_5)M(CO)_2N{=}NCH_2SiMe_3$ (M = Mo or W)[72].

Related reactions between ethyl diazoacetate and $[(\pi\text{-}C_5H_5)M(CO)_3]^-$ afforded unusual carbene complexes, the structures of which were unravelled by detailed X-ray studies[73, 74]. A number of equilibria involving these complexes were established (Scheme 8). It is uncertain whether these complexes should be regarded as nitrene–carbene complexes, containing the group $\ddot{\underset{\cdot\cdot}{N}}{-}NR_2$, or as true azo-carbene derivatives, in which the metal–carbon bond has some multiple-bond character.

A brief report describes the synthesis of $[(\pi\text{-}C_5H_5)_2W(N{=}NAr)]BF_4$ from $(\pi\text{-}C_5H_5)_2WH_2$ and the appropriate diazonium salt[75].

The tetranuclear complexes $[ArN{=}NMo(CO)_2(OH)]_4$ were obtained

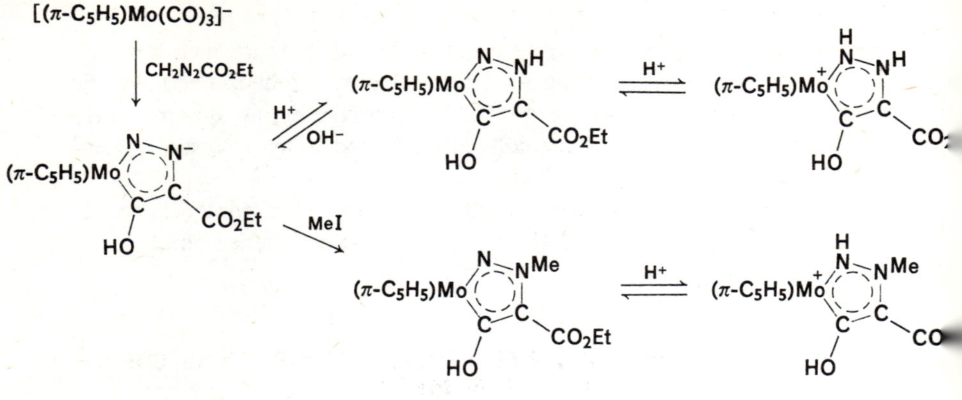

SCHEME 8.

from $Cs_4[Mo_4(CO)_{12}(OH)_{12}(OH)_4]$ and aryldiazonium cations; these complexes formed 1:4 adducts with oxygen donors at the facial OH groups. With 1,10-phenanthroline, [(phen)$Mo(CO)_2(OH)N{=}NAr$] was obtained.

B. Rhenium

Treatment of *mer*-$ReCl_3(PMe_2Ph)_3$ with phenylhydrazine has given $ReCl_2(N{=}NPh)(PMe_2Ph)_3$ as one product characterized by an X-ray structural study (see below)[77].

C. Iron

Aryldiazonium salts react with $Fe(CO)_3(PPh_3)_2$ affording $[Fe(N{=}NAr)(CO)_2(PPh_3)_2]BF_4$, which has a trigonal bipyramidal structure, with apical phosphine ligands[77a]. A similar reaction using $[Fe(CO)_3NO]^-$, followed by addition of PPh_3, gave $Fe(N{=}NAr)(CO)(NO)(PPh_3)$.

D. Ruthenium and Osmium

Cationic arylazo complexes $[RuCl(N{=}NAr)_2(PPh_3)]BF_4$ (Ar = *p*-MeC_6H_4, *p*-$MeOC_6H_4$) have been prepared by reactions between $RuHCl(PPh_3)_3$ and diazonium salts; with $RuCl_2(PPh_3)_3$, $[RuCl_2(N{=}NAr)(PPh_3)_2](BF_4)$, thought to be dimeric, was obtained[75]. Reduction (sodium amalgam in ethanol) of the bis-arylazo complexes gave $ArNH_2$ and NH_3, and chlorine afforded $[RuCl_3(N{=}NAr)(PPh_3)_2]$ and the diazonium chloride. In contrast, later reports[78,140] describe addition of diazonium salts, followed by lithium halide, to $RuX_2(PPh_3)_3$ or $OsX_2(PPh_3)_3$ (X = Cl, Br), to give the neutral arylazo complexes $MX_3(N{=}NAr)(PPh_3)_2$. The ruthenium chloro complex has also been obtained in a one-step reaction by addition of 1,3-diaryltriazenes and $RuCl_3$ to a boiling ethanolic solution of PPh_3.

E. Cobalt

The unstable complex $Co(N{=}NAr)(CO)_2(PPh_3)$ was obtained from the reaction between $[Co(CO)_4]^-$ and aryldiazonium cation, followed by addition of PPh_3[77a].

F. Rhodium

The initial report[79] of the reaction between $RhHCl_2(PPh_3)_2$ and p-$FC_6H_4N_2^+BF_4^-$ describes the product as $RhCl_2(N{=}NC_6H_4F\text{-}p)(PPh_3)_2$; using $RhCl(PPh_3)_3$, unstable $[RhCl(N{=}NC_6H_4F)(PPh_3)_3]BF_4$ is probably formed. Several related complexes, containing representative electron-withdrawing and -donating substituents, have also been described[80]. More recently however, these complexes have been assigned the aryldiimine structures $RhCl_3(NH{=}NAr)(PPh_3)_2$, having been obtained from true arylazo derivatives by addition of HCl[78,140]:

$$RhCl_2(N{=}NAr)(PPh_3)_2 \underset{NEt_3}{\overset{HCl}{\rightleftharpoons}} RhCl_3(NH{=}NAr)(PPh_3)_2$$

The arylazo complexes were obtained from $RhX(PPh_3)_3$ (X = Cl, Br) and the diazonium cation, followed by addition of LiX, or directly from 1,3-diaryltetrazenes, $RhCl_3$ and PPh_3.

G. Iridium

Diazonium salts react with $IrH_3(PPh_3)_3$ to give the aryl–diimino complexes $[IrH_2(NH{=}NC_6H_4R\text{-}p)(PPh_3)_3]BF_4$ (R = MeO, Me, NMe_2 or NO_2)[81,82].

The iridium complex $[IrCl(N{=}NAr)(PPh_3)_2]^+$ can be easily prepared by addition of an aroyl azide, followed by the aryldiazonium salt, to a chloroform solution of Vaska's complex, $IrCl(CO)(PPh_3)_2$[82a]. Addition of donor ligands affords the five-coordinate complexes $[IrCl(N{=}NAr)(PPh_3)_2L]^+$ (L = CO, RNC, PR_3, AsR_3, SbR_3). Six-coordinate complexes, $IrCl(X)(N{=}NC_6H_4Y\text{-}p)(PPh_3)_2L$, (X = Cl, Br, I, N_3, NO_2, NCO; Y = F, H, Me; L = CO, EtNC) have also been obtained. Reactions between $[IrCl(N{=}NAr)(PPh_3)_2]^+$ and HBF_4, $Et_3O^+BF_4^-$ or MeCOCl afford complexes containing HN=NAr, EtN=NAr or MeCON=NAr ligands. With $IrCl(CO)L_2$ (L = PMe_2Ph or $AsMe_2Ph$) and diazonium salts, the complexes $IrCl_2(N{=}NPh)(CO)L_2$ were obtained, in which the two Group V ligands are mutually *trans*, and the N=NPh group is *trans* to CO[83].

These reactions contrast with those between $IrCl(CO)(PPh_3)_2$ and p-$FC_6H_4N_2^+BF_4^-$, which afford the tetrazene complex and the o-metallated complexes **26** and **26a**, which have been described in Section IV, H[41].

H. Platinum

Several arylazoplatinum compounds, $ArN{=}NPtCl(PEt_3)_2$, were obtained from reactions between diazonium salts and $PtHCl(PEt_3)_2$[84]. These red or green complexes are probably formed via a cationic intermediate, $[ArN{=}NHPtCl(PEt_3)_2]^+$, which on treatment with base affords the neutral complex. Catalytic decomposition on alumina) affords the corresponding aryl complex, $PtArCl(PEt_3)_2$.

Reduction of the cationic complex with hydrogen (25°C/1 atm, Pt catalyst) gives the phenylhydrazine complex, $[ArNHNH_2PtCl(PEt_3)_2]^+$,

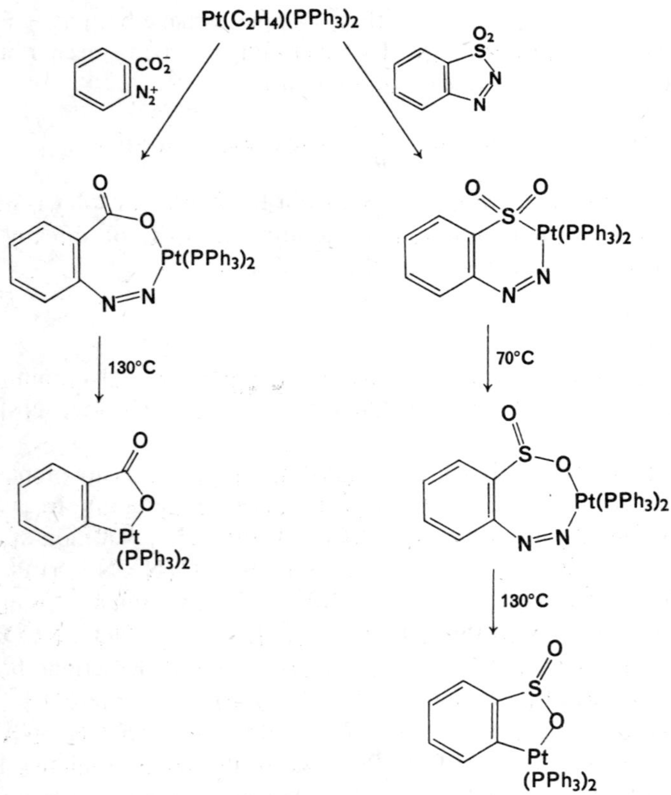

SCHEME 9.

which, on longer reaction, regenerates the original platinum hydride complex together with the phenylhydrazine. As with the iridium complex (**26**) discussed above, close analogies are found between these reductions and those occurring in nitrogen-fixing bacteria.

Cyclic azo compounds were isolated from attempts to trap benzyne as its platinum complex, by reactions between $Pt(C_2H_4)(PPh_3)_2$ and either benzenediazonium carboxylate or sulphonate. On heating, the complexes lose nitrogen, affording *o*-metallated products (Scheme 9)[85]. Sulphonyl azides react with $Pt(PPh_3)_4$ to give $Pt(N_2SO_2Ar)_2(PPh_3)_2$, thought to have the cyclic tetrazene structure $(Ph_3P)_2\overline{PtN(SO_2Ar){:}N{-}N}{=}N(SO_2Ar)$, although a bis-arenesulphonylazo formulation cannot be ruled out[85a]. Addition of diazonium salts to the platinum(0) precursor affords the cationic derivatives $[Pt(N{=}NC_6H_4R\text{-}p)(PPh_3)_3]^+$ ($R = NO_2$, F, H, OMe, Me, NMe_2), which readily lose dinitrogen in $CHCl_3$ to give a cationic σ-arylplatinum complex[85b]. Non-coordinating strong acids protonate the complexes to the dications $[Pt(NH{=}NAr)(PPh_3)_3]^{2+}$.

I. Azo Complexes, MN=NM′

Complexes containing an azo function bridging two metal atoms so far have not been prepared. Treatment of $RuH_2(N_2)(PAr_3)_3$ (Ar = Ph or *p*-tol) with inner diazonium salts derived from $B_{10}H_{10}^{2-}$, e.g. $B_{10}H_8(N_2)_2$ or $N_2B_{10}H_8SMe_2$, affords complexes containing RuN=NB groups, such as $B_{10}H_8[N_2Ru(H_2)(PPh_3)_3]_2$[86].

Bridged dinitrogen complexes are well-known, but fall outside the scope of this article. Compounds such $[(C_5H_5)_2Ti]_2N_2$, $[(arene)Cr(CO)_2]_2N_2$, $\{[Ru(NH_3)_5]_2N_2\}^{4+}$, $[(NH_3)_5RuN_2Os(NH_3)_5]^4$ and $[(PCy_3)_2Ni]_2N_2$ have been reviewed recently[86a]. The rhenium complexes $ReCl(N_2)(PR_3)_4$ react with other metal compounds [e.g. $ScCl_3$, $TiCl_4(THF)_3$, $CrCl_3(THF)_3$, $FeCl_2(THF)_{1.5}$ or $Pt_2Cl_4(PEt_3)_2$] to give binuclear derivatives, probably containing a dinitrogen bridge between two metals of different groups[86b].

J. Structures of Arylazo-Metal Complexes

Spectroscopic studies of the platinum complexes indicate substantial contributions of resonance structures such as:

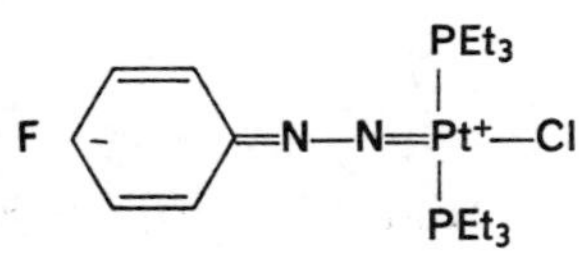

and the ν(CO) frequencies in the various molybdenum complexes indicated that the RN═N group has a very high π-acceptor ability. Comparison of ν(N═N) for the molybdenum and platinum complexes suggests that there is a greater degree of metal–nitrogen backbonding in the latter[84].

The isoelectronic nature of the NO and ArN_2 groups leads to the expectation that both linear and bent M—N—N geometries will be found in the arylazo–metal complexes, corresponding to the formal bonding schemes:

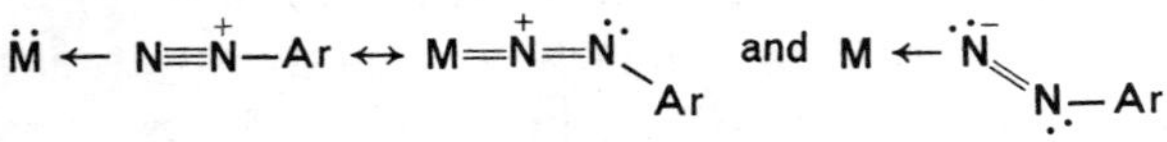

Structural data for the three complexes reported below indicate that only the linear M—N—N arrangement has been found at present. Structural studies have been reported for $HB(pz)_3Mo(CO)_2N{=}NPh$ (Figure 8)[87], for $ReCl_2(N{=}NPh)(PMe_2Ph)_3$ (Figure 9)[77], and $RuCl_3(N{=}NC_6H_4Me\text{-}p)$-

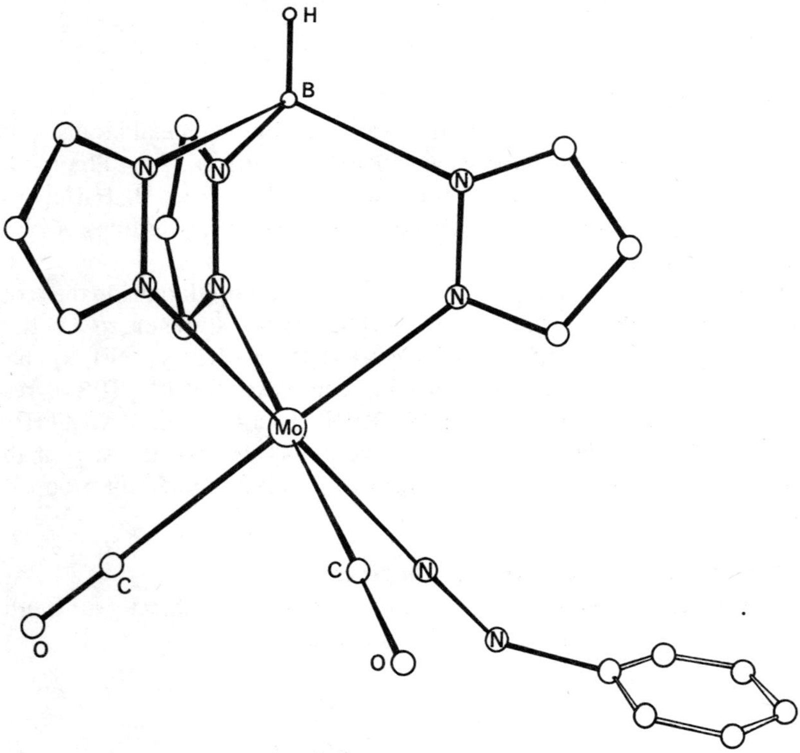

FIGURE 8. Structure of $HB(pz)_3Mo(CO)_2N{=}NPh$. (From ref. 87).

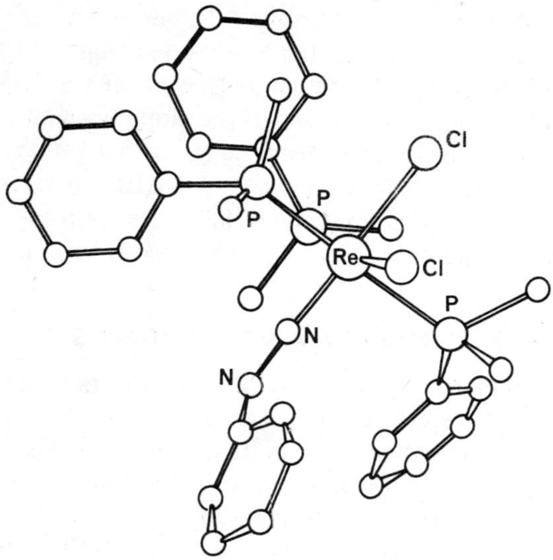

FIGURE 9. Structure of ReCl(N=NPh)(PMe$_2$Ph)$_3$. (From ref. 77).

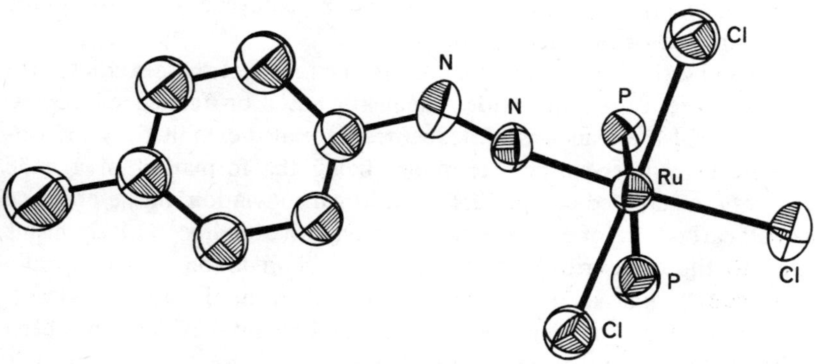

FIGURE 10. Structure of RuCl$_3$(N=NC$_6$H$_4$Me-p)(PPh$_3$)$_2$·Me$_2$CO. (From ref. 87a).

(PPh$_3$)$_2$.Me$_2$CO (Figure 10)[87a], and the important bond parameters are indicated below:

In the rhenium complex, the short Re—N bond indicates substantial multiple-bond character, although the N=N bond length is identical with that found in MeN=NMe. These data suggest that the PhN=N group acts as a σ- and π-donor, in contrast to the situation with the platinum complex, where other physical evidence suggests that this group behaves as a σ-donor and π-acceptor. Comparison of ν(CO) frequencies (where appropriate) for these complexes and the related nitrosyl derivatives, indicates a reduced π-acceptor capacity for the N=NAr ligand.

VIII. SUMMARY AND PROGNOSIS

It is evident from the survey given above that the transition metal chemistry of azobenzene is both extensive and varied, and within the confines of this chapter it has not been possible to do more than indicate the ramifications. Future developments in the organometallic field will no doubt result in the elucidation of more precise mechanisms of the *o*-metallation and rearrangement reactions, whereas organic chemists are likely to find reactions of the resulting complexes to be more exciting. The importance of using these metal derivatives as stoicheiometric reagents must not be underestimated in the present concern with catalytic applications of transition metals, especially if by their use unusual heterocyclic systems can be obtained in one- or two-step reactions.

It is also pertinent to draw attention to the fact that these studies form only one aspect of a much wider chemistry based on related molecules. Thus, many of the reactions of azobenzene are also exhibited by the isoelectronic benzylideneaniline, examples being the formation of a wide variety of *o*-metallated complexes[88], and its carbonylation, in the presence of cobalt carbonyls, for example, to *N*-phenylphthalimidine[36]. Mechanisms related to those described above, for the carbonylation of structurally similar compounds such as Schiff bases and aromatic ketone phenylhydrazones, have been suggested[89], but no experimental evidence has been put forward to support these theories.

At the time of writing, the area in which most progress is being made is that of the *o*-metallation reaction. Azoarenes are far from being the only *N*-donor ligands to undergo this reaction. Furthermore, the facile *o*-metallation of appropriately substituted benzenes has been reported for ligands containing other donor atoms. Other aromatic systems have also been metallated, including naphthalenes, quinolines, ferrocenes, and most recently, the 1-dimethylphosphino-1,2-carborane ligand. Many of these are summarized in Table 2, and it is evident that this reaction occurs even more commonly than has been recognized to date.

TABLE 2. Some other examples of ligands for which the *o*-metallation or related reactions have been reported

Ligand	Donor atom	Atom metallated	Metal	References
PhC_2Ph	C	C(aryl)	Fe, Ti	90, 91
Ph_2C:	C	C(aryl)	Ru	92
$C(NHC_6H_4Cl$-*p*)(NHEt)	C	C(aryl)	Pt	92a
ArCH=NR	N	C(aryl)	Mn, Re, Fe, Ru, Co, Rh, Ni, Pd, Pt	93
NapCH=NPh	N	C(nap)	Fe	94
ThienCH=NPh	N	C(thien)	Fe	94
$PhCH_2NR_2$	N	C(aryl)	Mn, Pd, Pt	95–97
$C_6H_4(CH_2NR_2)_2$	N	C(aryl)	Pd	98
$PhCH_2NHPh$	N	C(aryl)	Mn	99
$PhNH_2$	N	C(aryl)	Os	100
Nap-1-NMe_2	N	C(nap)	Pd	96
CH_2=CHCHMeNMe_2	N	C(allyl)	Pt	101
Benzaldoxime	N	C(aryl)	Pd	102
Acetophenone oxime	N	C(aryl)	Pd, Pt	102, 103
2-Acetonaphthone oxime	N	C(nap)	Pd	102
Ph_2C=N . NHPh	N	C(aryl)	Pd	104
N-Phenylosazones	N	C(aryl)	Pd	105
2-Phenylpyridine	N	C(aryl)	Mn, Re, Rh, Pd	106–108
2-Vinylpyridine	N	C(vinyl)	Mn, Re, Pd	106, 108
2-Phenylquinoline	N	C(aryl)	Pd	108
8-Ethylquinoline	N	C(sec)	Pd	109
1-Phenylpyrazole	N	C(aryl)	Mn, Re, Pd	106, 110
N-Benzylpiperidin-4-one	N	C(aryl)	Pd	104
2,3-Diphenylquinoxaline	N	C(aryl)	Mn	106
1,4-Diphenylphthalazine	N	C(aryl)	Mn	111
4-Phenylpyrimidine	N	C(aryl)	Mn	111
2,5-Diphenyloxazole	N	C(aryl)	Mn	111
Benzo[h]quinoline	N	C(aryl)	Mo, Mn, Re, Ru, Os, Rh, Pd	107, 112
Ar_2PPh	P	C(aryl)	Fe, Ru, Rh, Ir	113–115
$PhPR_2$	P	C(aryl)	Pt	116
$PhCH_2PMe_2$	P	C(aryl)	Mn	99
$Ph_2PCH_2CH_2PPh_2$	P	C(aryl)	Fe	117
$NapPR_2$	P	C(nap)	Rh, Ir	118
(*o*-tol)PBu^t_2 } (*o*-tol)$_2PBu^t$- }	P	C(prim)	Pd, Pt	119
(*o*-EtC_6H_4)PBu^t_2	P	C(sec)	Pd, Pt	120
(*o*-Pr-*i*C_6H_4)PBu^t_2	P	C(tert)	Pd	120
$Me_2PCH_2CH_2PMe_2$	P	C(prim)	Ru	121
$P(OAr)_3$	P	C(aryl)	Ru, Os, Rh, Ir, Pd, Pt	114, 122

TABLE 2 (*cont.*)

Ligand	Donor atom	Atom metallated	Metal	References
1-(Me_2P)-1,2-$C_2B_{10}H_{11}$	P	B	Ir	123
$AsPh_3$	As	C(aryl)	Ir	115
$PhCH_2AsMe_2$	As	C(aryl)	Mn	99
(*o*-tol)$AsBu^t_2$ } (*o*-tol)$_2AsBu^t$ }	As	C(prim)	Pt	115a
$SbPh_3$	Sb	C(aryl)	Ir	115
PhOCO (benzoate)	O	C(aryl)	Ti	91
PhCOR	O	C(aryl)	Mn, Re	123a
Anthraquinone	O	C(aryl)	Re	123a
$Ph_2\overline{PC_6H_4Mn}(CO)_4$	O	C(aryl)	Mn	124
$PhCH_2SMe$	S	C(aryl)	Mn	125
Ph_2CS	S	C(aryl)	Fe, Ru, Pd, Pt	126
PhCS(OEt) (thiobenzoate)	S	C(aryl)	Fe	125a

Abbreviations: Ar, aryl; R, alkyl or aryl; Nap, naphthyl; Thien, thienyl.

Finally, it can be expected that many new derivatives containing arylazo–metal moieties will be synthesized, and with them will come a greater understanding of the factors which influence their reactivity. For example, the contrast between the structures of complexes obtained from aryl-diazonium salts and either rhodium or iridium precursors is striking. Presumably these differences reflect the increasing nucleophilicity, or ease of oxidative addition, on going from rhodium to iridium, and they emphasize the close relationship that exists between azoarene and arylazo–metal complexes. Studies of the protonation of these complexes, and of analogous bi-metallic azo compounds yet to be synthesized, may lead to significant progress in the elucidation of the mechanism of nitrogen fixation.

IX. REFERENCES

1. D. Nicholls, B. A. Warburton and D. H. Wilkinson, *J. Inorg. Nuclear Chem.*, **32,** 1075 (1970).
2. H. Baumann and H. R. Henschel, *Fortschr. Chem. Forsch.*, **7,** 643 (1967).
3. R. Price, in *The Chemistry of Synthetic Dyes* (Ed. K. Venkataraman), Vol. 3, Chap. 8, Academic Press, New York, 1970, p. 303.
4. S. Otsuka, A. Nakamura and H. Minamida, *Chem. Comm.*, 1148 (1969).
5. M. Green, R. B. L. Osborn and F. G. A. Stone, *J. Chem. Soc.* (*A*), 3083 (1968).

6. J. P. Collman and J. W. Kang, *J. Amer. Chem. Soc.*, **89,** 844 (1967).
7. J. Ashley-Smith, M. Green, N. Mayne and F. G. A. Stone, *Chem. Comm.*, 409 (1969).
8. H. F. Klein and J. F. Nixon, *Chem. Comm.*, 42 (1971).
9. S. Otsuka, T. Yoshida and Y. Tatsuno, *Chem. Comm.*, 67 (1971).
10. R. S. Dickson, J. A. Ibers, S. Otsuka and Y. Tatsuno, *J. Amer. Chem. Soc.*, **93,** 4636 (1971); R. S. Dickson and J. A. Ibers, *J. Amer. Chem. Soc.*, **94,** 2988 (1972).
11. C. J. Brown, *Acta Cryst.*, **21,** 146 (1966).
11a. R. G. Denning and J. Thatcher, *J. Amer. Chem. Soc.*, **90,** 5917 (1968).
12. M. Kilner, *Adv. Organometallic Chem.*, **10,** 115 (1972).
13. A. J. Carty, *Organometallic Chem. Rev.*, **7A,** 191 (1971).
13a. R. J. Doedens, *Inorg. Chem.*, **9,** 429 (1970).
14. R. J. Doedens and J. A. Ibers, *Inorg. Chem.*, **8,** 2709 (1969).
15. S. Otsuka, T. Yoshida and A. Nakamura, *Inorg. Chem.*, **8,** 2514 (1969).
16. M. Dekker and G. R. Knox, *Chem. Comm.*, 1243 (1967).
17. C. D. Campbell and C. W. Rees, *Chem. Comm.*, 537 (1967).
18. R. P. Bennett, *Inorg. Chem.*, **9,** 2184 (1970).
18a. P. C. Ellgren and S. L. McMullin, *Inorg. Chem.*, **12,** 2004 (1973).
18b. M. Herberhold and W. Golla, *J. Organometallic Chem.*, **26,** C27 (1971).
19. L. G. Kuz'mina, N. G. Bokii, Y. T. Struchkov, A. V. Arutyunyan, L. V. Rybin and M. I. Rybinskaya, *Zh. strukt. Khim.*, **12,** 875 (1971).
20. H. Kisch, *J. Organometallic Chem.*, **30,** C25 (1971).
21. A. N. Nesmeyanov, L. V. Rybin, M. I. Rybinskaya, A. V. Arutyunyan, N. T. Gubenko and P. V. Petrovskii, *Izvest. AN SSSR, Ser. Khim.*, 1574 (1971).
21a. L. V. Rybin, A. V. Arutyunyan, P. V. Petrovskii and M. I. Rybinskaya, *Izvest. AN SSSR, Ser. Khim.*, 190 (1972).
22. H. A. Patel, A. J. Carty, M. Mathew and G. J. Palenik, *J. C. S. Chem.Comm.*, 810 (1972).
23. H. Alper, *J. Organometallic Chem.*, **50,** 209 (1973).
24. M. I. Bruce, R. C. F. Gardner and F. G. A. Stone, unpublished results.
25. R. F. Heck, *J. Amer. Chem. Soc.*, **90,** 313 (1968).
26. M. I. Bruce, M. Z. Iqbal and F. G. A. Stone, *J. Chem. Soc. (A)*, 3204 (1970).
27. M. I. Bruce, B. L. Goodall and F. G. A. Stone, *J. C. S. Chem. Comm.*, 558 (1973).
28. M. I. Bruce, B. L. Goodall and F. G. A. Stone, unpublished results.
29. M. M. Bagga, P. L. Pauson, F. J. Preston and R. I. Reed, *Chem. Comm.*, 544 (1965).
30. P. E. Baikie and O. S. Mills, *Chem. Comm.*, 707 (1966).
31. M. M. Bagga, W. T. Flannigan, G. R. Knox and P. L. Pauson, *J. Chem. Soc. (C)*, 1534 (1969).
32. M. I. Bruce, M. Z. Iqbal and F. G. A. Stone, *Chem. Comm.*, 1325 (1970).
33. M. I. Bruce, M. Z. Iqbal and F. G. A. Stone, *J. Chem. Soc. (A)*, 2820 (1971).
33a. M. I. Bruce, M. Z. Iqbal and F. G. A. Stone, *J. Organometallic Chem.*, **31,** 275 (1971).
34. J. D. Gilbert, D. Rose and G. Wilkinson, *J. Chem. Soc. (A)*, 2765 (1970).
35. M. I. Bruce, R. C. F. Gardner and F. G. A. Stone, *J. Organometallic Chem.*, **40,** C39 (1972).

35a. M. I. Bruce, R. C. F. Gardner, B. L. Goodall, F. G. A. Stone, R. J. Doedens and J. A. Moreland, *J. C. S. Chem. Comm.*, 185 (1974).
36. T. Joh, N. Hagihara and S. Murahashi, *J. Chem. Soc. Japan*, **88,** 786 (1967).
37. A. R. M. Craik, G. R. Knox, P. L. Pauson, R. J. Hoare and O. S. Mills, *Chem. Comm.*, 168 (1974).
38. R. J. Hoare and O. S. Mills, *J. C. S. Dalton*, 2138 (1972).
39. M. I. Bruce, B. L. Goodall, M. Z. Iqbal and F. G. A. Stone, *Chem. Comm.*, 661 (1971).
39a. R. J. Hare and O. S. Mills, *J. C. S. Dalton*, 2141 (1972).
40. M. I. Bruce, B. L. Goodall, F. G. A. Stone and B. J. Thomson, unpublished results.
41. F. W. B. Einstein, A. B. Gilchrist, G. W. Rayner-Canham and D. Sutton, *J. Amer. Chem. Soc.*, **93,** 1826 (1971).
42. A. B. Gilchrist, G. W. Rayner-Canham and D. Sutton, *Nature*, **235,** 42 (1972).
43. J. P. Kleiman and M. Dubeck, *J. Amer. Chem. Soc.*, **85,** 1544 (1963).
44. V. A. Semion, I. V. Barinov, Y. A. Ustynyuk and Y. T. Struchkov, *Zh. strukt. Khim.*, **13,** 543 (1972).
45. Y. A. Ustynyuk and I. V. Barinov, *J. Organometallic Chem.*, **23,** 551 (1970).
45a. Y. A. Ustynyuk, V. A. Chertkov and I. V. Barinov, *J. Organometallic Chem.*, **29,** C53 (1971).
46. I. V. Barinov, T. I. Voevodskaya and Y. A. Ustynyuk, *J. Organometallic Chem.*, **30,** C28 (1971).
47. A. C. Cope and R. W. Siekman, *J. Amer. Chem. Soc.*, **87,** 3272 (1965).
48. R. W. Siekman and D. L. Weaver, *Chem. Comm.*, 1021 (1968); D. L. Weaver, *Inorg. Chem.*, **9,** 2250 (1970).
49. B. Crociani, T. Boschi, R. Pietropaolo and U. Belluco, *J. Chem. Soc.* (*A*), 531 (1970).
50. A. L. Balch and D. Petridis, *Inorg. Chem.*, **8,** 2247 (1969).
51. M. I. Bruce, B. L. Goodall, M. B. Rees and F. G. A. Stone, unpublished results.
52. H. Takahashi and J. Tsuji, *J. Organometallic Chem.*, **10,** 511 (1967).
53. G. W. Parshall, *Accounts Chem. Res.*, **3,** 139 (1970).
54. M. I. Bruce, B. L. Goodall, G. L. Sheppard and F. G. A. Stone, unpublished results.
55. P. E. Baikie and O. S. Mills, *Inorg. Chim. Acta*, **1,** 55 (1967).
56. T. Joh, N. Hagihara and S. Murahashi, *Bull. Chem. Soc. Japan*, **40,** 661 (1967).
57. G. A. Olah, K. Dunne, D. P. Kelly and Y. K. Mo, *J. Amer. Chem. Soc.*, **94,** 7438 (1973).
58. S. Murahashi and S. Horiie, *J. Amer. Chem. Soc.*, **78,** 4816 (1956).
59. S. Murahashi and S. Horiie, *Bull. Chem. Soc. Japan*, **33,** 88 (1960).
60. J. Tsuji, J. Kiji, S. Isamura and M. Morikawa, *J. Amer. Chem. Soc.*, **86,** 4350 (1964).
61. J. Tsuji, M. Morikawa and J. Kiji, *J. Amer. Chem. Soc.*, **86,** 4851 (1964).
62. W. W. Prichard, U.S. Pat. 2769003 (1956); *Chem. Abstr.*, **51,** 7412 (1957).
63. M. I. Bruce, B. L. Goodall, A. D. Redhouse and F. G. A. Stone, *J. C. S. Chem. Comm.*, 1228 (1972).
64. I. Jardine and F. J. McQuillin, *Chem. Comm.*, 626 (1970); P. Abley, I. Jardine and F. J. McQuillin, *J. Chem. Soc.* (*A*), 840 (1971).

65. M. Green and G. Swinden, *Inorg. Chim. Acta*, **5**, 49 (1971).
66. D. R. Fahey, *Chem. Comm.*, 417 (1970); *J. Organometallic Chem.*, **27**, 283 (1971).
67. Y. A. Ustynyuk, I. V. Barinov and E. I. Sirotkina, *Doklady AN SSSR*, **187**, 112 (1969).
68. R. B. King and M. B. Bisnette, *J. Amer. Chem. Soc.*, **86**, 5694 (1964); *Inorg. Chem.*, **5**, 300 (1966).
69. M. L. H. Green, J. R. Sanders and R. N. Whiteley, *Z. Naturforsch.*, **23b**, 106 (1968).
70. A. N. Nesmeyanov, U. A. Chapovsky, N. A. Ustynyuk and L. G. Makarova, *Izvest. AN SSSR, Otdel. Khim. Nauk.*, 449 (1968).
71. S. Trofimenko, *Inorg. Chem.*, **8**, 2675 (1969); *J. Amer. Chem. Soc.*, **92**, 5118 (1970).
72. M. F. Lappert and J. S. Poland, *Chem. Comm.*, 1061 (1969).
73. M. L. H. Green and J. R. Sanders, *Chem. Comm.*, 956 (1967); *J. Chem. Soc. (A)*, 1947 (1971).
74. J. R. Knox and C. K. Prout, *Acta Cryst.*, **B25**, 1952 (1969).
75. J. A. McCleverty and R. N. Whiteley, *Chem. Comm.*, 1159 (1971).
76. F. J. Lalor and P. L. Pauson, *J. Organometallic Chem.*, **25**, C51 (1970).
77. V. F. Duckworth, P. G. Douglas, R. Mason and B. L. Shaw, *Chem. Comm.*, 1083 (1970).
77a. W. E. Carroll and F. J. Lalor, *J. Organometallic Chem.*, **54**, C37 (1973).
78. K. R. Laing, S. D. Robinson and M. F. Uttley, *J. C. S. Chem. Comm.*, 176 (1973).
79. M. C. Baird and G. Wilkinson, *J. Chem. Soc. (A)*, 865 (1967).
80. G. W. Rayner-Canham and D. Sutton, *Can. J. Chem.*, **49**, 3994 (1971).
81. L. Toniolo and R. Eisenberg, *Chem. Comm.*, 455 (1971).
82. G. Caglio and M. Angoletta, *Gazz. Chim. Ital.*, **102**, 462 (1972).
82a. B. L. Haymore and J. A. Ibers, *J. Amer. Chem. Soc.*, **95**, 3052 (1973).
83. A. J. Deeming and B. L. Shaw, *J. Chem. Soc. (A)*, 1128 (1969).
84. G. W. Parshall, *J. Amer. Chem. Soc.*, **87**, 2133 (1965); **89**, 1822 (1967).
85. C. D. Cook and G. S. Jauhal, *J. Amer. Chem. Soc.*, **90**, 1464 (1968).
85a. W. Beck, M. Bauder, G. La Monica, S. Cenini and R. Ugo, *J. Chem. Soc. (A)*, 113 (1971).
85b. S. Cenini, R. Ugo and G. La Monica, *J. Chem. Soc. (A)*, 3441 (1971).
86. W. R. Knoth, *J. Amer. Chem. Soc.*, **94**, 104 (1972).
86a. A. D. Allen, R. O. Harris, B. R. Loescher, J. R. Stevens and R. N. Whiteley, *Chem. Rev.*, **73**, 11 (1973).
86b. J. Chatt, J. R. Dilworth, R. L. Richards and J. R. Sanders, *Nature*, **224**, 1201 (1969); J. Chatt, J. R. Dilworth, G. J. Leigh and R. L. Richards, *Chem. Comm.*, 955 (1970); J. Chatt, R. C. Fay and R. L. Richards, *J. Chem. Soc. (A)*, 702 (1971).
87. G. Avitabile, P. Ganis and M. Nemiroff, *Acta Cryst.*, **B27**, 725 (1971).
87a. J. V. McArdle, A. J. Schultz, B. J. Corden and R. Eisenberg, *Inorg. Chem.*, **12**, 1676 (1973).
88. R. L. Bennett, M. I. Bruce, B. L. Goodall, M. Z. Iqbal and F. G. A. Stone, *J. C. S. Dalton*, 1787 (1972).
89. A. Rosenthal and I. Wender, in *Organic Syntheses via Metal Carbonyls* (Eds. I. Wender and P. Pino), Wiley, New York, 1968, pp. 405–466.

90. W. Hübel and E. H. Braye, *J. Inorg. Nuclear Chem.*, **10,** 250 (1959); M. van Meerssche, *Ind. Chim. Belge*, **28,** 507 (1963).
91. I. S. Kolomnikov, T. S. Lobeeva, V. V. Gorbacheskaya, G. G. Aleksandrov, Y. T. Struchkov and M. E. Volpin, *Chem. Comm.*, 972 (1971).
92. A. W. Parkins, E. O. Fischer, G. Huttner and D. Regler, *Angew. Chem.*, **82,** 635 (1970); *Int. Edit.*, **9,** 633 (1970).
92a. K. W. Muir, R. Walker, J. Chatt, R. L. Richards and G. H. D. Royston, *J. Organometallic Chem.*, **56,** C30 (1973).
93. R. L. Bennett, M. I. Bruce, B. L. Goodall, M. Z. Iqbal and F. G. A. Stone, *J. C. S. Dalton*, 1787 (1972) [Mn, Re, Ru, Rh, Pd]; M. M. Bagga, W. T. Flannigan, G. R. Knox, P. L. Pauson, F. J. Preston and R. I. Reed, *J. Chem. Soc.* (*A*), 36, 1968 [Fe]; Y. A. Ustynyuk, V. A. Chertkov and I. V. Barinov, *J. Organometallic Chem.*, **29,** C53 (1971) [Ni]; H. Onoue and I. Moritani, *J. Organometallic Chem.*, **43,** 431 (1972) [Pd].
94. W. T. Flannigan, G. R. Knox and P. L. Pauson, *J. Chem. Soc.* (*C*), 2077 (1969).
95. M. I. Bruce, M. Z. Iqbal and F. G. A. Stone, unpublished results; R. J. Doedens and R. G. Little, *Inorg. Chem.*, **12,** 844 (1973).
96. A. C. Cope and E. C. Friedrich, *J. Amer. Chem. Soc.*, **90,** 909 (1968) [Pd, Pt].
97. S. Trofimenko, *Inorg. Chem.*, **10,** 1372 (1971) [Pd].
98. S. Trofimenko, *J. Amer. Chem. Soc.*, **93,** 1808 (1971).
99. R. L. Bennett, M. I. Bruce and F. G. A. Stone, unpublished results.
100. C. C. Yin and A. J. Deeming, *J. C. S. Dalton*, 1013 (1974).
101. J. M. Kliegman and A. C. Cope, *J. Organometal. Chem.*, **16,** 309 (1969).
102. H. Onoue, K. Minami and K. Nakagawa, *Bull. Chem. Soc. Japan*, **43,** 3480 (1970) [Pd].
103. H. Onoue, K. Nakagawa and I. Moritani, *J. Organometal. Chem.*, **35,** 217 (1972) [Pd].
104. B. N. Cockburn, D. V. Howe, T. Keating, B. F. G. Johnson and J. Lewis, *J. C. S. Dalton*, 404 (1973).
105. G. Bombieri, L. Caglioti, L. Cattalini, E. Forsellini, F. Gasparrini, R. Graziani and P. A. Vigato, *Chem. Comm.*, 1415 (1971).
106. M. I. Bruce, B. L. Goodall and F. G. A. Stone, unpublished results [Mn, Re].
107. M. Nonoyama and K. Yamasaki, *Inorg. Nucl. Chem. Letters*, **7,** 943 (1971) [Rh].
108. A. Kasahara, *Bull. Chem. Soc. Japan*, **44,** 1272 (1968) [Pd].
109. V. I. Sokolov, T. A. Sorokina, L. L. Troitskaya, L. I. Solovieva and O. A. Reutov, *J. Organometal. Chem.*, **36,** 389 (1972).
110. S. Trofimenko, *Inorg. Chem.*, **12,** 1215 (1973) [Pd].
111. M. I. Bruce and I. Matsuda, unpublished results.
112. M. I. Bruce, B. L. Goodall and F. G. A. Stone, *J. Organometallic. Chem.*, **60,** 343 (1973) [Mo, Mn, Re, Ru, Os, Rh, Pd]; M. Nonoyama, K. Suzuki and K. Yamasaki, *Proc. Japan. Acad.*, **45,** 605 (1969) [Pd].
113. A. Sacco and M. Aresta, *Chem. Comm.*, 1223 (1968) [Fe]; W. Keim, *J. Organometal. Chem.*, **14,** 179 (1968); **19,** 161 (1969) [Rh].
114. M. I. Bruce, R. C. F. Gardner and F. G. A. Stone, *J. Organometallic Chem.*, **40,** C39 (1972) [Ru].
115. M. A. Bennett and D. L. Milner, *J. Amer. Chem. Soc.*, **91,** 6983 (1969) [Ir].
115a. B. L. Shaw and R. E. Stainbank, *J. C. S. Dalton*, 2394 (1973).

116. A. J. Cheney, B. E. Mann, B. L. Shaw and R. M. Slade, *J. Chem. Soc.* (*A*), 3833 (1971).
117. G. Hata, H. Kondo and A. Miyake, *J. Amer. Chem. Soc.*, **90,** 2278 (1968).
118. J. M. Duff and B. L. Shaw, *J. C. S. Dalton*, 2219 (1972).
119. A. J. Cheney and B. L. Shaw, *J. C. S. Dalton*, 754 [Pt], 860 [Pd] (1972); A. J. Cheney, W. S. McDonald, K. O'Flynn, B. L. Shaw and B. L. Turtle, *J. C. S. Chem. Comm.*, 128 (1973) [Pt].
120. D. F. Gill, B. E. Mann and B. L. Shaw, *J. C. S. Dalton*, 270 (1973).
121. J. Chatt and J. M. Davidson, *J. Chem. Soc.*, 843 (1965).
122. J. J. Levison and S. D. Robinson, *J. Chem. Soc.* (*A*), 639 (1970) [Ru]; M. I. Bruce, G. Shaw and F. G. A. Stone, *J. C. S. Dalton*, 1667 (1973) [Ru]; E. W. Ainscough, T. A. James, S. D. Robinson and J. N. Wingfield, *J. Organometallic Chem.*, **60,** C63 (1973) [Os]; E. K. Barefield and G. W. Parshall, *Inorg. Chem.*, **11,** 964 (1972) [Rh]; E. W. Ainscough, S. D. Robinson and J. J. Levison, *J. Chem. Soc.* (*A*), 3413 (1971) [Ir]; M. A. Bennett and R. Charles, *Aust. J. Chem.*, **24,** 427 (1971) [Ir]; N. Ahmad, E. W. Ainscough, T. A. James and S. D. Robinson, *J. C. S. Dalton*, 1151 (1973) [Pd, Pt].
123. E. L. Hoel and M. F. Hawthorne, *J. Amer. Chem. Soc.*, **95,** 2712 (1973).
123a. R. McKinney, G. Firestein and H. D. Kaesz, *J. Amer. Chem. Soc.*, **95,** 7910 (1973).
124. R. J. McKinney, B. T. Huie, C. B. Knobler and H. D. Kaesz, *J. Amer. Chem. Soc.*, **95,** 633 (1973).
125. R. L. Bennett, M. I. Bruce, I. Matsuda, R. J. Doedens, R. G. Little and J. T. Veal, *J. Organometallic Chem.*, **67,** C72 (1974).
125a. H. Alper and A. S. K. Chan, *J. C. S. Chem. Comm.*, 724 (1973).
126. H. Alper and A. S. K. Chan, *Chem. Comm.*, 1203 (1971); H. Alper and A. S. K. Chan, *J. Organometallic Chem.*, **61,** C59 (1973) [Ru]; H. Alper, *J. Organometallic Chem.*, **61,** C62 (1973) [Pd, Pt].

ADDENDUM (December 1973)

Since the preparation of this article, several important advances have been made on the topics covered. The salient features are summarized below, together with those of several papers inadvertently omitted from the main text.

II. COMPLEXES CONTAINING NO METAL–CARBON INTERACTION

Azobenzene reacts directly with $Ti(CO)_2(C_5H_5)_2$, $V(C_5H_5)_2$ or $Co(PMe_3)_4$ to give the complexes [**2**; M = $Ti(C_5H_5)_2$[127], $V(C_5H_5)_2$[127], and $Co(PMe_3)_2$[128], respectively]. The first two react with aqueous ethanol or iodine to give PhNHNHPh and PhN_2Ph, respectively.

IV. COMPLEXES CONTAINING *ORTHO*-METALLATED AZO COMPOUNDS AS LIGANDS

An alternative preparation of $[RhCl(azb)_2]_2$ (**22**) is from azobenzene and $[RhCl(C_2H_4)_2]_2$. Reaction of **22** with $[RhCl(PF_3)_2]_2$ affords [RhCl(azb)-$(PF_3)_2$], analogous to **24**; ready exchange of CO and PF_3 occurs, via the mixed complex $Rh_2Cl_2(azb)_2(CO)(PF_3)$[129]. $[RhCl(PF_3)_2]_2$ acts as a catalyst for the $LiAlH_4$-reduction of azobenzene to hydrazobenzene.

Syntheses of 2-(arylazo)phenylmercurials according to the route shown proceed in about 55% overall yield[130].

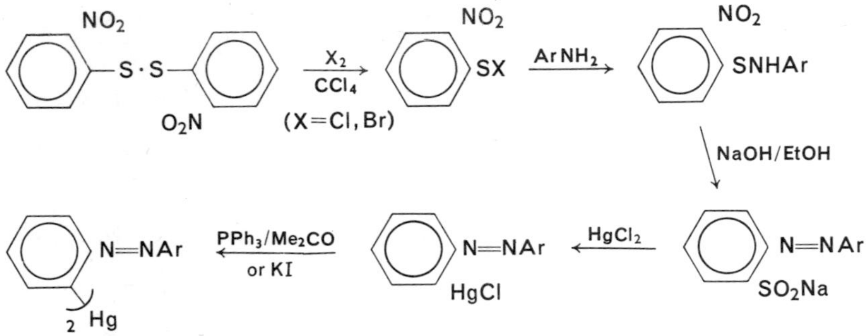

These reagents promise to be of considerable synthetic utility in the formation of new (phenylazo)phenyl derivatives of transition metals[144, 145].

VII. ARYLAZO AND ARYLDIIMINE COMPLEXES OF TRANSITION METALS

The relationship between arylazo-metal compounds and dinitrogen complexes has been emphasized by the discovery of the reactions[131]:

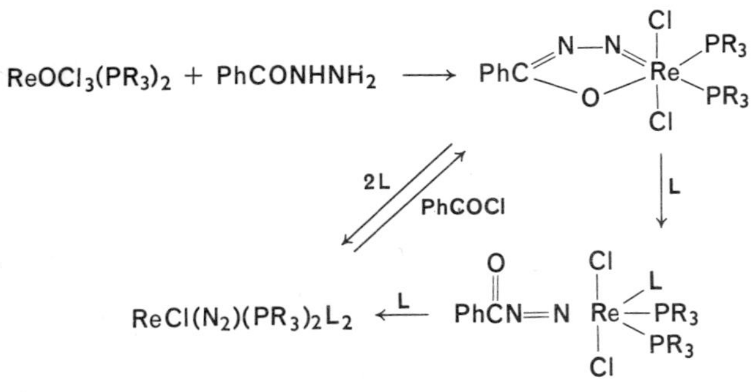

Acid halides react with *trans*-$[W(N_2)_2(dppe)_2]$ to give $WCl(N_2HCOR)$-$(dppe)_2$, which can lose HCl to give the chelated aroyl- or acylazo complex

$WCl(N{=}NCOR)(dppe)_2$[131a]. In the molybdenum and tungsten systems, arylhydrazines react with $MoOCl_2(PMe_2Ph)_3$ to give $MoCl_2(NAr)$-$(ArCON_2Ar)(PMe_2Ph)$, probably containing a five-membered chelate ring, while the bis-dinitrogen complexes $M(N_2)_2(PR_3)_4$ (M = Mo, W) react with aroyl halides to give $MX(N_2COAr)(PR_3)_4$[132]. Also known as a 1 : 1 adduct of $MoO(S_2CNR_2)_2$ and $N_2(CO_2Et)_2$ of undetermined structure[133].

Reactions between *cis*-$PtCl_2(PR_3)_2$ and $(R'CONH)_2$ (R′ = Me, Ph) afford $Pt(R'CON_2COR')(PR_3)_2$[134], in which the arylhydrazine is coordinated in the enolized form (as determined[135] by a crystal structure for R=R′=Ph):

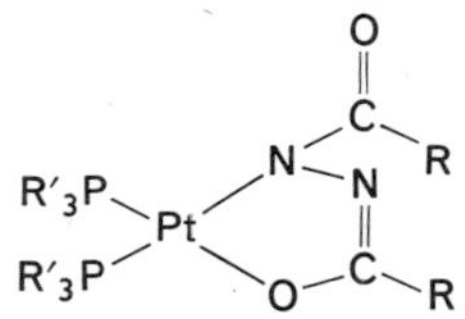

The dibenzoyl derivative (R′=Ph) can also be prepared directly from $Pt(PR_3)_4$ and $PhCON_2COPh$. Acids react with this complex to give dibenzoylhydrazine, while bromine displaces $PhCON_2COPh$ almost quantitatively. The palladium complexes are extremely unstable.

Oxidative addition of NOCl to $M(N{=}NAr)(CO)_2L$ ($L = C_5H_5$, M = Mo; $L = HBpz_3$, M = Mo or W) affords the nitrosyl complexes $MCl(N{=}NAr)(NO)L$[136].

The complex $Fe(CO)_3(PPh_3)_2$ reacts with *p*-$XC_6H_4N_2^+BF_4^-$ (X = F, Cl, Br, H, NO_2, OMe, OH) to give $[Fe(N{=}NC_6H_4X\text{-}p)(CO)_2(PPh_3)_2]BF_4$, in which the more effective π-acceptor arylazo ligand has replaced an equatorial CO group in the tricarbonyl[137].

A potentially important new synthetic route to arylazo-metal complexes is the diazotization of amines by coordinated nitrosyl groups, as exemplified by the reactions of aromatic amines with $[RuCl(NO)(bipy)_2]^{2+}$ to give $[RuCl(N{=}NAr)(bipy)_2]^{2+}$. The latter are considered to contain coordinated aryldiazonium cations, and with phenols, for example, the azo coupling product is obtained[138].

The five-coordinate cation $[RhCl(N{=}NPh)L]^+$ [$L = PhP(CH_2CH_2CH_2$-$PPh_2)_2$] is readily obtained from RhClL and $PhN_2^+PF_6^-$. The arylazo group is doubly-bent, corresponding to coordination as $PhN{=}N^-$ to Rh(III):

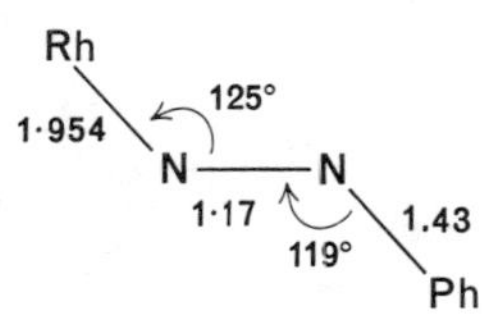

In an infrared study, 'modified' ν(NN) frequencies (see original paper) can be correlated with observed geometries[139].

Diazonium salts react with $RhHCl_2(PPh_3)_3$ to give diaryldiimine complexes $RhCl_3(NH{=}NAr)(PPh_3)_2$, previously formulated[79] as solvated $RhCl_2(N{=}NAr)(PPh_3)_2$ complexes. Other diaryldiimine complexes $[MX(NH{=}NAr)(CO)(PPh_3)_2]^+$ (M = Ru or Os) were obtained similarly from $MHCl(CO)(PPh_3)_3$ (X = Cl) or $MH_2(CO)(PPh_3)_3$ (X = H). Unlike the other arylazo complexes, the rhodium derivatives contain doubly bent Rh—N=N—Ar groups. The interrelationship of arylazo and aryldiimine complexes was studied in detail:

$$RhCl_2(N{=}NAr)(PPh_3)_2 \underset{HCl}{\overset{HCl}{\rightleftharpoons}} RhCl_3(NH{=}NAr)(PPh_3)_2 \xrightarrow{H_2} [RhCl_3(NH_2NHAr)(PPh_3)]_2$$

hydrogenation of the latter affords the insoluble hydrazine complexes[140].

Other papers have described the syntheses of $[RuCl(NH{=}NPh)(CO)_2(PPr^n_2Bu^t)_2]BF_4$[141] [from $RuHCl(CO)_2(PPr^n_2Bu^t)_2$ and $PhN_2^+BF_4^-$], and of $IrCl_2(N{=}NAr)(CO)(PMe_2Bu^t)_2$[142] [from $IrCl(CO)(PMe_2Bu^t)_2$ and $PhN_2^+Cl^-$].

REFERENCES

127. G. Fachinetti, G. Fochi and C. Floriani, *J. Organometallic Chem.*, **57,** C51 (1973).
128. H. F. Klein, *Angew. Chem.*, **83,** 363 (1971); *Internat. Edit.*, **10,** 343 (1971).
129. M. Kooti and J. F. Nixon, *J. Organometallic Chem.*, **63,** 415 (1973).
130. R. J. Cross and N. H. Tennent, *J. Organometallic Chem.*, **61,** 33 (1973).
131. J. Chatt, J. R. Dilworth and G. J. Leigh, *Chem. Comm.*, 687 (1969); J. Chatt, J. R. Dilworth and G. J. Leigh, *J. Organometallic Chem.*, **21,** P49 (1970); J. Chatt, J. R. Dilworth, G. J. Leigh and V. D. Gupta, *J. Chem. Soc. (A)*, 2631 (1971); J. Chatt, J. R. Dilworth and G. J. Leigh, *J. C. S. Dalton*, 612 (1973).
131a. J. Chatt, G. A. Heath and G. J. Leigh, *J. C. S. Chem. Comm.*, 444 (1972).
132. J. Chatt and J. R. Dilworth, *Chem. Comm.*, 549 (1972); J. Chatt, G. A. Heath, N. E. Hooper and G. J. Leigh, *J. Organometallic Chem.*, **57,** C67 (1973).
133. P. W. Schneider, D. C. Bravard, J. W. McDonald and W. E. Newton, *J. Amer. Chem. Soc.*, **94,** 8640 (1972).
134. J. R. Dilworth, A. S. Kasenally and F. M. Hussein, *J. Organometallic Chem.*, **60,** 203 (1973).
135. S. D. Ittel and J. A. Ibers, *Inorg. Chem.*, **12,** 2290 (1973).
136. M. Deane and F. J. Lalor, *J. Organometallic Chem.*, **57,** C61 (1973).
137. D. R. Fisher and D. Sutton, *Can. J. Chem.*, **51,** 1697 (1973); W. E. Carroll and F. J. Lalor, *J. C. S. Dalton*, 1754 (1973).

138. W. L. Bowden, W. F. Little and T. J. Meyer, *J. Amer. Chem. Soc.*, **95,** 5084 (1973).
139. T. E. Nappier, D. W. Meek, R. M. Kirchner and J. A. Ibers, *J. Amer. Chem. Soc.*, **95,** 4194 (1973); A. P. Gaughan, B. L. Haymore, J. A. Ibers, W. H. Myers, T. E. Nappier and D. W. Meek, *J. Amer. Chem. Soc.*, **95,** 6859 (1973).
140. K. R. Laing, S. D. Robinson and M. F. Uttley, *J. C. S. Dalton*, 2713 (1973).
141. D. F. Gill, B. E. Mann and B. L. Shaw, *J. C. S. Dalton*, 311 (1973).
142. B. L. Shaw and R. L. Stainbank, *J. C. S. Dalton*, 223 (1972).
143. A. Nakamura, M. Aotake and S. Otsuka, *J. Amer. Chem. Soc.*, **96,** 3456 (1974): interaction of $(\pi\text{-}C_2H_5)_2MH_2$ (M = Mo and W) with azo or diazo compounds.
144. R. J. Cross and N. H. Tennent, *J. Organometallic Chem.*, **72,** 21 (1974): synthesis of Mn, Ni, Pd and Pt derivatives using these mercurials.
145. Direct mercuration of azobenzene using $Ag(OAc)_2$ has been achieved: P. V. Roling, J. L. Dill and M. D. Rausch, *J. Organometallic Chem.*, **69,** C33 (1974); M. I. Bruce and D. Pilouras, unpublished work.

CHAPTER **10**

Radiation chemistry of hydrazo and azo compounds

A. HOROWITZ and L. A. RAJBENBACH

Soreq Nuclear Research Centre, Yavne, Israel

I. INTRODUCTION

Radiation chemistry is mainly concerned with the study of the chemical effects of high energy radiation. The sources of energy most commonly employed are radioactive sources emanating electromagnetic radiation (γ-rays) and accelerated particles. In recent years ^{60}Co sources and electron accelerators have become the most commonly used energy sources in radiolytic studies. It should be noted that, since the γ-rays emanating from ^{60}Co radioactive sources deposit their energy through the Compton effect, the agency which causes the chemical change is the same as that produced by electron beam generators, i.e., fast-moving electrons. These electrons lose most of their energy through ionization processes. The primary radiolytic species which result are thus ions and, to a smaller extent, excited molecules. The latter are formed partly in the direct energy degradation process but mostly following molecular cation electron recombination reactions. High-energy radiation is frequently referred to as ionizing radiation because charged species are the primary precursors of radiolytic products. Subsequent decomposition of electronically excited molecules and/or ions leads to the formation of free radicals. It is presently accepted that most of the substances produced by the radiolysis of pure organic compounds are formed through the intermediacy of free-radical reactions. It should be noted, however, that in polar media where charged species can be stabilized by solvation, or in solutions containing compounds which react easily with charged species, a considerable portion of the radiolytic products can be traced to reactions between ions, electrons and molecules. The pulse radiolysis technique lends itself particularly well to the study of transient species, which are produced by a pulse of ionizing radiation. These species are then monitored by means of their physical properties such as optical absorption, electron spin resonance spectra, electric conductivity, etc. This technique may, in fact, be considered as an analogue of flash photolysis. Conventional (non-pulsed) radiolytic studies deal mostly with the nature and overall yield of chemical change.

In the study of the effect of radiation on solutions and mixtures one should consider the difference in modes of energy absorption between radiolysis and photolysis. The contrast between the two systems is best appreciated from consideration of the net chemical effect. In u.v. irradiation, regardless of the dilution, energy is absorbed only by the component which can respond to the wave length of the illuminating source. In the case of ionizing radiation, energy is absorbed by all the molecules of the system and in very dilute solutions practically all of the energy is deposited in the solvent. Chemical reactions which occur are the result of reactions of the reactive species originating from the solvent.

The net chemical change following radiolysis is expressed in terms of the number of molecules which are formed or disappear per 100 eV energy absorbed. This yield, the *G*-value, is generally accepted as expressing the irradiation-induced chemical yield. Most *G*-values lie between 0·1 and 10. Higher *G*-values usually indicate secondary chain reactions. On the average about 20 eV are absorbed per chemical bond broken. Since chemical bond strengths of organic molecules usually vary between 3–4 eV it is evident that most of the energy absorbed in the course of radiolysis does not lead to chemical change but is dissipated in the form of heat.

Most radiation studies of organic compounds were carried out in the liquid state. In recent years, however, a growing number of publications were devoted to e.s.r. studies of ionizing radiation effects in rigid glasses and crystalline solids. In most of these studies the e.s.r. technique was used for the identification of radiolytically formed transient species.

The hydrazo, azo and azoxy family of compounds has not attracted the sustained interest of radiation chemists. There is no systematic effort towards the elucidation of the primary radiation effects. This review will be concerned only with the radiation chemistry of hydrazo and azo compounds. To the best of our knowledge, no work has been published on the radiolysis of azoxy compounds.

II. RADIATION CHEMISTRY OF HYDRAZO COMPOUNDS

A. Hydrazine and its Methyl Derivatives

Hydrazine (HD) and its monomethyl (MMHD) and 1,1-dimethyl (DMHD) derivatives are important rocket propellants. Thus the interaction of these compounds with ionizing radiation is of considerable interest to scientists engaged in space research. This is probably why these compounds are among the few hydrazo compounds which have attracted the attention of radiation chemists.

Lucien and Pinns[1] exposed liquid and gaseous HD and DMHD to X-ray radiation. The irradiations were carried out at room temperature to a total dose of approximately 1×10^6 rads. According to these authors the main identifiable products of the radiolysis of HD were N_2, H_2 and NH_3. The same products plus methylamine, dimethylamine and HD were observed in the radiolytic decomposition of DMHD. The following values for the decomposition of the starting materials are given: $G(\text{-HD})_{\text{liq}} \simeq 2 \times 10^2$,

TABLE 1. Gaseous products from the radiolysis of MMHD and DMHD

Substrate	Radical scavenger concentration by volume			G (products)		
	1,3-Pentadiene	Methyl methacrylate	CCl_4	H_2	CH_4	N_2
MMHD	—	—	—	4·7	3·4	4·0
	5%	—	—	3·6	3·1	3·6
	—	5%	—	4·4	4·1	4·6
	—	—	5%	3·3	25·6	30·4
DMHD	—	—	—	5·7	0·75	5·3
	—	5%	—	3·7	0·57	5·3

Total dose $8{\cdot}5 \times 10^6$ rads, irradiated at room temperature with a ^{60}Co source.

$G(\text{-HD})_{gas} \simeq 7 \times 10^7$, $G(\text{-DMHD})_{liq} \simeq 1 \times 10^3$ and $G(\text{-DMHD})_{gas} \simeq 2 \times 10^6$. The high G values of decomposition indicate that both HD and DMHD are removed from the systems by some unidentified chain reactions.

The radiolytic formation of non-condensed gases from MMHD and DMHD was studied by Shelberg[2,3]. The results on these studies are summarized in Table 1.

The observation that the yield of gaseous products in MMHD is not significantly affected by the presence of olefinic free-radical scavengers was considered as proof that these compounds are formed via molecular or ionic processes. The author claims that the radiolytic decomposition of MMDH apparently follows the simple stoicheiometric relationship: $(CH_3NHNH_2)_n \rightarrow n\ H_2 + n\ N_2 + n\ CH_4$. The fact that the addition of CCl_4 to MMHD caused a considerable increase in the yields of CH_4 and N_2 is rationalized in terms of a chain reaction induced by the products of the decomposition of CCl_4.

$$CCl_4 \rightsquigarrow Cl + CCl_3 \quad (1)$$

$$Cl + CH_3NHNH_2 \longrightarrow HCl + N_2 + CH_4 + H \quad (2)$$

$$CCl_3 + CH_3NHNH_2 \longrightarrow CHCl_3 + N_2 + CH_4 + H \quad (3)$$

$$H + CCl_4 \longrightarrow HCl + CCl_3 \quad (4)$$

In the case of DMHD it is considered that part of the product was formed via free-radical processes. This conclusion was reached because methyl methacrylate was found to cause a significant decrease in the yield of H_2 and CH_4.

On the strength of the available data it is rather difficult to form a clear picture of the chemical processes which occur in liquid and gas phase radiolysis of HD and its alkyl derivatives.

E.s.r. studies on the nature of radicals formed in the γ radiolysis of solid hydrazine[4] and its frozen aqueous solutions[5] yield evidence as to the importance of the scission of the N—H bond. No e.s.r. studies are available on the type of radicals formed in the radiolysis of MMHD and DMHD.

B. *Radiochemical Synthesis of Hydrazine*

Hydrazine is formed in the radiolysis of ammonia. Despite the relatively low yields $G(N_2H_4) < 1$ the possibility of industrial production of N_2H_4 by γ-radiation techniques has attracted considerable interest[6].

E.s.r. studies by Tupikov and co-workers[7] on γ-irradiation of solid NH_3 and its frozen aqueous solutions have proved the existence of NH_2 radicals. The radiolytic formation of N_2H_4 in condensed phases probably results from the dimerization reaction of NH_2 radicals. In the irradiation of solid

NH_3 a $G(N_2H_4)$ value of 0·1 was obtained[8]. The yield of N_2H_4 was found to be independent of the irradiation dose. In the radiolysis of liquid NH_3 a $G(N_2H_4)$ value of 0·2 was obtained at the irradiation dose of 2×10^{19} eV/g. However, a progressive decrease in the yield of hydrazine was observed with an increase in dose[8]. This decrease in the radiolytic yield is probably due to secondary reactions of N_2H_4 with reactive species such as hydrogen atoms and solvated electrons. This interpretation of dose effect is consistent with the observation that the addition of isopropanol to NH_3 stabilizes the yield of hydrazine. The effect is most probably the result of the ability of the alcohol to scavenge H atoms and electrons which would otherwise cause a reduction in the yield of hydrazine. In the radiolysis of liquid ammonia, in the presence of 5% isopropanol, a $G(N_2H_4)$ value of 0·59 was reported[8]. This represents the initial yield of hydrazine in γ irradiated liquid ammonia at 25°C, and is in agreement with the conclusions of Dainton and co-workers[9] who estimated, by a different method, an initial $G(N_2H_4)$ value of $1 \pm 0{\cdot}5$. Somewhat higher yields of hydrazine (G-value of 1·5 from liquid NH_3) are obtained under high-intensity irradiation with fission fragments[10]. Improved yields have also been reported in the γ-irradiation of NH_3 absorbed on sodium aluminium silicate. Under such conditions a $G(N_2H_4)$ value of 0·99 has been obtained[11].

The low yields of hydrazine and its sensitivity to attack by hydrogen atoms suggest that there is little hope for economical utilization of this process.

C. *Aqueous Solutions of Hydrazine and its Derivatives*

The radiation chemistry of aqueous solutions of hydrazine and methyl hydrazines was studied with the use of α-[12] and γ-radiation[13–15] as well as by pulse radiolytic techniques[16]. The major products formed are ammonia, hydrogen and nitrogen. It was found that over a pH range of 0·3–13 the yields of nitrogen and hydrogen are equal and the yield of ammonia is twice that of nitrogen and equal to the yield of hydrazine consumed: $G(\text{-}N_2H_4) = G(NH_3) = 2\,G(N_2) = 2\,G(H_2)$. In acidic and neutral solutions the yields of these three products are constant at hydrazine concentrations above 2×10^{-2}M but fall off markedly at lower concentrations. The maximum yields of the products at high hydrazine concentrations are: $G(NH_3) =$ 5·1–5·5 and $G(N_2) = 2{\cdot}5$ at pH 0·3–8. In strongly acidic solutions (pH = 0) $G(N_2)$ increases to 2·9–3·2. In basic solutions the yields of the products are considerably lower than in the acidic and neutral solutions[15].

In order to account for these results it is necessary to consider the reaction between hydrazine and the primary species formed in the radiolysis of water. The overall process can be described as follows:

$$H_2O \rightsquigarrow OH\ (2{\cdot}8),\ e^-_{aq}\ (2{\cdot}8),\ H\ (0{\cdot}55),\ H_2\ (0{\cdot}45)\ \text{and}\ H_2O_2\ (0{\cdot}71)$$

where the values given in parentheses are the 100 eV yields of the various products. The reactions between these initially formed species and hydrazine are strongly pH-dependent and therefore have to be considered separately in different media.

According to Belloni and Haissinsky[15] the main reactions in acidic media are:

$$N_2H_5^+ + H \longrightarrow N_2H_3 + H^+ + H_2 \qquad (5)$$

$$N_2H_5^+ + OH \longrightarrow N_2H_3 + H^+ + H_2O \qquad (6)$$

$$2\,N_2H_3 \longrightarrow [N_4H_6] \longrightarrow N_2 + 2\,NH_3 \qquad (7)$$

The tetrazane formed in reaction (7) is unstable and decomposes unimolecularly. It was suggested by Hayon and Simic[16] that the transient spectra of one of the intermediates observed by them could be ascribed to tetrazane formed in reaction (7). Furthermore, they were able to observe the spectra of the hydrazyl radical N_2H_3 and of the analogous species in the radiolysis of 1,1-dimethylhydrazine and 1,2-dimethylhydrazine.

In neutral solutions, in addition to reactions (5) to (7), the reaction between hydrated electrons and $N_2H_5^+$ also occurs:

$$N_2H_5^+ + e^-_{aq} \longrightarrow N_2H_3 + H_2 \qquad (8)$$

Further increase of the pH results in the conversion of $N_2H_5^+$ into N_2H_4. Therefore, in basic media, reactions analogous to (5) through (8) occur but the reacting species is hydrazine itself. Consequently, the rate constants of reactions such as (5), (6) and (8) also change. Also, since in basic media the reaction between the H atom and the hydroxyl ion results in the formation of hydrated electron (reaction 9), reaction (8) takes place instead of reaction (5).

$$H + OH^- \longrightarrow e^-_{aq} \qquad (9)$$

While the addition of hydrogen to acidic solutions of hydrazine did not affect the yields, addition of hydrogen peroxide markedly reduced them. At 10^{-3}M H_2O_2 $G(H_2)$ decreased to 0·42 but $G(N_2)$ was not affected. The decrease in hydrogen yield in acidic solution was attributed to the competition between $N_2H_5^+$ and H_2O_2 for the hydrogen atoms and in neutral media for the hydrated electron.

Using pulse radiolytic techniques Hayon and Simic[16] determined the rate constants for the reactions between various hydrazines and hydroxyl radicals. These rate constants are summarized in Table 2 (reproduced from reference 16). As can be seen, the rate constants show a strong dependence upon the state of protonation of the hydrazines and their dissociation constants.

TABLE 2. Rate constants for reaction of OH radicals and e^-_{aq} with hydrazines in aqueous solution[a]

Solute	p*K*	$k(e^-_{aq} + S)$, $M^{-1}\,s^{-1}$	$k(OH + S)$, $M^{-1}\,s^{-1}$
H_2NNH_2	8·07	$2{\cdot}2 \times 10^8$ (pH 6·0)	$1{\cdot}0 \times 10^9$ (pH 6·0)
		$2{\cdot}3 \times 10^6$ (pH 10·5)	$1{\cdot}4 \times 10^{10}$ (pH 10·0)
CH_3HNNH_2	7·87	$1{\cdot}4 \times 10^9$ (pH 5·5)	
		$6{\cdot}5 \times 10^6$ (pH 12·0)	
$CH_3HNNHCH_3$	7·52	$2{\cdot}3 \times 10^9$ (pH 5·6)	$7{\cdot}2 \times 10^8$ (pH 3·5)
		$6{\cdot}1 \times 10^6$ (pH 12·4)	$1{\cdot}4 \times 10^{10}$ (pH 10·1)
$(CH_3)_2NNH_2$	7·21	$5{\cdot}8 \times 10^9$ (pH 5·6)	$8{\cdot}1 \times 10^8$ (pH 3·5)
		$2{\cdot}4 \times 10^7$ (pH 12·0)	$1{\cdot}6 \times 10^{10}$ (pH 9·7)
$(CH_3)_2NNHCH_3$	6·56	$1{\cdot}3 \times 10^{10}$ (pH 5·4)	
		$\sim 10^8$ (pH 10·4)	

[a] Reproduced with permission from E. Hayon and M. Simic, *J. Amer. Chem. Soc.*, **94**, 42 (1972).

D. The Use of 1,1-Diphenyl-2-Picrylhydrazyl (DPPH) in Radiolytic Studies

The effect of DPPH on the formation of radiolytic products has been investigated in numerous systems. In early radiolytic studies DPPH was used to evaluate the yield of radiolytically formed radicals in various systems. Following the pioneering work of Chapiro[17] DPPH was used extensively as an additive to radiolytic systems. It was added to solutions of CCl_4 in benzene[18], to alkyl chlorides and bromides that were undergoing γ-radiation-induced isomerization[19] to 1,2-dichloroethane[20] and to methyl iodide[21]. DPPH addition was found to affect the distribution of the products in the ıadiolysis of ketones[22–24], sulphur compounds, such as H_2S[25], mercaptans, organic sulphides and disulphides dissolved in toluene[26], polysulphone[27], alcohols[28, 29] and esters[30]. It was used as an additive in studies of the radiation-induced polymerization of, for example, acetylene[31], isobutylene[32] and allyl alcohol[33]. It was also used in the study of the radiolysis of polymers such as polyvinylchloride[34] and polymethyl methacrylate[35].

Despite its widespread use the effect of DPPH in radiolytic systems is not fully understood. It probably acts mainly as a radical scavenger, as was suggested by early investigators[17], according to reaction (10)[36]. The fate

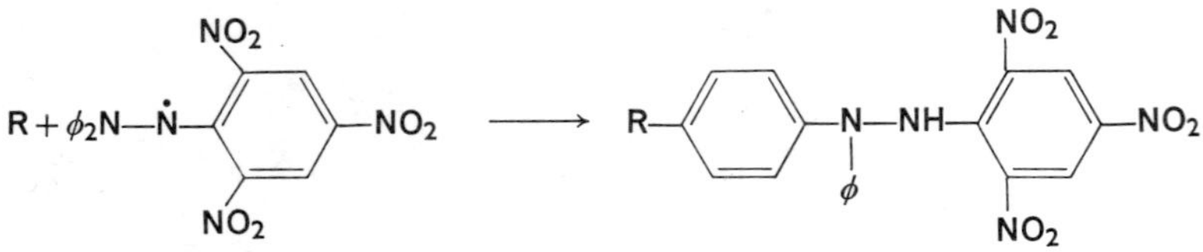
(10)

of the hydrazine derivative formed in reaction (10) is not known since there are practically no studies in which products formed from DPPH were determined. It should be noted that there is reason to believe that in many systems the effect of DPPH is solely due to its radical-scavenging ability. It is assumed that no free-radical mechanism is involved in those systems where the yield of the products is unaffected by the presence of DPPH. The opposite might not necessarily be the case in other systems. Electron scavenging and energy transfer to DPPH could be expected to be alternative modes through which DPPH could affect product formation, and on the grounds of currently available evidence these possibilities cannot be completely excluded.

It is worth mentioning that hydrazine derivatives are not the only products formed in radiolytic systems containing DPPH. In the radiolysis of DPPH in methyl iodide Kalus[21] found that the main products are 1,1-diphenyl-2-picrylhydrazine, *p*-nitrodiphenylamine, 2,6-dinitrophenylamine, 1-phenyl-1-(*p*-nitrophenyl)-2-picrylhydrazine and 2,4,6-trinitroaniline. These findings indicate that some of the products are formed by the splitting of the N—N bond in DPPH. Further evidence for N—N bond scission was obtained in e.s.r. studies of phenyl derivatives of hydrazine.

E. E.s.r. Studies of γ-Irradiated Hydrazo Compounds

It was suggested in the previous section that electron capture might be one of the processes by which DPPH affects the course of radiolysis. N—N bond scission could be envisaged as the outcome of this process. Indeed, the work of Shida and Kira[37], who carried out optical and e.s.r. studies of γ-radiolysed phenyl derivatives of hydrazine and dicarbazyl in rigid matrices at 77 K, indicates that those reactions are possible. From the absorption and e.s.r. spectra they concluded that in γ-irradiated tetraphenylhydrazine (TPH) in a methyltetrahydrofuran (MTHF) matrix dissociative electron capture by TPH results in the formation of a diphenylamine radical and a diphenylamide anion.

$$(C_6H_5)_2N{-}N(C_6H_5)_2 + e^- \longrightarrow (C_6H_5)_2N\cdot + (C_6H_5)_2N^- \qquad (11)$$

Upon changing the MTHF matrix to an alcoholic one, they observed a proton transfer reaction between the amide anion and the matrix:

$$(C_6H_5)_2N^- + ROH \longrightarrow (C_6H_5)_2NH + RO^- \quad (12)$$

N—N bond cleavage following electron capture also occurs in γ-irradiation of *asym*-diphenylhydrazine in the MTHF matrix. Shida and Kira[37] reported that electron capture by dicarbazyl in γ-irradiated MTHF matrix is non-dissociative. However, optical bleaching of the anion formed in this reaction results in the scission of the N—N bond similar to that in other hydrazine derivatives studied by them.

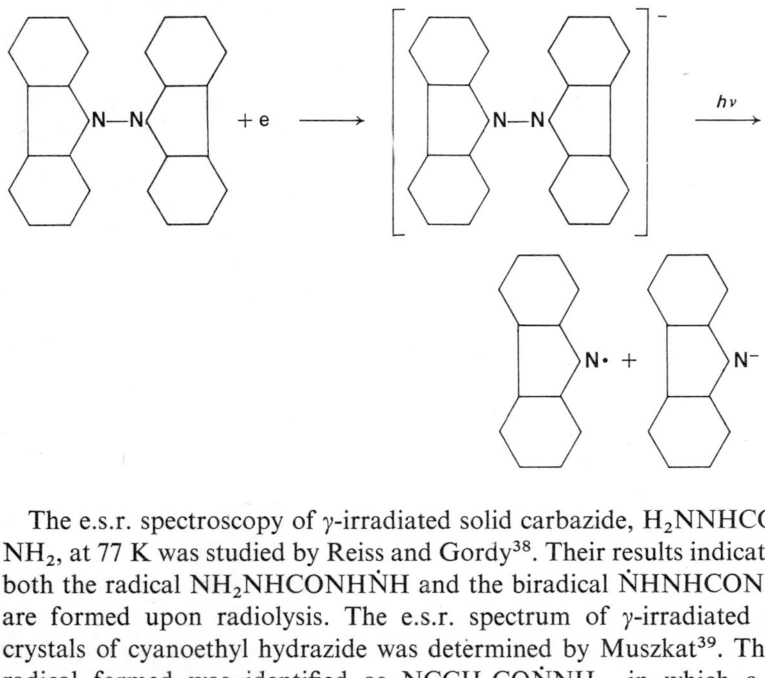

(13)

The e.s.r. spectroscopy of γ-irradiated solid carbazide, $H_2NNHCONH$-NH_2, at 77 K was studied by Reiss and Gordy[38]. Their results indicate that both the radical $NH_2NHCONH\dot{N}H$ and the biradical $\dot{N}HNHCONH\dot{N}H$ are formed upon radiolysis. The e.s.r. spectrum of γ-irradiated single crystals of cyanoethyl hydrazide was determined by Muszkat[39]. The free radical formed was identified as $NCCH_2CO\dot{N}NH_2$, in which a large fraction of the unpaired spin density is located on the α nitrogen atom.

III. RADIATION CHEMISTRY OF AZO COMPOUNDS

A. General Consideration

The radiation chemistry of azo compounds has been studied in aqueous and organic solutions and that of azomethane[40] in the vapour phase. Special attention was given to the radiolytic behaviour of azo dyes, because of their potential use in dosimetry[41–45]. Detailed product analysis was carried out only in the few cases where free-radical addition to the N—N bond was investigated and therefore the fate of the azo compounds in most of the radiolytic systems that were studied is not known.

In general, radiation-induced reactions of azo compounds in organic systems would be expected to be similar to those that are initiated photochemically. In cases where both radiolytic and photolytic studies of azo compounds were carried out, radical scavenging[46,47] and energy transfer processes[48] were indeed found to occur in a similar fashion. Reactions of charged species with azo solutes can be expected to occur in radiolytic systems, and should be favoured in polar solvents such as alcohols and water. However, reactions of this type have not been investigated.

B. Vapour-Phase Radiolysis of Azomethane

The vapour-phase radiolysis of azomethane was investigated by Stief and Ausloos[40]. The effect of dose, scavengers, pressure, temperature and inert gas (xenon) on the formation of the principal products was determined. These products are hydrogen, nitrogen, methane and ethane, with methylethyldimide, trimethylhydrazine and tetramethylhydrazine being formed at higher pressures. The authors proposed the following reaction scheme to account for the formation of methane, ethane and nitrogen.

$$CH_3N_2CH_3 \rightsquigarrow (CH_3N_2CH_3)^* \quad (14)$$

$$(CH_3N_2CH_3)^* \longrightarrow 2CH_3 + N_2 \quad (15)$$

$$CH_3 + CH_3N_2CH_3 \longrightarrow CH_4 + CH_2N_2CH_3 \quad (16)$$

$$CH_3 + CH_3 \longrightarrow C_2H_6 \quad (17)$$

$(CH_3N_2CH_3)^*$ represents an electronically excited molecule of azomethane which is formed either by direct excitation or following charge neutralization processes. Except for the primary step of excitation, the reactions are identical with those that occur in the photolysis of azomethane. This was further confirmed by the good agreement between the values of the ratio of rate constants $k_{16}/k_{17}^{1/2}$ as determined in the radiolysis and photolysis of azomethane. The products obtained at high pressure were ascribed to secondary reactions. However, the reaction sequence (14)–(17) does not account for the formation of hydrogen. Also, some of the nitrogen is apparently formed by an unidentified reaction other than (15). It was suggested that part of the observed radiolytic yield is due to the following reaction:

$$(CH_3N_2CH_3)^* \longrightarrow H + CH_2N_2CH_3 \qquad (18)$$

The finding of a residual yield in the presence of efficient free-radical scavengers led to a suggestion that part of the hydrogen yield is due to processes other than reaction (18). The postulated reactions are molecular hydrogen and/or hot hydrogen formation from excited azomethane and ion molecule reactions.

C. Radiochemical Reactions of Azo Esters

X-ray and photochemically induced reactions between azo esters (R—N=N—R; R = COOR′) and substrates (HAH) such as alcohols, ethers and esters, were studied by Schenck and Formanek[46] and Kopp[47]. Identical products were found in the radiolysed and photolysed systems. All these compounds, (HAH), were found to add to the azo esters by a free-radical chain mechanism. The mechanism of formation of the 1:1 adduct products is given by reactions (19)–(21).

$$R + HAH \longrightarrow RH + AH^{\bullet} \qquad (19)$$

(20)

(21)

In addition to the 1:1 adducts, tetrazanes formed in the biradical reaction between the R—Ṅ—N(AH)—R type radicals were also identified. The addition reactions of the following substrates were studied: dibenzyl ether, tetralin, isopropanol, *s*-butanol, isochromane, tetrahydrofuran, diethyl ether, dioxan and chloroform. In the last case $CHCl_2$ radicals add to the azo bond and reaction (22) then takes place.

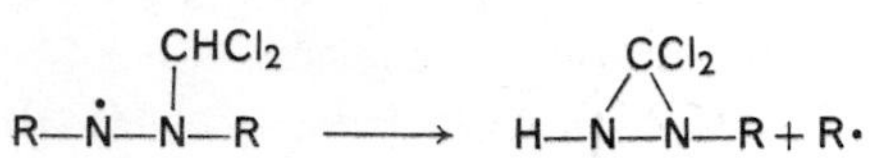

(22)

D. Aqueous Solutions of Azo Dyes

Irradiation of aqueous solutions of azo dyes induces decoloration that is enhanced by the presence of oxygen. According to Denio[44] the bleaching of air-free solutions of Orange IV is the result of the addition of hydrogen atoms and hydroxyl radicals to the dye. The author assumed that the hydrogen atoms probably add to the N=N bond while the hydroxyl radical adds to the aromatic nucleus. In the presence of oxygen both hydroxyl and hydroperoxyl radicals are the reactive species that add to the dye. Korgaonkar and Tarakanath[43] studied radiation-induced bleaching of Congo Red, Methyl Red and Methyl Orange in air-free aqueous solutions. *G*-values of bleaching not exceeding 2 were found under various experimental conditions.

E. Organic Solutions of Azo Dyes

Studies of organic solutions of azo dyes seem to indicate that radiolytic oxidation, rather than reduction, is favoured in these compounds and is responsible for their decoloration. However, as was pointed out by Grossweiner[45], the absence of sufficient evidence does not exclude the possibility that reduction may also occur under appropriate conditions.

Clark and Bierstedt[41] found that the *G*-values for the bleaching of the azo dyes, 2-hydroxyl-4-nitrophenylazo-β naphthol and Dithizone, in alcohols, were negligible in the presence of oxygen. (It should be noted that in radiolytic systems alcohols are considered to be reducing.) However, the results of the studies of Denio[44], where both aerated and deaerated methanolic and ethanolic solutions of Orange I and Orange IV were γ-radiolysed, indicate that decoloration occurs in the absence of oxygen but is suppressed in its presence. Those findings were attributed to the reoxidation of the reduced dye or the hydrazo radical by dissolved oxygen. According to Denio, in methanolic solutions of Orange IV, the azo dye is reduced to the hydrazo derivatives by the addition of hydrogen atoms and methylol, methyl and hydroxyl radicals. The azo bond appears to be the preferred site of attack by those radicals. Similar observations were made in that work in the case of Orange I solution in methanol and of the two azo dyes in ethanol.

The addition of chloroform or carbon tetrachloride to ethanolic solutions of azo dyes results in remarkably high *G*-values for the disappearance of the azo dye, which indicates that the dye is removed from the system by a chain mechanism. For example, in a mixture of 80% CCl_4 with ethanol the *G*(dye) was found to be 42·8 and 142 in solutions of Dithizone and Resazurin respectively while in a mixture of 50% CCl_4 with ethanol *G*(-2-hydroxy-4-nitrophenylazo-β-naphthol) of 190 was observed. Carbon tetrachloride appears to be a more effective sensitizing agent than chloroform in the X-ray-induced bleaching of azo dyes.

Clark and Bierstedt[41] postulated that oxidation of the azo dye is responsible for its disappearance in the above systems. In the case of Dithizone the following reaction occurs in CCl_4:

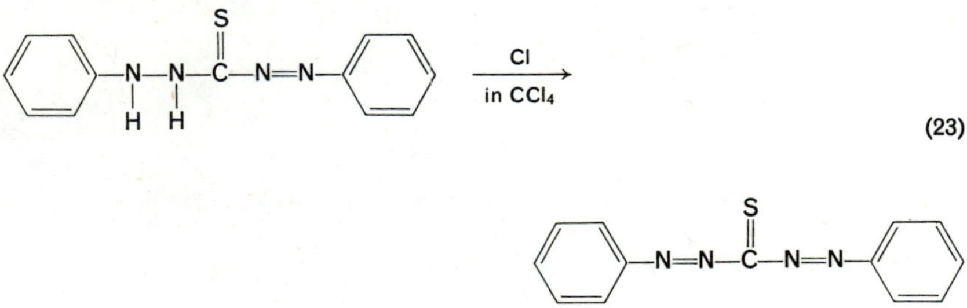

(23)

Irradiation of Methyl Yellow in chloroform results in the formation of a red product that is the salt of Methyl Yellow:

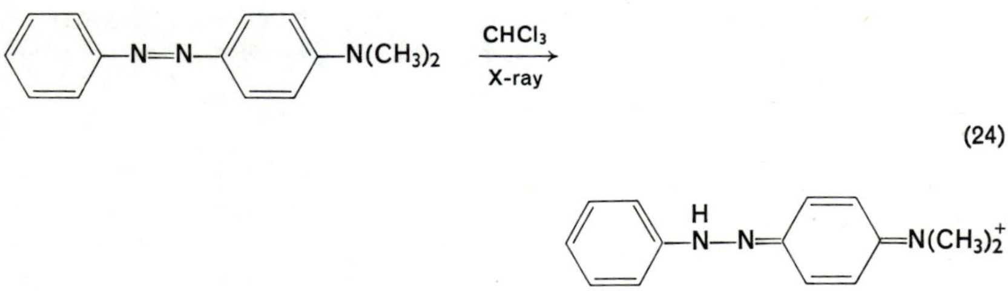

(24)

On the other hand in CCl_4 no red coloration is observed since, according to the authors, the following tautomeric rearrangement takes place:

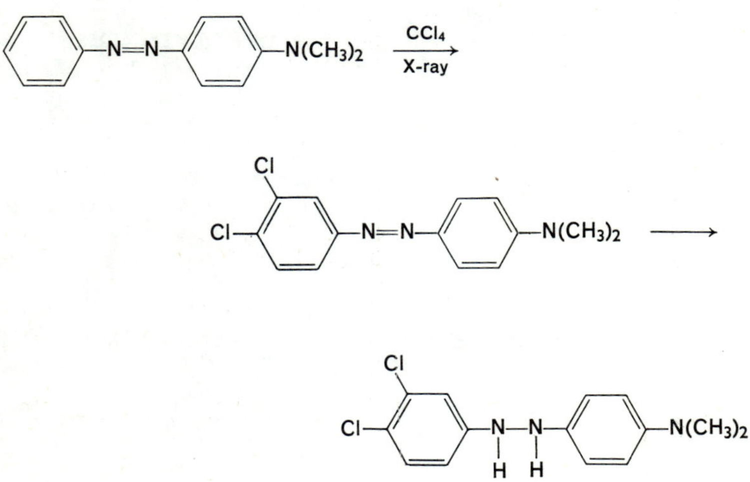

(25)

IV. ACKNOWLEDGMENT

The authors are indebted to the American Chemical Society and to Dr. E. Hayon for permission to reproduce Table 2.

V. REFERENCES

1. H. Lucien and M. L. Pinns, *NASA Technical Note*, TD-2452 (1964).
2. W. E. Shelberg, USNRDL—TR-843 (1965).
3. W. E. Shelberg, USNRDL—TR-896 (1965).
4. V. I. Tupikov and S. Ya. Pshezhetskii, *Russ. J. Phys. Chem.*, **37,** 1031 (1963).
5. V. I. Tupikov and S. Ya. Pshezhetskii, *Russ. J. Phys. Chem.*, **38,** 1364 (1964).
6. See, for example, F. W. Lampe, W. S. Koski and W. H. Johnston, *I.A.E.A. Vienna*, 1963, p. 41, and J. R. Puig and E. Schwartz, *ibid.*, p. 57.
7. V. I. Tupikov, V. I. Tsivenko, S. Ya. Pshezketskii, A. G. Kotov and V. K. Milinchuk, *Zh. Fiz. Khim.*, **37,** 138 (1963).
8. D. Schischkoff and D. Schulte-Frohlinde, *Z. Phys. Chem.*, **44,** 112 (1965).
9. F. S. Dainton, T. Skwarski, D. Smithies and E. Wezranowski, *Trans. Faraday Soc.*, **60,** 1068 (1964).
10. E. R. White, *Int. J. Appl. Radiat. Isotop.*, **16,** 419 (1965).
11. R. M. Black, *Rep. Prog. Appl. Chem.*, **48,** 312 (1963).
12. M. Lefort and M. Haissinsky, *J. Chem. Phys.*, **53,** 527 (1956).
13. H. A. Dewhurst and M. Burton, *J. Amer. Chem. Soc.*, **77,** 5781 (1955).
14. W. M. Garrison, *Radiat. Res., Suppl.*, **4,** 260 (1964).
15. J. Belloni and M. Haissinsky, *Int. J. Radiat. Phys. Chem.*, **1,** 519 (1969).
16. E. Hayon and M. Simic, *J. Amer. Chem. Soc.*, **94,** 42 (1972).
17. A. Chapiro, *Compt. Rend.*, **233,** 792 (1951).
18. N. Gilson-Leary and S. May, *J. Chem. Phys.*, **58,** 506 (1961).
19. M. Takehisa, G. Levey and J. E. Willard, *J. Amer. Chem. Soc.*, **88,** 5694 (1966).
20. H. Heda, *J. Phys. Chem.*, **71,** 3084 (1967).
21. H. W. Kalus, *Z. Naturforsch.*, **5,** 514 (1972).
22. P. Ausloos and J. F. Paulson, *J. Amer. Chem. Soc.*, **80,** 5117 (1958).
23. W. W. Bristowe, M. Kayayama and C. N. Trumbore, *J. Phys. Chem.*, **69,** 807 (1965).
24. P. Dunion and C. N. Trumbore, *J. Amer. Chem. Soc.*, **87,** 4211 (1965).
25. G. S. McNaughton, *Trans. Faraday Soc.*, **62,** 1812 (1966).
26. W. Audo, K. Sugimoto and S. Oae, *Bull. Soc. Japan*, **37,** 357 (1964).
27. A. Davis, M. H. Gleaves, H. J. Golden and M. B. Huglin, *Makromol. Chem.*, **129,** 63 (1969).
28. A. Prevost-Bernas, A. Chapiro, C. Cousin, Y. Landler and M. Magat, *Discuss. Faraday Soc.*, **12,** 98 (1952).
29. I. Zwiebel and R. H. Bretton, *A.I.Ch.E.*, **10,** 339 (1964).
30. T. Isoya, K. Hirota and M. Hatada, *Bull. Chem. Soc. Japan*, **38,** 1722 (1965).
31. Y. Tabata, B. Saito, H. Shibano, H. Solene and K. Ostiwa, *Makromol. Chem.*, **76,** 89 (1964).
32. A. I. Popova, A. P. Sheinker and A. D. Abkin, *Vysokomol. Soedin*, **7,** 1597 (1965).
33. S. A. Dolmatov and L. S. Polak, *Kinet. Katal.*, **7,** 27 (1966).

34. H. Ueda and D. Yashiro, *Bull. Chem. Soc. Japan*, **44,** 595 (1971).
35. H. Ueda and D. Yashiro, *Bull. Inst. Chem. Res., Kyoto Univ.*, **41,** 39 (1963).
36. A. Henglein, *Makromol. Chem.*, **15,** 188 (1955).
37. T. Shida and A. Kira, *Bull. Chem. Soc. Japan*, **73,** 4315 (1969).
38. K. Reiss and W. Gordy, *J. Chem. Phys.*, **55,** 5329 (1971).
39. L. Muszkat, *Chem. Phys. Letters*, **18,** 14 (1973).
40. L. J. Stief and P. Ausloos, *J. Phys. Chem.*, **65,** 877 (1961).
41. G. L. Clark and P. E. Bierstedt, *Radiat. Res.*, **2,** 199 (1955).
42. G. L. Clark and P. E. Bierstedt, *Radiat. Res.*, **2,** 295 (1955).
43. K. S. Korgaonkar and M. S. Tarakanath, *Int. J. Appl. Radiat. Isotop.*, **14,** 405 (1963).
44. A. A. Denio, Ph.D. thesis, Univ. of New Hampshire (1963).
45. L. I. Grossweiner, *Radiat. Res. Rev.*, **2,** 345 (1970).
46. G. O. Schenck and H. Formanek, *Angew. Chem.*, **70,** 505 (1958).
47. H. S. Kopp, Ph.D. thesis, Georg-August Universität Göttingen (1962).
48. V. A. Krongraus, *Int. J. Radiat. Phys. Chem.*, **1,** 465 (1969).

CHAPTER **11**

Formation and fragmentation of cyclic azo compounds

KENNETH MACKENZIE

University of Bristol, Bristol, England

I. INTRODUCTION

Information on the synthesis and decomposition of cyclic azo compounds is scattered throughout the literature, but with the exception of an excellent small volume dealing more generally with the mainly thermal extrusion of small molecules such as CO, CO_2, SO_2 and N_2 from appropriate structures[1], there appear to be few comprehensive single sources of information in this important area*. The past decade or so has, however, seen a considerable expansion in activity with carbocyclic azo compounds, especially with regard to the mechanistic features of the thermal and photochemical extrusion of nitrogen, a process which we have found it convenient to refer to as 'deazetation'[7] in analogy to the chelotropic[8] elimination of carbon monoxide from bridged carbonyl compounds, a process which is frequently called decarbonylation[9, 10].

In this chapter the sources of cyclic azo compounds—as strictly defined for the present purpose by the presence of the azo function in an alicyclic structure—are documented, and their mechanistically significant thermal and photochemical deazetation reactions are discussed. The chemistry of cyclic compounds containing the azo function adjacent to other hetero atoms is only incidentally described.

Material which has been reviewed in depth elsewhere is given only in summary form, attention being focused as far as possible, consistent with clarity, on the more cogent developments of the past decade.

II. MONOCYCLIC AZO COMPOUNDS

A. Diazirines

1. Preparation and general properties

It is some time since the results of electron diffraction studies[11] and more recently microwave[12] and i.r.[13] spectrometry established the structure of ground-state diazomethane unambiguously as a linear planar molecule containing sp^2 hybridized carbon. Thus the much earlier speculative discussions embracing the possibility of diazomethane having a cyclic structure such as **1** were rendered quite obsolete. Interest in the possible independent existence of such small-ring azo compounds later revived, however[14–16], and shortly afterwards the first examples of such compounds were more fully described[17, 18].

* References to specialized reviews on individual topics are given under the main subject headings in the text. In addition, certain other sources of information especially on mechanistic aspects, have been particularly useful[2–6].

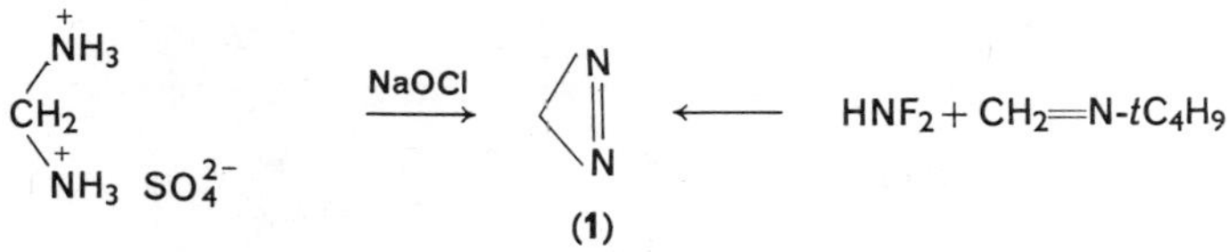

The parent member of the class of diazirines (**1**) can be made by oxidation of methylene diammonium sulphate with sodium hypochlorite, or by treating difluoramine with *t*-butylazomethine[19].

Diazirine can also be made from methylenebistrichloroacetamide by treatment with aqueous alkali followed by sodium hypochlorite[20].

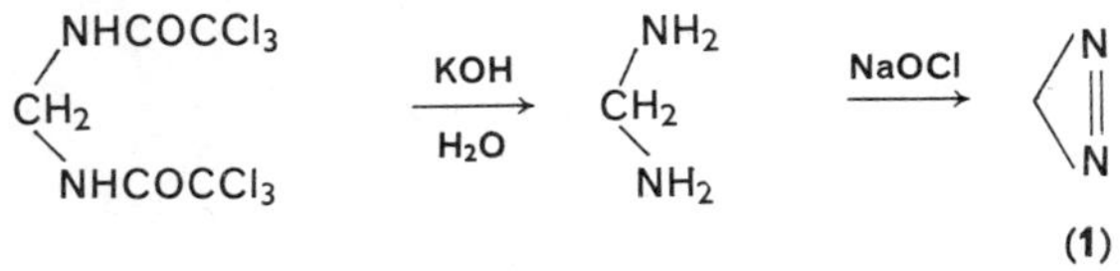

The cyclic structure of the molecule follows from bond lengths and internal angles derived from microwave spectrometry[21]. Similar data for methyl- and dimethyldiazirine, and their deuteriomethyl analogues yield dipole moments in the range 1·96–2·3D, e.g. Me_2CN_2 2·19 ± 0·07D[21a]. The equivalence of the two nitrogen atoms has been elegantly established by chemical incorporation of ^{15}N into the diazairine ring, reaction with cyclohexylmagnesium bromide, and exhaustive catalytic reduction of the resulting cyclohexyldiaziridine. The equal sharing of the nitrogen isotopes in the reduction products leaves no doubt as to the symmetrical nature of the azo function in the ring[22].

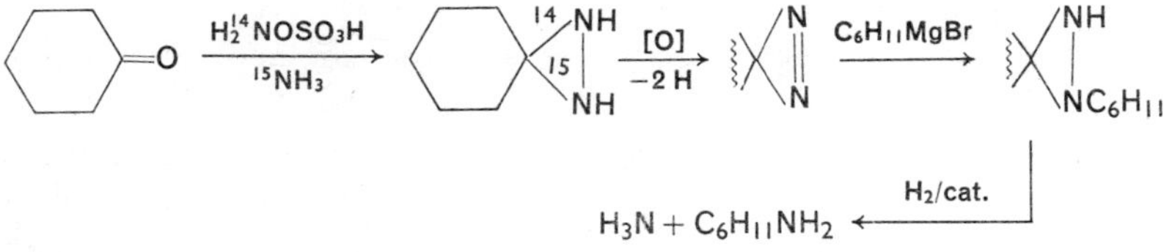

Substituted diazirines are accessible from the corresponding diaziridines, which are themselves formed from suitable carbonyl compounds, e.g. ketones in the presence of ammonia and chlorine[23]. Diaziridines may be oxidized to diazirines with yellow mercuric oxide in neutral solution, or with *alkaline* permanganate; in acidic oxidants hydrolytic ring opening can occur[14]. Polarographic studies indicate that reduction of diazirines takes place by a two-electron process in bases, but by a four-electron process in acids, in accord with the observed cleavage reactions on attempted oxidation of diaziridines with acidic permanganate[24].

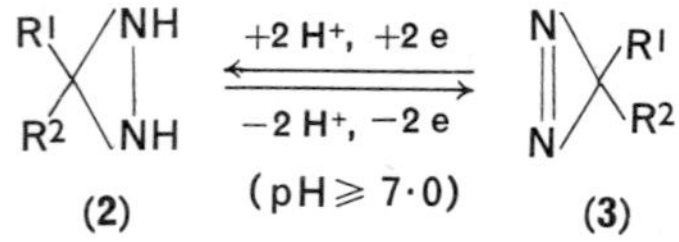

(2) **(3)**

An alternative method of preparation of diazirines consists[19] of the reaction of an aldehyde with ammonia and an oxidant to give a bicyclic diaziridine, e.g. **4**, which is then hydrolysed and the product oxidized with dichromate to give the substituted diazirine (**5**).

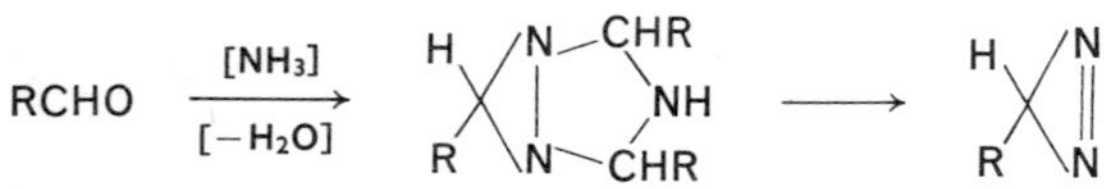

(4) **(5)**

A range of diazirines corresponding to structures **3** and **5** has now been made, e.g. with $R^1 = H$ and $R^2 = Me$, *n*-Pr, *i*-Pr and *t*-Bu; and with $R^1 = Me$ with $R^2 = Me$, Et and Ph. Spirocyclic compounds, e.g. **6** and **7**, have also been made as well as other simple aryl and alkyl analogues[25].

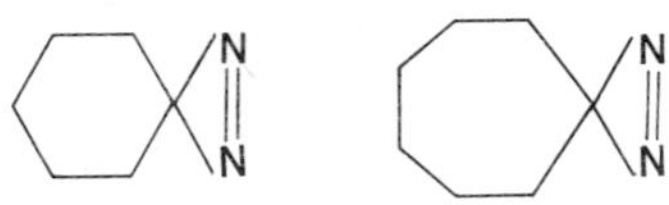

(6) **(7)**

The physical properties of a number of these compounds have been tabulated[25], including b.p., n_D^{20} and u.v. maxima.

The diazirines contrast markedly with the isomeric diazomethanes both as regards physical and chemical properties. Whilst diazomethane is a highly reactive yellow gas (λ_{max} 412 nm) diazirine (**1**) is relatively stable to phenols and acids and absorbs only in the u.v. region (λ_{max} 321 nm). The u.v. spectrum of, e.g., 3-methyldiazirine, is unusual, showing considerable vibrational fine structure in the 300–370 nm region, with at least 14 sharp bands. The resolution of fine structure in the electronic spectra of these molecules appears to be quite general, and those of diazirine, diazirine-d_1 and -d_2 and of diazirine-$^{14}N^{15}N$ have been analysed in some detail. Two overlapping sets of transitions result from splitting of the degeneracy of the nitrogen lone-pair levels[25a]. Similarly, phenylmethyldiazomethane is a red liquid, isolable at 0–10°C but decomposing within a few minutes at 25°C, whereas the isomeric diazirine is colourless and fairly stable at ambient temperatures. As with diazoalkanes, however, the diazirines should be handled with care since detonations have occurred on heating and, insidiously, merely on condensing the vapour into the liquid phase.

Chemically, the diazirines are remarkably stable to strong acids (and bases), requiring 80% sulphuric acid to decompose them. When reactions do occur, it is almost always at nitrogen. Thus diazirines are reduced either by lithium[14] or by sodium amalgam in aqueous alcohol, and react with Grignard reagents[19] to give 1-substituted diaziridines, e.g. **8**.

(8)

In general, diazirines exhibit electrophilic character, in contrast to the nucleophilic behaviour of diazoalkanes seen, e.g., in their reactions with carbonyl compounds. Correspondingly, electronegative substituents stabilize diazoalkanes (e.g., diazoketones where there is charge stabilization in the canonicals $>\overset{-}{C}-\overset{+}{N}\equiv N$) but the opposite effect is observed with diazirines. The chlorodiazirines (**10**) accessible by halogenation of alkyl or arylamidines (**9**), are, for example, noticeably less stable than the simple alkyl and aryldiazirines, slowly decomposing near 25°C. A number of analogues (**12–17**) of the chlorodiazirines (**10**) have been described[26]. The electronegativity of the vinyl group might similarly be a factor in the slow, room-temperature rearrangement of 3-methyl-3-vinyldiazirine into 3-methylpyrazole[26a].

Perhaps not surprisingly, methoxy compound **17** is very unstable, decomposing at ambient temperatures into diazine (**18**); difluorodiazirine however, is rather stable, decomposing only at elevated temperatures into perfluorocyclopropane and bisdifluoromethylenediazine and perhaps a factor in the stability here is back donation of electronic charge through the fluorine *p*-orbitals, resulting in a more uniform electron distribution in the molecule.

Other fluorinated diazirines recently reported include the difluoroamino compounds **19** and **20** and the bistrifluoromethyl compound (**21**), accessible from semicarbazide hydrochloride and suitable *N*-fluoroazomethines[27].

What is the origin of the marked contrast in properties of diazoalkanes

(9) (10) (11)

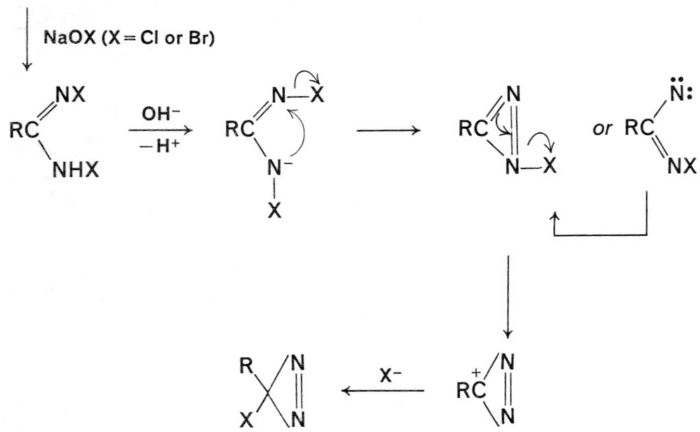

(12; R = Me, X = Br) (13; R = *t*-Bu X = Cl)

(14; R = cycloprop., X = Cl) (15; R = *p*-$MeOC_6H_4$, X = Cl)

(16; R = $CH_2{=}CH$, X = Cl)

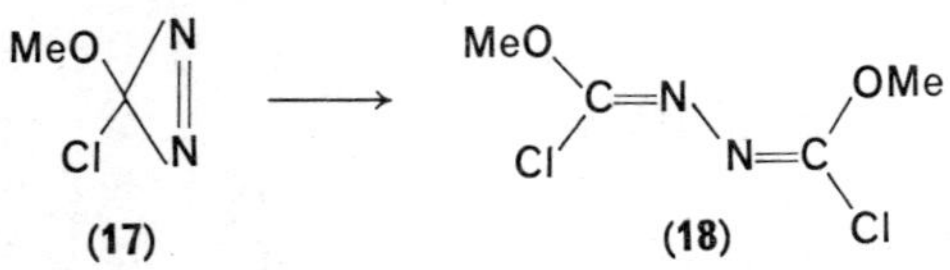

(17) (18)

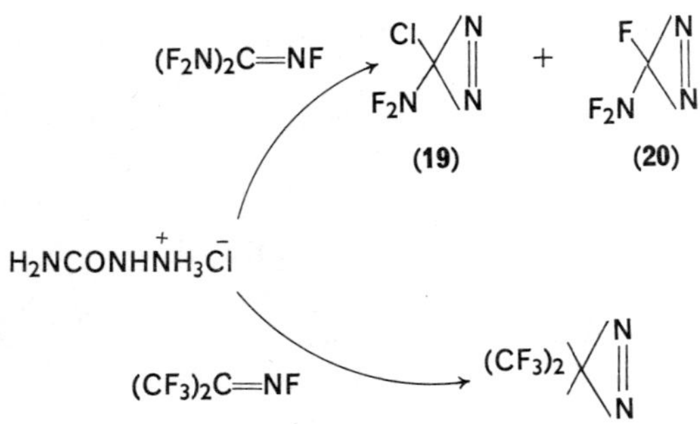

(19) (20) (21)

and diazirines? The introduction of the azo function into a three-membered ring introduces mixing involving the in-phase and out-of-phase nitrogen lone-pair orbital combinations; the result is electron delocalization. Bonding in the molecule can be regarded as akin to that in cyclopropane (as constructed from Walsh orbitals), with additional interactions involving, for example, the antisymmetric lone-pair combination and the methylene carbon $2p$ orbital[28]. Photoelectron spectroscopy confirms that the nitrogen lone-pair electrons are indeed delocalized: 56% on nitrogen in the $b_2(n_-)$ orbital and 43% in the $a_1(n_+)$ orbital[28a]. (Computer plot diagrams for diazirine M.O.s are available, together with relevant energies[28b].) The fact that diazirines do not form complexes with transition metals, unlike other azo compounds, e.g. 2,3-diazanorbornene, which do (especially with Cu^I salts) can therefore be understood.

FIGURE 1. Orbital mixing in diazirine involving nitrogen lone pairs.

2. Thermal and photochemical fragmentation

Because of the electronegativity of nitrogen, electron delocalization in diazirines is not as complete as in the carbocyclic analogue, cyclopropene, and the resemblance of the occupied M.O.s to those in nitrogen gives some basis in theory for expecting clean fragmentation of diazirines into carbenes[28b], rather than rearrangement typical of cyclopropene. Prompted by, among other reasons, their potential as sources of carbenes in situations where the diazoalkanes are inaccessible or dangerous, detailed mechanistic study of the thermolysis and photolysis of diazirines has been an area of recent activity.

At an elementary level, dropwise addition of diazirines to boiling nitrobenzene (210°C) results in the expected loss of nitrogen and formation of cycloalkane and olefinic products[29] as with diazirines **22** and **25**; the spriocyclic diazirines, e.g. **7**, behave similarly.

Photolysis of diazirine in solution can result in the formation of singlet and triplet methylene, depending on the conditions. Direct irradiation of a deuteriochloroform solution of diazirine whilst scanning the ^1H n.m.r. spectrum reveals the dynamic nuclear polarization emission spectrum of

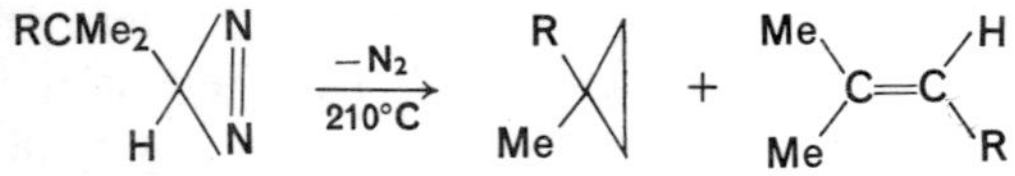

(22; R = Me) (23; R = Me) (24; R = Me)

(25; R = H) (26; R = H) (27; R = H)

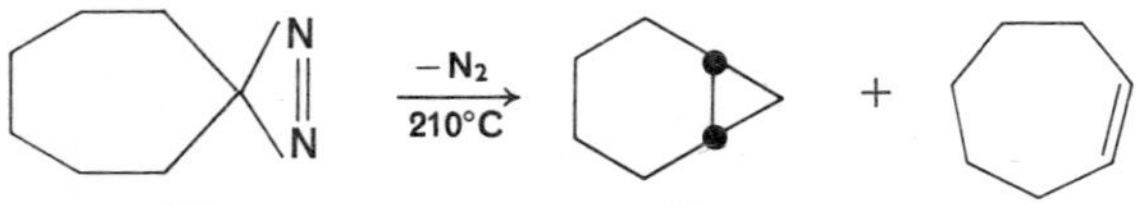

(7) (28) (29)

$DCCl_2CH_2Cl$ but *no* Cl_3CCH_2D is seen; in the photosensitized fragmentation, however, absorption lines due to Cl_3CCDH_2 appear; chlorine abstraction by singlet methylene and radical dimerization in the direct photolysis, as opposed to deuterium abstraction and radical combination in the sensitized process, points to triplet methylene being generated in the presence of the sensitizer. The results are an interesting example of the multiplicity specificity for reactions of methylene[30].

In the thermal decompositions of simple diazirines, however, the nature of the solvent appears to have little effect, and here it is suggested that a carbene formed as an intermediate very rapidly inserts, or rearranges to

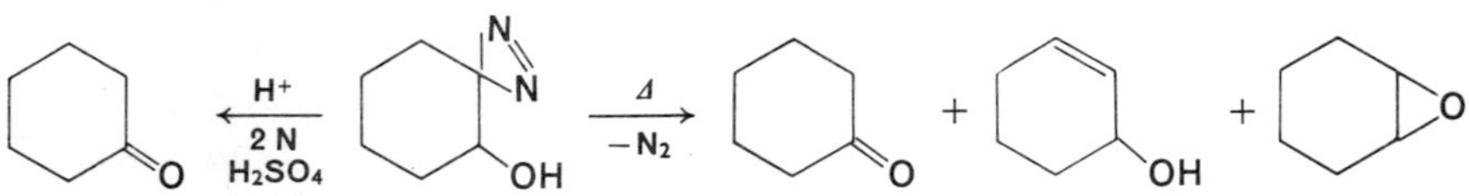

(31) (100%) (30) (31) (75%) (32) (20%) (33) (5%)

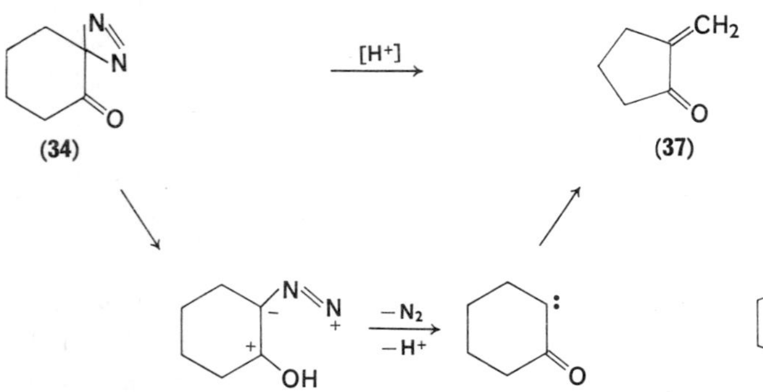

(34) (37) (35) (36)

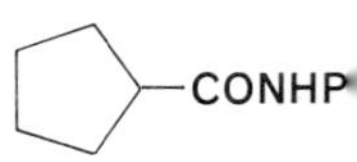

(38)

form olefin. Nevertheless, in some cases solvent effects can be seen, and neighbouring groups can also exert a considerable influence[29], e.g. in the decomposition of the hydroxydiazirine (**30**) catalysed by acids; the keto-diazirine (**34**) is even more acid-sensitive, fragmenting in media whose protonating power is at least 10^6-fold less effective than the acids required to decompose simple alkylated diazirines.

Probably, nonsynchronous electronic movement occurs in these decompositions, leading to an intermediate stabilized by the electronegative carbonyl group (in its protonated form). Decomposition of ketodiazirine (**34**) in aniline by way of contrast gives amide (**38**).

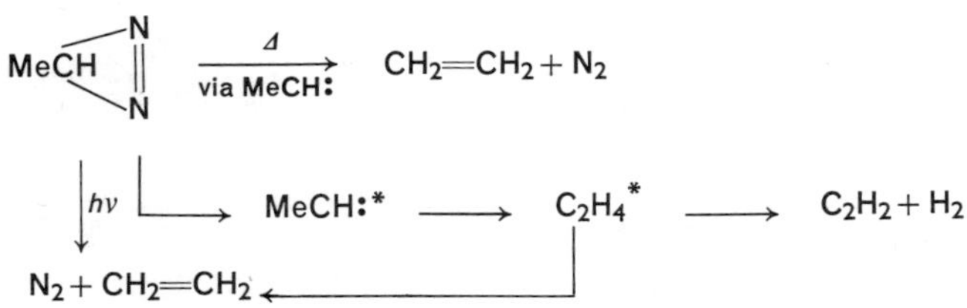

* ≡ vibrationally excited.

Detailed low-pressure, gas-phase kinetic studies of diazirine decompositions have revealed some interesting chemistry. At 150°C the decomposition of methyldiazirine is a smooth, clean unimolecular process giving only ethylene and nitrogen, and a carbene is presumed as the intermediate[31]. However, because of (*i*) the possibility of isomerization of diazirines into azoalkanes and (*ii*) the production of vibrationally excited carbenes, the photolysis of diazirines is likely to be much more complex than the thermal decomposition reactions, but correspondingly very much more informative. Thus, photolysis of methyldiazirine through Pyrex gives acetylene, ethylene, hydrogen and nitrogen[31]. The former two products are pressure-dependent and the main process appears to involve methylcarbene as the initial product, although up to 40% of the ethylene is formed directly in too low an energy state to decompose to acetylene; there appears to be no evidence for diazoethane being formed here, although in the liquid-phase photolysis of diazirine itself diazomethane and methylene are both observed.

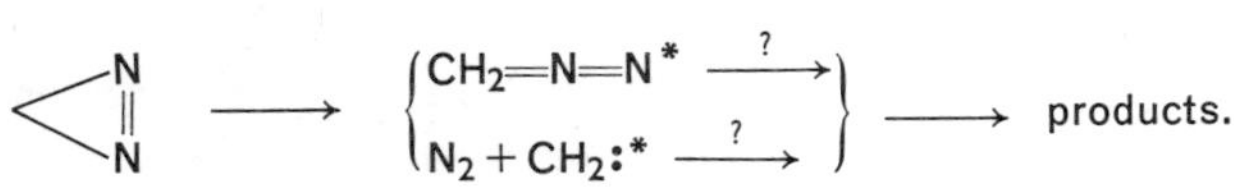

The three carbon nuclei in dimethyldiazirine allow potentially for a much wider spread of products from an indiscriminate photo-intermediate; those actually observed can be rationalized as deriving from some undefined state of dimethylcarbene which rearranges into 'hot' propylene as the initial step in the main decomposition pathway[32]. The vibrationally excited propylene disproportionates into various radical species which yield a number of dimeric products. Significantly, cyclopropane is also found in the products, and it too is probably derived by rearrangement of 'hot' propylene[33]. The fact that $^3P_1$ mercury-sensitized vapour photolysis of propylene also gives some cyclopropane suggests that a triplet species may also be involved in the diazirine photolysis, and the results generally support the decomposition pathways proposed for the latter in that the primary photoproducts appear to be the same.

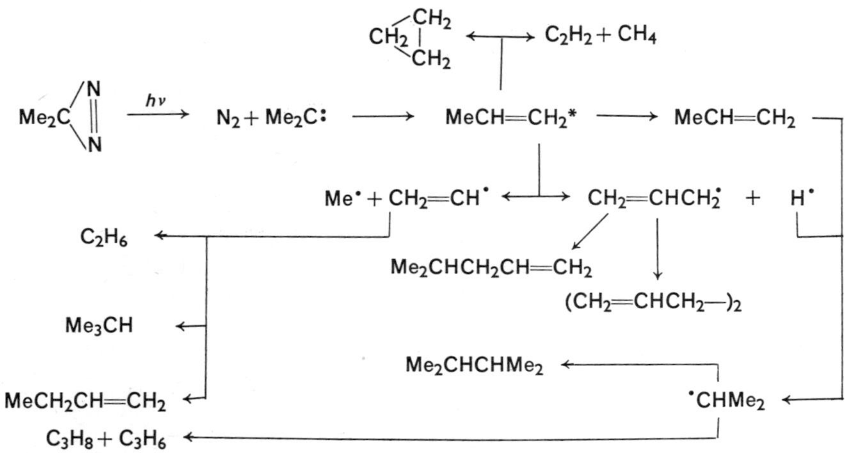

In none of these studies, however, has much direct evidence for the intermediacy of carbenes been adduced; no adducts of dimethylcarbene with cyclohexene can be found, and either the activation energy for carbene addition to olefins is *relatively* too high, or else rearrangement to 'hot' propylene must be extremely fast. On the other hand, in the thermal decomposition of dimethyldiazirine, giving nitrogen and propylene, special techniques[34] have allowed observation of the unimolecular reaction at as low a temperature as 50°C (where the half-life is three years) and the kinetic parameters here (E_a 33·2 kcal/mole) *are* consistent with fragmentation to dimethylcarbene, although the rearrangement to the lower-energy diazo-alkane cannot be rejected. Comparison of the products of diazirine thermolyses with an appropriate carbene chemically generated should, however, indicate whether it is a reasonable intermediate; for ethylmethyldiazirine the thermal products are 1-butene, the stereoisomeric 2-butenes and methylcyclopropane in essentially the same yields as found for thermolysis of 2-butanone tosylhydrazone sodium salt. It seems a fair supposition that the same carbene is involved in both decompositions.

In the photolysis of ethylmethyldiazirine, however, increased proportions of the thermodynamically least stable products are observed, once again suggestive of a 'hot' carbene intermediate; there are precedents for the intermediacy of vibrationally excited carbenes[35] but later evidence has cast doubt on the validity of this interpretation. For instance, the photolysis of diazirine in the presence of a large excess of *trans*-2-butene gives trans-*dimethylcyclopropane*, suggesting a stereospecific *singlet* methylene. Moreover, in the presence of saturated hydrocarbons, methylene produced in similar photolyses shows the same discriminative behaviour towards insertion into primary and secondary C—H bonds as methylene produced in the photolysis of ketene. Production of methylene by both these photoreactions and kinetic data for its reaction with cyclobutane (to give 'hot' methylcyclobutane as the precursor of ethylene and propylene) indicates that the carbene is produced in essentially the same energy state[33], probably the singlet excited state.

Turning to the question of the possible intermediacy of diazoalkanes in the photolysis of diazirines, it has been claimed that low-pressure irradiation of diazirine, with or without large partial pressures of nitrogen, results in up to 20% of the initial decomposition passing through the isomeric diazomethane[36]. However, it has been pointed out that the heat of formation of the strained ring diazirine is 60% higher than that of diazomethane; the latter very efficiently produces methylene by excitation in its absorption range (436) nm), whereas diazirine absorbs at 320 nm. Consequently, diazomethane produced by photolysis of diazirine must obviously be far more energetic than that initially produced by direct irradiation of diazomethane, and might therefore be expected to be in a higher electronic state. Since the diazomethane produced by the isomerization process must also be vibrationally excited to a much greater degree than when formed directly from the ground-state azoalkane, the chance of collisional deactivation can scarcely be the same for the two species; some effect of added nitrogen would be expected unless the possibility of a long-lived triplet excited diazomethane is admitted[33].

The formation of diazomethane from diazirine *has*, however, been observed by irradiation of a nitrogen matrix of the cyclic azoalkane at extremely low temperatures; but experiments with ^{15}N-enriched matrices indicate that its source is reaction of methylene with nitrogen. No diazirine is formed by irradition of diazomethane, nor is any diazomethane formed when 3,3-dideuteriodiazirine is irradiated[37]. However, the conversion of diazoacetyl compounds into diazirines when exposed to visible light has been observed comparatively recently[37a].

Decompositions of other alkylated diazirines have also been studied in detail and the results are, by and large, similar to those already described. The tendency to form larger amounts of the thermodynamically least stable products in photolyses is again evident in the comparative results from

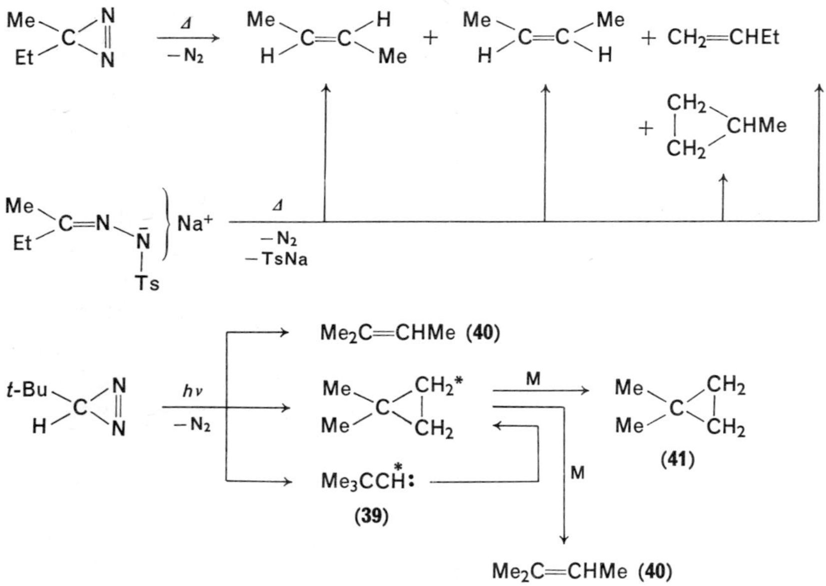

$Me_2CHCH{=}CH_2$
(42)

$MeCH_2CMe{=}CH_2$
(43)

* ≡ vibrationally excited.

thermolysis and photolysis of diethyldiazirine; the results can be interpreted in terms of a 'hot' carbene and since the products are pressure-insensitive secondary isomerizations are unimportant; similar considerations apply to 3-methyl-3-isopropyldiazirine[33].

The results for relatively high-pressure photolysis of *t*-butyldiazirine as a source of *t*-butyl carbene are very interesting when compared with the chemical preparation of the intermediate[38]. Under the photolysis conditions used here secondary isomerizations are suppressed and the differences observed in the product compositions from the two sources can again best be interpreted as due to 'hot' carbenes being formed in the photochemical route; irradiation of the diazirine gives 2-methyl-2-butene (**40**) (49%) and 1,1-dimethylcyclopropane (**41**) (51%) via intermediate **39**, but reaction of 1,1-di-iodo-2,2-dimethylpropane with methyl lithium[39] gives a much higher yield of the cycloalkane (95–97%) and only 3–5% of the methylbutene (**40**). Rather a similar result obtains with the products of thermolysis of the sodium salt of pivalic aldehyde tosylhydrazone, but with 3-methyl-1-butene (**42**) (1%) also appearing. Detailed kinetic analysis of the diazirine photolysis, and consideration of the effect of pressure on the relative yields of the olefins compared to those from dimethylcyclopropane thermolysis, suggests that the olefinic products derived from the 'hot' cyclopropane are also vibrationally excited. In agreement with the finding that *s*-butylcarbene gives cyclopropanes and 2-methyl-1-butene (but no 2-pentene) the experimental results show that E_a for methyl shift in the intermediate (**39**) is larger than for insertion into a C—H bond. The results of these experiments could also be interpreted in terms of two different excited states of the diazirine which give rise to the cyclopropane and the carbene selectively. Alternatively, a carbene might always be the intermediate, but again in two different states leading to cyclopropane or olefins.

Photolyses of spirocyclic diazirines have also been the subject of detailed study. In the gas phase the photolysis products from pentamethylenediazirine (**6**) are cyclohexene, methylenecyclopentane and bicyclo [3.1.0] hexane (97·0, 0·4, 2·5) and at low enough pressures (< 10 mm) the energetic cyclohexene is sufficiently insulated from vibrational damping to undergo cycloreversion to ethylene and butadiene. Very small amounts of methylcyclopentene have also been detected at even lower pressures, suggestive of 'hot' bicyclohexane as its source. The data support a carbene intermediate, but the pressure-composition relationships reveal that some cyclohexene is being formed with insufficient activation to decompose, although derived from electronically excited diazirine. Probably here nitrogen departs with the bulk of the excess energy, although the possibility that two excited states of the diazirine are involved must be considered, the higher-energy state presumably giving rise to the carbene, and the lower yielding unactivated cyclohexene. It is also possible that the result reflects the rather broad spread of energies possible among the hydrocarbon fragments on statistical thermodynamic grounds (see Section II. C. 2).

In the liquid-phase photolysis of pentamethylenediazirine there is definite infrared evidence for the formation of diazocyclohexane in the initial phases of the reaction and it seems possible that up to 40% of the diazirine is isomerized; it suggests that diazocyclohexane could also play a part in the gas-phase process. It is not known whether the isomerization occurs from the excited state, or from a vibrationally excited ground state. The formation of bicyclo[3.1.0]hexane in photolysis of pentamethylenediazirine is one of the few pieces of direct evidence for carbene intermediates in these fragmentations[32].

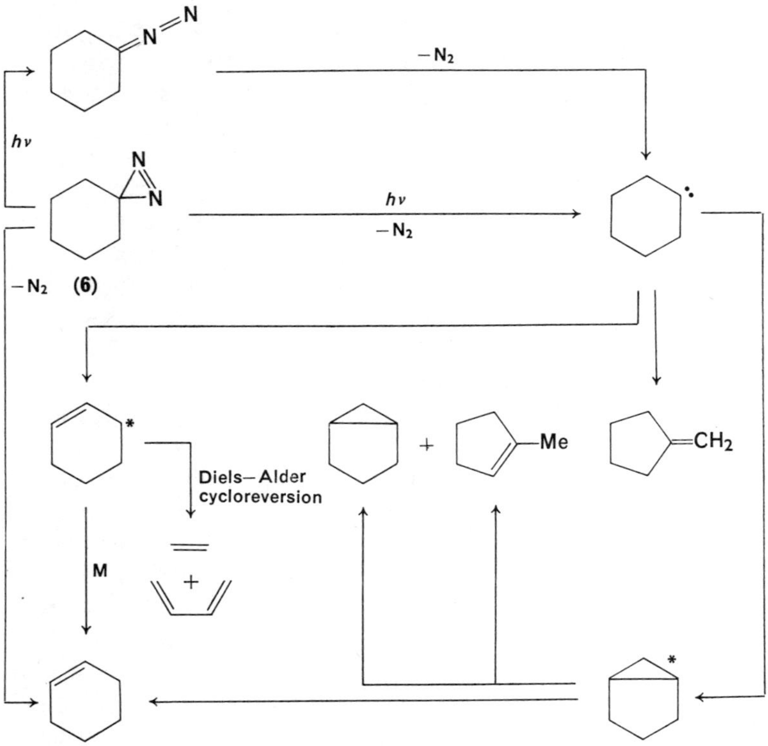

The photochemical fragmentation of pentamethylenediazirine.

Synthetically useful carbenes can also be made from diazirines; thus irradiation of a dilute solution of 3-bromo-3-phenyldiazirine (**44**) in an olefin gives an essentially quantiative yield of the cyclopropane (**45**); alkenes employed here include stereoisomeric 2-butenes, isobutene, and tetra- and trimethylethylenes. Competitive experiments in mixed olefins indicate that the carbene is similar, though not identical, to the species chemically derived from benzal bromide. The photochemically derived carbene is more electrophilic and it seems possible that chemical preparations lead to a carbene complexed with salts[40]. Halocarbenes are also formed in the gas-phase decomposition of 3-chloro-3-alkyldiazirines at 100–150°C;

3-chloro-3-methyldiazirine gives vinyl chloride and nitrogen by rearrangement of the intermediate (E_a's range from 29·5 to 31·1 kcal/mole for the high-pressure rates[40a].)

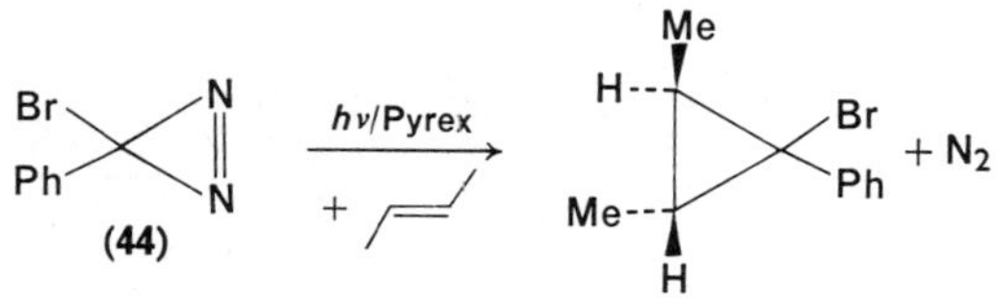

(44) (45)

Photolysis of 3,3-difluorodiazirine in olefins gives stereospecific adducts of the difluorocarbene liberated which, in common with other halocarbenes, is formed in the singlet state[41].

B. *Diazetines*

Very few references appear in the literature to four-membered rings containing the azo function, and the parent member of the class of diazetines remains unknown, although the tetramethyl derivative (**47**) has been postulated[42] as an intermediate in the photochemical decomposition of the bis-azide (**46**).

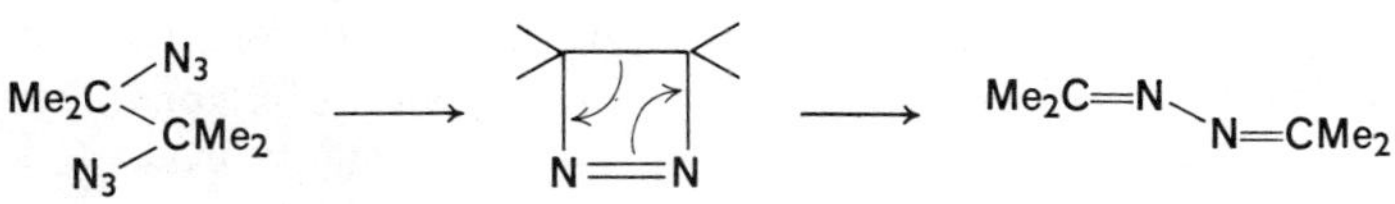

(46) (47) (48)

Tetrafluorodiazetine (**49**) has been made[43] by fluorination of cyanogen with argentic fluoride at 110–115°C; as with the diazirines the u.v. spectrum characteristically exhibits considerable fine structure. Although diazetine (**49**) is stable to water at low temperatures, and even to hot, dry hydrogen choride, it is dangerously hazardous, having been known to detonate spontaneously.

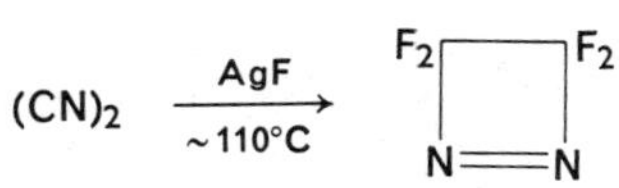

(49)

The interesting fused-ring diazetine (**52**) has been synthesized[44] from quadricyclane (**50**) by cycloaddition, and appropriate manipulation of the intermediate adduct (**51**); the thermochemistry of diazetine (**52**) is discussed later (see Section III. B. 1).

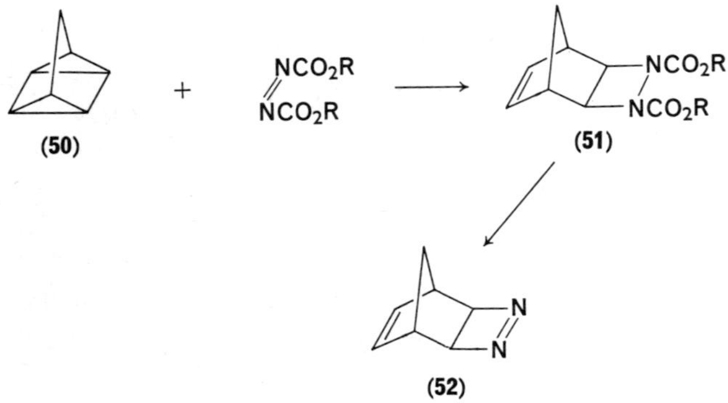

C. Pyrazolines
1. Synthesis

Routes to pyrazolines of the general structure (**54**) have been known since 1888 when Buchner recognized that the reaction product of the then relatively novel diazoacetic ester with α,β-unsaturated carbonyl compounds contained the hydropyrazole ring[45]. The tautomers (**54**) are regarded as 2-pyrazolines but 1-pyrazolines, e.g. **53**, clearly constitute the initial adducts and, by careful control of the reaction conditions, these may be isolated. Earlier literature on the Buchner synthesis of pyrazolines from azoalkanes has been reviewed[46] and later synthetic work has been summarized[47].

In the absence of special factors such as, for example, ring strain, azoalkanes fail to form adducts with compounds having isolated double bonds, and indirect methods must be used to obtain simple alkylated

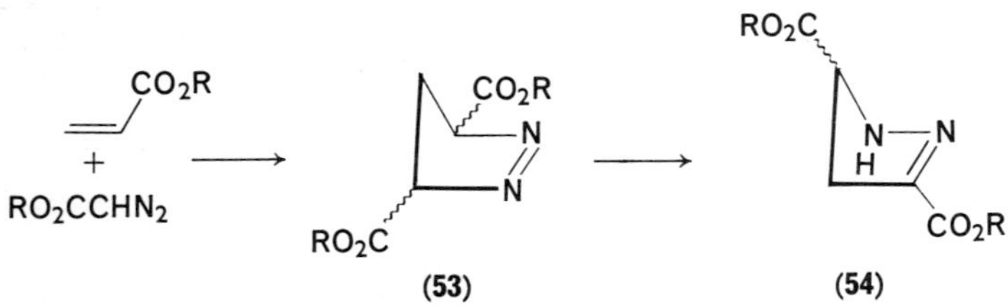

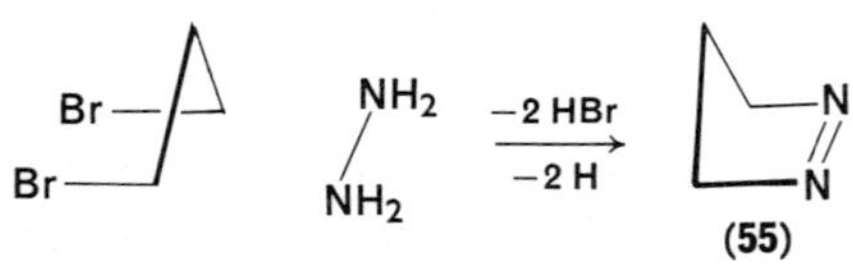

pyrazolines. 1-Pyrazoline (**55**) itself has been made by hydrazinolysis of 1,3-dibromopropane, followed by cupric-ion-catalysed air oxidation of the cyclic hydrazo compound first formed.

The procedure is specifically designed to exclude acids and bases, and a range of methylated and polymethylated 1-pyrazolines has been made by this method[48].

An interesting property of some of the simpler alkylated 1-pyrazolines is manifest in the ^{1}H n.m.r. spectra, where the 4-methylene hydrogens may or may not be equivalent due to conformational effects; rapid equilibrium between conformers **55a** and **55b** in 1-pyrazoline, for instance, renders the 4,4-protons equivalent, but in the 3-methyl derivative conformer **56a** is preferred on account of 1,3-diaxial nonbonded interactions and the 4,4-protons are no longer magnetically equivalent[48]. In such cases the chemical shift difference can be rather large ($\Delta\tau$ 0·75 in **56a**, 1·60 p.p.m. in 3,5-dimethyl-1-pyrazoline) due to the anisotropy of the azo linkage. Conformational effects of this kind can of course be significant mechanistically in the fragmentation of these compounds.

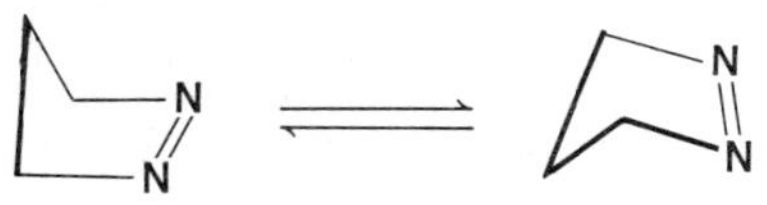

(55a) **(55b)**

(56a) **(56b)**

Returning to the more general synthesis of pyrazolines via cycloaddition of diazoalkanes to 'activated' olefins, considerable theoretical and experimental progress has been made in recent years with regard to the mechanistic principles involved, and it is useful to consider in broad outline the evidence bearing on this question. Apart from observing the stereochemistry of the addition three main experimental variables have been investigated: (*i*) the structure of the azoalkane, (*ii*) the structure of the activated olefin and (*iii*) the dielectric constant of the solvent. In some cases investigations of this kind have been complicated by the decomposition of the pyrazoline adduct into cyclopropane (which is sometimes more rapid than adduction, e.g. for additions involving diazofluorene), but provided the decomposition of the diazoalkane to a carbene is not an important competitive pathway, the initial additions can be regarded as 1,3-cycloadditions, and no change in addition mechanism need be considered, even though a pyrazoline is not isolated.

The relatively small effect of changes in the solvent dielectric constant on the rates of addition of a typical diazoalkane (e.g., diphenyldiazomethane) to a range of olefins (dipolarophiles) immediately points to an addition mechanism not involving polar intermediates. For example, the rate of addition of diphenyldiazomethane to dimethyl maleate is only doubled in passing from ethyl acetate (dielectric constant $\varepsilon = 6{\cdot}0$) to acetonitrile ($\varepsilon = 37{\cdot}5$) or dimethylformamide ($\varepsilon = 37{\cdot}6$) as solvent. With certain other olefins, omitting reactions in ethanol which seem exceptional, the rate of addition of the same diazoalkane scarcely changes at all over this range of solvents[49]. Such results suggest either a concerted addition mechanism or possibly a diradical intermediate in a stepwise sequence; but the large negative entropy of activation observed in many of these cycloadditions (typically −30 to −40 cal/mol/deg. C) serves to support the former possibility in indicating a transition state more constrained than the addenda. Concertedness here receives further powerful support from the stereospecific nature of these cycloadditions, and in the relative rates of addition of a given diazoalkane to stereoisomeric olefins. In the addition of diazomethane to tiglic ester (**57**) for example, the relative stereochemistry of the methyl groups is retained in the adduct (**58**) whereas for a diradical (or dipolar) intermediate internal rotation could, of course, lead to some loss of stereospecificity[50].

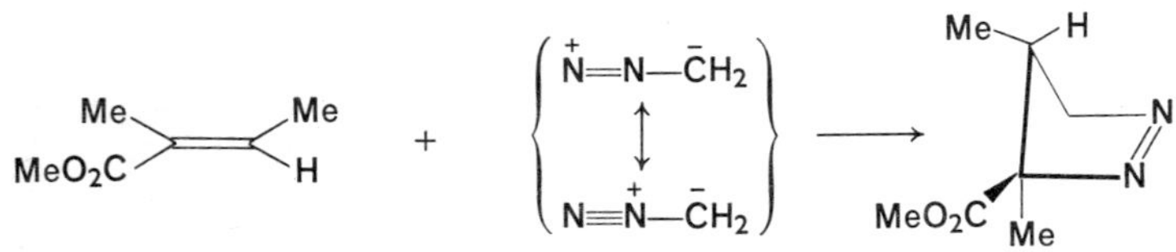

(57) (58)

Further, in the addition of diphenyldiazomethane to maleic and fumaric esters in dimethyl formamide (40°C) the *trans*-diester reacts roughly 40-fold faster than the *cis*-compound, consonant with a steric retardation due to the proximity of the ester groups in the latter, which move closer together as hybridization changes occur at carbon in the reaction centres during the course of the concerted addition. An even more marked steric effect is evident in the similar additions to α-methylacrylic ester and the *trans-β*-methyl analogue; the addition rates are retarded 14-fold and 280-fold respectively compared to cycloadditions with ethyl acrylate itself. These steric effects are fully consistent with the concept of a relatively rigid constrained multicentre transition state and the negative entropy factor[51]; they also clearly suggest that there is adequate driving force for steric relaxation by rotation in hypothetical diradical intermediates, which should therefore lead to loss of stereospecificity.

In a provocative but useful critique of the theory of concerted cycloaddition as applied to 1,3-dipolar reagents including diazoalkanes, it has been suggested that the features of these additions could be satisfactorily explained if the reactions involved addition of a spin-paired diradical to the olefin, giving a 1,5-diradical intermediate capable of reverting to starting materials or closing to a cyclic product faster than rotation about a σ-bond, i.e. with a rate-determining addition[52]. The activation energies for the cyclization and reversion steps must then be very small, with the intermediate lying in an exceedingly shallow depression at the apex of the reaction coordinate. Furthermore, it is argued that, of the potential 1,5-diradicals that could arise in these processes, the 'best diradical' i.e., that which is expected to be the most stable species, can be discerned in many instances as intermediates, e.g. structures like **59–62**.

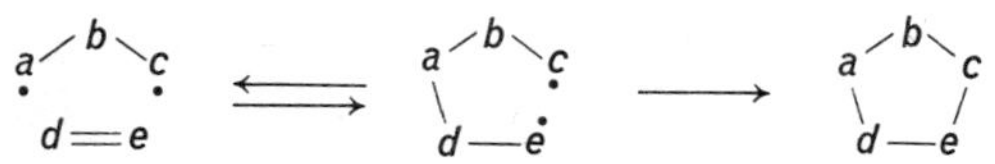

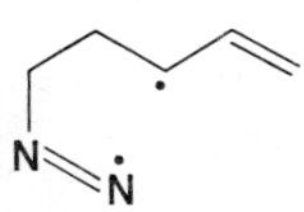

(59)

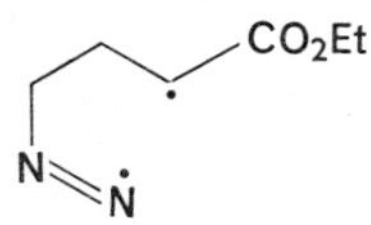

(60)

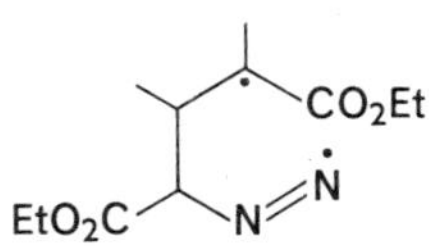

(61)

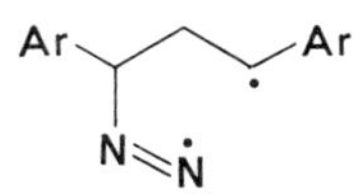

(62)

However, such a reaction scheme implies that the rate of reversion of the intermediate diradical to reactants must be at least 30-fold faster than rotation about the σ-bond for the stereospecificity to be maintained (i.e., $\leqslant 3\%$ stereoisomeric adduct) and $\Delta G^{\neq}$ for regression must then be ~2·1 kcal smaller than the rotation barrier, which itself is not larger than 3·4 kcal. On these calculations the barrier to reversion must be < 1·3 kcal, an unreasonably small barrier if the diradical intermediate is to have any real significance[53]. With regard to the question of the intermediacy of stabilized diradicals, it is cogently pointed out that 1,1-dichloro-2,2-difluoroethylene combines 5.5-fold faster with the methylated olefin unit in isoprene than with the nonmethylated double bond because the diradical (**63**) which forms the intermediate here is stabilized both by the vinyl group (allylic stabilization ca. 11 kcal) and the methyl group. However, by contrast, in the addition of diphenyldiazomethane to isoprene at 20°C addition occurs *preferentially* at the nonmethylated double bond. Similarly, in the addition of diazomethane to 2-phenylbutadiene, a diradical such as **64**—if it is an intermediate—should be stabilized even better than **63**; in fact, however, the product of the rather slow addition is the isolable pyrazoline (**65**).

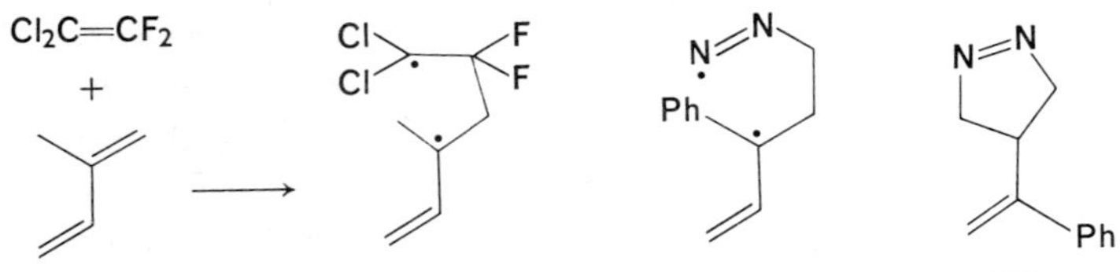

(63) **(64)** **(65)**

Here, then, is reasonable convincing evidence that the 'best diradicals' which might be postulated as intermediates in these additions do not in fact play any significant role.

It has been argued[52] that if additions of diazoalkanes represented by the resonance canonicals **66** ↔ **67** are to be considered 'dipolar cycloadditions' then there are cases where the 'polarity' of the dipole leads to the 'wrong' prediction with regard to the product. Examples that might be cited are the addition of diphenyldiazomethane to 1-*p*-anisyl-2-nitroethylene to give the

$$\bar{C}R_2{-}\overset{+}{N}{\equiv}N \quad \longleftrightarrow \quad \overset{+}{N}{=}N{-}\bar{C}R_2$$

(66) **(67)**

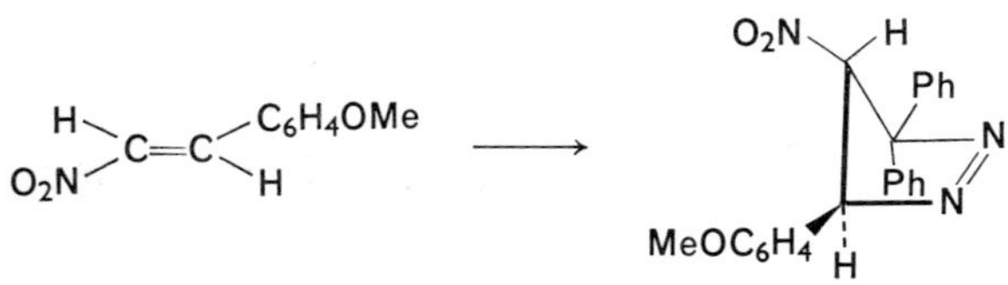

(68)

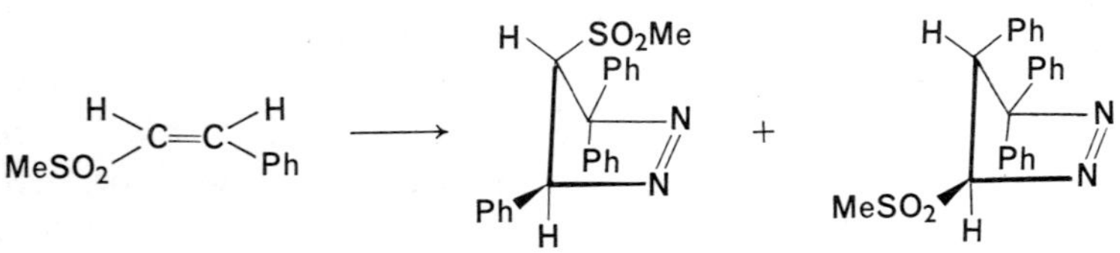

(69) (60%) **(70)** (40%)

adduct (**68**)[54] and the similar addition to sulphones where the favoured adduct is, for example, the pyrazoline (**69**). It could be said in both these cases that the 'negative end' of the dipole is adding to the electron-rich end of the dipolarophile.

The possibility that cycloadditions involving diazoalkanes proceed through either a spin-paired diradical or a dipolar species represented by a resonance canonical is, however, founded on a misconception about the nature of bonding and the distribution of electrons in 1,3-dipolar reagents.

Diazoalkanes can be regarded as members of the diazonium betaines[51] which can sometimes be usefully represented by the valence bond canonicals **66** ↔ **67**, **71** ↔ **72** and **73** ↔ **74**. Their relative importance can be judged from predicted and actual dipole moments, e.g. for linear diphenyldiazomethane represented as **67**. Here the expected dipole moment is 6·4D but the measured value (benzene at 25°C) is only 1·42D, and is reduced to 0·62D in *p,p′*-dichlorodiphenyldiazomethane, revealing the importance of structures such as $R_2C{=}\overset{+}{N}{=}\overset{-}{N} \leftrightarrow R_2\overset{-}{C}{-}\overset{+}{N}{\equiv}N$. The molecular orbital picture of these reagents, however, provides a better insight into their ability to react with olefins[55].

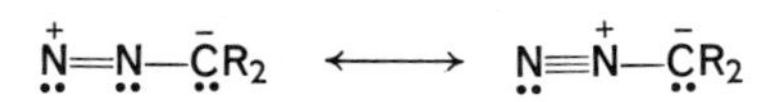

(**66**) (**67**)

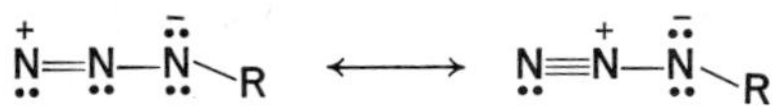

(**71**) (**72**)

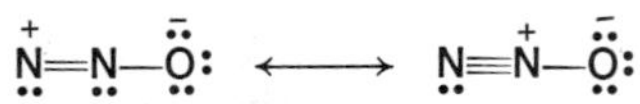

(**73**) (**74**)

All 1,3-dipoles are isoelectronic with ozone or nitrous oxide and have a 4-π electron system (**76**) analogous to the allyl anion, the highest occupied M.O. (HOMO) being antisymmetric with respect to a mirror plane bisecting the molecule in a normal plane; this is precisely the symmetry required for interaction with the lowest vacant M.O. (LVMO) in an olefinic bond (eventually leading to the antisymmetric σ-orbital component in the adduct, filled by two of the available electrons). The interaction of HOMO and LVMO for the reacting species stabilizes the transition state for the cycloaddition and the closer the relevant HOMO/LVMO energies the stronger the interaction. The idea can be illustrated by reference to an orbital energy diagram[56] for allyl anion–olefin cycloaddition.

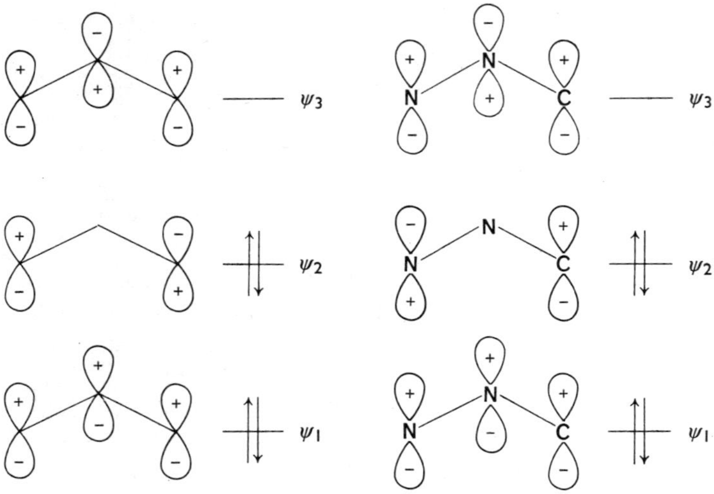

$C_3H_5^-$ **(75)** CH_2N_2 **(76)**

Calculations indicate that, for small deviations from linearity in **(76)**, the allyl-type nonbonding ψ_2-orbital is the HOMO. See Appendix.

Electronic effects will be reflected in the relative positions of the M.O.s; electron-attracting substituents in the olefinic dipolarophile will lower both ψ_A and ψ_B with the results that the interaction of ψ_A with φ_N will quickly get stronger; although ψ_B and φ_A will interact less strongly because they separate in this process, the relative effect will be small by comparison with the large orbital energy difference in the unsubstituted molecule, and the interaction between the ethylenic LVMO and the allyl (or dipole) HOMO will dominate. This is in fact what is observed in practice as the nature of the dipolarophile in a 1,3-dipolar cycloaddition is varied. For similar reasons good dipolarophiles are, in general, good dienophiles.

In this picture of dipolar cycloadditions it is seen that the governing factors may not necessarily be electrostatic in character; the selectivity for one of the alternative directions in an addition may depend on the relative efficiency for suitable orbital interactions to develop as the assembly proceeds along the reaction coordinate. However, in most cases it is reasonable to expect electronic distribution and, especially, polarizability to be important factors. Such a viewpoint shows that 1,3-dipolar cycloadditions are quite naturally stereospecific; the 1,3-dipole and dipolarophile approach each other in separate but parallel planes to achieve the transition state,

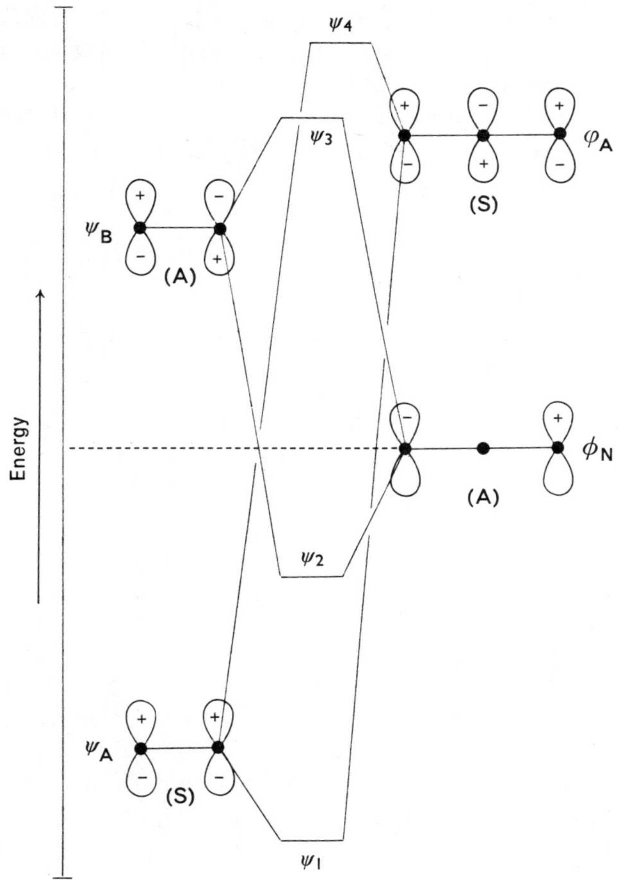

FIGURE 2. HOMO–LVMO interactions for an ethylenic π-bond and π-allyl anion. The situation will be similar for a π-bond, and diazomethane at deviations up to ~ 30° from linearity.

orbital symmetry is conserved as the maximum degree of orbital interaction develops, and these cycloadditions satisfy the simplified Woodward–Hoffmann rules for thermal concerted cycloadditions in involving $(4n + 2)\pi$-electrons in the transition state[55]. (The problem of regiospecificity in dipolar cycloadditions (i.e., the orientation of the addenda when there is more than one possible mode of *cis*-addition) has been more generally studied both from the experimental[56a] and theoretical[56b] point of view.)

In cases where solvent effects on addition rates are small and activation entropies are negative or at least very small, it seems safe to assume that the concerted process is in operation, but of course there could be exceptions where a dipolar intermediate is involved. Such appears to be the case in the addition of *p*-anisyldiazomethane to *p*-methoxystyrene; a mixture of *cis*- and *trans*-3,5-di-*p*-anisylpyrazolines results. This result can be rationalized in terms of a zwitterionic species in which the *p*-methoxy group stabilizes a benzylic carbocation[57].

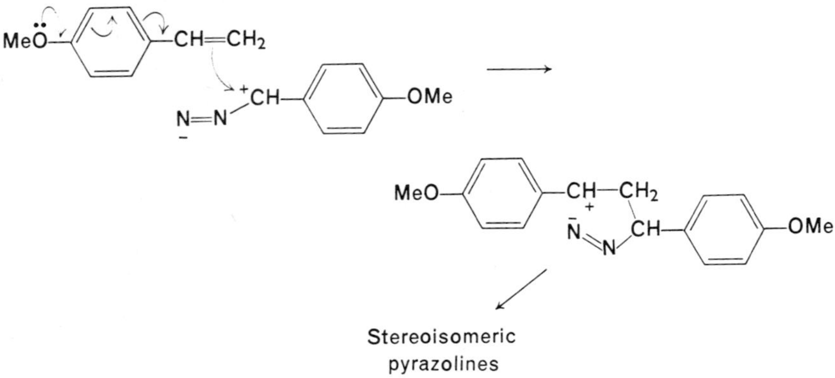

It is interesting that, whilst the addition of an allyl anion to a dipolarophile has not been observed, the reverse orbital symmetry-allowed process has been realized; thus the anion (**78**) made from pyrazoline (**77**) decomposes at room temperature releasing the stable allylic anion (**79**), with virtually quantitative evolution of nitrogen. The relative rate of fragmentation of pyrazoline (**77**) compared to electrocyclic ring opening of the anion derived

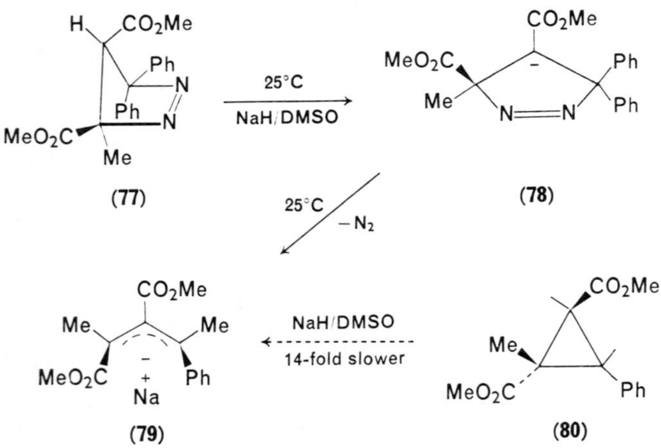

from cyclopropane (**80**) under similar conditions is 14:1, and the cyclopropane is clearly not an intermediate in the decomposition of the azo compound. Calculations show that E_a for the fragmentation is ca. 12 kcal/mole less than for formation of a diradical species by loss of nitrogen[58].

Before passing on to a general discussion of the thermal and photochemical decomposition of pyrazolines it is useful to summarize briefly factors affecting their practical synthesis from activated olefins and diazoalkanes[49, 50]. With regard to the structure of the latter, conjugating substituents reduce their activity in cycloaddition, the order of reactivity being: $CH_2N_2 > Ph_2CN_2 > N_2CHCO_2Me > —COCN_2 > —COCN_2CO$. The additions always give 1-pyrazolines as the initial products, but these may rapidly tautomerize if there is a suitably acidic hydrogen in the adduct; if a suitable proton source is absent, this tautomerism can only be accomplished in acidic media. The structure of the dipolarophile is widely variable but, in general, conjugating substituents on the olefin activate in the following order: $CO_2Ph > CO_2Et > C{\equiv}N > COMe > p\text{-}C_6H_4NO_2 >$ Ph. As already indicated, *trans*-olefins react faster than *cis*-olefins, and the maximum relative rate range among the commoner dipolarophiles exceeds 20,000, e.g., with maleic anhydride compared to 1,1-diphenylethylene. (On the same scale, incidentally, acetylenes are rather more reactive; phenylacetylene reacts with diphenyldiazomethane over four times as fast as diphenylethylene, whilst a maximum relative rate exceeding 35,000 is seen with acetylene dicarboxylic esters). In special cases unactivated olefinic bonds can react with 1,3-dipolar reagents including diazoalkanes, usually as a result of ring strain, and norbornene, for example, reacts with diphenyldiazomethane at roughly the same rate as *trans*-ethyl crotonate (dimethylformamide at 40°C, rate ratio 1·15).

As in the case of Diels–Alder cycloadditions, it should be possible to observe additions between components exhibiting inverse electron demand. For example, trifluordiazoethane (**81**) slowly forms an adduct (**82**) with ethylene if the components are kept in a sealed tube, and simple alkylated ethylenes give similar stereoisomeric mixtures of pyrazolines[59]. (The same pyrazolines are rapidly formed in the more conventional addition of diazoethane to 3,3,3-trifluoropropene.) For these additions involving diazoalkane (**81**) vinylic alkyl substituents reduce olefin reactivity, but trifluoroalkylvinyl substituents increase it, and seem about as effective as the alkoxycarbonyl group. These are precisely the results to be expected for the more usual cycloaddition process[56] in which the dipolarophile LVMO 1,3-dipole HOMO interaction is the dominant effect. Consequently, a clearcut example of an inverse electron demand dipolar cycloaddition remains obscure.

It is worth noting in connexion with fluorinated dipolarophiles that perfluorobicyclohexadiene (**84**) readily forms a mono-adduct (**85**) and isomeric bis-adducts (**86** and **87**) with diazomethane; in contrast there is no reaction with vinyl fluoride and for the compound **84** ring strain is an obvious factor[60].

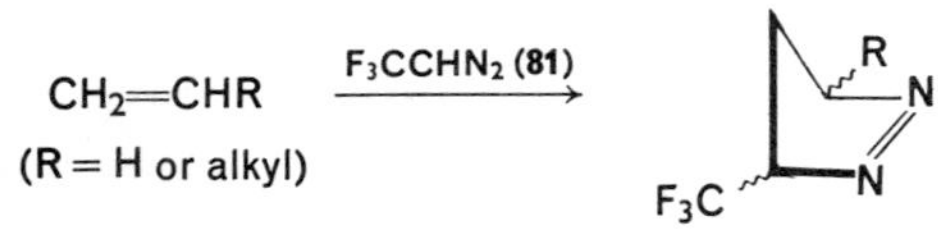

(**82**; R = H)
(**83**; R = Alkyl)

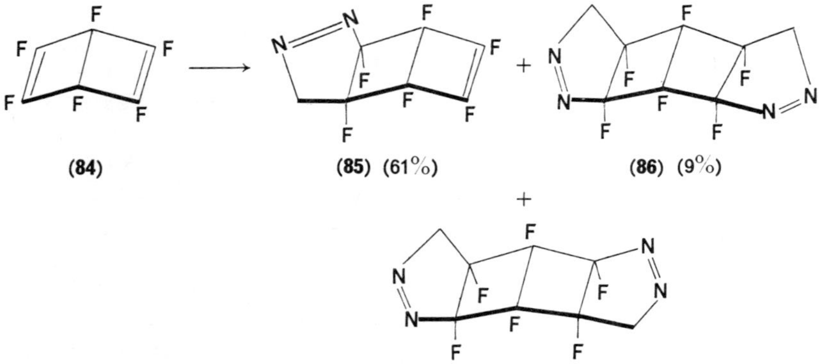

(**84**) (**85**) (61%) (**86**) (9%) (**87**)

2. Thermal and photochemical fragmentation

The ease with which 1- and 2-pyrazolines fragment with ejection of molecular nitrogen has been appreciated since the time of their discovery[45] and in some instances provides useful syntheses of cyclopropanes[61–64], especially where thermolysis is carried out in the vapour phase with alkylated 1-pyrazolines having an alkoxycarbonyl group in the 3-position[65]. The stereoelectronic features of their decompositions have been effectively probed only comparatively recently however, and since during its development the subject has passed fairly quickly through several phases, it is relevant to give a rather detailed account here.

Two main decomposition mechanisms can be visualized in outline, the first involving stepwise rupture of the C—N bonds leading to a zwitterionic intermediate such as **88** and/or **89**, and the second implying extrusion of nitrogen via diradical species such as **90** and/or **91**. Clearly, the nature of the substituents R^1 and R^2 and others which could be present can be expected to effect whichever of these pathways might be followed, besides providing experimentally useful stereochemical labels.

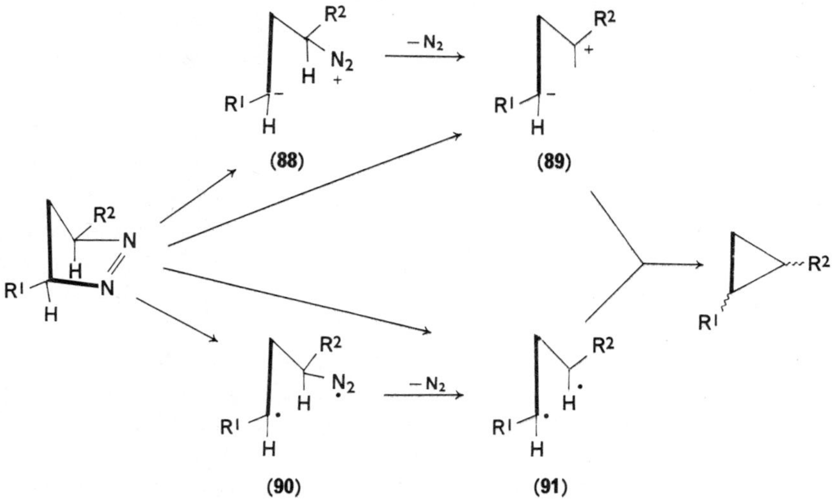

It was established in one of the few earlier studies[61] that thermolysis of pyrazoline (**92**) gave both stereoisomeric cyclopropanes **93** and **94** together with olefinic products, and a careful reinvestigation 30 years later confirmed this result and extended the study to the isomeric pyrazoline (**97**), which gave precisely the same *four* products **93–96**, although in different relative proportions[66].

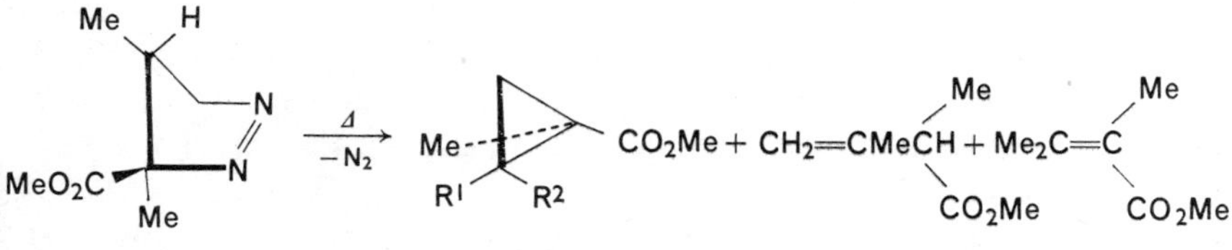

(92) **(93)** $R^1 = H$; $R^2 = Me$ (0·7) **(95)** (0·24) **(96)** (3·73)
(97) (*trans*-diMe) **(94)** $R^1 = Me$; $R^2 = H$ (1·0)

$\xrightarrow[-N_2]{\Delta}$ **(93)** (1·22); **(94)** (1·00), **(95)** (0·15), **(96)** (1·16)

These experiments showed slight stereoselectivity in favour of retained geometry in the products, and it was suggested, following an earlier observation[67], that the intermediate was a zwitterionic species in which nitrogen loss is favoured over σ-bond rotation.

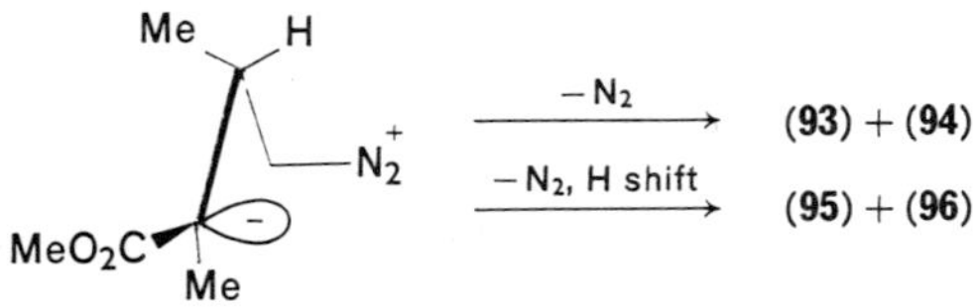

It was also suggested that a diradical analogous to **91** formed by simultaneous C—N cleavage could be involved or that loss of nitrogen might give a zwitterionic species which could be regarded as a canonical of the singlet trimethylene diradical. The distinctly different triplet state diradical was also discussed as a possible intermediate. The photolysis of pyrazoline (**92**), however, stereospecifically gave the *cis*-dimethylcyclopropane carboxylic ester (**94**) and some methyl tiglate (by formal loss of diazomethane) and in order to account for these results it was proposed that photoexcitation led to an intermediate or transition state (**98**) capable of extruding either nitrogen or diazomethane or both, thus accounting for the stereochemistry observed. The possibility that a singlet state diradical might be involved in the decomposition was discarded on the grounds that the

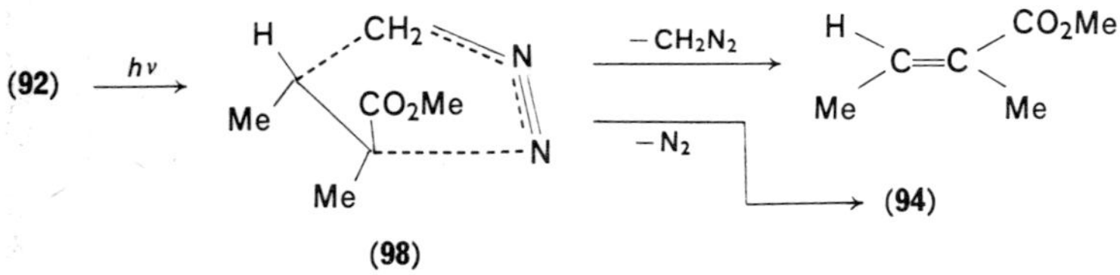

species would be in the excited state and unlikely to cyclize to stable products[50]. This summary serves to illustrate the state of knowledge and ideas current just over a decade ago. In the years since then much more has been discovered.

Exploring the possibility that zwitterionic species are involved in pyrazoline fragmentations, evidence was sought among the products for hydride and alkyl shifts to the cationic site supposedly developed[68]. No examples of alkyl or aryl migrations from C-4 in analogues of pyrazoline (**92**) had previously been observed, although 1,2-hydrogen shift must be postulated to account for olefinic products such as **96** formed from compound **92**. Clearly a useful test was to decompose 4,4-dialkylated pyrazolines such as **99**, **103** and **106**, and search for products of 1,2-alkyl shift. The results of these experiments were rewarding and seemed to support the zwitterion postulate, since alkyl migration products *were* observed, and the reaction rates and proportions of olefins increased with solvent polarity, without much effect on the cyclopropane stereochemistry. However, the rate increases did not correlate well when compared to solvent acceleration of other carbonium ion processes and the stereoselectivity observed in olefin formation seemed to rule out the formation of a completely free carbonium ion which should partition more equally among the stereoisomeric olefins.

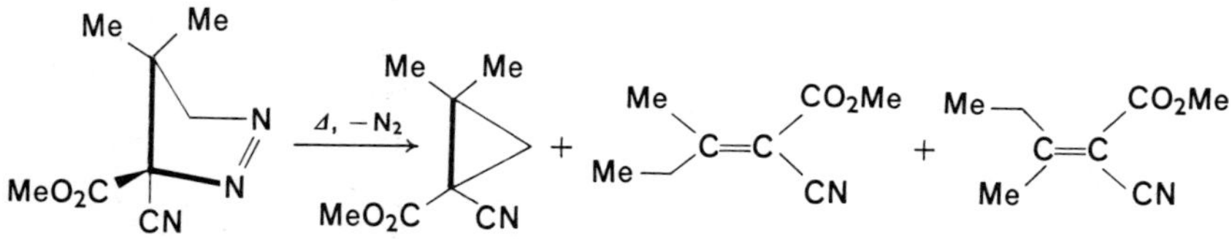

(99) **(100)** **(101)** (5%) **(102)** (85%)

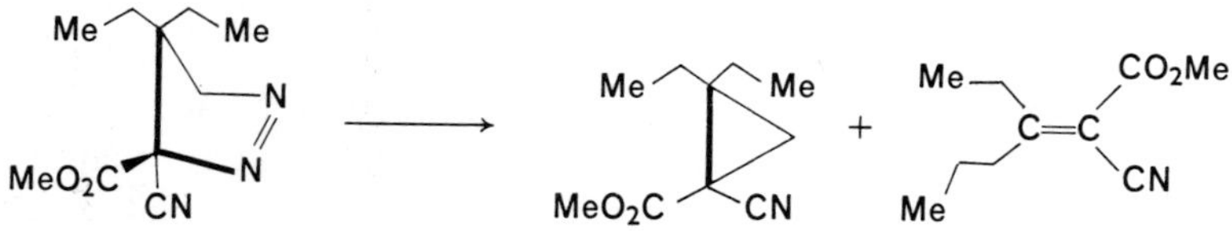

(103) **(104)** (15%) **(105)** (85%)

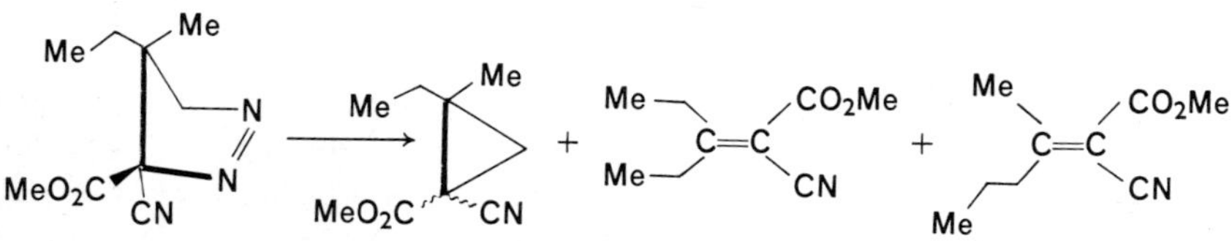

(106) **(107)** (stereoisomers, 25%) **(108)** (35%) **(109)** (isomers, 40%)

It was therefore proposed that alkyl shifts are either concerted with C—N bond scission or that considerable carbonium ion character develops at carbon before the C—N bond is completely broken[68].

Other evidence bearing on the possibility of zwitterionic species as

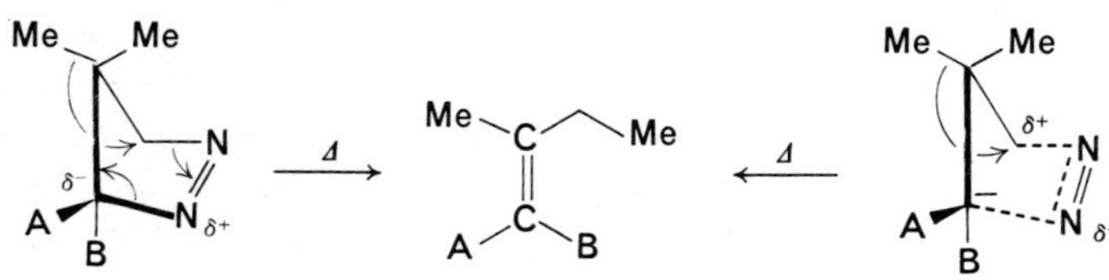

intermediates included the observation that dihydrofurans are formed in the vapour-phase thermolysis of 3-acetyl pyrazolines, as well as the usual cyclopropanes and olefins, suggesting some similarity to polar chemical processes known to give dihydrofurans from carbonyl compounds. The yields of dihydrofurans are strongly sensitive to the reaction conditions; e.g. thermolysis of pyrazoline (**110**) in cyclohexane gives 34% dihydrofuran, (compared to 13% in the vapour-phase decomposition, see scheme) but the yield falls to nearly zero in acetonitrile. The formation of the dihydrofuran appears to be at the expense of *trans*-cyclopropane specifically, and none is formed in thermolysis of the neat pyrazoline. In this connexion boron

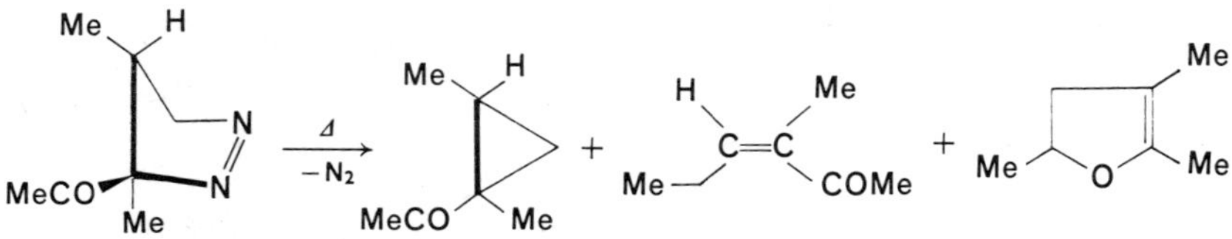

(110) **(111)** (48%, 27% *cis*-diMe) **(112)** (7% + 3% *trans*) **(113)** (13%)

trifluoride strongly catalyses acetylpyrazoline decompositions, reaction ensuing even at 5°C. The product composition is, of course, modified in comparison to that from thermolysis, and there is evidence that the complexes involved here show conformational preferences, e.g., for the *cis*- and *trans*-isomers of acetylpyrazoline (**110**): the former gives both *cis*- and *trans*-ketone (**112**), whilst the latter gives only the *trans*-isomer[69].

With regard to cyclopropane stereoselectivity in pyrazoline thermolyses, the data available at the time (1965) seemed to be at variance. Whereas the thermolysis of pyrazolines (**92**) and (**97**) resulted in only slight stereochemical preference[50] it had meanwhile been found[70] that heating diphenylpyrazoline (**114**) gave *trans*-diphenylcyclopropane stereospecifically. Comparison of this result with thermolyses of 3,5-dialkylpyrazoline carboxylates, e.g. **116** and **122**, invited itself and under a variety of conditions moderately stereoselective formation of *trans*-cyclopropane (**118**) from *cis*-pyrazoline (**116**), and *cis*-cyclopropane (**117**) from *trans*-pyrazoline (**122**) was observed, together with essentially stereospecific formation of *cis*-olefin (**120**) from *cis*-pyrazoline (**116**), and the *trans*-olefin (**119**) from *trans*-compound (**122**). (The yields of cyclopropanes in these decompositions ranged from 65% of **118**, to 76–79% of *cis*-compound **117**.) These results further disfavoured the zwitterionic intermediate; a net inversion of configuration at one of the carbon termini seems to be implied and if σ-bond rotation or pyramidal carbanion inversion within a zwitterion is possible one might then expect the same (or at least closely similar) products from both pyrazolines **116** and **122**, and even the possibility that reclosure of the intermediate to a pyrazoline should result in stereoisomerism in partially decomposed material; no isomerization of *trans*-pyrazoline (**122**) was, however, observed.

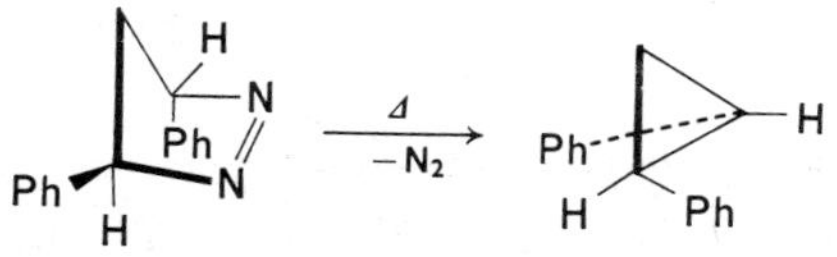

(114) **(115)**

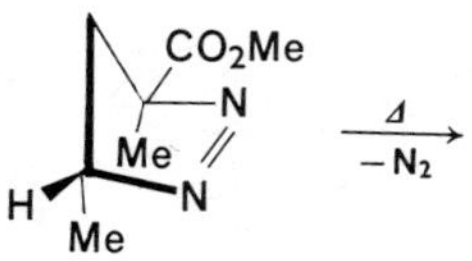

(116)
(122) (*trans*-diMe)

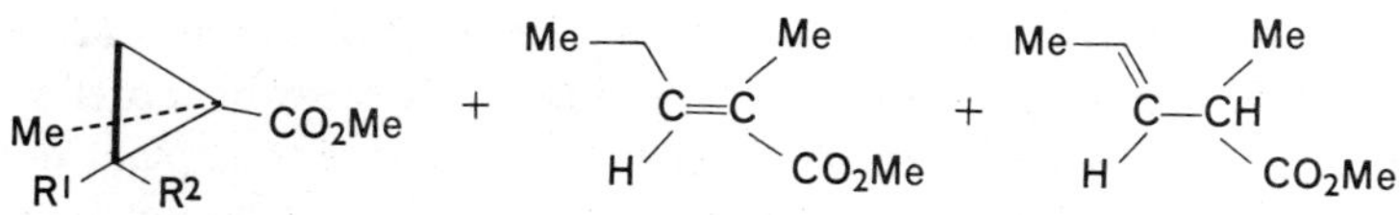

(117; R^1 = Me, R^2 = H (≡**93**)) **(119)** **(121)**
(118; R^2 = Me, R^1 = H (≡**94**)) **(120)** (*cis*)

Synchronous cleavage of both C—N bonds in the transition state as the alternative appears to be supported by the fact that in 3,5-dialkylpyrazolines both terminal substituents enhance the rate of decomposition. The process could be visualized as C—N stretching motions perpendicular to the C-3/C-5 plane with development of two carbon *p*-orbitals more or less parallel to each other but whose precise orientation would depend on the substituents present. It was suggested that such a process could conceivably lead to a twisted transition state with *p*-orbitals generated in such a way as to allow inversion at one centre as the σ-bond formed. Further experiments with more highly substituted pyrazolines had indicated that the actual mechanism need not require total inversion or retention of configuration and, for example, pyrazoline (**123**) gives cyclopropane with predominantly retained configuration[71].

Another possible rationalization of the predominant inversion at one of the C—N termini in a pyrazoline decomposition envisaged stepwise loss of

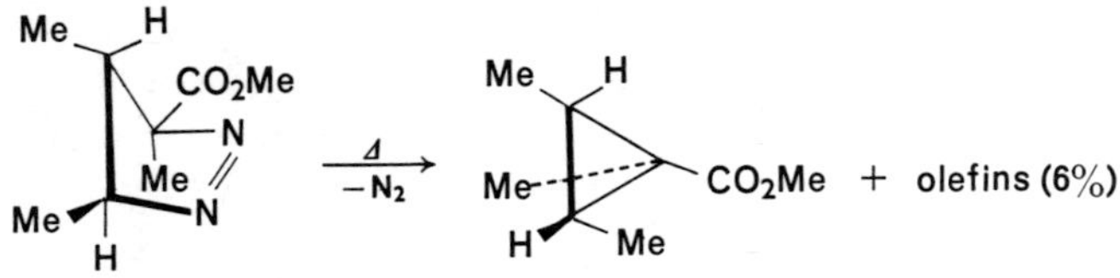

(123) **(124)** (61%) **(125)** (all-*cis*-Me 32%)

nitrogen with backside displacement in a rotamer of the initial intermediate[72], but this scheme also suffers the drawbacks inherent in any mechanism involving polar intermediates.

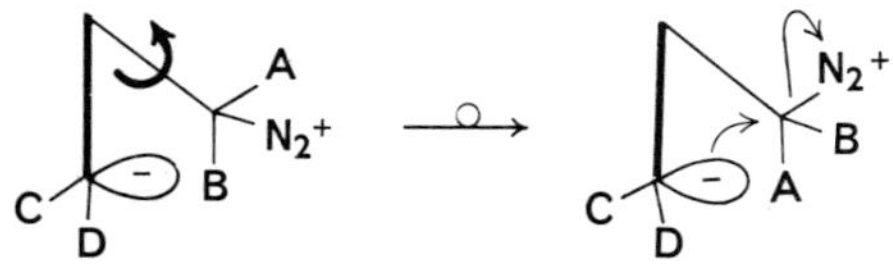

With regard to the stereochemistry of olefin formation, examination of conformers of *cis*- and *trans*-dimethylpyrazolines (**116** and **122**) shows that shift of *trans*-4 hydrogen to C-5 during C—N scission is preferred in each case, and it is this hydrogen which must migrate to explain olefin stereospecificity. Similarly, hydrogen migration to C-3 would give *trans*-β,γ-

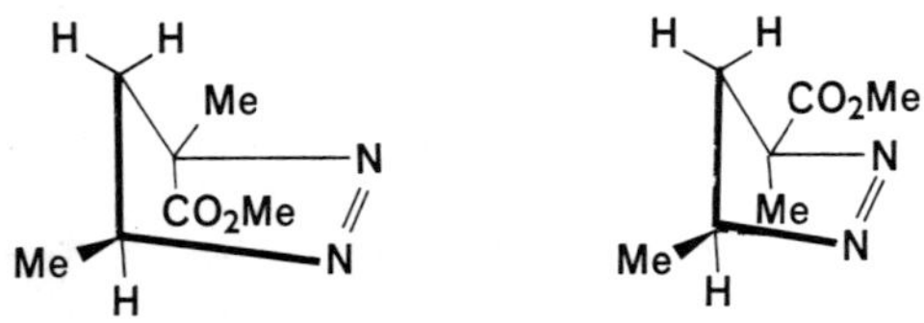

(116) **(122)**

unsaturated ester, and it is significant that *only trans*-β,γ-unsaturated ester is observed. A difficulty in using a zwitterionic species to explain these results is that the ratio of hydride to proton shifts to give α,β- and β,γ-unsaturated esters would have to be about 14:1; whilst hydride shifts in carbonium ions are well-known and rapid processes (if the relevant orbitals are parallel) proton shift to the developing carbanion site might be rather slow.

The mechanism of olefin formation was further explored using the 4-ethyl derivatives (**126** and **127**) of 3-methylpyrazoline-3-carboxylic ester[73]. The results indicated that the general trend is the same as with the dimethyl analogues (**92** and **97**), except that there is more stereoselective cyclopropane formation. Clear stereospecificity for the α,β-unsaturated ester is observed and the results can be accounted for in each case if the C-4 hydrogen atom is *trans* to the cleaving C—N bond in the transition state. The ground-state conformation as determined by ^{1}H n.m.r. suggests that the 4-ethyl group is pseudo-equatorial, so the advantage gained in the transition state must compensate for any adverse effects of the inverted conformation. These studies interestingly suggested that C-4-alkylated analogues of pyrazoline (**126**) which, on the basis of ^{3}J H-4/H-5 n.m.r. couplings, mainly prefer an axial conformation, also give the highest yield of olefins.

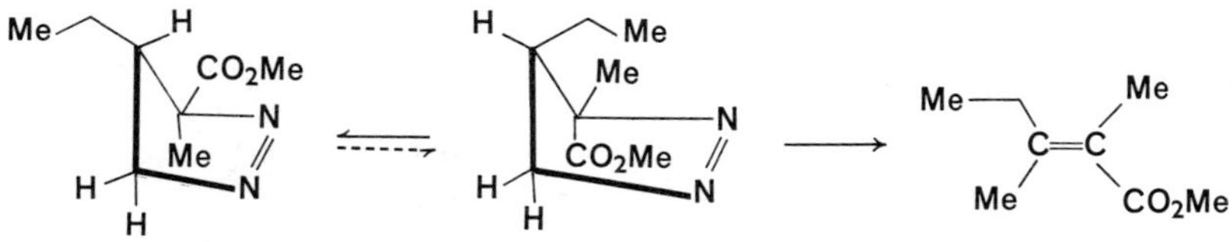

(**126**) (**126a**)

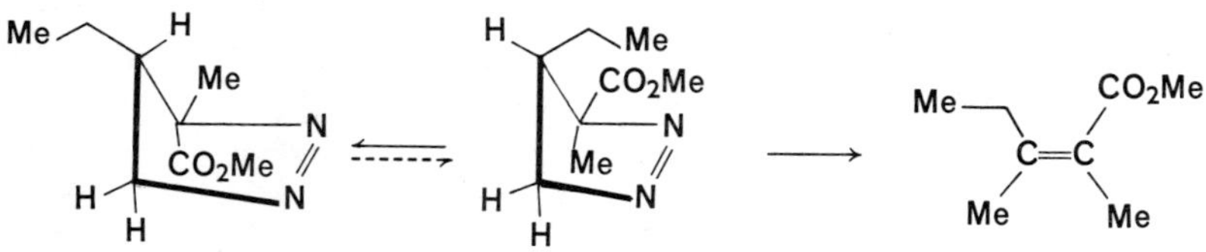

(**127**) (**127a**)

The results of photolysis of a number of pyrazolines have in general borne out the earlier finding that more cyclopropane is formed than in the thermal reactions, frequently stereospecifically but, when not, with a tendency to maintain the relative stereochemistry of the pyrazoline[74]. To rationalize these findings it was suggested[71] that fragmentation was from a vibrationally excited ground state derived by radiation emission; the general similarity of the photolysis products to those from thermolyses seemed to require no completely new mechanism.

The really significant alternative to these various reaction schemes in order to account for the products of pyrazoline decomposition is that all the products, thermal and photochemical, cyclopropane and olefinic, derive from a single common intermediate, i.e., a species like **91**. Kinetic data and product analysis seemed to suggest that the intermediate is a

π-type diradical, but it is better described as [0,0]-trimethylene with its terminal orbitals necessarily antisymmetric about a mirror plane bisecting the methylene group, which must therefore usually close to a cyclopropane in a conrotatory manner[75]. Depending on the C_1-C_2-C_3 angle, other related structures for the intermediate, e.g., with the terminal p-orbitals mutually orthogonal, or σ-parallel in the plane of the carbon atoms (designated '0,90' and '90,90') may be of higher energy than the '0,0' species, but at angles $\ll 120°C$ the '90,90' intermediate is by far the most stable, and easily cyclizes to a cyclopropane.

[0,0] [0,90] [90,90]

Considerable evidence has accumulated to support this viewpoint. For example, comparison of the kinetic parameters for thermolysis of a series of alkylated pyrazolines[76] (Table 1) reveals that successive methylation at the

TABLE 1. Activation parameters for thermolysis of substituted pyrazolines

Pyrazoline	T(°C)	E_a (kcal/mole)	$\Delta S^{\neq}$(e.u.)	Log A
Unsubstituted	202–237	42·4 ± 0·3	11.2 ± 0·6	15·93 ± 0·13
3-Methyl	190–230	41·0 ± 0·3	10·1 ± 0·7	15·7 ± 0·15
4-Methyl	199–234	42·2 ± 0·2	10·8 ± 0·3	15·85 ± 0·05
4,4-Dimethyl	247–286	42·8 ± 0·2	2·9 ± 0·3	14·10 ± 0·07
3,3-Dimethyl	184–211	40·0 ± 0·2	10·8 ± 0·2	15·85 ± 0·03
cis-3,5-Dimethyl	191–224	40·3 ± 0·3	9·4 ± 0·5	15·54 ± 0·11
trans-3,5-Dimethyl	191–224	40·2 ± 0·2	10·0 ± 0·5	15·67 ± 0·11
3,3,5-Trimethyl	183–215	39·0 ± 0·4	8·4 ± 0·9	15·42 ± 0·21
3,3,5,5-Tetramethyl	176–219	37·7 ± 0·4	4·6 ± 0·8	14·49 ± 0·12
2,3-Diazanorbornene	171–202	36·9 ± 0·2	5·8 ± 0·5	14·7 ± 0·10
(Cf. 2,3-Diazanorbornene[153]	130–180	37·3 ± 0·3	8·7 ± 0·4	14·9 ± 0·1)

3- or 5-position brings about a decrease in E_a. Furthermore, since the activation energies for *cis*- and *trans*-3,5-dimethylpyrazoline and for 3,3,5-trimethyl- and 3,3,5,5-tetramethylpyrazolines are less than for 3,3-dimethylpyrazoline, cleavage of the C—N bonds at both termini is important in the transition state. However, it might be objected quite legitimately that such small changes in activation energy could be due to rising ground-state energies as methyl substitution increases, lowering E_a progressively relative to pyrazoline. Fortunately, replacement of hydrogen by deuterium rather than by methyl groups removes this ambiguity[77]. The effect on $\Delta G^{\ddagger}$ by α-deuterium substitution in azoalkanes is known ($\Delta\Delta G^{\ddagger}$ 80–110 cal/mol/D, due to force constant effects[78]) and the observed differences in activation free energies for 3,3-bisdeuterio- and 3,3,5,5-tetradeuteriopyrazoline as derived from the progressively smaller unimolecular rate constants for thermolysis are 86 ± 6 and 84 ± 6 cal/mol/D respectively. The consonance of these results leaves little doubt that concerted bond cleavage is involved in the transition state. In this connexion the rather larger than normal β-deuterium isotope effect is consistent with the hyperconjugative effect of the hydrogen atoms on the central carbon of the [0,0]trimethylene species[75]. The rate constants for cyclization of the β-deuteriated species are also smaller than for the protio analogues, whereas α-deuteriation has no kinetic effect on this process; however, slightly more propylene-d_2 is produced from pyrazoline-3,3-d_2. Careful analysis of the ^{1}H n.m.r. spectra of the olefinic products indicates a product ratio of 52:48 for 3,3-bisdeuterio- and 1,1-bisdeuteriopropylenes, consistent with an inverse kinetic isotope effect k_H/k_D 0·92 as expected for the sp^2 to sp^3 hybridization change during rearrangement.

In all these alkylated pyrazoline decompositions cyclopropanes strongly predominate (89–99%) except for 4-methylpyrazoline; the products here, methylcyclopropane and 2-methylpropene, are of comparable importance, and such a striking difference in behaviour suggests a changed mechanism. To check this 4-deuterio-4-methylpyrazoline was decomposed; the proportions of cyclopropane and alkene *do* change (from 52 to 66% and 47·7 to 34% respectively) but only to the extent expected for a primary kinetic isotope effect on the necessary deuterium migration in the common intermediate, and hence the increased yield of cycloalkane.

The relative rate constants for decomposition of pyrazoline, 4-methyl- and 4,4-dimethylpyrazoline show that the latter decomposes at ca. 1/130th of the rate for pyrazoline, but the rate for the monomethyl compound is almost the same as the parent structure; this suggests that the 'flap' angle in the pyrazoline *decreases* as the C—N bonds break and the terminal carbon atoms approach sp^2 hybridization, causing transition state steric compression, which becomes rather more important when it cannot be relieved in either of the conformations. Alternatively, the effect could be due to the increased C-3, C-4, C-5 angle resulting in a decrease in the geminal methyl separation. Whichever of these effects is actually operative, both are satisfyingly consistent with the proposed [0,0]trimethylene intermediate.

Further information is secured by comparing the olefinic products (and the kinetic parameters) for decomposition of *cis*- and *trans*-3,4-dimethylpyrazolines (**128** and **129**) with those for the 5,5-bis-deuterio analogues; the kinetic isotope effect, k_H/k_D 1·19 and 1·21 respectively, indicates that at least the primary C—N bond is cleaving in the transition state. Two discrete intermediates could be produced from each stereoisomer, depending on conformational preferences, but careful ^1H n.m.r. integration for the olefinic products shows that the olefin derived from *cis*-isomer (**128**) contains >94% deuteriomethylbutene (**130**), and that from *trans*-compound (**129**) is more than 96% deuteriomethylbutene (**131**). Cyclopropane formation here is not stereospecific. This result is also consistent with the steric crowding effect of a methyl group at C4 *syn* to the azo function in conformer **128a**[79].

The stereochemistry of the olefins formed in these decompositions is also more generally consistent with a trimethylene intermediate. For example, *cis*-3,5-dimethyl- and *trans*-3,5-dimethylpyrazoline should give inter-

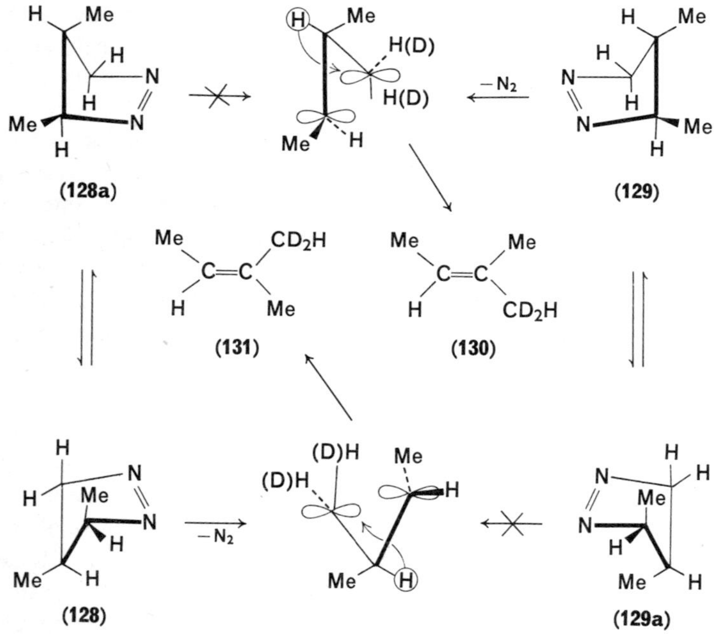

(128a) (129) (131) (130) (128) (129a)

(132) (133)

mediates **132** and **133**; migration of hydrogen from C4 to C3 *or* C5 gives only *trans*-2-pentene from **132** but for **133** similar migrations give respectively *trans*- and *cis*-2-pentene. This is precisely the difference observed in the products from the two pyrazolines; but what is the origin of isomeric cyclopropanes so often found as products of thermolysis of pure pyrazolines? Perhaps the answer lies in the fact that species like **132** and **133** can exist in symmetric (S) and antisymmetric (A) singlet states which may not be too energetically different. (There is also the possibility of a triplet state with one electron in each of the (S) and (A) M.O.s. Due to electron interaction this state is actually of lowest energy at equilibrium geometry, and is the true ground state.) The orbital symmetry and relative energy of these planar states of the intermediate depend on the C3-C4-C5 angle (due to interaction of the terminal p-orbitals) and on interaction with substituents at C4. In trimethylene itself the effect at C4 is due to an antibonding interaction between the terminal p-orbitals and the localized σ-orbitals of π-symmetry which contribute to the CH_2 group; this has the effect of raising the symmetric M.O. whilst the antisymmetric orbital cannot, by symmetry, interact[75,76]. Obviously substitution at C4 is liable to change this energy order for the trimethylene M.O.s and, therefore, either conrotatory or disrotatory ring closure to a cyclopropane will result, there

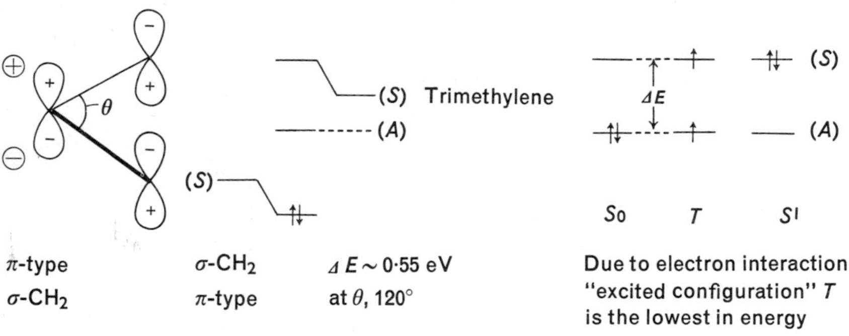

being appreciable barriers to other internal rotations in these intermediates. The conrotatory or disrotatory cyclization of antisymmetric and symmetric π-species like **132** and **133**, whose accessibility might then be a function of the substituents present, could thus explain rather well the diverse stereochemical results which have accumulated.*

* Recent calculations show that terminal substituents reduce the stabilization for the conrotatory mode relative to the disrotatory mode of cyclization of [0,0]trimethylene[79a].

Conformational effects ensure that two geometrically different trimethylene species are involved in the decompositions of pyrazolines **128** and **129**; such effects are absent, however, for *cis*- and *trans*-isomers of 3-methyl-4-deuteriopyrazolines **134** and **135**. The results of their decomposition clearly show that a common intermediate trimethylene is involved[80]. The ratio of stereoisomeric methyldeuteriocyclopropanes is 1:1 in each case and, on the basis of product data for 3-methylpyrazoline, assuming a kinetic isotope effect k_H/k_D of 2, the predicted and experimentally observed yields of deuterated methylcyclopropane, 1-butene and stereoisomeric 2-butenes correlate quite well, and are closely similar for the two isomers. Calculation of the actual kinetic isotope effect for the two olefins from the ratio of allylic to vinylic protons in the ^{1}H n.m.r. spectrum shows that its values are virtually independent of the source, at least within experimental error (Table 2) as required by a common intermediate. Certainly the data are inconsistent with a concerted stereospecific migration of hydrogen *trans* to the cleaving C—N bond in the transition state.

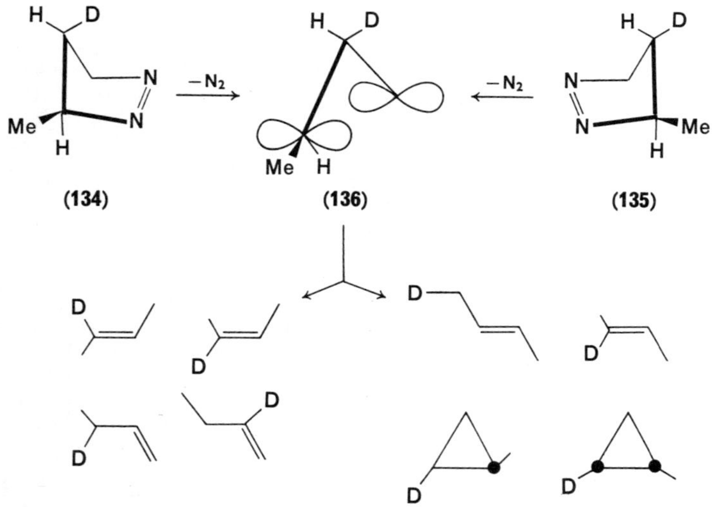

The intermediacy of [0,0]trimethylene in thermolyses of simple pyrazolines therefore seems very convincing, and could be viewed in a simple way as the natural consequence of concerted cycloreversion of the azo compound into asymmetric trimethylene and nitrogen. However, contemporaneous observations with bicyclic azo compounds (see Section III. A) suggested the possibility that inversion occurred at both carbon termini during departure

TABLE 2. Thermolysis of 3-methylpyrazoline and the *cis*- and *trans*-4-deuterio derivatives **134** and **135**

Products:			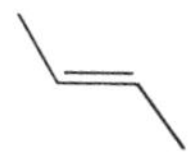	
3-Methylpyrazoline	93·3 ± 0·6	1·9 ± 0·2	1·16 ± 0·12	3·7 ± 0·3
Predicted for **134** and **135**	94·9	1·4	0·9	2·8
Found, **134**	95·1 ± 0·5	1·4 ± 0·2	1·0 ± 0·1	2·5 ± 0·3
mole % H shift product in total		72 (d ~28)	63 (d ~37)	78 (d ~22)
k_H/k_D		2·7 ± 1	1·7 ± 0·6	3·5 ± 1
Found, **135**	94·7 ± 0·5	1·5 ± 0·2	1·1 ± 0·2	2·7 ± 0·3
mole % H shift product in total		63 (d ~37)	65 (d ~35)	75 (d ~25)
k_H/k_D		3·0 ± 0·6	1·7 ± 0·6	1·9 ± 0·6

of nitrogen (as in **137**) or that an initially pyramidal biradical underwent a similar double inversion, e.g. into **138**, before converting into products. According to both these ideas *trans*-3,5-dimethylpyrazoline ought to give stereospecifically *trans*-dimethyl cyclopropane, but 73 % *cis*-dimethylcyclopropane and only 25 % *trans*-isomer are actually observed, together with 1 % each of stereoisomeric 2-pentenes[76]. More significantly thermolysis of

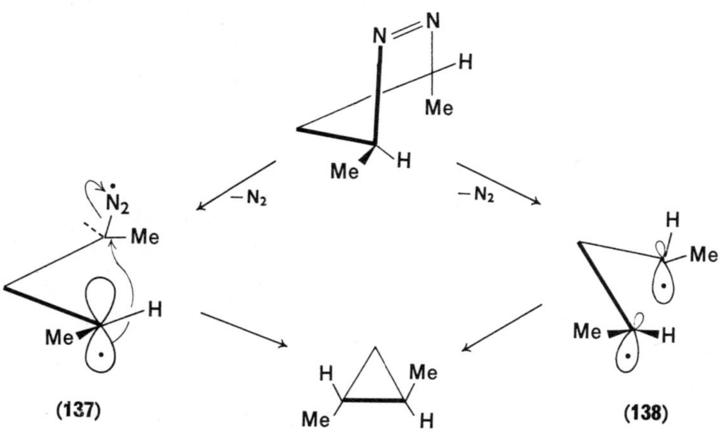

3R:5R(+)-*trans*-3,5-dimethylpyrazoline gives 6 % of optically active *trans*-dimethylcyclopropane and this *does* suggest that at least some of the

molecules decompose through intermediates which undergo double inversions at the chiral centres, e.g., in the pyramidal species, which closes to optically active product; but it is not clear what sequence of events occurs. Does the trimethylene intermediate precede a non-planar 1,3-diradical species or vice versa, or is the pyramidal species formed in a separate pathway? Orbital symmetry conservation would suggest that the former is likely, but these questions have certainly not been answered[81].

Calculations indicate that the lowest singlet S^0 state of [0,0]trimethylene is the planar antisymmetric π-species, but there is also the excited singlet state S^1 which has no barriers to internal rotation. Whilst the thermally accessible S^0 species is not related to the concerted addition of methylene to olefins[75] it does seem reasonable that the nonstereoselective intermediate involved in additions of triplet carbenes to olefins is related to the triplet trimethylene which could be formed in ketone-sensitized decompositions of pyrazolines.

In the direct vapour-phase photolysis of 3- and 4-methylpyrazolines, the products—olefins and methylcyclopropane—very likely arise from S^1 trimethylene cyclizing to vibrationally excited cyclopropanes which partially fragment, since product formation is pressure-sensitive; for *cis*- and *trans*-3,5-dimethylpyrazolines comparative experiments show that fragmentation to olefins occurs *only* in the direct photolysis, and appears to be independent of pressure or phase, whilst in the sensitized decomposition there is very little fragmentation, the products being *cis*- and *trans*-dimethyl cyclopropanes (39% and 60% in each case)[82]. The *trans*-pyrazoline decomposes with similar stereoselectivity in thermolysis *and* sensitized photolysis, but for the *cis*-isomer, opposite stereoselectivity is observed in the two processes. The latter stereochemical results are virtually reversed in liquid phase photolyses. In these reactions involving triplet trimethylene, internal rotation and/or spin inversion and ring closure could involve a collisionally stabilized diradical so that a 'cold' cyclopropane having no tendency to fragment results. The lack of competing olefin formation can be understood on the basis of the triplet intermediate involved; here there is a parallel to the addition of triplet methylene to olefins at relatively high pressures where there is also very little fragmentation of the adduct. Regarding the stereochemistry of ring closure in the direct S^1 photolysis of *cis*- and *trans*-3,5-dimethylpyrazolines, the resulting products resemble the stereochemistry of the reactants as the stereoselectivity increases with pressure; in this respect the *cis*-isomer resembles its behaviour in the thermal reaction, but the *trans*-compound behaves in the opposite manner. Broadly, these results can be understood in terms of the 'floppy' S^1 state of the trimethylene species; the final ring-closure step must obviously be symmetry-controlled, but in contrast to the S^0 state the stereochemistry in the S^1 species is mobile and sensitive to external influences.

Vibrationally excited triplet species are proposed to explain the products of $^3P_1$ mercury-sensitized photolysis of the stereoisomers of 3,4-dimethylpyrazoline[83]. Among the products, *cis*- and *trans*-2-butenes, *cis*- and *trans*-2-pentenes, 2-methyl-(1 and 2)-butenes and ethane, there is also some 3-methyl-1-butene. This product does not appear in the direct photolysis mixture, but is a component in the products of reaction of triplet methylene with the 2-butenes, and 1,2-dimethyl-1,3-trimethylene triplet diradical is the intermediate suggested. The formation of the olefins in this photosensitized reaction can be ascribed to the hydrogen and methyl migrations within the diradical, and further evidence for this is the suppression of olefin formation under partial pressures of nitrogen where collisional deactivation of the 'hot' triplet supervenes[84]. This result appears to be one of the first examples where 'hot' triplets are involved in pyrazoline decompositions.

Any detailed analysis of the mechanism of photolysis of pyrazolines must consider the fundamental question of how the excess energy absorbed by the chromophore is utilized in the reaction, e.g. how much is carried away by the ejected nitrogen or retained by the hydrocarbon fragments as vibrational energy. Experimental information is available from the variation of product composition with pressure, since at low or 'zero' pressure fragmentation or olefin formation from an energetic intermediate will be at a maximum. For example[85], the direct photolysis of 1-pyrazoline in the gas phase gives cyclopropane, propylene and ethylene; the propylene/cyclopropane ratio is pressure-sensitive, whereas ethylene formation is not very dependent on pressure changes. Propylene appears to be formed by two pathways, only one of which is pressure-sensitive. A reaction scheme accommodating the data includes a 'hot' cyclopropane formed by cyclization of singlet trimethylene as a source of pressure-sensitive propylene, the pressure-insensitive source being decomposition of triplet pyrazoline formed from the initial S^1 state. A statistical mechanical treatment of the data uses, among others, a Gaussian energy distribution function in evaluating the theoretically expected propylene/cyclopropane ratio at various energies, k_a/w, (where k_a is a function of the rate of deactivation, the distribution function f_e and the rate constant for isomerization k_e, at energy E integrated over the energy range, and w is the rate of deactivation). The results of using this function to fit the experimental propylene/cyclopropane ratios give an average peak energy of 75 kcal/mole for the 'hot' cyclopropane formed (with a peak width of $\pm$ 40 kcal/mole). The fraction of cyclopropane molecules formed at any energy is assumed to be proportional to the energy level density at that energy, which is in turn a function of the energy level density for the 'cyclopropane' part of the pyrazoline, the remaining degrees of freedom of the pyrazoline and, of course, the total available energy. The energy level densities can be evaluated from the vibrational modes for which suitable models may exist (here cyclobutanone is used, having many frequencies similar to those in cyclopropane, the remaining

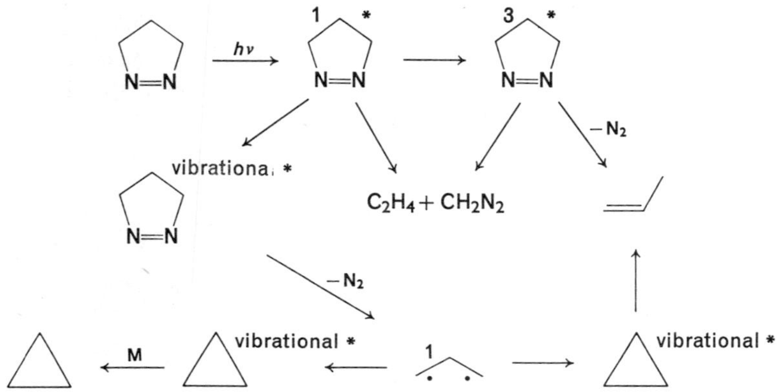

vibrational modes being fitted to the nitrogen part of the ring). On this basis the total available energy term fits for the experimental product ratio when it is 105 kcal/mole although the total available energy from radiation absorbtion is theoretically ca. 130 kcal/mole. The conclusion from this and similar analyses is that the total calculated radiation energy is *not* available to the hydrocarbon fragments, possibly because of poor vibrational coupling between the absorbing entity and the rest of the structure. (The experimental data show that at 'zero' pressure the propylene/cyclopropane ratio is 0·9–1·0, showing that only half of the cyclopropane molecules isomerize, which is substantiated by the figure of 75 kcal/mole as the average peak energy; further manipulation of the data with thermochemical values gives a maximum of 105 kcal/mole for the cyclopropane formed by ring closure of S^1 trimethylene—close to the maximum energy predicted by fitting the calculated product ratio versus $(\text{pressure})^{-1}$ curve to the experimental curve[85].) A very similar analysis has been carried out for photolysis of 4-methyl and 3-vinyl 1-pyrazolines, with essentially the same conclusions regarding energy availability and, interestingly, the suggestion that the relatively small amount of energy apparently in the hydrocarbon fragment is due to the lack of coupling between the reaction coordinate and the responsible N—N vibrational mode during concerted C—N scission, a result which had been anticipated for the 4-methylpyrazoline, but which appears to be more generally observed. The only significant effect of the vinyl substituent appears to be in stabilizing the intermediate trimethylene, causing a reduction of E_a; the substituent does *not* promote a stepwise radical decomposition involving the diaza biradical $\dot{}RN{=}N\dot{}$ as had been expected[86].

This last result incidentally shows that further information is potentially available with regard to the nature of the trimethylene intermediate, especially by varying the 3,5-substituents in the pyrazoline. For example, thermolysis of *trans*-3,5-di-*p*-anisylpyrazoline gives 93·3% *trans*-cyclopropane (+ 6·7% *cis*-isomer), whereas photolysis gives essentially pure *trans*-compound, no *cis*-isomer or olefinic products being detected. Similarly, thermolysis of the *cis*-pyrazoline gives predominantly *trans*-cycloalkane, but comparable amounts of *cis*-isomer appear as well (57:43%), whilst in photolysis the proportions of products are exactly reversed. The results have been interpreted in terms of rotational effects in a conventional *biradical*, the increased temperature enabling C—C rotation to compete with cyclization of the biradical, hence the lack of stereospecificity for the thermolysis of the internally strained *cis*-compound (which is much faster than the *trans*-isomer)[87].

However, if these decompositions are considered to be concerted, computation of the rates by comparison with those known for methylated pyrazolines should be possible on the basis of similar comparative experimental rate data for methylated and arylated azoalkanes. The calculated kinetic parameters for thermolysis of 3,5-diphenyl- and 3,5-di-*p*-anisylpyrazoline on this basis are in fact in good agreement with the redetermined experimental rates for these compounds (diphenylpyrazoline: E_a calc. 26·8 kcal/mole, E_a found 27·5 ± 0·5 kcal/mole, log A 14·1; di-*p*-anisylpyrazoline: E_a calc. 26·5 kcal/mole, E_a found 26·1 ± 0·6 kcal/mole, log A 14·9) and the data show that the aryl groups exert the same stabilization effect on the intermediate in both cyclic and open-chain azo compounds[88]. The earlier, different rate data for the aryl pyrazolines seem to have been due to stereochemical impurity[87]. In agreement with the earlier work, however, the stereochemistry of cyclopropane formation here is temperature-sensitive; a change in temperature from 85 to 50°C for the decomposition of the diphenylpyrazoline results in ca. 3% increase in the proportion of the mainly *trans*-diphenylcyclopropane produced; clearly σ-bond rotational effects within the intermediate must be important, unless there is only a very small separation between a symmetric and antisymmetric [0,0]trimethylene species, assuming that these descriptions remain meaningful for the terminally arylated trimethylene whose properties might be expected to be profoundly modified by heavy substitution. In this connexion low-temperature photolysis of 3,5-diphenylpyrazoline in a nujol or fluorolube matrix at 77 K results in the appearance of an e.s.r. signal due to a free radical which is *not*, however, a triplet state; the result at least seems consistent with the stereospecific formation of *trans*-diphenylcyclopropane since a triplet species should not be so stereospecific[89].

To summarize, the thermal and photochemical behaviour of alkylated and arylated pyrazolines seems best interpreted in terms of a 1,3-trimethylene intermediate capable of existing in stereochemically different singlet and triplet states and modified, especially by arylation, when the intermediate behaves more like a bis-benzylic biradical. The trimethylene species can be regarded as being formed by the concerted extrusion of nitrogen, more especially in the alkylated pyrazolines. Where the pyrazoline is heavily substituted with electronegative groups, it is possible that zwitterionic-like intermediates are involved, although solvent effects are rather smaller than expected. More likely concerted processes with unsymmetrical transition states occur here.

Amusingly, recent calculations[90] indicate that stabilization within the 1,3-trimethylene species can be regarded as due either to electron-pairing effects, or to a strong 1,3-zwitterionic contribution to the wave function: an idea expressed in much earlier work[50]. The calculations[90] surprisingly suggest that the most stable conformation for the 1,3-trimethylene species is actually [90,90] with the terminal methylenes folded inwards as in **139**.

(139)

Very little work appears to have been done in probing the chemistry of ionic species derived from pyrazolines by solvolytic reactions of appropriate derivatives, where the ionization of the leaving group is potentially subject to neighbouring-group effects involving the azo function. However, acetolysis of the *p*-bromobenzenesulphonate **140**-OBs results in the elimination of nitrogen to give olefinic products **141**–**143**; no acetate **140**-OAc is formed. By contrast, the thermal decomposition of the corresponding tetramethylpyrazoline is over a hundredfold slower at the same temperature, giving almost entirely tetramethylcyclopropane as the product, with a very little 2,4-dimethyl-2-pentene. Kinetic studies with the bis-trideuteriomethyl analogue of **140**-OBs show k_H/k_D is close to unity, implying that the rate-limiting step does not involve rehybridization about C3. However, there is an effective kinetic isotope effect in the product-

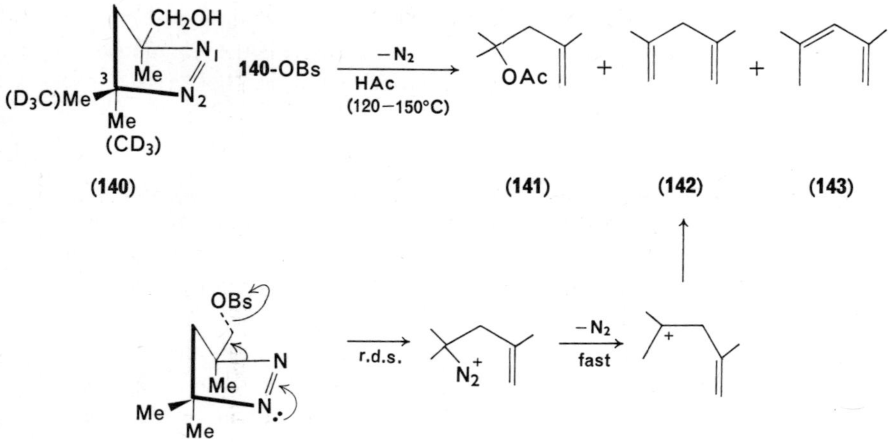

(140) (141) (142) (143)

partitioning step (k_H/k_D 1·95) showing that an intermediate is involved, and a scheme utilizing azo group participation leading to a diazonium cation, whose fast decomposition to dimethylpentenyl carbocation as the source of the olefinic products rationalizes the results[91]. Under somewhat similar conditions, 2-*exo*-5,6-diazanorborn-5-enyl brosylate solvolyses 117-fold faster than the *endo*-epimer, and a similar reaction scheme involving a diazonium ion precursor of cyclopent-3-enyl carbocation nicely accounts

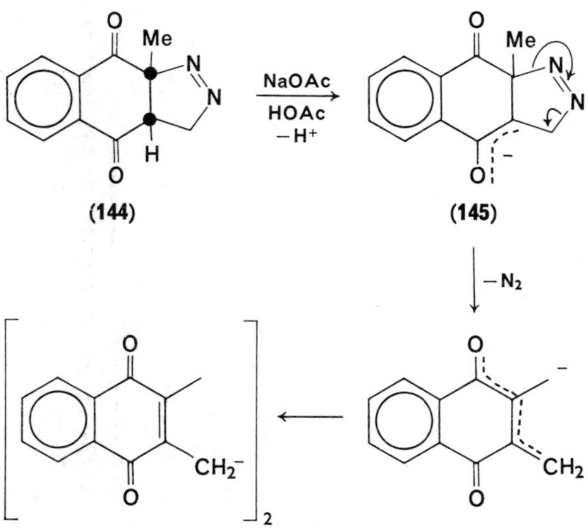

(144) (145) (147) (146)

for the products[91a]. Carbanions can also be implicated in the decomposition of pyrazolines, as already seen[58]; a further example is the fragmentation of the diazomethane adduct of 2-methylnaphthaquinone (**144**)[92].

D. *Methylenepyrazolines, Pyrazolenines and Vinylpyrazolines*

1. Methylenepyrazolines: synthesis and fragmentation

Substitution of an exocyclic methylene group into a pyrazoline ring is readily effected by making use of allenes in cycloadditions with diazoalkanes. The synthesis of methylenepyrazolines has been prompted from a desire to study thermal, and especially photochemical, decompositions where the theoretically interesting triplet state trimethylenemethane[93] is most likely involved; their decomposition is also a source of methylene cyclopropanes whose synthesis and structure[94], thermal rearrangement[95], and carbonium-ion chemistry[96] have attracted attention in recent years.

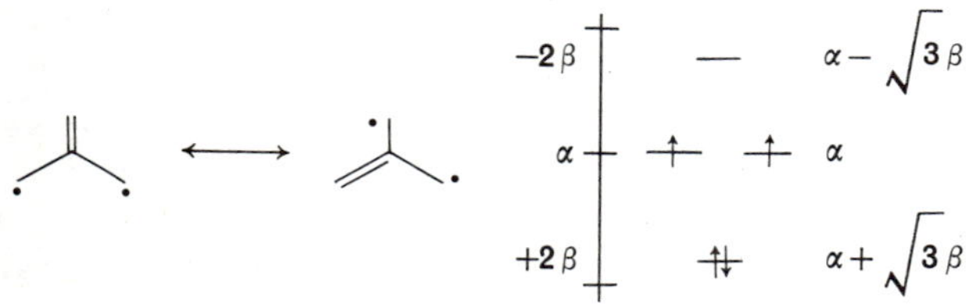

Trimethylenemethane: Hückel energy levels

The incorrect proposal, for instance, that the sex attractant of the American cockroach had structure **152** led to synthetic studies[97]; addition of 2-diazopropane (**149**) to the allenic ester (**150**) gives the pyrazoline (**151**), whose photolysis affords methylenecyclopropanes (**152** and **153**) in roughly 1:2 ratio. It is presumed that a trimethylenemethane (**154a** or **154b**) is involved here. That the initial pyrazoline is **151** rather than the alternatives **155** or **156** follows from spectroscopic evidence, and the formation of two products in photolysis, whereas structure **156** should give only one product if a diradical is involved.

The nature of the trimethylenemethane formed in these photolysis reactions has been probed experimentally with the aid of sensitizers and triplet quenchers[98]. The benzophenone-sensitized photolyses of methylenepyrazolines (**157** and **160**), for example, gives a higher proportion of the rearranged products **158** and **161** than statistically expected from the possible modes of ring closure.

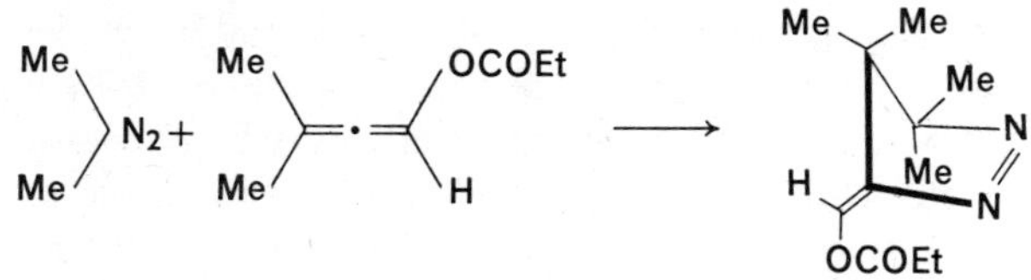

(149) **(150)** **(151)**

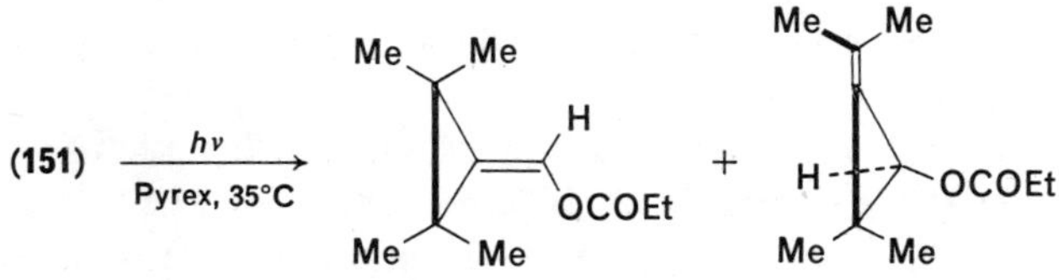

(152) (33%) **(153)** (66%)

(154a) **(154b)**

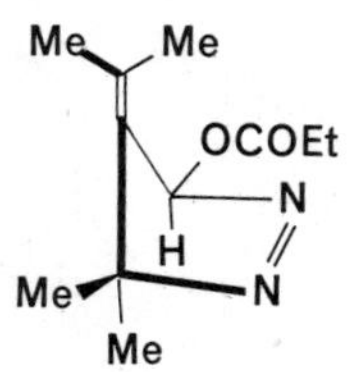

(155)

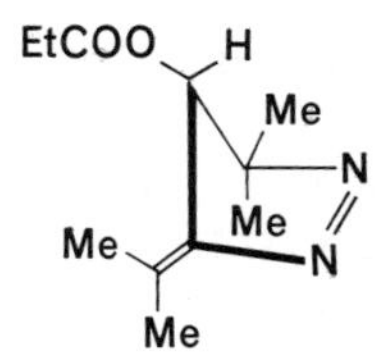

(156)

The intersystem crossing rate for $S^1 \rightarrow T^1$ benzophenone should be fast enough to ensure efficient triplet activation of the pyrazoline, which gives ground-state nitrogen and triplet trimethylenemethane as the intermediate, steric or electronic factors increasing the relative amounts of **158** and **161**.

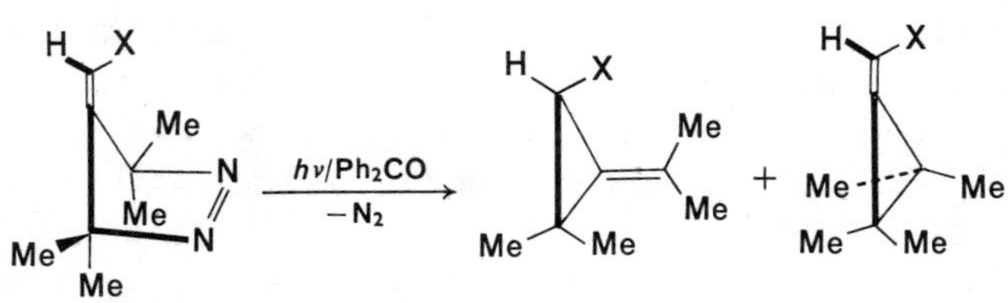

(**157**; X = CO_2Et) ⟶ (**158**) + (**159**)
(**160**; X = Cl) ⟶ (**161**) + (**162**)

The direct-photolysis product ratio is not affected by added triplet quenchers (e.g., piperylene) and in this case singlet-excited pyrazoline is the precursor of a proposed non-planar trimethylenemethane where closure to cyclopropane competes effectively with processes required to achieve a planar singlet or triplet state. Under a variety of conditions irradiation at low temperature fails, however, to produce an e.s.r.-active species and although, due to symmetry and for practical reasons, this does not *prove* that planar trimethylenemethane fails to form in the direct photolyses, it is at least evidence in favour of singlet processes.

Other similar product analysis studies include the photolysis of the alkylidenepyrazoline carboxylates (**163** and **168**), where the benzophenone-sensitized process produces approximately the *same* ratios of alkylidene cyclopropanes (**164** and **165**) (ca. 83:10) in each case, unlike the direct photolysis which gives a ratio for these products of 78·6:6·8 from **163**, and almost the inverse ratio 6·1:84·6 from **168**. Of the two major products the

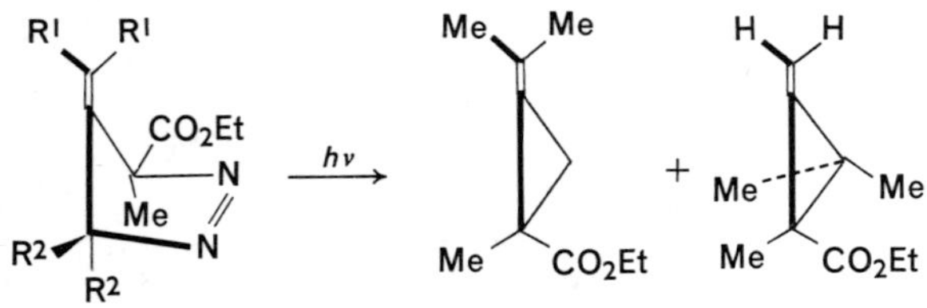

(**163**; $R^1 = Me$, $R^2 = H$) (**164**) (**165**)
(**168**; $R^1 = H$, $R^2 = Me$)

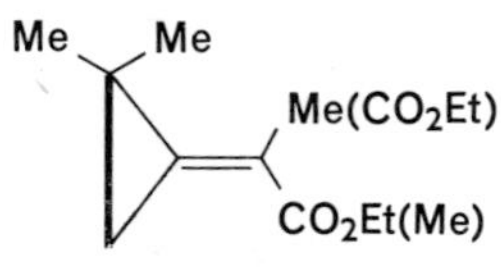

(**166**)
(**167**)

isopropylidene compound is thermodynamically the most stable, and clearly kinetically controlled ring closure of the trimethylenemethane biradical must in each case control the direct photolysis to the photostable products, whereas for the sensitized photolyses a common species seems implicated; again the lower energy triplet trimethylenemethane seems the most reasonable candidate[99].

There is *direct* evidence for the existence of trimethylenemethane in a ground-state triplet configuration (as predicted by elementary Hückel theory[93]). Adduction of diazomethane and carefully dried allene (with the latter as solvent) is complete in 24 h in a sealed tube; the regiospecific reaction gives only 4-methylene-1-pyrazoline. Photolysis of dilute fluorocarbon solutions of the azo compound at –185°C (λ 315–335 nm) gives a solution whose e.s.r. spectrum indicates a triplet species stable at –196°C up to a month. The e.s.r. signal disappears on warming the solution to –150°C, in itself a strong indication that a triplet species is present. Small discrepancies between the experimental e.s.r. spectrum and that calculated can be resolved by assuming that there is some negative spin density at the central carbon atom[100].

The thermolysis of 4-methylene-1-pyrazoline has E_a 9·6 kcal/mole lower than the parent pyrazoline, whilst $\Delta S^{\neq}$ is decreased[101], consistent with the lower probability of thermal formation of the planar ground-state triplet intermediate, if this *is* indeed the species involved. Theoretically, the singlet ground-state of methylenepyrazoline and methylenecyclopropane can be orbital symmetry correlated with a planar-singlet trimethylenemethane (by either a conrotatory or disrotatory motion) but the ground-state triplet intermediate directly correlates only with the triplet-excited ($n \rightarrow \pi^*$) state of methylenepyrazoline[102]. As a check on the symmetrical nature of the intermediate, 3,3-bisdeuterio-4-methylenepyrazoline (**169**), decomposed into methylenecyclopropane products, should show two-thirds of the deuterium in the ring and one-third in the exocyclic methylene group; but in fact a nonstatistical deuterium distribution appears, with excess in the vinyl positions. The explanation could be either an isotopic induced slower

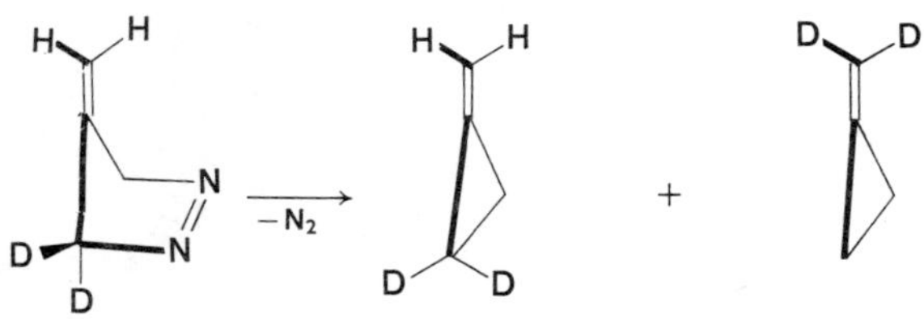

(169) **(170)** **(171)**

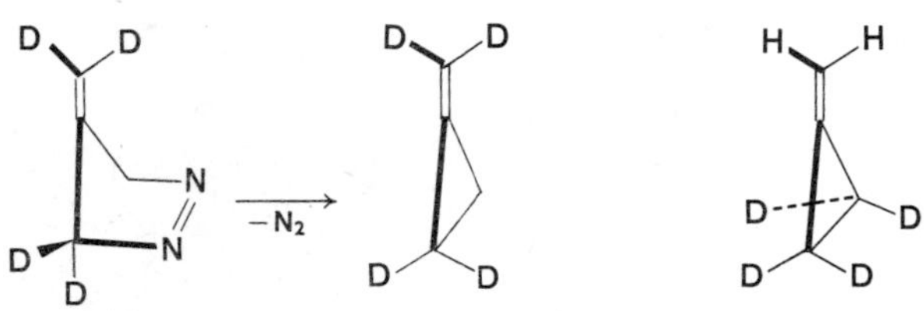

(172) **(173)** **(174)**

Observed: 73·8% 26·2% (±2%)

Calculated: 73·2% 26·8% (±0·8%)

rotation into the conformation necessary for ring closure in a planar electronically symmetrical species, or that such a planar symmetrical species is not produced. The former seems much the most likely explanation and this is brilliantly demonstrated by observing the product partitioning in the decomposition of the double-labelled pyrazoline (**172**); use of the secondary kinetic isotope ratio k_H/k_D (1·37 ± 0·05) obtained from the kinetic data for thermolysis of methylenepyrazoline-d_2 (**169**) correctly predicts the ratio of the products **173** and **174**. The required secondary kinetic isotope effect is considered rather large, perhaps because of the relative moments of inertia of the CD_2 and CH_2 groups (ratio: 2:1).

The nature of trimethylenemethane has also been probed by generating the species in solution in butadiene (e.g., from cyclobutanone) and observing the products[103]. More relevant in the present context is the generation of a trimethylenemethane (**176**) from 7-isopropylidenediazanorbornene (**175**) and the formation of four dimers of unspecified stereochemistry (**177–180**), derived from the thermal intermediate. Very significantly, recording the ^1H n.m.r. spectrum during the decomposition of the azo compound (**175**)

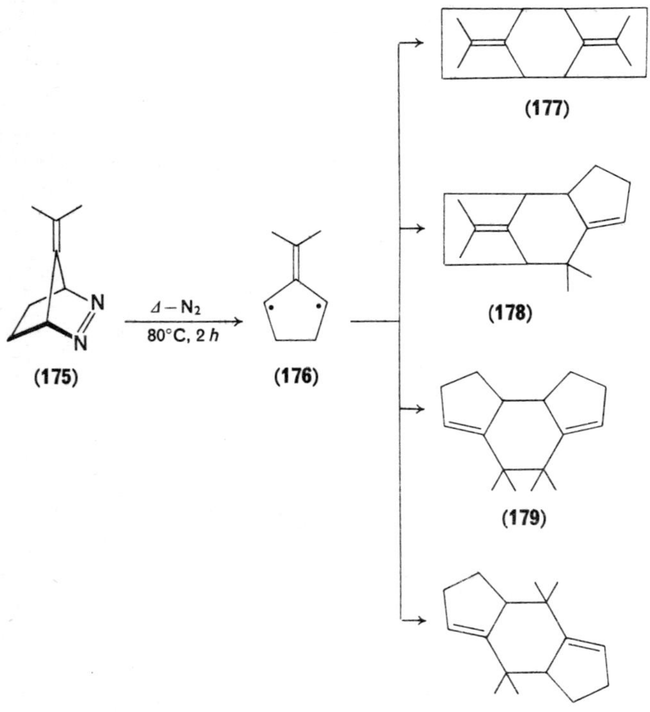

(175) (176) (177) (178) (179) (180)

shows strong nuclear polarization emission signals at positions corresponding to the nonpolarized resonance lines in the dimers[104]. Recent theory shows that only triplet–triplet dimerization can give products with nuclear polarization emission[105]. These experiments therefore establish unambiguously that the triplet ground-state of trimethylenemethane is readily accessible thermally, although clearly a singlet–triplet transition is required[102].

The vapour-phase photolysis of 4-methylenepyrazoline at 25°C gives only methylenecyclopropane and nitrogen. Deuterium-labelling studies again indicate a symmetrical intermediate, and the effect of added nitrogen and piperylene show that it is formed in the singlet state in the direct photolysis. In the (more efficient) benzene-sensitized decomposition both singlet and triplet trimethylenemethane appear to be implicated, methylenecyclopropane and 1,4-dimethylenecyclohexane appearing as products, but added oxygen suppresses formation of the dimeric dimethylenecyclohexane, which is therefore believed to arise from the triplet species. There seems a good chance that singlet states of benzene are the sensitizing species here[106]; other examples of benzene singlet-sensitization are, however, known[107].

2. Pyrazolenines

The spontaneous, light-accelerated unimolecular decomposition of diazoalkenes, e.g. 3-diazopropene (**181**), has been known for thirty years[108, 109] and recently it has been shown that the activating wavelength (310–380 nm) is outside the region of absorption of the diazo compound. Cyclization of the diazoalkene gives a pyrazolenine (**183**), and the latter chromophoric species, which is present in only small concentration, is the source of the final product, pyrazole (**184**)[110].

This proposal has received strong support from kinetic data for a series of *trans*-3-diazo-3-alkyl-1-phenylpropenes with a *p*-Me or *p*-NO_2 substi-

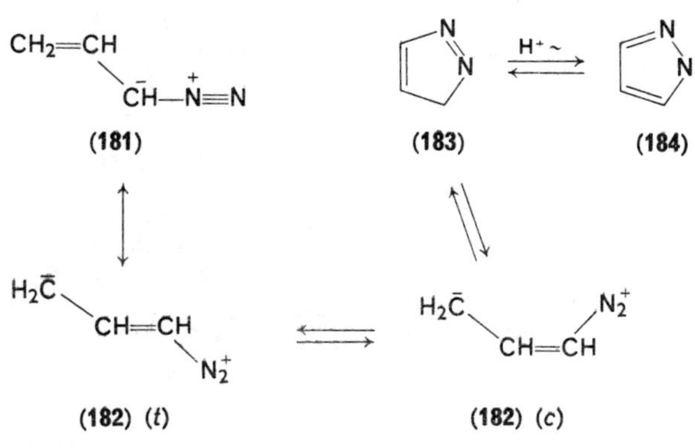

tuent; the Hammett ρ value is only $-0{\cdot}4$ for the unimolecular dark reaction leading to the intermediate pyrazolenine. The overall cyclization process is actually six-fold faster than, for example, 3-diazopropene, which could be due to stabilization of the partial charge in a nonsymmetrical transition state for an otherwise synchronous dipolar cycloaddition. Substituents on the α-carbon of the diazo group have a pronounced effect, the relative cyclization rate being accelerated 13-fold for α-methylation, possibly due to destabilization of the diazo compound by the presence of the substituent on what could be regarded as a carbanion carbon[111].

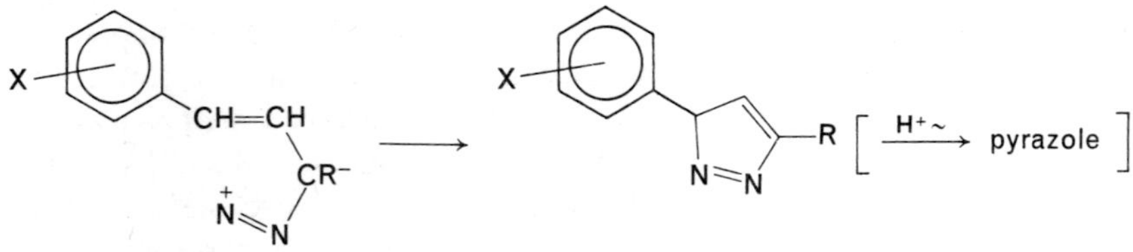

(185) (186)

X = *p*-Me or *m*-NO_2
R = e.g. Me

Pyrazolenines have also been implicated in the base-catalysed thermolyses of α,β-unsaturated aldehyde and ketone tosylhydrazones, and alkylated analogues, e.g. **189**, have been isolated. These azo compounds are thermally stable under the reaction conditions, and cyclopropenes concomitantly formed arise from azoalkenes which give carbenes as the penultimate intermediates in a separate pathway[112].

Pyrazolenines do, however, provide good sources of cyclopropenes by photofragmentation[113]. Interestingly, photolysis of the bicyclic pyrazolenine

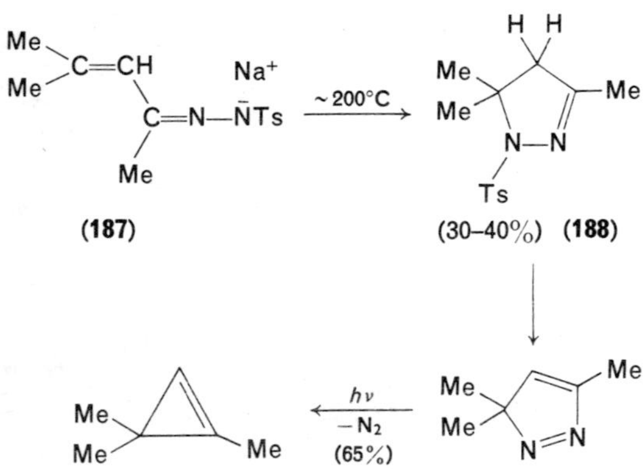

(187) (188) (189) (190)

(**191**) in pentane at 13°C gives the expected cyclopropene (**192**) but at low temperatures in methylene chloride and other polar solvents reaction is diverted, giving the tricyclic compound (**193**) which, however, reverts to **191** on warming to 0°C[114]. This valence tautomerism is perfectly general and also occurs with alkylated pyrazolenines[115].

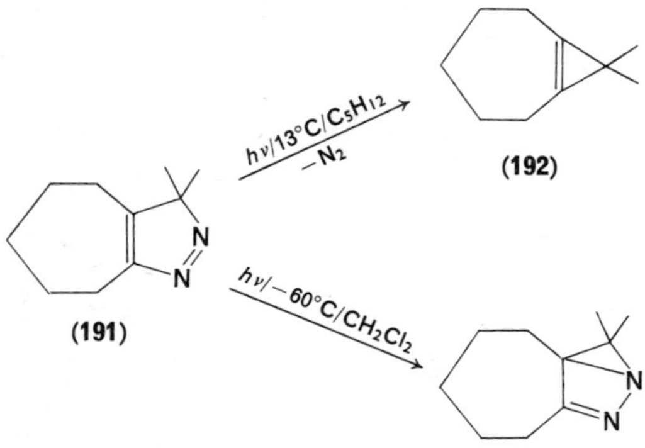

(**191**) (**192**) (**193**)

The observation of a transient coloration during photolytic deazetation of pyrazolenines suggested that an intermediate diazoalkene might be the cyclopropene precursor, loss of nitrogen leading to a vinylcarbene as the penultimate intermediate. By varying the irradiation wavelength, using filters and working in neutral media it is possible to prepare and characterize the intermediate diazoalkenes [e.g., λ_{max} 279 and 522 nm, ε_{max} 7000 and 23, ν_{max} 2035 cm^{-1} for 4-methyl-2-diazopent-3-ene derived from pyrazolenine (**189**)]; these, however, are rapidly decomposed in acids. The absence of any effect of senistizers on the reaction course suggests that the same excited

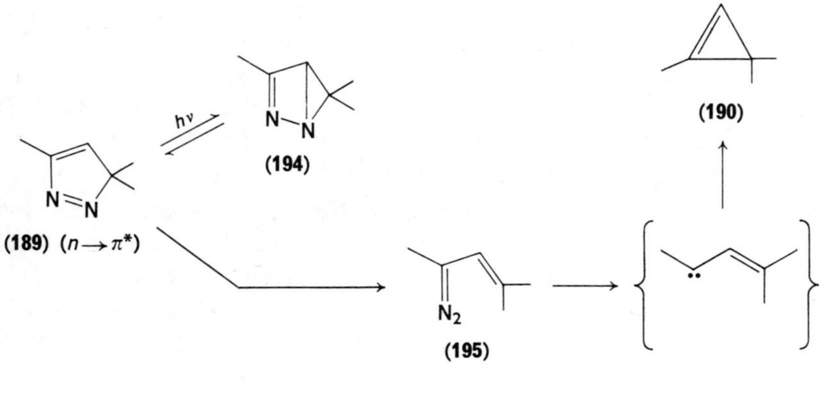

(**189**) ($n \rightarrow \pi^*$) (**194**) (**190**) (**195**)

species is involved for conversion to the bicyclic tautomer and/or the diazoalkene[115].

The involvement of diazoalkenes in photolyses of this kind has also been adduced by trapping experiments with carboxylic acids; e.g. whilst photolysis of the diazopropene adduct of methylbutynol (**196**) in boiling ether gives a deep-red solution which quickly fades, yielding cyclopropene (**197**)

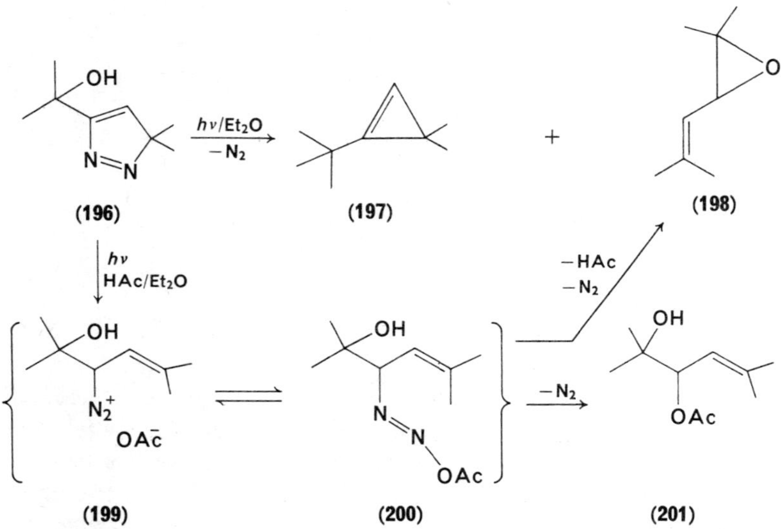

and epoxyolefin (**198**), similar experiments in the presence of 2% acetic acid give only ester **201** and a small amount of epoxide (**198**), with no colour developing[116]. By way of contrast similar experiments with simple esters of pyrazoleninol (**196**) give the dienyl and allenic esters **206** and **207** (in ratio 2:3), the course of reaction being unaffected by acids, but photolysis of mixtures of different esters of differently substituted pyrazolenines gives cross-over products, indicating *intermolecular* processes. The required rearrangement via diazoalkene/diazotate/diazonium carboxylate (**203–205**) is visualized as too fast to be affected by acids, and in these reactions no direct evidence for diazoalkenes is apparent[117].

Aromatic fused-ring pyrazolenines, e.g. **208**, have been used as sources of benzocyclopropenes[118]; these are also isolated from the photolysis products of spiro-fused pyrazolenines, e.g. analogues of compound **211**. Rearrangement gives benzo-fused pyrazolenines (**213**) which lose nitrogen. The precise mechanism here is uncertain, either a zwitterionic species

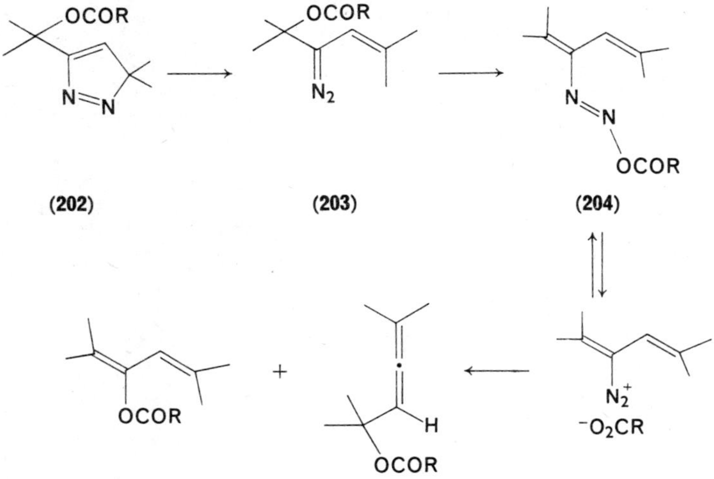

(202) (203) (204)

(206) (207) (205)

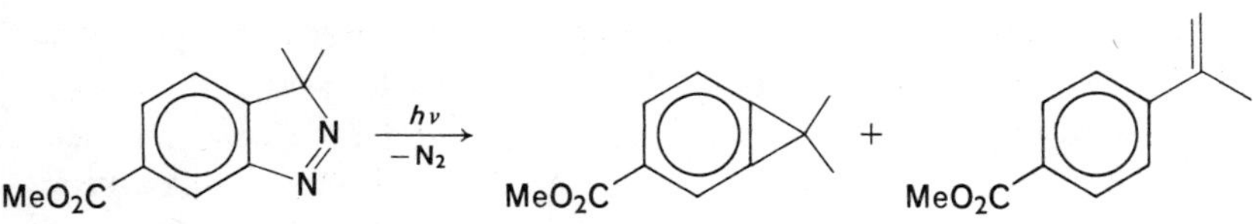

(208) (209) (210)

(t/2 15 min/80°C)

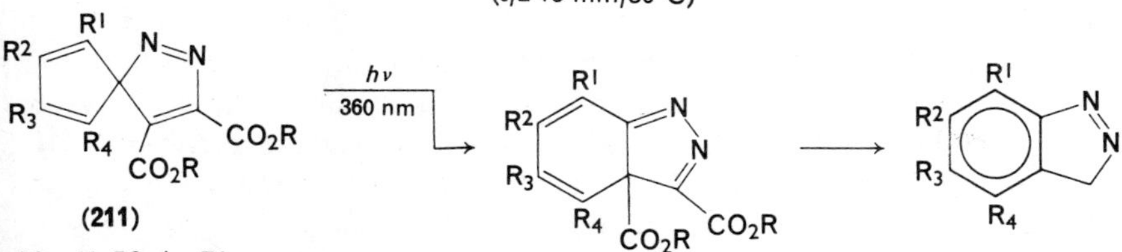

(211) (212) (213)

R¹ = H, R²⁻⁴ = Ph
R¹ = Ph, R²,³ = H, R⁴ = Ph
R¹⁻⁴ = Ph

formed by loss of nitrogen from **213** ring closes to the cyclopropene, or rearrangement to a diazoalkene as the precursor of a carbene could rationalize the products[119]. In certain analogous reactions of spiro-fused pyrazolenines, a spiro-fused cyclopropene results, e.g. cyclopropene (**215**) from pyrazolenine (**214**). The base-promoted conversion of *N*-*p*-tosyl-2-pyrazolenines into pyrazoles is also believed to involve transient pyrazolenines[120].

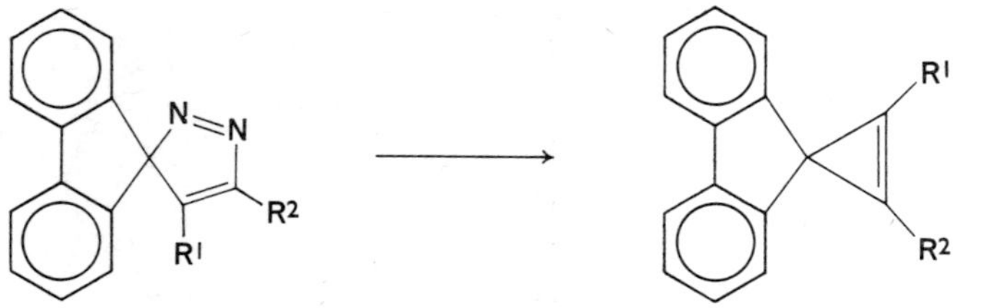

(**214**; $R^{1,2} = CO_2Me$ or $R^1 = H$, $R^2 = CO_2Me$)

(**215**; $R^{1,2} = CO_2Me$ or $R^1 = H$, $R^2 = CO_2Me$)

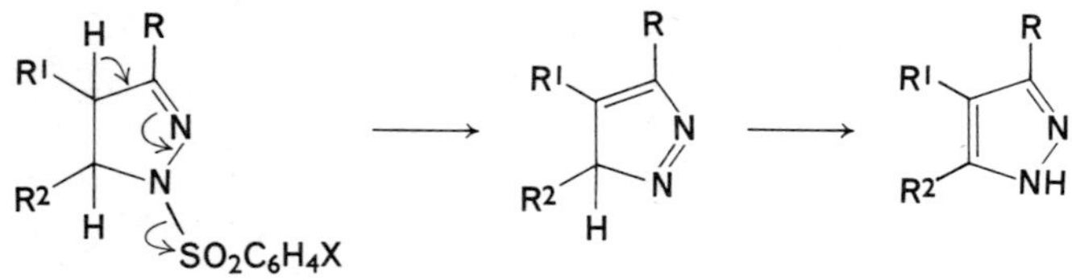

(216) **(217)** **(218)**

3. Vinylpyrazolines and miscellaneous pyrazoline rearrangements

Although the ready cyclization of diazoalkenes to pyrazolenines and thence to pyrazoles has long been known[108–110] very little work has been reported on the properties of adducts of the relatively novel diazoalkenes with dipolarophiles. However, vinyldiazomethane forms 5-vinyl-2-pyrazolines with methyl maleate and methyl fumarate via initial 1-pyrazolines which readily tautomerize. The initial cycloadditions may be nonstereospecific, although this does not appear to have been rigorously established. For example, methyl maleate with vinyldiazomethane gives a single product, **220**, which is thought to arise from 1-pyrazoline (**219**); methyl fumarate also gives diester (**220**) together with the 4,5-*cis*-stereoisomer. Nonstereospecific cycloaddition would imply polar intermediates, and it is suggested these may be involved[121]. (Vinyldiazomethane also undergoes cycloaddition with acetylenic dipolarophiles, yielding vinylpyrazoles[122].)

The tautomerism of 2-pyrazolines has been discussed[123] and more recently a kinetic study has shown that, under conditions where the 1-pyrazoline extrudes nitrogen, the tautomeric rearrangement is rate-determining, e.g. in the partial decomposition of the two isomeric pyrazolines **221** and **222** no evidence for their interconversion is found in either[124]. The decomposition of substituted 2-pyrazolines similar to **221**, e.g. **224**, usually gives stereoisomeric mixtures of cyclopropanes but, by contrast, decomposition of the analogous 3,4-dicarbomethoxy-5-phenyl compounds

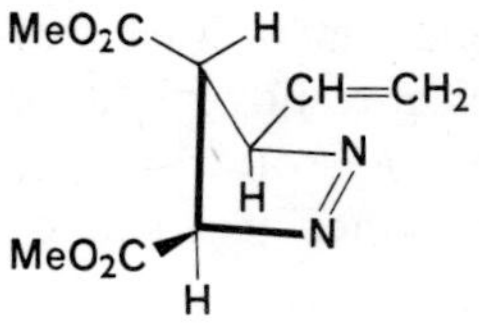

(219)

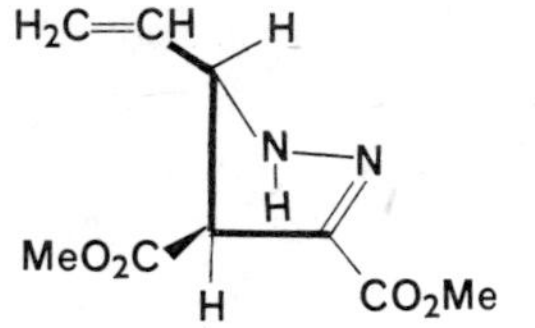

(220)

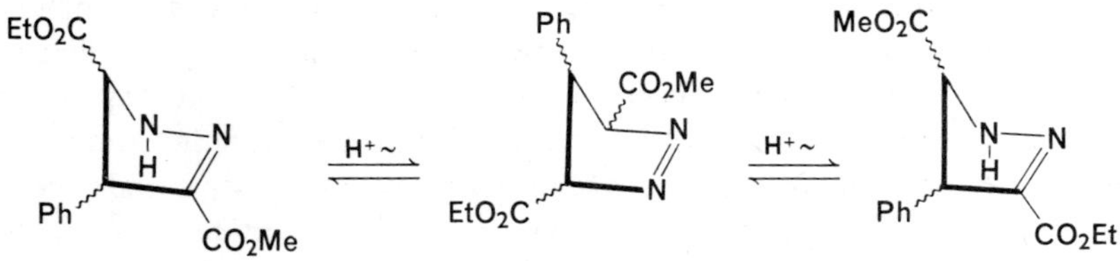

(221) (223) (222)

227 and **228** gives only the *trans*-dicarbomethoxy isomer in each case, with just detectable traces of the *cis*-isomer[125]. Prior to this work, it had been suggested that the stereochemical course in 2-pyrazoline decompositions was generally determined by the relative thermodynamic stability of the intermediate 1-pyrazoline. Clearly, however, if the rates of decomposition

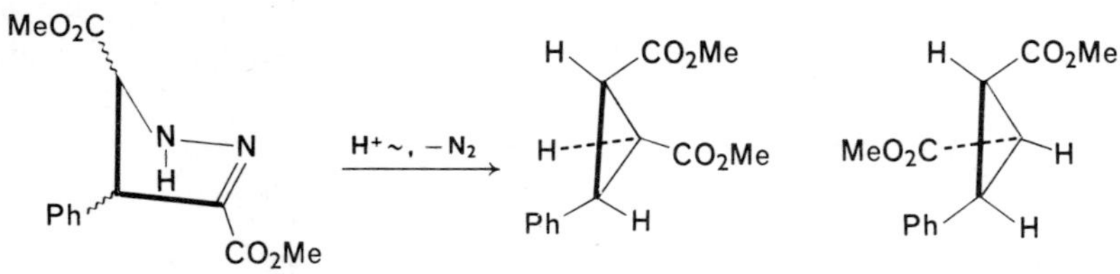

(224) (225) (69%) (226) (31%)

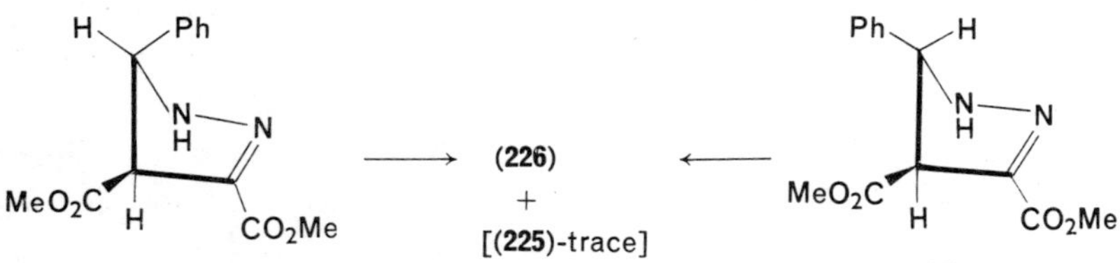

(227) (228)

of the intermediate stereoisomeric 1-pyrazolines are comparable, and in all cases stereospecific and faster than the tautomerism of the 2-isomers, the ultimate product composition is kinetically determined by the rates of protonation to the alternative 1-pyrazoline stereoisomers in the tautomeric rearrangement. If the 1-pyrazoline decomposition is not stereospecific, the degree of stereoselectivity observed will then depend on at least two independent factors, and a wide range of stereochemical results might be expected. In this connexion it is worth recalling that, in prototropic rearrangements, the *least* thermodynamically stable tautomer of a pair of isomers is frequently favoured kinetically[126].

The synthesis of cyclopropanes by the decomposition of pyrazolines potentially introduces chiral centres and a study of the absolute configurational consequences in synthesis and fragmentation of pyrazolines is obviously relevant to the mechanism. The result of partial asymmetric synthesis depends on the Prelog–Cram model of the transition state[126a] to predict the absolute configuration of the product. For example, the addition of diphenyldiazomethane to (–)-menthyl acrylate and to (–)-menthylmethacrylate gives (–)-2,2-diphenylcyclopropane carboxylic acid and the (+)-1-methyl-2,2-diphenyl analogue of the same relative configuration. The Prelog–Cram model assumes that the acrylic esters will be in the *transoidal* configuration with the menthyl group flanking the carbonyl group, and predicts that the approach of the diazoalkane from below the plane will be favoured and lead to an excess of configuration **229a** over isomer **229b**. However, if the menthyl group flanks the olefinic bond, or the conformation of the ester is *cisoidal*, the major product will have the opposite configuration, **229b**. In fact, the configuration **229b** is actually formed in excess and, in order to rationalize this fact, assuming that the Prelog–Cram model is valid,

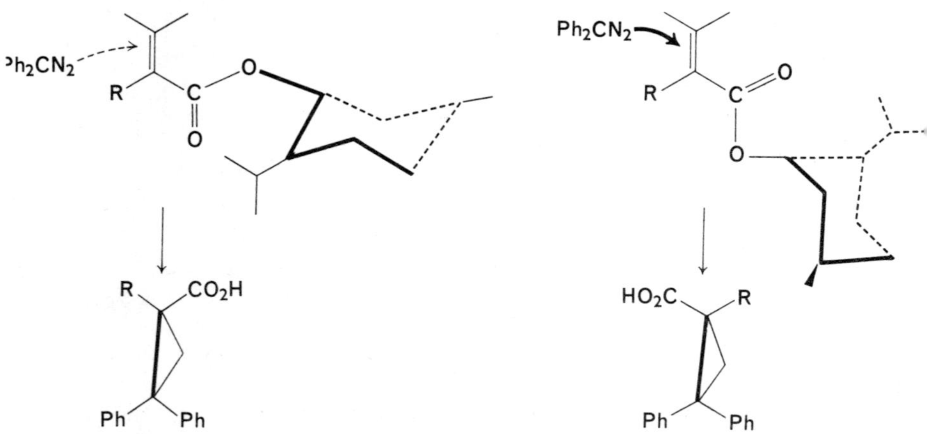

(229a) (229b)

diazo group transfer and rotation around a σ-bond at the chiral centre is suggested. In the initial phase of the cycloaddition, diazoalkane will approach the acrylic ester to give the transition state **A** which can either form pyrazoline **B**, or transform into dipolar intermediates **C** and **D** whose relative proportions will determine the outcome of asymmetric synthesis, as indicated in the reaction scheme[72]. The result is an important one which

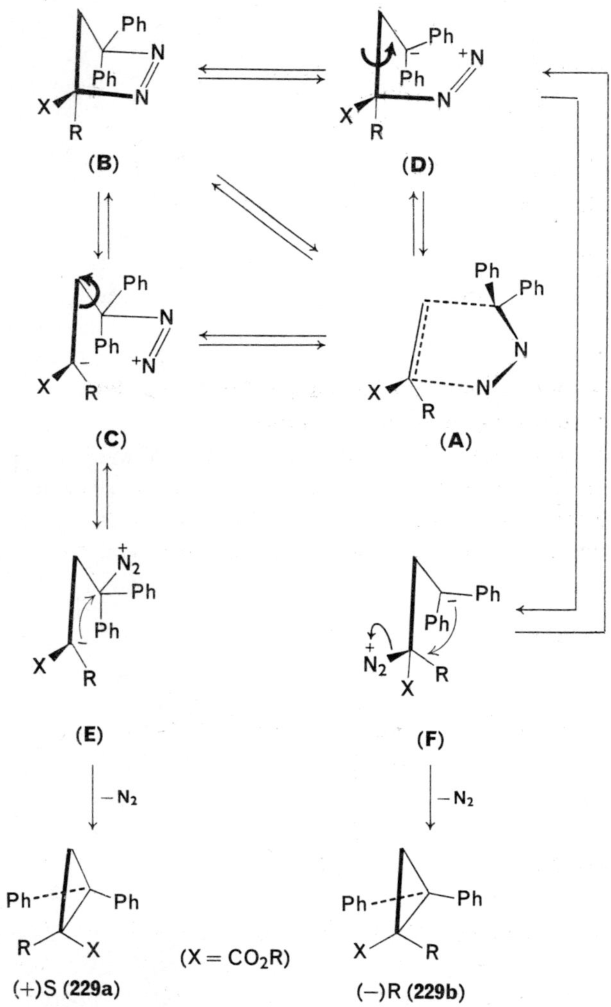

would seem to imply, if a trimethylene π diradical-like intermediate is preferred to a dipolar species, that one of the conrotatory modes of ring closure is faster than the other in the stereochemically preferred $[0,0]\pi$-

species. This can be understood if there is a residual interaction between the trimethylene orbitals and the departing nitrogen which stereochemically ensures an excess of one conrotatory mode over the other. It has been suggested that such residual interactions play a role in other, fused-ring pyrazoline deazetations (see below).

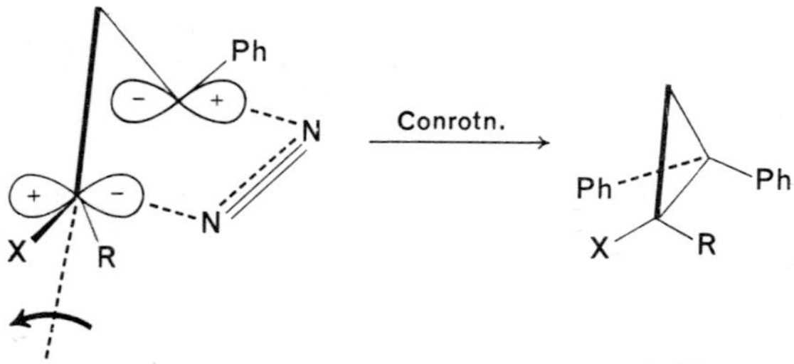

(230) (−)R

E. *Medium- and Large-Ring Azo and Bis-Azo Compounds*

I. Medium- and large-ring compounds

The addition of diazo compounds to a variety of olefins constitutes a versatile synthesis of pyrazolines, and Diels–Alder addition of azodicarboxylic esters (**232**) to acyclic dienes affords access to the homologous tetrahydropyridazines, e.g. **234** (although not always in good yields)[127].

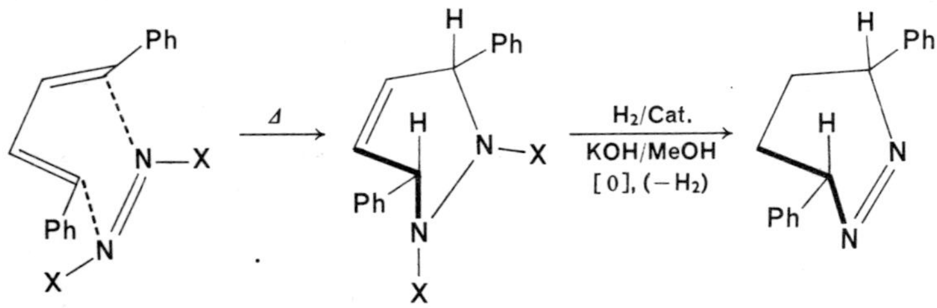

(**232**; X = CO_2R) (**233**) (**234**)

Seven-membered ring homologues of tetrahydropyridazine (**234**), e.g. **235**, can be made from 1,5-diketones by condensation with hydrazine (note the double reduction step)[128]. The corresponding eight-membered ring compounds are accessible from 1,6-diol ditosylates which condense with hydrazodiformic esters in the presence of sodium hydride to give the required intermediate, e.g. diester (**240**) from diol derivative (**239**); *cis*- and *trans*-isomers of 3,8-dimethyl-*cis*-1,2-diazacyclo-1-octene (**241**; R = Me)

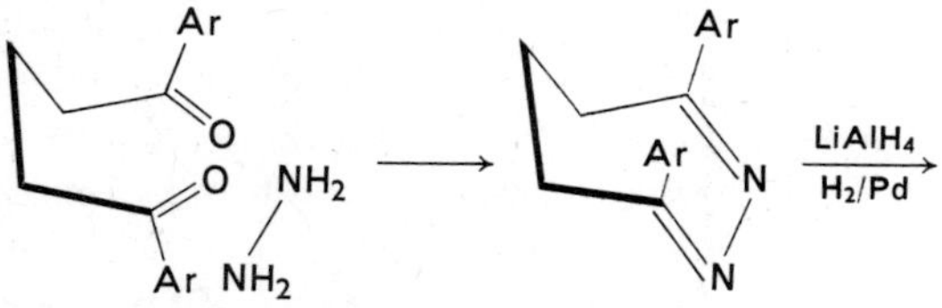

(235) (236)

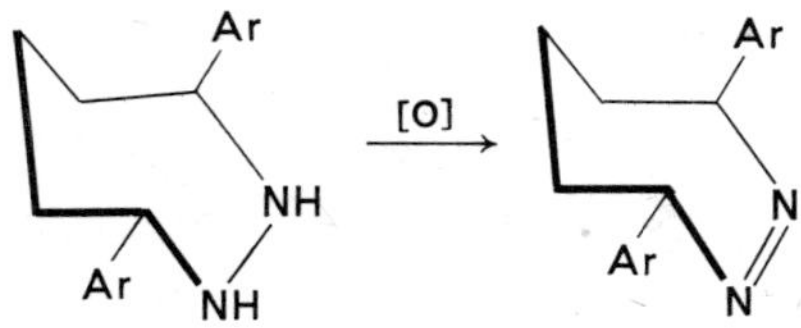

(237) (238)

and the 3,8-diphenyl analogue have been made in this way, along with the parent compound *cis*-1,2-diazacyclooctene[129, 130].

Thermolysis of some of these rather labile azo compounds is complicated by concomitant tautomerism into cyclic hydrazones, especially in the solid state. However, thermal deazetation of tetrahydropyridazine (**234**) is complete, and 100-fold faster in solution than the acyclic analogue, giving

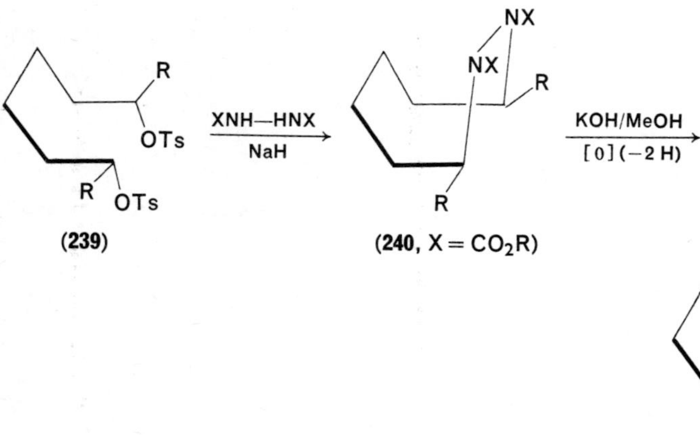

(239) (240, X = CO_2R)

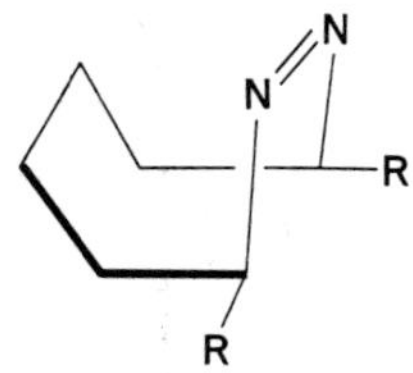

(241)

styrene; similarly, the homologues (**238**) give *cis*- and *trans*-1,2-diphenylcyclopentane (65%) and 1,5-diphenylpentenes (23–27%) with an E_a 29·7 kcal/mole whilst the somewhat slower decomposition of diphenyldiazocyclooctenes (**241**) (E_a 36·7 kcal/mole) yields stereoisomeric 1,2-diphenylcyclohexanes (43%) and 1,6-diphenylhexenes (50%). These results can all be interpreted in terms of a diradical intermediate which either cyclizes or disproportionates into olefin. The rather slower decomposition of the eight-membered ring compounds is understandable, since in stable conformations the benzyl groups are not co-planar with the azo function and cannot incipiently stabilize the transition state as, for example, occurs in the decomposition of azo-1-phenyl-*n*-propane (E_a 32·3 kcal/mole). Compared to thermolysis, the photolysis of *trans*-3,8-diphenyl-*cis*-1,2-diazacyclooctene shows an increased tendency to form the disproportionation products (steroisomeric diphenylcyclohexanes, 15% *cis* and 20% *trans*), and in the benzophenone-sensitized photolysis 95% of the product is *trans*-diphenylhexene. This result is rationalized on the grounds that in the twisted boat conformation for the diazacyclooctene the hydrogen atom which must transannularly migrate to the appropriate radical centre in order to form olefin is particularly well-orientated[130, 131]. In the photosensitized reaction

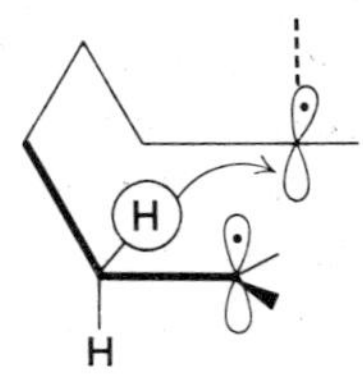

the triplet diradicals are longer-lived, additionally favouring the disproportionation. A similar result is seen with *cis*-1,2-diazacyclooctene, the disproportionation product, 1-hexene, rising to 84% in the sensitized decomposition*.

The stereochemical course and propensity for olefin formation in fragmentation of medium-ring azo compounds has been probed[132] with

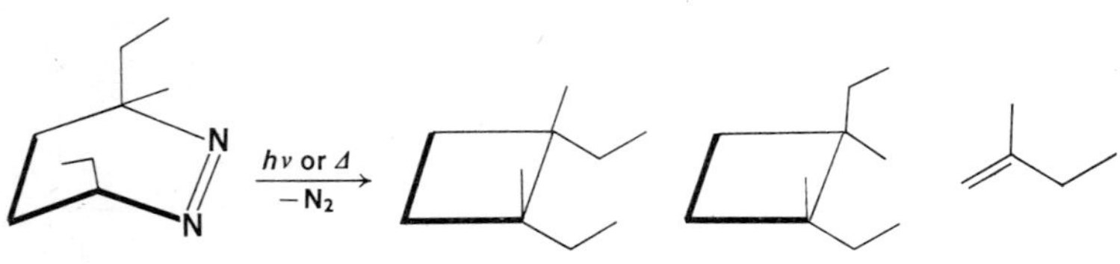

(242) **(244)** **(245)** **(243)**

* Recently, the photostereomutation of the *cis*-azo linkage in medium rings[131a] has been described; the resulting *trans*-azo compounds are characterized by their smaller dipole moments, lower intensity ultraviolet absorption, and reduced tendency to form complexes with chemical shift reagents, e.g. Eu(fod)$_3$, especially under competitive conditions, the *cis*-isomers complexing strongly.

the *meso* and *dl* forms of 3,6-diethyl-3,6-dimethyltetrahydropyridazines (**242**). The products of thermolysis, direct photolysis and thioxanthene-sensitized photolysis of the reactants are summarized below. The results indicate a degree of retention of configuration in the cyclobutane products

Isomer		% **243**	% **244**	% **245**
meso	145–148°C	49	43	2·5
dl	145–148°C	51	3·5	42
meso	*hv*	61	35	3·5
dl	*hv*	60	4·0	33
meso	sensit. *hv*	77	11·5	8
dl	sensit. *hv*	75	8	12

of > 98 % for the thermal process for both isomers, 95 and 97 % for the direct photolysis of the *meso* and *dl* forms but a considerable reduction in stereoselectivity for the sensitized reactions, (61 and 65 %). These differences reflect the longer lifetimes of the triplet diradicals formed in the sensitized process. Concerted radical-pair formation is the only efficient mode of energy release from the triplet excited state when geometrical/conformational changes are stereochemically disfavoured[132]. In the singlet-state reactions closure to cyclic products with predominant retention of configuration may very well reflect the differences in energy and stereochemical properties between the required 1,4-tetramethylene species and the planar [0,0]π-trimethylene characteristic of pyrazoline decompositions.

Incorporation of a π-bond into analogues of diazacyclohexene (**242**) has a dramatic effect both on the rate of deazetation and on the products. Oxidation of the unsaturated hydrazo compounds **246** and **249** results in rapid evolution of nitrogen from the intermediate azo compounds even at low temperatures (–50°C), giving stereospecifically the *trans–trans* and *cis–trans* hexadienes **248** and **251** respectively[133]. It seems highly likely that these reactions are examples of orbital, symmetry-allowed, concerted ($\pi_s^4 + \pi_s^2$) cycloreversions where the highest occupied product orbital interacts with the lowest vacant antisymmetric orbital of the nitrogen molecule in the transition state, ensuring a low-energy reaction profile. This is equivalent to π-bond participation in the decomposition because the olefinic LVMO has the correct symmetry to interact with the antisymmetric σ-bond component to nitrogen[134]. Other examples of the same effect are found in certain bicyclic and tricyclic azo compounds, especially where a cyclopropane ring replaces the double bond; examples are discussed below.

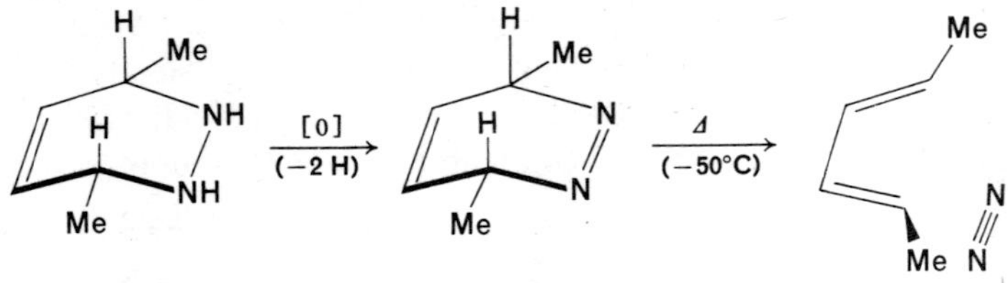

(246) (247) (248)

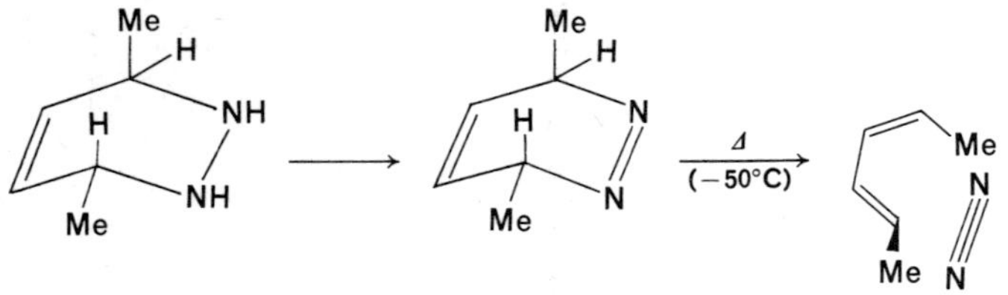

(249) (250) (251)

2. Cyclic bis-azo compounds

It has been reported that diazomethane forms a dimer in sunlight, 1,2:4,5-tetrazene[46] but cyclic bis-azo compounds are otherwise comparatively rare. They can be made from cyclic azines derived by condensing suitable diketones with hydrazine where the product consists of the mono- and bis-azines, e.g. **253** and **254**. Reduction of these azines under very specific conditions followed by oxidation (HgO) affords the

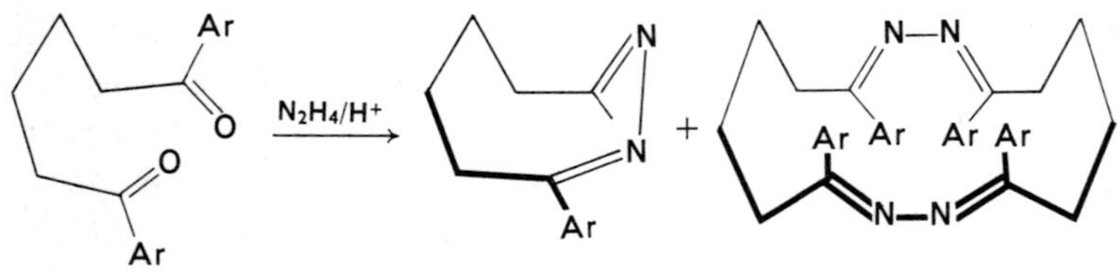

(**252**; Ar = *p*-MeOC$_6$H$_4$) (**253**; λ_{max} 286 nm) (**254**; λ_{max} 312 nm)

corresponding azo compounds. A number of analogues of compound **255** have been made in this way[135–137], e.g. **256–258**. Simpler members of this series of compounds are catalytically decomposed by acids at 80°C, e.g. **255**, but the pure substance loses nitrogen[136] only at elevated temperatures (> 150°C) and gives mainly olefin, 1,6-di-*p*-anisylhexene, presumed formed by disproportionation of the intermediate 1,6-diradical. However, compounds **257** and **258** decompose at 120°C in xylene mainly to give products ascribable to the dimerization of the intermediate α,ω-diradicals, and bis-azo compound **257** gives tetraphenylcycloeicosane (**259**), as well as up to 11% olefinic product. Attempts to demonstrate the formation of radicals by decomposing these azo compounds in polymerizable solvents are inconclusive although some adducts are obtained (with stilbene or with maleic anhydride as additives). The results may indicate a cage effect, operating to ensure ring closure rather than polymerization chain initiation.

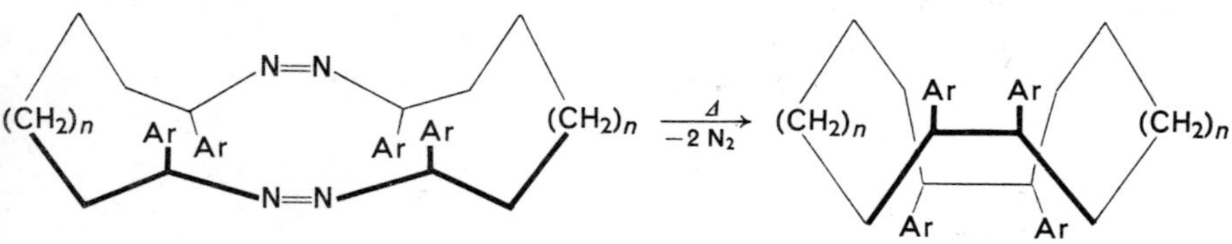

(**255**; $n = 1$) (**256**; $n = 3$)
(**257**; $n = 5$) (**258**; $n = 7$)
Ar = Ph (**255**; Ar = p-MeOC_6H_4)
λ_{max} 360 nm

(**259**; $n = 5$; Ar = Ph) (72%)
(**260**; $n = 7$; Ar = Ph) (66%)

Kinetic data for the thermolysis of azo compounds **256–258** gives an average E_a of ~30 kcal/mole, a similar value to the open-chain azo compounds, but the rate plots consistently fail to pass through the origin and this could be evidence of stepwise loss of each nitrogen molecule, ring closure intervening between each separate activation[135,137]. The question has been investigated in greater detail as part of the more general problem of the transmission of 'activation energy' since these bis-azo compounds provide good models for probing how far apart reaction centres may be separated before the efficiency for transmission of excess vibrational energy released at one point to another reaction centre falls off. (Collisional deactivation is inefficient enough to allow part of the energy released to transfer to another point in the molecule; for compound **257** transmission through the methylene chain is unlikely.) Kinetic data for compound **257** and its benzylic tetradeuterio analogue show that concerted loss of both nitrogen molecules is (as expected) ruled out (k_H/k_D is 1·20 instead of 1·57 calculated) and cross-over experiments using the azo compound and the labelled isotopic isomer **257**-d_4 show that cyclization of the initially formed biradical resulting from loss of one of the azo groups as nitrogen is $\nless$100-fold more rapid than loss of the second nitrogen molecule[138].

III. POLYCYCLIC AZO COMPOUNDS

A. Bicyclic, Bridged Bicyclic and Related Polycyclic Azo Compounds

1. Bicyclic pyrazolines and other non-bridged systems

The steric constraint inherent in fusion of a small ring onto the 1-pyrazoline system is expected to modify the ability of the molecule to decompose through the intermediacy of a [0,0]trimethylene formed by a concerted deazetation, and several recent investigations have explored fragmentation mechanisms in bicyclic structures. For example, thermolysis (by glc injection), and photolysis of the stereoisomeric diazabicyclohexenes **261** and **264** shows a marked stereoselectivity, the former *exo*-methyl compound giving 98% *trans*-pentadiene (**262**) and 2% of the stereoisomer (**263**), whilst for compound **264** the proportions of the two products are

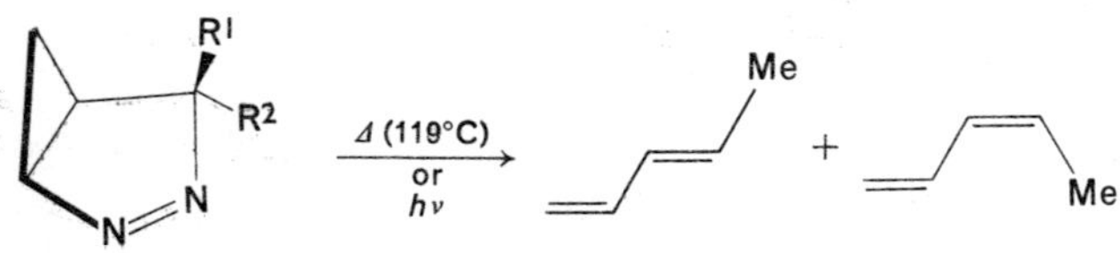

(**261**; R¹ = H, R² = Me) (**262**) (**263**)

(**264**; R¹ = Me, R² = H)

exactly reversed. The introduction of an *exo*-methyl group at the cyclopropane methylene increases the complexity of the product, the proportions of the various components depending on the orientation of the methyl group in the azo ring. The results can be accommodated in part by the involvement of carbenes **271** and **272** formed by nitrogen loss from intermediate diazoalkenes resulting from dipolar cycloreversion.

The products of base-catalysed decomposition of the tosylhydrazone of 4-methylpent-3-enal show a similar distribution but not, however, the very strong temperature dependence of the proportion of propenylcyclopropanes (*trans*- and *cis*-**266**) stereospecifically formed from azo compounds **264a** and **264b**. It is believed that higher temperatures favour a direct route to the carbene, bypassing the azoalkene, and that the relatively large amounts of *cis–trans* and *cis–cis* hexadiene (**268** and **269**) indicate competitive pathways involving a 1,3-biradical species formed by deazetation of the azo compounds[139].

A similar complexity and distribution is found in the products of (*in situ* glc) thermolysis, and photolysis of compounds in the homologous diazabicyclo[3.2.0]heptene series made by addition of diazoalkanes to appropriate cyclobutenes. The parent compound **273** affords bicyclo[2.1.0]pentane (**274**)

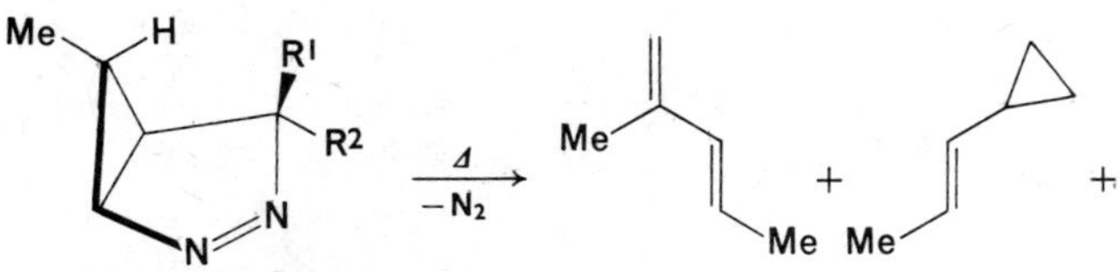

(264a; R1 = H, R1 = Me)
(264b; R1 = Me, R2 = H)

(265) (266)

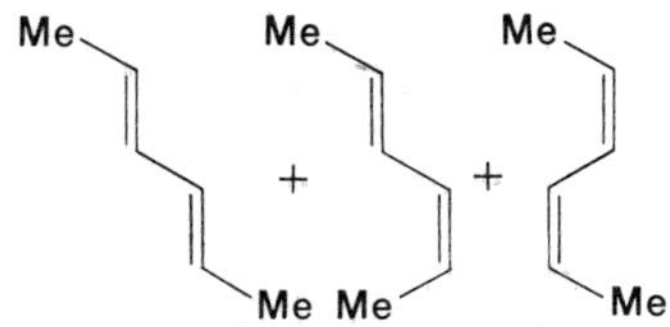

(267) (268) (269)

(264a)

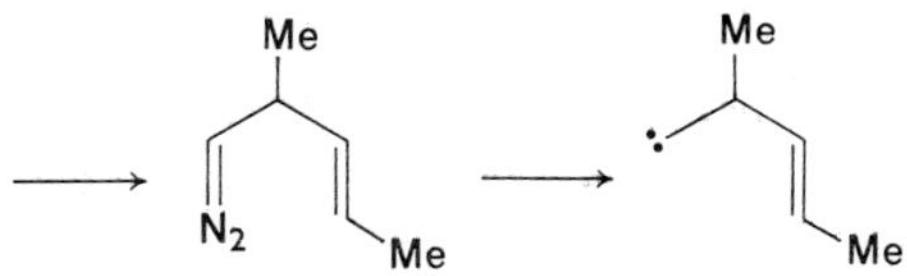

(270a) (271)

(264b)

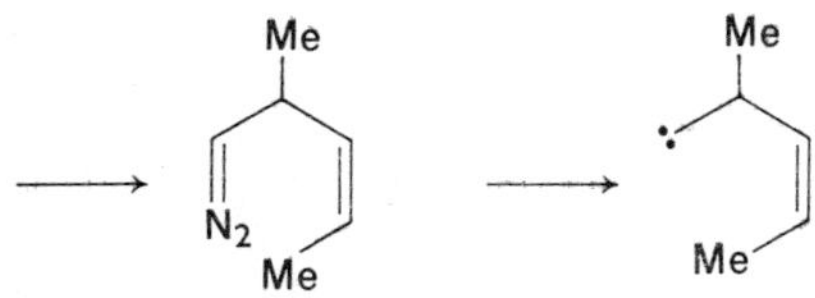

(270b) (272)

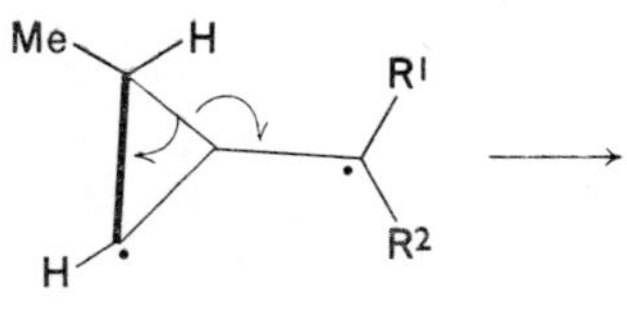

(267), (268), (269)

and five olefinic products, and again competitive pathways involving carbenes and 1,3-diradicals could account for the products, especially when the bicyclopentane yield rises to 90% in sensitized photolysis; since the interesting possibility arises that 1,4-pentadiene (**275**) derives from an equilibrium involving the obviously important diradicals **280** and **281**, ring methylation is used as a labelling technique as a test for this possibility.

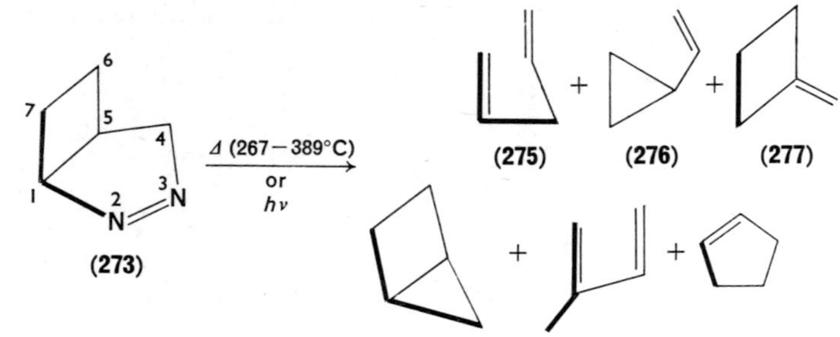

(273) (275) (276) (277) (274) (278) (279)

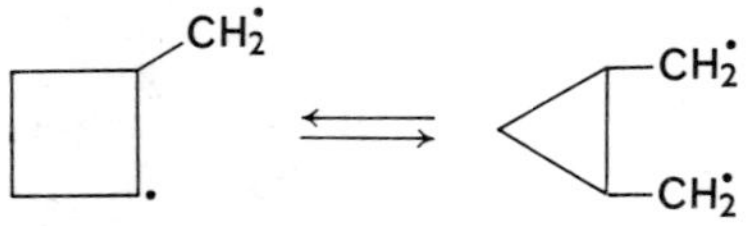

(280) (281)

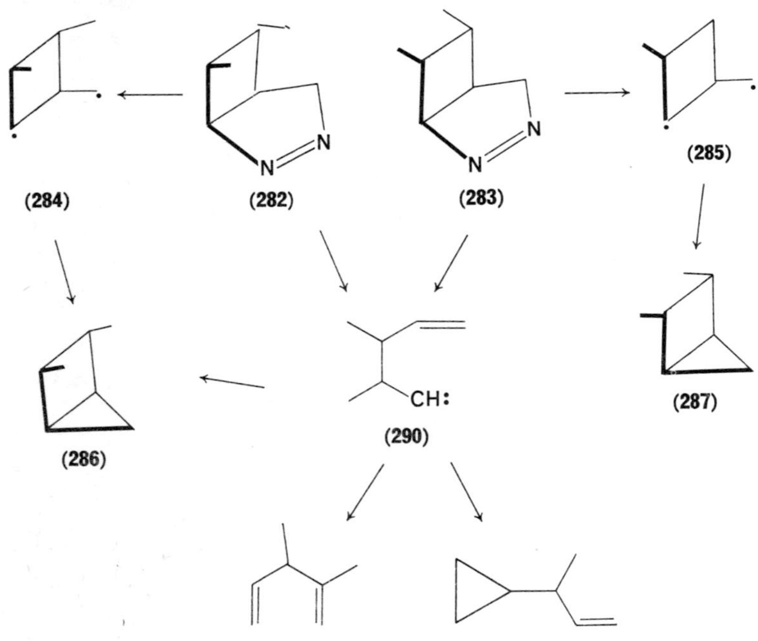

(284) (282) (283) (285) (286) (290) (287) (288) (289)

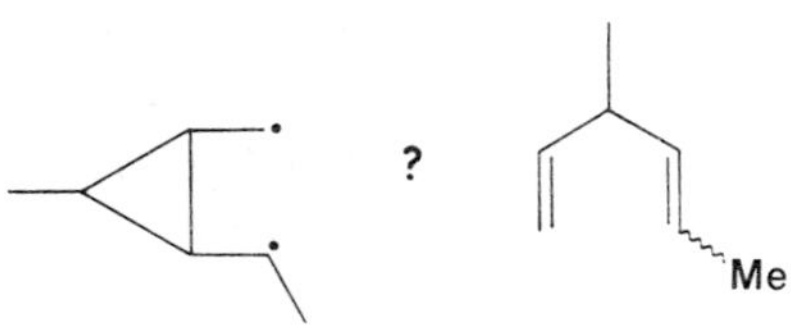

(291) (292)

A reaction scheme employing diradical **281** as the precursor of diene **275** implies that a similar scheme for the decomposition of the *syn*- and *anti*-6,7-dimethyldiazabicyclo[3.2.0]heptenes **282** and **283** should lead to *cis*- or *trans*-4-methylhexa-2,5-diene (**292**) through a diradical such as **291** formed from 1,3-diradicals **284** and **285**. However, diene **292** is completely absent and therefore diradical **291** is not an intermediate. Again the relative product yields are noticeably temperature-dependent, as would be expected if vibrationally excited carbenes are among the intermediates[140]. In similar studies with the *exo*- and *endo*-4-methyl derivatives of diazabicycloheptenes **293** and **294**, the products are similarly complex (nine components). Significantly, the stereochemistry of the 5-methylbicyclopentanes which result corresponds to predominant inversion of configuration at the methylated carbon and since steric restraints clearly preclude a planar [0,0]trimethylene intermediate it seems probable that σ-rotation within a diazabiradical **295**, and backside displacement of nitrogen, might account for this observation, the diazabiradical also serving as a source of 1,3-diradicals analogous to species **284** and **285**. Again, dipolar cycloreversion to an azoalkene which produces a carbene provides a competitive pathway[141].

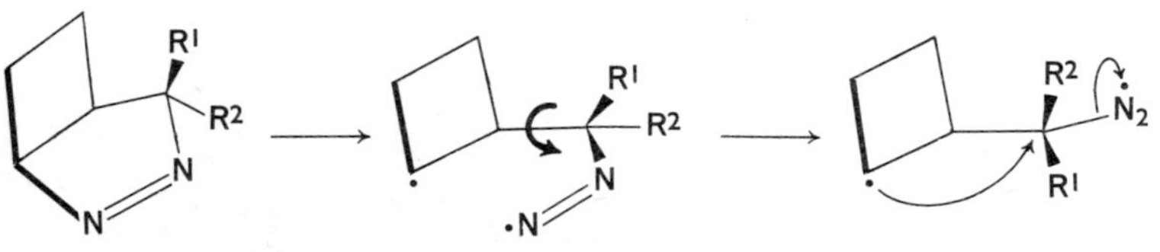

(**293**; R1 = H, R2 = Me)
(**294**; R1 = Me, R2 = H)
(**295**)

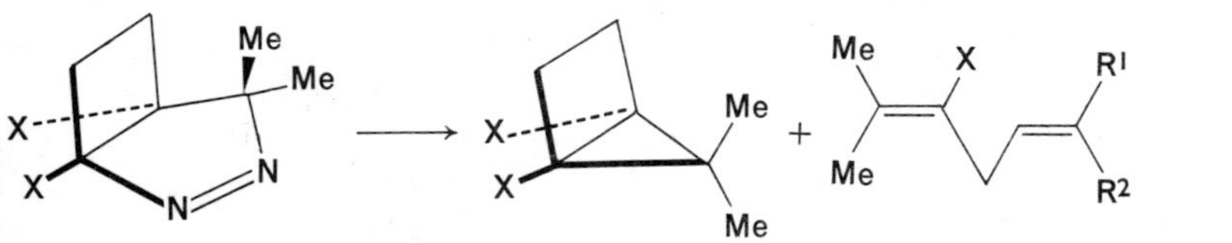

(**296**) (**297**) (**298**; R1 = H, R2 = X = CO_2Me)
(**299**; R2 = H, R1 = X = CO_2Me)

Carbene species are also invoked to rationalize the unsaturated products from singlet-excited decomposition of the more heavily substituted diazabicyclo[3.2.0]heptene (**296**), which gives bicyclopentane (**297**) and the stereoisomeric dienes **298** and **299**. In the benzophenone triplet-sensitized photolysis, the bicyclopentane accounts for 88% of the product and the dienes **298** and **299** only 4 and 8%, respectively. The triplet-sensitized photolysis of the *syn-* and *anti*-dichlorodiazabicycloheptenes **300** and **302** likewise leads to high yields of bicyclopentanes **301** and **303** with retained stereochemistry; probably in these reactions the low energy triplet species and the ability of the intermediates and products to dissipate vibrational energy facilitates the product specificities observed[142]. Photolysis of analogues of diazabicyclics **296** and **300** therefore provides a generally useful entry to the bicyclo[2.1.0]pentane series, especially if sensitizers are used, whereas thermolysis almost always gives olefinic compounds in

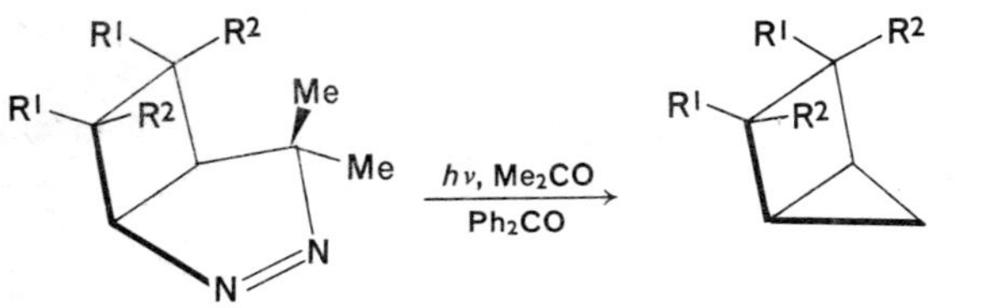

(**300**; $R^1 = H$, $R^2 = Cl$) (**301**; $R^1 = H$, $R^2 = Cl$)

(**302**; $R^1 = Cl$, $R^2 = H$) (**303**; $R^1 = Cl$, $R^2 = H$)

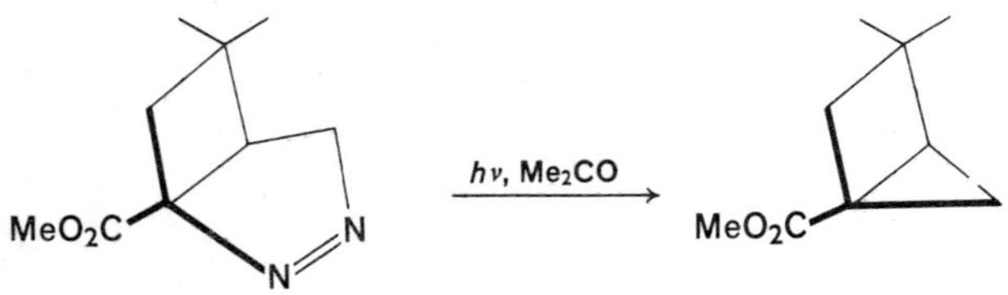

(**304**) (**305**)

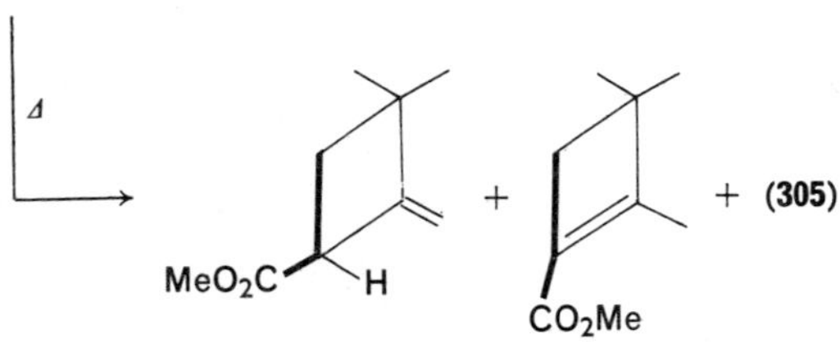

(**306**) (**307**)

competitive reactions. A further striking example is the photolysis of azo compound **304** which gives[143] bicyclopentane (**305**); but on thermolysis the major products are the olefins **306** and **307**.

As the size of the ring fused to the pyrazoline nucleus is made larger, fragmentations involving [0,0]trimethylene should become more accessible. Certainly, the dramatic decline in the complexity of the products formed in the decomposition of 2,3-diazabicyclo[3.3.0]octenes points to a change in mechanism, if only a reduction in the number of competing pathways; diazoalkenes, for example, appear to play no part here. The steric restraints imposed by a three-carbon bridge vicinally fused onto a trimethylene intermediate would seem significant, but kinetic data for thermolysis of diazabicyclooctene (**308**) and its 4,4-bisdeuterio analogue give $\Delta\Delta G^{\neq}$ ($= -RT \ln k_H/k_D$) 93 cal/mole/deuterium atom. For the unsaturated analogues, the corresponding value is 83 cal/mole/deuterium atom. Both these results are in line with the expected values for concerted extrusion of nitrogen[78]. The products formed from compound **308**, bicyclo[3.1.0]hexane, methylcyclopentene and methylenecyclopentane (**309–311**) accord with a trimethylene pathway insofar as simple olefinic products correlate with this intermediate[144]. Similar results are found for the thermolysis of the 4-*exo* and 4-*endo* methyl derivatives of pyrazoline (**308**); in this case also, unsaturated products arise, viz. ethylcyclopentene and ethylidenecyclopentane, whilst the bicyclic products, *endo*- and *exo*-6-methyl derivatives of bicyclo-[3.1.0]hexane (**309**) show predominant inversion of stereochemistry in each case[145]. The stereoselectivity here is not too different from that found in 3,4-dimethylpyrazolines and although a bridged trimethylene species (**319**) must be strained, simple models indicate that for approximate co-planarity with the terminal *p*-orbitals parallel, the strain involved is not too severe; more significantly, only one of the possible conrotations seems stereochemically feasible from this intermediate; that leading to inversion of configuration at the methylated centre, as is indeed observed for the major product. Ring-closure modes involving localized diazabiradicals or pyramidally inverting 1,3-diradicals (cf. **295** and **318** below) could also explain the data.

Pyrazolines **312** and **314** also produce bicyclics (e.g., **313–316**), with predominant inversion of configuration at the methylated centre in the latter case, but here the complete lack of monocyclic olefinic products suggests that a different mechanism is involved[144]; either a diazabiradical (**317**) which undergoes σ-rotation and backside displacement of nitrogen, or a pyramidal diradical (**318**) which, for reasons which are as yet obscure, inverts and rotates around the σ-bond before ring closing, would accord with the observation. Clearly, the considerable allylic stabilization could be important in an intermediate such as **317** and **318**; but the analogous species formed from pyrazoline **308** and its methyl derivatives on the other hand cannot be so favourable. The saturated and unsaturated bicyclic azo compounds **308** and **312** contain similar structural elements to 3-methyl- and 3-vinyl-pyrazolines and, interestingly, the free-energy change $\Delta\Delta G^{\neq}$ relative to the kinetic parameters for these compounds does resemble that

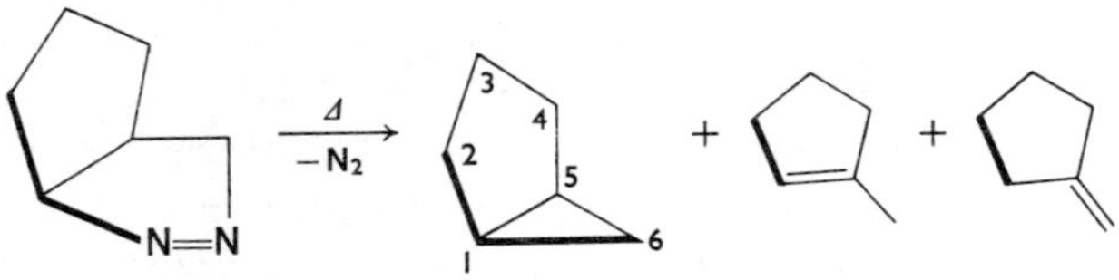

(**308**) (**309**) (**310**) (**311**)

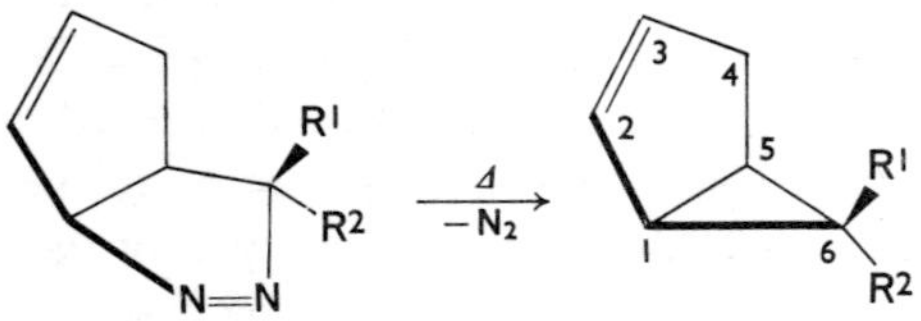

(**312**; R1 = R2 = H) ⟶ (**313**; R1 = R2 = H)

(**314**; R1 = H, R2 = Me) ⟶ (**315**; R1 = Me, R2 = H (62%) + (**316**; R1 = H, R2 = Me) 38%)

(**317**)

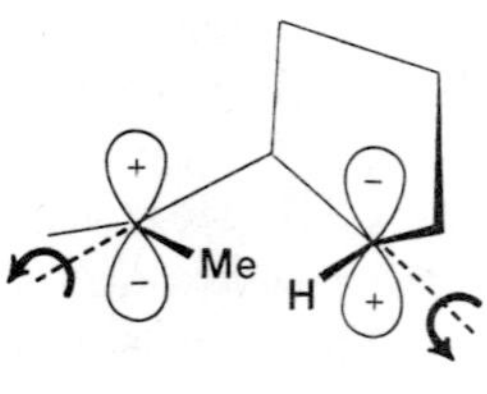

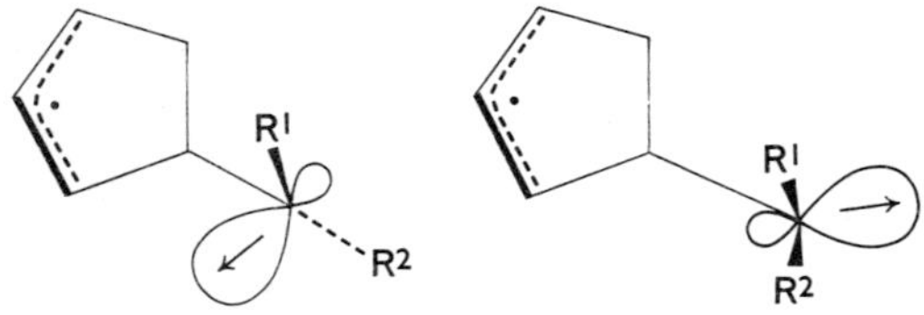

(**319**) (**318**)

for comparison of 3-methyl- with 3-vinyl-pyrazoline; however, the differences are in the probability term for the bicyclic molecules whereas in comparing the methyl and vinyl pyrazolines the difference appears in the entropy term[144]. This suggests that, for the latter pair, the difference lies only in the degree of stabilization of the trimethylene intermediate (as deduced from molecular dynamics[86]) and it is a fair supposition that such an intermediate is less probable for the bicyclic pyrazolines.

Photolysis of the 4-*exo*-methyl derivative of pyrazoline **308**, on the other hand, proceeds with predominantly retained configuration in the methylated bicyclic products. A kinetic analysis capable of providing ratios for the σ-bond rotation/ring closure rates shows that as in other types of reaction, diradicals best explain the products. As expected, sensitized photolysis leads to triplet diradicals which cyclize more slowly[145].

The manifestation in product composition of steric restraints in fragmentations of bicyclic azo compounds originates largely in the nitrogen-free component, but has a counterpart in stereochemical control by the azo function in certain special cases. Oxidation of the cyclic hydrazo compound **320** at low temperature and photolysis of the product **321** at –70°C, or warming to –10°C, results in the *identical* product, *trans–trans*-hepta-2,5-diene in clear 'violation' of the orbital symmetry conservation rules for concerted cycloreversions. Similar results are found for the other stereoisomers of azo compound **321**; viz. the *cis–syn* compound yields *cis–cis*-hepta-2,5-diene. Disrotation observed in these homocycloreversions in the excited state could be occurring from a *trans*-azo linkage present in the activated molecule[133] (cf. ref. 131a.)

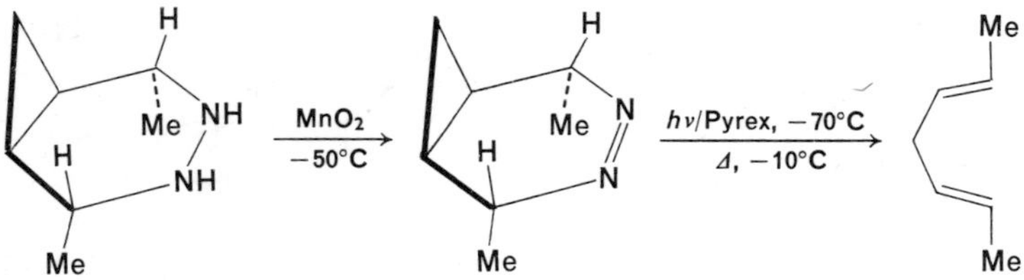

(320) **(321)**

In this connexion, the exclusive formation of *cis–cis*-hepta-2,5-diene (**325**) from the *cis–syn* azo compound **322** contrasts with the formation of *trans–trans* diene (**248**) from the olefinic azo compound **247**, and suggests that cycloreversion must be from transition state **323** rather than from the conformational isomer **324**, even though there is considerable methyl–methyl repulsion not evident in the latter[133]. The lower energy reaction coordinate concomitant on maintaining bonding interactions between the terminal orbitals and the opening transannular bond as nitrogen departs ensures a concerted pathway for conformer **323** which is not available to isomer **324**. Similar effects are to be found in other polycyclic azo compounds discussed below.

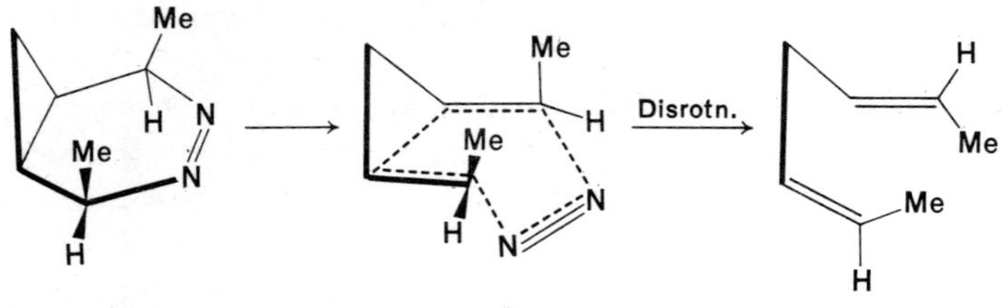

(322) (323) (325)

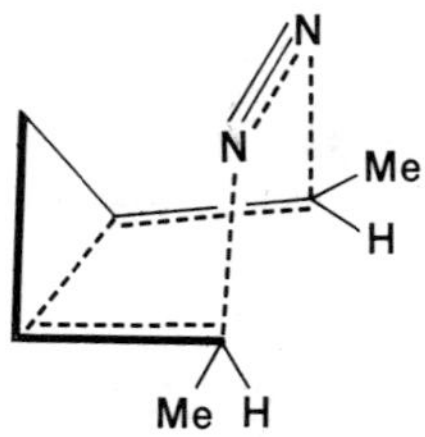

(324)

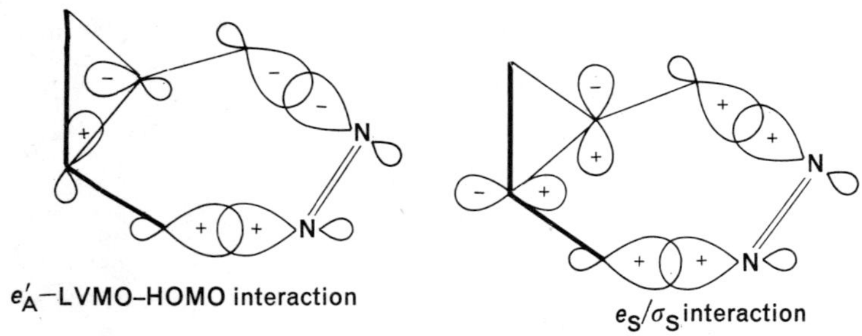

FIGURE 3. Simplified Walsh orbital viewpoint of component M.O.s in transition state **323**.

2. Diazabicyclo[2.2.1]heptenes and related bridged bicyclic azo compounds

The addition of cyclopentadiene to ethyl azodicarboxylate to form a 1:1 adduct (**325**), was the first clearly defined example of the Diels–Alder diene synthesis whose more general application to the synthesis of cyclic com-

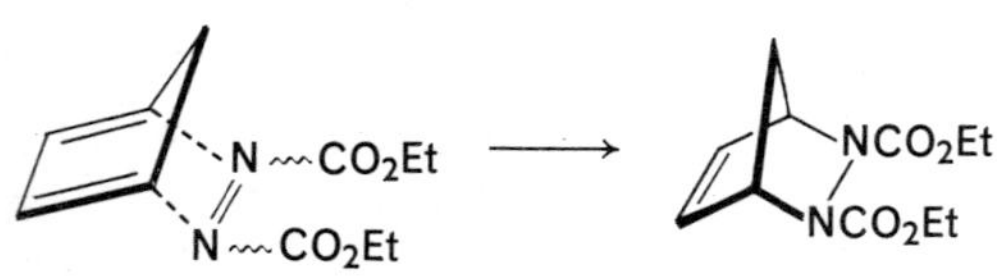

(325)

pounds developed from an acute appreciation of the basic principles[146] anticipated in connexion with the structure of adducts formed between benzoquinone and isoprene[146, 147].

The more general chemical[148, 149] and physical[150] properties of analogues of ester **325** have been investigated, but perhaps the most important reaction sequence is reduction, hydrolysis and concomitant decarboxylation to an intermediate hydrazo compound (**326**) whose air-autoxidation or dehydrogenation with mercuric oxide gives the important azo compound **327**. Oxidation of the (frequently nonisolated) intermediate is also conveniently accomplished with cupric chloride, and under these conditions the azo compound is isolated as its acid-dissociable, reddish-brown Cu^I complex, $(C_5H_8N_2)_2Cu^{+}$[151].

Moderate thermolysis of azo compound **327** gives bicyclo[2.1.0]pentane (**328**) as the only product[152] and similar fragmentations constitute an important general source of bicyclopentanes of this class and of other fused ring alicyclic compounds.

The preparation of azo diene **329** from adduct **325** has also been investigated[153]. Hydrolysis of ester **325** and immediate work-up, however, gives only hydrazo compound **326**, azo compound **327** and cyclopentadiene. Intermolecular disproportionation (cf. diimide[154]), together with fragmentation of diene **329** under conditions where the adduct **325** is stable, accounts for the products.

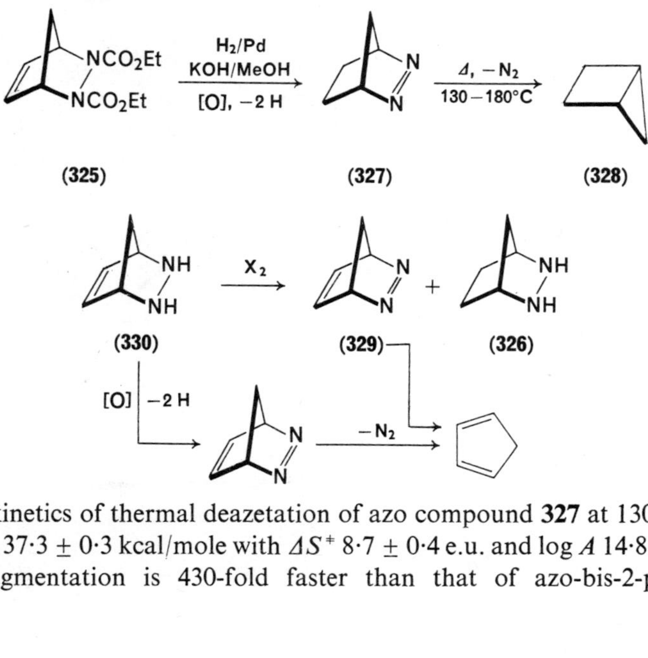

(325) (327) (328)

(330) (329) (326)

The kinetics of thermal deazetation of azo compound **327** at 130–180°C gives E_a 37·3 ± 0·3 kcal/mole with $\Delta S^{\neq}$ 8·7 ± 0·4 e.u. and log A 14·86 ± 0·1; this fragmentation is 430-fold faster than that of azo-bis-2-propane

($Me_2CHN{=}NCHMe_2$) at 250°C. For the noncyclic azoalkanes the entropy and frequency factors ($\Delta S^{\neq}$ 12–15 e.u., log A 15) suggest simultaneous rupture of both C—N bonds in the transition state, and the values observed for **327** are also consistent with concerted elimination of nitrogen[153].

Azo compound **327** shows u.v. light absorption ($n \rightarrow \pi^*$) similar in appearance to that of diazirines in having several sharp bands, with λ_{max} 341·5 nm (isooctane); this vibrational fine structure disappears in solutions in ethanol. Excitation of the $n \rightarrow \pi^*$ mode in the gas-phase photolysis of 2,3-diazabicyclo[2.2.1]hept-2-ene (diazanorbornene) gives bicyclopentane, cyclopentene, cyclopentadiene and penta-1,4-diene; the effect of pressure changes on relative product composition interpreted by a statistical mechanical treatment of the energy distribution within the intermediates, and the effect of added nitrogen as vibrational quencher indicate that *both* primary photoproducts, cyclopentene and bicyclo[2.1.0]pentane are formed sufficiently 'hot' to undergo further fragmentation[155]. Solution photochemistry of azo compound **327** indicates a triplet excited state 20 kcal/mole lower in energy than the S^1 state (84 kcal/mole)[156] but, perhaps surprisingly, both species are accessible using aromatic sensitizers; thus 2-acetonaphthone and acenaphthylene have quite adequate triplet energies to sensitize decomposition of **327**, but the quantum yields for the two sensitized photolyses differ so much that the high quantum yield for acenaphthylene can only be ascribed to singlet sensitization, (S^1 acenaphthylene, 89·1 kcal/mole) under conditions where the excited sensitizer lifetime is sufficient to ensure that diffusion-controlled energy transfer is an order of magnitude faster than sensitizer decay. This important observation is confirmed by the activity of **327** as a fluorescence quencher; its efficiency drops off sharply for fluorescers having singlet excitation energy <~82 kcal/mole. Experi-

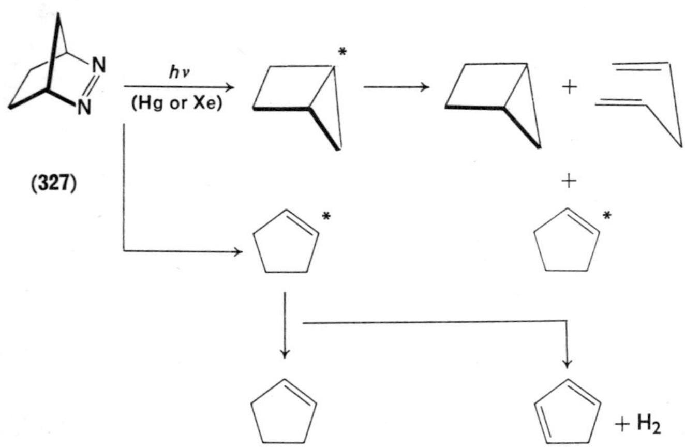

(327)

ments with oxygen as triplet quencher show that $S^1 \rightarrow T^1$ is not an important process, and the direct photolysis is unaffected; the test of sensitivity of the photolysis product ratio, [bicyclopentane/cyclopentene], to the excited state mutiplicity is of no value here, perhaps because of insensitivity of hydrogen atom migration to radical multiplicity[157].

Azo-2-methylpropane photolysis is also both singlet and triplet sensitized in the presence of a variety of aromatic species (anthracene, pyrene, phenanthrene, triphenylene). Addition of triplet quenchers to photolyses of the azoalkane has no effect on the decomposition, however, and in the presence of benzophenone and triphenylene, when only triplet species *can* be present, there is *no* decomposition. Clearly, there is a spin correlation effect, the triplet species undergoing geometrical changes, but the singlet intermediate fragmenting. For diazabicycloheptene (**327**), on the other hand, both the singlet and triplet intermediates decompose, presumably because of the impossibility of geometrical changes in the triplet state[158].

Kinetic data for the thermolysis of diazabicycloheptene (**327**) suggest concerted loss of nitrogen[153] and, in a search for a possible intermediate, the *exo*-bis-5,6-deuterio derivative **330** was thermolysed; in the gas phase at low pressure the product consists of bicyclopentanes **331** and **332** in ca. 1:1 proportion, but in the condensed phase *anti*-isomer **331** predominates (75 %) with net inversion of the bridge. The result could be rationalized by a double inversion of a pyramidal diradical (**333**), although it is not clear why several such inversions do not occur before cyclization, giving a random distribution (1:1) of bicyclopentanes. Alternatively, a biradical (**334**) could necessarily and more convincingly account for the product[159]; on the other hand, the orientation of the relevant orbitals involved in generating the transannular bond in structure **334** is not ideal[160]. However, some support for an intermediate like **334** comes from the stereospecific addition of the highly reactive dienophile *N*-phenyltriazolindione (**336**) to bicyclopentane

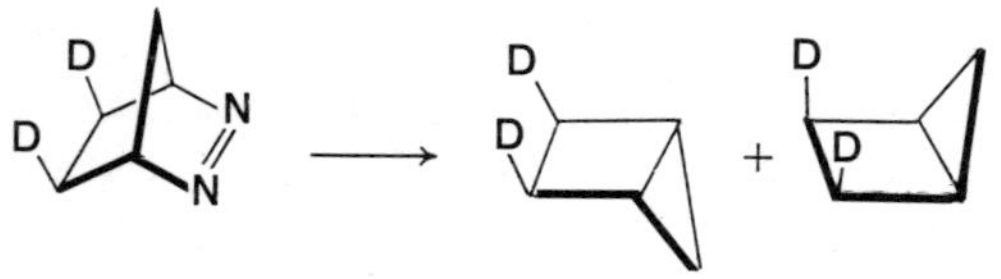

(330) (331) (332)

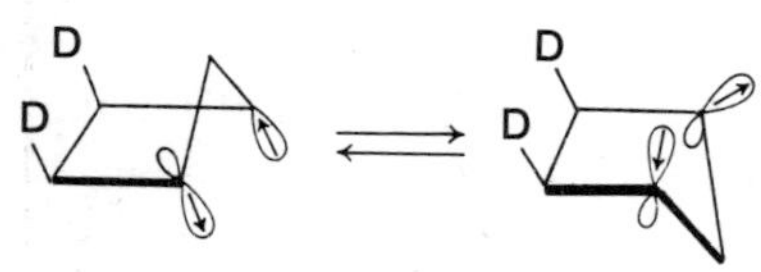

(333)

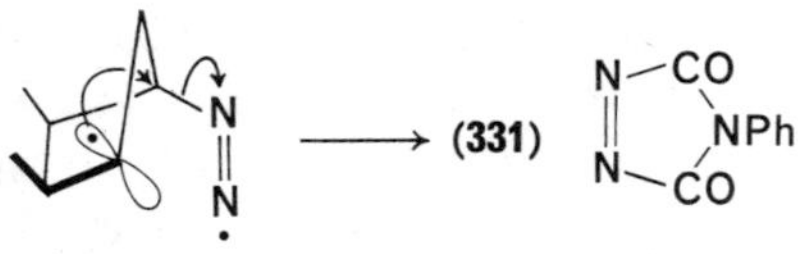

(334) (336)

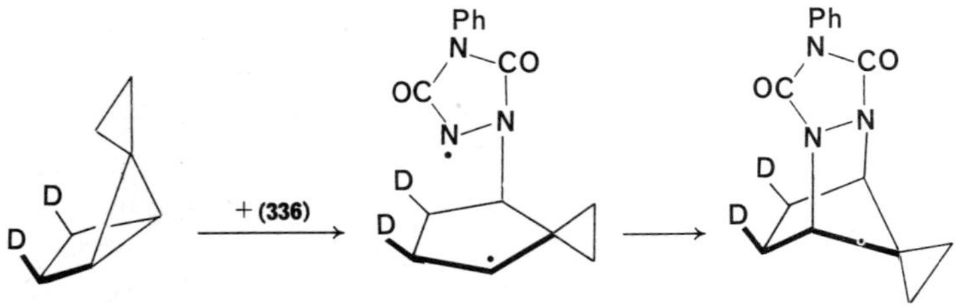

(335) (337) (338)

(**335**) to give adduct **338**. Although the principle of microscopic reversibility can, strictly speaking, only be applied to observable equilibria[161] it is reasonable to assume that if, in general, cycloadditions to cyclopentanes follow the same *unusual* stereochemical course as cycloreversions, then the mechanisms for the forward and reverse examples will most likely be identical. On this basis, since the addition of dienophile **336** and **335** must on steric grounds involve a biradical (**337**), structure **334** could represent a reasonable intermediate for cycloreversion of diazanorbornenes.

Somewhat similar results have been observed in the products of relatively high-pressure, gas-phase thermolysis of the *exo*-5-methoxydiazanorbornene (**339**), which gives methoxybicyclopentanes, again with predominant inversion of the methylene bridge (if allowance be made for the thermal equilibration of **340** and **341** under the reaction conditions) and net inversion of the product is also found with the *endo*-5-methoxy analogue of **339**, but the composition of the product from the two epimers differs widely. Clearly, a common intermediate such as a pair of rapidly interconverting pyramidal biradicals like **333** cannot account for these observations, but if pyramidal inversion occurs followed by ring-closure before complete equilibration, the results could be understood[160]. Methylene bridge inversion is also seen in the direct solution photolysis of the epimeric pair of methoxydiazanorbornenes, but in triplet-sensitized reactions the same product mixture arises from each epimer; in the solid state, however, direct irradiation results in a high degree of retained configuration in the product bicyclopentanes. Non-equilibrating singlet biradicals formed in the direct photolysis and longer-lived, lower-energy triplet species which equilibrate could rationalize these observations. In the solid state, crystal lattice inhibition to inversion of the biradical intermediate, and presumably rapid vibrational quenching of the product, could account for the high degree of retained stereochemistry[160].

All the proposals outlined have some merit in rationalizing the observational data, but they may actually be different aspects of a more subtle scheme, as apparent when the problem is analysed from a molecular orbital viewpoint[162]. Charge-density contour calculations show that there is a substantial antibonding component of the molecular charge distribution located beneath the 'flap' of the bicyclopentane molecule. If the antisymmetric component of the C—N σ-bond combination is considered as the nitrogen molecule departs from the diazanorbornene to form bicyclopentane, the development of the antibonding region in the 'under flap' region of the product molecule can be accommodated by the motion of the C_5 ring towards planarity as the nitrogen departs. As the nitrogen molecule becomes completely free, the C_5 fragment can cyclize in either of the two possible directions, thus rationalizing the observation of a 1:1 mixture of stereoisomers from bisdeuteriodiazanorbornene (**330**) at *low* pressures in the gas phase. However, in the condensed phase, or at relatively high pressures in the gas phase, the departing nitrogen molecule is likely to remain in the vicinity of the C_5 fragment; a small interaction between the nitrogen molecule and the antibonding component of charge developing beneath the bridge as cyclization proceeds towards bicyclopentane ensures that there is a predominance of bridge inversion in the product. The calculations show that the attractive force is worth up to 10 kcal/mole and is maintained even at relatively large distances between the components. A similar conclusion can be reached from a consideration of the theoretical result that the easiest motion for any system will have its greatest amplitude where the interaction of the HOMO/LVMO for the components is greatest. This criterion predicts two possible motions for the C—N bonds in diazanorbornenes as they fragment; a concerted symmetric motion, with inversion at the carbon termini (*A*) and an antisymmetric motion (*B*) leading to stepwise fragmentation. The kinetic[153] data clearly support the symmetric mode. To account for the observed *partial* inversion of the methylene bridge in these decompositions requires either that both of these processes occur, or that the interaction with the departing nitrogen molecule in the alternative picture is only partly effective, some of the nitrogen becoming free, so that there is a distribution between the alternative bicyclopentanes.

Low-temperature photolysis of the diazanorbornene derivative **342** gives a good yield of the *syn*-bicyclopentane (**343**) (79%) which, however, equilibrates with the *anti*-isomer **344** at 70°C (E_a 29·0 kcal/mole)[161]. In this

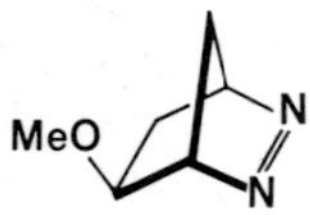

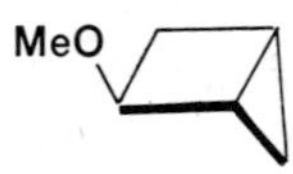

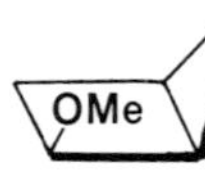

(339) **(340)** **(341)**

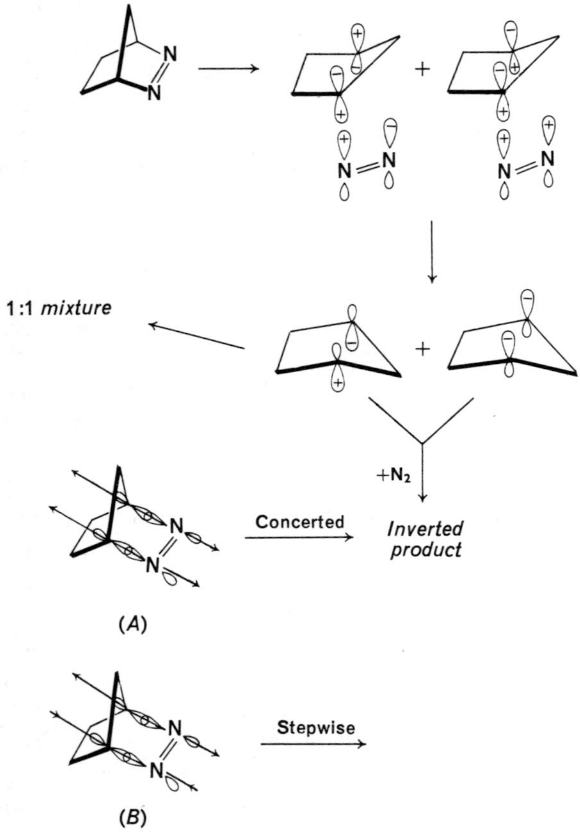

connexion, thermolysis of azo compound **345** and its analogues **348** and **351** gives entirely the products of inverted methylene configuration (**347**, **350** and **353**); brief direct solution photolysis (Pyrex), however, gives mixtures containing up to 30% of the retained configuration *endo–syn* compounds **346**, **349** and **352**, recognized from their unusual ^{1}H n.m.r. spectra (strong deshielding of the *endo*-cyclopropyl methylene proton) and their conversion at *room temperature* into the *endo–anti* isomers (E_a 22–23 kcal/mole)[163].

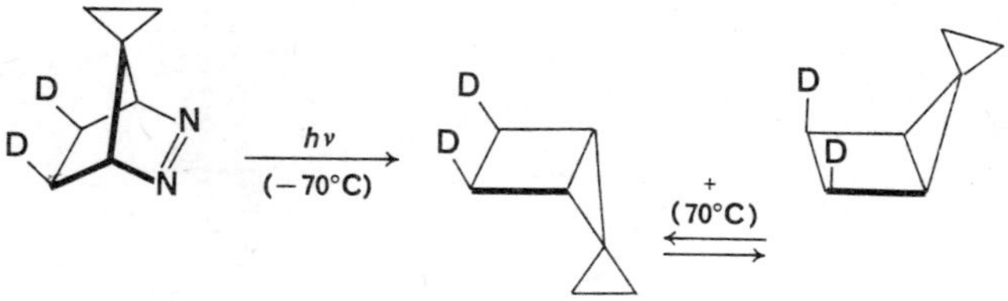

(342) (344) (343) (79%)

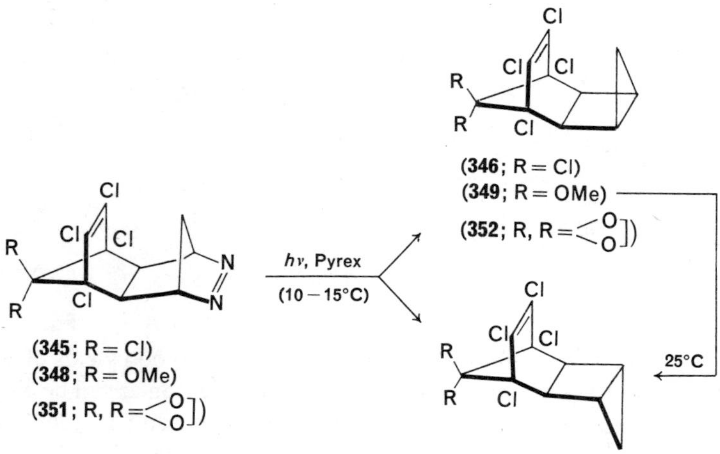

(345; R = Cl)
(348; R = OMe)
(351; R, R = $<^{O}_{O}]$)

(346; R = Cl)
(349; R = OMe)
(352; R, R = $<^{O}_{O}]$)

(347; R = Cl)
(350; R = OMe)
(353; R, R = $<^{O}_{O}]$)

One of the very few *endo* fused diazanorbornenes to have received attention is the structure **355** made from isopyrazole (**354**) and cyclobutadiene. Photolysis gives a quantitative yield of the product with retained bridge configuration **356** and in analogy to, for example, **346** it seems very unlikely that any inverted bridge product could survive for very long on account of the severe compression which would be present in the molecule. Interestingly, mild thermolysis of tricycloheptene (**356**) in C_2Cl_4 gives only quadricyclane (**357**) (hence no diradical here), which cycloreverts to norbornadiene (**358**) at 175°C. The reverse of this reaction is accomplished photochemically, but higher temperature thermolysis converts diene (**358**) into a tropilidene[164].

Trimethylenemethane has already been discussed in connexion with 4-methylenepyrazoline decompositions (Section II, D 1), and generation of a cyclic analogue (**148a**) by mild thermal deazetation of azo compound **359**

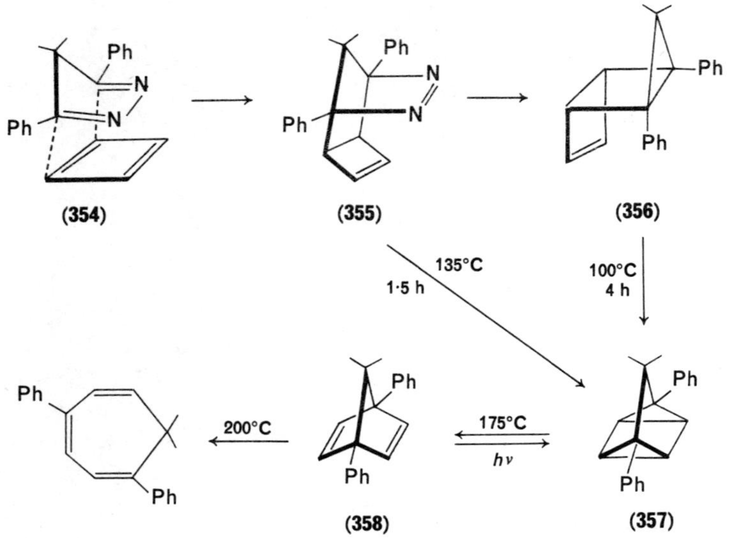

in a large excess of olefin supresses its dimerization; good yields of adducts with, for example, 1,3-hexadiene, and the 2,4-diene isomer can be obtained, whilst maleic and fumaric esters give bridged adducts (**360**), as do dichloroethylene and acrylonitrile, but azodienophiles give adducts such as **361**[165].

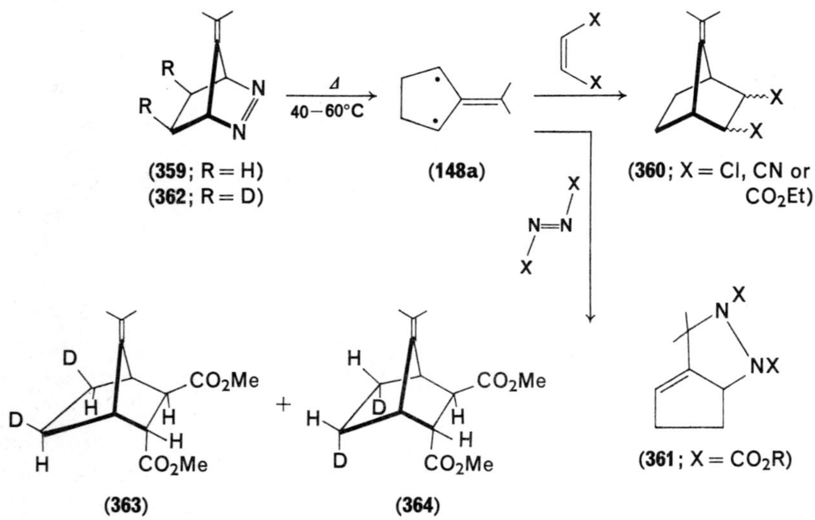

Decomposition of the deuterium-labelled diaza-7-isopropylidenenorbornene (**362**) and trapping with maleic ester gives a 1:1 mixture of the *trans*-ester adducts (**363** and **364**), which must clearly result from trimethylenemethane approaching the π-face of the olefin in a parallel plane with equal probability of deuterium being *syn* or *anti*. However, only a final stepwise ring closure can account for the *trans*-cycloadduct, but olefin dilution experiments suggest that initially addition is stereospecific with a singlet intermediate, and only subsequently is a non-stereospecific triplet species formed (in agreement with theoretical expectation). If a two-step addition process is involved, the fact that incompletely consumed olefin is recovered quite unchanged shows that it is irreversible[166].

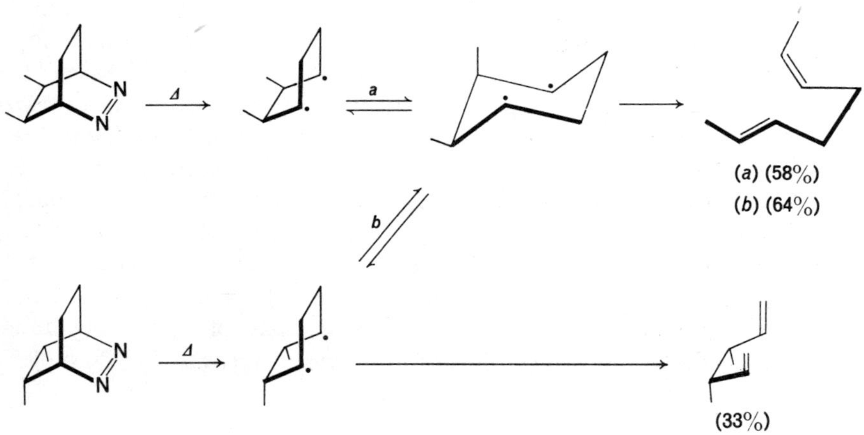

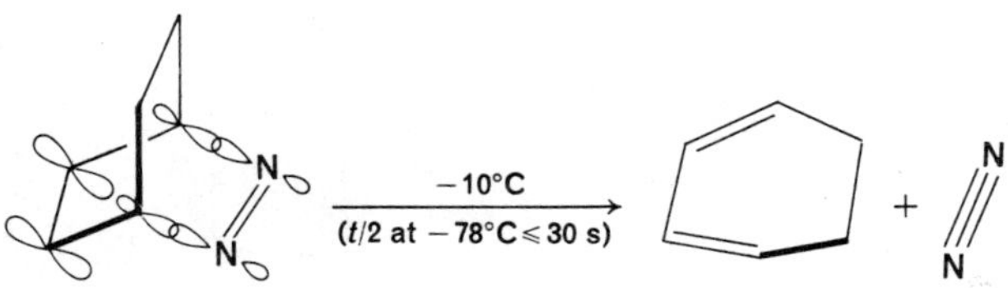

(365)

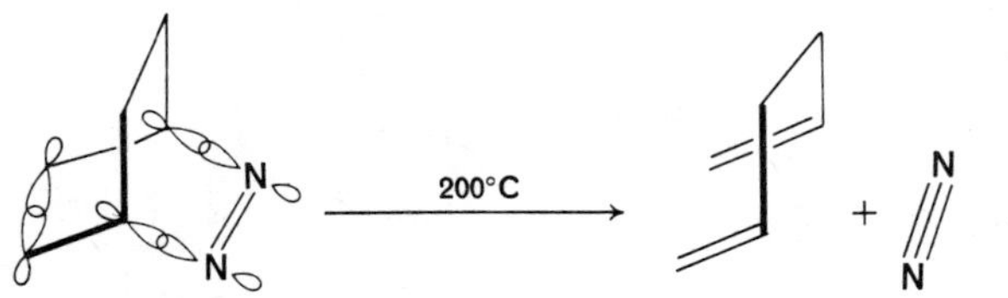

(366)

B. Bridged Tri-, Tetra- and Pentacyclic Azo Compounds and Related Caged Structures

1. Bridged systems related to diazabicyclo[2.2.1]heptene and -[2.2.2]octene

Among the bridged polycyclic azo compounds groups of structures can be discerned which sharply divide into those characterized by unusually rapid extrusion of nitrogen, typically at or below room temperature, and those whose properties resemble the 'normal' pyrazolines or pyridazines in decomposing only at elevated temperatures. One example has already been cited, pyridazine (**321**), and another is the diazabicyclo[2.2.2]octa-2,5-diene (**365**); this compound loses nitrogen readily at –78°C, but the dihydro-derivative (**366**) by contrast fragments rapidly only at 200°C. The difference can be rationalized by the pericyclic participation of the π-bond in **365**, whose orbitals lie practically parallel to the reorganizing σ-bonds, whereas the relevant σ-orbitals in the dihydro-derivative (**366**) are orthogonal to the C—N σ-bond orbitals, and fragmentation is therefore through a biradical[167]. This is further illustrated by the thermolysis of the *endo–endo* and *exo–exo* 5,6-dimethyl analogues of **366**, which give similar (though not identical) mixtures of *trans–trans*, *cis–trans* and *cis–cis* octa-2,6-dienes rich in the *cis–trans* isomer, together with 33% of 3,4-dimethylhexa-1,5-diene; these results accord with equilibration of the initially formed boat-like diradicals and a chair conformer as the precursor of the *cis–trans* diene[168]. This observation suggests that any suitably orientated π or π-type orbitals with the correct symmetry properties are capable of interacting and accelerating C—N σ-bond scission and this is indeed the case. For example,

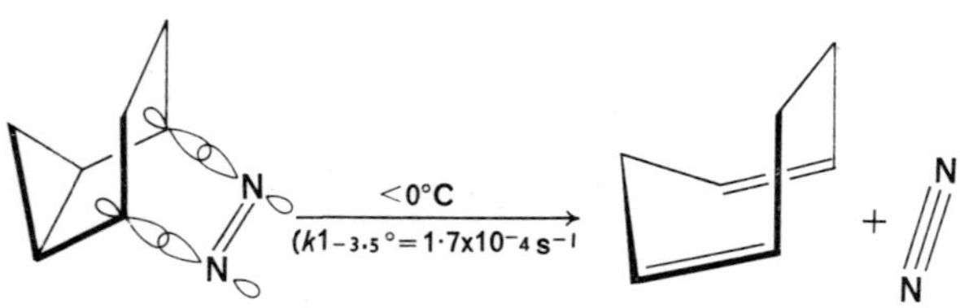

(**367**)

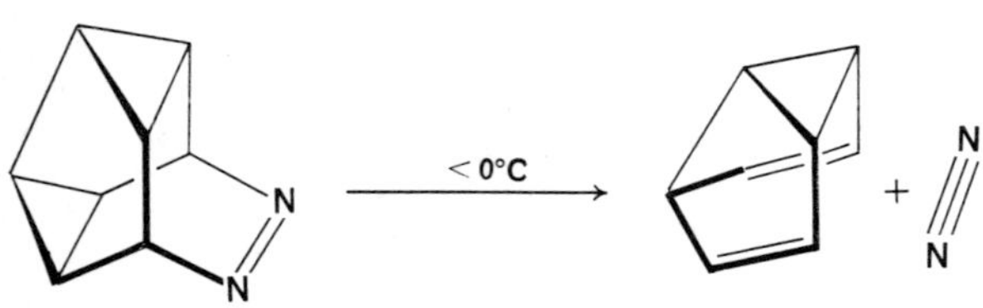

(**368**)

a cyclopropane fused anti-parallel to the C—N bonds as in **367** and **368** renders the molecule labile at below 0°C[169,170].

The fragmentation of **368** to simibullvalene is accompanied by a secondary process giving up to 20% of cyclooctatetraene, depending on the source of the azo compound. In the presence of transition metals, e.g. Ag^+, semibullvalene is known to rearrange into cyclooctatetraene, but there is also the possibility that the observed decomposition of the Cu^I complex of azo compound **368** entirely into cyclooctatetraene could be an example of a competing orbital symmetry allowed cycloreversion (best visualized from the point of view of the reverse, addition of nitrogen to the tetraene)[171].

The relative unimolecular rate parameters for decomposition of these tricyclic azo compounds reveal the remarkable differences apparent when compared to monocyclic pyrazolines; e.g. the large negative $\Delta S^{\neq}$ term in the case of compound **367** (–21 e.u.) whilst very likely subject to a large error, nevertheless if compared to the modest positive value for $\Delta S^{\neq}$ for compound **366** (10·5 e.u.) suggests a change in mechanism from a concerted cycloreversion for **367** to something like a biradical process for **366**[172]. On the basis of similar kinetic data decomposition of compound **369** is also more obviously a concerted process compared to loss of nitrogen from diazanorbornene **327**. In this connexion it is interesting to compare compound **365** with diazetine (**52**); at –78°C symmetry-allowed cycloreversion of the former is at least 10^{22}-fold faster than the symmetry-forbidden (and therefore probably stepwise) fragmentation of the diazetine. The rate difference is a conservative measure of the energy requirements for

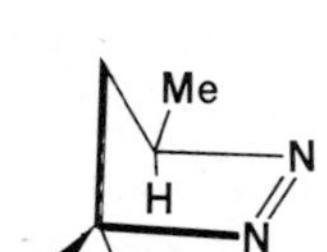

(**137a**) (0·01)
(40 ± 0·3)

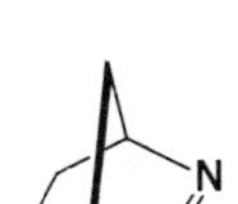

(**327**) (1·00)
(37·3 ± 0·3)

(**369**) (1·6 x 10^{11})
(18·9 ± 1·5)

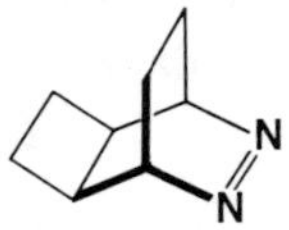

(**370**) (10^5)
(39·2 ± 0·3)
(Rel. rates)
(*E*a kcal/mole)

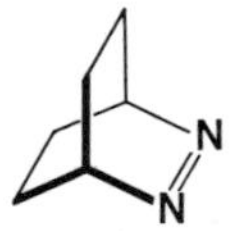

(**366**) (1·00)
(44·6 ± 0·2)

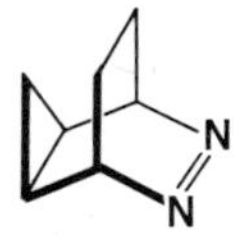

(**367**) (10^{17})
(14·9 ± 1·5)

symmetry-allowed versus 'disallowed' processes, since strain release factors for diazetine (**52**) tend to diminish the rate contrast. Kinetic parameters here show that the energetic cost of symmetry forbiddeness may be as high as ca. 18 kcal/mole, underlining the fact that for symmetry conserved and nonconserved processes the differences likely to be found will usually depend on the type of reaction selected for study; in the present case, the very large rate difference may be due to a considerable separation in the energy levels for the relevant reactant and product M.O.s for the diazetine[173]. Interestingly, diazetines **52** and **371** decompose at comparable rates and have similar activation parameters; there must, on the evidence, be little possibility of remote transannular orbital participation in **52** and it is therefore perhaps not surprising that norbornadiene forms the major fragmentation product of the azo compound when the formation of quadricyclane is otherwise symmetry-allowed.

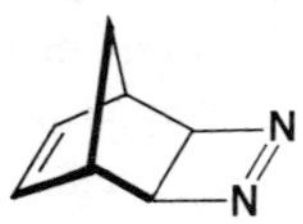

(52)

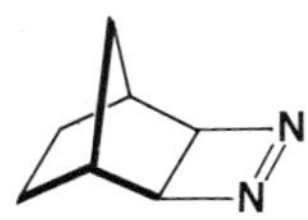

(371)

In the compounds **367**, **368** and **369** the cyclopropane ring is more or less ideally situated for participation in the fragmentation transition state. The question naturally arises as to the effect of graded structural changes in this class of compounds which might result in less ideal geometry, and has been probed with the series of compounds **369**, **372**, **373** and **374**. The striking data similarity for diazanorbornene (**327**) and compound **372** (Table 3) strongly suggests a quite different mechanism for deazetation here, compared to the concerted processes signalled for compounds **373** and **374** in analogy to **369**. In **372** the rigidity imposed by the one-carbon bridge precludes the motion necessary for maintenance of favourable interaction throughout the reaction coordinate involving the cyclopropane Walsh e_S and e_A' orbitals and the cleaving S and A σ C—N bond combinations, the two- and three-carbon bridges in **373** and **374** being much more elastic in this respect[174]. (See the relevant photoelectron spectroscopic data.[174a])

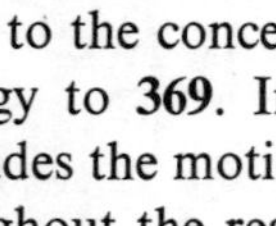

(372)

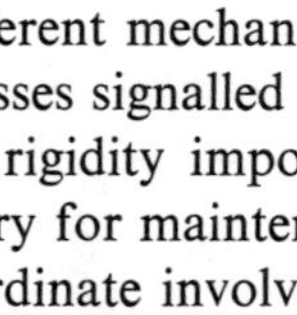

(373)

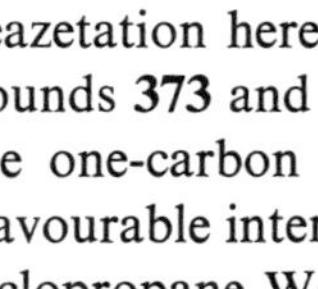

(374)

TABLE 3. Kinetic parameters for decomposition of azo compounds

Compound:	**327**	**369**	**372**	**373**	**374**
Rel. rate	1	$2{\cdot}2 \times 10^{11}$	9·2	$5{\cdot}2 \times 10^{8}$	$1{\cdot}9 \times 10^{11}$
E_a, kcal/mole	$37{\cdot}3 \mp 0{\cdot}3$	$17{\cdot}7 \mp 1{\cdot}6$	$36{\cdot}5 \mp 0{\cdot}3$	$23{\cdot}3 \mp 0{\cdot}4$	$19{\cdot}6 \mp 0{\cdot}9$
$\Delta S^{\ddagger}$, e.u.	8·7	−13	8·3	−5	−6

Other variations in the structure of polycyclic azo compounds can also exert a profound effect on the relative ease of decomposition. The azoxy compounds **375–377**, for example, are perfectly stable compounds which are decomposed only when heated to elevated temperatures (200–400°C). The result can be visualized as a reflection of the lack of energy-lowering HOMO/LVMO orbital interactions in the reaction coordinate due to the different orbital volumes at the nitrogen termini of the nitrous oxide extruded[170]. Since these azoxy compounds may be reduced to the corresponding azo compounds by hexachlorodisilane, their stability could be exploited as a method of storing labile azo compounds.

The synthesis of polycyclic azo compounds **372–374** makes use of the uncommon homodiene addition of *N*-phenyltriazolindione (**336**) to norbornadiene[175] and its bridge-expanded homologues. However, the

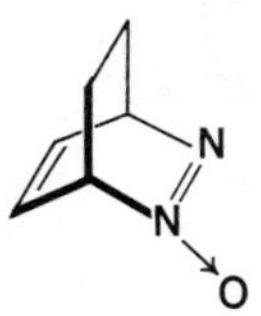

(375)

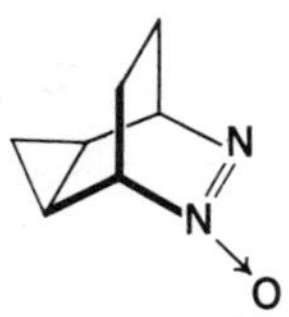

(376)

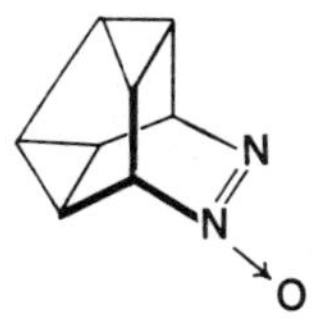

(377)

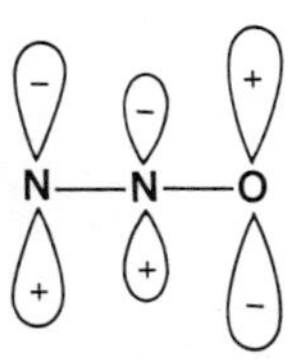

HOMO

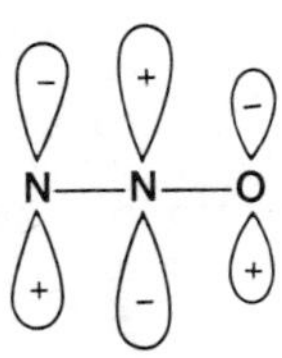

LVMO

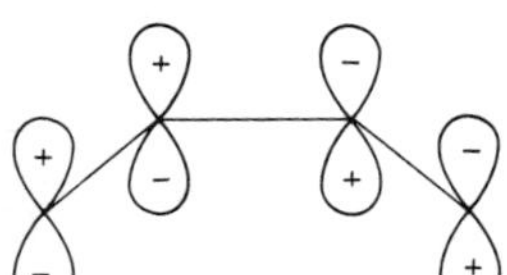

HOMO (φ_2)

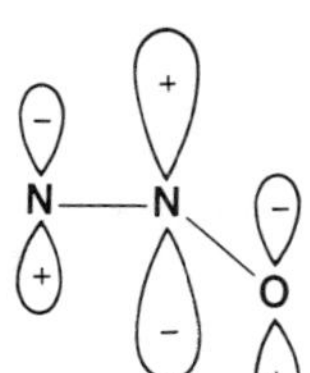

LVMO

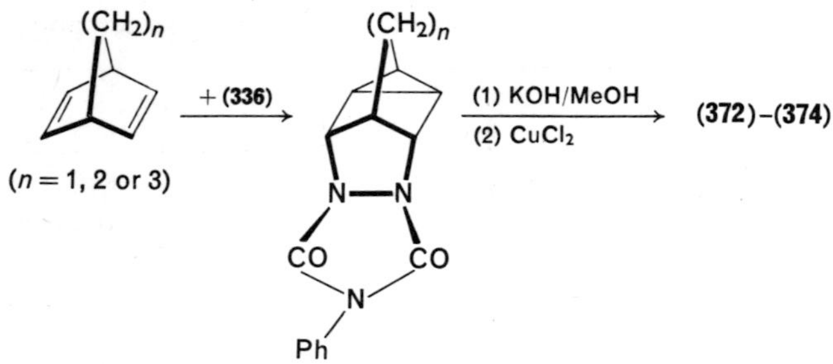

synthesis of tricyclics **367** and **369** employs addition of diazomethane to a bicyclic azo ester, e.g. **325**, and the intermediate adduct may either be thermolysed to give subsequently the desired tricyclic azo compound **369**, or be converted to the tricyclic bis-azo compound **378** which itself has interesting properties[176]. For example, mild thermolysis of **378** gives the bicyclopentane derivatives **379** and **380**, together with pyrazole (**381**), via isopyrazole derived from cycloreversion of **378**. An additional product, tricyclohexane (**382**), appears in the analogous photolysis. The cycloreversion of bis-azo compound **378** into pyrazole is one of the few known cases of heterocycle extrusion, since the propensity for nitrogen loss from

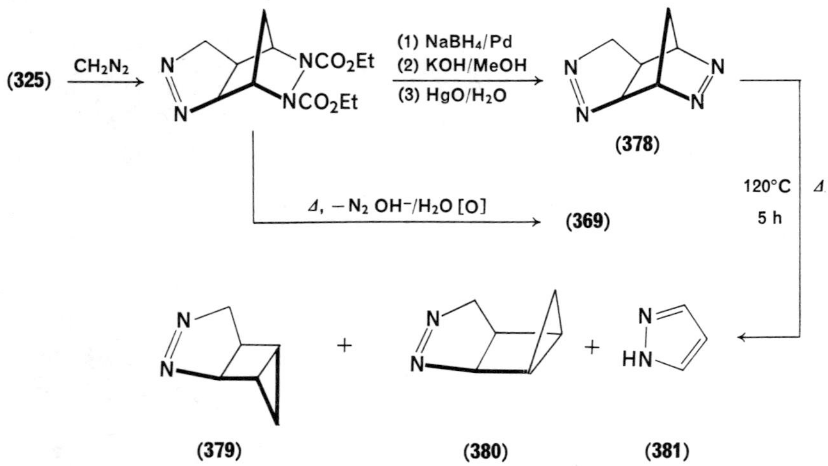

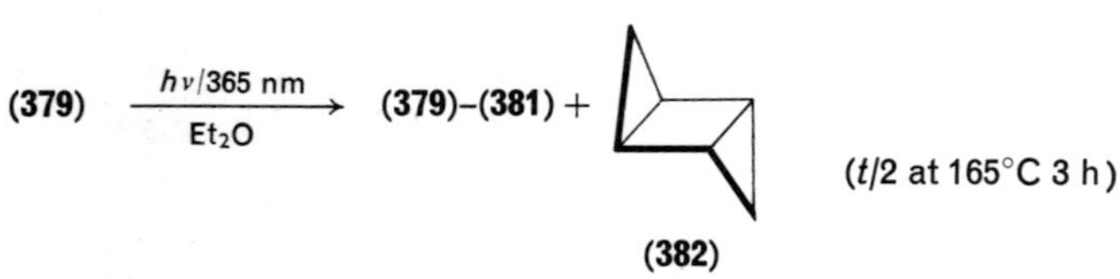

polycyclic azo compounds dominates, as is shown by the deazetation of the 3,3-diphenyl-derivative of **378**, which gives only diphenylcyclohexa-2,5-diene when heated[7]. In the latter case, nitrogen loss from the diphenylpyrazoline ring must, of course, be especially rapid on account of the stabilizing effect of the phenyl groups on the intermediate (which is presumably a biradical, since steric effects would seem to preclude any resemblance to a trimethylene species). However, another example of heterocycle extrusion is found with the diazatricycloundecatriene (**383**); this compound cycloreverts exclusively at room temperature into tetrachlorobenzene and pyrazole (E_a 22·29 $\pm$ 0·46 kcal/mole $\Delta S^{\neq}$ −7·3 $\pm$ 1·4 e.u. log A 11·63 $\pm$ 0·31) and the kinetic parameters indicate a close relationship in thermal behaviour to that of the carbocyclic analogue (E_a 22·74 $\pm$ 0·69 kcal/mole $\Delta S^{\neq}$ −4·8 $\pm$ 2·2 log A 12·18 $\pm$ 0·47). Photolysis of compound **383**, on the other hand, gives the tricyclic compounds **384** and **385**, tetrachlorinated derivatives of the otherwise unknown stereoisomeric tricyclo-[4.3.0.0^{1,7}]nona-2,4-trienes[7,178]. An interesting contrast in this connexion is the photolysis of azo compound **386**, *with nitrogen retention*, to give mainly diazacyclooctatetraene (**388**), 'Dewar' benzene and benzvalene, whereas thermolysis gives benzene exclusively[179]. (More recently it

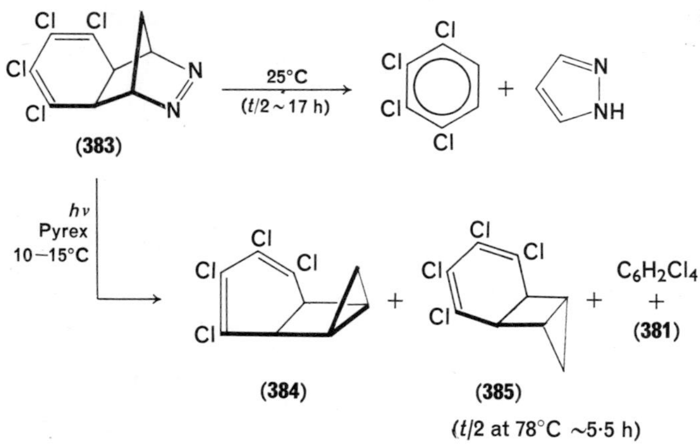

has been shown[179a] that photolysis of compound **386** gives prismane, half-life 11 h at 90°C.) The methylene bridged homologue of compound **386** on the other hand (**372**) precludes photocycloreversion and gives norbornadiene when irradiated or heated[175].

Azo compound **392** is accessible from the dibromocyclooctatriene adduct of ethyl azodicarboxylate[180] (**391**) and it thermolyses exclusively to

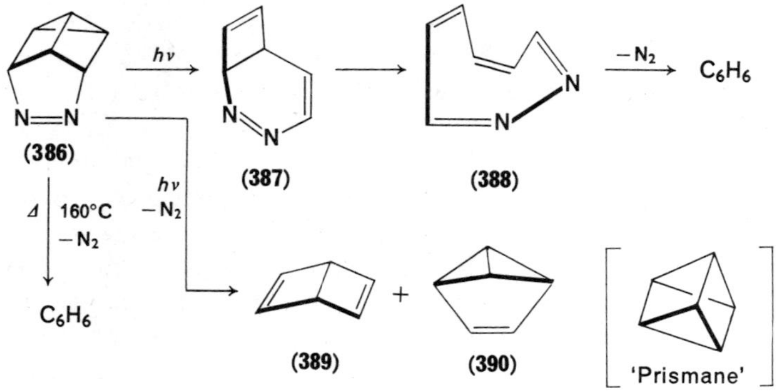

cyclooctadiene (**394**). If the tricyclic compound **393** is involved here, it rearranges into **394** under the reaction conditions, but it can be isolated by low temperature photolysis of **392**. Interestingly, none of the *syn*-stereoisomer of tricyclooctane (**393**) is observed suggesting stereoselective elimination of nitrogen with inversion at the carbon termini[181], unlike the photolysis of diazanorbornenes which proceeds with predominantly retained configuration[161]. Since the dienes **394** and **395** are formed in comparable amounts it seems unlikely that the cyclobutane ring has any appreciable influence on the photochemical reaction course which seems best explained in terms of 1,4-biradical intermediates.

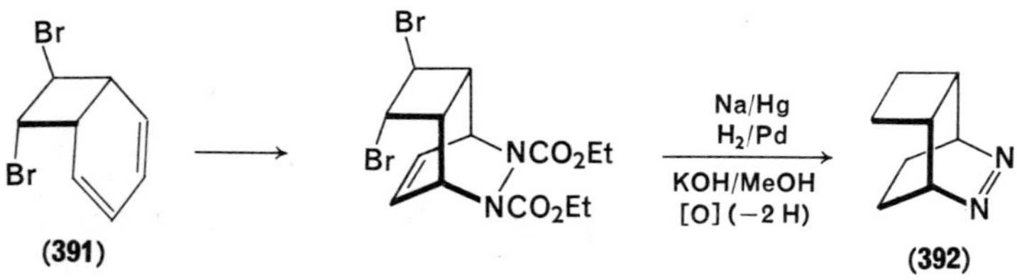

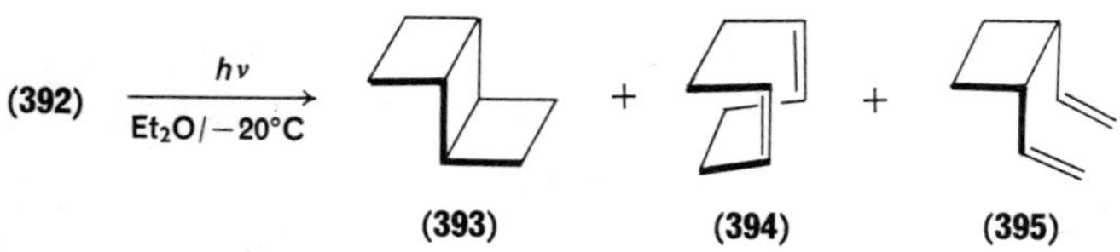

Similarly, the lower homologue of compound **392**, diazatricyclononene (**396**) gives exclusively cycloheptadiene (**398**) when heated, but photolysis again results in inversion and ring closure at the C—N termini when nitrogen is eliminated, giving tricycloheptane (**397**), in addition to diene

(**398**) and possibly the unstable *cis*-divinylcyclopropane (**399**)[182]. It seems likely that the thermal reactions leading exclusively to dienes **394** and **398** are homo- and bis-homocycloreversions involving transition states like **400**.

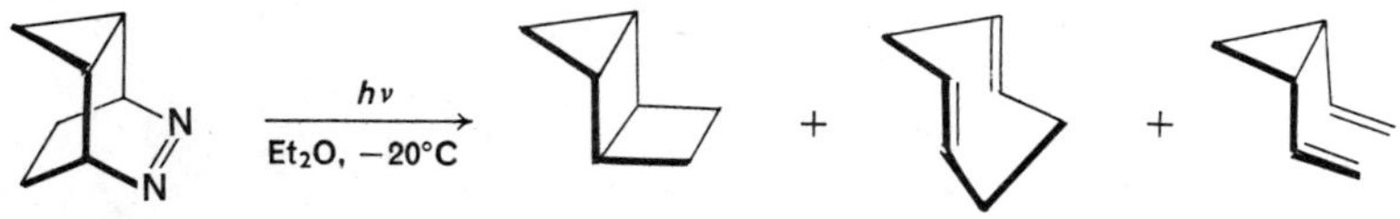

(**396**) (**397**) (**398**) (**399**)

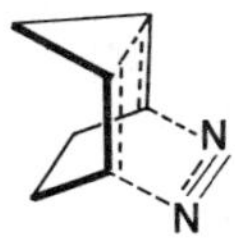

(**400**)

2. Bridged polycyclic azo compounds derived from triazines and tetrazines

Triazines such as **401** exhibit inverse electron demand in $(\pi_s^4 + \pi_s^2)$ cycloadditions, and they rapidly react with strained ring olefins to give bridged azo compounds such as **402**; nitrogen is readily lost from these and

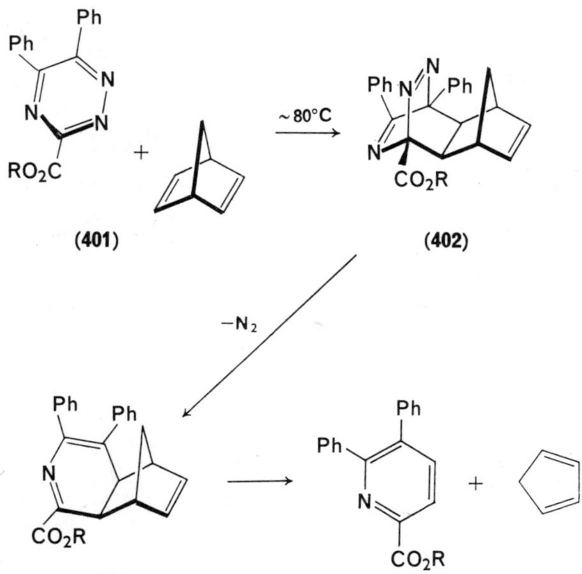

(**401**) (**402**) (**403**)

arylated pyridine carboxylic esters result from cycloreversion[183] of the intermediate compound **403** in precise analogy to the carbocyclic counterparts[184].

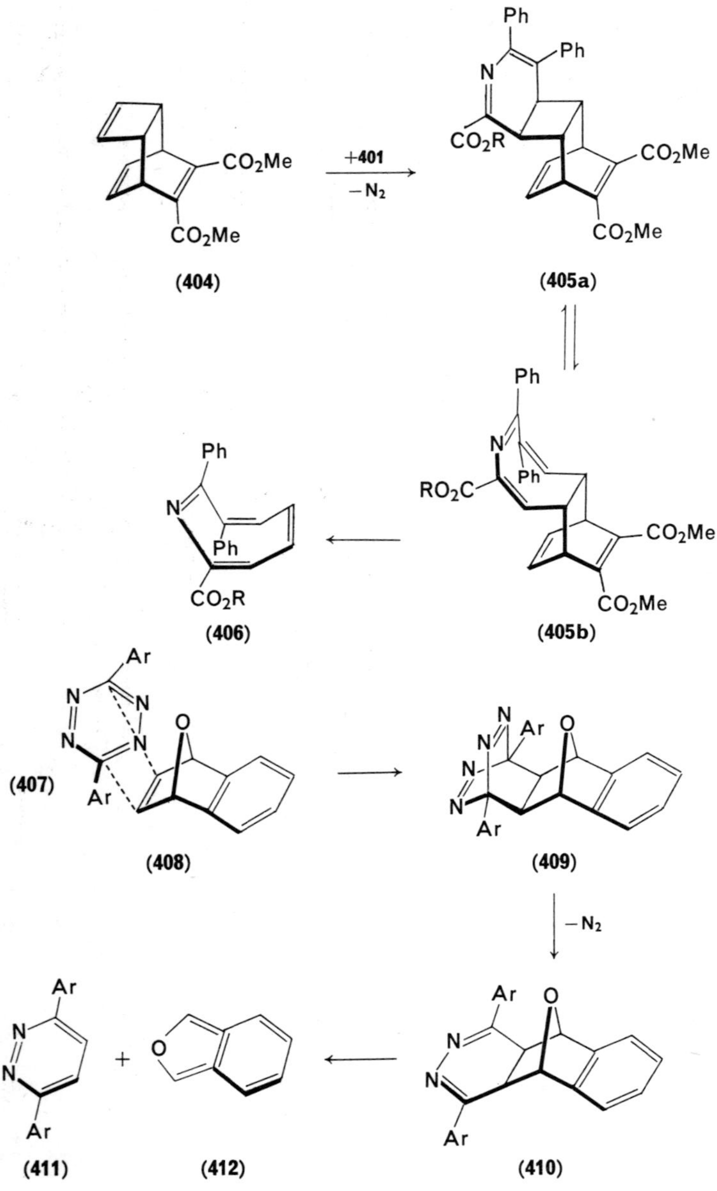

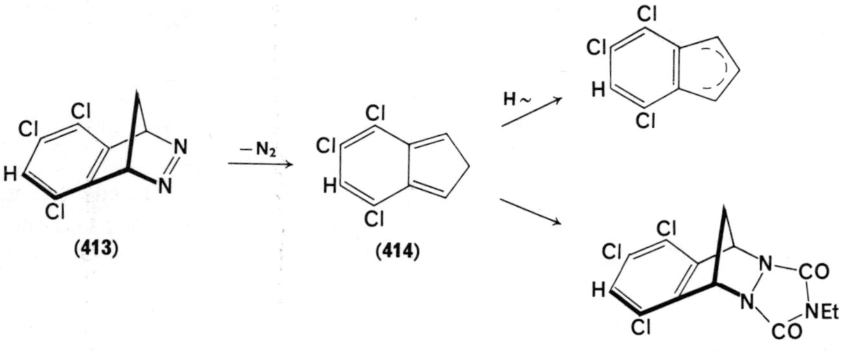

(413) (414) (415)

Similar cycloreversion of the deazetation product **405** formed from the adduct of triazine (**401**) and cyclooctatetraene-ynoic ester adduct (**404**) provides[183] a synthesis of azacyclooctatetraene (**406**). Tetrazines such as 3,6-diphenyl and 3,6-dipyridyltetrazine (**407**) can also form adducts[185] with strained ring dienophiles, e.g. **408**; the primary adducts (**409**) easily extrude nitrogen, and the products (**410**) behave like the carbocyclic structures[184] and fragment into the interesting isobenzfuran (**412**) and diarylpyridazines (**411**)[186].

In this connexion trichloroisoindene (**414**), a carbocyclic analogue of isobenzfuran (**412**), is liberated by mild thermolysis of diazabenzonorbornadiene (**413**), and is trapped by the highly reactive dienophile *N*-ethyltriazolindione to give adduct **415**[7].

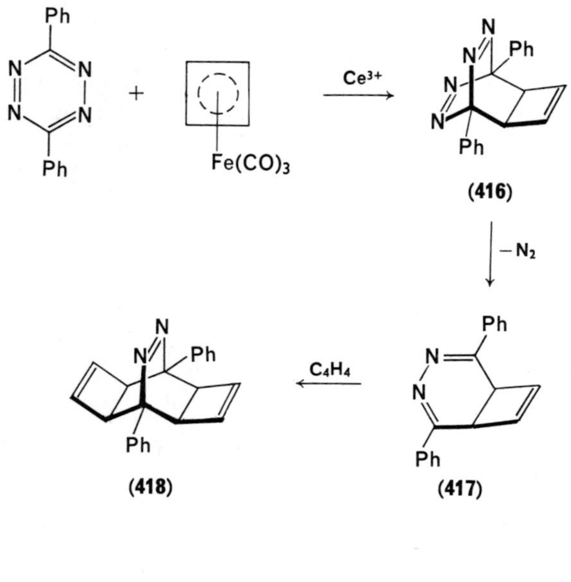

(416) (417) (418)

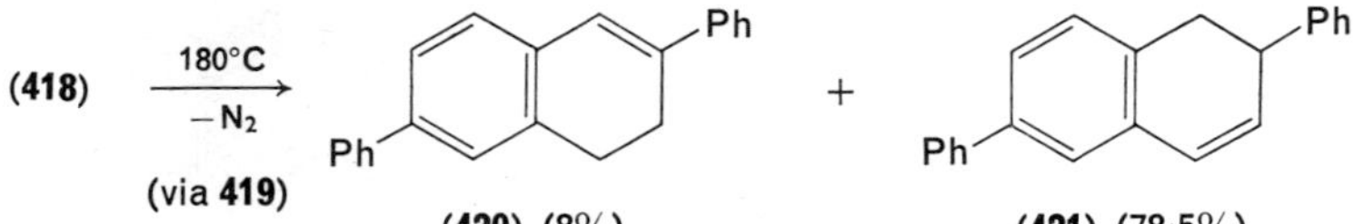

(420) (8%) (421) (78·5%)

Diaryltetrazines also form adducts, e.g. **416** with cyclobutadiene; these likewise easily lose nitrogen and the resulting diazacyclooctatetraene tautomers (**417**) react with a further molecule of cyclobutadiene to give the all-*syn* fused adduct **418**.

The decomposition of the azo-bridged adduct **418** at 180°C gives diphenyldihydronaphthalenes (**420** and **421**) via a transient cyclodecapentaene which electrocyclizes to 9,10-dihydronaphthalene (**419**); the latter undergoes thermal 1,5-sigmatropic hydrogen transfers, preferentially by way of the most stable conjugated intermediate, hence the higher yield of isomer **421**[187].

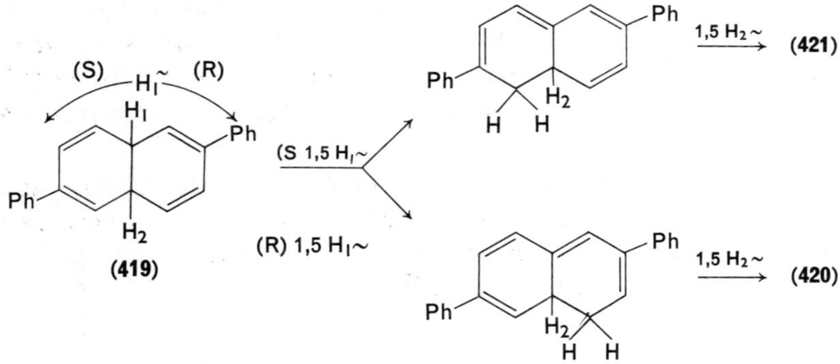

Tetrazines have also been used to make adducts from cyclopropenes and even diphenylcyclopropene will function as dienophile; once again nitrogen is lost from the initial adducts forming diazanorcaradienes, which either tautomerize into diazapines or can be trapped with suitable dienophiles[188].

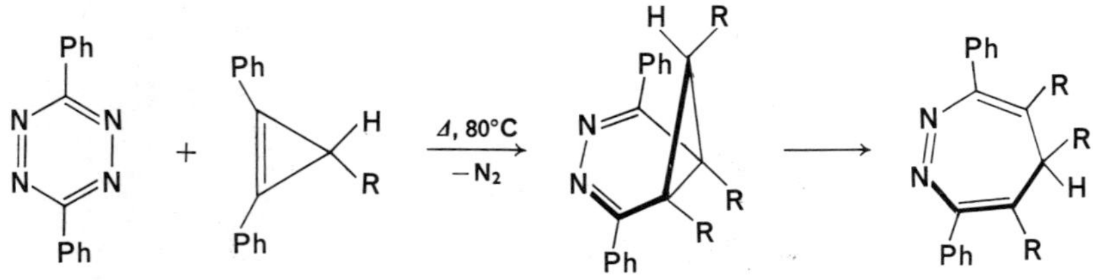

(422; R = Ph) (424; R = H)

(423; R = Ph) (425; R = H)

483 349

^{1}H n.m.r. studies show that diazanorcaradiene (**424**) equilibrates with diazapine (**425**) at 110°C and the equilibrium favours the diazapine at 180°C. By way of contrast, the diazanorcaradiene (**422**) is not observed; instead the diazapine formed **423** rearranges above 110°C into a tautomer at first thought to be **423a** (since it decomposed into tetraphenylpyrrole and and benzonitrile at 235°C), but an X-ray[189] structural determination has shown that the structure of the tautomer is **423b**, formally the product of consecutive 1,5-sigmatropic hydrogen shifts. The result does not preclude compound **423a** being involved in the equilibria leading to fragmentation into tetraphenylpyrrole and benzonitrile.

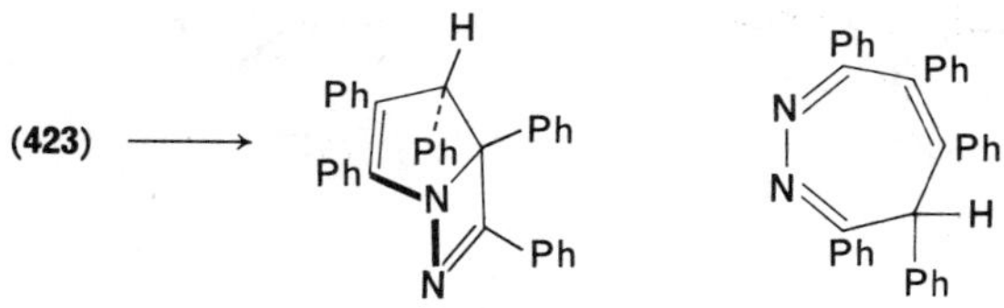

(423a) (423b)

Azo bridged compounds are also accessible from cycloadditions involving diphenyl-3,5-diazacyclopentadienones; the adducts, e.g. with norbornene, lose carbon monoxide and hydrogen to give pyradazinonorbornenes[190].

3. Caged structures containing the azo function

Entry into caged systems containing the azo function is rather easily accomplished, e.g. by photocyclization of the (indirectly prepared) cyclooctatetraene adduct of ethyl azodicarboxylate (**426**). The adduct itself is readily converted into the corresponding azo compound, which as expected expels nitrogen and readily reverts to cyclooctatetraene through the intermediate valence tautomer (**428**). By way of contrast the diazabishomocubene (**430**) (diazabasketene) is thermally rather stable[180], vigorous pyrolysis yielding intractable tars, and photolysis giving only a low yield of cyclooctatetraene. Flash vacuum pyrolysis, however, and low-temperature trapping gives the parent structure of compound **406**, the highly unstable azocine (**431**); this, whilst stable at –196°C decomposes on warming to –50°C[191].

An ingenious synthesis of a bis-azo cage structure starts from basketene (**432**); this undergoes multiple cycloaddition with ethyl azodicarboxylate to give a high yield of the bis-adduct **433**, which photocyclizes to a cage structure (**434**), and the usual steps transform this heptacyclic product into the bis-azo cage compound **435**. Flash vacuum pyrolysis of the bis-azo

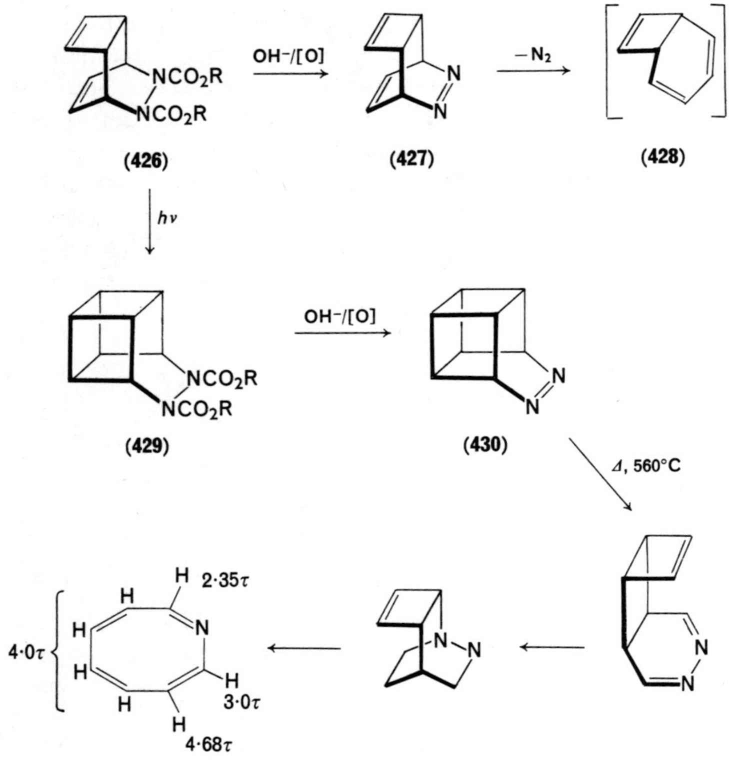

(426) (427) (428) (429) (430) (431) (C_8H_8, τ4·3)

compound gives dihydrofulvalene (**440**) as the only product, no pentaprismane (the anticipated product) nor any mono-azo compounds are detectable in the pyrolysate, and the mass spectrum supports the view that both nitrogen molecules are rapidly lost. Thermolysis kinetics yield E_a

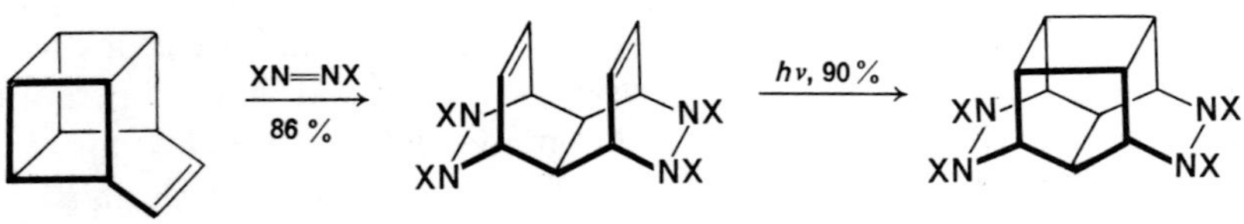

(432) (433; X = CO_2R) (434)

45·1 ± 0·4 kcal/mole and $\Delta S^{\neq}$ 6·6 e.u., the molecule decomposing 10^{-4}-fold slower than, for example, azo compound **392**, but almost an order of magnitude faster than diazabicyclooctene (**366**), suggesting cyclobutane participation in the fragmentation of diazatricyclodecene (**392**), but not

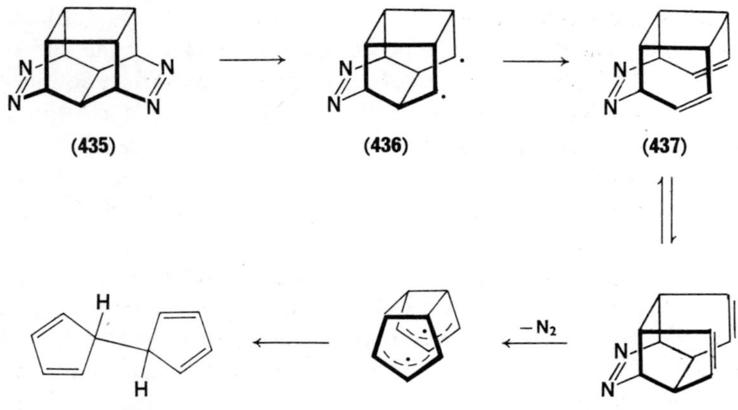

(435) (436) (437) (438) (439) (440)

for **435**. A possible mechanism involves rapid sequential nitrogen extrusion to give the bis-allylic biradical (**439**)[192]. In this connexion, bis-adduct **433** deuterium labelled at both termini of one of the olefinic bonds has been used as a source of *cis*-2,3-bisdeuterio-9,10-dihydronaphthalene, valuable in probing the degenerate rearrangements of the hydrocarbon. The required intermediate bis-azo compound is readily made by the usual hydrolysis–oxidation procedure[193].

The mechanistically fascinating rearrangements of cubane and allied structures catalysed by silver tetrafluoroborate (and Pd^{II} salts)[194] find a

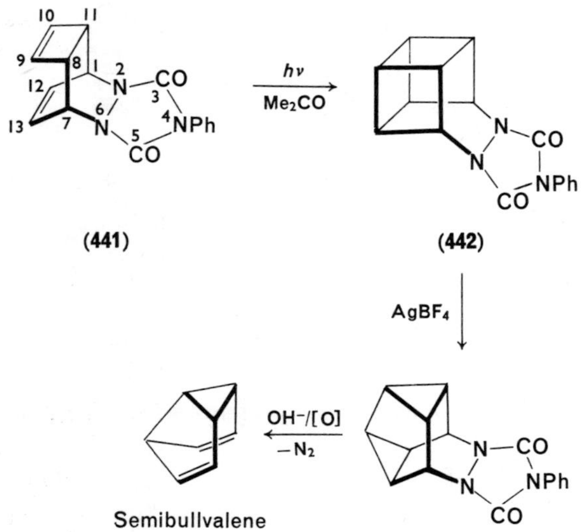

(441) (442) (443)

parallel in the similar (though *not necessarily* related) transformation of cyclooctatetraene-4-phenyltriazolindione adduct (**442**) into the isomer **443**, a source, by the usual operations, of azo compound **368**. Since adduct **441** is potentially available in large quantities, the thermolysis of azo compound **368** is an extremely valuable practical route to semibullvalene[195].

Attempts to bring about a similar rearrangement of diazabasketene (**430**) result in decomposition, and cycloreversion/rearrangement into azocine (**431**) seems a distinct possibility.

Other complex rearrangements involving adduct **426** have been uncovered[196]; thermolysis of the adduct results in a complex sequence of events which could involve electrocyclic and sigmatropic shifts, and/or intramolecular cycloaddition. Some of the steps which would appear necessary to rationalize the data do not seem to follow precisely the orbital symmetry selection rules, even though electrocyclic processes for diazacyclooctatetraene derivatives resemble those in the hydrocarbon[197]. The propensity for rearrangements in the $C_{10}H_{10}$ hydrocarbon series is well known[198], and it seems likely that similar complexity attends the behaviour of the diaza analogues. In this connexion bullvalene (**446**) is produced by photolysis of azo compound **444**, possibly via the intriguing bis-allylic biradical (**445**); in the thermolysis of pyrazoline (**444**), on the other hand, only bullvalene and its isomer (**448**) are produced[198].

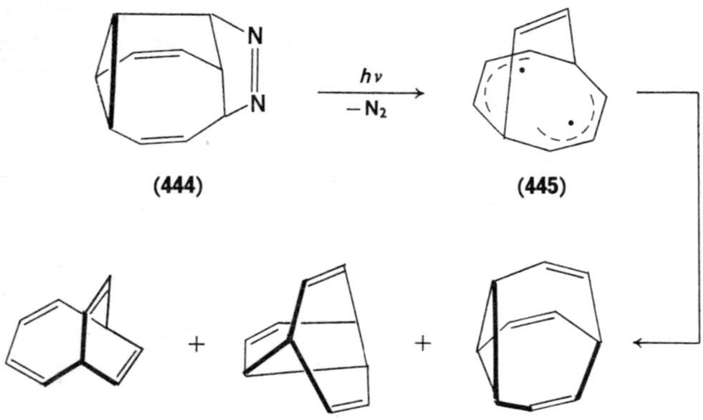

(**444**) (**445**)

(**448**) (**447**) (**446**)

C. *4-Phenyl-1,2,4-triazolindione as a Potential Source of Polycyclic Azo Compounds*

Early syntheses[199] of 4-phenyl-1,2,4-triazolindione (**336**) have proved unreliable, but its accessibility through phenylisocyanate[200] has rendered a particularly powerful and useful dienophile readily available; more recently the parent triazolindione has been reported[201]. The cerise azo compound (**336**) is best prepared in solution by oxidation of the hydrazo precursor with *t*-butyl hypochlorite[175] or lead tetraacetate[200] and the crystalline compound can be obtained by vacuum sublimation of the solution evaporate; *note that some hazard attends handling of any substantial quantities of the solid compound*[202]. The azo compound forms normal Diels–Alder adducts, even with those dienes which frequently undergo thermal 'ene' addition, as for example cycloheptatriene[203] and cyclohexa-1,3-diene with ethyl azodicarboxylate (the latter do however react photochemically Diels–Alder wise, perhaps an example of the rare excited state $\pi_s^4 + \pi_s^2$ concerted cycloaddition[204]). The azo compound reacts instantaneously with cyclopentadiene at –78°C and with cycloheptatriene at –50°C; rapid low-temperature addition also ensues with butadiene and isoprene, and at ambient temperatures with anthracene, cyclooctatetraene and norbornadiene (cf. compound **372**). The reaction with cycloheptatriene and cyclooctatetraene can involve their bicyclic valence tautomers norcaradiene and bicyclo[4.2.0]octadiene giving, for example, from the latter, adduct **441** (100°C dioxan[174]) but a change in reaction conditions (25°C acetone) can lead to an alternative adduct **449**[205].

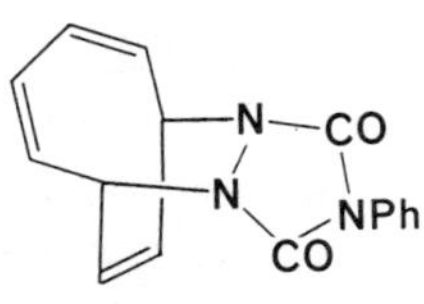

(449)

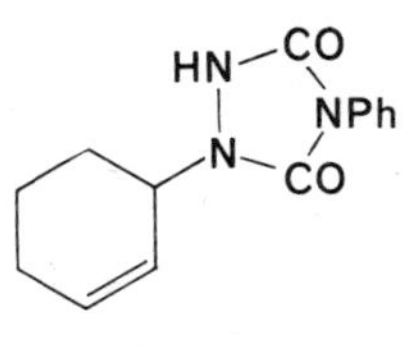

(450)

Valence tautomerism in substituted cyclooctatetraenes can lead to isomeric bicyclo[4.2.0]octatrienes, and equilibria in chloro- and bromocyclooctatetraene have been investigated by trapping the tautomers with *N*-phenyltriazolindione; chlorocycloctatetraene gives 8-chloro-**441** together with the vinylic-substituted isomer 9-chloro-**441**; similar bromo-adducts are formed from bromocyclooctatetraene with, in addition, the bromo-analogue of adduct **449**[206]. In principle, triazolindione adducts can be hydrolysed, e.g. by heating with methanolic potassium hydroxide or using similar bases in ethanediol[151], and the resulting hydrazo compounds are readily oxidized to the polycyclic azo compounds.

Ene-type additions do, however, occur with *N*-phenyltriazolindione[207] and, for example, cyclohexene reacts with **336** 3×10^4-fold faster than with ethyl azodicarboxylate to give adduct **450**. The lack of a marked solvent effect in this reaction points to a concerted addition, uncatalysed by added peroxides; other simple olefins, e.g. cyclopentene and the isomeric 2-butenes, react similarly.

In cycloadditions, **336** typically reacts two orders of magnitude faster than tetracycanoethylene, and will even form an adduct with the very unreactive hexachlorocyclopentadiene, with which the latter fails to react; kinetic data for various additions are compared in Table 4[208].

TABLE 4. Relative reactivities of triazolindione (**336**) and tetracyanoethylene

Diene	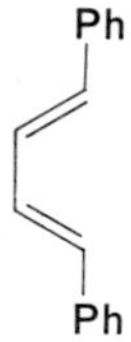	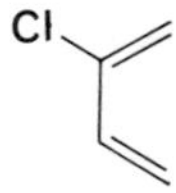	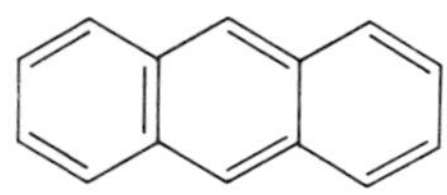	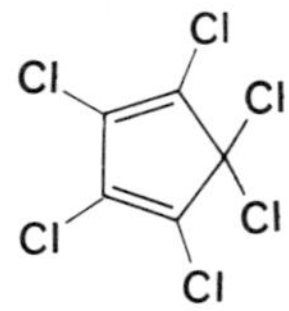
Relative rate with				
(*a*) azo dienophile (**336**)	$1{\cdot}13 \times 10^5$	$2{\cdot}3 \times 10^4$	$6{\cdot}9 \times 10^3$	27
(*b*) tetracyanoethylene	$8{\cdot}2 \times 10^2$	1·00	$2{\cdot}7 \times 10^4$	No reaction

The inability of hexachlorocyclopentadiene to form an adduct with tetracyanoethylene, and the slow reaction with **336**, could be considered due to 'inverse electron demand'[209, 56]. In accord with this, although dimethoxytetrachlorocyclopentadiene (**451**) slowly forms an adduct with triazolindione (**336**)[210] the trimethoxy (and tetramethoxy) dienes **452** and **453** do not react appreciably faster, if at all[211]. Interestingly, the reaction of

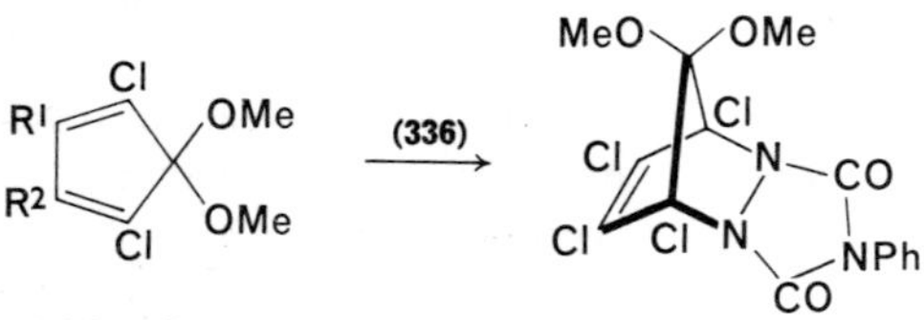

(**451**; R¹,² = Cl)
(**452**; R¹ = Cl, R² = OMe)
(**453**; R¹,² = OMe)

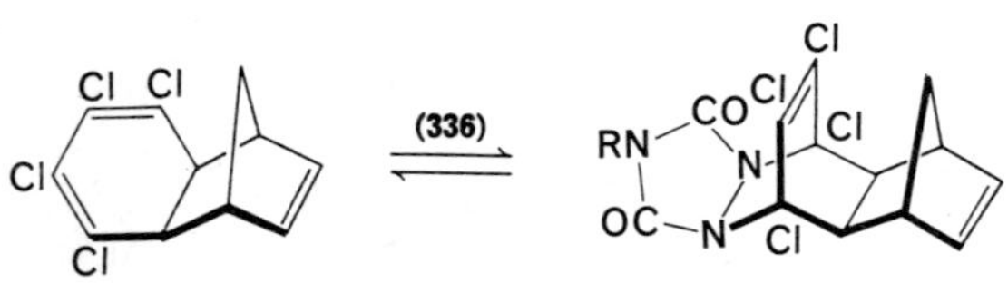

(**454**)

(**455**; R = Ph)
(**456**; R = Et)

the azo dienophile with tetrachlorodiene (**454**) is *reversible*; dissolution of the crystalline adduct **455** in the presence of 4-ethyl-1,2,4-triazolindione at room temperature gives the more stable adduct **456** by equilibrium displacement[7,178], and possibly the steric requirement of the somewhat bulky phenyl group is a factor here. Cycloreversions of adducts of **336** whilst rare are not without precedent; e.g. adducts with 9,10-dialkoxyanthracene, whilst being especially sensitive to acids, also revert to the addenda when heated in *neutral* media[208].

Reference has been made (Section III, B, 2) to the use of azo compounds to trap the reactive diene isoindene (**414**), and the 'Dewar' isomer of *o*-xylylene (**457**) also forms an adduct (**458**) which easily aromatizes when warmed in methanol[212].

Triazolindione **336** forms a normal ($\pi_s^4 + \pi_s^2$) transannular adduct with tropone and similar compounds are formed from azepines and diazepines, but *not* with tropolone which instead reacts by substitution addition[213]. Somewhat similar cycloadditions are seen with cyclic tetraenes **460** which give unsymmetrical adducts **461**[214]. Reactions with methylenecyclopropanes on the other hand give the unusual four-membered ring compounds **463**[215],

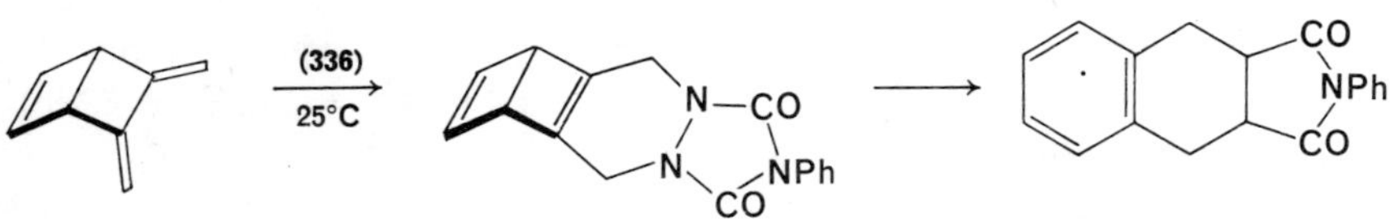

(457) **(458)** **(459)**

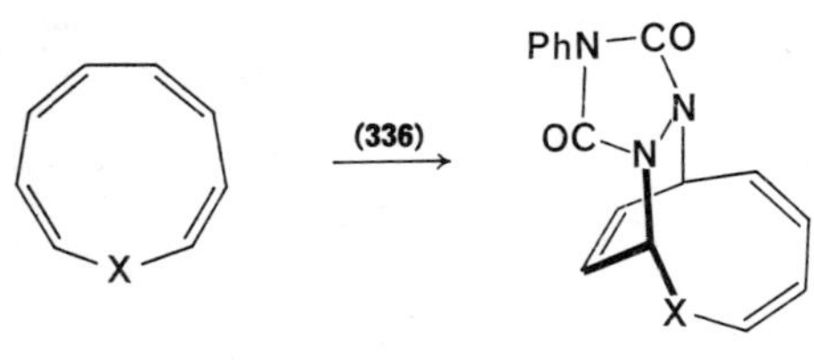

(460; X = NCONMe$_2$ or NAc) **(461)**

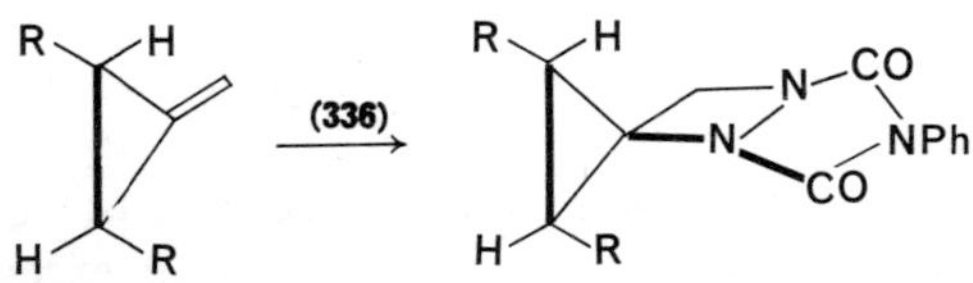

(462) **(463)**

among the few known reactions where a four-membered diaza ring is synthesized.

In analogy to the homo-diene addition with norbornadiene[175] (a source of azo compound **372**) and related bridged dienes[174], **336** forms a homodiene adduct with hexamethylbicyclo[2.2.0]hexadiene (cf. compound **386**) and with its photo-tautomer hexamethylprismane[216]. (More recently the mechanisms of these additions have been investigated[179a].)

Triazolindione (**336**) also readily forms adducts with isopyrazoles[217] and isopyrazole *N*-oxides[218]. The reaction with isopyrazoles is especially interesting since these dienes do not react with maleic anhydride, tetracyanoethylene or ethyl azodicarboxylate; the isopyrazole adducts thermally or photochemically extrude nitrogen, e.g. from adduct **464** to give diazabicyclopentane **465** or a disproportionation product **466** (depending on the detailed structure of the adduct). In the thermolysis of azo compound **464** ($R^1 = Ph$, $R^2 = CD_3$, $R^{3,4} = Me$) all the deuterium is retained in the product **466** and there is a strong preference for deuterium over hydrogen migration ($k_D/k_H = 2$). Photolysis of adducts **464** gives the bicyclopentane analogues **465** which may be isolated for substituents, $R^{1-4} = Me$, and $R^{1-3} = Me$, $R^4 = CH_2Ph$, but in other cases these compounds are unstable and their presence is inferred from ^1H n.m.r. observations; amusingly these diazabicyclopentanes behave like their carbocyclic analogues and readily react with triazolindione **336** (even at $-50°C$ for **465** $R^{1,2} = Ph$, $R^{3,4} = Me$) to give adducts **467**.

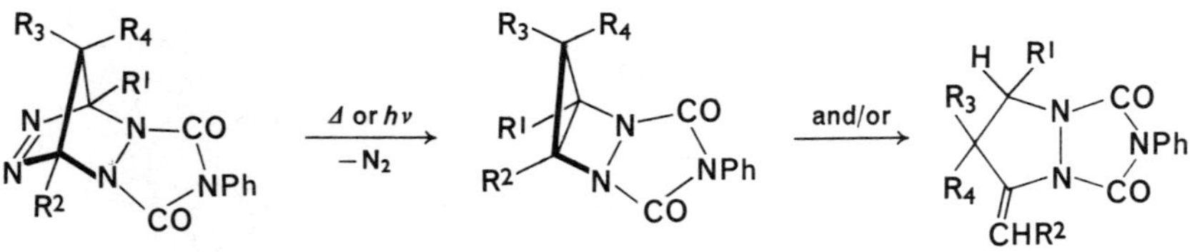

(464) **(465)** **(466)**

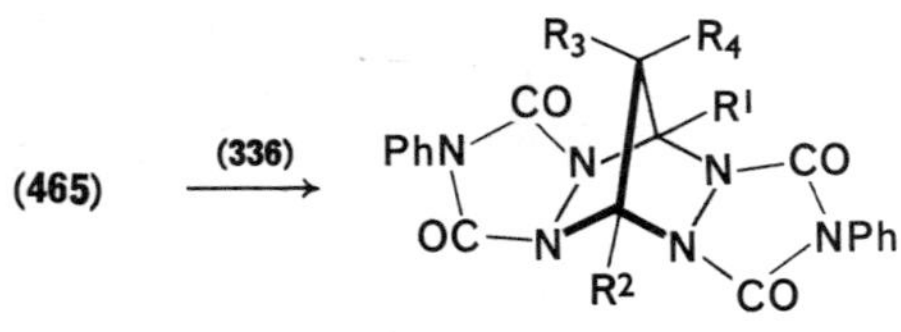

(467)

The ^1H n.m.r. spectra of compounds **465** at ambient temperatures shows that there is rapid inversion of the cyclopropane ring (except for $R^{1,3} = Me$, $R^4 = CH_2Ph$); for the tetramethyl compound (**465**; $R^{1-4} = Me$) k_1, the unimolecular rate constant for inversion, is 0·3 s^{-1} at 28°C whilst for **465** (R^1 = Ph, R^{2-4} = Me) its value is 0·5 s^{-1} at −8°C (higher temperatures causing ring opening to compounds **466**). If irradiation of the azo compounds **464** is carried out at low enough temperatures (e.g., −60°C) mixtures of the stereoisomeric 'bicyclopentanes' are observed and e.s.r. monitoring indicates that a singlet biradical is an intermediate here. At even lower tempera-

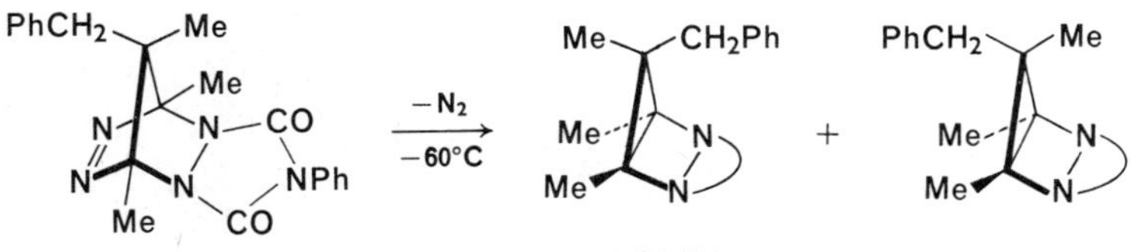

tures (matrix at 4–77 K) mainly the *syn* compound (**A**) is formed and thus greater stereoselectivity is observed here and, interestingly, with inversion of the methylene bridge. At these low temperatures, triplet intermediate diradicals are involved; an e.s.r. signal may be observed for many hours at 77 K but disappears on warming to 190 K. Triplet e.s.r. signals are not observed for **464** (R^{1-4} = Me or R^1 = Ph, R^{2-4} = Me) until the temperature is reduced to 4 K. This reflects the ability of aromatic substituents to stabilize the triplet biradicals formed in arylated derivatives of **464**. Clearly, the triplet species involved here represents the ground state for the biradical intermediate and the higher energy singlet species involved in the remarkable stereomutations of the cyclopropane ring in examples of compound **465** must clearly be stabilized by the adjacent hetero atoms[217]. The singlet species must also be involved in the disproportionation leading to compounds **466**.

Addition of isopyrazoles to dialkylpyrazolindiones[219] gives adducts **468** similar to **464** which do not, however, form triplet intermediates by photochemical nitrogen extrusion[217]; it seems possible that their conversion into monocyclic products by transannular N—N scission reflects the greater stability of the triazolidindione ring present in all triazolindione adducts (which, incidentally, are degraded with bases with rather variable ease[220]).

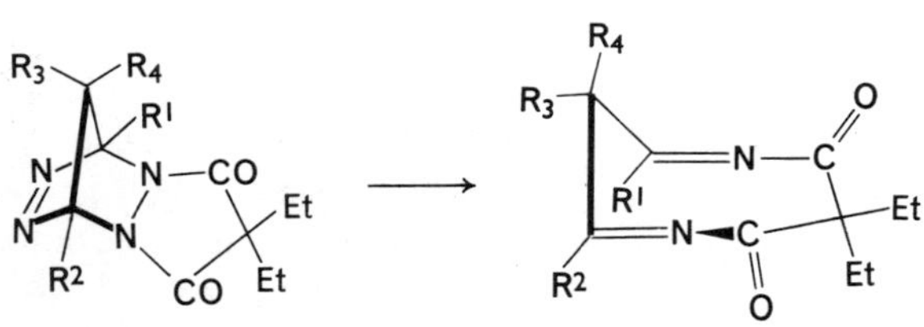

(468) **(469)**

The azoxy analogues **470** of adducts **464** lose nitrogen on photolysis to give hydroxy analogues of the compounds **466**, e.g. **471**; these are catalytically dehydrated by acids and the same bis-exomethylene product is formed by simple thermolysis of the initial adduct in, for example ethanol or chloroform, but not, it is claimed, in dimethyl sulphoxide[218].

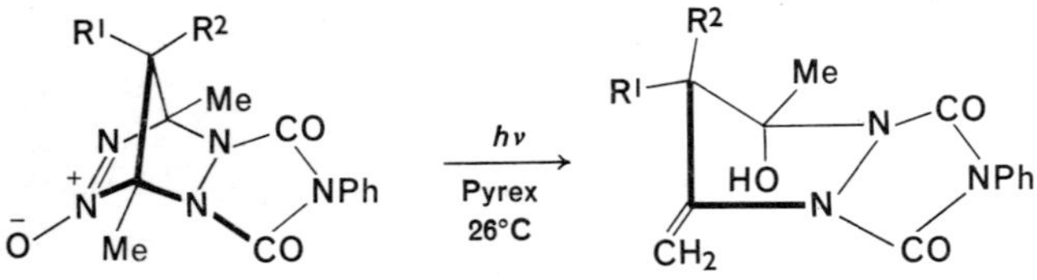

(470) (471)

Finally, mention should be made of 2,3-diazaquinones (3,6-pyridazinediones and phthalazine-1,4-diones) which exhibit powerful dienophilic reactivity, giving fused-ring heterocyclic structures; these have not generally found application for regenerating cyclic azo compounds, but their thermochemical properties have been investigated; e.g. the cyclopentadiene adduct cycloreverts and, e.g. phthalazindione loses nitrogen to form a bis-ketene, which yields benzocyclobutenedione as the final product[222].

Azodicarbononitrile formed by pyrolysis of cyanogen azide is also a very reactive dienophile potentially useful for introducing the azo function; its use is mitigated by its sensitivity to heat and mechanical shock[223]. Other cyclic azo compounds which have been investigated recently include diazacyclopentadienones[224, 225].

IV. APPENDIX

The Addition of Diazomethane to Olefines to form Pyrazolines (Theoretical)

We have already seen how the cycloaddition of a 1,3-dipolar reagent such as diazomethane to a dipolarophile viewed in molecular orbital terms uses the isoelectronic allyl anion as a model for diazomethane. This approach gives a valuable insight into the cycloaddition mechanism and indicates the effect of substituents in both 1,3-dipolar reagent and dipolarophile[56]. It has recently been stressed[226] that further insight into the mechanism of cycloaddition reactions may be obtained by a more detailed analysis of the electronic structure of both reactants and products. An example of this is the extrusion of nitrous oxide from 2,3-diazabicyclo[2.2.2]octa-2,3-diene *N*-oxide discussed above, (Section III. B. 1.) where the slower extrusion of nitrous oxide relative to nitrogen from the parent diaza- compound is explained in terms of the interaction of the diene HOMO and the LVMO of the bent nitrous oxide molecule[170]. The reduction in symmetry caused by the strong oxygen perturbation diminishes orbital overlap and raises the interaction energy. The reaction is then classified as partly forbidden. As such conclusions can only be reached by considering the precise electronic configuration of the species involved some MINDO/2 S.C.F. calculations have been carried out on diazomethane[227]. The MINDO method[228] appears at the present time to give a quite reliable estimate of ground-state properties and these sanguine expectations are further justified by the results for the ionization potentials shown below for diazomethane, allene and ketene, by way of comparison.

Molecule	Experimental	Ionization Potential, eV		
		MINDO/2	CNDO/2[229]	*Ab initio*[230]
Diazomethane	8·99	9·18	9·67	10·15
Allene	10·19	9·75	11·27	12·50
Ketene	9·60	9·62	10·28	11·51

In the experimentally determined[12] linear C_{2v} geometry, the HOMO for diazomethane is indeed analogous to the nonbonding orbital of the allyl anion having the form $(-0{\cdot}67C_12p_z - 0{\cdot}21N_12p_z + 0{\cdot}71N_22p_z)$ with energy $-9{\cdot}18$ eV. The central node is no longer present due to the heteroatom perturbation. However, as diazomethane approaches the dipolarophile a considerable distortion away from the linear geometry is expected and calculations show the nature of the electronic configuration at a few selected conformations.

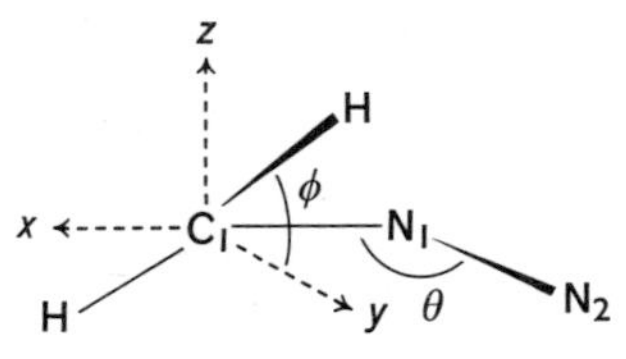

θ	Heats of formation, kcal/mole		
	$\phi = 0°$	$\phi = 90°$	$\phi = 45°$
120°	136·53	116·59	113·05
150°	89·49	90·92	90·28
180°	83·73	83·73	

Predictably the linear geometry is found to be the most stable of the configurations investigated. As θ is decreased, the all-planar conformation with $\theta = 120°$ and $\phi = 0°$ is strongly destabilized with respect to the conformation with the methylene groups rotated through 45° or 90°. This instability is due to the increased nuclear and electronic repulsions between the hydrogen H_1 and nitrogen N_2 on bending. That these factors are not manifest for $\theta = 150°$ would be expected from the distance dependence of the nuclear and electronic repulsion terms. It would seem unlikely that diazomethane would retain the all-planar conformation as the dipolarophile approached, indeed in this conformation the HOMO is no longer the nonbonding allyl type, which becomes unoccupied due to the severe repulsive forces that would operate if it were filled.

If the conformer with $\theta = 120°$ and $\phi = 45°$ is considered a more realistic model for the construction of the transition state then the form of its HOMO is useful in determining the extent of interaction. Indeed, this MO is of the nonbonding allyl type in both xy and xz planes though the amplitude is greater in the xz plane:

$$C_1(-0{\cdot}43p_y - 0{\cdot}40p_z) + N_1(-0{\cdot}08p_y - 0{\cdot}22p_z) + N_2(0{\cdot}19p_y + 0{\cdot}51p_z)$$

However, for the only slightly less stable conformation with $\phi = 90°$ the nonbonding allyl type orbital is strongly stabilized by a hyperconjugative interaction with an antisymmetric combination $(H_1 - H_2)$. The hyperconjugative effect stabilizes all conformers to some extent but when $\phi = 0°$ it is only affecting the $2p_y$ orbitals which are destabilized on bending more severely than the $2p_z$ orbitals.

It may be said that whatever precise geometry diazomethane adopts in the transition state of a 1,3-dipolar addition, the HOMO–LVMO interactions are going to be markedly dependent on a delicate balance of energy factors.

V. ACKNOWLEDGMENT

I should like to express my thanks to my student and colleague Dr. Woo Pok Lay for his stimulating contributions to our work with cyclic azo compounds, and to Dr. John R. Telford for valuable initial contributions.

VI. REFERENCES

1. B. P. Stark and A. J. Duke, *Extrusion Reactions*, Pergamon Press, Oxford, 1969.
2. *Chemical Society Annual Reports*, **57–69** (1960–1972).
3. A. Ledwith and P. J. Russell, in *Organic Reaction Mechanisms* (Ed. B. Capon and C. W. Rees), Vol. 7, 1971, p. 275; A. Ledwith, in *Organic Reaction Mechanisms*, (Ed. B. Capon and C. W. Rees), Vol. 6, 1970, p. 305.

4. A. Ledwith and I. Cowell, *Quart. Rev. (London)*, **24,** 119 (1970).
5. J. B. Aylward, *Quart. Rev. (London)*, **25,** 407 (1971).
6. R. Huisgen, R. Grashey and J. Sauer, in *The Chemistry of the Alkenes* (Ed. S. Patai), Wiley–Interscience, London, 1964.
7. K. Mackenzie, W. P. Lay and J. R. Telford, *J. Chem. Soc.*, 3199 (1971).
8. R. B. Woodward and R. Hoffmann, *The Conservation of Orbital Symmetry*, Verlag Chemie–Academic Press, 1970.
9. E. I. Becker, M. A. Ogliaruso and G. M. Romaneli, *Chem. Rev.*, **65,** 261 (1965).
10. C. H. F. Allen, *Chem. Rev.*, **62,** 653 (1962); **37,** 209 (1945).
11. H. Boersch, *Monatsch.*, **65,** 331 (1935).
12. A. P. Cox, L. F. Thomas and J. Sheridan, *Nature*, **181,** 1000 (1958).
13. C. Bradley-Moore and G. C. Pimental, *J. Chem. Phys.*, **40,** 329, 340, 1529 (1964).
14. S. R. Pauson, *Angew. Chem.*, **72,** 781 (1960).
15. E. Schmitz and R. Ohme, *Chem. Ber.*, **94,** 2166 (1961).
16. E. Schmitz and R. Ohme, *Tetrahedron Letters*, 612 (1961).
17. E. Schmitz, R. Ohme and R. D. Schmidt, *Chem. Ber.*, **95,** 2714 (1962).
18. E. Schmitz and R. Ohme, *Kernenergie*, **5,** 357 (1962).
19. E. Schmitz and R. Ohme, *Chem. Ber.*, **95,** 795 (1962); *Angew. Chem. Internat. Ed.*, **3,** 333 (1964); *Angew. Chem.*, **73,** 115, (1961).
20. E. Schmitz and R. Ohme, *Z. Chem.*, **7,** 226 (1967).
21. L. Pierce and V. Dobyns, *J. Amer. Chem. Soc.*, **84,** 2651 (1962).
21a. J. E. Wollrab, LeRoy H. Sharpen and P. D. Ames, *J. Chem. Phys.*, **49,** 2405 (1968).
22. E. Schmitz, *Adv. Heterocyclic Chem.*, **2,** 122 (1963).
23. E. Schmitz, *Adv. Heterocyclic Chem.*, **2,** 104 (1963).
24. H. Lund, *Adv. Heterocyclic Chem.*, **12,** 270 (1970).
25. R. A. Reed, *Chem. Ind. (London)*, 529 (1966) and references cited therein.
25a. Cf. J. A. Merritt, *J. Ala. Acad. Sci.*, **38,** 138 (1967).
26. W. H. Graham, *J. Amer. Chem. Soc.*, **87,** 4396 (1965).
26a. E. Schmitz, C. Hoerig and C. Greundemann, *Chem. Ber.*, **100,** 2093 (1967).
27. J. L. Zollinger, C. D. Wright, J. J. McBrady, D. H. Dytvig, F. A. Fleming, C. A. Kurhajee, R. A. Mitsch and E. W. Neuvar, *J. Org. Chem.*, **38,** 1065 (1973); see also R. A. Mitsch, *J. Heterocyclic Chem.*, **1,** 59 (1964); R. A. Mitsch *et al.*, *J. Heterocyclic Chem.*, **1,** 233 (1964) and R. A. Mitsch *et al.*, *J. Amer. Chem. Soc.*, **87,** 758 (1965).
28. R. Hoffmann quoted in H. M. Frey, *Adv. Photochem.*, **4,** 225 (1966); cf. R. Bonaccorsi, E. Scrocco and J. Tomasi, *J. Chem. Phys.*, **52,** 5270 (1970).
28a. E. Haselbach, E. Heilbronner, A. Mannschreck and W. Seitz, *Angew. Chem., Internat. Ed.*, **9,** 902 (1970).
28b. W. L. Jorgenson and L. Salem, *The Organic Chemists Book of Orbitals*, Academic Press, New York and London 1973.
29. E. Schmitz, D. Habisch and A. Stark, *Angew. Chem.*, **75,** 723 (1963) [*Internat. Ed.*, **2,** 548 (1963)].
30. H. D. Roth, *J. Amer. Chem. Soc.*, **93,** 4935 (1971).
31. H. M. Frey and I. D. R. Stevens, *J. Chem. Soc.*, 1700 (1965).
32. H. M. Frey, *Pure and Appl. Chem.*, **9,** 527 (1964).
33. H. M. Frey, *Adv. Photochem.*, **4,** 225 (1966).

34. G. A. Bottomley and G. L. Nyberg, *Australian J. Chem.*, **17,** 406 (1964).
35. B. S. Rabinowitch and D. W. Setser, *Can. J. Chem.*, **40,** 1425 (1962).
36. M. J. Amrich and J. A. Bell, *J. Amer. Chem. Soc.*, **86,** 292 (1964).
37. C. B. Moore and G. C. Pimental, *J. Chem. Phys.*, **41,** 3504 (1964).
37a. G. Lowe and J. Parker, *Chem. Comm.*, 1135 (1971).
38. H. M. Frey and I. D. R. Stevens, *J. Chem. Soc.*, 3101 (1965).
39. W. Kirmse and W.v. Wedel, *Annal.*, **666,** 1 (1963).
40. R. A. Moss, *Tetrahedron Letters*, 4905 (1967).
40a. M. R. Bridge, H. M. Frey and M. T. H. Lui, *J. Chem. Soc.*(*A*), 91 (1969); H. M. Frey and M. T. H. Lui, *J. Chem. Soc.*(*A*), 1916 (1970).
41. R. A. Mitsch, *J. Amer. Chem. Soc.*, **87,** 758 (1965).
41a. W. J. Baron, M. R. De Camp, M. E. Hendrick, M. Jones, Jr., R. H. Levin and M. B. Sohn in M. Jones and R. A. Moss, *Carbenes*, Vol. 1, Chap. 1, John Wiley, New York, 1973; G. W. Griffen and N. R. Bertoniere, *Carbenes*, p. 205; R. A. Moss, *Carbenes.*, p. 153.
42. P. Margaretha and S. Solar, *Angew. Chem.*, *Internat. Ed.*, **11,** 1024 (1972).
43. H. J. Emeleus and G. L. Hurst, *J. Chem. Soc.*, 3276 (1962).
44. D. M. Lemal, N. Rieber, J. Alberts and J. A. Lipsky, *J. Amer. Chem. Soc.*, **91,** 5668 (1969).
45. E. Buchner, *Ber.*, **21,** 2637 (1888); **23,** 701 (1890); E. Buchner and A. Papendieck, *Annal.*, **273,** 232 (1893).
46. L. I. Smith, *Chem. Rev.*, **23,** 193 (1938).
47. R. Huisgen, *Angew. Chem.*, **67,** 439 (1955); T. L. Jacobs, in *Heterocyclic Compounds* (Ed. R. C. Elderfield), Vol. 5, 1952. p. 72.
48. R. J. Crawford, A. Mishra and R. J. Dimmel, *J. Amer. Chem. Soc.*, **88,** 3959 (1966).
49. R. Huisgen, H. Stangl, H. J. Sturm and H. Wagenhofer, *Angew. Chem.*, **73,** 170 (1961).
50. R. Huisgen, *Angew. Chem.*, *Internat. Ed.*, **2,** 566, 633 (1963).
51. R. Huisgen, *Proc. Chem. Soc.*, 357 (1961).
52. R. A. Firestone, *J. Org. Chem.*, **33,** 2285 (1968).
53. R. Huisgen, *J. Org. Chem.*, **33,** 2291 (1968).
54. W. E. Parham, C. Serres, P. R. O'Connor, *J. Amer. Chem. Soc.*, **80,** 588 (1958); W. E. Parham, H. G. Braxton and P. R. O'Connor, *J. Org. Chem.*, **26,** 1805 (1961).
55. R. Hoffmann and R. B. Woodward, *Acc. Chem. Res.*, **1,** 17 (1968).
56. R. Sustmann, *Tetrahedron Letters*, 2717 (1971).
56a. A. Eckell, R. Huisgen, R. Sustmann, G. Wallbillich, D. Grashey and E. Spindler, *Chem. Ber.*, **100,** 2192 (1967); G. Bianchi, C. De Micheli, R. Gandolfi, P. Grünanger, P. V. Finzi and O. V. de Pava, *J. Chem. Soc. Perk.*, I, 1148 (1973).
56b. A. Eckell, R. Huisgen and R. Sustmann, G. Wallbillich, D. Grashey and E. Spindler, *Chem. Ber.*, **100,** 2192 (1967), *cf.* K. N. Houk, *J. Amer. Chem. Soc.*, **95,** 4092 (1973).
57. C. G. Overberger, N. Weinschanker and J- P. Anselme, *J. Amer. Chem. Soc.*, **87,** 4119 (1965).
58. P. Eberhard and R. Huisgen, *J. Amer. Chem. Soc.*, **94,** 1345, 1346 (1972).
59. J. H. Atherton and R. Fields, *J. Chem. Soc.* (*C*), 1507 (1968).
60. N. G. Barlow and R. N. Haszeldine, *Chem. Comm.*, 931 (1969).

61. K. von Auers and F. König, *Annal.*, **496,** 252 (1932).
62. J. Smith and K. L. Howard, *J. Amer. Chem. Soc.*, **65,** 159 (1943).
63. L. Staudinger, E. Anthes and F. P. Fenninger, *Ber.*, **49,** 1928 (1916).
64. J. B. Cloke, *J. Amer. Chem. Soc.*, **56,** 2710 (1934).
65. D. E. McGreer, P. Morris and G. Carmichael, *Can. J. Chem.*, **41,** 726 (1963); D. E. McGreer, W. Wai and G. Carmichael, *Can. J. Chem.*, **38,** 2410 (1960).
66. T. V. van Auken and K. L. Rinehart, *J. Amer. Chem. Soc.*, **84,** 3736 (1962).
67. N. G. Young, L. J. Andrews, S. L. Lindenbaum and S. J. Cristol, *J. Amer. Chem. Soc.*, **66,** 810 (1944).
68. D. E. McGreer, R. S. McDaniel and M. G. Vinje, *Can. J. Chem.*, **43,** 1398 (1965); cf. J. Buss, H. Steinberg and T. J. de Boer, *Monatsh.*, **98,** 1155 (1967).
69. D. E. McGreer, N. W. K. Chiu and M. G. Vinje, *Can. J. Chem.*, **43,** 1398 (1965).
70. C. G. Overberger and J-P. Anselme, *J. Amer. Chem. Soc.*, **86,** 658 (1964).
71. D. E. McGreer, N. W. K. Chiu, M. G. Vinje and K. C. K. Wong, *Can. J. Chem.*, **43,** 1407 (1965).
72. H. M. Walborsky and C. G. Pitt, *J. Amer. Chem. Soc.*, **84,** 4831 (1962).
73. D. E. McGreer and Weh-Sai Wu, *Can. J. Chem.*, **45,** 461 (1967).
74. Cf. D. E. McGreer and J. W. McKinley, *Can. J. Chem.*, **49,** 105 (1971).
75. R. Hoffmann, *J. Amer. Chem. Soc.*, **90,** 1475 (1968); cf. T. C. Chambers and G. B. Kistiakowsky, *J. Amer. Chem. Soc.*, **56,** 399 (1934); A. K. Q. Sim, W. M. St. John and E. F Hayes, *J. Amer. Chem. Soc.*, **92,** 7249 (1970).
76. R. J. Crawford and A. Mishra, *J. Amer. Chem. Soc.*, **88,** 3963 (1966).
77. R. J. Crawford and B. H. Al-Sader, *Can. J. Chem.*, **46,** 3301 (1968).
78. S. Seltzer and S. G. Mylonakis, *J. Amer. Chem. Soc.*, **89,** 6584 (1967); cf. reference 77.
79. R. J. Crawford and L. H. Ali, *J. Amer. Chem. Soc.*, **89,** 3908 (1967).
79a. N. D. Epiotis, *J. Amer. Chem. Soc.*, **95,** 1200 (1973).
80. R. J. Crawford and G. L. Erikson, *J. Amer. Chem. Soc.*, **89,** 3907 (1967).
81. R. J. Crawford and A. Mishra, *Can. J. Chem.*, **47,** 1515 (1969).
82. R. J. Crawford, A. Mishra and R. Moore, *Can. J. Chem.*, **46,** 3305 (1968).
83. E. B. Klunder and R. W. Carr, *Chem. Comm.*, 742 (1971).
84. B. S. Rabinowitch, *J. Phys. Chem.*, **72,** 191 (1968).
85. P. Cadman, H. M. Mennier and A. F. Trotman-Dickenson, *J. Amer. Chem Soc.*, **91,** 7640 (1969).
86. F. H. Dorer, E. Brown, J. Do and R. Rees, *J. Phys. Chem.*, **65,** 1640 (1971).
86a. H. Andrist, *J. Org. Chem.*, **38,** 1772 (1973).
87. C. G. Overberger, N. Weinshenker and J-P. Anselme, *J. Amer. Chem. Soc.*, **87,** 4119 (1965).
88. J. W. Timberlake and B. K. Bandlish, *Tetrahedron Letters*, 1393 (1971).
89. C. G. Overberger and J-P. Anselme, *J. Amer. Chem. Soc.*, **86,** 658 (1964).
90. Y. Jean and L. Salem, *Chem. Comm.*, 382 (1971).
91. E. L. Allred and C. R. Flynn, *J. Amer. Chem. Soc.*, **94,** 5891 (1972).
91a. E. L. Allred and C. R. Flynn, *J. Amer. Chem. Soc.*, **92,** 1064 (1970).
92. F. M. Dean, L. E. Houghton and R. B. Morton, *J. Chem. Soc.* (*C*), 1980 (1967); 2065 (1968).
93. R. Hoffmann, *Chem. Comm.*, 240 (1969); A. Streitwieser, *Molecular Orbital Theory for Organic Chemists*, John Wiley, New York and London 1961, p. 43; cf. W. W. Schoeller, *Tetrahedron Letters*, 2043, 2047 (1973).

94. R. E. Erickson, *Tetrahedron Letters*, 1753 (1966); N. Wakabayashi, *J. Org. Chem.*, **32,** 489 (1967); A. N. Herriott and W. M. Jones, *Tetrahedron Letters*, 2387 (1967).
95. G. Köbrich and B. Rösner, *Tetrahedron Letters*, 2031 (1973) and references cited therein; W. v. E. Doering and H. D. Roth, *Tetrahedron*, **26,** 2825 (1970) and references cited therein; A. Bezaguet and M. Bertrand, *Compt. Rend.*, **362C,** 1679 (1966); J. K. Crandall and D. R. Paulson, *J. Amer. Chem. Soc.*, **88,** 4302 (1966).
96. T. C. Shields, B. A. Shoulders and P. D. Gardner, *J. Org. Chem.*, **31,** 1303 (1966).
97. A. C. Day and M. C. Whiting, *J. Chem. Soc.*, 464 (1966).
98. S. D. Andrews and A. C. Day, *Chem. Comm.*, 667 (1966).
99. T. Janjiki, H. Kato and M. Ohta, *Chem. Comm.*, 496 (1968).
100. P. Dowd, *J. Amer. Chem. Soc.*, **88,** 2587 (1966).
101. R. J. Crawford and D. M. Cameron, *J. Amer. Chem. Soc.*, **88,** 2589 (1966).
102. W. T. Borden, *Tetrahedron Letters*, 259 (1967).
103. P. Dowd and K. Sachder, *J. Amer. Chem. Soc.*, **90,** 2715 (1968); P. Dowd, G. Sengupta and K. Sachder, *J. Amer. Chem. Soc.*, **92,** 5726 (1970).
104. T. A. Berson, R. J. Bushby, J. M. McBride and M. Tremelling, *J. Amer. Chem. Soc.*, **93,** 1544 (1971).
105. G. L. Closs, *J. Amer. Chem. Soc.*, **93,** 1546 (1971).
106. T. J. Gajewski, A. Yeshurum and E. J. Blair, *J. Amer. Chem. Soc.*, **94,** 2138 (1972).
107. See for example S. L. Murov, R. S. Cole and G. S. Hammond, *J. Amer. Chem. Soc.*, **90,** 2957 (1968).
108. J. W. Kenner and D. W. Adamson, *J. Chem. Soc.*, 286 (1935).
109. C. D. Hurd and S. C. Liu, *J. Amer. Chem. Soc.*, **57,** 2656 (1935).
110. A. Ledwith and D. Parry, *J. Chem. Soc.*, 41 (1967); cf. D. Y. Curtin and S. M. Gerber, *J. Amer. Chem. Soc.*, **74,** 4052 (1952).
111. J. L. Brewster and H. Hart, *J. Amer. Chem. Soc.*, **91,** 711 (1969).
112. G. L. Closs, L. E. Closs and W. A. Böll, *J. Amer. Chem. Soc.*, **85,** 3796 (1963).
113. G. L. Closs and W. A. Böll, *Angew. Chem., Internat. Ed.*, **2,** 399 (1963); cf. G. Ege, *Tetrahedron Letters*, 1667 (1963); D. P. G. Hamon, and L. J. Holding, *Chem. Comm.*, 1330 (1970).
114. G. L. Closs and W. A. Böll, *J. Amer. Chem. Soc.*, **85,** 3904 (1963).
115. G. L. Closs, W. A. Böll, H. Heyn and V. Dev, *J. Amer. Chem. Soc.*, **90,** 173 (1968).
116. A. C. Day and M. C. Whiting, *J. Chem. Soc. (C)*, 1719 (1966).
117. A. C. Day and M. C. Whiting, *Chem. Comm.*, 292 (1965).
118. R. Anet and F. A. L. Anet, *J. Amer. Chem. Soc.*, **86,** 525 (1964); E. Vogel, S. Korte, W. Grimme and H. Günther, *Angew. Chem., Internat. Ed.*, **7,** 289 (1968).
119. H. Dürr and L. Schrader, *Chem. Ber.*, **103,** 1334 (1970).
120. G. Ege, *Tetrahedron Letters*, 1665 (1963).
121. I. Tabushi, K. Takagi, M. Okano and R. Oda, *Tetrahedron*, **23,** 2621 (1967).
122. J. Manecke and H. V. Schenck, *Tetrahedron Letters*, 2061 (1968).
123. S. G. Beech, J. H. Turnbull and W. Wilson, *J. Chem. Soc.*, 4686 (1952).
124. W. M. Jones, *J. Amer. Chem. Soc.*, **82,** 3136 (1960).
125. W. M. Jones, *J. Amer. Chem. Soc.*, **81,** 5153 (1959).

126. C. K. Ingold, *Structure and Mechanism in Organic Chemistry* (2nd. ed.), Chap. XI, Bell, London, 1969; D. J. Cram and R. T. Uyeda, *J. Amer. Chem. Soc.*, **86,** 5466 (1964).
126a. V. Prelog, *Helv. Chim. Acta*, **36,** 308 (1953); D. J. Cram and F. A. Abd Elhafez, *J. Amer. Chem. Soc.*, **74,** 5828 (1952); cf. J. A. Mills and W. Klyne, *Progress in Stereochemistry*, Vol. 1, Academic Press, New York, 1954, p. 198; W. Klyne and P. M. Scopes, *Progress in Stereochemistry*, Vol. 4, Academic Press, New York, 1969, p. 133; E. Eliel, *Stereochemistry of Carbon Compounds*, McGraw-Hill, New York, 1962.
127. S. G. Cohen, S. Hsiao, E. Saklad and C. H. Wang, *J. Amer. Chem. Soc.*, **79,** 2661, 4400 (1957).
128. C. G. Overberger and J. G. Lombardino, *J. Amer. Chem. Soc.*, **80,** 2317 (1958).
129. C. G. Overberger and I. Tashlick, *J. Amer. Chem. Soc.*, **81,** 217 (1959).
130. C. G. Overberger and J. W. Stoddard, *J. Amer. Chem. Soc.*, **92,** 4922 (1970).
131. A. C. Cope, S. W. Fenton and C. F. Spencer, *J. Amer. Chem. Soc.*, **74,** 5884 (1952).
131a. C. G. Overberger, M. S. Chi, D. G. Pucci and J. A. Barry, *Tetrahedron Letters*, 4565 (1972).
132. P. D. Bartlett and N. A. Porter, *J. Amer. Chem. Soc.*, **90,** 5317 (1968).
133. J. A. Berson and S. S. Olin, *J. Amer. Chem. Soc.*, **91,** 777 (1969).
134. R. B. Woodward and R. Hoffmann, *Acc. Chem. Res.*, **1,** 17 (1968).
135. C. G. Overberger and M. Lapkin, *J. Amer. Chem. Soc.*, **77,** 4651 (1955).
136. J. M. Zanden and G. De Vries, *Rec. Trav. Chim.*, **75,** 1159 (1956).
137. C. G. Overberger, I. Tashlick, M. Bernstein and R. G. Hiskey, *J. Amer. Chem. Soc.*, **80,** 6556 (1958).
138. S. G. Mylonakis and V. Seltzer, *J. Amer. Chem. Soc.*, **90,** 5487 (1968).
139. D. F. Eaton, R. G. Bergman and G. S. Hammond, *J. Amer. Chem. Soc.*, **94,** 1351 (1972).
140. D. H. White, P. D. Condit and R. G. Bergman, *J. Amer. Chem. Soc.*, **94,** 1348 (1972); R. A. Keppel and R. G. Bergman, *J. Amer. Chem. Soc.*, **94,** 1350 (1972).
141. D. H. White, P. D. Condit and R. G. Bergman, *J. Amer. Chem. Soc.*, **94,** 7931 (1972).
142. M. Franck-Neumann, *Tetrahedron Letters*, 2979 (1968).
143. T. H. Kinstle, R. L. Welch and R. W. Exley, *J. Amer. Chem. Soc.*, **89,** 3660 (1967).
144. M. P. Schneider and R. J. Crawford, *Can. J. Chem.*, **48,** 628 (1970).
145. P. D. Condit and R. G. Bergman, *Chem. Comm.*, 4 (1971).
146. O. J. Diels, J. H. Blum and W. Kroll, *Annal.*, **443,** 242 (1925).
147. H. v. Euler and K. O. Josephson, *Ber.*, **53,** 822 (1920).
148. E. L. Allred, C. L. Anderson and R. L. Smith, *J. Org. Chem.*, **31,** 3493, 3498 (1966).
149. J. K. Stille and T. Anyos, *J. Org. Chem.*, **27,** 3352 (1962).
150. J. E. Anderson and M. J. Lehn, *J. Amer. Chem. Soc.*, **89,** 81 (1967).
151. P. G. Gassman, *Org. Synth.*, **49,** 1 (1969).
152. R. Criegee and A. Rimmelin, *Chem. Ber.*, **90,** 414 (1957).
153. S. Cohen, R. Zand and C. Steel, *J. Amer. Chem. Soc.*, **83,** 2895 (1961); cf. ref. 76.

154. cf. S. Hünig, H. R. Müller and W. Thier, *Angew. Chem., Internat. Ed.*, **4,** 271 (1965); R. Stolle and A. Benrath, *J. Prakt. Chem.*, **70,** 263 (1904).
155. T. F. Thomas, C. I. Sutin and C. Steel, *J. Amer. Chem. Soc.*, **89,** 5107 (1967); cf. T. F. Thomas and C. Steel, *J. Amer. Chem. Soc.*, **87,** 5290 (1965).
156. P. S. Engel, *J. Amer. Chem. Soc.*, **90,** 5731 (1968); cf. I. I. Abram, G. S. Milne, B. S. Solomon and C. Steel, *J. Amer. Chem. Soc.*, **91,** 1220 (1969).
157. P. S. Engel, *J. Amer. Chem. Soc.*, **91,** 6903 (1969).
158. P. D. Bartlett and P. S. Engel, *J. Amer. Chem. Soc.*, **90,** 2960 (1968).
159. W. R. Roth and M. Martin, *Tetrahedron Letters*, 4695 (1967); W. R. Roth and M. Martin, *Annal.*, **702,** 1 (1967).
160. E. L. Allred and R. L. Smith, *J. Amer. Chem. Soc.*, **91,** 6766 (1969).
161. W. R. Roth and K. Enderer, *Annal.*, **730,** 82 (1969).
162. F. S. Collins, J. K. George and C. Trindle, *J. Amer. Chem. Soc.*, **94,** 3732 (1972).
163. W. P. Lay, K. Mackenzie, J. R. Telford and D. L. Williams-Smith, *Chem. Comm.*, 761 (1969).
164. L. Paquette and L. M. Leichter, *J. Amer. Chem. Soc.*, **93,** 4922 (1971).
165. J. A. Berson, D. M. McDaniel, L. R. Corwin and J. H. Davis, *J. Amer. Chem. Soc.*, **94,** 5508 (1972).
166. J. A. Berson, D. M. McDaniel and L. R. Corwin, *J. Amer. Chem. Soc.*, **94,** 5509 (1972).
167. M. Martin and W. R. Roth, *Chem. Ber.*, **102,** 811 (1969).
168. W. R. Roth and M. Martin, *Tetrahedron Letters*, 3865 (1967).
169. R. Askani, *Tetrahedron Letters*, 3349 (1970); cf. reference 195.
170. J. P. Snyder, L. Tee, V. T. Banduro, C. Y. Yu and R. J. Boyd, *J. Amer. Chem. Soc.*, **94,** 3260 (1972).
171. R. M. Moriarty, C-L. Yeh and N. Ishibi, *J. Amer. Chem. Soc.*, **93,** 3085 (1971).
172. E. L. Allred and J. C. Hinshaw, *Chem. Comm.*, 1021 (1969); E. L. Allred, J. C. Hinshaw and A. L. Johnson, *J. Amer. Chem. Soc.*, **91,** 3382 (1969).
173. N. Rieber, J. Alberts, J. A. Lipsky and D. M. Lemal, *J. Amer. Chem. Soc.*, **91,** 5668 (1969).
174. E. L. Allred and A. L. Johnson, *J. Amer. Chem. Soc.*, **93,** 1300 (1971).
174a. *cf.* A. Schweig, B. Trost, *et al.*, *J. Amer. Chem. Soc.*, **96,** 622 (1974).
175. R. C. Cookson, S. S. H. Gilani and I. D. R. Stevens, *J. Chem. Soc.* (*C*), 1905 (1967).
176. E. L. Allred and J. C. Hinshaw, *J. Amer. Chem. Soc.*, **90,** 6885 (1968).
177. J. C. Hinshaw and E. L. Allred, *Chem. Comm.*, 72 (1969).
178. K. Mackenzie and W. P. Lay, *Tetrahedron Letters*, 3241 (1970).
179. B. M. Trost and R. M. Cory, *J. Amer. Chem. Soc.*, **93,** 5573 (1971).
179a. T. J. Katz and N. Acton, *J. Amer. Chem. Soc.*, **95,** 2738 (1973).
180. R. Askani, *Chem. Ber.*, **102,** 3304 (1969).
181. H. Tanida, S. Teratake, Y. Hata and M. Watanabe, *Tetrahedron Letters*, 5341 (1969).
182. H. Tanida, S. Teratake, T. Hata and M. Watanabe, *Tetrahedron Letters*, 5345 (1969). *cf.* J. A. Berson *et al.*, *Tetrahedron*, **30,** 1639 (1964).
183. J. A. Elix, W. S. Wilson and R. N. Warrener, *Tetrahedron Letters*, 1837 (1970).
184. K. Mackenzie, *J. Chem. Soc.*, 473 (1960); cf. reference 178.

185. R. A. Carboni and R. V. Lindsey, *J. Amer. Chem. Soc.*, **81,** 4342 (1959); cf. J. Sauer, *Angew. Chem., Internat. Ed.*, **5,** 211 (1966).
186. R. N. Warrener, *J. Amer. Chem. Soc.*, **93,** 2346 (1971).
187. L. Paquette and J. F. Kelly, *Tetrahedron Letters*, 4509 (1969).
188. M. A. Battiste and T. J. Barton, *Tetrahedron Letters*, 1227 (1967); cf. J. Sauer and E. Heinrich, *Tetrahedron Letters*, 4979 (1966).
189. J. N. Brown, R. L. Towns and L. M. Trefonas, *J. Amer. Chem. Soc.*, **92,** 7436 (1970).
190. P. J. Fagen and M. J. Nye, *J. Amer. Chem. Soc.*, **93,** 537 (1971).
191. D. W. McNiel, M. E. Kent, E. Hadaya, P. F. D'Angelo and P. O. Schissel, *J. Amer. Chem. Soc.*, **93,** 3817 (1971).
192. Kei-Wei Shen, *J. Amer. Chem. Soc.*, **93,** 3064 (1971).
193. L. Paquette, *J. Amer. Chem. Soc.*, **93,** 7110 (1971).
194. L. Cassar, P. E. Eaton and J. Halpern, *J. Amer. Chem. Soc.*, **92,** 6366 (1970).
195. L. Paquette, *J. Amer. Chem. Soc.*, **92,** 5765 (1970).
196. J. Maier, *Chem. Ber.*, **102,** 3310 (1969).
197. J. Maier, U. Heep, M. Wiessler and M. Strausser, *Chem. Ber.*, **102,** 1928 (1969).
198. S. Masamune, H. Zenda, M. Wiesel, N. Nakatsuka and G. Bigam, *J. Amer. Chem. Soc.*, **90,** 2727 (1968).
199. J. Thiele and O. Stange, *Annal.*, **283,** 1 (1894); J. Stolle, *Ber.*, **45,** 273 (1912); F. Arndt, L. Loewe and A. Tarlan-Akön, *Chem. Abstr.*, **42,** 8190 (1948); *Rev. faculté Sci. Univ. Istanbul*, **13A,** 127 (1948).
200. B. T. Gillis and J. D. Hagarty, *J. Org. Chem.*, **32,** 330 (1967).
201. M. G. De Amazua, M. Lora-Tamayo and J. L. Soto, *Tetrahedron Letters*, 2407 (1970).
202. M. A. Battiste, personal communication.
203. J. M. Cinnamon and K. Weiss, *J. Org. Chem.*, **26,** 2644 (1961).
204. R. Askani, *Chem. Ber.*, **98,** 2551 (1965).
205. A. B. Evnin, R. D. Miller and G. R. Evanega, *Tetrahedron Letters*, 5863 (1968).
206. R. Huisgen, W. E. Konz and G. E. Gream, *J. Amer. Chem. Soc.*, **92,** 4105 (1970).
207. W. H. Pirkle and J. C. Stikler, *Chem. Comm.*, 760 (1967).
208. J. Sauer and B. Schröder, *Chem. Ber.*, **100,** 678 (1967).
209. J. Sauer and H. Wiest, *Angew. Chem., Internat. Ed.*, **1,** 269 (1962).
210. C. M. Anderson, J. B. Bremner, I. W. McCay and R. N. Warrener, *Tetrahedron Letters*, 1255 (1968).
211. K. Mackenzie, P. R. Young, P. C. Billot and J. G. Barker, unpublished observations.
212. F. R. Farr and N. Bauld, *J. Amer. Chem. Soc.*, **92,** 6695 (1970).
213. T. Sasaki, K. Kanematsu and K. Hayakawa, *Chem. Comm.*, 82 (1970).
214. A. G. Anastassiou, R. P. Cellura, J. M. Spence and S. W. Eachus, *Chem. Comm.*, 325 (1972).
215. D. J. Pasto and A. Fu-Tai Chen, *Tetrahedron Letters*, 2995 (1972).
216. D. M. Lemal and J. P. Lokensgard, *J. Amer. Chem. Soc.*, **88,** 5934 (1966).
217. A. B. Evnin, D. R. Arnold, L. A. Karnischky and E. Strom, *J. Amer. Chem. Soc.*, **92,** 6218 (1970).
218. W. R. Dolbier, W. D. Loehle and W. M. Williams, *Chem. Comm.*, 867 (1972).

219. H. Stetter and P. Woermle, *Annal.*, **724**, 150 (1969); cf. B. T. Gillis and R. A. Izydore, *J. Org. Chem.*, **34**, 3181 (1969).
220. Several polychlorinated diene adducts and related compounds we have handled resist hydrolysis, K. Mackenzie, J. R. Telford and W. P. Lay, unpublished observations; cf. also M. Heyman, V. T. Bandurco and J. P. Snyder, *Chem. Comm.*, 297 (1971).
221. R. A. Clement, *J. Org. Chem.*, **25**, 1724 (1960); **27**, 1115 (1962); T. J. Kealy, *J. Amer. Chem. Soc.*, **84**, 966 (1962).
222. D. L. Forster, T. L. Gilchrist, C. W. Rees and E. Stanton, *Chem. Comm.*, 695 (1971).
223. F. D. Marsh and M. E. Hermes, *J. Amer. Chem. Soc.*, **87**, 1819 (1965).
224. B. T. Gillis and R. Weinkam, *J. Org. Chem.*, **32**, 3321 (1967); J. P. Freeman, T. J. Gannon and D. L. Surbey, *J. Org. Chem.*, **34**, 187 (1969).
225. C. W. Rees and M. Yelland, *Chem. Comm.*, 221 (1973).
226. N. D. Epiotis, *J. Amer. Chem. Soc.*, **95**, 1191 (1973); **94**, 1924 (1972) and accompanying papers.
227. B. Astin, University of Bristol, private communication of previously unpublished work.
228. M. J. S. Dewar and E. Hasselbach, *J. Amer. Chem. Soc.*, **92**, 590 (1970).
229. J. Del Bere and H. H. Jaffé, *J. Chem. Phys.*, **50**, 1126 (1969).
230. J. M. André, M. Ch. André, G. Le Roy and J. Weiler, *Internat. J. Quantum Chem.*, **3**, 1013 (1969).

CHAPTER **12**

The electrochemistry of azoxy, azo and hydrazo compounds

F. G. THOMAS

Chemistry Department, James Cook University of North Queensland, Townsville, Australia, 4811

and

K. G. BOTO

Electrochemistry Group, Defence Standards Laboratories, Ascot Vale, Australia, 3032

I. INTRODUCTION

Electrochemically, the azoxy, azo and hydrazo groups are closely related as shown in equation (1). These steps usually occur as two- or four-electron transfers and the reversibility of a particular step will depend on the experi-

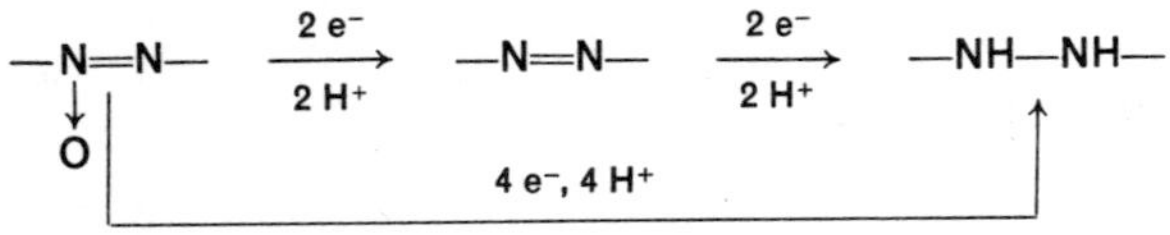

(1)

mental conditions and on the nature of the molecule in which the electroactive group is situated. In aprotic solvents, however, many intermediate radical species, such as the azobenzene radical ion, may be stabilized. In

$$(C_6H_5—N—\dot{N}—C_6H_5)^-$$

these cases, the two-electron transfer steps of equation (1) can be resolved into two one-electron steps. This usually leads to a more detailed understanding of the mechanism of the electrode reaction.

As equation (1) indicates, the electrochemistry of any one of these groups may involve one or both of the other two. Because of this, it has been found convenient in the subsequent discussion to consider the electrochemistry in terms of the end product of the electrode reactions (Section II on electrochemical preparation) or of the starting material (Section III on electrode reactions). This enables a more complete description of the electrochemical reactions of one functional group in a particular experimental situation to be given in one section of the text.

Most of the earlier investigations of the electrochemistry of these groups was concerned with the electrolytic preparation of aromatic azo and hydrazo compounds. With the advent of polarography these investigations rapidly expanded and the emphasis shifted from preparative methodology to a study of the electrode reactions and to the development of electroanalytical methods for determining these groups. These studies were mainly concerned with aromatic azo compounds because of (*i*) the importance of this group in the dyestuff industry, (*ii*) the interest in the carcinogenic properties of azo compounds, and (*iii*) the use of azo compounds for the indirect determination of non-electroactive metals. While this work led to a reasonable understanding of the electrode processes involved, many questions remained unanswered because of the almost universal use of solvent systems containing water. The major problem encountered when water is present is the adsorption of many of these compounds at the electrode–solution interface. Such adsorption phenomena often modify the electrode reactions, and in polarographic studies may cause the appearance of prewaves, etc., which complicate the interpretation of the results.

In recent years, two factors have lead to a much better understanding of these reactions. The first has been the use of anhydrous, and particularly aprotic, solvent systems to study the electrochemistry of these groups. This has minimized, and in many cases removed, the problem of adsorption. The second factor has been the tremendous expansion in the range of electrochemical methodology and availability of instrumentation. The application of techniques such as cyclic voltammetry to the study of the electrode reactions of these groups has enabled detailed redox mechanisms to be established in many cases.

II. ELECTROLYTIC PREPARATION OF AZOXY, AZO AND HYDRAZO COMPOUNDS

The most common starting materials for the electrolytic preparation of azoxy, azo or hydrazo compounds are nitrobenzene and its derivatives. These electrolytic reductions have been well established and many patents exist which detail the optimum conditions for obtaining the best yields in terms of both current consumption and chemical conversion. Nitroso compounds have also been used as source material. In addition, azoxy compounds may be reduced to azo or hydrazo compounds and azo compounds are often a convenient source for hydrazo compounds. In fact, if unsymmetrically substituted hydrazo compounds are sought it is necessary to use the corresponding azo or azoxy compound since reduction of the nitro compounds leads to symmetrically substituted azoxy, azo or hydrazo compounds.

A. *Azoxy Compounds*

The electrolytic reduction of nitrobenzene and its derivatives is usually carried through to the azo or hydrazo stage. However, by careful control of the conditions and the amount of current passed through the cell it is possible to stop the electrolysis at the azoxy stage[1,2].

The reduction of nitrobenzene to azoxybenzene is a combination of electrolytic and chemical steps[2], namely

$$2\,C_6H_5NO_2 + 4\,e^- + 2\,H_2O \longrightarrow 2\,C_6H_5NO + 4\,OH^- \qquad (2a)$$

$$C_6H_5NO + 2\,e^- + 2\,H_2O \longrightarrow C_6H_5NHOH + 2\,OH^- \qquad (2b)$$

$$C_6H_5NO + C_6H_5NHOH \longrightarrow C_6H_5\underset{\displaystyle\overset{\downarrow}{O}}{N}{=}NC_6H_5 + H_2O \qquad (2c)$$

which give the overall reaction:

$$2\,C_6H_5NO_2 + 6\,e^- + 3\,H_2O \longrightarrow C_6H_5\underset{\displaystyle\overset{\downarrow}{O}}{N}{=}NC_6H_5 + 6\,OH^- \qquad (2d)$$

In addition to reaction (2c), phenylhydroxylamine may react to give azobenzene as follows:

$$2\ C_6H_5NHOH \longrightarrow C_6H_5N{=}NC_6H_5 + 2\ H_2O \qquad (2e)$$

In order to obtain the maximum yield of the azoxy compound, conditions must be controlled so that reaction (2c) predominates over reaction (2e). The relative rates of these two reactions depend on the alkalinity of the aqueous phase[3]; with weakly alkaline conditions favouring production of the azoxy compound as reaction (2c) is then much faster than reaction (2e).

In practice, the electrolysis is carried out using a well stirred suspension of nitrobenzene (8 %) in 2·5 % sodium hydroxide solution. Further control of the reduction is achieved by passing only the theoretical amount of electricity through the cell as required by equation (2d); i.e. three faradays of charge per mole of nitrobenzene. Low current densities (ca. 2 A dm^{-2}) and the use of low overvoltage cathode materials such as nickel also increase the yield of azoxy compound and minimize the reduction to azo and hydrazo compounds. Temperature is also found to effect the yields: 75°C is the optimum working temperature for the maximum yield of azoxy compound. Control of the current and the use of a low overvoltage material for the cathode most probably effect control of the products via reaction (2b). For maximum yield of azoxy compound only half of the nitrosobenzene produced initially at the electrode surface should be further reduced to phenylhydroxylamine. This should then react with the remainder of the nitrosobenzene rather than with itself. Too high a current density and too high an electrode potential would result in a high 'concentration' of electrons at the electrode surface. This would be expected to increase the rate of production of hydroxylamine and hence its concentration at the electrode surface and so increase the probability of reaction (2e) in the vicinity of the electrode. In the extreme case one expects the reduction of nitrosobenzene to go beyond hydroxylamine to aniline when sufficiently high electron 'concentrations' occur at the electrode.

B. Azo Compounds

The electrolytic reduction of aromatic nitro compounds dissolved or suspended in aqueous sodium hydroxide can lead directly to the production of azobenzene derivatives according to the overall reaction (equation 3).

$$2\ XC_6H_4NO_2 + 4\ H_2O + 8\ e^- \longrightarrow XC_6H_4N{=}NC_6H_4X + 8\ OH^- \qquad (3)$$

Higher current densities and more alkaline solutions are used for this reduction compared with those used for the preparation of azoxy compounds since the reduction is easier to control. However, in too drastic conditions complete reduction of the nitro compound to the amine may occur. Further reduction of the azo compound to the hydrazo compound under the alkaline conditions used for this preparation appears to require a metal or metal-ion catalyst[4] (Section II, C). In the absence of such catalysts, reduction to the hydrazo compound is minimal.

Experimental conditions are similar to those employed for the production of azoxy compounds except that higher current densities (>10 A dm^{-2}) are used and four faradays of charge per mole of nitro compound are passed through the cell. Both electrodes are usually made of iron. In addition to azobenzene[5,6], 1,1′-dichloroazobenzene and 4,4′-dimethylazobenzene have been prepared in large quantities by this method[7]. Laboratory scale preparations of many other disubstituted azobenzenes have also been carried out electrolytically.

C. Hydrazo Compounds

The electrolytic production of aromatic hydrazo compounds has received much more detailed study than the production of azoxy and azo compounds. The hydrazo compounds have been prepared by the electrolytic reduction of alkaline suspensions of nitro compounds[5,7–11], azoxy compounds[7] and azo compounds[5,7,12]. Because of their availability, nitrobenzene derivatives are usually used as starting materials. If unsymmetrical compounds are required, it is necessary to use the corresponding azoxy or azo compound as starting material.

Although the overall reduction of nitro compounds to hydrazo compounds may be represented by equation (4), it appears that the reduction

$$2\,XC_6H_4NO_2 + 6\,H_2O + 10\,e^- \longrightarrow XC_6H_4NHNHC_6H_4X + 10\,OH^- \qquad (4)$$

of these suspensions occurs in two distinct stages[4]. The first stage is the multi-step reduction to the azo compound and the second stage is a chemical reduction of the azo compound to the hydrazo compound by a metal. The metal ion so produced in this stage is then reduced back to the metal electrolytically. This second stage is shown in equation (5) and is effectively

$$XC_6H_4N{=}NC_6H_4X + 2\,H_2O + M \longrightarrow XC_6H_4NHNHC_6H_4X + M^{2+} + 2\,OH^- \qquad (5a)$$

$$M^{2+} + 2\,e^- \longrightarrow M \qquad (5b)$$

a catalytic reduction with the metal acting as catalyst. Since both the free metal (usually plated on the cathode) and the metal ion catalyse this reduction[11,12], equations (5a) and (5b) can occur in the order given when the metal is catalyst or in the reverse order when the metal ion is the catalyst. Metals that have been used as catalysts are cadmium, lead, tin and zinc and the metal ions that have been used are lead(II) and zinc(II).

TABLE 1. Cathodes used for the electrolytic preparation of hydrazo compounds in alkaline media

Base metal	Spongy metal coating	Reference
Pb	Sn, Zn, Cd or Pb	9
Fe	Sn, Zn or Cd	9
	Pb	9, 11
Stainless steel	Pb	10, 13
	Sn, Zn or Cd	13

Since azobenzene is readily reduced in alkaline media at the dropping mercury electrode to hydrazobenzene without the need of metal catalysts, it would appear that there is no need for the catalyst in the electrolytic preparation of hydrazo compounds. However, the polarographic results apply to *solutions* of azo compounds (alcohol being present to effect solution) whereas the preparative methods described here use *suspensions* of the azo compounds in aqueous alkaline media. Under such conditions the electrolytic reduction of the azo to the hydrazo compound would require a considerably more cathodic potential than needed for solutions of the azo compound. Thus the role of the metal is to enable the reduction to occur at a relatively low cathodic potential, namely at the potential of the metal–ion reduction (equation 5b). This results in a saving in power and an efficient and rapid conversion of the azo to the hydrazo compound.

Considerable effort has been expended in the development of suitable cathodes for the large scale preparation of hydrazo compounds. The more common cathodes are listed in Table 1. With the aid of these cathodes almost quantitative yields of hydrazo compounds can be obtained[9]. The alternative way of catalysing this reduction, by adding lead(II), tin(II) or zinc(II) ions to the alkaline media[5,11,12] and using nickel, steel or lead cathodes does not appear to be as suitable as using the spongy metal cathodes.

The experimental conditions for the large scale electrolytic preparation of aromatic hydrazo compounds are not very critical. A two-compartment cell with an alkali-resistant porous membrane is used. The anode compartment contains 20–30% aqueous sodium hydroxide solution and the cathode compartment 2–20% aqueous sodium hydroxide solution with 6–10% of the starting material suspended in it. This suspension is agitated and the electrolysis carried out at ca. 90°C with current densities in the range 10 to 30 A dm^{-2}. Some examples of the electrolytic preparation of hydrazo compounds are listed in Table 2.

TABLE 2. Some hydrazo compounds prepared by large scale electrolyses

Precursor	Products	Chemical yield (%)	Current efficiency (%)	Reference
$C_6H_5NO_2$	$C_6H_5NHNHC_6H_5$	85–95	ca. 100	5
$C_6H_5NO_2$	$C_6H_5NHNHC_6H_5$	90		8
	$C_6H_5N{=}NC_6H_5$	4		
	$C_6H_5NH_2$	2		
$C_6H_5NO_2$	$C_6H_5NHNHC_6H_5$	90–95[a]	80–88[a]	9
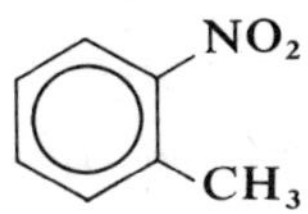	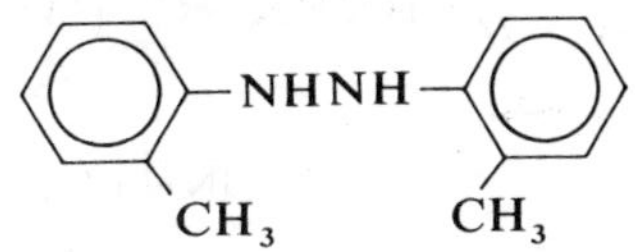	100[b]	ca. 100[b]	5
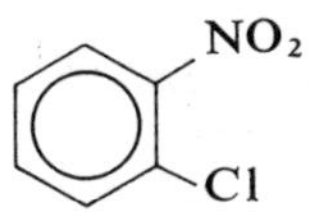	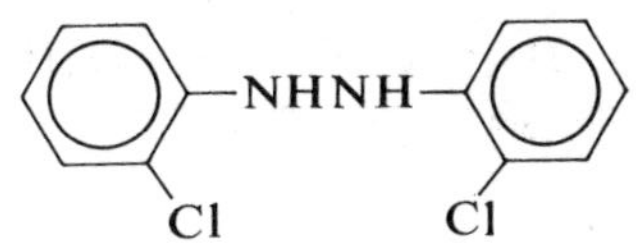	60	33	7
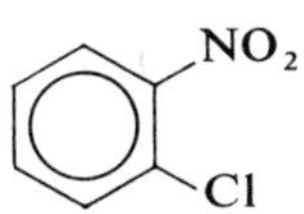	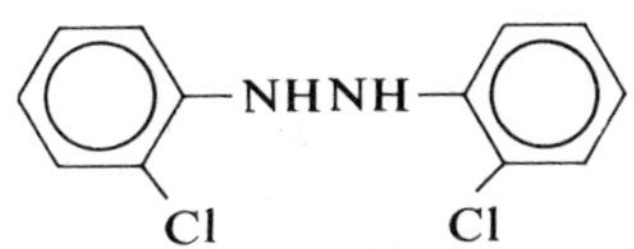	81		10
	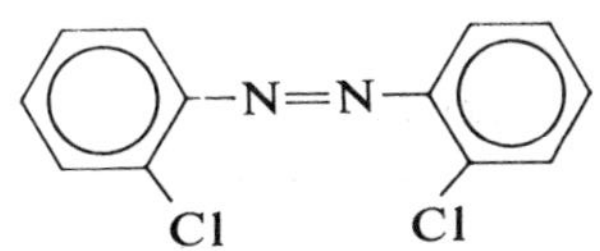	13		

TABLE 2 (*cont.*)

Precursor	Products	Chemical yield (%)	Current efficiency (%)	Reference
	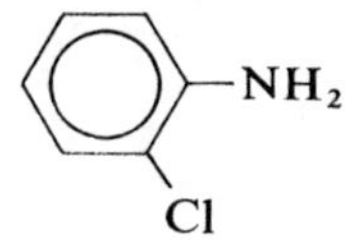	6		
		72	42	7
$C_6H_5N{=}NC_6H_5$	$C_6H_5NHNHC_6H_5$		ca. 100	5
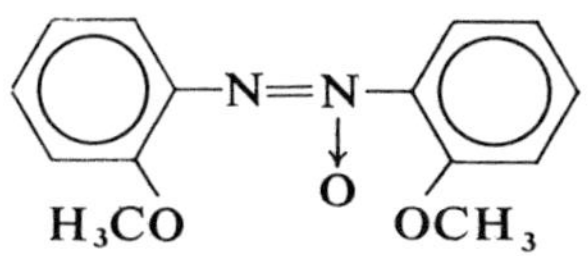	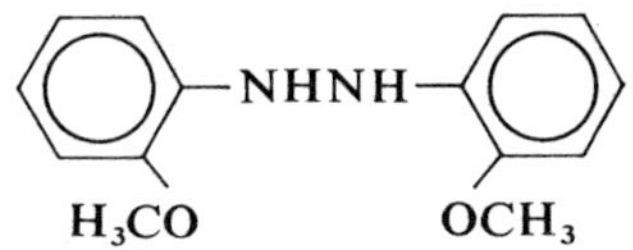	92	77	7
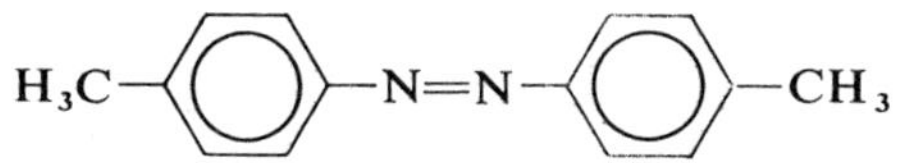		90	66	7

[a] Yield and current efficiency higher at larger current densities.
[b] Only report of quantitative conversion in large scale preparations. Method used Pb(II), Sn(II) or Zn(II) ions as catalysts with an iron cathode.

Attempts to carry out large scale preparations of aromatic hydrazo compounds in acid solution have not been successful because the reduction usually goes through to the corresponding aniline or the hydrazo compound undergoes a benzidine type rearrangement. A successful laboratory preparation of hydrazobenzene in neutral, alcoholic (30%) solution has been reported[14]. In this method a two-compartment cell with mercury pool electrodes was used for the electrolysis. The anode compartment was filled with 0·5M-tetraethylammonium bromide in 30% aqueous alcohol and the cathode compartment contained the same solution saturated with azobenzene. The product, hydrazobenzene, precipitated from solution thus facilitating separation. In liquid ammonia, containing ammonium salts as electrolytes, hydrazobenzene has been prepared by the electrolytic reduction of nitrosobenzene, azoxybenzene and azobenzene[15,16] at a mercury cathode.

Future development of the electrolytic preparation of these compounds could well be directed towards a much greater use of non-aqueous solvents, particularly the aprotic ones such as dimethylformamide and acetonitrile. Such solvents offer the advantage of controlling the availability of protons required in the electrode reactions. The stoichiometric quantity of protons needed to convert the nitro group, etc., to water or the azo group to the hydrazo group can be added in the form of an acid, rather than have an unlimited supply of protons as is the case with the protolytic solvents such as water. This should result in easier control of the production of the intermediate products in the reduction sequence [equation (1)]. The main disadvantage of such solvents is that the products of the electrolysis will be soluble in these media so that some extraction process will be necessary for their recovery.

One other area that has not received much attention in the large scale electrolytic preparations is that of potential control. In general, controlled potential electrolysis offers a much greater degree of control over the nature of the products than does controlled current electrolysis. Unfortunately, potential control results in a much slower rate of electrolysis. While this is no serious problem in laboratory scale preparations it is a serious disadvantage in any large scale operation.

III. ELECTROCHEMICAL REACTIONS

The use of protolytic solvents such as water and aqueous alcohols results in electrode reactions involving an even number of electrons [equation (1)] and protons because of the ready availability of protons in these solvent systems. On the other hand, when many of these electrochemical reactions are studied in aprotic media, single electron steps can be observed and radical ions can be produced in solution. These aprotic solvents therefore offer a means of studying the electrode processes in more detail, usually without the complications arising from associated adsorption phenomena. It is convenient, therefore, to consider the electrochemistry of these functional groups in the two types of solvents separately.

A. Electrochemistry in Protolytic Solvents

1. Azoxy compounds

The polarographic reduction of azoxybenzene in a variety of aqueous alcoholic mixtures ranging from 20 to 100% alcohol (ethanol or methanol) occurs in a single four-electron step[17-20]. The half-wave potential for the reduction depends on the solvent composition and on the pH of the solution, becoming more negative as the pH and/or the alcohol concentration increase. This move to more negative reduction potentials is consistent with the decrease in proton availability as the pH increases and the decrease in surface concentration of the azoxybenzene (due to decreasing adsorption at the electrode surface) as the alcohol content increases. No simple relationship between the half-wave potential and pH in a given solvent mixture has been observed, probably because of the irreversible nature of the reduction process. The reduction product is hydrazobenzene, produced according to the reaction

$$C_6H_5\underset{\underset{O}{\downarrow}}{N}{=}NC_6H_5 + 4\,e^- + 4\,H^+ \longrightarrow C_6H_5NHNHC_6H_5 + H_2O \qquad (6)$$

In aqueous methanolic solution the polarographic waves exhibit maxima[18]. Addition of surfactants to help suppress the maxima results also in a shift of the half-wave potential[17]. Carboxymethyl cellulose and triphenylphosphine oxide cause a shift to more negative potentials while the addition of gelatin results in the half-wave potential shifting to more positive values. These effects, coupled with the fact that azoxybenzene is strongly adsorbed on mercury from these solutions indicate the influence of surface adsorption phenomena on these electrode reactions.

Holleck[17, 18] has studied the polarographic reduction of azoxybenzene in aqueous methanol, pure anhydrous methanol and anhydrous methanol containing known amounts of a range of phenols and carboxylic acids. In anhydrous methanol, the addition of weak acids, causes the reduction wave to shift slightly to more positive potentials while the addition of stronger acids results in the development of a prewave. This prewave eventually reaches the same height as the original azoxybenzene wave after excess of strong acid has been added. The mechanism proposed to account for these observations is a two-stage process

$$\underset{\displaystyle\downarrow\atop\displaystyle O}{C_6H_5N{=}}NC_6H_5 + 2\,AH + 2\,e^- \longrightarrow C_6H_5N{=}NC_6H_5 + H_2O + 2\,A^- \quad (7a)$$

$$C_6H_5N{=}NC_6H_5 + 2\,AH + 2\,e^- \longrightarrow C_6H_5NHNHC_6H_5 + 2\,A^- \quad (7b)$$

In proton-poor media, e.g. aqueous alkaline methanol or pure methanol, AH is the solvent (water or methanol) and A^- the solvo base (OH^- or OCH_3^-) while in the presence of added acid, HA is the acid and A^- its conjugate base. It should be noted that in pure methanol solutions two of the products of reaction (7a), namely H_2O and OCH_3^-, are expected to react to produce OH^- and CH_3OH.

The electrochemistry of substituted azoxybenzenes[19] is generally similar to that of the parent compound when the substituent groups are not themselves electroactive. However, the presence of electroactive substituents, particularly in the 2 position, may greatly modify the electrochemistry. Also the relative basicities of the substituents can effect the extent of the reduction. The more basic the substituent, the greater the possibility of the reduction proceeding beyond the hydrazo compound to the aniline derivative, particularly if the reduction is carried out in acidic solution.

The polarographic reduction of a series of symmetrically disubstituted azoxybenzenes in 80% ethanol containing (*i*) 1M-acetic acid and 1M-ammonium acetate buffer or (*ii*) 0·1M-ammonia and 0·1M-ammonium acetate buffer has been studied by Hazard and Tallec[19]. When non-reducible groups such as CN, Cl, CH_3, SCH_3 and OCH_3 are symmetrically substituted in the 4 and 4′ positions, the half-wave potentials ($E_{1/2}$) vary approximately according to a Hammett type relation (8) in both buffer solutions.

$$E_{1/2}(X) = E_{1/2}(H) + k\sigma_p(X) \quad (8)$$

where $E_{1/2}(H)$ is the half-wave potential of the parent azoxybenzene, $E_{1/2}(X)$ the half-wave potential of the X substituted azoxybenzene and $\sigma_p(X)$ the Hammett *para* substituent constant.

In acetate buffer $E_{1/2}(H) = -0{\cdot}64$ V vs. saturated calomel electrode (s.c.e.) and in the ammonia buffer $E_{1/2}(H) = -0{\cdot}84$ V vs. s.c.e. The values of k were found to be 0·55 and 0·66, respectively. The plot of $E_{1/2}(X)$ vs. $\sigma_m(X)$, the Hammett *meta* substituent constant, for these groups substituted in the 3 and 3′ positions gives a straight line of slope 0·41 in acetate buffer and a slope of 0·46 in ammonia buffer. However there is a fair amount of scatter of the experimental results around the line of best fit for both the symmetrical 4,4′ and 3,3′ substituted series.

When the substituents are in the 2 and 2′ positions, no simple relationship between $E_{1/2}(X)$ and $\sigma_0(X)$, the Hammett *ortho* substituent constant, emerges. However, as with the 3,3′ and 4,4′ derivatives, the half-wave potential becomes more negative (i.e. the azoxy compound is more difficult to reduce) as the electron donating properties of the substituent increase.

Constant potential coulometric and cyclic voltammetric studies show that the reduction of the parent azoxybenzene and all the dicyano, dichloro, dimethyl, dimethoxy and dimethylthio derivatives are quantitatively reduced in ammoniacal buffer solution according to equation (6). All these disubstituted derivatives with the exception of the 4,4′-dimethoxy and the 4,4′-dimethylthio azoxybenzenes are likewise reduced to the corresponding hydrazo compounds in acetate buffers. In the case of the 4,4′-dimethoxy and 4,4′-dimethylthio derivatives approximately 5·4 faradays of charge are required for the reduction compared with the 4 faradays per mole required by equation (6).

With these two derivatives in acetate buffers the hydrazo compounds produced according to equation (6) undergo two further competing reactions. The first of these competing reactions is a chemical auto oxidation–reduction reaction (e.g. for the 4,4′-dimethoxy derivative):

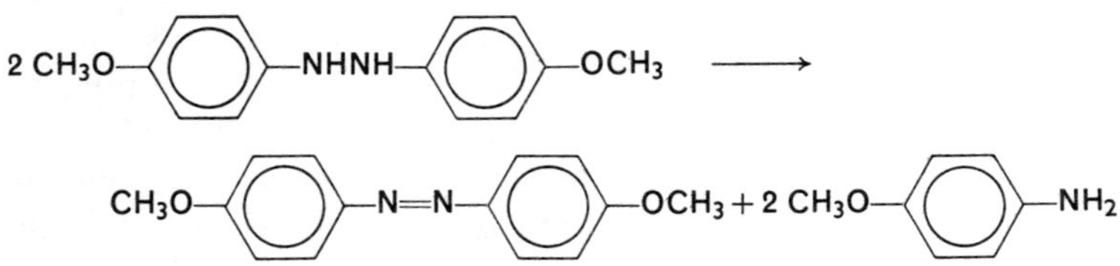

(9a)

followed by an electrochemical reduction of the azo compound

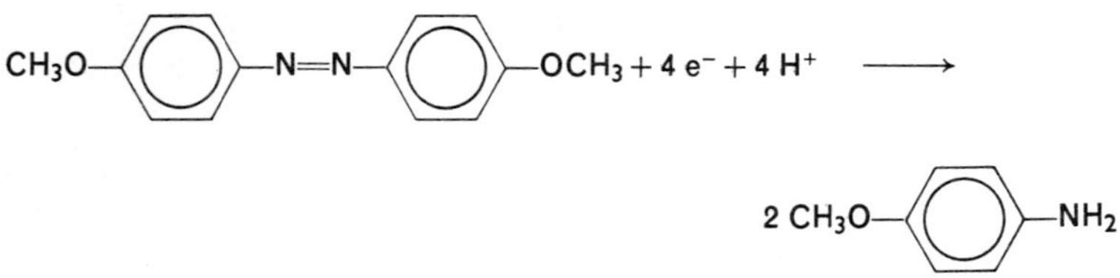

(9b)

which, in combination with equation (6) gives the overall reaction for the reduction of 4,4′-dimethoxyazoxybenzene:

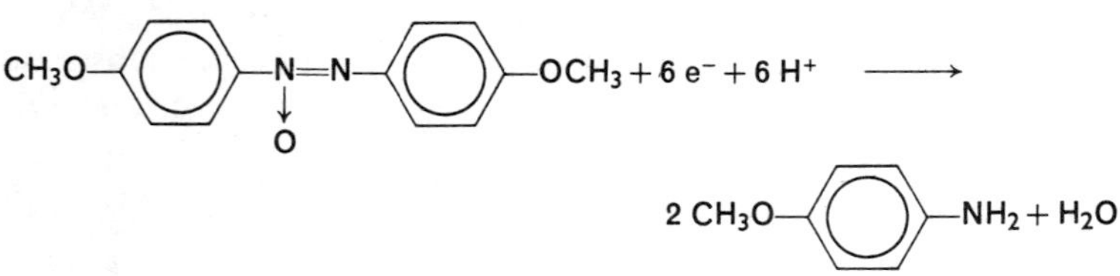

(10)

The second of the competing reactions is a semidine rearrangement

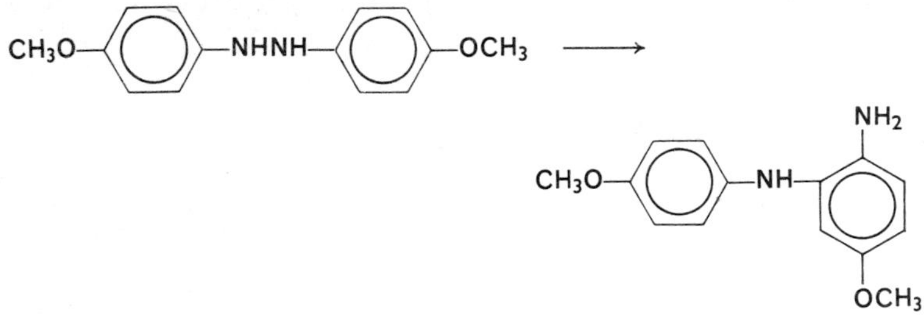

(11)

to give the electroinactive *o*-semidine.

Analysis of the products shows that 80% of the 4,4′-dimethoxyazoxybenzene is converted to *p*-anisidine [reactions (9) and (10)], showing that the auto oxidation reduction reaction is favoured in the acetate buffers. On the basis of an 80% conversion to *p*-anisidine and 20% conversion to the *o*-semidine, the electrolysis is calculated to require 5·6 faradays of charge per mole of 4,4′-dimethoxyazoxybenzene. This compares favourably with the experimentally observed value of 5·4 faradays per mole. The 4,4′-dimethylthio derivative undergoes an analogous reaction sequence. The behaviour of these two electron donating substituents shows that the electrode reactions depend on three factors—the electron donating ability of the substituent (i.e. its basicity), the pH of the medium and the position of the substituent in the aromatic ring.

The presence of an electroactive substituent such as the nitro group in one or both of the aromatic rings introduces the question of which group is reduced first, particularly in this case where the two electroactive groups are reduced at similar potentials. The effect of each group on the other is important in this regard and as will be seen from the subsequent discussion, the relative positions of the nitro and azoxy groups plays an important role in deciding which group is reduced first. The relative positions of these two groups also effect the course of the electrode reaction and the nature of the final products. Details of the polarographic behaviour of several mono- and di-nitroazoxybenzenes in 80% ethanol containing either acetate or ammonia buffers are listed in Table 3. In addition to the polarographic results, constant potential coulometric data were used to interpret the electrode reaction mechanisms of these compounds[19].

In the reduction of 4-nitroazoxybenzene, a total of 12 electrons are consumed per molecule to give aniline and *p*-phenylenediamine as the final products in both acidic and basic solutions. In both buffers the azoxy group is first reduced to the hydrazo group. In acetate buffer, this is followed by a four-electron reduction of the nitro group to give 4-hydroxylaminohydrazobenzene which then undergoes a series of autooxidation–reduction reactions and electrode reductions to give the final products. The sequence is shown in the following reaction scheme.

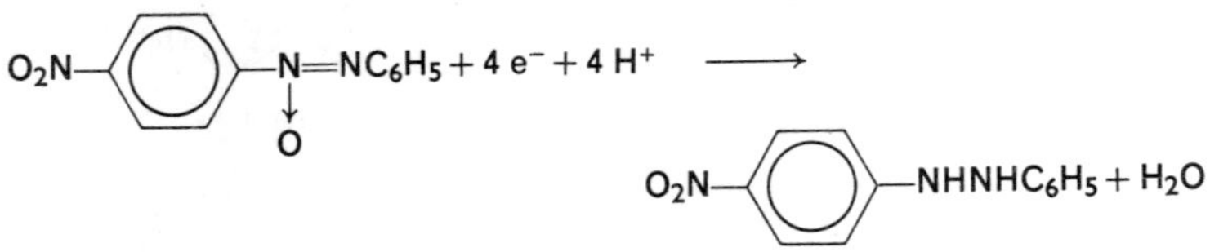

(12a)

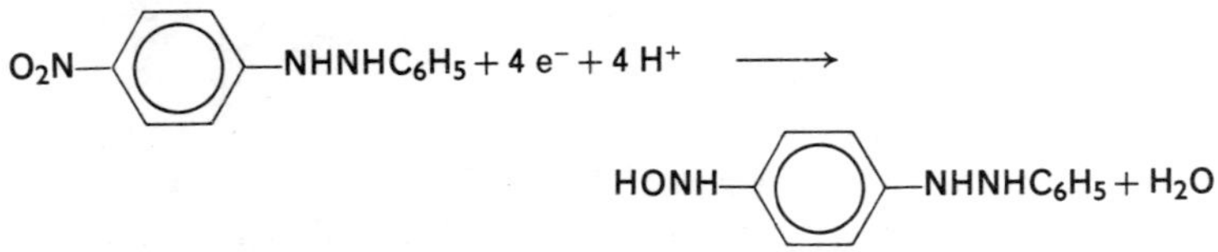

(12b)

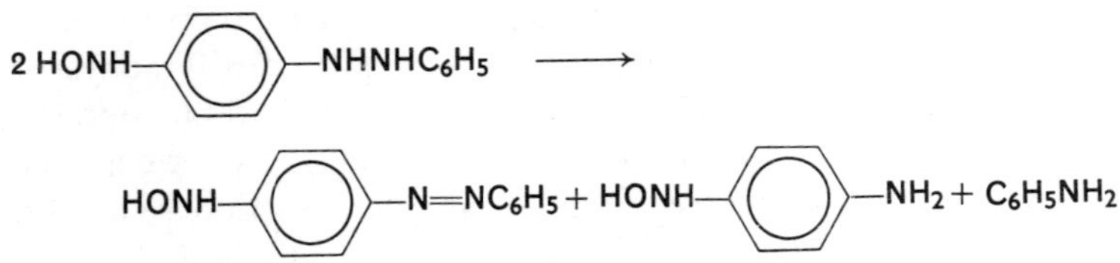

(12c)

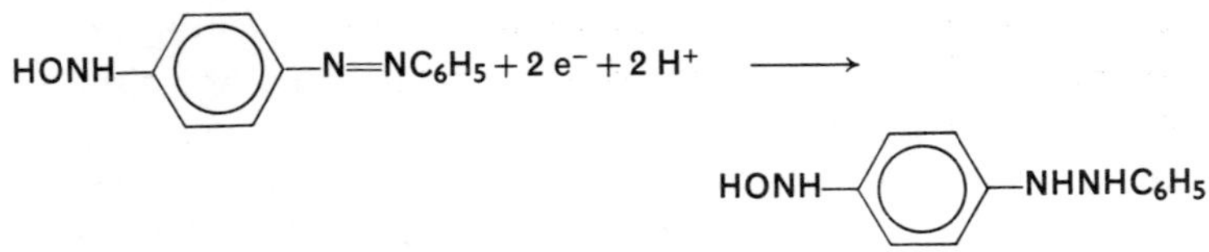

(12d)

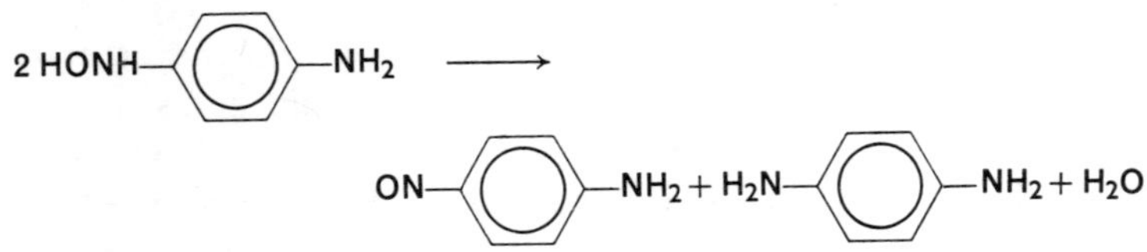

(12e)

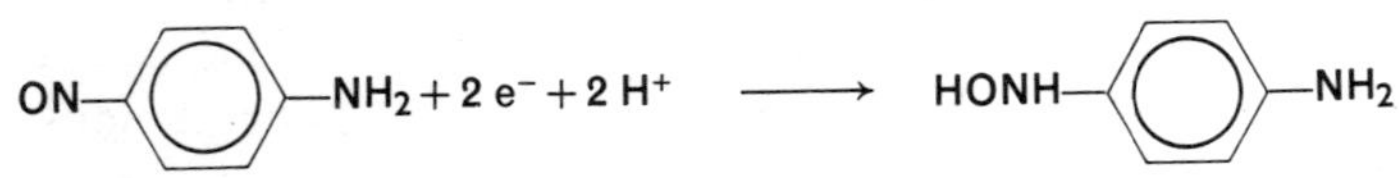

(12f)

This gives the overall reaction

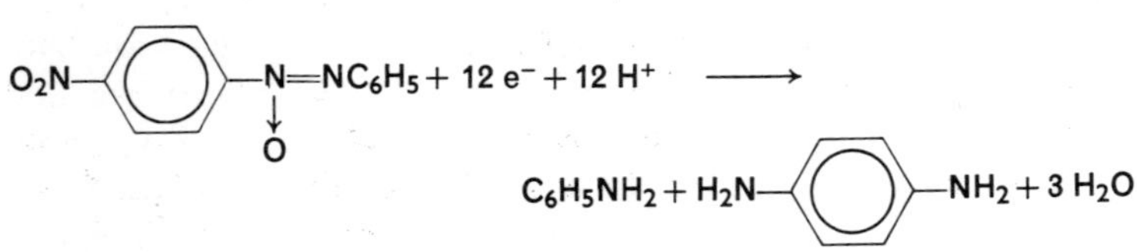

(12g)

TABLE 3. Polarographic characteristics of various mono- and di-nitroazoxybenzenes in 80% ethanol–water

Compound	1M-acetate buffer $E_{1/2}$[a]	n[b]	0·1M-ammonia buffer $E_{1/2}$[a]	n[b]
2-Nitroazoxybenzene	–0·43	4	–0·59	4
	–0·65	5·2	–0·86	4·8
4-Nitroazoxybenzene	–0·29	4	–0·45	4
	–0·69	8	–0·92	6 (8)
2,2′-Dinitroazoxybenzene	–0·3[c]	ca. 6·2 (15)	–0·4[c]	ca. 7·5 (15)
	–0·6[c]	ca. 6·8	–0·8[c]	ca. 5·5
3,3′-Dinitroazoxybenzene	–0·34	5·8 (16)	–0·47	4·1 (16)
	–0·64	7·1	–0·77	6·6
	–1·06	0·7		
4,4′-Dinitroazoxybenzene	–0·11	4	–0·25	4
	–0·65	14	ca. –0·83[d]	12 + 2

[a] Volts vs. saturated calomel electrode.
[b] Determined by comparison of wave height with that for the four-electron wave of azoxybenzene. Values in brackets from coulometric measurements.
[c] Poorly separated waves.
[d] Wave appears to be split into two parts corresponding to $n = 12$ and $n = 2$.

In ammonia buffer, an additional, alternate route for the reduction of 4-nitrohydrazobenzene [reaction (12b)] also occurs, namely

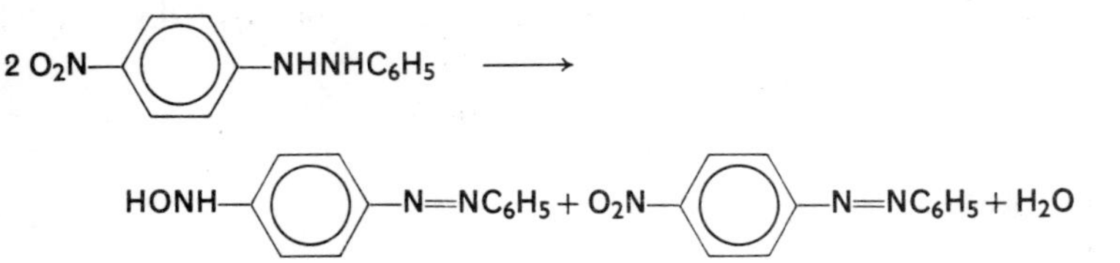
(12h)

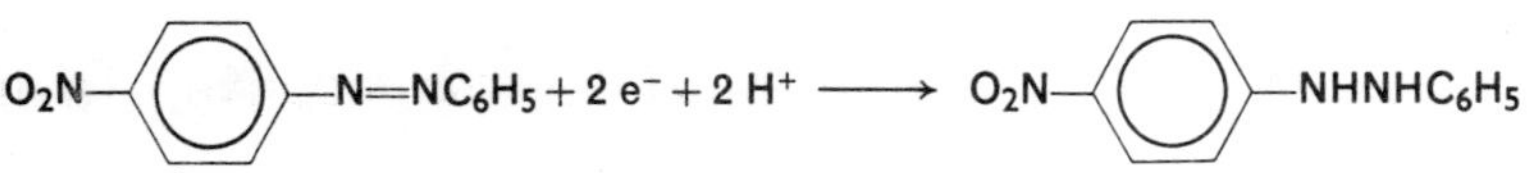
(12i)

In contrast to the 4-nitro derivative, 2-nitroazoxybenzene only requires approximately nine electrons per molecule for complete reduction in both buffer systems. The products of electrolysis are aniline, *o*-phenylenediamine and 2-phenylbenzotriazole. Since the reduction to the triazole requires only six electrons per molecule, cf. the 12 electrons per molecule required for the reduction to the two amines, the observed values of 9·2 electrons per molecule (acetate buffer) and 8·8 electrons per molecule (ammonia buffer) mean that 53% and 47% of the 2-nitroazoxybenzene is reduced to the amines in acetate and ammonia buffer, respectively. Thus the reduction to the amines is slightly favoured in acidic solutions while the reduction to the triazole is slightly favoured in basic solution.

In the first four-electron step in the reduction of the 2-nitro derivative the reduction is non-selective in that either of the two electroactive groups, —NO_2 or —N(O)=N—, is reduced. Thus the first polarographic wave corresponds to the two electrode reactions

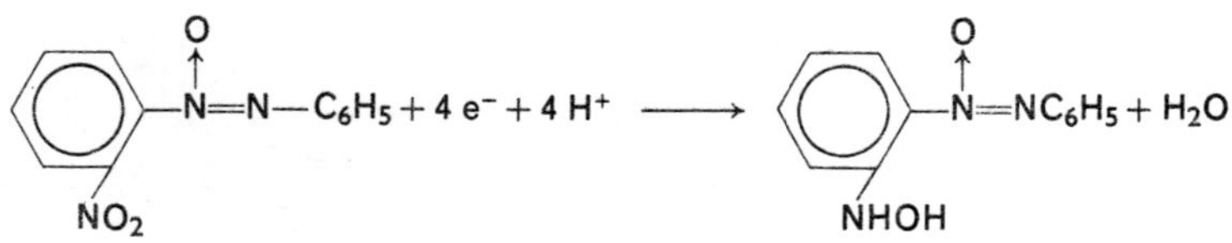
(13a)

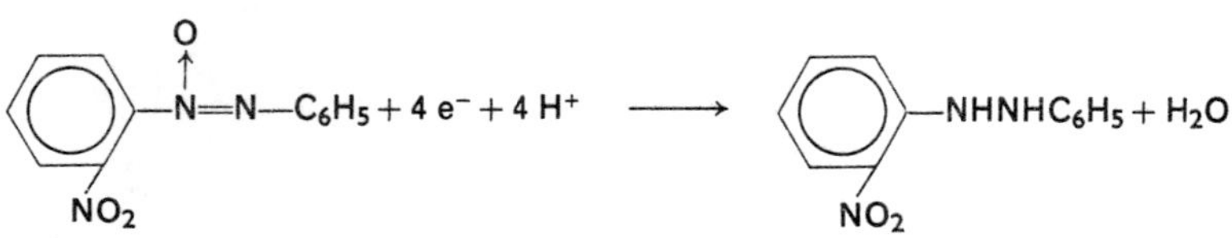
(13b)

occurring at the same potential. The 2-hydroxylaminoazoxy benzene then undergoes the following sequence of reactions to produce the amines.

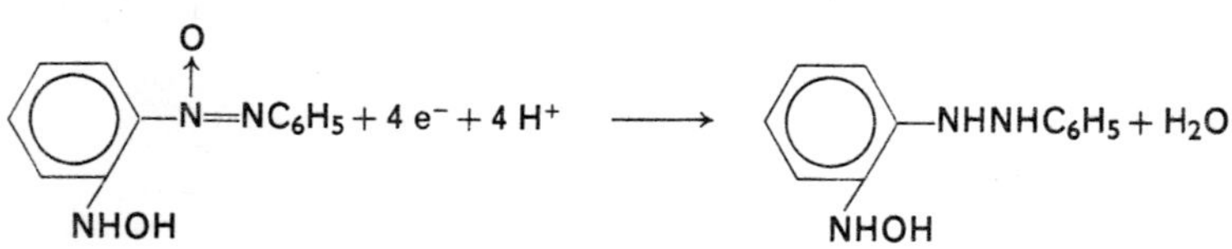
(13c)

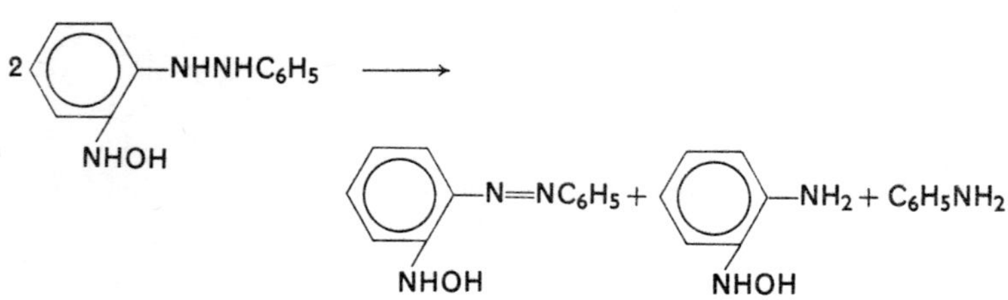
(13d)

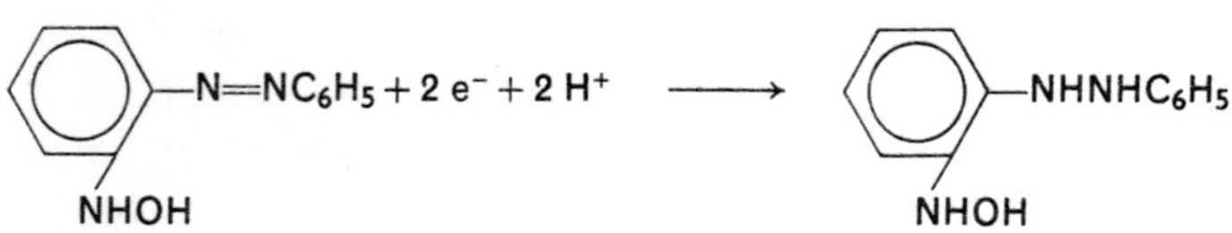
(13e)

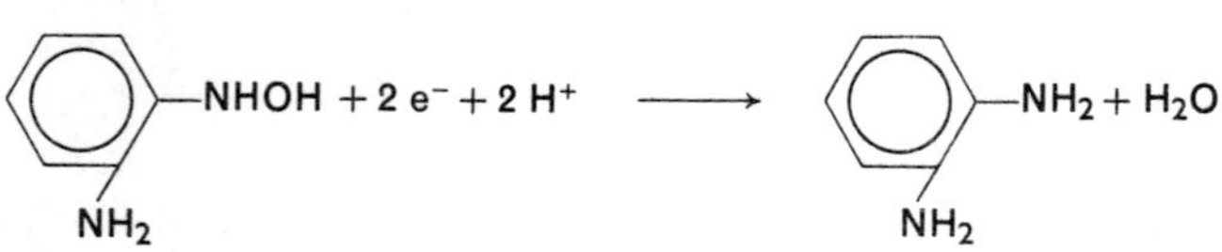
(13f)

The 2-nitrohydrazobenzene produced in equation (13b) undergoes the following reactions to produce 2-phenyltriazole.

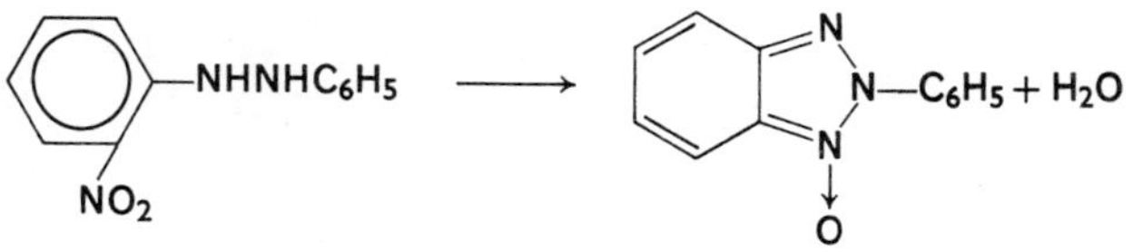

(13g)

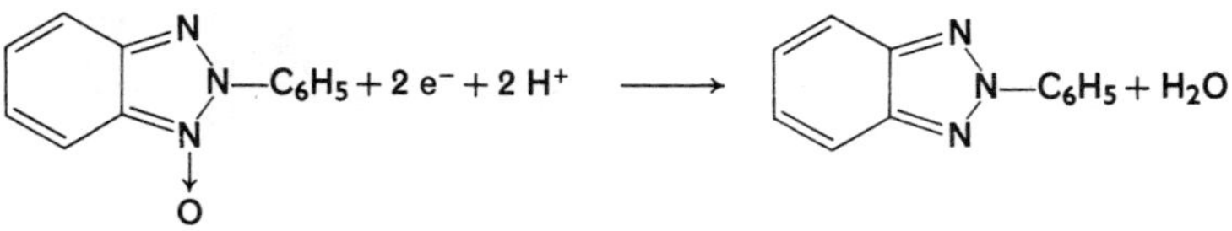

(13h)

Since 2-nitrohydrazobenzene can also be reduced to 2-hydroxylamino-hydrazobenzene

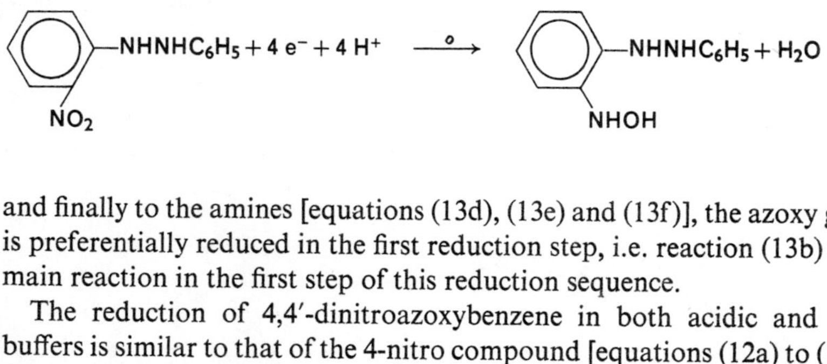

(13i)

and finally to the amines [equations (13d), (13e) and (13f)], the azoxy group is preferentially reduced in the first reduction step, i.e. reaction (13b) is the main reaction in the first step of this reduction sequence.

The reduction of 4,4′-dinitroazoxybenzene in both acidic and basic buffers is similar to that of the 4-nitro compound [equations (12a) to (12g)]. However, no evidence for an alternative route in basic solutions [analogous to equations (12h) and (12i)] was found. The complete reduction requires 18 electrons per molecule and the end product is the expected *p*-phenylene-diamine.

The presence of the second nitro group in 2,2′-dinitroazoxybenzene results in the selective reduction of the azoxy group to the hydrazo group being the first step in the reduction of this compound. This is in contrast to the 2-mono-nitro compound where both groups can be reduced. Although the polarographic results are difficult to interpret due to the merging of the two waves, constant potential coulometry provides enough information for the following reaction mechanism to be established.

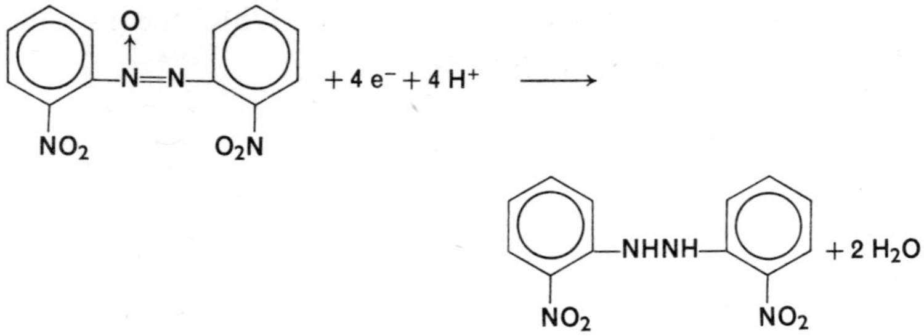

(14a)

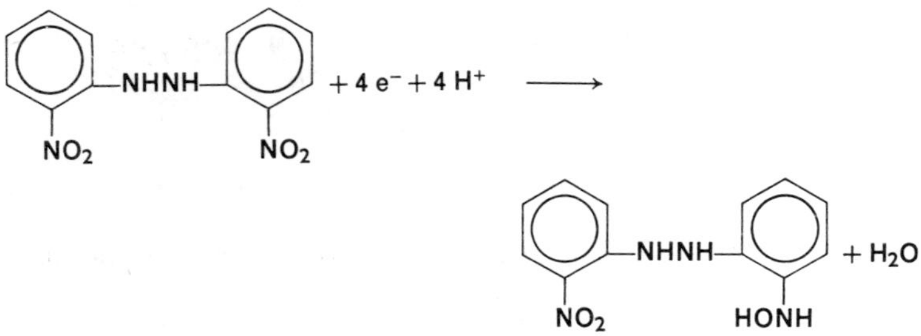

(14b)

The 2-nitro-2′-hydroxylaminohydrazobenzene now undergoes further

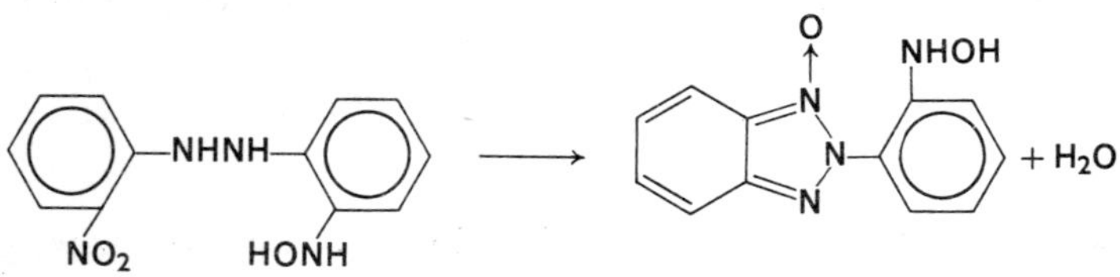

(14c)

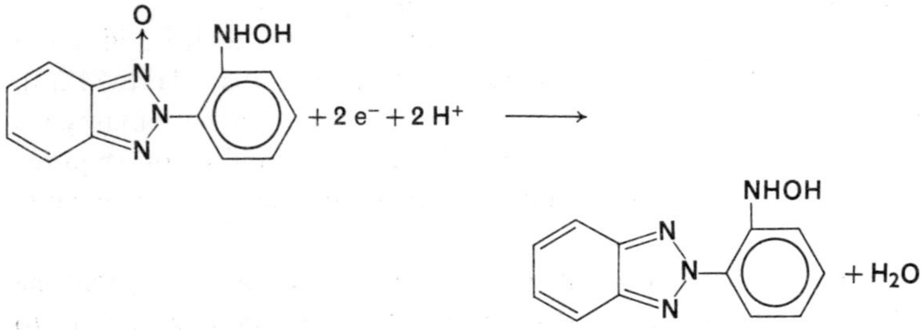

(14d)

reduction by either of two routes. The first gives 2-(2′-hydroxylaminophenyl)benzotriazole in 40% yield based on starting material. The second route which accounts for the remaining 60% of the starting material gives *o*-phenylenediamine as follows:

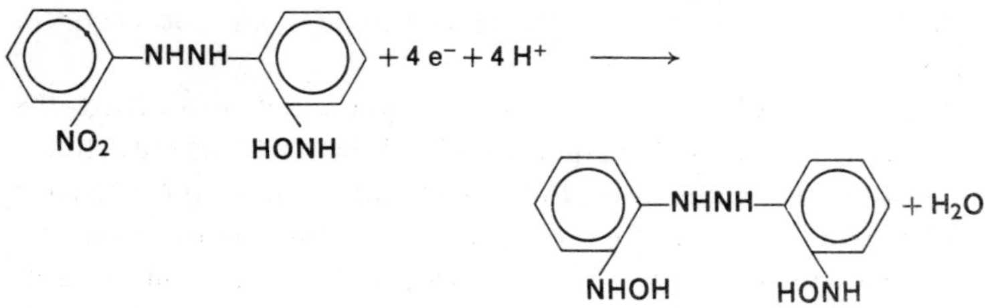

(14e)

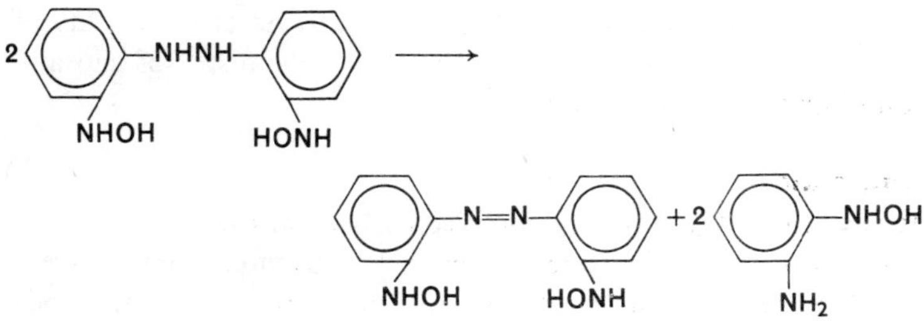

(14f)

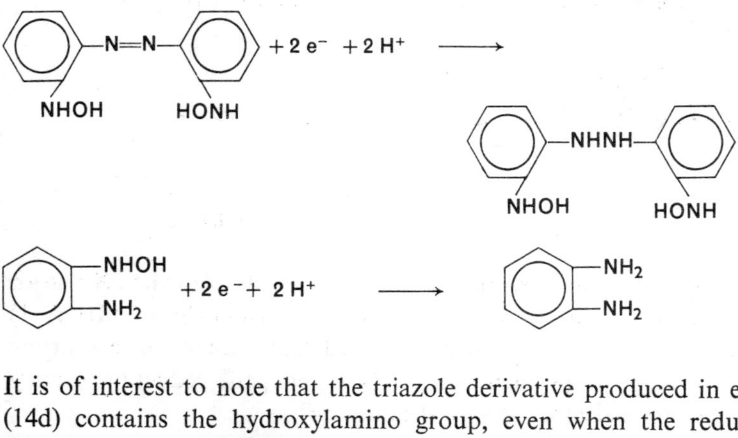

(14g)

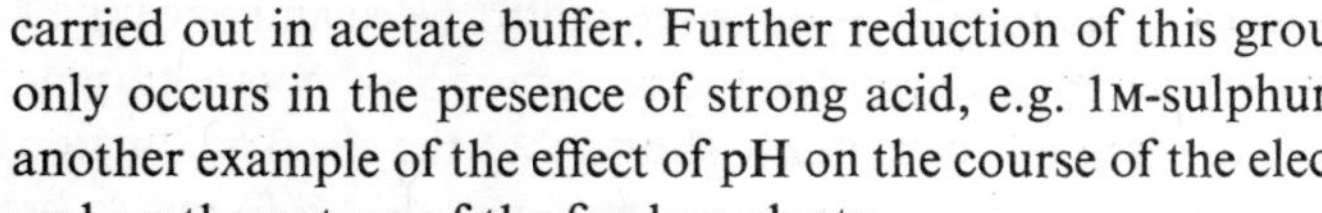

(14h)

It is of interest to note that the triazole derivative produced in equation (14d) contains the hydroxylamino group, even when the reduction is carried out in acetate buffer. Further reduction of this group to the amine only occurs in the presence of strong acid, e.g. 1M-sulphuric acid. This is another example of the effect of pH on the course of the electrode reactions and on the nature of the final products.

In the reduction of 3,3′-dinitroazoxybenzene, the polarographic results indicate complex behaviour. On the other hand, coulometric studies show that the reduction is quite straightforward. Both nitro groups are first reduced in an eight-electron transfer. This is in marked contrast to all the other nitro derivatives studied where the azoxy group is the most easily reduced group. The 3,3′-dihydroxylaminoazoxybenzene so produced is then further reduced in two four-electron steps to 3,3′-diaminohydrazobenzene. Further reduction to *m*-phenylenediamine does not occur in either buffer solution.

The electrochemistry of these substituted azoxybenzenes shows how the substituents can modify the course and products of the electrode reductions. The three major factors which affect the course and extent of the electrode reaction under given solution conditions (pH, solvent composition and supporting electrolyte) in the reduction of azoxybenzene derivatives are: (*i*) the position of the substituent; (*ii*) the basicity of the substituent (i.e. its electron donating or attracting ability), and (*iii*) the electroreactivity of the substituent. Similar factors influence the electrochemistry of azo and hydrazo compounds to be discussed below.

2. Azo compounds

The extensive literature covering the electrochemistry of the aromatic azo–hydrazo redox system in aqueous media has been comprehensively reviewed by Che-Man Chang[21]. In general, aromatic azo compounds undergo the following two-electron reduction at the dropping mercury electrode[22, 23]:

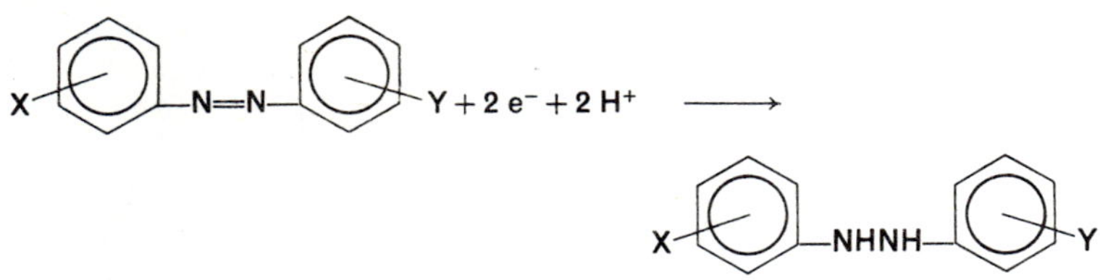

(15)

There are, however, some exceptions to this. For example, when X and/or Y are electroactive groups (cf. azoxybenzene derivatives in Section III, A, 1) complex electrode behaviour may be observed[19], and when they are strong electron donating groups such as OH, NH_2, etc., the reduction usually proceeds via a four-electron step to the amines[24].

Although the overall reaction (15) is quite straightforward and one might anticipate a fairly simple mechanism for the reduction, the polarographic behaviour of these compounds commonly shows that the electroreduction reactions are very complex[25–29]. The half-wave potentials are found to be dependent on the concentration of the azo compound[26, 30], solvent composition[31], the presence of surfactants[27] and the pH of the solution (Table 4). If reaction (15) is assumed to be reversible, the shape of the polarographic wave will be described by the Nernst equation and the half-wave potential at 25°C is given by[37a]

$$E_{1/2} = \text{const.} - \frac{0{\cdot}059q}{n}.\text{pH} \qquad (15a)$$

where q and n are the number of protons and electrons, respectively, involved in the electrode reaction. Since $q = n = 2$ for reaction (15) the half-wave potential for the reduction of azo compounds should change by –0·059 V for every unit increase in the pH of the solution. From the data in Table 4 it can be seen that for many azo compounds the relationship between half-wave potential and pH differs considerably from the theoretical value of –0·059 V per pH unit. In many cases, non-linear relationships are reported[27, 33, 35], or two linear portions are noted at different pH ranges[24, 37]. These observations are due to irreversibility of the overall reduction process and/or the influence of surface adsorption processes. The latter effects are shown also by the observation that surfactants affect the slope of the pH–half-wave potential plot[25–27, 33].

Most workers report that the limiting polarographic current is diffusion controlled and proportional to the concentration of the azo compound at low concentrations[25, 33, 34]. The linear region of the concentration–current relationship has been used for the quantitative determination of aromatic azo compounds[37, 38]. The reversibility of reaction (15) was found to depend not only on the pH of the solution—being more reversible at the extremes of the pH scale—but also on the type and concentration of the buffer used[25–29, 32]. In addition, the type of electrode material used[39] and the nature of any added surfactant affected the reversibility of the electrode reaction[27, 40, 41]. Despite all these complications, considerable headway has been made towards an understanding of the electrode reduction of aromatic azo compounds. The more significant aspects of the electrochemistry of certain types of aromatic azo compounds will now be considered in more detail.

a. The effects of adsorption. The adsorption of azo compounds and their reduction products at the mercury–solution interface and the effect this has on the electrochemistry of these compounds has been one of the major difficulties encountered in the study of these systems. Adsorption of azo and hydrazo compounds at the dropping mercury electrode is particularly pronounced in solvents with a high water content and is one of the major factors affecting the reversibility of the electrode reactions[26, 28, 45]. For example, oscillographic polarographic studies have shown that the rate of the electrode reaction in 30% methanol in the pH range 1 to 7 is determined by the adsorption kinetics of the azo and hydrazo species involved[26].

Tensammetric waves have been observed in the a.c. polarographic studies of azobenzene derivatives[25]. Electrocapillary studies have also demonstrated the strong adsorption of azo compounds at the mercury–solution interface. Determinations of the surface concentrations on mercury electrodes of methyl orange and of 4-aminoazobenzene in 30% methanol[46] show that these compounds are readily adsorbed to give mono-layer coverage. The results indicate that 4-aminoazobenzene is adsorbed with the molecules flat on the surface (i.e. the plane of the benzene rings is parallel to the surface) whereas methyl orange, which contains the strongly hydrophilic sulphonic acid group, is adsorbed 'side-on' to the surface (i.e. the plane of the benzene rings is perpendicular to the surface). Similar results were obtained from a study of the adsorption of methyl red at a mercury electrode in aqueous solutions in the pH range 7–12·3[47]. In this work it was found that the anion of methyl red was more strongly adsorbed than the acid form and that monolayer coverage of the anion is achieved at solution concentrations of 10^{-4}M or greater. Another interesting feature was that methyl red anion exhibits two position adsorption. At positive potentials the molecules lie flat on the surface whereas at potentials slightly positive to and in the region of the onset of electrode reduction, the molecules reorientate themselves so that they lie side-on to the surface. What role this re-orientation has in the reduction process is as yet unknown.

The studies of Holleck and co-workers[40,41] on the effect of surfactants which are more strongly adsorbed at the mercury–solution interface than the azo and hydrazo compounds show that these materials usually inhibit the electrode reaction. An apparent anomaly is the case of gelatin. In this case there are two competing effects—the retarding effect due to the surface layer of gelatin blocking the electrode reaction and the catalytic effect of the gelatin in promoting the protonation of the azo compound prior to reduction, and hence catalysing the overall electrode reaction. These effects are summarized as follows:

$$C_6H_5N{=}NC_6H_5 \underset{-H^+}{\overset{+H^+}{\rightleftharpoons}} C_6H_5N{=}NHC_6H_5^+ \quad \text{(electrode surface)} \qquad (16a)$$

$$C_6H_5N{=}NC_6H_5 \underset{-H^+}{\overset{+H^+}{\rightleftharpoons}} C_6H_5N{=}NHC_6H_5^+ \quad \text{(in solution)} \qquad (16b)$$

$$C_6H_5N{=}NHC_6H_5^+ + 2\,e^- \longrightarrow C_6H_5\bar{N}{-}NHC_6H_5 \qquad (16c)$$

$$C_6H_5\bar{N}NHC_6H_5 \underset{-H^+}{\overset{+H^+}{\rightleftharpoons}} C_6H_5NHNHC_6H_5 \quad \text{(electrode surface)} \qquad (16d)$$

$$C_6H_5\bar{N}NHC_6H_5 \underset{-H^+}{\overset{+H^+}{\rightleftharpoons}} C_6H_5NHNHC_6H_5 \quad \text{(in solution)} \qquad (16e)$$

The catalytic effect operates in equations (16a) and (16d) while the blocking effect interferes with (16c). Which of these two effects predominates depends on the gelatin concentration. At high gelatin concentrations (0·1 %) the blocking effect becomes important. At low concentrations, (up to 0·1 %),

TABLE 4. Variation of half-wave potential ($E_{1/2}$) with pH for some azobenzenes in several hydroxylic solvents

Compound	Solvent	pH Range	Concentration (mM)	$E_{1/2}$ dependence on pH (V vs. s.c.e.)	Comment	Ref.
cis-Azobenzene	10% EtOH	2·85–12·50	0·1	$E_{1/2} = +0{\cdot}06 - 0{\cdot}062$ pH	—	32
trans-Azobenzene	10% EtOH	2·85–12·50	0·1	$E_{1/2} = +0{\cdot}06 - 0{\cdot}062$ pH	—	32
cis-Azobenzene	30% MeOH	2·0–10·9	0·25	$E_{1/2} = +0{\cdot}064 - 0{\cdot}058$ pH	—	29
trans-Azobenzene	30% MeOH	2·0–10·9	0·25	$E_{1/2} = +0{\cdot}054 - 0{\cdot}059$ pH	—	29
Azobenzene	30% MeOH	1–9	0·5	Non-linear	—	27
Azobenzene	30% MeOH	1–9	0·5	0·078 V slope	0·01% gelatin	27
Azobenzene	30% MeOH	1–13	0·04	0·060 V slope	0·005% gelatin	26
Azobenzene	5–20% EtOH	1–13	0·05–0·2	$E_{1/2} = +0{\cdot}068 - 0{\cdot}060$ pH	—	28
Azobenzene	25–100% EtOH	0–12·4	1	0·060 V slope	—	31
Monosubstituted azobenzenes						
4-Sulphonic acid	H_2O	1–12	0·05–0·2	$E_{1/2} = +0{\cdot}125 - 0{\cdot}061$ pH	—	28
4-*N*,*N*-Dimethylamino	50% EtOH	1·5–13·5	0·1–1·0	Non-linear	0·01% methyl cellulose	33
4-*N*,*N*-Dimethylamino	20% EtOH	1–6	0·08	$E_{1/2} = +0{\cdot}24 - 0{\cdot}095$ pH	$n = 4$	24
		8–13		$E_{1/2} = -0{\cdot}05 - 0{\cdot}058$ pH	$n = 2$	
4-*N*,*N*-Dimethylamino	30% iPrOH	2–12	0·03–0·1	0·085 V slope	1% gelatin	34
4-Amino	10% EtOH	3·5–9·8	—	$E_{1/2} = +0{\cdot}05 - 0{\cdot}072$ pH	—	35
4-Amino	20% EtOH	1·5–5	0·08	$E_{1/2} = +0{\cdot}27 - 0{\cdot}11$ pH	$n = 4$	24
		8–13		$E_{1/2} = -0{\cdot}04 - 0{\cdot}060$ pH	$n = 2$	
4-Nitro	45% MeOH	1·9–12	0·1	$E_{1/2} = +0{\cdot}20 - 0{\cdot}059$ pH	Several surfactants	36
4-Carboxylic acid	20% EtOH	1·5–13	0·08	$E_{1/2} = +0{\cdot}105 - 0{\cdot}060$ pH	—	24
4-Methoxy	20% EtOH	1·5–13	0·08	$E_{1/2} = +0{\cdot}025 - 0{\cdot}062$ pH	—	24
4-Acetyl	20% EtOH	1·5–13	0·08	$E_{1/2} = +0{\cdot}125 - 0{\cdot}061$ pH	—	24
Disubstituted azobenzenes						
4,4′-Disulphonic acid	H_2O	1–12	0·05–0·2	$E_{1/2} = +0{\cdot}158 - 0{\cdot}059$ pH	—	28

TABLE 4 (*cont.*)

Compound	Solvent	pH Range	Concentration (mM)	$E_{1/2}$ dependence on pH (V vs. s.c.e.)	Comment	Ref.
4-*N*,*N*-Dimethylamino-4′-sulphonic acid	10% EtOH	3·5–9·8	—	$E_{1/2} = +0{\cdot}060–0{\cdot}069$ pH	—	35
4-*N*,*N*-Dimethylamino-4′-sulphonic acid	H_2O	3·5–13	0·08	$E_{1/2} = 0–0{\cdot}058$ pH	$2 < n < 4$ (pH 3·5–6) $n = 2$ (pH 8–13)	24
4-*N*-Phenylamino-3′-sulphonic acid	10% EtOH	3·5–9·8	—	Non-linear	—	35
4-Amino-4′-sulphonic acid	H_2O	3–6	0·08	$E_{1/2} = +0{\cdot}18–0{\cdot}088$ pH	$2 < n < 4$	24
		7·5–13		$E_{1/2} = +0{\cdot}02–0{\cdot}058$ pH	$n = 2$	
2,4-Diamino	10% EtOH	3·5–9·8	—	$E_{1/2} = -0{\cdot}03–0{\cdot}072$ pH	—	35
4-Dimethylamino-2′-carboxylic acid	10% EtOH	3·5–9·8	—	$E_{1/2} = +0{\cdot}09–0{\cdot}064$ pH	—	35
4-Dimethylamino-2′-carboxylic acid	20% EtOH	5–13	0·08	$E_{1/2} = +0{\cdot}04–0{\cdot}061$ pH	$n = 4$ pH 5–8 $n = 2$ pH 9–13	24
Multisubstituted azobenzenes						
2,4-Dihydroxy-4′-sulphonic acid	H_2O	3–13	0·08	$E_{1/2} = +0{\cdot}105–0{\cdot}080$ pH	$n = 4$	24
3,3′-Dimethyl-4-amino	10% EtOH	3·5–9·8	—	Non-linear	—	35
2-Hydroxy-2′,4′-diamino-5-sulphonic acid	10% EtOH	3·5–9·8	—	$E_{1/2} = -0{\cdot}05–0{\cdot}066$ pH	—	35

Phenylazonaphthol derivatives					
1-(4-Hydroxynaphthylazo)-benzene-4-sulphonic acid	H_2O	1·5–6 7–13	0·08	$E_{1/2} = +0{\cdot}145 - 0{\cdot}082$ pH $n = 4$ $E_{1/2} = +0{\cdot}010 - 0{\cdot}060$ pH $2 < n < 4$	24
1-(2-Hydroxynaphthylazo)-benzene-4-sulphonic acid	H_2O	3–13	0·08	$E_{1/2} = +0{\cdot}005 - 0{\cdot}060$ pH $n = 4$	24
1-(2-Hydroxynaphthylazo)-2-phenol-4-sulphonic acid	H_2O	3–13	0·08	$E_{1/2} = -0{\cdot}020 - 0{\cdot}060$ pH $n = 4$	24
Heterocyclic azo compounds					
4-(2-Pyridylazo)-phenol	50% EtOH	3·8–13	0·2	$E_{1/2} = +0{\cdot}295 - 0{\cdot}094$ pH $n = 2$	37
4-(2-Pyridylazo)-3-hydroxy-phenol	H_2O	1–5·6 5·6–12·3 >12·3	0·2	$E_{1/2} = +0{\cdot}183 - 0{\cdot}098$ pH $n = 2$ (pH 1–9) $E_{1/2} = +0{\cdot}007 - 0{\cdot}070$ pH $E_{1/2} = -0{\cdot}783$ (indep. of pH) $n = 3{\cdot}5$ (pH >12)	37
2-[(5-Bromo-2-pyridyl)azo]-5-(*N*,*N*-diethyl)amino-phenol	50% EtOH	2–11·2 11·2–13·5	0·2	$E_{1/2} = +0{\cdot}086 - 0{\cdot}076$ pH $E_{1/2} = -0{\cdot}770$ (indep. of pH) $n = 2$ (1–8) $n = 3$ (13)	37
1-(2-Pyridylazo)-2-naphthol	50% EtOH	2–11·3 11·3–13·5	0·2	$E_{1/2} = +0{\cdot}114 - 0{\cdot}075$ pH $n = 2$ (1–8) $E_{1/2} = -0{\cdot}740$ (indep. of pH) $n = 3{\cdot}1$ (13)	37

where only partial replacement of azobenzene by gelatin on the electrode surface occurs, the catalytic effect is evident.

b. The effects of pH and solution composition. In addition to the effects of pH expected from equation (15) and noted in Table 4, it has been found that the reduction of azo compounds is more reversible at the extremes of the pH range[25, 26, 28]. Because of the importance of the protonation steps [e.g. equations (16b) and (16e)] in the reduction of azo compunds, factors which affect the supply of protons at the electrode surface may influence the mechanism of the reduction and introduce unusual features in the polarograms. In particular, when the reduction is carried out in poorly buffered media, or in the presence of buffer systems composed of acids which have slow dissociation steps, multi-step polarographic waves are observed. Such multi-step waves have been observed in the reduction of azobenzene[49, 50], methyl orange and methyl yellow[38] and methyl red[51, 52]. These multi-step waves are caused by the depletion of protons at the electrode surface during the course of the reduction of the azo compound.

The depletion of hydrogen ions at the electrode surface during the reduction of azobenzene in dilute solutions of strong acids and in buffer solutions of low buffer capacity has been studied quantitatively[50]. The two polarographic waves observed between pH 4·5 and 7·0 have been adequately accounted for by means of theoretically-derived equations which describe the shape and limiting currents of the waves. An interesting application of this multi-step reduction is due to Delahay and co-workers[49]. They studied the kinetics of dissociation of monochloroacetic acid by using the first of the two waves observed in the reduction of azobenzene in the presence of monochloroacetic acid–monochloroacetate buffers. This first wave was a kinetic wave with the current being controlled by the rate of dissociation of the acid.

A further effect due to the nature of the buffer used to control pH is the ‘specific buffer effect’ observed in the polarographic reduction of azobenzene-4-sulphonic acid and azobenzene-4,4′disulphonic acid and in the polarographic oxidation of the corresponding hydrazo compounds[28]. In the vicinity of pH 9, buffers containing ammonia or ethanolamine caused an increase in reversibility of the electrode process as compared with any other of a wide range of buffer systems. The presence of these two species also resulted in an increase in the height of the current peaks in the a.c. and Kalousek[53] polarograms—an observation consistent with the increase in the reversibility of the electrode reaction. The specific buffer effect was not observed when the reduction occurred on the positive side of the electrocapillary maximum, i.e. when the electrode surface is positively charged. It appears to be caused by association between the buffer cations and the sulphonate anions. The increase in reversibility is probably due either to weaker adsorption of the ion pair at the electrode surface or to an increased rate of electron transfer through ammonium or ethanolammonium ion 'electron bridges'.

The pH of the solution may also effect the nature of the final products of the reduction of azo compounds. For example, the product of the 2-electron reduction of azobenzene, hydrazobenzene, is stable and can be re-oxidized to azobenzene in alkaline solutions. However, in acid solution, hydrazobenzene slowly rearranges to form, mainly, benzidine and diphenyline

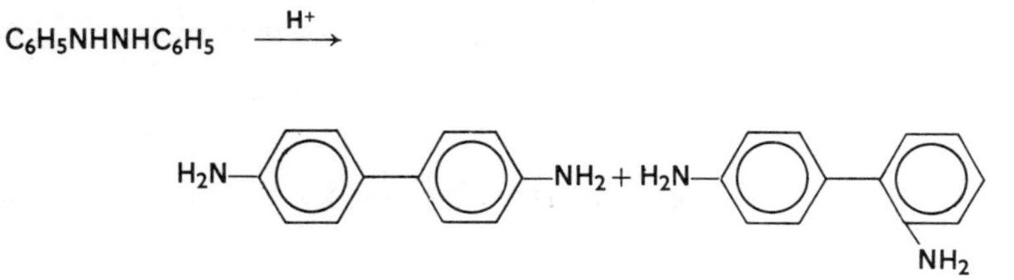

(17)

[equation (17)]. The rate of this rearrangement depends on pH, being faster in more acidic solutions[21,48]. In strongly acidic aqueous or alcoholic solution, the end products of the reduction of azobenzene were benzidine (70%) and diphenyline (30%)[48]. Since these compounds are not electrolytically re-oxidized back to azobenzene, their formation results in the electroreduction of azobenzene being irreversible.

Whether the observed polarographic waves are irreversible or not will, however, depend on the relative rates of the rearrangement and electrode reactions [equations (15) and (17)]. If the rearrangement is slow compared with the electrode reaction, it will occur mainly away from the electrode surface. Under these circumstances, the reduction may give a reversible polarographic wave since the concentration of the hydrazo compound at the electrode surface is virtually that in the absence of any rearrangement reaction. On the other hand, a very rapid rearrangement results in complete irreversibility of the polarographic wave.

c. Cis- *and* trans- *isomers of azobenzene.* The difference in polarographic behaviour of the *cis*- and *trans*- isomers of azobenzene and azobenzene disulphonic acids was first studied by Winkel and Siebert[42]. They reported that the *trans* isomers were always reduced at more cathodic potentials than the corresponding *cis* isomers. These results were confirmed by later workers[29,43], although one report[29] indicated no significant difference between the reduction potentials of the isomers of azobenzene in aqueous methanol, possibly due to lack of sensitivity in the potential measurements. At pH values less than 4 the half-wave potentials of the *cis*- and *trans*-azobenzene were close together but at higher pH, the separation increases[44]. In most studies of aromatic azo compounds *cis–trans* isomeric effects have not been studied and, in general, studies have concentrated on the more stable *trans*-isomers.

d. Substituted azobenzenes. The electrochemistry of substituted azobenzenes is basically the same as that of the parent compound except when the substituents are electroactive or strong electron donating groups. In the absence of electroactive groups, good correlations have been found between the polarographic half-wave potentials and Hammett σ constants for substituents in the *m*- and *p*-positions of the benzene rings[24, 38]. For example, the relationship

$$E_{1/2}(\mathrm{X}) = E_{1/2}(\mathrm{H}) + 0{\cdot}136\sigma_p(\mathrm{X}) \tag{18}$$

where $E_{1/2}(\mathrm{X})$ is the half-wave potential for the reduction of the substituted azobenzene, $E_{1/2}(\mathrm{H})$ is the same value for azobenzene and $\sigma_p(\mathrm{X})$ is the Hammett *para* σ constant; was found to hold for eight mono-substituted azobenzenes in 20% ethanol at pH 0. Correlation between the Hammett *ortho* σ constants and $E_{1/2}$ values is not so good because of steric factors.

e. Strong electron donating substituents. As indicated in equation (15), the reduction of aromatic azo to hydrazo compounds involves the addition of only two electrons. However, a number substituted azobenzenes, such as methyl orange and methyl red[24], and the 4-hydroxy[24, 54], 4-amino[24], 4-*N*,*N*-dimethylamino[24, 33, 34], 4-hydroxy-4′-sulphonic acid[24], 2,4-dihydroxy-4′-sulphonic acid[24] and 2,2′,4-trihydroxy-4′-sulphonic acid derivatives[24] give four-electron polarographic reduction waves. The products of these reductions are the corresponding amines resulting from the splitting of the azo *N-N* bond.

Using the techniques of a.c., d.c., Kalousek, and single sweep polarography, coulometry[24], cyclic voltammetry and chronoamperometry[21] the mechanism of the reduction of these compounds has been elucidated. The expected hydrazo compound, formed by the two-electron reduction of the azo compound, is involved as an unstable intermediate. The reaction sequence given below for 4-*N*,*N*-dimethylaminoazobenzene is typical of the reduction mechanism of substituted azobenzenes containing strong electron donating substituents such as hydroxy, amino and *N*-substituted amino groups.

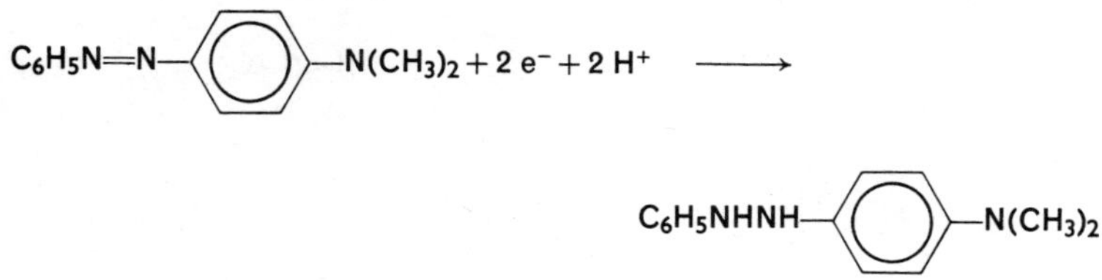

(19a)

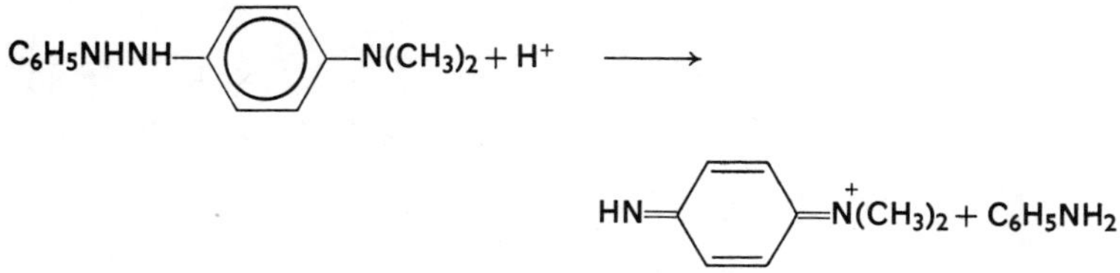

(19b)

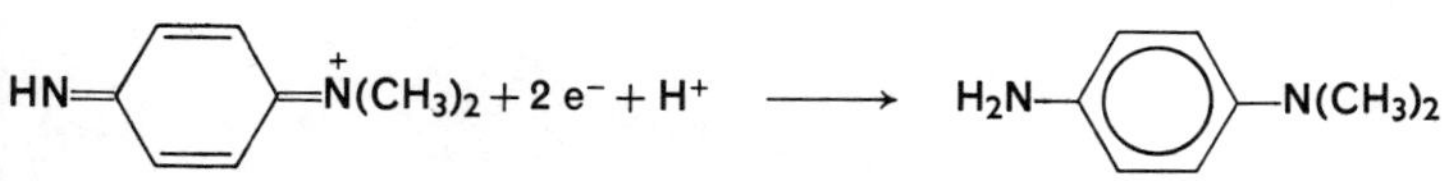

(19c)

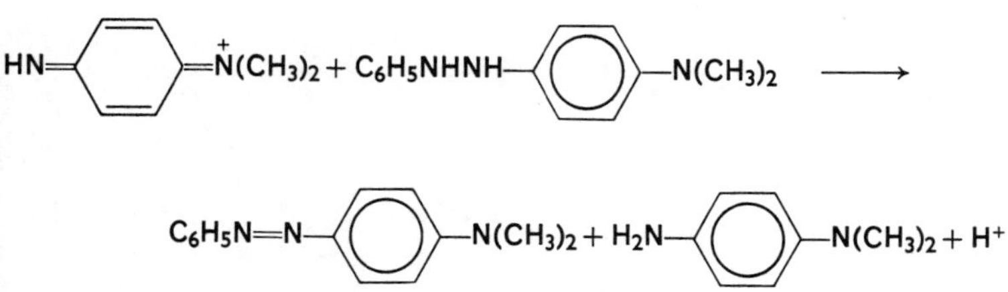

(19d)

overall

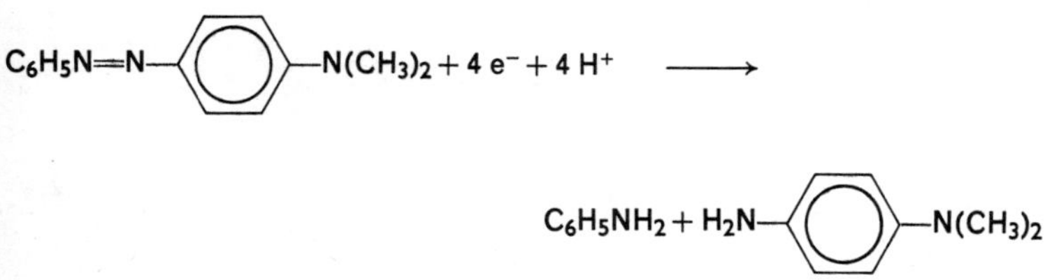

(19e)

Since (19b) and (19d) are fast and (19c) occurs at the same potential as (19a), only a single four-electron polarographic wave is observed.

f. Summary. The factors discussed in the above sections III, A, 2a–e can be summarized in the following mechanism for the reduction of azobenzenes at a mercury electrode.

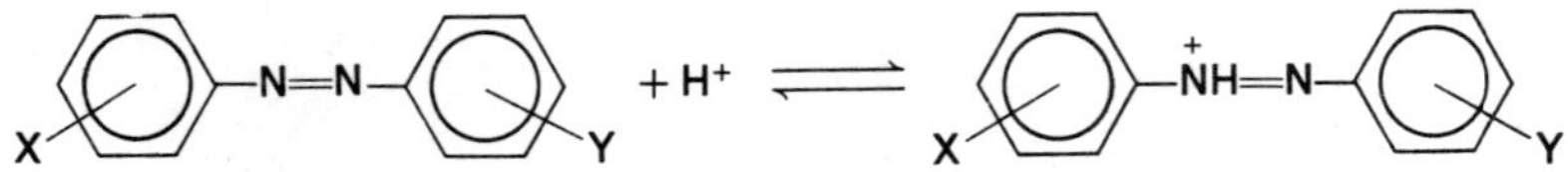

(20a)

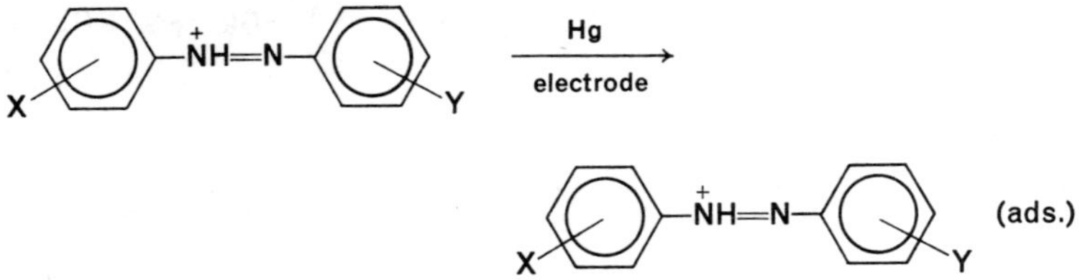

(20b)

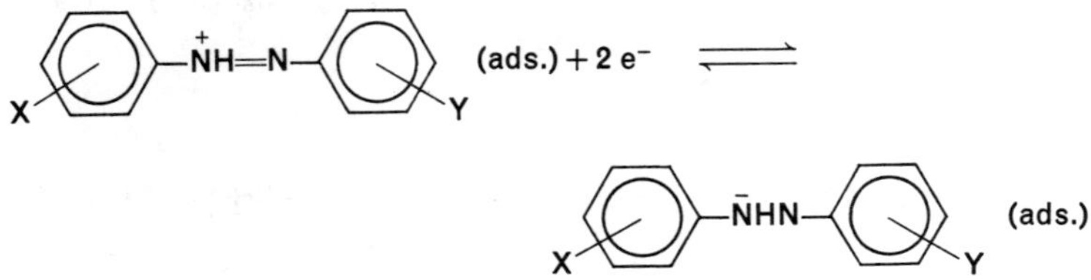

(20c)

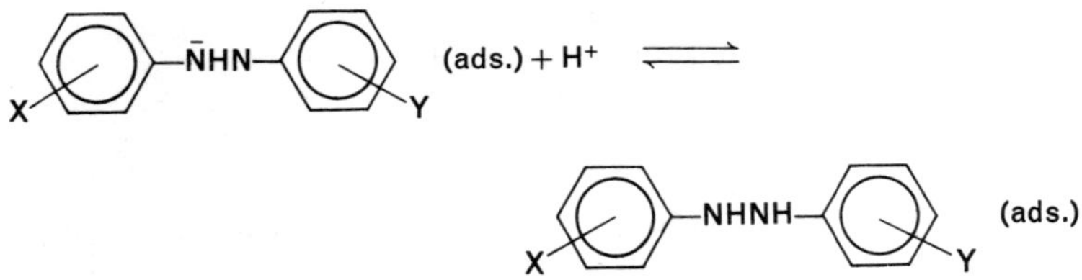

(20d)

or

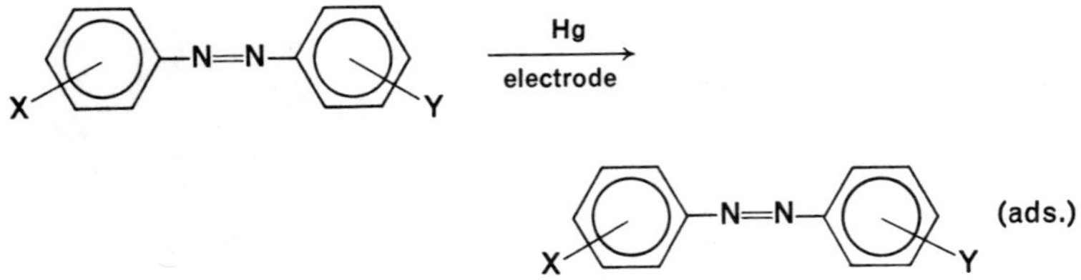

(20e)

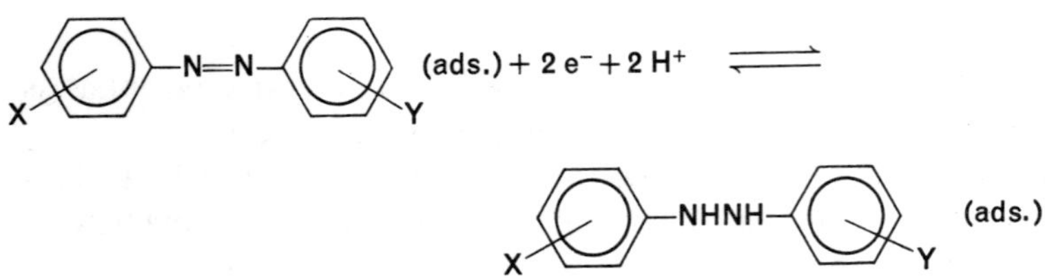

(20f)

Then

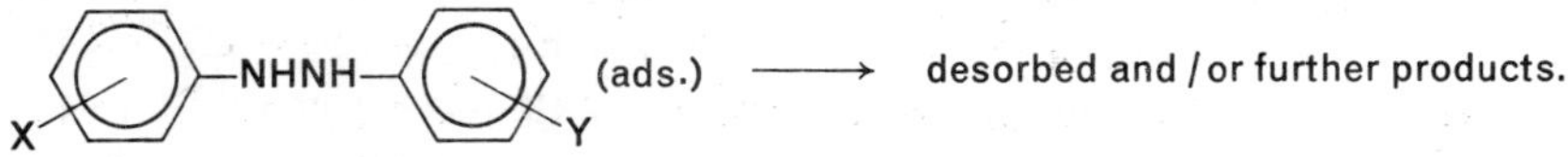

In the above scheme, the azobenzene can be reduced via the adsorbed azonium form [(20c) and (20d)] produced in the pre-protonation step (20a) or via the adsorbed, unprotonated form (20f). The hydrazo compound produced is generally strongly adsorbed and may undergo additional chemical (17) or electrode reactions (19), depending on the pH and/or the substituent groups X and Y. The overall rate of the reduction is determined by the relative rates of the pre-protonation and adsorption reactions[18, 28] since most authors consider that the actual electron transfer steps are rapid and reversible. When the system is poorly buffered the rates of reactions (20a) and (20d) or of (20f) may be slow and controlled by the rate of production of the proton needed in these reactions[49, 50]. Inhibition of the electrode reaction by strongly adsorbed products and added surfactants may also occur. The reactions of the reduction products, if any, also affect the overall reversibility of the electrode reaction.

Thus the overall reduction scheme in protolytic media can be very complex. It is difficult to make any predictions about the reduction mechanism of a particular compound unless conditions such as the composition of the solvent, pH, the type and concentration of buffer used and the nature of any surfactant that may be present are carefully specified.

g. Azobenzenes with electroactive substituents. As in the case of azoxy compounds with other electroactive groups in the molecule (Section III, A, 1), azobenzenes containing such groups often exhibit complex electrode behaviour. For example, mono- and di-nitroazobenzenes exhibit similar electrochemical behaviour[19] (other than the first step which is a two-electron step rather than the four-electron step for azoxybenzenes) and give the same reduction products as the corresponding azoxybenzenes described in Section III, A, 1.

h. Other aromatic azo compounds. The electrochemistry of phenylazonaphthalenes is essentially the same as that of the azobenzenes, i.e. in the absence of strong electron donating groups or other electroactive groups the azo compound is reduced to the corresponding hydrazo compound. As with the azobenzenes, the presence of strong electron donating groups results in a four-electron reduction with cleavage of the N–N bond to give the corresponding amines[55, 56].

In 48% ethanolic solution, the d.c. polarographic reduction of 1-phenylazo-2-naphthol and a series of 1-(X-phenylazo)-2-naphthols, where X is

Cl, CH_3 or OCH_3 in the 2′, 3′ or 4′ positions, proceeds via a single, four-electron step in the pH range 4 to 12[55]. Oscillographic polarographic studies of these compounds indicate that the hydrazo compound is an unstable intermediate in this reduction. The half-wave potentials, extrapolated to pH = 0, vary linearly with the Hammett substituent constants of the three groups. With the exception of the 2′-methoxy- and 2′-chloro-derivatives, the half-wave potentials at pH 0 are given by

$$E_{1/2}(X) = -0{\cdot}105 + 0{\cdot}13\sigma(X) \tag{21}$$

The plot of half-wave potential vs. pH for these compounds shows two distinct linear portions. Below pH 10 the slope is ca. −66 mV per pH unit while above pH 10 the slope falls to about −40 mV per pH unit. Since the change in slope occurs at a pH close to the value of the pK_a of the azonaphthol, it is considered to be caused by a change in the electroactive species from the acid form to the anion. The acid form is also expected to be internally hydrogen bonded (**1**) whereas this is not possible in the anion (**2**) and this feature may also be an important factor in the change of slope which may reflect a change in mechanism.

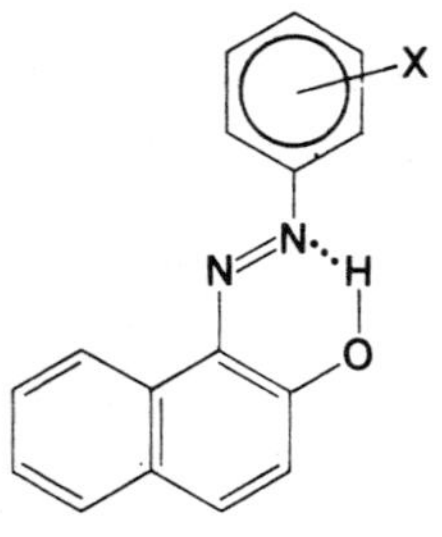

(1)

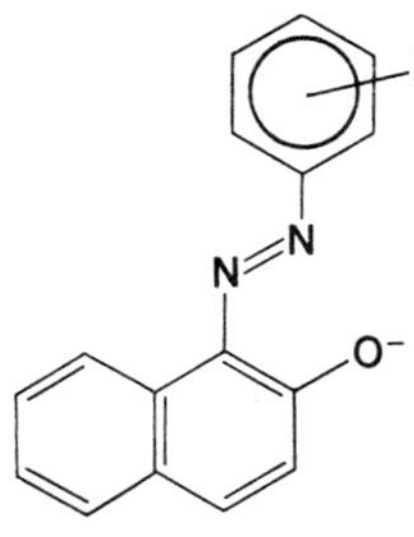

(2)

X = H; 2′, 3′ or 4′, Cl, CH_3 or OCH_3

Similar results were obtained for a series of 4-phenylazo-1-naphthols, mono-substituted in all three positions of the phenyl ring with Cl, CH_3 and OCH_3[56]. In this case, however, the slope of the pH vs. half-wave potential was greater at pH values above 10 than the −66 mV per pH unit slope at the lower pH values. Again the break in these plots occurs at a pH value close to that of the pK_a of the azonaphthol. The main difference between the 4-phenylazo-1-naphthols and the 1-phenylazo-2-naphthols is that the former exists as a zwitterion (**3**) at low pH in equilibrium with a quinonoid

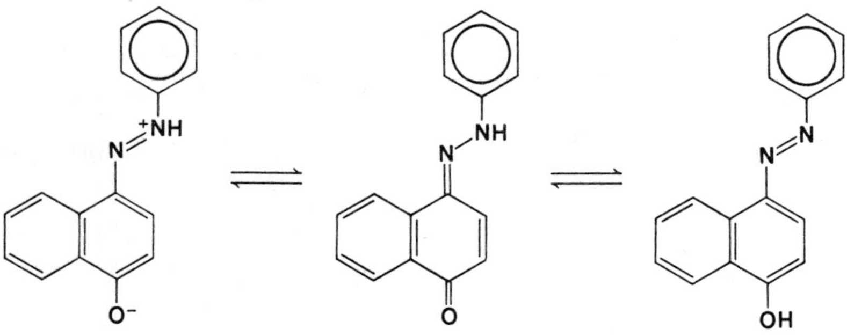

(3) (4) (5)

(22)

form (**4**) as well as the azonaphthol form (**5**). In aqueous alcohol, the zwitterion appears to be the major form, although in solvents of low dielectric constant one of the two uncharged forms is expected to be the major species in solution.

The role of acid dissociation in the electrode reactions of the phenylazo-naphthols is further evidenced by the unbroken linear, half-wave potential vs. pH plot obtained at all pH values for 4-phenylazo-1-naphthyl methyl ether. In this case the same electroactive species exists at all pH values since there is no acidic hydrogen in this molecule.

The mechanism for the electrode reduction of the 4-phenylazo-1-naphthol is considered to involve the zwitterion as follows:

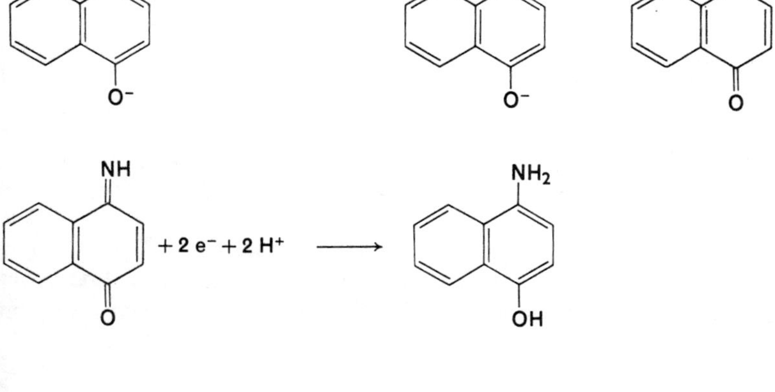

(23a)

(23b)

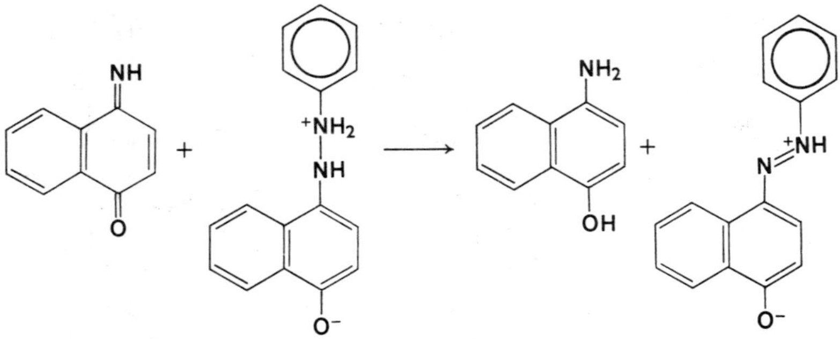

(23c)

The hydrazo compound produced initially is very unstable and dissociates to produce aniline and 1,4-naphthaquinone-4-imide. The latter is reduced to 4-amino-1-naphthol either electrolytically (23b) or by the hydrazo compound in the vicinity of the electrode surface (23c). All these reactions proceed essentially to completion during the lifetime of a mercury drop (ca. 5 sec) and so the polarographic currents are not kinetic in nature.

In a recent study of the polarography of heterocyclic azo compounds[37] some 25 pyridylazo, thiazolylazo and quinolylazo phenols and naphthols were investigated. It was found that these heterocyclic hydroxy azo compounds yield two-electron reduction waves in acid media and four-electron waves only in strongly alkaline media (Table 4). This is in contrast to the hydroxy-azobenzenes which give four-electron waves only in acid solution. This stabilization of the heterocyclic hydrazo compound formed in the first stage of the reduction is due to the electron attracting properties of the pyridyl, thiazolyl and quinolyl groups tending to cancel the electron donating effect of the hydroxyl or amino substituents.

Constant potential electrolysis of the pyridylazo compounds shows that the hydrazo compounds produced disproportionate to give the amines and the azo compounds [e.g. for 4-(4-pyridylazo)-phenol]:

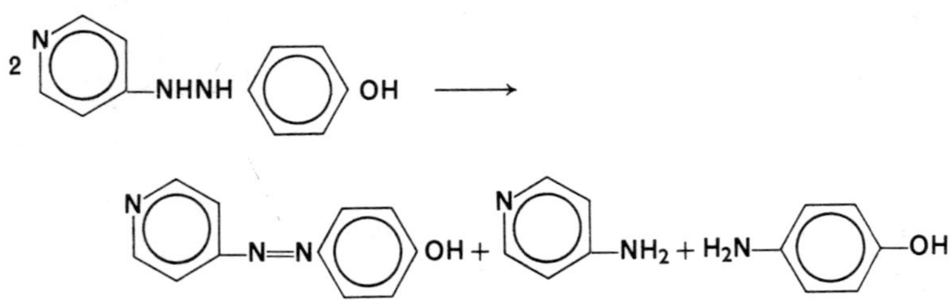

(24)

As in the case of the azobenzene derivatives[24], this reaction is first order with respect to the hydrazo compound. However, the disproportionation of hydrazobenzenes was found to be first order with respect to hydrogen ion concentration, whereas the disproportionation of the pyridylhydrazo compounds shows a small inverse hydrogen ion dependence, which suggests that in this case the disproportionation is base catalysed. Although disproportionation via reaction (24) results in an overall four-electron reduction to the amines on exhaustive electrolysis, the rate of this reaction is sufficiently slow so that during the lifetime of a drop (ca. 5 sec) no appreciable decrease in the concentration of the hydrazo compound at the electrode surface occurs. Thus only a two-electron polarographic wave is observed.

The plots of half-wave potential vs. pH for these compounds show several linear sections. The points of discontinuity are distinct and occur at pH values close to the pK_a values of the azo compound. For example, with 4-(2-pyridylazo)resorcinol the breaks occur at pH values of 5·75 and 12·3.

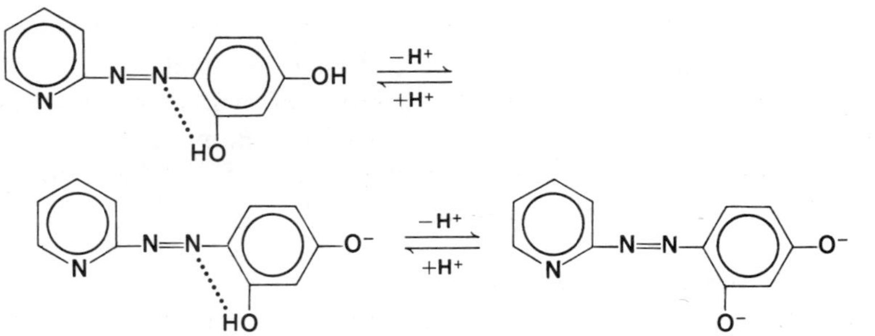

(25)

The values of pK_a for the acid dissociation reactions are 5·50 to 5·83 and 11·9 to 12·5, respectively[57]. This change of slope in the half-wave potential vs. pH plot is analogous to that observed with the azonaphthols and arises from the change in the nature of the electroactive species. This in turn may lead to a change in adsorption behaviour of the solute, a change in the detail of the electrode reaction mechanism or changes in the physical parameters of the electrode reaction.

The naphthalene and heterocyclic azo compounds, like the azobenzenes, often exhibit complex electrode behaviour. While some predictions may be made about the course of a particular electrode reaction, the nature of the particular groups attached to the azo linkage can have quite specific effects in determining the actual course and products of an electrode reaction.

3. Hydrazo compounds

The electrochemistry of the reduction of hydrazo compounds has been covered in the above sections (III, A, 1) and (III, A, 2) since hydrazo compounds are the immediate products of the reduction of azo and azoxy compounds in protolytic solvents. The electro-oxidation of hydrazo compounds is usually the reverse of the electro-reduction of azo compounds[58], i.e. the reverse of reaction (15). In many instances, hydrazo compounds have been admixed with azo compounds in, or their oxidation studied concurrently with, the study of the reduction of azo compounds in order to help elucidate the electrode processes[19, 24, 29, 40, 45].

Generally, hydrazo compounds are much more unstable than azo or azoxy compounds and as a result, their reduction to the amines commonly involves a combination of chemical auto-oxidation–reduction steps and electron exchange reactions; quinonoid intermediates are often formed and the azo compound produced as an intermediate is rapidly reduced back to the hydrazo compound. Examples of these reactions are given in reaction sequences (9), (12), (13), (14) and (19). The charge-transfer steps in the reduction of hydrazo compounds often involve the intermediates (azo or quinone compounds) and not the hydrazo compound.

Another complicating feature of the reduction of hydrazo compounds is the possibility of a benzidine type rearrangement in acid solution. Because of this, the electrode reaction is often less than 100% efficient in reducing hydrazo compounds to amines.

B. *Electrochemistry in Aprotic Solvents*

Despite the extensive literature on the electrochemistry of azoxy, azo and hydrazo compounds in aqueous and other protolytic solvents, relatively few studies have been carried out on the electrochemical behaviour of these compounds in aprotic media. The publications to date have concentrated on the azo–hydrazo redox system using dimethylformamide (DMF)[59–64], acetonitrile (AN)[14, 65–70], propylene carbonate (PC)[71] and pyridine (PY)[72] as solvents.

In contrast to the results in protolytic solvents, where a two-electron reduction of the azo to the hydrazo compound is observed, the reduction of azo compounds in DMF, AN and PY at the dropping mercury electrode proceeds via two one-electron steps (Table 5). The only exceptions to this occur when proton donors are added to the system[61, 62, 65, 72–75] or when proton releasing substituents such as —SO_3H, or —OH are present in the molecule[72]. Both a.c. and d.c. polarography and cyclic voltammetry show that the first of these steps is reversible and the second is usually irreversible.

The variation in half-wave potential of the first wave has been shown to be a linear function of the Hammet substituent constant for 3- and 4-substituted azobenzenes and for 2-substituted 5-phenylazopyridines. Reported relationships are listed in Table 6. However, 2-substituted azo-

TABLE 5. Polarographic data for selected azo and hydrazo compounds in aprotic solvents

Compound	Solvent	Supporting electrolyte	Reference electrode	Half wave potential (V) (Mercury electrode)	No. of electrons	Slope E vs $\log[(i_d - i)/i]$ (mV)	Ref.
Azobenzene	DMF	TBAP[a]	SCE	−1·36	1	61	62
				−2·03	1	82	
	DMF	TEAP[b]	Ag/Ag$^+$ (0·1M) in DMF	−1·81	1	59	60
				−2·29	1	90	
	AN	TEAP	SCE	−1·405	1	59	70
				−1·755	1	79	
	PC	TBAP	SCE	−1·40	2		71
	PY	TEAP	SCE	−1·350	1	61	72
				−1·805	1	70	
1-Phenylazonaphthalene	AN	TBAI[c]	SCE	−1·26	1	55[d]	68
				−1·74	1	100[d]	
1,1′-Azonaphthalene	DMF	TBAP	SCE	−1·13	1	58	62
				−1·69	1	103	
	AN	TBAI	SCE	−1·16	1	58[d]	68
				−1·62	1	100[d]	
Substituted azobenzenes							
4-Chloro	DMF	$NaNO_3$	SCE	−1·22	1	55[d]	64
	AN	TEAP	SCE	−1·325	1	64	70
				−1·660	1	79	
4-Sulphonate (TEA salt)[e]	AN	TEAP	SCE	−1·350	1	55	70
				−1·655	1	96	
4-*N*,*N*-Dimethylamino	DMF	$NaNO_3$	SCE	−1·56	1	55[d]	64
	AN	TEAP	SCE	−1·610	1	67	70
				−1·890	1	85	

TABLE 5 (*cont.*)

Compound	Solvent	Supporting electrolyte	Reference electrode	Half wave potential (V) (Mercury electrode)	No. of electrons	Slope E vs $\log[(i_d - i)/i]$ (mV)	Ref.
2-Methyl	AN	TEAI[f]	SCE	−1·376	1		69
				−1·97	1		
3-Methyl	AN	TEAI	SCE	−1·406	1		69
				−1·98	1		
4-Methyl	AN	TEAI	SCE	−1·48	1		69
				−1·97	1		
3,3′-Dimethyl	DMF	TBAP	SCE	−1·38	1	63	62
				−2·01	1	110	
4,4′-Diphenyl	DMF	TBAP	SCE	−1·22	1	63	62
				−1·77	1	109	
Heterocyclic azocompounds							
4,4′-Azopyridine	DMF	TBAP	SCE	−0·80	1	60	62
				−1·53	1	67	
9,9′-Azoacridine	AN	$LiClO_4$ + $HClO_4$	Ag/Ag^+ (0·10M) in AN	+0·35[g]	2		65
				−0·65[g]	2		
3-Phenylazopyridine	DMF	$NaNO_3$	SCE	−1·17	1	50[d]	64
2-Chloro-5-phenylazo-pyridine	DMF	$NaNO_3$	SCE	−1·09	1	55[d]	64
2-Acetylamino-5-phenyl-azopyridine	DMF	$NaNO_3$	SCE	−1·42	1	50[d]	64
2-Amino-5-phenylazo-pyridine	DMF	$NaNO_3$	SCE	−1·43	1	55[d]	64

						Comment	
Hydrazo compounds (oxidation)							
Hydrazobenzene	DMF	TBAP	SCE	ca. +0·2	2		62
	AN	TEAP	SCE	ca. +0·3	2	Large maxima	14
9,9′-Hydrazoacridine	AN	$LiClO_4$	Ag/Ag (0·01M)	−0·350	1	Base added	65
				−0·050[g]	1		
				+0·100[g]	1	Acid added	
				+0·600	1	Split wave	
				+1·500			
Hydrazo compounds (reduction)							
9,9′-Hydrazoacridine	AN	$LiClO_4$ + $HClO_4$	Ag/Ag (0·01M)	−0·420	2		65

[a] TBAP = Tetra-n-butylammonium perchlorate.
[b] TEAP = Tetraethylammonium perchlorate.
[c] TBAI = Tetra-n-butylammonium iodide.
[d] $E_{1/4}$–$E_{3/4}$ values (55 mV for reversible electrode reaction).
[e] TEA = Tetraethylammonium ion.
[f] TEAI = Tetraethylammonium iodide.
[g] Platinum electrode.

TABLE 6. Correlation between half-wave potential and Hammett σ functions for first reduction wave of substituted azobenzenes and 5-phenylazopyridines

Compounds	Solvent	Supporting electrolyte	$E_{1/2}$-σ relationship	Ref.
Azobenzenes				
4-Cl, 4-$N(CH_3)_2$, 4-SO_3^-, 4-$N(CH_3)_3^+$, 4-CH_3O	AN	TEAP	$E_{1/2}(X) = E_{1/2}(H) + 0{\cdot}26\sigma_p$	70
2-CH_3, 3-CH_3, 4-CH_3, 2-Cl, 3-Cl, 4-Cl, 2-OCH_3, 3-OCH_3, 4-OCH_3, 2-NH_2, 4-NH_2	AN	TEAI	$E_{1/2}(X) = E_{1/2}(H) + 0{\cdot}33\sigma$	69
4-I, 4-Br, 4-Cl, 4-F, 4-SCH_3, 4-CH_3, 4-OCH_3, 4-NH_2, 4-$N(CH_3)_2$	DMF	$NaNO_3$	$E_{1/2}(X) = E_{1/2}(H) + 0{\cdot}36\sigma_p$	64
2-*X*-5 *phenylazopyridines*				
X = I, Br, Cl, F, $NHCOCH_3$, CH_3, OH, NH_2	DMF	$NaNO_3$	$E_{1/2}(X) = E_{1/2}(H) + 0{\cdot}36\sigma_p$ (using σ_p values for benzene)	64

benzenes do not show such good correlations[69] due to steric effects. Good correlations have also been noted between the polarographic half-wave potential of the first wave for substituted azobenzenes and (*a*) the $n \rightarrow \pi^*$ transition energy ($E_{1/2} = 12{\cdot}33 - 4{\cdot}84\bar{\nu}$)[69], (*b*) the energy of the lowest vacant π molecular orbital[63, 69]. The latter correlation has also been reported to be good for the 2-substituted 5-phenylazopyridines[63] and for azo-naphthalenes[68]. Although the second one-electron reduction step is irreversible, the half-wave potentials for this reduction for several 4-substituted azobenzenes in AN are given reasonably well by the relation[70] $E_{1/2}(\text{X}) = E_{1/2}(\text{H}) + 0{\cdot}25\sigma_p$.

One notable exception to the above is 4-nitroazobenzene[70]. The reduction potentials of both waves are much more positive than expected from the $E_{1/2} - \sigma_p$ relationship and both reduction steps are reversible. This is undoubtedly due to the presence of the electro-active nitro group which may be the site of electron transfer rather than the azo group.

The electro-reduction of azobenzene[59–62] and of several azo aromatic hydrocarbons[62] has been studied thoroughly in DMF using polarography, cyclic voltammetry, coulometry and e.s.r. spectroscopy. A similar study of the electro-reduction of azobenzene and several of its 4-substituted derivatives has been carried out in AN[14, 70] using a.c. and d.c. polarography, coulometry and u.v.-visible spectrophotometry, and of azobenzene in PY[72]. As a result of these studies the mechanism of the reduction of aromatic azo compounds in DMF and AN is considered to be that given in reaction sequence (26) while in PY it appears to be the same although there is as yet no certainty about the final step and products.

$$\text{R}^1\text{—N=N—R}^2 + \text{e}^- \rightleftharpoons \{\text{R}^1\text{—N—N—R}^2\}^{\overline{\cdot}} \qquad (26\text{a})$$

$$\{\text{R}^1\text{—N—N—R}^2\}^{\overline{\cdot}} + \text{e}^- \rightleftharpoons \{\text{R}^1\text{—N—N—R}^2\}^{2-} \qquad (26\text{b})$$

$$\{\text{R}^1\text{—N—N—R}^2\}^{2-} + \text{SH} \longrightarrow [\{\text{R}^1\text{—N—N—R}^2\}^{2-}]\text{H}^+ + \text{S}^- \text{ (fast)} \qquad (26\text{c})$$

where SH = solvent

In this mechanism (reaction sequence (26)) both charge transfer steps are reversible and the observed irreversibility of the second polarographic wave is attributed to the very fast, irreversible protonation of the dianion produced in the second charge transfer step (26b). The dianion is considered to be a strong enough base, except in the case of the 4-nitro derivative, to abstract a proton from the 'so-called' aprotic solvents to produce the protonated dianion (26c).

The proton of the protonated dianion is considered to be bound initially to the hydrocarbon part of the molecule when DMF is used as solvent[62].

This species is then considered to undergo a slow second protonation and rearrangement eventually to produce hydrazobenzene.

On the other hand, the studies in AN indicate that the final product of electrolysis is the protonated dianion, and that the proton is attached to one of the nitrogen atoms. Solutions of the protonated dianion are stable in air-free AN for long periods and although the u.v.-visible spectrum has absorption peaks at the same wavelengths as hydrazobenzene, the extinction coefficients are many times larger. Also, the polarographic behaviour of the reduction product is quite different from that of hydrazobenzene—the former is oxidized at the dropping mercury electrode in a single two-electron step at $-1{\cdot}1$ V vs. s.c.e., whereas the latter is oxidized in a single two-electron step at $+0{\cdot}3$ V vs. s.c.e. This polarographic behaviour shows that the reduction product is not the dianion (26b) as the dianion is expected to be re-oxidized in two one-electron steps at similar potentials to the one-electron reduction steps of the parent azobenzene. Also the spectrum is quite different to that reported for chemically produced solutions of the dianion[76].

Thus it is concluded that in both solvents the final product is a protonated form of the azobenzene dianion. This is quite stable in AN but in DMF it slowly undergoes further protonation in the bulk solvent to produce eventually hydrazobenzene. This minor difference in the two solvents reflects the slight difference in the very weak proton donating properties of these two solvents.

In pyridine, the polarographic behaviour is very similar to that in AN and DMF. However, the second wave is only slightly irreversible, cf. the complete irreversibility observed in DMF and AN. While it is considered that the mechanism in PY involves reactions (26a) and (26b) some doubt exists as to whether or not reaction (26c) is applicable. PY is a much weaker proton donor than either AN or DMF and it seems unlikely that the protonation reaction will take place in this solvent, although Elving[77] postulates the possibility of this type of reaction occurring in the reduction of benzophenone in PY. Until a product analysis of this system in PY is completed no definite conclusion about the final product of electro-reduction of azobenzene in PY can be made.

The monoanion produced in reaction (26a) undergoes an auto-oxidation–reduction reaction in both AN and DMF[14, 61, 62]. The rate of this reaction depends on the size of the cation of the supporting electrolyte, being fastest with the smallest ions (Li^+, Be^{2+}, etc.). In the presence of tetraethylammonium ions however it is so slow that it does not affect the polarographic or cyclic voltammetric measurements in either AN or DMF[14, 62]. It does, however, affect the coulometric results, giving values of $1{\cdot}06 \pm 0{\cdot}02$ and

0·96 ± 0·02 for the number of electrons required for the coulometric reduction of azobenzene to the monoanion and for the reduction of the monoanion to the dianion respectively. The mechanism of this reaction is considered to involve ion pairs[61]:

$$\{C_6H_5NNC_6H_5\}^{\bar{\cdot}} + (C_2H_5)_4N^+ \underset{}{\overset{K}{\rightleftharpoons}} \{C_6H_5NNC_6H_5\}^{\bar{\cdot}}(C_2H_5)_4N^+ \quad (27a)$$

$$\{C_6H_5NNC_6H_5\}^{\bar{\cdot}} + \{C_6H_5NNC_6H_5\}^{\bar{\cdot}}(C_2H_5)_4N^+ \xrightarrow{k} C_6H_5N{=}NC_6H_5 + \{C_6H_5NNC_6H_5\}^{2-}(C_2H_5)_4N^+ \quad (27b)$$

$$\{C_6H_5NNC_6H_5\}^{2-}(C_2H_5)_4N^+ + HS \longrightarrow \{C_6H_5NHNC_6H_5\}^- + (C_2H_5)_4N^+ + S^- \quad (27c)$$

This reaction was studied spectrophotometrically in 0·1M-tetraethylammonium perchlorate solutions in AN and found to be second order with respect to the monoanion[14]. The apparent second order rate constant, $k_2 = Kk[(C_2H_5)_4N^+]$, was found to be 0·14 dm³ mol⁻¹ sec⁻¹. When small cations are present, e.g. Li^+, reactions (27a), (27b) and (27c) are much faster and compete with reactions (26b) and (26c) in the overall electrode process leading to the formation of $\{C_6H_5NHNC_6H_5\}^-$ from $\{C_6H_5NNC_6H_5\}^{\bar{\cdot}}$.

The reduction of 4-nitroazobenzene in AN[14] proceeds as in equations (26a) and (26b). However, the dianion of 4-nitroazobenzene so formed does not appear to be a strong enough base to abstract a proton from the solvent to form the protonated dianion (26c). Polarograms of the reduction product show two reversible one-electron oxidation waves at the same potentials as the reversible one-electron reduction waves of the initial 4-nitroazobenzene. This, together with the similarity between the spectrum of the reduction product and the spectra of chemically produced azobenzene dianions supports the view that the dianion of 4-nitroazobenzene is the product of electrolysis in air-free AN solutions.

The addition of proton donors to DMF, AN or PY solutions of azo compounds alters the mechanism of the electrode reaction considerably[61, 62, 72–75]. Excess proton donor results in a single two-electron reduction to the hydrazo compound which usually occurs at more positive potentials than the first one-electron reduction wave observed in the absence of proton donors. With very weak proton donors, however, the two-electron reduction occurs at essentially the same potential as the first one-electron step in the absence of proton donors[62, 75].

A comprehensive study of the effects of proton donors on the electrode reactions of azobenzene and several of its 4-substituted derivatives in AN has recently been carried out by the authors[72–75]. The effects of proton donor strength (pK_a) and concentration were investigated and it was found that the reduction potentials and reaction mechanisms depended mainly on the pK_a of the proton donor and to a lesser extent on the Hammett σ function of the substituent. The presence of basic substituents such as —OCH_3 and —$N(CH_3)_2$ also modified the reaction mechanism.

The d.c. polarograms of azobenzene, 4-nitroazobenzene and 4-trimethylammonioazobenzene perchlorate in the presence of strong proton donors such as sulphuric acid (first dissociation only), hydrogen chloride and hydrogen bromide, all showed the development of a new irreversible reduction wave at ca. 1·5 V more positive than the proton reduction wave of the added donor and up to 2 V more positive than the first one-electron wave of the azo compound. This wave was diffusion controlled and its height was proportional to the proton donor concentration but independent of the azo compound concentration when the azo compound was in excess (i.e. when [azo]:[proton donor] > 1:2). When the acid is in excess, the height is proportional to the azo compound concentration and corresponds to a two-electron reduction.

The following mechanism accounts for the observed polarographic behaviour:

$$RC_6H_4N{=}NC_6H_5 \xrightarrow{\text{diffusion}} \text{electrode surface} \quad (28a)$$

$$HA \xrightarrow{\text{diffusion}} \text{electrode surface} \quad (28b)$$

$$HA \rightleftharpoons H^+{}_{(solv.)} + A^- \quad (28c)$$

$$RC_6H_4N{=}NC_6H_5 + H^+{}_{(solv.)} \rightleftharpoons RC_6H_4N{=}\overset{+}{N}HC_6H_5 \quad (28d)$$

$$RC_6H_4N{=}\overset{+}{N}HC_6H_5 + e^- \longrightarrow RC_6H_4\dot{N}NHC_6H_5 \text{ (slow)} \quad (28e)$$

$$RC_6H_4\dot{N}NHC_6H_5 + e^- \longrightarrow \{RC_6H_4NNHC_6H_5\}^- \text{(fast)} \quad (28f)$$

$$\{RC_6H_4NNHC_6H_5\}^- + HA \longrightarrow RC_6H_4NHNHC_6H_5 + A^- \text{ (fast)} \quad (28g)$$

overall:

$$RC_6H_4N{=}NC_6H_5 + 2\,HA + 2\,e^- \longrightarrow RC_6H_4NHNHC_6H_5 + 2\,A^- \quad (28h)$$

The azonium species formed in reaction (28d) is much more easily reduced than either the azo compound or HA ($H^+_{(solv)}$) and accounts for the azonium wave observed at very positive potentials. The azonium species is reduced in a slow, one-electron step to form the protonated radical species $RC_6H_4\dot{N}NHC_6H_5$ (28e), which is rapidly reduced at the same potential to eventually produce the hydrazo compound [reactions (28f) and (28g)]. That sulphuric acid behaves as a monobasic acid in this sequence is in agreement with its observed behaviour in AN[78]. The HSO_4^- ion is a very weak acid in AN and its effect on the reduction of azo compounds is similar to that of other very weak acids discussed below.

The polarography of 4-*N*,*N*-dimethylamino- and 4-methoxy-azobenzene in the presence of sulphuric acid is complicated by the basic nature of the substituent. In these cases, the azonium wave is split, the first portion being partly kinetically controlled. The total current of the two parts of the split azonium wave is diffusion controlled and corresponds to the two-electron reduction to the hydrazo compound. The split wave arises because of the presence of several protonated species in these solutions. For example, in the case of the 4-*N*,*N*-dimethylamino- compound the following species exist in equilibrium in solution:

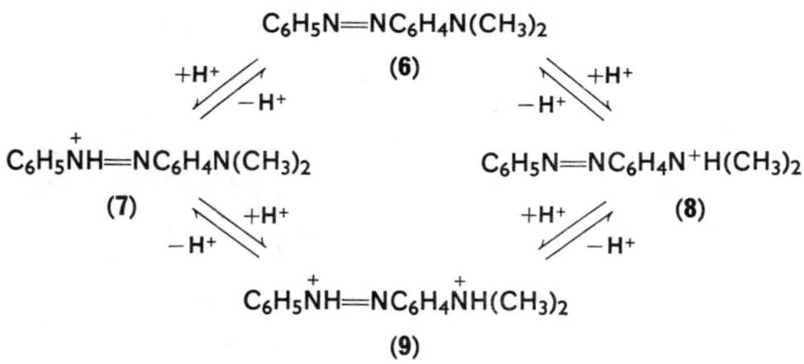

(29)

The dimethylammonium species (**8**) is expected to be reduced at ca. −1·0 V[78, 79] and therefore cannot be the species being reduced. The two parts of the split azonium wave (half-wave potentials of +0·1 V and −0·45 V) thus arise from the reduction of species (**7**) and (**9**). Because of the strong electron attracting properties of the $\overset{+}{\text{—NH}}(CH_3)_2$ group, species (**9**) is expected to be more easily reduced than species (**7**) and so the first wave at +0·1 V is attributed to the reduction of (**9**) and the second part of the split wave is attributed to the reduction of (**7**). Although the concentration of species (**9**) is expected to be very small, the equilibrium between (**9**) and (**7**) is considered to be fast so that as (**9**) is consumed at the electrode surface, more is formed from both (**7**) and (**8**) during the life of a drop resulting in an appreciable current. The formation of (**9**) from the other species is however, not fast enough for all of the azo compound to be converted to (**9**) during the life of a drop so that a second wave, due to the reduction of (**7**), is observed. A similar explanation accounts for the split azonium waves observed in the polarograms of 4-methoxyazobenzene–sulphuric acid system.

The addition of weak proton donors, such as acetic acid and 3,3,3-triphenylpropanoic acid, phenol and methoxyphenol to azobenzene in

AN solutions results in the appearance of two azonium waves in the polarogram[74]. In the case of acetic acid, the two waves appear at –0·9 V and –1·2 V vs. s.c.e. The total current of the two waves corresponds to a two-electron reduction to the hydrazo compound and it is proportional to the acetic acid concentration while the azo compound is in excess; the excess azobenzene being reduced in two one-electron steps at the normal potentials (Table 5). Since the first azonium wave is depressed by the addition of acetate ions to the system and is no longer evident when the mole ratio of acetate to acetic acid exceeds 15:1, it is considered that the two azonium waves are due to the reduction of the same species formed by two different routes—a specific protonation of azobenzene by acetic acid and a general protonation by the solvated proton (28d). The overall reaction mechanism is the same as that for the strong acids with an additional step

$$RC_6H_4N{=}NC_6H_5 + HA \rightleftharpoons RC_6H_4N{=}\overset{+}{N}HC_6H_5 + A^- \qquad (30)$$

for the formation of the azobenzene azonium ion (30). This specific protonation step (30) is an alternate route to the general protonation reactions (28c) and (28d) and in acetic acid solutions both routes are of similar importance. Since excess acetate depresses the first wave, this wave is accounted for by the general protonation sequence (28). The second wave incorporates the specific protonation step (30) in reaction sequence (28) instead of steps (28c) and (28d).

The splitting of the azonium waves occurs with the weak acids because the $H^+_{(solv)}$ concentration is many orders of magnitudes lower for these acids than it is for a similar concentration of sulphuric acid. The rate of reaction (28d) is correspondingly slower and so the specific protonation step (30) now becomes important, particularly at more negative potentials.

Apart from the complications caused by the mixed general and specific protonation steps, it is noted that the half-wave potentials of the azonium waves with both classes of acids are related to the pK_a of the acid. With sulphuric acid, the half-wave potential for the azobenzene azonium ion is +0·28 V and for acetic acid it is –0·9 V. The relationship[80]

$$E_{1/2} = (RT/2F)\ln K_a + \text{const.}$$

only holds approximately in this case because of the irreversibility of the waves.

The effect of weak proton donors on the polarography of 4-nitroazobenzene is similar to that noted for the effects of very weak proton donors on azobenzene, i.e. no azonium wave is formed. This is because the 4-nitro derivative is more easily reduced and is a much weaker base than azobenzene.

When very weak proton donors such as water and diethyl malonate are added to solutions of azobenzenes in AN, no azonium waves are observed in the polarograms[75]. However, on addition of water, for example, the

TABLE 7. Mole ratios of water:azo compound required to produce a single two-electron wave

Compound	Mole ratio
Azobenzene	368:1
4-Nitroazobenzene	900:1
4-*N*,*N*-Dimethylaminoazobenzene	119:1
4-Methoxyazobenzene	310:1

height of the first one-electron reduction wave increases with water concentration while the sum of the currents of both waves remains constant. Eventually, only a single two-electron reduction wave results when a large excess of water is present (Table 7). When the slightly stronger proton donor, diethyl malonate, is added to azobenzene only a 4:1 mole ratio of diethyl malonate to azobenzene is required to give the two electron wave. These results are similar to those reported for the 4-azopyridine–hydroquinone system in DMF[62]. The most probable mechanism for the electrode reactions in this case is:

$$C_6H_5N{=}NC_6H_5 + e^- \rightleftharpoons \{C_6H_5N{-}NC_6H_5\}^{\bar{\cdot}} \qquad (31a)$$

$$\{C_6H_5N{-}NC_6H_5\}^{\bar{\cdot}} + RH \longrightarrow C_6H_5NH\dot{N}C_6H_5 + R^- \text{ (slow)} \qquad (31b)$$

$$2\,C_6H_5NH\dot{N}C_6H_5 \longrightarrow C_6H_5N{=}NC_6H_5 + C_6H_5NHNHC_6H_5 \qquad (31c)$$

An alternate mechanism[81] which could well describe the results involves step (31a) and (31b) followed by:

$$C_6H_5NH\dot{N}C_6H_5 + e^- \longrightarrow \{C_6H_5NHNC_6H_5\}^- \text{ (fast)} \qquad (31d)$$

$$\{C_6H_5NHNC_6H_5\}^- + RH \longrightarrow C_6H_5NHNHC_6H_5 + R^- \qquad (31e)$$

where $RH = H_2O$, etc. and the reduction potential of step (31d) is more positive than that of step (31a).

The essential difference between reaction sequence (31) and reaction sequence (28) is that the first electron transfer step *precedes* protonation in the former because of the very weak proton donating properties of the donor. Evidence for reactions (31b) and (31c) comes from studies on bulk solutions of azobenzene radical ions containing known amounts of water. Azobenzene and hydrazobenzene are the products of the reaction with water and the rate of the reaction is first order with respect to both water and azobenzene radical ion concentrations. The rate of this reaction is given by

$$\text{Rate} = k[H_2O][\{C_6H_5NNC_6H_5\}^{\cdot -}] \qquad (32)$$

with $k = 8{\cdot}5 \times 10^{-3}$ mol^{-1} dm^3 sec^{-1} at 298 K. Although the radical ion slowly disproportionates to azobenzene and hydrazobenzene in the absence of water [reaction sequence (27)], addition of water speeds up the formation of products and when $[H_2O] > 25$ mM, reactions (31b) and (31c) are the only significant route to the production of azobenzene and hydrazobenzene from the radical anion.

Thus the addition of proton donors markedly alters the course of the electrode reactions of azo compounds in aprotic solvents. As the proton donating strength of the added proton donor decreases, the reduction potential of the azonium species produced becomes more negative until a stage is reached where the azo compound is not protonated to a sufficient degree for the azonium wave to appear before the onset of the one-electron reduction of the azo compound itself. When this happens it is the radical anion so produced that is protonated.

By studying the electrochemistry of azo compounds in aprotic solvents, the role of the proton can be better understood because of the great degree of control that can be achieved over the amount and availability of protons in the solution. Furthermore, complications due to adsorption phenomena are not apparent in most aprotic systems. The ability to prepare reasonably stable solutions of radical anions of azo compounds means that the chemistry of such systems can be studied in detail. One interesting application of the use of these radical anion solutions is in polymer chemistry[82]. Acrylonitrile was distilled into a solution of azobenzene radical anions in pyridine and found to polymerize, the radical anion acting as an initiator. This immediately raises the possibility of electro-initiation of polymerization by the *in situ* production of azo radical anions by direct electrolysis of a monomer containing small amounts of a suitable azo compound.

IV. REFERENCES

1. C. Kerns, *Trans. Electrochem. Soc.*, **62,** preprint, 10 pp. (1932); *Chem. Abstr.*, **26,** 4759² (1932).
2. E. Bamberger, *Chem. Ber.*, **33,** 274 (1900).
3. F. Haber and H. Schmidt, *Z. phys. Chem.*, **32,** 271 (1900).

4. T. Sekine, *J. Chem. Soc. Japan*, **76,** 694 (1955).
5. A. Piguet, *U.S. Patent* 1,225,052, May 8 (1916); *Chem. Abstr.*, **11,** 1932[1] (1917).
6. *Ges. für Chem. Ind. Swed.*, 41675, Dec. 13 (1916); *Chem. Abstr.*, **11,** 1932[9] (1917).
7. K. Sugino, K. Odo, T. Sekine, K. Shirai, and E. Ichikawa, *J. Electrochem. Soc., Japan*, **30,** E167 (1962).
8. T. Sekine, *Kogyo Kagaku Zasshi*, **60,** 918 (1957); *Chem. Abstr.*, **53,** 8883f (1959).
9. K. Sugino and T. Sekine, *J. Electrochem. Soc.*, **104,** 497 (1957).
10. C. E. Van Winkel, R. V. Smith and E. Hurtz, *U.S. Patent* 2925371, Feb. 16 (1960).
11. K. S. Udepa, G. S. Subramanian and H. V. K. Udupa, *J. Electrochem. Soc.*, **108,** 373 (1961).
12. E. Brunner, *Brit. Patent* 147541, May 17 (1920).
13. K. Sugino and T. Sekine, *Jap. Patent* 4962, June 13 (1959).
14. K. G. Boto and F. G. Thomas, *Aust. J. Chem.*, **26,** 1251 (1973).
15. W. S. Harris and G. B. Matson, *U.S. Department of Commerce, Office of Technical Services*, AD293446 (1962).
16. H. Siba, T. Inouc and R. Mujasaka, *Sci. Papers Inst. Phys. Chem. Research (Tokyo)*, **35,** 455 (1939); *Chem. Abstr.*, **33,** 6170 (1939).
17. L. Holleck and G. Holleck, *Z. Naturforsch.*, **19B,** 162 (1964).
18. L. Holleck and A. M. Shams-el-Din, *Electrochim. Acta*, **13,** 199 (1968).
19. R. Hazard and A. Tallec, *Bull. Soc. Chim. France*, 2917 (1971).
20. D. Lester and L. A. Greenberg, *J. Am. Chem. Soc.*, **66,** 496 (1944).
21. J. Che-Man Chang, *Ph.D. Thesis*, University of Illinois (1966); *Univ. Microfilms*, Ann Arbor, 66–12,302.
22. M. Shikata and I. Tachi, *Coll. Czech. Chem. Comm.*, **10,** 368 (1938).
23. T. N. Hoang, *J. Chim. Phys.*, **35,** 345 (1938); **36,** 164 (1939).
24. T. M. Florence, *Aust. J. Chem.*, **18,** 609, 619 (1965).
25. T. M. Florence and G. H. Aylward, *Aust. J. Chem.*, **15,** 65, 416 (1962).
26. B. Nygard, *Arkiv. Kemi*, **20,** 163 (1962).
27. L. Holleck, A. M. Shams-el-Din, R. M. Saleh and G. Holleck, *Z. Naturforsch.*, **19B,** 161 (1964).
28. T. M. Florence and Y. J. Farrer, *Aust. J. Chem.*, **17,** 1085 (1964).
29. S. Wawzonek and J. D. Fredrickson, *J. Am. Chem. Soc.*, **77,** 3985, 3988 (1955).
30. T. N. Hoang, *Compt. Rend.*, **208,** 1979 (1939).
31. A. L. Markman and E. V. Zinkova, *J. Gen. Chem. U.S.S.R.*, **29,** 3058 (1959).
32. C. R. Castor and J. H. Saylor, *J. Am. Chem. Soc.*, **75,** 1427 (1953).
33. H. A. Laitinen and T. J. Kneip, *J. Am. Chem. Soc.*, **78,** 736 (1956).
34. G. Costa and A. Puxeddu, *Ric. Sci.*, 27*A*, *Polarografia*, **3,** 94 (1957).
35. I. Rusznak, F. Peter and G. Palyi, *Acta Chim. Acad. Sci. Hung.*, **35,** 199 (1963).
36. L. Holleck and G. Holleck, *Monatsh.*, **95,** 990 (1964).
37. T. M. Florence, D. A. Johnson and G. E. Batley, *J. Electroanal. Chem. Interfacial Phenomena* (1973). In press.
37a. L. Meites, *Polarographic Techniques*, 2nd Ed., Interscience, New York, 1965, p. 217.

38. A. Pattoni, *Atti. Inst. Veneto Sci. Venezia*, **106,** 29 (1948); *Atti. Soc. Med.-Chir. Padova*, **25,** 125 (1947).
39. P. J. Elving, L. Chuang and I. Fried, *Anal. Chem.*, **37,** 1528 (1965).
40. G. Holleck and L. Holleck, *Naturwissenschaften*, **51,** 433 (1964).
41. L. Holleck and G. Holleck, *Polarography*, 1964, Vol. II, Macmillan, London, 1966, pp. 825–834.
42. A. Winkel and H. Siebert, *Chem. Ber.*, **74B,** 670 (1941).
43. F. Wessely and J. Wratil, *Mikrochemie ver. Mikrochim. Acta*, **33,** 248 (1947).
44. P. J. Hillson and P. P. Birnbaum, *Trans. Farad. Soc.*, **48,** 478 (1952).
45. G. Holleck and L. Holleck, *Naturwissenschaften*, **51,** 212 (1964).
46. B. Kastening, *Ber. Bunsen Ges. Phys. Chem.*, **68,** 979 (1964).
47. K. G. Boto and F. G. Thomas, *Aust. J. Chem.*, **23,** 43 (1970).
48. R. B. Carlin, R. G. Nelb and R. C. Odioso, *J. Am. Chem. Soc.*, **73,** 1002 (1951).
49. P. Delahay, S. Oka and H. Matsuda, *J. Am. Chem. Soc.*, **82,** 329 (1960).
50. P. Rüetschi and G. Trümpler, *Helv. Chim. Acta*, **35,** 1021 (1952).
51. L. Faith, *Farmacia*, **29,** 286 (1960); *Chem. Abstr.*, **55,** 9791h (1961).
52. A. Volpi, *Gazz. Chim. Ital.*, **77,** 473 (1947).
53. M. Kalousek, *Coll. Czech. Chem. Comm.*, **13,** 105 (1948); **19,** 1099 (1954).
54. G. Costa, *Gazz. Chim. Ital.*, **83,** 875 (1953).
55. S. Millefiori and G. Campo, *Ann. Chim. (Rome)*, **59,** 128 (1969).
56. S. Millefiori, *Ann. Chim. (Rome)*, **59,** 138 (1969).
57. S. Shibata, *Chelates in Analytical Chemistry*, Vol. IV, Marcel Dekker, New York, 1972.
58. L. Holleck and S. Vavřička, *Z. Naturforschung*, **B23,** 748 (1968).
59. G. H. Aylward, J. L. Garnett and J. H. Sharp, *Chem. Commun.*, 137 (1966).
60. G. H. Aylward, J. L. Garnett and J. H. Sharp, *Anal. Chem.*, **39,** 457 (1967).
61. G. H. Aylward, J. L. Garnett and J. H. Sharp, *Rev. Polarog. (Japan)*, **14,** 322 (1967).
62. J. L. Sadler and A. J. Bard, *J. Am. Chem. Soc.*, **90,** 1979 (1968).
63. T. M. Krygowski and P. Tomasik, *Bull. Acad. Polon. Sci. Ser. Sci. Chim.*, **18,** 303 (1970).
64. P. Tomasik, *Roczniki Chem.*, **4,** 1211 (1970).
65. G. Cauquis and G. Fauvelot, *Polarography* 1964, Vol. II, Macmillan, London, 1966, pp. 847–861.
66. G. Cauquis and G. Fauvelot, *Bull. Soc. Chim. France*, 2014 (1964).
67. T. A. Gough and M. E. Peover, *Polarography* 1964, Vol. II, Macmillan, London, 1966, pp. 1017–1033.
68. S. Millefiori, *Ricerca Sci.*, **39,** 616 (1969).
69. S. Millefiori and G. Favini, *Z. physik. Chem. N.F.*, **75,** S23 (1971).
70. K. G. Boto and F. G. Thomas, *Aust. J. Chem.*, **24,** 975 (1971).
71. J. E. McClure and D. L. Maricle, *Anal. Chem.*, **39,** 236 (1967).
72. K. G. Boto, and F. G. Thomas, *Aust. J. Chem.*, in press.
73. K. G. Boto and F. G. Thomas, *Aust. J. Chem.*, **26,** 1669 (1973).
74. K. G. Boto and F. G. Thomas, *Aust. J. Chem.*, **26,** 2409 (1973).
75. K. G. Boto and F. G. Thomas, *Aust. J. Chem.*, **26,** 2417 (1973).
76. A. G. Evans, J. C. Evans, P. J. Emes, C. L. James and P. J. Pomeroy, *J. Chem. Soc. (B)*, 1484 (1971).
77. R. F. Mitchielli and P. J. Elving, *J. Am. Chem. Soc.*, **90,** 1989 (1968).

78. J. F. Coetzee and I. M. Kolthoff, *J. Am. Chem. Soc.*, **79,** 6110 (1957).
79. M. Kilpatrick, Jr. and M. L. Kilpatrick, *Chem. Rev.*, **13,** 131 (1933).
80. K. Takamura and Y. Hayakawa, *J. Electroanal. Chem. Interfacial Electrochem.*, **31,** 225 (1971).
81. G. J. Hoitjink, J. van Schooten, E. de Boer and W. Y. Aaldersberg, *Rec. Trav. Chim.*, **73,** 355 (1954).
82. K. G. Boto and E. Senogles, Private communication.

CHAPTER 13

Biological formation and reactions of hydrazo, azo and azoxy groups

TETSUO MIYADERA

Central Research Laboratories, Sankyo Co., Ltd., Tokyo, Japan

ABBREVIATIONS

AB	4-Aminoazobenzene
ATP	Adenosine triphosphate
AY	Acid yellow
CoA	Coenzyme A
DAB	4-Dimethylaminoazobenzene
DCA	3,4-Dichloroaniline
DNA	Deoxyribonucleic acid
$FADH_2$	Reduced form of flavine-adenine dinucleotide
FMN	Flavine mononucleotide
GSH	Glutathione(γ-L-glutamyl-L-cysteinylglycine)
GSSG	Oxidized glutathione
HPUra	6-(*p*-Hydroxyphenylazo)uracil
INH	Isoniazid, isonicotinic acid hydrazide
MAB	4-Methylaminoazobenzene
MAM	Methylazoxymethanol
MAMA	Methylazoxymethyl acetate
MAO	Monoamine oxidase
NADH	Reduced form of NAD(nicotinamide-adenine dinucleotide)
NADP	Nicotinamide-adenine dinucleotide phosphate
NADPH	Reduced form of NADP
OAT	Orthoaminoazotoluene, 2′,3-dimethyl-4-aminoazobenzene
RNA	Ribonucleic acid
TCAB	3,3′,4,4′-Tetrachloroazobenzene

I. INTRODUCTION

The understanding of the biological action of a molecule requires a chemical and physicochemical knowledge of the molecule. Even if a functional group plays the major role in the biological action, the remaining part of the molecule cannot be ignored, since it may be essential for transportation to the site of action and there may be a steric requirement for the biological reaction or interaction of the functional group with biological constituents. The majority of biologically active compounds reveal their activities by weak and reversible bonding such as hydrogen bonding, hydrophobic bonding, electrostatic and Van der Waals interactions with biological constituents. There are, however, many cases where a functional group is involved through covalent bond formation. It is, therefore, of considerable interest to make a survey of correlation between biological reactions of a certain functional group and their biological responses. Covalent bond formation has been frequently observed with hydrazo, azo and azoxy compounds, although in some cases the functional groups themselves are not directly linked in their biological reactions. These compounds are often pharmacologically weak or inactive in the intact forms and are activated only by metabolic reactions. This chapter is intended to give chemical explanations for the biological action of hydrazo, azo and azoxy compounds. The three functional groups to be discussed together are closely related in a biological sense.

Hydrazo, azo and azoxy groups are concerned with a number of important biological reactions such as protein synthesis inhibition, carcinogenesis, thiol-disulphide interchange, azo reduction, monoamine oxidase inhibition, immunochemical affinity labelling and nitrogen fixation.

Biological studies on azo compounds came much earlier than on the other two classes of compounds because of their use as colouring materials. Azo dyes containing the N=N linkage are one of the most common chemical types in the modern dye industry. Early in the 1900s Sisley and Porcher found that some azo dyes, when orally administered, gave rise to the aromatic amines through reductive fission of the azo linkage. With respect to azo reductions, in the 1930s an important discovery led to extensive development of sulphanilamides. The dye prontosil, also known as an antibacterial agent, was found to be active following the metabolic azo reduction to *p*-aminosulphanilamide. Azo dyes available today are characterized by the water-solubilizing substituent —SO_3Na on the azobenzene ring and are susceptible to azo reduction as the principal metabolic step. In general, the introduction of a water-soluble substituent makes it easier to detoxify azo compounds. For the purpose of safe usage, it is essential to study the metabolic reactions as well as the biological effects of the dyes and their metabolites.

The widespread use of synthetic dyes resulted in a crucial problem, when some dyes were found to be carcinogenic. For instance, Kinoshita found that the dye once used for colouring margarine and butter, 4-dimethylaminoazobenzene (DAB, butter yellow) induces tumours in rat livers. Miller and co-workers have elaborated the metabolic reactions of DAB and found an important metabolic pathway to a potent carcinogen capable of binding to proteins and nucleic acids.

Unfortunately the biogenesis of naturally occurring hydrazo and azoxy compounds is unknown. The occurrence of these functional groups is rather rare and it was only in the 1950s that macrozamin and cycasin were found as the first natural azoxy compounds. The finding of cycasin, the toxic and carcinogenic constituent of cycad plants, motivated mechanistic investigations of the biological alkylation of hydrazo, azo and azoxyalkanes including methylazoxymethanol, the aglycone of cycasin. It has now been recognized that the carcinogenesis of cycasin is possibly associated with the methylation of guanine residues of nucleic acids by an ultimate carcinogen generated from the aglycone. The mode of action of the carcinogenic aliphatic compounds is completely different from that of azobenzenes. Although no azo compound occurs in nature to the best of our knowledge, recently Bartha and co-workers have found that azobenzenes are enzymically formed from anilines in soils. This enzymic transformation is the reverse of the reduction of azobenzenes, which is considered as a detoxification reaction. Biological studies on azobenzenes are of considerable importance, since chloroaniline derivatives have been widely used for agricultural use as anilides, carbamates and ureas.

An application of moderately reactive azo compounds to biological studies has recently been successfully attempted by Kosower and Kosower and co-workers. The azo reagents designated as azoester and diamide have been employed for the selective oxidation of intracellular glutathione (GSH), a cofactor of several enzymes, to the disulphide(GSSG) in order to probe the biological roles of the cofactor. The two reagents are capable of oxidizing GSH without affecting cellular function, so that the normal function of the cells is resumed after regeneration or addition of GSH. The use of the thiol-oxidizing agents can be of great benefit in elucidating biological changes arising from the deficiency of GSH or from the perturbation of GSH–GSSG ratio. A variety of reversible biological responses have been observed, including protein synthesis inhibition, growth inhibition of bacteria and fungi and many other biological effects.

Biological formation and reactions of hydrazo compounds became of great concern to chemists and pharmacologists after the beginning of clinical use of hydrazo monoamine oxidase (MAO) inhibitors as antidepressants and of isoniazid as an anti-tubercular agent. During treatment of tubercular patients with the hydrazo compound iproniazid, the drug was observed to stimulate the central nervous system and inhibit the activity of MAO both *in vivo* and *in vitro*. These findings stimulated energetic searches for a great variety of amines and hydrazine derivatives in the search for drugs for pharmacotherapy of depression. Isoniazid and iproniazid are closely related compounds from the viewpoint of pharmacology as well as of chemistry, and they release hydrazine and an alkylhydrazine *in vivo* and *in vitro*. Generally, hydrazine derivatives resulting from enzymic or non-enzymic hydrolysis are chemically reactive, displaying a variety of biological effects and frequently presenting hazards. Accordingly, it is essential to study metabolic reactions and the resulting metabolites from such hydrazine derivatives.

The medical, agricultural and industrial uses of hydrazo, azo and azoxy compounds will be further developed in an effort to provide useful substances with no undesirable hazards. In this light it becomes imperative to understand various biological reactions and the resulting biological effects.

II. NATURAL OCCURRENCE OF HYDRAZO, AZO AND AZOXY COMPOUNDS

A. Azoxy Compounds

1. Macrozamin and cycasin

Up to 1950 no hydrazo, azo or azoxy compounds had been found in nature, but in 1951 the toxic macrozamin, first isolated by Cooper[1] from the Australian cycad (*Macrozamia spiralis*), was characterized as primeverosyloxyazoxymethane[2]. Several years later Nishida's group (1955)[3] and Riggs (1956)[4] had isolated the aliphatic azoxy glycoside cycasin from seeds, roots and leaves of cycad plants which grew in Guam and Japan, respectively. Cycad seeds were utilized as flour in tropical and subtropical areas after being repeatedly washed and dried in order to remove the toxic principle and the high incidence of diseases in these areas is due to the eating of cycad seeds[5]. The aliphatic azoxy glycoside displays no toxicity to germfree[6] or intraperitoneally injected rats, while toxic signs are observed in rats which have been fed cycasin[7]. These interesting results can be understood when it is realized that cycasin is hydrolysed to the toxic aglycone by intestinal bacteria[8]. The resulting aglycone, methylazoxymethanol (MAM) is hepatotoxic, carcinogenic, teratogenic and neurotoxic[9] irrespective of the route of administration. Enzymic hydrolysis of cycasin by cycad glucosidase *in vitro*[10] also gives rise to MAM which can be synthesized as the acetate[11] methylazoxymethyl acetate (MAMA) with similar biological activities.

$$\underset{\substack{\downarrow\\ O}}{CH_3N}{=}NCH_2O\text{—primeverose}$$

Macrozamin

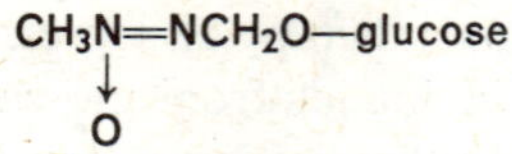

Cycasin

2. Elaiomycin and LL-BH 872a

Elaiomycin is an azoxy-containing antibiotic isolated as a metabolite of *Streptomyces hepaticus*[12,13]. The isolation and the structural determination[14] have been made by Haskell and co-workers. The structurally related LL-BH 872a is isolated from the fermentation beer of *Streptomyces hinnulinus*[15]. The two antibiotics share an α,β-unsaturated azoxy group, but show a marked difference in biological effects. Elaiomycin induces a variety of tumours[16] and is very effective *in vitro*, but not *in vivo* against *Mycobacterium tuberculosis*[17]. It would be of interest to investigate some correlation between the carcinogenesis by elaiomycin and its metabolic reactions. The other antibiotic LL-BH 872a is reported as a new antifungal agent.

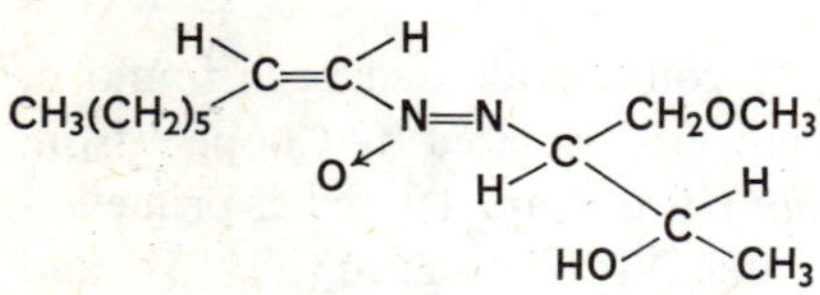

Elaiomycin

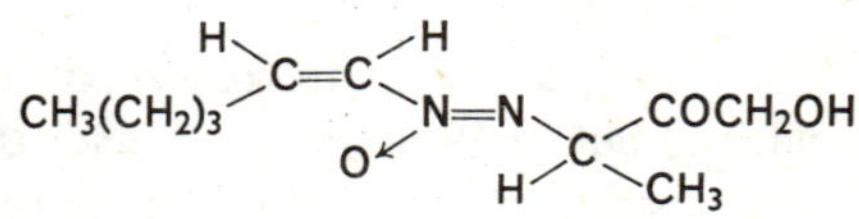

LL-BH 872a

B. *Hydrazo Compounds: Spinamycin*

Recently an antibiotic hydrazo compound has been found to occur in nature. Spinamycin has been isolated by Umezawa and co-workers from *Streptomyces albospinus* and characterized as a novel hydrazo compound[18]. The antibiotic inhibits the growth of some fungi and Yoshida rat sarcoma cells. The hydrazo group appears to be of essential importance in the antifungal action which is discussed in Section V. The occurrence of hydrazo compounds is very rare, but another natural product, negamycin with a hydrazine group, has been discovered in the same laboratory[19].

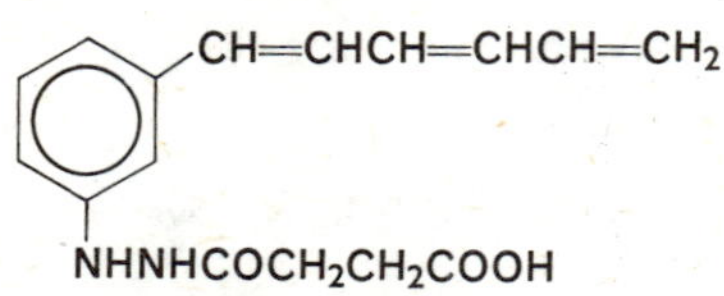

Spinamycin

Negamycin is not a hydrazo compound, but possesses a disubstituted hydrazide structure and has inhibitory activities against Gram-positive and especially Gram-negative bacteria. Its mode of action is the inhibition of protein synthesis with miscoding activity[20].

The biogenesis of the functional groups of these natural products is of considerable biogenetic interest and should be elucidated in the future.

III. METABOLIC REACTIONS OF HYDRAZO, AZO AND AZOXY COMPOUNDS

A. *Acetylation of Hydrazides and Hydrazines to Hydrazo Compounds*

Clinical use of a number of hydrazines as hydrazides and hydrazo derivatives has resulted in extensive research on the metabolism of these drugs. The toxicities frequently observed during treatment of patients with hydrazides and hydrazo compounds appear to be attributable to hydrazines arising from the metabolic reactions. The enzymic acetylation of hydrazines and hydrazides which can be designated as a biological hydrazo formation is an important metabolic pathway for detoxication in a literal sense. From the viewpoint of detoxication, the C—N bond formation involved in the acetylations is preferable to the C—N bond cleavage relating to the enhancement of the toxicity of original compounds.

Isoniazid (INH), a common drug used for the treatment of tuberculosis, provides a good example with which to describe the metabolic meanings in further detail. In dogs deficient in capacity for acetylation, INH is principally hydrolysed to isonicotinic acid and toxic hydrazine[21], so that the toxicity appears to a greater extent in the dog than it does in man and many other animals. In contrast, both in a man and in a monkey capable of acetylating hydrazines and hydrazides, INH is acetylated to yield acetyl-INH amounting to 91% and 50–100%, respectively, of the excreted metabolites[22]. The toxicity of hydrazine is diminished by acetylation to monoacetylhydrazine and intensively by further acetylation to diacetylhydrazine[23]. The acetyl groups are thought to be transferred from acetylCoA

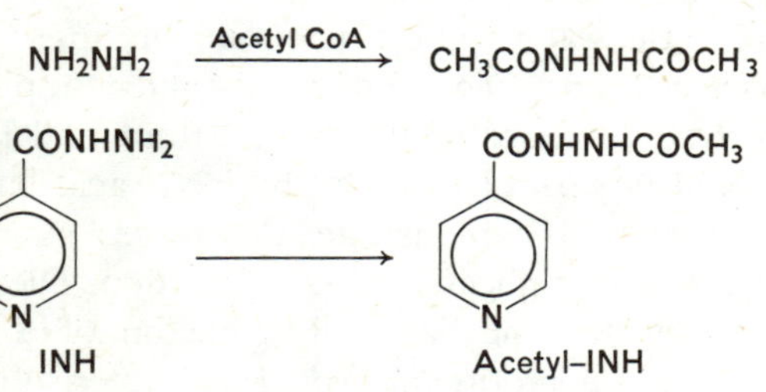

in the presence of acetyltransferases[24] and the ability to acetylate hydrazines and hydrazides depends upon individuals and species. Acetylation has also been observed with hydrazines such as monoamine oxidase inhibitors.

B. Azo Reduction

The *in vivo* azo reduction with the cleavage of the aromatic azo linkage is one of the earliest explored biological reactions of the functional groups. This metabolic reduction is considered as an important detoxication route together with that of aromatic hydroxylation. The two-stage azo reduction proceeding by way of hydrazo intermediates can be regarded as the reverse reaction of non-enzymic and enzymic azo formation from hydrazo compounds and aromatic amines.

Sisley and Porcher have obtained sulphanilic acid ($NH_2C_6H_4SO_3H$) as a metabolite of orange I (a phenylazonaphthalene dye) after oral administration to dogs[25]. Following Domagk's discovery in 1932 of the antibacterial activity of the dye prontosil, the biotransformation to the active sulphanilamide was hypothesized by Tréfouël and co-workers[26], and thereafter demonstrated by Fuller[27]. That is, prontosil undergoes azo reduction to form sulphanilamide (**1**), inhibitory against bacteria, and

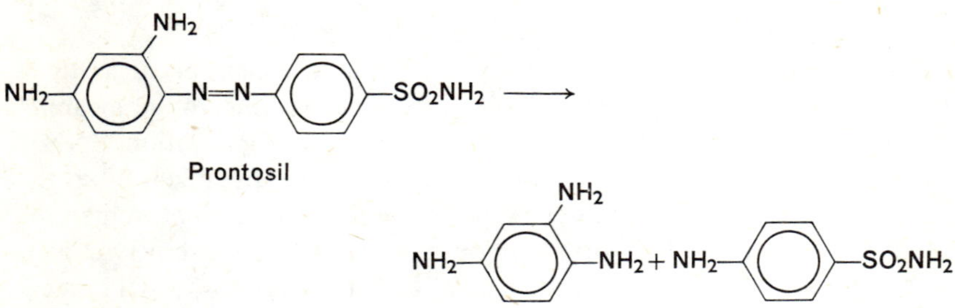

(**2**) (**1**)

1,2,4-triaminobenzene (**2**). Azo compounds are reduced mainly in liver and intestine[28]. Azo dyes commonly used as food colours ranging from red to yellow have the water-soluble functional group, sodium sulphonate SO_3Na. The bacterial azo reduction of water-soluble sulphonated azo dyes has been extensively investigated following Brohm and Frohwein's work on decolourization *in vitro* of azo dyes by lactic acid bacteria[29]. During the following decade it was demonstrated that the gut microflora, intestinal tissue and contents are implicated in azo reduction. Dieckhues has comprehensively investigated the bacterial reduction of a number of azo compounds by various microorganisms[30]. Most water-soluble sulphonated azo dyes can be reduced *in vitro* by intestinal contents and various bacteria, including the mammalian intestinal microorganisms, to the corresponding aromatic amines. Daniel[31] has studied the metabolism and excretion of a number of sulphonated azo dyes and found that the azo reductions occur mostly with the formation of amines and their derivatives when the dyes are orally administered to rabbits. Amaranth, ponceau SX and sunset yellow undergo similar azo reductions in rats[32]. Tartrazine is also found to be reduced by the gastrointestinal flora giving rise to sulphanilic acid in the urine of rat, rabbit and man[33]. Tartrazine and some other water-soluble azo dyes can be reduced by rat liver homogenate supernatant, cultures of *B. proteus* isolated from the rat intestinal flora and bacteria such as *P. vulgaris* and *E. coli*[34]. The requirement of bacteria for azo reduction can account for the failure of rats to reduce the azo linkage of intraperitoneally injected tartrazine. Acid yellow (AY) is readily reduced to sulphanilic acid and *p*-phenylenediamine sulphonic acid by the flora of the intestinal tract. In the *in vivo* reduction, *N*-acetylsulphanilic acid and 5-acetylamino-2-aminobenzene sulphonic acid are obtained as the conjugates of the AY metabolites[35]. With respect to the intestinal azo-reductase, Roxon, Ryan and Wright have concluded that an NADPH-dependent FMN-flavoprotein is involved in intestinal azo reduction[36a].

In contrast to water-soluble azo dyes, water-insoluble azo dyes are not susceptible to azo reduction by intestinal organisms. For instance, the water-insoluble azo dyes sudan III and IV are not reduced by intestinal bacteria in spite of high lipid solubility[36b]. Similarly, the lipid-soluble 4-dimethylaminoazobenzene (DAB) derivatives resist reduction by *P. vulgaris* or rat intestinal contents[34a].

The lipid-soluble azobenzene (**3a**) undergoes hepatic azo reduction leading to hydrazobenzene (**4a**) and aniline, together with hydroxylation products, when orally administered to rats and rabbits[37]. As stated earlier, the azo reduction of azobenzenes proceeds by way of hydrazobenzene (**4**) which can be further reduced to anilines and the corresponding aminophenols. The N—N single bond cleavage in the second step corresponds to the reductive fission of non-aromatic hydrazines such as hydrazine and methylhydrazine giving rise to ammonia and methylamine, respectively.

Stevenson and co-workers[38] have obtained *p*-aminophenol, *p*-phenylenediamine (**5b**, $R = NH_2$) and their *N*-acetyl derivatives as azo reduction products from the urine of rats fed DAB (**3b**). DAB is also reduced *in vitro* by rat liver slices[39] and more rapidly by rat liver homogenates[40] to *N*,*N*-dimethyl-*p*-phenylenediamine (**5b**, $R = (CH_3)_2N—$) and aniline. Mueller and Miller[40] have demonstrated that the reduction of DAB by liver homogenates is most favoured by an anaerobic system fortified with triphospho-

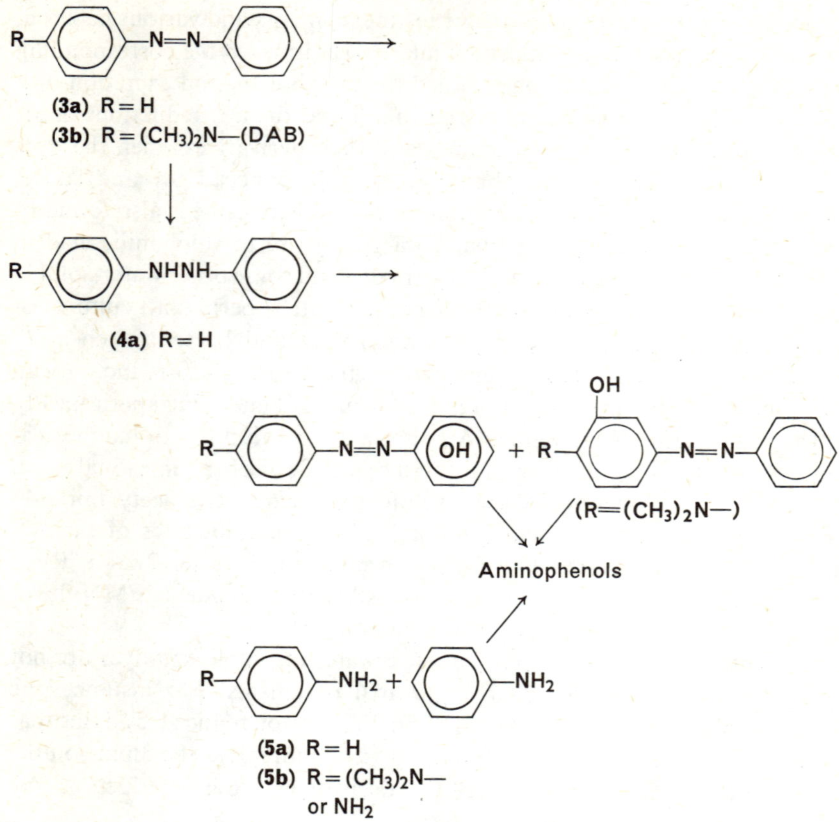

pyridine nucleotide, diphosphopyridine nucleotide, etc. The presence of atmospheric oxygen permits the reverse reaction, autoxidation of the hydrazo intermediate to the azo form. A riboflavin coenzyme is also implicated in the liver slice system. The major metabolites in the early stage of the DAB metabolism *in vivo* are shown to be the ethereal sulphates of 4′-OH—DAB and 4′-OH—MAB(4-methylaminoazobenzene)[41] which are also found in rat bile[42].

Hepatic azo reduction has been shown to be effected by microsomal azoreductase which appears identical with NADPH-cytochrome *c* reductase[43], but some other enzymic (i.e., cytochrome P_{450}) and non-enzymic (i.e., NADPH, NADH, $FADH_2$) processes are involved in reductive fission of azo groups[28].

The lipid-soluble prontosil can be reduced in the gut as well as in the liver following absorption, while the water-soluble neoprontosil (sulphonated azonaphthalene dye) is poorly absorbed due to its high polarity and reduced *in vivo* mainly by the gut flora[44]. When incubated with everted intestinal sacs from rats the lipid-soluble prontosil gives rise to the *N*-glucuronide (59%) and traces of sulphanilamide with the recovery of the unchanged dye (32%). The mucosal conjugation is considered to be necessary for the absorption of the dye. Most of the sulphanilamide in the urine resulting from the oral administration of prontosil is produced in the large intestine probably after enterohepatic circulation, accompanied by some azo reduction in the duodenum[44].

Azo reduction also occurs effectively in human placenta[45] in the presence of NADPH.

C. *Oxidative and Hydrolytic Cleavage of N—C Bonds*

Since the discovery of anti-depressive[46] and monoamine oxidase (MAO) inhibitory activity[47] of iproniazid, a number of hydrazo compounds have been prepared in an effort to provide more effective drugs for treatment of depression. Monoamine oxidases occur in various organs, but the enzymes from different origins are not identical in that they display differences in activity. The oxidases play a role in the oxidative deamination of monoamines such as serotonin, epinephrine (adrenaline), norepinephrine (noradrenaline), dopamine and tryptamine, etc. A MAO inhibitor inhibits the metabolism of the amines thus increasing the biogenic amine levels in brain and other tissues and thereby causing an elevation of mood. MAO inhibitors so far synthesized also non-selectively affect enzymes other than MAO, such that they have other pharmacological effects[48].

Isocarboxazid and nialamide are more effective inhibitors of the hydrazo type and have been used for the treatment of depressed patients. Davison[49] has suggested that the MAO inhibitory activity of iproniazid might be the

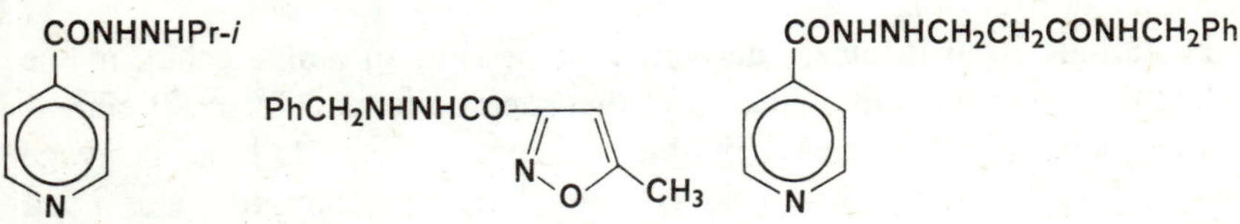

Iproniazid Isocarboxazid Nialamide

result of metabolic conversion to a potent MAO inhibitor, isopropylhydrazine. On the basis of the structure–activity relationship Barsky and co-workers[50] have concluded that alkylhydrazines are essential moieties for MAO inhibition of the hydrazo compounds. In reality, iproniazid undergoes hydrolytic cleavage in rats to isonicotinic acid and possibly isopropylhydrazine or oxidative cleavage to isoniazid (INH) and acetone[51].

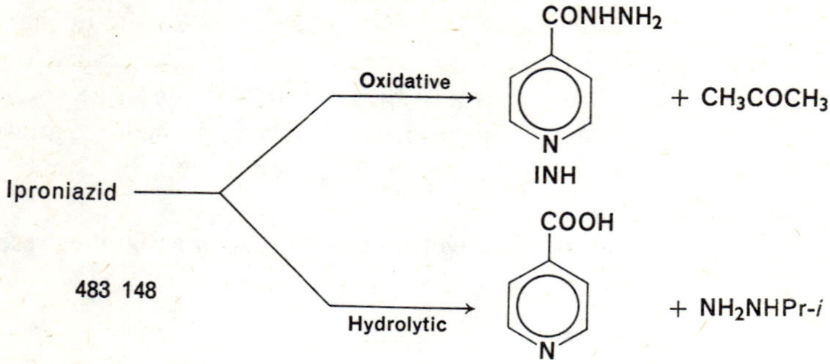

The resulting isopropylhydrazine is known as a long lasting inhibitor as the result of irreversible complexation with MAO as in the case of the majority of MAO inhibitors. Once the enzyme–isopropylhydrazine complex is formed, the enzyme cannot effect the oxidative deamination of the substrates[52]. The formation of acetone becomes possible following the α-C-hydroxylation. Hydroxylations at the carbon atom adjacent to nitrogen are very common metabolic reactions and a number of examples are given in this chapter. Further support for hydrolytic and oxidative cleavage has been obtained by the detection of benzylhydrazine when isocarboxazid is treated with rat liver homogenates[53]. Isocarboxazid undergoes hydrolytic and oxidative cleavage giving rise to benzylhydrazine and benzoic acid, respectively[54]. Recently it was reported that benzylhydrazine concentrations in tissues can be measured[55] and the benzyl hydrazine levels in rat brains reach a maximum shortly after administration of isocarboxazid[56]. The metabolic conversion proceeds by the catalytic action of a hydrolysing enzyme which is mainly present in microsomal fractions of liver.

Nialamide is an isoniazid derivative possessing an amide group in the substituent, so that it may undergo other metabolic reactions[57]. Preferential degradation of the MAO inhibitor occurs with amide bond cleavage and as a result isonicotinic acid is formed from nialamide to a much lesser extent than from iproniazid.

As is the case with the hydrazo MAO inhibitors, diacyl derivatives of hydrazine are readily hydrolysed enzymically or non-enzymically *in vivo* and *in vitro*. For instance, acetyl-INH gives rise to acetylhydrazine and isonicotinic acid on treatment with human serum[58] and affords 1,2-diacetylhydrazine when administered to humans and rabbits[59]. It is evident that the acetyl–nitrogen bond is more stable than the aroyl–nitrogen bond. Thus, the much less toxic diacetylhydrazine ($CH_3CONHNHCOCH_3$) is stable *in vivo* and is recovered unchanged. From the viewpoint of detoxication, it is significant that the acetyl–nitrogen bond is generally not susceptible to hydrolytic cleavage.

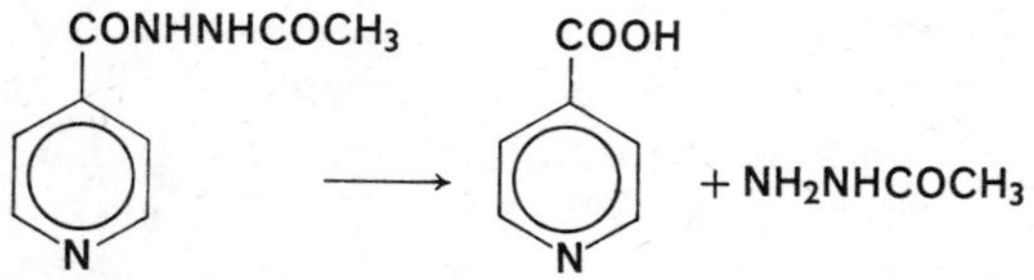

Acetyl-INH

D. Carcinogenic Alkylation of Aliphatic Compounds

1. Cycasin

Riggs[60] has found that cycasin may act as an alkylating agent capable of converting phenol to anisole in the presence of a catalytic amount of concentrated sulphuric acid. The chemical alkylation suggests the possibility of the *in vivo* methylation of biological constituents through similar processes. The occurrence of the *in vitro*[61] and *in vivo* methylation[62] has been demonstrated unambiguously by the isolation of methylated guanine base. Cycasin or the aglycone, methylazoxymethanol (MAM) undergoes preferential reaction with the guanine residues of DNA and RNA giving rise to 7-methylguanine (**6**) after hydrolysis. The carcinogenesis of MAM is thought to result from the biological decomposition to the methyl-containing molecular fragment responsible for methylation. The experiment using tritium-labelled MAM has also indicated the incorporation of the

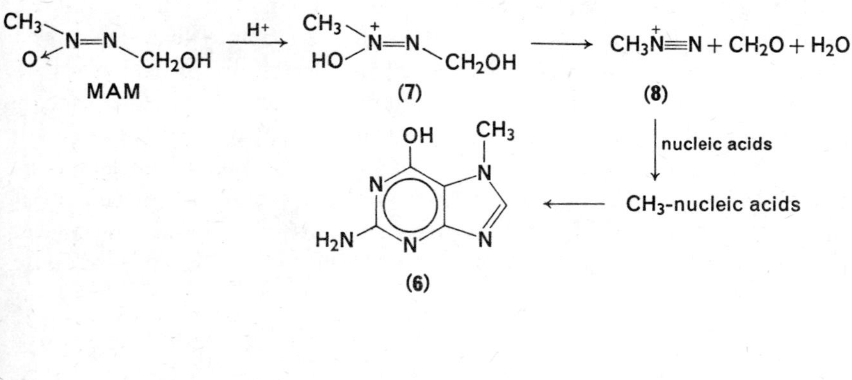

methyl group into the guanine residues of foetal brain DNA and RNA and to some extent into the proteins[61].

The chemical methylation may be initiated by protonation of the azoxy group, possibly at the *N*-oxide oxygen. The resulting ion (**7**) decomposes with loss of formaldehyde and water to yield methyldiazonium ion (**8**) or the potent carcinogen diazomethane (CH_2N_2). Diazomethane and diazoethane are known as powerful alkylating agents not only of carboxylic acids, phenols and enols, but also of purine and pyrimidine bases. The guanine and purine residues of nucleic acids are alkylated with diazomethane and diazoethane to form 7-methyl (or ethyl) guanine and 3-methyl (or ethyl) adenine in a ratio of 3–4:1[63]. The formation of diazomethane has also been postulated to be associated with the carcinogenesis of dimethylnitrosamine ($(CH_3)_2NN{=}O$) structurally and biologically related to MAM. Lijinsky and co-workers[64] have, however, precluded the possibility that diazomethane is involved in the methylation of nucleic acids with dimethylnitrosamine. The hexadeuterated dimethylnitrosamine ($(CD_3)_2NN{=}O$) when administered to rats gives rise to 7-CD_3-guanine, but not 7-CD_2H-guanine after the hydrolysis of the rat liver nucleic acids. If diazomethane is the ultimate carcinogen arising from the deuterated dimethylnitrosamine, 7-CD_2H-guanine should have been formed as the major methylation product. In view of the analogy of the metabolic reactions of the nitrosamine and MAM, the methyldiazonium ion appears to be most probably the ultimate carcinogen of MAM. In support of the intermediacy of the methyldiazonium ion, Benn and Kazmaier have obtained a trideuterated cresol ($CH_3C_6H_4OCD_3$) in the acid-catalysed reaction of *p*-cresol with the deuterated MAM ($CD_3N{=}NCD_2OH$)[65]. Concerning the mechanism of the metabolic reaction leading to the ultimate carcinogen, a six-membered transition state has been proposed (see Section III, D, 2). In any case it is likely that the driving force of the profound decomposition to the methyl unit is due to the formation of thermodynamically stable molecular nitrogen and formaldehyde.

The acetyl derivative (MAMA) of MAM, a similarly potent carcinogen, also inhibits the incorporation of thymidine-2-^{14}C into liver DNA and leucine-1-^{14}C into proteins[66]. Because of this inhibition, protein synthesis reaches 70% below control levels 2 h after a single injection, but the incorporation of the labelled amino acid returns to control levels within 24 h after the injection[67]. It is assumed that the alkylation of nucleic acids affects the function of the cellular macromolecules involving the control of protein synthesis, although it is not conclusive whether the biological effects of the aglycone MAM are attributable to its ability to alkylate nucleic acids.

2. Hydrazo-, azo- and azoxyalkanes

Some hydrazo-, azo- and azoxyalkanes are known to be alkylating agents with carcinogenic properties, together with many other types of alkylating agents. The simplest hydrazoalkane, 1,2-dimethylhydrazine, induces tumours[68] in rats after repeated subcutaneous or oral administration, while the isomeric 1,1-dimethylhydrazine is not carcinogenic. 1,2-Diethylhydrazine and some other higher homologues such as 1-methyl-2-*n*-propylhydrazine are also potent carcinogens[69]. The azo- and azoxyalkanes[70], the oxidized forms of these carcinogenic hydrazoalkanes, have slightly higher carcinogenic activities than the corresponding hydrazoalkanes. The three classes of carcinogenic compounds are, however, unable to induce tumours without undergoing profound chemical transformations, as can be realized by the observation that no tumour is produced at the sites of administration.

In general, hydrazoalkanes are readily oxidized to the azo derivatives in the presence of air oxygen and more rapidly by the catalytic action of heavy metal ions (i.e., cupric ion)[71]. The ease of oxidation of hydrazoalkanes suggests the involvement of *in vivo* transformation to the azo forms which make it possible to undergo metabolic conversion to carcinogens. This idea can account for the non-carcinogenicity of 1,1-dimethylhydrazine which is incapable of oxidation to the azo derivative. It is unlikely that the resultant azoalkanes are isomerized *in vivo* to the isomeric alkylhydrazones, since the ethylhydrazone of acetaldehyde is non-carcinogenic[69]. Druckrey and Lange[72] have postulated that azomethane (**9**) may undergo enzymic hydroxylation by α-C-hydroxylase after oxidation to azoxymethane (**10**). The resulting MAM is the potent carcinogen which is obtained by hydrolysis of cycasin. They have proposed a concerted reaction mechanism involving a six-membered transition state for the conversion of MAM to the

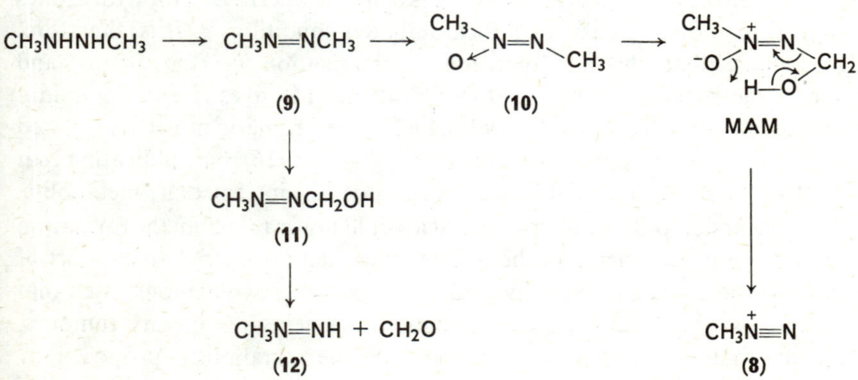
(9) **(10)** MAM **(11)** **(12)** **(8)**

ultimate carcinogen, **8**. The formation of aldehydes has been demonstrated unambiguously by their isolation from the incubation mixture of azo- and azoxyalkanes with the microsomal fraction of rat liver and NADPH[69]. The α-C-hydroxylation appears reasonable, since it is known that the α-carbon atom adjacent to nitrogen or oxygen is dealkylated following enzymic hydroxylation[73]. An alternative path to a possible ultimate carcinogen involves the formation of methyldiazene (**12**) via a hydroxymethyl intermediate (**11**) from **9**. Methyldiazene has been shown by Kosower and Tsuji[74] to react bimolecularly through a radical mechanism in the absence of oxygen to form methane. The diazene is also capable of introducing methyl and hydrogen units into double bonds, possibly by a radical process[75]. It remains, however, uncertain whether methyldiazene is involved in the methylation of genetic materials.

E. Carcinogenesis of 1-Methyl-2-benzylhydrazine Derivative (Procarbazine)

Procarbazine (also known as natulan) is cytostatic[76], but mutagenic[77] and carcinogenic[78] inducing a variety of tumours in mice and rats, and it has inhibitory effects on the biosynthesis of DNA[79], RNA[80] and proteins[81].

The benzylhydrazine is rapidly metabolized, as may be seen from the half-life of less than 15 min in the serum of man, dog and rat[82]. By analogy with hydrazoalkanes, procarbazine and the parent 1-methyl-2-benzylhydrazine are readily autoxidized to the azo derivatives, especially in the presence of metal ions[71]. The oxidation of procarbazine to azo-procarbazine proceeds rapidly in animals such as dogs, rats and man[83]. Both cytostatic and carcinogenic activity are also observed with the resulting azo-procarbazine (**13**)[84] being consistent with the biotransformation. Benzylazoalkanes are generally thermodynamically less stable than the isomeric alkylhydrazones, so that the azo isomers can be isomerized to the hydrazones under either acidic, basic or radical-initiated conditions[85]. It is, therefore, reasonable that the azo-hydrazone isomerization *in vivo* occurs and subsequently the hydrolysis of the hydrazone (**14**) gives rise to an amidobenzaldehyde (**15**) and methylhydrazine[86]. The aromatic moiety is isolated from urine as *N*-terephthalic acid isopropylamide (**16**)[83,87], indicating that methylhydrazine or a related species is a possible intermediate metabolite. The conversion of **13** to **16** proceeds at a 4-fold lower rate than the formation of **13** from procarbazine in the isolated perfused rat liver[88]. In support of the metabolic pathway, Kelly and co-workers[84] have demonstrated that **13** and **14** are as active as procarbazine in inducing pulmonary tumours. An alternative pathway for the procarbazine metabolism proposed by

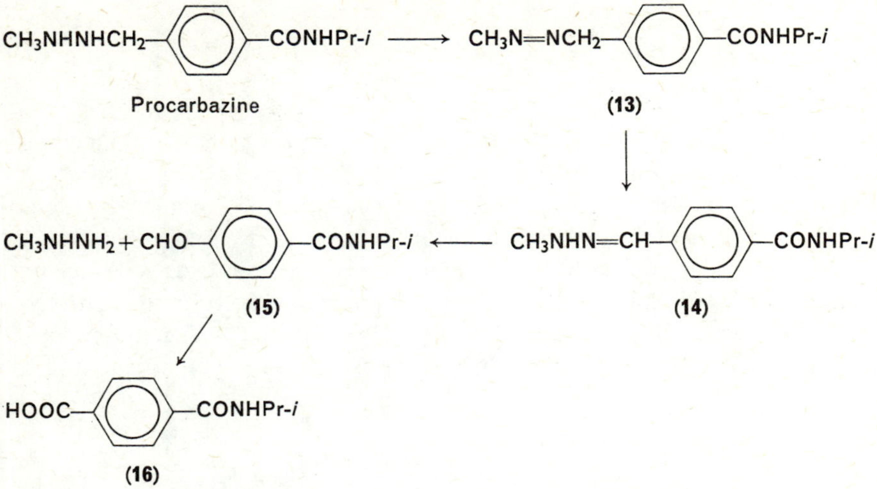

Baggiolini and co-workers[88] involves oxidative cleavage of the benzyl and methyl moieties of azo-procarbazine implying the formation of methyldiazene (**12**) and the aldehyde (**15**).

$$\underset{(\mathbf{13})}{\text{Azo-procarbazine}} \longrightarrow \underset{(\mathbf{12})}{CH_3N{=}NH} + \mathbf{15}$$

The methyl moiety of procarbazine is, to some extent, metabolized to methane and carbon dioxide[89]. Methane is thought to result from methyldiazene which arises either from methylhydrazine or directly from **13**. The carcinogenesis of procarbazine has been explained in terms of two different types of reactions: (*i*) methylation of nucleic acids; (*ii*) degradation of DNA by hydroxyl radicals arising from the autoxidation reactions. Kreis and co-workers[90] have shown that procarbazine when intraperitoneally injected to mice effected methylation of guanine residues of nucleic acids. Another postulate is that autoxidations involved in the metabolic reactions generate hydrogen peroxide and thence OH radicals capable of causing degradation of DNA[86].

F. Formation of an Azo Carcinogen from Aryldimethyltriazenes

Aryldimethyltriazenes are potent carcinogens capable of methylating nucleic acids by an azo intermediate resulting from the metabolic reaction. The dimethyltriazene moiety of aryldimethyltriazenes is closely related to dimethylnitrosamine (**17**) with regard to its electronic structure and the

(17)

carcinogenic action. It is, therefore, not surprising that aryldimethyltriazenes generate methyldiazonium ion or the related reactive species as does the carcinogen, **17**. The metabolic fate of 1-phenyl-3,3-dimethyltriazene (**18**) has been investigated *in vitro* and *in vivo* by Druckrey and co-workers[69]. The methyl group of the triazene undergoes enzymic hydroxylation on incubation with the complete system of rat liver and the resulting hydroxymethyltriazene (**19**) decomposes to formaldehyde and 1-phenyl-3-methyltriazene (**20**). The monomethyltriazene (**20**) is also a potent carcinogen

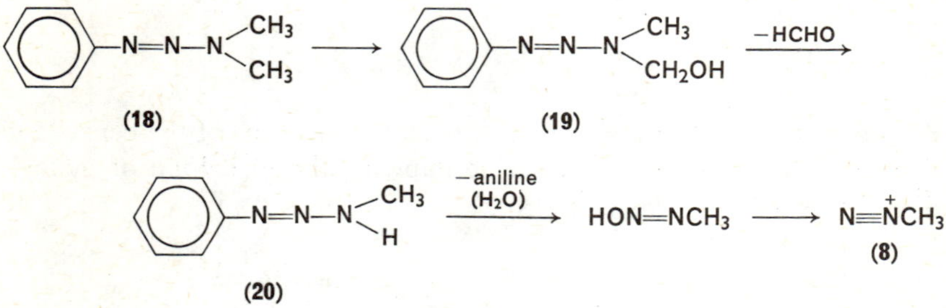

and chemically so reactive that it can be utilized as a methylating agent of carboxylic acids and phenols[91]. The *in vitro* treatment of DNA of low molecular weight from herring sperm (*Serva*) and DNA of high molecular weight from salmon sperm with **20** affords 7-methylguanine following hydrolysis of the methylated nucleic acids[69]. The metabolic activation of the nitrosamine **17** to the ultimate carcinogen is thought to proceed through the α-C-hydroxylation and subsequent release of formaldehyde[92]. The N—O bond cleavage of the resulting methylnitrosamine leads to the same ultimate carcinogen.

The same metabolic conversion mechanism has been proposed with 4(5)-(3,3-dimethyl-1-triazeno)imidazole-5(4)-carboxamide which induces tumours in Sprague–Dawley rats[93].

G. Carcinogenic Metabolism of Azobenzene Derivatives

Among numerous approaches to the elucidation of mechanisms of carcinogenesis, the most common investigation of chemical interest attempts to find carcinogenic reactions of chemical carcinogens with biological constituents. Chemical explanations are still limited to the biological reactions in early stages of carcinogenesis, but probably make it possible to design non-carcinogenic compounds of the same or related classes. Carcinogenic hydrazo, azo and azoxy compounds may assist better understanding of the mechanism of the biological reactions responsible for carcinogenesis. Azobenzenes give rise to ultimate carcinogens following small chemical changes with retention of the azobenzene structure, while hydrazo-, azo- and azoxyalkanes undergo degradative reactions to yield ultimate carcinogens. Although the azo linkage of azobenzenes is not directly involved, a brief description is attempted in view of its biological importance.

Yoshida's discovery[94] of the hepato-carcinogenesis using orthoamino-azotoluene (OAT) stimulated studies on carcinogenesis of azobenzenes. Butter yellow (4-dimethylaminoazobenzene, DAB) once used for colouring butter and margarine was found to be a potent carcinogen in rat liver[95] and has since been used as the most common compound for carcinogenic studies. DAB undergoes metabolic reactions including azo reduction and aromatic hydroxylations, but these reactions have been proved not to be associated with the carcinogenic reactions. The metabolism of DAB responsible for carcinogenesis involves the three following important reactions: oxidative *N*-demethylation[96]; *N*-hydroxylation of the resulting 4-methylaminoazobenzene (MAB); binding to macromolecules such as

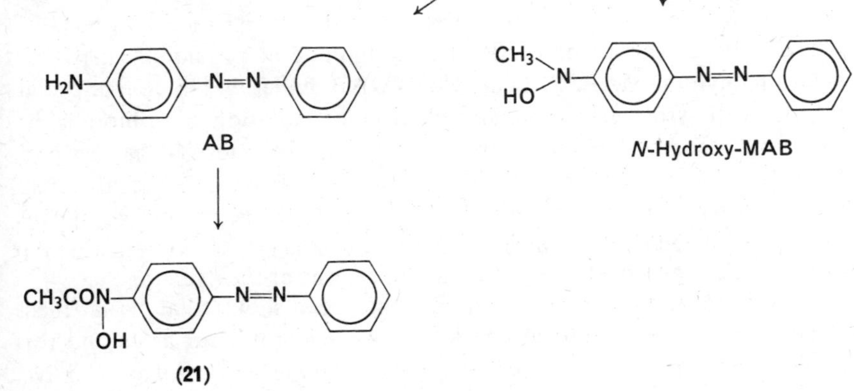

(21)

proteins and nucleic acids. The oxidative *N*-demethylation is catalysed by liver microsomal enzymes which require NADPH and oxygen[97]. MAB is also a potent carcinogen and follows a metabolic pathway similar to that of DAB when fed to rats.

Further demethylation to 4-aminoazobenzene (AB) results in a remarkable decrease of the carcinogenic activity. At least one *N*-methyl appears to be essential for the potent carcinogenicity of the azobenzene derivatives, since substitution of the two methyl groups by other substituents leads to disappearance of the carcinogenic activity[98]. The *N*-hydroxy-MAB which possibly results from MAB would be carcinogenic in the intact form inducing tumours at sites of administration[99]. Although there is no conclusive evidence for the possible intermediate, *N*-hydroxy-*N*-acetyl-*p*-aminoazobenzene (**21**) is isolated from the urine of rats injected with AB[100]. An important observation was made by Miller and co-workers[101] who pointed out that only carcinogenic azobenzenes are bound to the target organ (liver) proteins. No DAB-bound dye is found in hamsters, rabbits, etc., which are not susceptible to DAB. Although the position of the bonding of the protein-bound dye is still in dispute[102], methylthio-MAB (**22**)[103] is isolated after alkaline treatment of the protein-bound dye. The

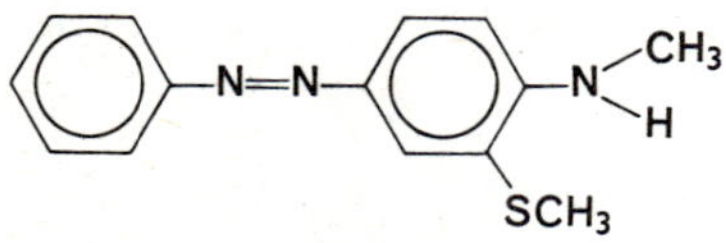

(22)

methylthio group is possibly transferred to MAB from methionine residues of proteins. The protein-bound dye is thought to be formed from the reaction of the *N*-hydroxy-MAB intermediate or its ester with the liver proteins. Model reactions[104] have been carried out with the *N*-acyloxy- or *N*-aroyloxy derivatives of MAB and methionine at physiological pH to form methylthio-MAB. *N*-Benzoyloxy-MAB undergoes a non-enzymic reaction with tyrosine, tryptophan and cysteine in addition to methionine[105]. The results suggest that *N*-hydroxy-MAB may be esterified *in vivo* and thereafter bind to the liver proteins, although no *N*-acyloxy derivative has been detected. An azo carcinogen (i.e., 3′-CH_3-DAB)-protein conjugate has been isolated from rat liver cytosol[106], but it is not known yet whether or not it plays a critical role in the carcinogenesis of the dye.

In addition to azo-protein binding, it has been shown that ^3H-labelled DAB is incorporated into DNA, RNA and acidic nuclear acids and the basic nuclear proteins, histone. Roberts, Warwick and Albert[107] have shown that the ^3H-labelled DAB is incorporated into DNA, RNA and acidic nuclear proteins in addition to proteins. The binding to nucleic acids is thought to be attained by the reaction with 4-hydroxymethylaminoazobenzene ($C_6H_5N{=}NC_6H_4NHCH_2OH$) forming cyclic compounds. This type of cyclic compound can be produced on treatment of AB with formaldehyde and amino group-containing nucleosides. *N*-Benzoyloxy-MAB also reacts with guanine, favouring it greatly over adenine and pyrimidine bases. However, Borenborm[108] and some other workers have either ruled out or regarded as less important the binding of the dye to nucleic acids.

OAT is a chemical carcinogen of the primary amino type which is frequently employed for studies on carcinogenesis. Unlike DAB, OAT does not require the presence of one *N*-methyl group to undergo carcinogenic reaction, but the mode of action is principally not different from that of DAB. The primary amino group appears to be oxidized to the *N*-hydroxy-OAT[102] which may be responsible for the successive carcinogenic reactions with proteins and nucleic acids.

DAB and some related carcinogenic azobenzenes have been shown to influence the biosynthesis of nucleic acids and proteins, as in the case of other types of carcinogens. Carcinogenic studies on DAB and a number of more or less potent carcinogenic azobenzenes have been extensive, but they are beyond the scope of this chapter. Although many problems remain unanswered, it is conceivable that the binding of the active azo dye to proteins and nucleic acids is somehow involved in the early stage of the carcinogenic processes.

IV. OXIDATION OF THIOLS TO DISULPHIDES BY AZO COMPOUNDS AND THE RESULTING BIOLOGICAL EFFECTS

A. General

In spite of extensive studies on azo compounds, addition reactions to the azo linkage excepting azo reductions are little known. The azo linkage becomes more activated and reactive to nucleophiles when substituted with electron-withdrawing groups on one or both nitrogen atoms. Azo compounds for biological use should be moderately activated by attaching groups so that a given azo compound can react selectively with a specific functional group in a biological system. If it is too reactive, the selectivity of the azo reagent is lost and undesired reactions may take place with less reactive functional groups. Thiols have higher reactivity towards azo groups than amino, hydroxyl groups or carboxylic acids present in biological systems. The greater nucleophilicity of thiols has also been observed with many other reagents.

It is of considerable importance to investigate biological changes resulting from the deficiency of a certain biological constituent, since it pinpoints the biological importance of the material. Kosower and Kosower and co-workers have employed moderately reactive azo compounds with a view to evaluating the biological role of glutathione (GSH)[109] by selective oxidation to the disulphide (GSSG) without affecting cellular function and found that azoester and diamide are the best oxidizing agents of intracellular GSH. The moderately reactive azo reagents undergo selective reactions with intracellular GSH in various biological systems, giving rise to the disulphide and the corresponding hydrazo derivatives. GSH exists in many parts of animals and plants and acts as cofactor of some enzymes catalysing disulphide–thiol interchanges. The main biological significance of GSH has been previously considered to be the trapping of potentially harmful free radicals, redox buffering and chelating with heavy metals. Hitherto unknown roles of biological importance have become apparent as the result of the oxidative conversion of GSH to GSSG.

The oxidation of a thiol compound (R″SH) to the corresponding disulphide (R″SSR″) by an azo compound (R—N═N—R′) involves addition reaction of R″SH and successive nucleophilic attack of another molecule of R″SH at the sulphur atom of the resultant 1–1 adduct. There is evidence that an azo-thiol adduct is formed on treatment of ethyl azodicarboxylate with *n*-pentylmercaptan in a non-polar solvent[110]. The intermediacy of azo-thiol adducts has also been indicated in the reactions of azoester and diamide with GSH. It is noteworthy that a C═N bond also permits addition reaction with a thiol, quickly leading to an equilibrium mixture[111], but the corresponding disulphide formation is only made possible by a radical

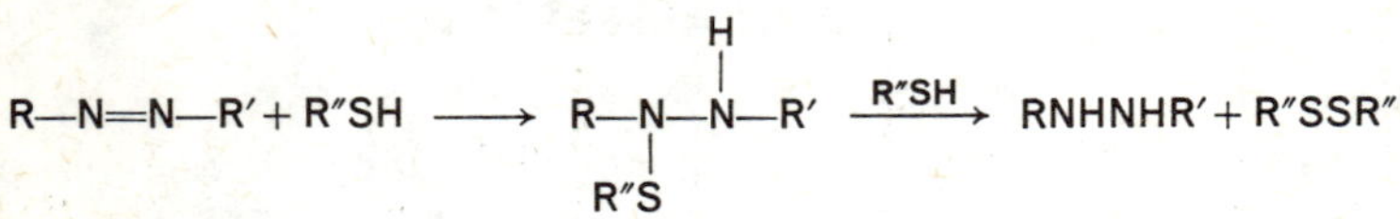

mechanism at higher temperatures[112]. As the two reagents, azoester and diamide, are suitable for biological studies as compared with the initially used acetylphenylhydrazine and others, details of these reactions are given in the following paragraphs.

B. *Azoester(methyl phenyldiazenecarboxylate)*

GSH is rapidly oxidized bimolecularly by azoester inside red blood cells[113] without requiring oxygen[114]. Although the azo-GSH adduct has not been isolated, the intermediacy of some element, possibly the GSH adduct (**23**), is recognized by the observation that, after a delay, the formation of hydrazoester (**24**) follows the disappearance of azoester[115]. On the basis of this time lag and its stoichiometry, the azoester–GSH reaction is represented as follows:

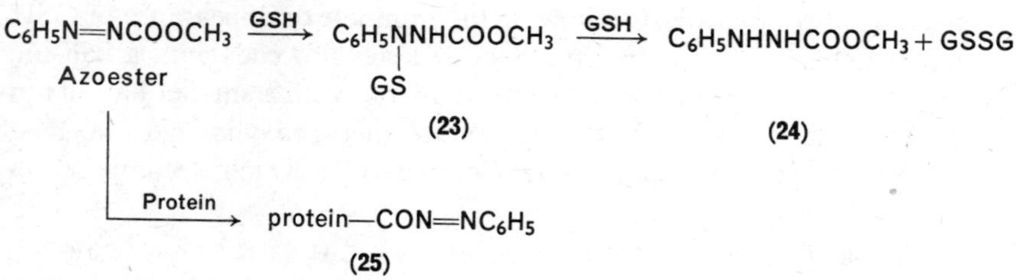

Azoester also undergoes a similar reaction with cysteine leading to the formation of hydrazoester and the disulphide cystine, but there is no delay after the disappearance of the azoester. The rate constants of the reactions of azoester with GSH and cysteine in buffer solutions are given in Table 1[115].

TABLE 1. Rate constants of azoester in the buffer solutions containing 0·001M EDTA at 25°C (ionic strength, 0·1)

Azoester (mole)	GSH (mole)	pH (buffer)	k (M^{-1} s^{-1})
$7·96 \times 10^{-5}$	$3·76 \times 10^{-4}$	6·59 (phosphate)	2·1
$7·96 \times 10^{-5}$	$3·83 \times 10^{-4}$	7·62 (phosphate)	10·1
$7·96 \times 10^{-5}$	$3·76 \times 10^{-4}$	8·71 (tetraborate)	140·1
$0·775 \times 10^{-5}$	$0·387 \times 10^{-4}$	9·65 (tetraborate)	400 (approx.)
Azoester (mole)	**L-Cysteine (mole)**	**pH (buffer)**	**k (M^{-1} s^{-1})**
$8·0 \times 10^{-5}$	$4·36 \times 10^{-4}$	6·6 (phosphate)	2·2
$8·0 \times 10^{-5}$	$4·40 \times 10^{-4}$	7·6 (phosphate)	3·4
$8·0 \times 10^{-5}$	$4·43 \times 10^{-4}$	8·76 (tetraborate)	31·2
$8·0 \times 10^{-5}$	$4·27 \times 10^{-4}$	9·62 (tetraborate)	80 (approx.)

The kinetic measurements indicate that GSH reacts with azoester 3–5 times faster than cysteine in the range of pH = 7·6–9·6 and that thiolate anion may be involved in the addition reaction. Azoester when reacted in buffer solutions (pH = 6·56–9·65) consumes two mole-equivalent of GSH to give GSSG in a quantitative yield, but the use of the ester in red blood cells requires about an equivalent quantity, possibly due to the reaction with protein amino groups (azoester → **25**).

The intracellular reaction of azoester with GSH can be carried out without detectable intracellular damage, giving rise to hydrazoester and GSSG in high yields. Azoester when used in an excess amount is, however, subjected to slow hydrolysis following the complete disappearance of GSH in red blood cells[114]. The hydrolysis of azoester proceeds with a half-life of about 20 min in neutral aqueous solution, but with a smaller half-life in biological systems. The resulting phenyldiazenecarboxylic acid (**26**) may be decarboxylated to unstable phenyldiazene (**27**) which decomposes in

$$C_6H_5N{=}NCOOCH_3 \longrightarrow \underset{(\mathbf{26})}{C_6H_5N{=}NCOOH} \longrightarrow \underset{(\mathbf{27})}{C_6H_5N{=}NH} \longrightarrow C_6H_5\cdot \downarrow \text{membrane damage}$$

the presence of oxygen to generate phenyl radical[116]. The phenyl radical causes membrane damage leading to haemolysis if there is a deficiency of GSH. This illustrates that azoester can act as a chemical agent causing membrane damage in addition to being a thiol-oxidizing agent, and that GSH plays an important role in preventing membrane damage by chemical agents. The other effects resulting from very excessive use of azoester are an increase of methaemoglobin formation and Heinz body formation, possibly through the following chemical processes[114].

$$\text{Protein-S}\cdot + \text{Protein-S}\cdot \longrightarrow \text{Protein-S—S-Protein}$$

$$C_6H_5\cdot + \underset{\text{Oxyhaemoglobin}}{Fe^{2+}O_2} \longrightarrow C_6H_5OO^- + \underset{\text{Methaemoglobin}}{Fe^{3+}}$$

$$\text{or } C_6H_5OO\cdot + Fe^{2+}O_2 \longrightarrow C_6H_5OO^- + O_2 + Fe^{3+}$$

Incubation of the GSH-deficient red blood cells with GSSG reductase and NADPH completely regenerates GSH from the disulphide. The recovery of GSH falls from over 90% to about 60% in the 4-fold excess use of azoester, but is not lowered below 60% even with much greater addition of azoester[114].

It has been shown that haemoglobin thiols do not undergo reaction with azoester added to convert all the GSH to GSSG and there is no influence of the chemical state of human erythrocyte haemoglobin (oxy-, deoxy-, carbon monoxide or met-) on the ease of the interconversion between GSH and GSSG[114].

C. Diamide

The reagent is water-soluble and stable in water, while azoester is sparingly soluble in water and susceptible to hydrolysis. The azo linkage of diamide is moderately activated by the two adjacent carbamoyl groups and reacts selectively with GSH[117]. Diamide oxidizes GSH stoichiometrically and very rapidly to form GSSG and dihydrodiamide (**28**) in excellent yields. Diamide has advantages over azoester if excess reagent is employed and membrane damage is undesirable. About 0·6 mole of diamide is sufficient to oxidize one mole of GSH in red blood cells and no effect has been observed on haemoglobin, osmotic fragility and cell density after regeneration of GSH. GSH is regenerated from GSSG in red blood cells within a short period after incubation with glucose.

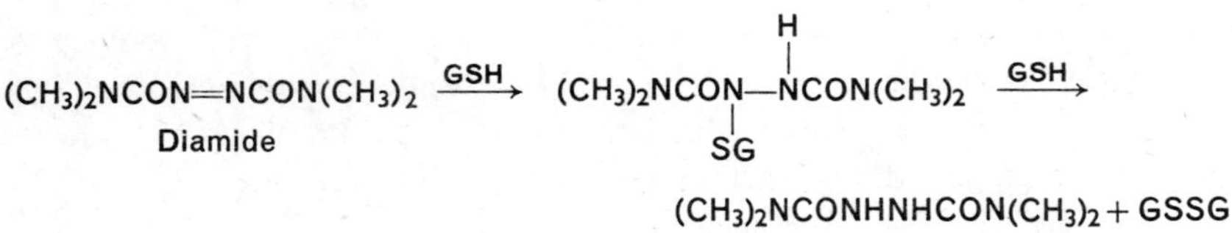

(28)

D. Biological Effects of Azoester, Diamide and some Azo Compounds on Bacterial Growth

High concentrations of GSH in *Escherichia coli* (ca. 10^{-4}M)[118] imply some biological importance of the tripeptide. Kosower and co-workers[119] have demonstrated by using diamide that the growth of the bacteria depends upon GSH. Addition of sufficient amount of diamide to a culture of *E. coli* B growing exponentially in M-9 media at 37°C results in growth inhibition, but not death of the cells. Growth is resumed after excess diamide (10^{-4}–6×10^{-4}M) is converted to the dihydro derivative (**28**) by reduction with slowly regenerated GSH. The use of an excess of diamide for the oxidation of thiols causes no damage to the bacteria, as is realized from the resumption of normal growth after the excess diamide has been completely consumed. The length of the time lag depends upon the concentrations of diamide added in the culture. Normal growth is also resumed after addition of excess GSH to consume the remaining diamide. The growth rate is not enhanced by the addition of cysteine and methionine, each at 20 μg/ml, together with GSH. Even when a higher concentration of diamide (3×10^{-3}M) is used, the growth of *E. coli* B is resumed after washing out the reagent on a membrane filter or adding excess GSH. Rose and co-workers[120] have recently investigated the effects of diamide and some related derivatives on a number of bacteria including *E. coli* and found that high concentration of diamide (3×10^{-3}M) is, after several hours, bactericidal for pure cultures of *E. coli*, *Proteus* sp. and *Salmonella enteritidis*, but not for *Pseudomonas aeruginosa* whose growth is only delayed for 1–2 h. Although the mechanism of the bactericidal action has not been explained, the growth inhibition is thought to be mainly associated with the temporary loss of GSH and other low molecular weight thiols such as cysteine[119]. The growth inhibition of bacteria is correlated with protein synthesis inhibition in the case of deficiency of GSH or of a high GSSG/GSH ratio (see Section IV, E).

The mode of action differs from that of the antibacterial 6-(*p*-hydroxyphenylazo)uracil (HPUra) and the closely related arylazopyrimidines. Brown and co-workers[121] have found that HPUra selectively inhibits DNA replication of Gram-positive bacteria without affecting the synthesis of protein and RNA. There is evidence that the antibacterial activity is associated with the inhibition of ATP-dependent polymerization of deoxyribonucleotides. The mechanism of the inhibition of DNA synthesis is not fully explained, and further work is needed to interpret the selective inhibition of Gram-positive bacteria.

The bacteriostatic effects of a number of arylazo derivatives of malononitriles and some others have been reported[122].

E. Effects of Diamide on Protein Synthesis in Rabbit Reticulocytes

In connection with the temporary inhibition of the growth of *E. coli*, Kosower and co-workers[123] have investigated the effects of the thiol-oxidizing agent diamide on protein synthesis. Addition of diamide (1·5–1·7 moles per mole of intracellular GSH) to protein-synthesizing rabbit reticulocytes leads to the immediate inhibition of incorporation of ^{14}C-labelled leucine into soluble proteins. The leucine incorporation is, however, resumed about 10 min after the addition of diamide. The period of delay depends upon the amounts of diamide as is seen with the growth inhibition of *E. coli*. The rate of protein synthesis is dependent on the concentrations of the intracellular GSH. The GSH oxidized to GSSG is rapidly regenerated after complete consumption of the diamide and protein synthesis starts again when the regeneration of GSH reaches about 70%. The rate of the leucine incorporation is recovered to that of control at the time of over 80% GSH regeneration. Systems which have been treated with large amounts of diamide cannot, however, resume the normal rate even after all the intracellular GSH is regenerated. Profound changes have been observed in the pattern of the ribosomal aggregates and in the radioactivity of ribosomes. The ribosomal pattern and the distribution of the radioactivities change gradually after diamide addition, but become similar to those of control by the time that regeneration of GSH is complete (25 min after diamide addition). Furthermore, it has been shown that initiation is more GSH-dependent than translation and release of protein from ribosomes. Although the preceding description is focused on GSH disappearance, Kosower and co-workers[124] have found that the concentration of GSSG affects protein synthesis. Addition of GSSG ($1–2 \times 10^{-4}$ M) instead of a thiol-oxidizing agent to a rabbit reticulocyte lysate (GSH content, $7–10 \times 10^{-4}$ M) also results in profound inhibition of initiation of protein synthesis. The incorporation of ^{14}C-alanine 6–8 min after addition slows down over 2–4 min, then creases. Protein synthesis can be resumed by adding glucose and NADP to the lysate at the beginning of incubation, but not after inhibition of protein synthesis has started. The change of ribosomal pattern (conversion of polysomes to monosomes) is also observed together with loss of the labelled protein from ribosomal fraction. There is no doubt that the biological changes arising from the addition of a thiol-oxidizing agent or GSSG are associated with the inhibition of the protein synthesis and thiols are intimately linked to some of the stages of protein synthesis.

It would be of great interest to investigate a possible physiological regulatory role of GSSG in normal and genetically-deficient systems (i.e., glucose-6-phosphate dehydrogenase deficiency, GSSG-reductase deficiency)[124].

F. Other Biological Effects of Diamide and Azoester

Azoester also influences the feeding behaviour of blood-sucking invertebrates such as ticks (*Ornithodoros tholozani*), leeches (*Hirudo medicinalis*), Tsetse flies (*Glossina austeni*) and mosquitoes (*Aëdes aegypti*) in spite of the presence of feeding stimulants[125]. For instance, long-starved ticks do not feed on 10^{-3} M azoester solutions containing glucose ($5{\cdot}5 \times 10^{-4}$ M)–ATP (10^{-3} M) and glucose ($5{\cdot}5 \times 10^{-4}$ M)–leucine (10^{-2} M) during 60 min, while they are readily stimulated to feed by GSH, ATP and a number of amino acids in the presence of glucose[126]. After removal of the azoester, over 50% of the ticks feed on the glucose ($5{\cdot}5 \times 10^{-4}$) and GSH (10^{-3} M) solution during 1 h. Mosquitoes are repelled by the addition of azoester to the stimulant saline–ATP solution with a potency comparable to pyrethrins and other mosquito repellents. Taking into account the high GSH content of neurones[127], it is conceivable that intracellular GSH may be involved in the detection of chemical stimuli. Another remarkable biological response is that of hydra following treatment with azoester[128]. The resulting major behavioural consequences are: (*i*) inhibition of capture of shrimps used as the prey; (*ii*) inhibition of tentacle contractions; (*iii*) inhibition of mouth opening. The sensitivities to the reagent are in the order shown above, the functioning of nematocysts virtually disappearing after the hydra are treated for 5 min with $1\text{–}3 \times 10^{-5}$M azoester. At this concentration the volvent nematocysts are not paralysed, whereas no nematocysts function after being treated with higher concentrations of azoester. Treatment with 10^{-3}M azoester solution causes lysis and death in 10 min.

V. ANTIFUNGAL ACTION OF HYDRAZO AND AZO COMPOUNDS

The chemical bases of antifungal actions have been extensively studied with a variety of compounds and considered to depend on any of the usual interactions: Van der Waals, hydrogen, ionic, dative and covalent bondings[129]. The major sites of antifungal action may be coenzymes, apoenzymes and substrates of enzymes[130]. The following description is mainly concerned with the antifungal actions of hydrazo and azo compounds through covalent bond formation with coenzymes or apoenzymes.

Recently, the mode of action of antifungal hydrazo and azo compounds has been investigated in detail. A number of hydrazo and azo, but not many azoxy, compounds are known to have antifungal activities. Antifungal actions of hydrazo and azo compounds are of biological interest, making it possible to understand the biological significance of thiols of coenzymes and apoenzymes in the germination, growth and sporulation of fungi. Elucidation of the correlation of antifungal activities with the biochemical reactions would be of importance in order to prove the mode of action and to develop more effective antifungal agents.

Kosower and associates[131] have investigated the effect of azoester and some azo compounds on the germination of the fungus *Trichoderma viride* and found that germination is either inhibited or severely delayed by the reagent. This strongly indicates the important role of the coenzyme GSH or of other thiols in maintaining the life of the fungus. Richmond and Somers have observed that the germination of the fungus *Neurospora crassa* accompanies an increase in the SH content[132] and the rate of spore germination decreases on treatment with the irreversible alkylating agent iodoacetic acid which diminishes the soluble SH content of the conidia[133]. GSH has been shown to be present in mycelium and conidia of the fungus *Penicillium griseofulvum* as the major component of a hot-water-soluble fraction[134]. GSH is generally regarded as being the major non-protein component of the soluble thiol pool in living cells[135]. Azoester reacts selectively with GSH in the presence of protein thiols, so that the intracellular oxidation of GSH to GSSG may be associated with the reduction of germination, growth and sporulation of the fungus[131]. The perturbation of the GSH–GSSG balance in fungi is thought to influence the initiation of protein synthesis. The importance of thiols in living organisms is also reflected by the observations that the reduction of GSSG to GSH takes place in germinating pea seeds[136] and cell division is dependent upon SH–disulphide transformation[137].

Kosower and co-workers[138] have made an attempt to systematize the structure–activity relationship of antifungal azo compounds in the belief that antifungal activities may be correlated with the reactivities of azo compounds towards thiol groups. A model study has been carried out with hexahydrospinamycin (**29**)[139]. This hexahydro derivative of the antifungal antibiotic spinamycin is much more convenient to use than spinamycin itself which is too sensitive to air because of the presence of the conjugated triene. If the hydrazo group plays a central role in the antifungal action, the reduction of the unsaturated side-chain will not abolish the activity of the

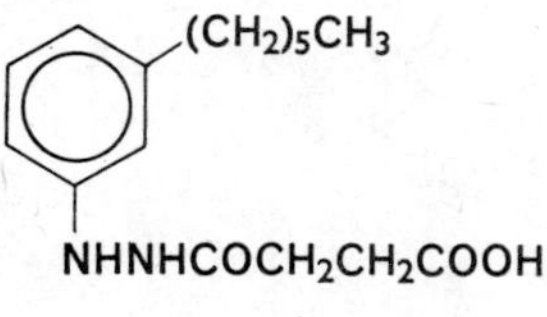

(29)

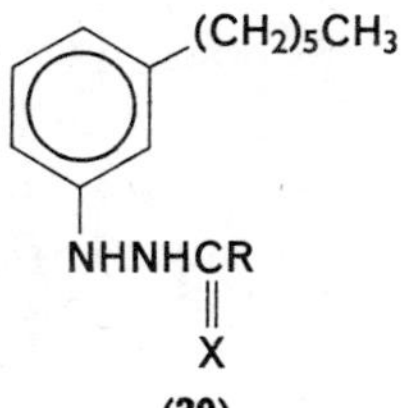

(30)

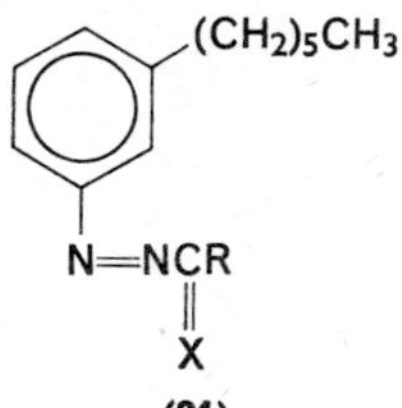

(31)

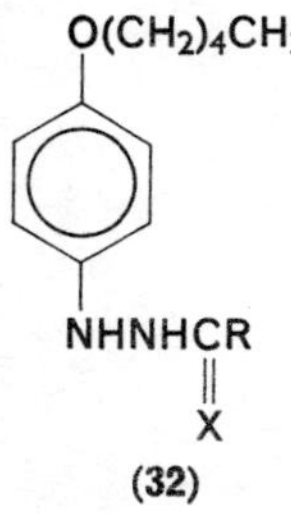

(32)

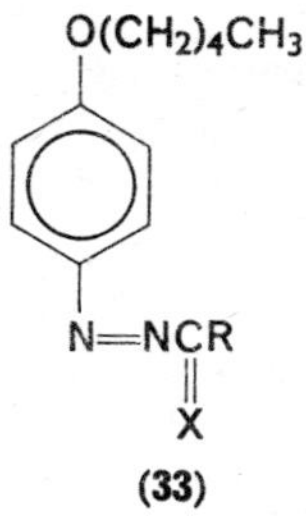

(33)

X = O or S

R = NH_2, $N(CH_3)_2$

antibiotic. Besides the side-chain reduction, the replacement of the hydrocarbon chain with any other substituent of similar size and lipid-solubility would not profoundly affect the antifungal activity. From this point of view, a number of the related hydrazo and azo compounds in addition to hexahydrospinamycin have been synthesized for kinetic measurements and antifungal tests.

The azo form of hexahydrospinamycin is rather unstable, yielding succinic anhydride and *n*-hexylbenzene. It is very likely that the decomposition proceeds by way of the corresponding reactive phenyldiazene derivative (*m*-hexyl—$C_6H_4N{=}NH \rightarrow$ *n*-hexylbenzene + N_2). Highly active antifungal azo compounds have been shown to react rapidly with GSH, the latter acting as a nucleophile towards the azo linkage. Potent antifungal hydrazo compounds (**30**, **32**, X = S) become coloured in the culture solution, and are comparable (or a little weaker) to the corresponding azo derivatives (**31**, **33**, X = S) in activity. It is conceivable that the hydrazo compounds are readily oxidized to the azo derivatives, thereafter undergoing reactions with reactive thiols (mainly with most reactive GSH). Pluijgers and coworkers[140] have obtained coloured phenyldiazenethiocarboxamide after allowing an alkaline solution of the antifungal hydrazo derivative to stand. The hydrazo compound seems to be activated by oxidation to the azo form, since both compounds have similar inhibitory activities against the spore germination of a number of fungi. The introduction of an electron-donating group into the benzene ring of the phenylhydrazine derivatives is expected to facilitate autoxidation of the hydrazo group to the azo derivative. For this reason, the pentyloxy derivatives (**32**) have been prepared and found to be readily oxidized to the azo derivative (**33**)[138]. Such hydrazo or azo compounds can be employed *in vivo* for intracellular GSH oxidation following regeneration of the azo derivatives by autoxidation. The oxidation–reduction mechanism is depicted schematically as follows[138]:

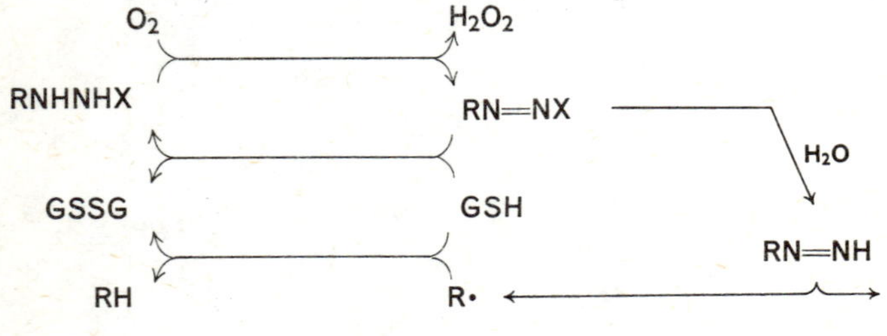

Chain autoxidation in the absence of sufficient GSH.

However, the rate of autoxidation may not reflect the potency of the hydrazo or the resulting azo compounds, since electron-donating groups make the azo linkage less reactive to the nucleophilic GSH.

Pluijgers and co-workers[141] have demonstrated that the antifungal activities of phenyldiazene cyanide and its *p*-substituted derivatives (**34**) can be correlated with the rates of their reactions with cysteine, the fungitoxicity increasing with the reactivities towards cysteine. *para*-Substitution of the benzene ring with an electron-donating group such as dimethylamino or methoxy results in lower rates of reaction with cysteine. The mode of action of the phenyldiazenecyanides is thought to involve the cyanide and azo groups in covalent bond formation with fungal cell thiols in a non-specific way.

The reaction of the azocyanides with cysteine proceeds through the initial addition of the thiol to the cyano group and the successive reduction of resultant adduct (**35**) with two moles of cysteine to give the disulphide cystine and the hydrazo derivative (**36**). Zsolnai[122] has tested a number of

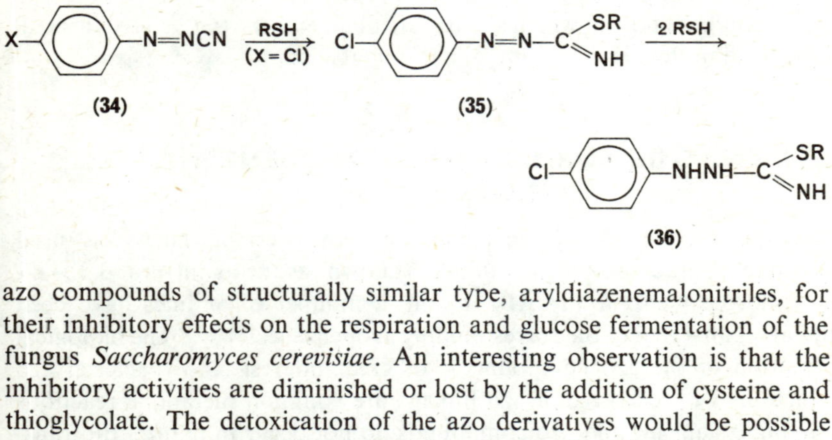

(**34**) (**35**) (**36**)

azo compounds of structurally similar type, aryldiazenemalonitriles, for their inhibitory effects on the respiration and glucose fermentation of the fungus *Saccharomyces cerevisiae*. An interesting observation is that the inhibitory activities are diminished or lost by the addition of cysteine and thioglycolate. The detoxication of the azo derivatives would be possible with glutathione instead of the thiol reagents. The biological reaction involved in the detoxication process is assumed to be similar to the possible inactivation mechanism of sulphhydryl enzymes responsible for the growth inhibition.

The antifungal agent, Dexon (sodium 4-dimethylaminophenyldiazene sulphonate), appears to have a different mode of action, since there is no indication of the reaction with GSH in buffer solutions[138]. The low reactivity of the azo linkage can be readily understood as the result of the attachment of sulphonate ion and the electron-donating group. Dexon readily undergoes photochemical decomposition to 4-dimethylaminophenol, but the photochemical reaction is not responsible for the antifungal action. It is suggested that Dexon inhibits phosphorylation.

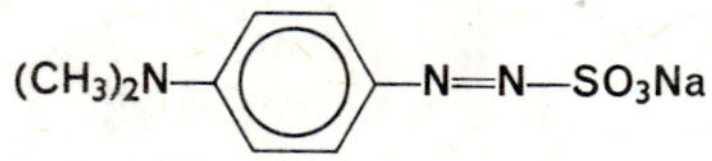

Dexon

As is the case with deacylation reactions of the acylated hydrazo compounds (see Section III, C), some antifungal hydrazo compounds might undergo hydrolytic cleavage to the corresponding monosubstituted hydrazines responsible for the antifungal activities. In view of the antifungal activities of a number of phenylhydrazines, there is a possibility that spinamycin and hexahydrospinamycin become active following enzymic or non-enzymic hydrolysis to the phenylhydrazine derivatives. Horsfall and Lukens[142] have assumed that the inhibition by phenylhydrazines is the result of trapping glyoxylic acid, essential in sporulation, as the phenylhydrazones. The antifungal action of the phenylhydrazines can be regarded as an additional type in which a substrate of an enzyme (i.e., glyoxylic acid) is involved in the reaction with an antifungal agent.

VI. ENZYMIC FORMATION OF AZOBENZENES FROM ANILINES IN SOILS

Since the discovery[143] of the metabolic conversion of the widely used herbicide 3,4-dichloropropionanilide (known as propanil) to 3,3′,4,4′-tetrachloroazobenzene (TCAB) in soil, a number of anilines have been similarly converted to the corresponding azobenzenes[144, 145]. The biological formation of the azo compound is of great interest since bacterial azo reductions with cleavage of azo linkage are common metabolic reactions in living organisms and azo compounds do not occur in nature. Biotransformation of an aromatic amine to the corresponding azo compound in rats is, however, known as a minor metabolic reaction, although it is very rare[146]. The unusual metabolic reactions of propanil have been extensively studied because of the importance of environmental contamination with azobenzenes and because of the mechanistic interest of the reaction. The formation of azobenzene derivatives is not desirable because of their stability and biological effects, although no tumours have been found in rats fed TCAB[147]. Biological azo formation is of great concern to mankind, since a number of aniline derivatives have been used in agriculture (carbamates, ureas and other anilides).

The *N*-propionyl substituent of propanil is removed by microorganisms to give 2,4-dichloroaniline (DCA)[148, 149] with the evolution of carbon dioxide[150]. Soil organisms also effect the condensation reaction of the resulting substituted aniline[151] as is evident from the failure to detect TCAB in propanil- or DCA-treated soils when sterilized or poisoned[150]. The herbicide also undergoes enzymic conversion to DCA in plants[152], but TCAB has not been found in plant tissues[153]. Since two molecules are involved in the formation of the azo compound, cross-reactions can occur with two different anilines or anilide herbicides giving rise to a mixed azobenzene[154, 155]. The formation of TCAB depends upon the concentration of DCA in soils[156]. The most common enzymes in microorganisms for catalysing the transformation of DCA to TCAB are peroxidases, which are known to effect the oxidation of substituted anilines with the formation of appreciable amounts of azobenzenes[157]. Bartha and co-workers have extensively investigated the mechanism of the transformation of DCA to TCAB and shown the involvement of peroxidase-catalysed oxidation[144, 158] and of labile intermediates[159] in the condensation reaction. The presence of a labile intermediate in the azo formation has been indirectly indicated by incubation of a mixture of enzymically reactive and unreactive anilines[159]. The incubation mixture gives rise to a mixed azobenzene, while the latter aniline alone remains unchanged in soils. In the absence of either peroxidase or hydrogen peroxide, no azobenzenes are produced[144, 160]. The peroxidase-catalysed conversion of DCA to TCAB appears to proceed by way of 3,4-dichlorophenylhydroxylamine (**37**), since the addition of trisodium pentacyanoaminoferroate gives rise to a complex with this intermediate, **37**[160]. Hydroxylamine formation is thought to involve an enzymic hydrogen abstraction from DCA followed by hydroxylation by the peroxidase–hydrogen peroxide complex. The resulting hydroxylamine is capable of undergoing non-enzymic reaction with DCA to form 3,3′,4,4′-tetrachlorohydrazobenzene (**38**). The intermediacy of **38**, although detection failed, can be rationalized by indirect evidence that it is very susceptible to autoxidation in phosphate buffer solutions and, when incubated in soil, rapidly gives rise to TCAB.

Moreover, generation of the 3,4-dichloroanilino radical $Cl_2C_6H_3\dot{N}H$ is postulated[160] on the basis of the result that the antioxidant 2-(*t*-butyl)-4-methylphenol when added undergoes the oxidative dimerization which is known to occur in the presence of a radical. The anilino radical may afford the hydroxylamine **37** and the hydrazobenzene **38** by reacting with hydroxyl radical and another anilino radical, respectively, but these paths are regarded as being less important.

Bordeleau and Bartha[145] have performed the *in vitro* biotransformation

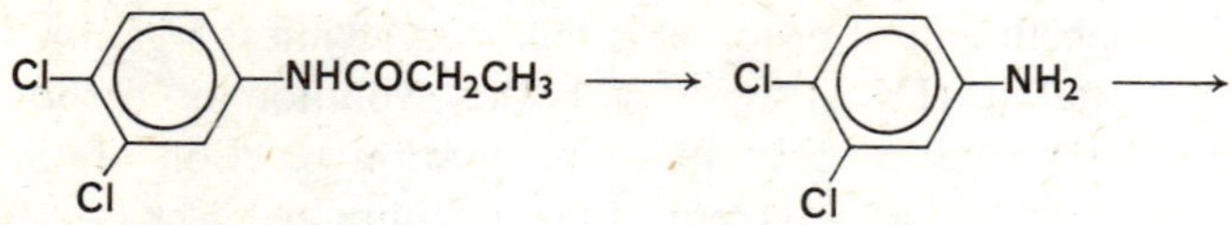

Propanil DCA

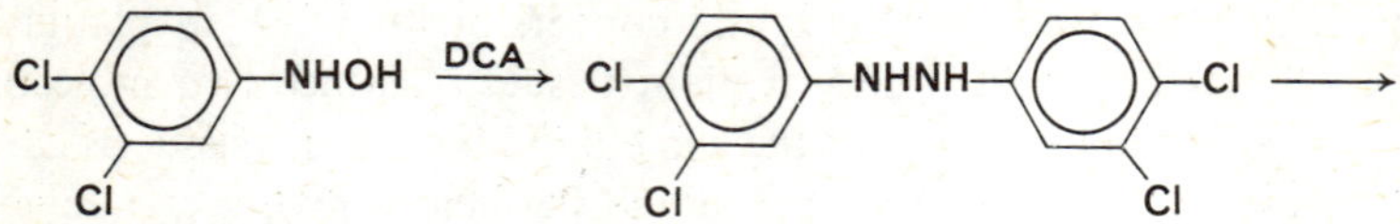

(37) (38)

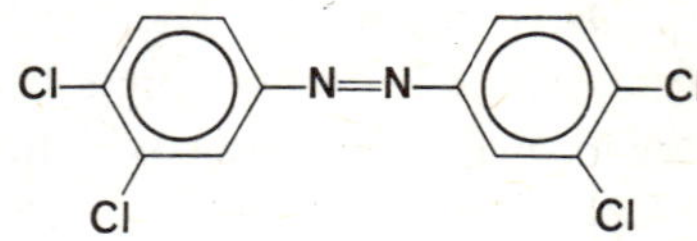

TCAB

of anilines, including aniline itself, to the corresponding azobenzenes and higher polymers using partially purified *G-candidum* L-3 peroxidase and aniline oxidase. Electron-releasing substituents enhance the reactivity of the aniline, favouring the formation of polymers but decreasing the yields of azobenzenes, the order of reactivity being: $NO_2 < F < Cl < Br < I < H < CH_3 < OCH_3$. The position of substituents also influences the reactivity of anilines and, for monosubstituted anilines with peroxidase, the order has been demonstrated to be *meta* < *ortho* < *para*.

In the metabolic reaction of propanil, two other azo derivatives have been isolated and characterized as 4-(3,4-dichloroanilino)-3,3′,4′-trichloroazobenzene (**39**)[161, 162] and 1,3-bis(3,4-dichlorophenyl)triazene (**40**)[150, 163]. The transformation of DCA to TCAB is also possible photochemically in the presence of riboflavin-5′-phosphate[164] or the photosensitizer benzophenone[165] giving rise to 4-(3,4-dichloroanilino)-3,3′,4′-trichloroazo-

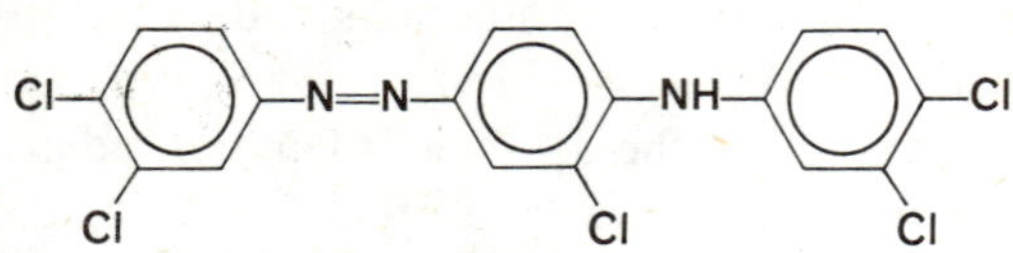

(39)

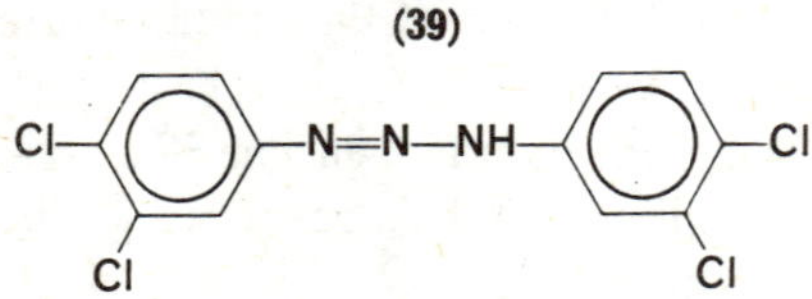

(40)

benzene and 3,3′,4,4′-tetrachloroazoxybenzene, respectively. A herbicide 4-nitrodiphenylether derivative is photochemically converted to the corresponding azo compound together with some other products[166].

VII. BIOLOGICAL NITROGEN FIXATION POSSIBLY INVOLVING DIIMIDE AND HYDRAZINE INTERMEDIATES

Nitrogen in food protein can be supplied from air nitrogen through biological nitrogen fixation, but modern agriculture has utilized inorganic fertilizers as the most important source of nitrogen. One of the starting materials, ammonia, has been catalytically manufactured from nitrogen and hydrogen by the Haber process at high temperatures and pressures. Biological nitrogen fixation permits the production of ammonia under much milder conditions by means of the enzyme nitrogenases. The nitrogenases from different sources are not identical, but commonly appear to be a complex of a Mo–Fe protein and a Fe protein which have been separated[167]. Neither purified Mo–Fe proteins nor Fe proteins alone show any enzymic activities, but the complex of the two shows catalysed nitrogen fixation. Since the first discovery of a N_2 complex of a transition metal[168] a variety of such complexes has been prepared for the purpose of elucidating the nitrogenase-catalysed nitrogen reduction. The model studies have succeeded in the facile reduction of molecular nitrogen to ammonia under mild conditions.

Nitrogenase can reduce some substrates such as acetylene, nitriles, isonitriles, azide and nitrous oxide in addition to nitrogen[167]. Acetylene, which is similar to nitrogen in size, readily undergoes two-electron reduction by nitrogenase, giving rise to ethylene.

Two mechanisms have been proposed for the biological nitrogen fixation and the related transition metal-catalysed reductions. Hardy and coworkers[167] have proposed the stepwise reduction mechanism for the biological nitrogen fixation involving enzyme-bound diimide and hydrazine as the reduction intermediates. Although the intermediates have not been detected, they have provided evidence supporting a metal-bound intermediate capable of leading to hydrazine[169]. Molecular nitrogen is thought

$$N_2 \longrightarrow \underset{\text{Diimide}}{HN{=}NH} \longrightarrow H_2N{-}NH_2 \longrightarrow 2\,NH_3$$

to undergo the stepwise reduction by the cooperation of Mo and Fe following the initial complex formation with Fe in nitrogenase[167]. A good model of biological nitrogen fixation by cooperation of Mo^{III} and Ti^{III}

has been reported by Shilov and co-workers[170]. The two-electron reduction of hydrazine to ammonia by a binuclear molybdenum(III) compound resembles the last step of the reduction pathway, supporting the feasibility of the stepwise mechanism[171]. Van Tamelen's group[172] has suggested an alternative mechanism involving the formation of a nitride intermediate on the basis of the experiments using titanium–nitrogen complexes, and they have obtained hydrazine by the transition metal-catalysed reduction of molecular nitrogen. Since the first production of hydrazine from molecular nitrogen, there have appeared a number of papers concerning the formation of hydrazine from different transition metal–N_2 complexes. Biogenetic reactions using metal–N_2 complexes will be extensively developed in order to elucidate the mechanism of biological nitrogen fixation.

VIII. INTRODUCTION OF PHENYLAZO GROUP INTO PROTEINS AND AMINO ACIDS

A. Chromophoric Labelling

Chemical modifications of a specific functional group in proteins provide a useful means for investigating the active sites of enzymes. A number of reagents have been devised for such specific chemical modification by which the thiol group can be modified most readily and selectively because of its highest reactivity. The azo group is a suitable chromophore to attach to a moiety carrying a sufficiently reactive group for specific chemical modification. This sort of modification may enable a quantitative determination of thiol groups in enzymes and proteins by spectrophotometric measurements in visible regions without interference by biological constituents. This section is not intended to review the chromophoric reagents, but to give only a few examples of recent reagents for labelling protein thiols.

1. 4-Phenylazo-2-chloromercuriphenol and 4-(*p*-nitrophenylazo)-2-chloromercuriphenol. A number of organic mercurials[173] have been prepared for probing the biological roles of thiols in biological constituents. The two azo reagents[174] which were devised as improvements of earlier azo mercurials react selectively and stoichiometrically with the thiols of

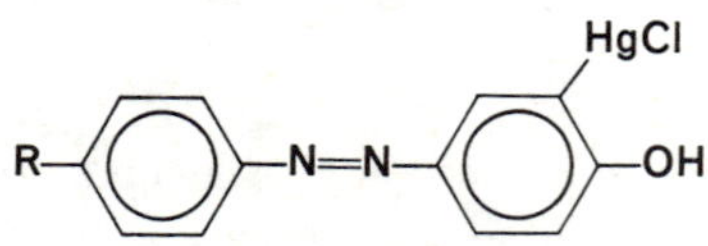

R = H, NO_2

cysteine residues in proteins. Approximately equivalent amounts of the reagents undergo reaction with one thiol group of ficin and inhibit the enzyme activity. Both reagents have advantages over the thiol-labelling reagent, *N*-(4-dimethylamino-3,5-dinitrophenyl)maleimide owing to their intense colouration especially in acids and bases and their higher solubility in aqueous ethanol (15%).

2. Arylsulphenyl halides. Arylsulphenyl halides are useful reagents for the formation of mixed disulphides of proteins and related compounds[175].

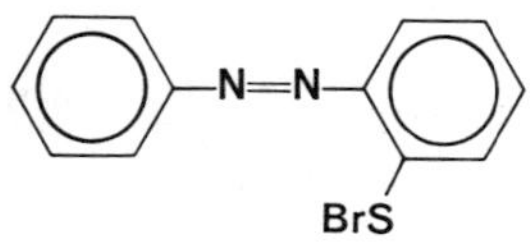

Studies on whether disulphide formation causes a biological change or not may permit the estimation of the biological function of protein thiols. The nitro derivatives, 2-nitro-, 4-nitro- and 2,4-dinitrophenylsulphenyl chloride undergo a selective reaction with tryptophan and cysteine residues in proteins[176]. The introduction of a phenylazo group increases the stability and water solubility[177] of the phenylsulphenylbromide reagent. The reagent is selective and reacts only with cysteinyl residues in proteins, but not with tryptophan residues.

3. 4-Phenylazodiphenylcarbamoyl chloride or fluoride. The reagent inactivates enzymes such as trypsin, chymotrypsin and acetylated trypsin,

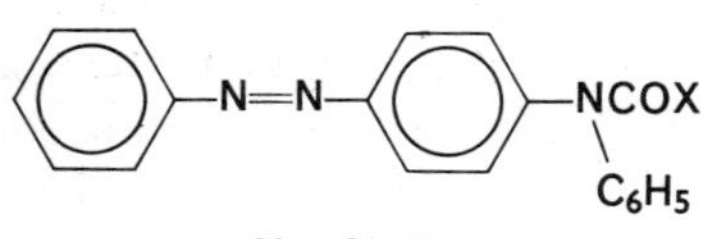

X = Cl, F

giving rise to yellow derivatives[178], and can be used for chromophoric labelling of these enzymes. It can be also used to determine the concentration of active enzymes; the inactivated trypsin can be reactivated.

B. *Immunochemical Reaction of Proteins with Diazotized Amines with the Formation of an Azo Linkage*

The reaction of aromatic diazonium ions with amino acids or proteins possessing phenyl and imidazole rings has long been known as a method of introducing an azo linkage. The coupling reactions are of use in immunochemistry, colorimetry for identification of tyrosine and histidine residues, and conformational studies of proteins.

The phenyl rings in the tyrosyl residues in proteins couple with diazotized aromatic amines such as arsanilic acid, sulphanilic acid or *p*-aminobenzoic acid[179]. Exclusive substitution by one azo group at the *o*-position can be achieved by the action of one equivalent of a diazotized amine. The second azo group can be introduced at the other *o*-position on treatment with two equivalents of the diazonium ion. Some time ago it was found that a protein

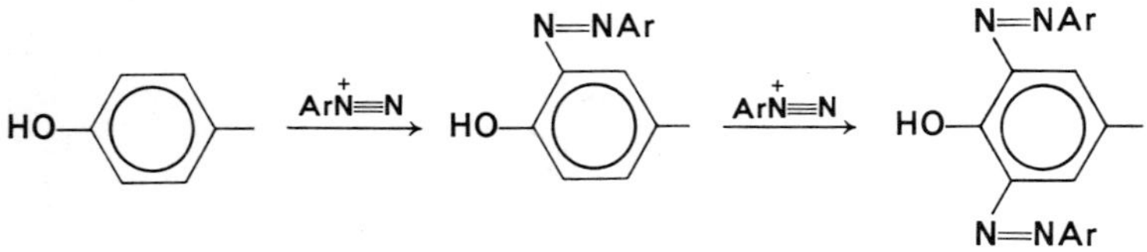

coupled with diazotized arsanilate was capable of producing antibodies on injection into rabbits[180]. Since then a number of proteins (synthetic and

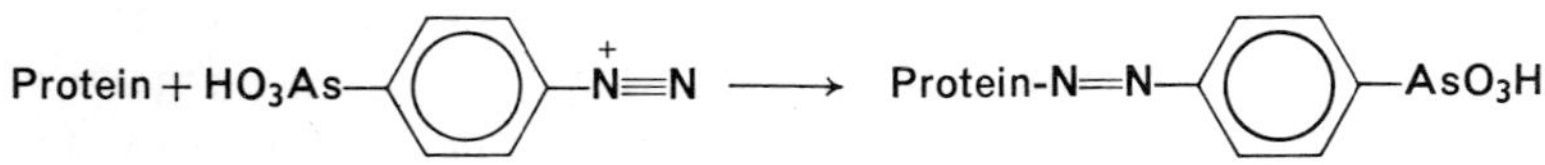

natural) have been submitted to coupling reactions with a variety of small molecules designated as haptens. The antibodies thus formed have specificity for the hapten. A small molecule, i.e. *p*-nitrobenzenearsonic acid, resembling the haptenic moiety inhibits the hapten–antibody complexation, suggesting the presence of sites with affinity for the grouping in the antibody. Singer and co-workers[181] have proposed and extensively studied the method of labelling the active sites of proteins. Aryldiazonium salts have been utilized to form irreversibly a covalent azo bond between the two aromatic rings of the reagent and the antibody for affinity labelling. Affinity labelling of antibodies is of use for investigating the nature of the complexing regions of antibodies. Thus, the presence of tyrosine residues in the antibody–hapten complexing regions has been demonstrated on the basis of an absorption maximum due to azo tyrosine.

IX. REFERENCES

1. J. N. Cooper, *Proc. Roy. Soc. New South Wales*, **74,** 450 (1940).
2. B. W. Langley, B. Lythgoe and N. V. Riggs, *Chem. Ind.*, 75 (1951); B. W. Langley, B. Lythgoe and N. V. Riggs, *J. Chem. Soc.*, 2309 (1951); B. W. Langley, B. Lythgoe and L. S. Rayner, *J. Chem. Soc.*, 4192 (1952).

3. K. Nishida, A. Kobayashi and T. Nagahama, *Bull. Agr. Chem. Soc. Japan*, **17,** 77 (1955).
4. N. V. Riggs, *Chem. Ind.*, 926 (1956).
5. M. G. Whiting, *Econ. Botany*, **17,** 270 (1963).
6. G. L. Laqueur, *Fed. Proc.*, **23,** 1386 (1964).
7. A. Kobayashi and H. Matsumoto, *Fed. Proc.*, **23,** 1354 (1964); *Arch. Biochem. Biophys.*, **110,** 373 (1965).
8. M. Spatz, E. G. McDaniel and G. L. Laqueur, *Proc. Soc. Exp. Biol. Med.*, **121,** 417 (1966); M. Spatz, D. W. E. Smith, E. G. McDaniel and G. L. Laqueur, *Proc. Soc. Exp. Biol. Med.*, **124,** 691 (1967).
9. Reviewed by M. Spatz, *Ann. N.Y. Acad. Sci.*, **163,** 848 (1969), and references cited therein.
10. A. Kobayashi, *Agr. Biol. Chem.*, **26,** 203 (1962).
11. H. Matsumoto, T. Nagahama and H. O. Larson, *Biochem. J.*, **95,** 13c (1965).
12a. J. Ehrlich, L. E. Andersen, G. L. Coffey, W. H. Feldman, M. W. Fischer, A. B. Hillegas, A. G. Karlson, M. P. Kundsen, J. K. Weston, A. S. Youmans and G. P. Youmans, *Antibiotics Chemotherap.*, **4,** 338 (1954).
12b. T. H. Haskell, A. Ryder and Q. R. Bartz, *Antibiotics Chemotherap.*, **4,** 141 (1954).
13. K. Okuma, G. Nakamura and S. Yamashita, *J. Antibiot., Ser. A*, **10,** 224 (1957).
14. C. L. Stevens, B. T. Gillis, J. C. French and T. H. Haskell, *J. Amer. Chem. Soc.*, **80,** 6088 (1958).
15. W. J. McGahren and M. P. Kunstmann, *J. Amer. Chem. Soc.*, **91,** 2808 (1969); **92,** 1587 (1970).
16. R. Schoental, *Nature*, **221,** 765 (1969).
17. A. G. Karlson, *Antibiotics Chemotherap.*, **12,** 446 (1962).
18. E. L. Wang, M. Hamada, Y. Okami and H. Umezawa, *J. Antibiot., Ser. A*, **19,** 216 (1966); H. Naganawa, T. Takita, K. Maeda and H. Umezawa, *J. Antibiot.*, **21,** 241 (1968).
19. M. Hamada, T. Takeuchi, S. Kondo, Y. Ikeda, H. Naganawa, K. Maeda, Y. Okami and H. Umezawa, *J. Antibiot.*, **23,** 170 (1970).
20. S. Mizuno, K. Nitta and H. Umezawa, *J. Antibiot.*, **23,** 581 (1970); **23,** 589 (1970); Y. Uehara, S. Kondo, H. Umezawa, K. Suzukake and M. Hori, *J. Antibiot.*, **25,** 685 (1972).
21. A. S. Yard, H. McKennis, Jr. and J. H. Weatherby, *J. Pharmacol. Exptl. Therap.*, **119,** 195 (1957).
22. J. N. Vivien and J. Grosset, *Advan. Tuberc. Res.*, **11,** 45 (1961); J. B. Hughes, *J. Pharmacol. Exptl. Therap.*, **109,** 444 (1953).
23. H. McKennis, A. S. Yard, J. H. Weatherby and J. A. Hagy, *J. Pharmacol. Exptl. Therap.*, **126,** 109 (1959) and references cited therein.
24. T. C. Chou and F. Lipmann, *J. Biol. Chem.*, **196,** 89 (1952). Metabolism of hydrazines and hydrazides is reviewed by L. B. Colvin, *J. Pharm. Sci.*, **58,** 1433 (1969); M. R. Juchau and A. Horita, *Drug Metab. Revs.*, **1,** 71 (1972).
25. P. Sisley and C. Porcher, *Compt. Rend.*, **152,** 1062 (1911).
26. J. Tréfouël, Mme. J. Tréfouël, F. Nitti and D. Bovet, *C.R. Soc. Biol. Paris*, **120,** 756 (1935).
27. A. T. Fuller, *Lancet*, **I,** 194 (1937).
28. Reviewed by R. Walker, *Food Cosmet. Toxicol.*, **8,** 659 (1970).

29. K. Brohm and E. Frohwein, *Z. Lebensmittelunters. u.-Forsch.*, **73,** 30 (1937).
30. B. Dieckhues, *Zentr. Bakteriol. Parasitenk.* (*Abt. I*), **180,** 244 (1960).
31. J. W. Daniel, *Toxicol. Appl. Pharmacol.*, **4,** 572 (1962).
32. J. L. Radonski and T. J. Mellinger, *J. Pharmacol.*, **136,** 259 (1962).
33. R. Jones, A. J. Ryan and S. E. Wright, *Food Cosmet. Toxicol.*, **2,** 447 (1964).
34a. A. J. Ryan, J. J. Roxon and A. Sivayavirojana, *Nature*, **219,** 854 (1968).
34b. R. Jones, A. J. Ryan and S. E. Wright, *Food Cosmet. Toxicol.*, **4,** 411 (1966).
34c. J. J. Roxon, A. J. Ryan and S. E. Wright, *Food Cosmet. Toxicol.*, **4,** 419 (1966).
35. R. R. Scheline and B. Longberg, *Acta Pharmacol. Toxicol.*, **23,** 1 (1965).
36a. A. J. Ryan and P. G. Welling, *Food Cosmet. Toxicol.*, **5,** 645 (1967).
36b. A. J. Ryan and P. G. Welling, *Food Cosmet. Toxicol.*, **5,** 755 (1967).
37a. L. A. Elson and F. L. Warren, *Biochem. J.*, **38,** 217 (1944).
37b. H. G. Bray, R. C. Clowes and W. V. Thorpe, *Biochem. J.*, **49,** lxv (1951).
37c. R. T. Williams in *The Metabolism and Detoxication of Drugs*, 2nd ed., Chapman and Hall, London, 1959, p. 478.
37d. J. A. Miller and E. C. Miller, in *Advances in Cancer Research* (Ed. J. P. Greenstein and A. Haddow), Vol. 1, Academic Press, New York, 1953, p. 339.
38. E. S. Stevenson, K. Dobriner and C. P. Rhoads, *Cancer Res.*, **2,** 160 (1942).
39. C. J. Kensler, *Ann. N.Y. Acad. Sci.*, **49,** Art. **1,** 29 (1947); C. J. Kensler, *Cancer*, **1,** 483 (1948); C. J. Kensler and W. C. Chu, *Arch. Biochem.*, **25,** 66 (1950); C. J. Kensler, *J. Biol. Chem.*, **179,** 1079 (1949).
40. G. C. Mueller and J. A. Miller, *J. Biol. Chem.*, **176,** 535 (1948); **180,** 1125 (1949); **185,** 145 (1950).
41. M. DuPlooy and J. Dijkstra, *Chem.-Biol. Interactions*, **4,** 163 (1972).
42. M. Watanabe and M. Ishidate, *Chem. Pharm. Bull.* (*Tokyo*), **15,** 1461 (1967).
43. J. J. Kamm, Ph.D. Dissertation. Georgetown University, Washington, D.C. (1963); P. H. Hernandez, J. R. Gillette and P. Mazel, *Biochem. Pharmacol.*, **16,** 1859 (1967); P. H. Hernandez, P. Mazel and J. R. Gillette, *Biochem. Pharmacol.*, **16,** 1877 (1967).
44. R. Gingell, J. Bridges and R. T. Williams, *Biochem. J.*, **114,** 5p (1969); R. Gingell and J. W. Bridges, *Biochem. J.*, **125,** 24p (1971).
45. M. R. Juchau, *Federation Proc.*, **31,** 48 (1972) and references cited therein.
46. I. J. Selikoff, E. H. Robitzek and G. G. Ornstein, *Quart. Bull. of Sea View Hosp.*, 17 (1952); *Am. Rev. Tuberc.*, **67,** 212 (1953).
47. E. A. Zeller, J. Barsky, J. R. Fouts, W. F. Kircheimer and L. S. Van Orden, *Experientia*, **8,** 349 (1952); E. A. Zeller and J. Barsky, *Proc. Soc. Exptl. Biol. Med.*, **81,** 459 (1952).
48. E. A. Zeller and J. R. Fouts, in *Annual Review of Pharmacology* (Ed. W. C. Cutting), Annual Reviews, Inc. (Palo Alto), p. 13, 1963 and references cited therein.
49. A. N. Davison, *Biochem. J.*, **67,** 316 (1957).
50. J. Barsky, W. L. Pacha, S. Sarker and E. A. Zeller, *J. Biol. Chem.*, **234,** 389 (1959).
51. B. A. Koechlin and V. Iliev, *Ann. N.Y. Acad. Sci.*, **80,** 864 (1959).
52. L. S. Seiden and J. Westley, *Federation Proc.*, **21,** 416 (1962).
53. M. A. Schwartz, *Proc. Soc. Exp. Biol., N.Y.*, **107,** 613 (1961).

54. M. A. Schwartz, *J. Pharmacol. Exptl. Therap.*, **130,** 157 (1960); M. A. Schwartz, *Proc. Soc. Exp. Biol., N.Y.*, **107,** 613 (1961); B. A. Koechlin, M. A. Schwartz and W. E. Oberhaensli, *J. Pharmacol. Exptl. Therap.*, **138,** 11 (1962).
55. T. Satoh and K. Moroi, *Analytica Chim. Acta*, **51,** 449 (1970).
56. T. Satoh, S. Yamamoto and K. Moroi, *Japan J. Pharmacol.*, **22,** 629 (1972).
57. J. A. Schneider, B. M. Bloom, C. Delahunt, K. F. Finger, M. Finkelstein, R. P. Rowe, R. Stebbins and A. Weissman, *J. Soc. Das Ciencias Med. DeLisboa, Suppl.*, **123,** 43 (1959).
58. M. Wenzel, *Naturwissenschaften*, **42,** 370 (1955).
59. A. S. Yard and H. McKennis, Jr., *J. Med. Pharm. Chem.*, **5,** 196 (1962); A. S. Yard and H. McKennis, Jr., *Federation Proc.*, **17,** 421 (1958).
60. N. V. Riggs, *Nature*, **207,** 632 (1965).
61. H. Matsumoto and H. Higa, *Biochem. J.*, **98,** 20c (1966); Y. Nagata and H. Matsumoto, *Proc. Soc. Exp. Biol. Med.*, **132,** 383 (1969).
62. P. N. Magee and E. Farber, *Biochem. J.*, **83,** 114 (1962); P. N. Magee and T. Hultin, *Biochem. J.*, **83,** 106 (1962); R. C. Shank and P. N. Magee, *Biochem. J.*, **100,** 35p (1966); P. N. Magee and J. M. Barnes, *Advan. Cancer Res.*, **10,** 163 (1967).
63. E. Kriek and P. Emmelot, *Biochem. Biophys. Acta*, **91,** 59 (1964).
64. W. Lijinsky, J. Loo and A. E. Ross, *Nature*, **218,** 1174 (1968).
65. M. H. Benn and P. Kazmaier, *Chem. Comm.*, 887 (1972).
66. M. S. Zedeck, S. S. Sternberg, R. W. Poynter and J. McGowan, *Cancer Res.*, **30,** 801 (1970).
67. P. B. Lundeen, G. S. Banks and R. W. Ruddon, *Biochem. Pharmacol.*, **20,** 2522 (1971).
68. H. Druckrey, R. Preussmann, F. Matzkies and S. Ivankovic, *Naturwissenschaften*, **53,** 557 (1966); B. Toth, *Cancer Res.*, **32,** 804 (1972) and references cited therein.
69. H. Druckrey, R. Preussmann, S. Ivankovic and A. V. Hodenberg, *Ann. N.Y. Acad. Sci.*, **163,** 697 (1969) and references cited therein.
70. H. Druckrey, R. Preussmann, F. Matzkies, S. Ivankovic, C. H. Schmidt, B. T. So and C. Thomas, *Z. Krebsforsch.*, **67,** 31 (1965).
71. H. Aebi, B. Dewald and H. Suter, *Helv. Chim. Acta*, **48,** 656 (1965).
72. H. Druckrey and A. Lange, *Federation Proc.*, **31,** 1482 (1972).
73. R. E. McMahon, *J. Pharm. Sci.*, **55,** 457 (1966).
74. E. M. Kosower and T. Tsuji, *J. Amer. Chem. Soc.*, **93,** 1992 (1971).
75. B. Akermark, *Acta Chem. Scand.*, **21,** 2909 (1967).
76. W. Bollag and E. Grünberg, *Experientia*, **19,** 130 (1963); W. Bollag, A. Kaiser, A. Langemann and P. Zeller, *Experientia*, **20,** 503 (1964).
77. D. W. E. Smith, *Science*, **152,** 1273 (1966).
78. M. G. Kelly, R. W. O'Gara, K. Gadekar, S. T. Yancey and V. T. Oliverio, *Cancer Chemother. Rep.*, **39,** 77 (1964); J. C. Heuson and R. Heimann, *European J. Cancer*, **2,** 385 (1966).
79. G. Weitzel, F. Schneider, D. Kummer and H. Ochs, *Z. Krebsforsch.*, **70,** 354 (1968).
80. G. Weitzel, F. Schneider, W. Herschmann, W. Durst, J. Thauer, H. Ochs and D. Kummer, *Hoppe-Seyler's Z. Physiol. Chem.*, **348,** 443 (1967).
81. H. Koblet and H. Diggelmann, *Eur. J. Cancer*, **4,** 55 (1968).

82. F. N. Dost and D. J. Reed, *Biochem. Pharmacol.*, **16,** 1741 (1967).
83. J. Raaflaub and D. E. Schwartz, *Experientia*, **21,** 44 (1965).
84. M. G. Kelly, R. W. O'Gara, S. T. Yancy, K. Gadekar, C. Botkin and V. T. Oliverio, *J. Natl. Cancer Inst.*, **42,** 337 (1969).
85. A. J. Bellamy and R. D. Guthrie, *J. Chem. Soc.*, 3528 (1965).
86. K. Berneis, M. Kofler, W. Bollag, P. Zeller, A. Kaiser and A. Langemann, *Helv. Chim. Acta*, **46,** 2157 (1963).
87. V. T. Oliverio, C. Denham, V. T. Vita and M. G. Kelly, *Cancer Chemother. Rep.*, **42,** 1 (1964).
88. M. Baggiolini, B. Dewald and H. Aebi, *Biochem. Pharmacol.*, **18,** 2187 (1969).
89. J. Roaflaub and D. E. Schwartz, *Experientia*, **21,** 44 (1965).
90. W. Kreis, S. B. Piepho and H. V. Bernhard, *Experientia*, **22,** 431 (1966); W. Kreis, J. H. Burchenal and D. J. Hutchison, *Proc. Am. Assoc. Cancer Res.*, **9,** 38 (1966); W. Kreis, *Cancer Res.*, **30,** 82 (1970).
91. O. Dimroth, *Ber.*, **36,** 909 (1903).
92. H. Druckrey, R. Pressman and S. Ivankovic, *Ann. N.Y. Acad. Sci.*, **163,** 676 (1969).
93. J. L. Skibba, R. O. Johnson and G. T. Bryan, *Proc. Am. Assoc. Cancer Res.*, **11,** 73 (1970).
94. T. Yoshida, *Nippon Byori Gakkai Kaishi*, **23,** 636 (1933).
95. R. Kinoshita, *Gann*, **30,** 423 (1936).
96. G. C. Mueller and J. A. Miller, *J. Biol. Chem.*, **202,** 579 (1953); M. Matsumoto and H. Terayama, *Gann*, **52,** 239 (1961).
97. A. H. Conney, R. R. Brown, J. A. Miller and E. C. Miller, *Cancer Res.*, **17,** 628 (1957); T. Kurihara and H. Terayama, *Seikagaku*, **36,** 198 (1964).
98. J. A. Miller and E. C. Miller, *J. Exptl. Med.*, **87,** 139 (1948).
99. J. A. Miller and E. C. Miller, *Progr. Exp. Tumor Res.*, **11,** 273 (1969).
100. J. A. Miller, K. Sato, L. A. Poirier and E. C. Miller, *Proc. Am. Assoc. Cancer Res.*, **5,** 45 (1964).
101. E. C. Miller and C. A. Baumann, *Cancer Res.*, **6,** 289 (1946); E. C. Miller, J. A. Miller, R. W. Sapp and G. M. Weber, *Cancer Res.*, **9,** 336 (1949).
102. H. Terayama, in *Methods in Cancer Research*, Vol. 1, Academic Press, New York, London, 1967, pp. 399–449 and references cited therein.
103. J. D. Scribner, J. A. Miller and E. C. Miller, *Biochem. Biophys. Res. Commun.*, **20,** 560 (1965) and references cited therein.
104. P. D. Lotliker, J. D. Scribner, J. A. Miller and E. C. Miller, *Life Sciences*, **5,** 1263 (1966); L. A. Poirier, K. Sato, E. C. Miller and J. A. Miller, *Proc. Am. Assoc. Cancer Res.*, **7,** 56 (1966); L. A. Poirier, J. A. Miller, E. C. Miller and K. Sato, *Cancer Res.*, **27,** 1600 (1967); E. C. Miller and J. A. Miller, *Ann. N.Y. Acad. Sci.*, **163,** 731 (1969).
105. J.-K. Lin, J. A. Miller and E. C. Miller, *Biochemistry*, **7,** 1889 (1968); **8,** 1573 (1969).
106. S. Sorof, V. M. Kish and B. Sani, *Biochem. Biophys. Res. Commun.*, **48,** 860 (1972).
107. J. J. Roberts and G. P. Warwick, *Biochem. J.*, **93,** 18p (1964); J. J. Roberts and G. P. Warwick, *Nature*, **197,** 87 (1963); A. E. Albert and G. P. Warwick, *Chem.-Biol. Interactions*, **5,** 61 (1972) and references cited therein.
108. M. Borenborm, *Cancer Res.*, **19,** 1045 (1959).

109. E. M. Kosower and N. S. Kosower, *Nature*, **224,** 117 (1969).
110. T. Mukaiyama and K. Takahashi, *Tetrahedron Letters*, 5907 (1968).
111. G. W. Stacy, R. I. Day and R. J. Morath, *J. Amer. Chem. Soc.*, **77,** 3869 (1955).
112. T. R. Oakes and G. W. Stacy, *J. Amer. Chem. Soc.*, **94,** 1594 (1972).
113. N. S. Kosower, K-R. Song and E. M. Kosower, *Biochem. Biophys. Acta*, **192,** 1 (1969); N. S. Kosower, G. A. Vanderhoff, E. M. Kosower, P-K. C. Huang, *Biochem. Biophys. Res. Commun.*, **20,** 469 (1965).
114. N. S. Kosower, K-R. Song and E. M. Kosower, *Biochem. Biophys. Acta*, **192,** 15 (1969).
115. N. S. Kosower, K-R. Song, E. M. Kosower and W. Correa, *Biochem. Biophys. Acta*, **192,** 8 (1969).
116. P.-K. C. Huang and E. M. Kosower, *J. Amer. Chem. Soc.*, **90,** 2354 (1968); **90,** 2362 (1968); **90,** 2367 (1968); P.-K. C. Huang and E. M. Kosower, *Biochem. Biophys. Acta*, **165,** 483 (1968).
117. N. S. Kosower, E. M. Kosower, B. Wertheim and W. Correa, *Biochem. Biophys. Res. Commun.*, **37,** 593 (1969).
118. R. B. Roberts, P. H. Abelson, D. B. Cowie, E. T. Bolton and R. J. Butler, *Carnegie Inst. Wash. Publ.*, **607,** 318 (1955).
119. R. Wax, E. Rosenberg, N. S. Kosower and E. M. Kosower, *J. Bacteriol.*, **101,** 1092 (1970).
120. M. J. Rose, N. K. Enkiri and W. L. Sulzbacher, *Appl. Microbiol.*, **22,** 1141 (1971).
121. N. C. Brown and R. E. Handschumacher, *J. Biol. Chem.*, **241,** 3083 (1966); N. C. Brown, *Proc. U.S. Nat. Acad. Sci.*, **67,** 1454 (1970); N. C. Brown, C. L. Wisseman and T. Matsushita, *Nature New Biology*, **237,** 72 (1972).
122. T. Zsolnai, *Biochem. Pharmacol.*, **14,** 1325 (1965).
123. T. Zehavi-Willner, N. S. Kosower, T. Hunt and E. M. Kosower, *Biochem. Biophys. Res. Commun.*, **40,** 37 (1970); T. Zehavi-Willner, E. M. Kosower, T. Hunt and N. S. Kosower, *Biochem. Biophys. Acta*, **228,** 245 (1971).
124. N. S. Kosower, G. A. Vanderhoff, B. Bonerofe, T. Hunt and E. M. Kosower, *Biochem. Biophys. Res. Commun.*, **45,** 816 (1971).
125. R. Galun, E. M. Kosower and N. S. Kosower, *Nature*, **224,** 181 (1969).
126. R. Galun and S. H. Kindler, *J. Insect. Physiol.*, **14,** 1409 (1968).
127. H. McIlwain, *Biochemistry and the Central Nervous System*, J. and A. Churchill, London, 1966, p. 412.
128. H. M. Lenhoff, E. M. Kosower and N. S. Kosower, *Nature*, **224,** 717 (1969).
129. H. P. Burchfield and E. E. Storrs, in *Fungicides* (Ed. D. C. Torgeson), Vol. 2, Chap. 2, Academic Press, New York and London, 1969, pp. 61–100.
130. R. J. Lukens, *Chemistry of Antifungal Action*, Springer-Verlag, Berlin, 1971.
131. D. Weinman, E. Galun, N. S. Kosower and E. M. Kosower, *Experientia*, **26,** 40 (1970).
132. D. V. Richmond and E. Somers, *Biochem. Biophys. Res. Commun.*, **25,** 657 (1966).
133. D. V. Richmond and E. Somers, *Ann. Appl. Biol.*, **57,** 231 (1966).
134. K. J. Bent and A. G. Morton, *Biochem. J.*, **92,** 260 (1964), references cited therein concerning the presence of GSH.
135. H. Stern, in *Moderne Methoden der Pflanzenanalyse*, Vol. 6, Springer-Verlag, Berlin, 1963, p. 21.

136. S. P. Spragg, P. M. Lievesley and K. M. Wilson, *Biochem. J.*, **83,** 314 (1962).
137. D. Mazia, in *Sulfur in Proteins* (Ed. R. Benesch), Academic Press, New York, 1959, p. 367; H. Stern, in *Sulfur in Proteins* (Ed. R. Benesch), Academic Press, New York, 1959, p. 391.
138. E. M. Kosower and T. Miyadera, *J. Med. Chem.*, **15,** 307 (1972).
139. T. Miyadera and E. M. Kosower, *J. Med. Chem.*, **15,** 339 (1972).
140. C. W. Pluijgers, J. Berg, A. K. Sijpesteijn, A. Tempel and A. Verloop, *Rec. Trav. Chim. Pays-Bas*, **87,** 833 (1968); A. K. Sijpesteijn, C. W. Pluijgers, A. Verloop and A. Tempel, *Ann. Appl. Biol.*, **61,** 473 (1968).
141. F. Haverkate, A. Tempel and C. W. Pluijgers, *Rec. Trav. Chim.*, **90,** 641 (1971).
142. J. G. Horsfall and R. J. Lukens, *Conn. Agr. Expt. Sta., New Haven, Bull.*, **694,** 1 (1968).
143. R. Bartha and D. Pramer, *Science*, **156,** 1617 (1967).
144. R. Bartha, H. A. B. Linke and D. Pramer, *Science*, **161,** 582 (1968).
145. L. M. Bordeleau and R. Bartha, *Can. J. Microbiol.*, **18,** 1873 (1972).
146. M. Matsumoto and H. Terayama, *Gann.*, **56,** 339 (1965).
147. R. Bartha and D. Pramer, *Advan. Appl. Microbiol.*, **13,** 317 (1970).
148. P. C. Kearney, *J. Agr. Food Chem.*, **13,** 561 (1965).
149. R. Bartha, *J. Agr. Food Chem.*, **16,** 602 (1968).
150. H. Chisaka and P. C. Kearney, *J. Agr. Food Chem.*, **18,** 854 (1970).
151. R. Bartha, H. A. B. Linke and D. Pramer, *Umschau*, **69,** 182 (1969); *Chem. Abstr.*, **71,** 2407c (1969).
152. M. Adachi, K. Tonegawa and T. Ueshima, *Noyaku Seisan Gijutsu*, **14,** 19 (1966); D. S. Frear and G. G. Still, *Phytochemistry*, **7,** 913 (1968).
153. H. B. Lieb and C. C. Still, *Plant Physiol.*, **44,** 1672 (1969); G. G. Still, *Weed Res.*, **9,** 211 (1969).
154. P. C. Kearney, J. R. Plimmer and F. B. Guardia, *J. Agr. Food Chem.*, **17,** 1418 (1969).
155. L. M. Bordeleau, H. A. B. Linke and R. Bartha, *Bacteriol. Proc.*, 4 (1969); R. Bartha, *Science*, **166,** 1299 (1969).
156. P. C. Kearney and J. R. Plimmer, *J. Agr. Food Chem.*, **20,** 584 (1972).
157. D. G. H. Daniels and B. C. Saunders, *J. Chem. Soc.*, 822 (1953).
158. R. Bartha and L. M. Bordeleau, *Soil Biol. Biochem.*, **1,** 139 (1969); *Chem. Abstr.*, **71,** 90256f (1969).
159. L. M. Bordeleau and R. Bartha, *Bull. Environ. Contam. Toxicol.*, **5,** 34 (1970).
160. L. M. Bordeleau, J. D. Rosen and R. Bartha, *J. Agr. Food Chem.*, **20,** 573 (1972).
161. H. A. B. Linke and R. Bartha, *Bacterial Proc.*, **70,** 9 (1970).
162. H. A. B. Linke, *Naturwissenschaften*, **57,** 307 (1970).
163. J. R. Plimmer, P. C. Kearney, H. Chisaka, J. B. Yount and U. I. Klingebiel, *J. Agr. Food Chem.*, **18,** 859 (1970).
164. J. D. Rosen and M. J. Siewierski, *J. Agr. Food Chem.*, **19,** 50 (1971).
165. J. R. Plimmer and P. C. Kearney, 158*th A.C.S. Meeting*, New York, September, 1969, Pest-29.
166. D. G. Crosby and M. Nakagawa, 162*nd A.C.S. Meeting*, Washington, D.C., September, 1971, Pest-30.

167. R. W. F. Hardy, R. C. Burns and G. W. Parshall, in *Bioinorganic Chemistry* (Ed. R. E. Gould), American Chemical Society, Washington, D.C., 1971, p. 219 and references cited therein.
168. A. D. Allen and C. V. Senoff, *Chem. Comm.*, 621 (1966).
169. E. K. Jackson, G. W. Parshall and R. W. F. Hardy, *J. Biol. Chem.*, **243,** 4952 (1968).
170. A. Shilov, N. Shuvalov, N. Denisov, N. Shuvalova, O. Efimov and A. Shilova, *Nature*, **231,** 460 (1971).
171. P. C. H. Mitchell and R. D. Scarle, *Nature*, **240,** 417 (1972).
172. E. E. Van Tamelen, *Accounts Chem. Res.*, **3,** 361 (1970); E. E. Van Tamelen, in *Bioinorganic Chemistry* (Ed. R. E. Gould), American Chemical Society, Washington, D.C., 1971, p. 95.
173. H. S. Bennett and R. M. Watts, in *General Cytochemical Methods* (Ed. J. F. Danielli), Vol. 1, Academic Press, New York, 1958, p. 317 and references cited therein.
174. S. F. Chang and I. E. Liener, *Nature*, **203,** 1065 (1964).
175. A. N. Glazer, in *Annual Review of Biochemistry* (Ed. E. E. Snell), Vol. 39, Annual Reviews, Inc., Palo Alto, 1970, p. 105 and references cited therein.
176. A. Fontana, F. M. Veronese and E. Scoffone, *Biochemistry*, **7,** 3901 (1968).
177. A. Burawoy, F. Liversedge and C. E. Vellins, *J. Chem. Soc.*, 4481 (1954).
178. H. Kaufman, S. M. Vratsanos and B. F. Erlanger, *Abstr. Papers, Am. Chem. Soc.*, No. 152, 174c (1966); September, New York.
179. M. Tabachinick and H. Sobatka, *J. Biol. Chem.*, **234,** 1726 (1959) and references cited therein.
180. K. Landsteiner, in *The Specificity of Serological Reactions*, 2nd ed., Harvard University Press, Cambridge, Mass., 1945.
181. L. Wofsy, H. Metzger and S. J. Singer, *Biochemistry*, **1,** 1031 (1962); **2,** 979 (1963); H. Metzger, L. Wofsy and S. J. Singer, *Proc. Natl. Acad. Sci. U.S.*, **51,** 612 (1964); S. J. Singer, L. Wofsy and A. H. Good in *Molecular and Cellular Basis of Antibody Formation* (Ed. J. Sterzl), Czech. Acad. Sci., Prague, 1965, p. 135.

CHAPTER **14**

Oxidation and synthetic uses of hydrazo, azo and azoxy compounds

BRIAN T. NEWBOLD

Département de Chimie, Université de Moncton, Moncton, New Brunswick, Canada

I. INTRODUCTION

The oxidation reactions of both aliphatic and aromatic hydrazo and azo compounds have attracted considerable interest in recent years apparently because of their utility for the synthesis of azo and azoxy derivatives, which are useful starting materials for a number of reactions. The literature up to about 1960 concerning the above oxidations has been reviewed in detail by Zollinger[1]. A number of pertinent general treatises are also available[2–10]. Recently, Sandler and Karo[11] have critically reviewed the available methods of synthesis, including oxidative routes, for both aliphatic and aromatic azo and azoxy compounds. Other pertinent reviews relating to more specific aspects are mentioned in the appropriate sections in this chapter. The oxidation reactions of azo dyes have not been included here since they have already been reviewed elsewhere[1, 12].

An emphasis has been placed on reaction conditions because of their importance, particularly with regard to oxidations of the rather sensitive aliphatic hydrazo and azo compounds; and in view of the synthesis aspect product yields have been frequently cited. Information on the separation, purification, and determination of structure of azo and azoxy products has also been given occasionally since there have been some important recent advances in techniques available for these purposes. Studies on mechanisms and the effects of substituents on the course of reaction have been given particular attention wherever possible since relatively little has been reported on these aspects in the earlier literature.

Unfortunately, most investigations published to date on oxidation reactions leading to the formation of azo or azoxy derivatives have not indicated the geometric configurations of the products obtained, however in this chapter, where possible, information in this regard has been included, either by naming the substances involved or by showing their structures. In addition, the structures of the positional isomers of unsymmetrically substituted azoxy compounds have been frequently shown.

In general, the IUPAC system of nomenclature has been employed, especially for unsymmetrical derivatives, although in some cases other names cited in the original references have been used. The reader is referred to useful summaries of the naming systems currently in use for azo and azoxy compounds that have appeared recently[11].

II. OXIDATION OF ALIPHATIC HYDRAZINES

Since relatively few preparative routes to aliphatic azo compounds are available, oxidation (dehydrogenation) of hydrazines is an important method for the synthesis of these substances. The oxidation reactions of hydrazines have led to a wide variety of products ranging from simple azoalkanes to complex cyclic azo derivatives. In general, aliphatic hydrazines are quite unstable to oxidation and hence readily converted to azo compounds by numerous reagents, such as halogens, hypohalites, peroxides, mercuric oxide, molecular oxygen or air, nitric acid, and *N*-bromosuccinimide. The parent substance, hydrazine, is easily oxidized to unstable di-imide (HN=NH), which undergoes disproportionation to hydrazine and nitrogen.

Aliphatic azo compounds are good sources of free radicals and have therefore proved popular as starting materials for studies on the latter, and this has stimulated interest in their synthesis via oxidation of hydrazines. However, some of these azo derivatives are unstable, being sensitive to decomposition, such that their preparation has necessitated precautions and careful control of reaction conditions. In some cases attempts to synthesize azo derivatives from hydrazines have failed due to inappropriate conditions being used and/or the instability of the desired product. The importance of reaction conditions, and the need for control of the amount of oxidant and for chromatographic analysis of the product, with regard to oxidations of hydrazines have been underlined in a recent review[11].

Some aspects of the oxidation reactions of hydrazines with periodates and lead tetra-acetate; and the electrochemical oxidation of dialkylhydrazines, have recently been briefly reviewed elsewhere[13–15].

A. Symmetrically Substituted Hydrazines to Azo Derivatives

Symmetrical aliphatic azo compounds have been prepared from the corresponding *N,N'*-disubstituted hydrazines using various oxidizing agents and especially mercuric oxide. Cohen and Wang[16] have oxidized *N,N'*-bis-benzhydrylhydrazine (**1**) to azo-bis-(diphenylmethane) (**2**) in 76% yield with 30% hydrogen peroxide in ethanol; and showed that a previously reported failure to convert **1** to **2** with the mild oxidant mercuric oxide was due to the use of too high a temperature (equation 1). They oxidized **1** to **2** with mercuric oxide in benzene suspension at 10–15°C and

$$\underset{(\mathbf{1})}{Ph_2CH{-}NH{-}NH{-}CHPh_2} \xrightarrow[\text{or HgO}]{H_2O_2} \underset{(\mathbf{2})}{Ph_2CH{-}N{=}N{-}CHPh_2} \qquad (1)$$

thus increased the yield of **2** to 90%. However, they also cited occasional difficulties with these preparations, rapid recrystallization at low temperature being necessary to obtain a high yield of pure product. Compound **2** decomposed on melting to give nitrogen and 1,1,2,2-tetraphenylethane.

Overberger and Gainer[17] have prepared neophyl-type azo compounds of general formula (**3**), where R is —H, —OMe, or —NHCOMe, by direct oxidation of the corresponding non-isolated hydrazo derivatives (made by reduction of appropriate azines) with 30% hydrogen peroxide in buffered

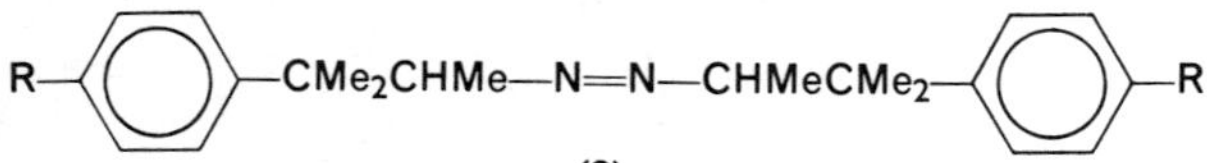

(**3**)

media. Seltzer and Hamilton[18,19] used yellow mercuric oxide to obtain azobis-α-phenylethane (**4**) and a hexadeutero derivative of **4** from the

$$\underset{(\mathbf{4})}{PhMeCH{-}N{=}N{-}CHMePh}$$

corresponding hydrazines. An instance of an unsuccessful oxidation with mercuric oxide was reported by Gerson and co-workers[20] who found that a dibenzylhydrazine derivative on treatment with the above reagent was converted to a 1,2-diphenylethane instead of the appropriate azoalkane.

Simple azoalkanes, e.g. azopropane, azo-n-butane, and azocyclohexane have been synthesized by oxidizing the hydrazines with sodium hypochlorite[21], and azomethane was obtained via oxidation with fuming nitric acid[22].

Mercuric oxide seems to have become a preferred reagent for oxidation of aliphatic hydrazines since it often leads to good yields of azoalkanes, but it is not always satisfactory for this purpose. The literature cites the use of both yellow and red mercuric oxide in this connection, the former variety being more frequently employed; however in some papers the species of mercuric oxide used has not been indicated. Both the *meso* and *dl* forms of 1,1′,2,2′-tetraphenylazoethane (**5**) have been made using yellow mercuric oxide[23]. Mercuric oxide oxidation of the symmetrical di-isopropylhydrazine

$$\underset{(\mathbf{5})}{PhCH_2CHPh{-}N{=}N{-}CHPhCH_2Ph}$$

(**6**) affords *cis*-azoisopropane in 4% and *trans*-azoisopropane (**7**) in 96% yield; the *cis*-isomer being isolable by distillations from large scale prepara-

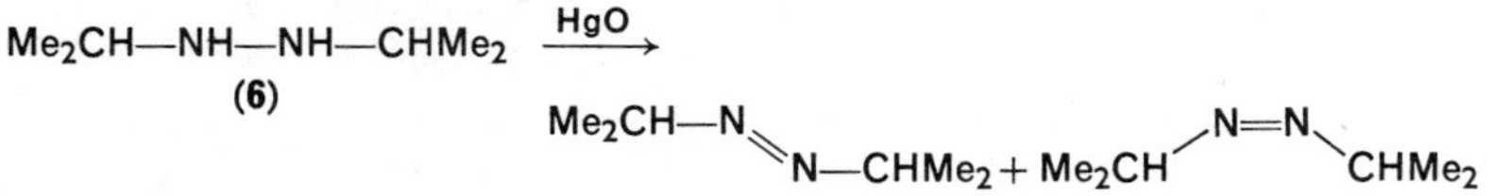

(6) (7) (2)

tions[24], (equation 2). On the other hand, Gibian and Corley[25] recently reported that they had encountered considerable difficulty in reproducing literature methods (including bromine water and yellow mercuric oxide) for oxidation of hydrazines to azo compounds. However, they found that symmetrical azo derivatives of general formula $R^1C_6H_4CHR$—N=N—$CHRC_6H_4R^1$ where R is —Me, i-Bu, i-Am, or —CH_2Ph and R^1 is —H, —Me, —OMe or —Cl, could be obtained in fair to good yield (43–82%) by air oxidation in the dark, even when using crude hydrazine oils. Shelton and Liang[26, 27] have prepared a series of symmetrically disubstituted 1,1′-diphenyl-1,1′-azoethanes (RC_6H_4CHMe—N=N—$CHMeC_6H_4R$), where R is for example —H, *o*-OMe, *m*-Me, *m*-Cl, *m*-CF_3, *p*-Et, *p*-Cl or *p*-OMe, by oxidation of the corresponding hydrazines with freshly made mercuric oxide.

Many symmetrical aliphatic azo compounds having other functional groups besides the —N=N— linkage have also been synthesized by oxidation of the corresponding hydrazines using suitable reagents. For example, α-carbonylazo derivatives[28] may be obtained from hydrazines by treatment with nitric acid[29], and halogens[30]. Thus, exposure of diethyl hydrazodicarboxylate (**8**) to yellow fuming nitric acid at 0–5°C affords diethyl

$$\underset{(\mathbf{8})}{\text{EtOOC—NH—NH—COOEt}} \xrightarrow{HNO_3} \underset{(\mathbf{9})}{\text{EtOOC—N=N—COOEt}} \qquad (3)$$

azodicarboxylate (**9**) in 80% yield[21], (equation 3). Compound **9** was previously prepared in lower yield from **8** via oxidation with bromine water. *N,N′*-Dibenzoylhydrazine is oxidized by chlorine in alkali to give azodibenzoyl in 40% yield[21].

Acylazo compounds may also be synthesized by the action of mercuric chloride or silver nitrate on the appropriate diacylhydrazines to give the metal derivatives, followed by addition of iodine or bromine[21]. For instance,

$$\underset{(\mathbf{10})}{Ph_3CCH_2CO\text{—N=N—}COCH_2CPh_3}$$

mercury *N,N'*-di(3,3,3-triphenylpropionyl)hydrazine is converted to di-(3,3,3-triphenylpropionyl) di-imide (**10**) in 78% yield by addition of iodine at 25°C. Campbell and co-workers[31] have recently obtained symmetrical azodiacyl compounds, such as azodi-*p*-bromobenzoyl and azodi-*p*-nitrobenzoyl, by oxidation of the mercury(II) salts of the dicarbonylhydrazines with bromine in appropriate solvents.

N-Bromosuccinimide has proved to be a most useful reagent for the oxidation of aliphatic hydrazines, having been applied with success to a wide range of substrates, including α-carbonylhydrazo derivatives. For example, t-butyl azodiformate (**11**) was obtained in 90% yield from t-butyl hydrazodiformate (**12**) by the action of this reagent in pyridine[32] (equation 4). Bock and co-workers[33, 34] have used *N*-bromosuccinimide to prepare numerous derivatives of azodicarboxylic acid from hydrazines. Azodibenzoyl, azodi-*p*-chlorobenzoyl, and diphenyl azodicarboxylate (**13**) were obtained in this manner, (equation 5). These workers also found that this reagent could be advantageously applied to easily solvolysed hydrazines, such as $R_2OP{-}NH{-}NH{-}POR_2$, where R is —Ph, —OPh, or $-NMe_2$, to obtain the corresponding azobisphosphonic acid derivatives. *N*-Bromosuccinimide has been shown to be suitable for the selective oxidation of hydrazines in inert solvents, the formally +1-valent bromine atom being reduced to the element or, in the presence of pyridine, to hydrogen bromide[33].

$$\underset{\textbf{(12)}}{Me_3COOC{-}NH{-}NH{-}COOCMe_3} \xrightarrow{NBS} \underset{\textbf{(11)}}{Me_3COOC{-}N{=}N{-}COOCMe_3} \quad (4)$$

$$PhOOC{-}NH{-}NH{-}COOPh \xrightarrow{NBS} \underset{\textbf{(13)}}{PhOOC{-}N{=}N{-}COOPh} \quad (5)$$

Recently, the simplest aliphatic diacylazo compound, azodiacetyl, which had been previously reported but not adequately characterized, was obtained in modest yield (30–35%) by stirring equimolar amounts of the hydrazine, *N*-bromosuccinimide, and quinoline in carbon tetrachloride at room temperature[31]. Azodipivaloyl (**14**) was similarly prepared in 50% yield using *N*-bromosuccinimide and pyridine.

$$\underset{\textbf{(14)}}{Me_3CCO{-}N{=}N{-}COCMe_3}$$

Symmetrical azonitriles, such as 1,1′-azobis-1-cyclohexanenitrile (**15**) and 2,2′-azobis-isobutyronitrile (**16**) have been made by treating appropriate hydrazines with halogens[35]. Thus, **15** was produced in 90% yield by the action of bromine on the hydrazine in ethanol/HCl at room temperature

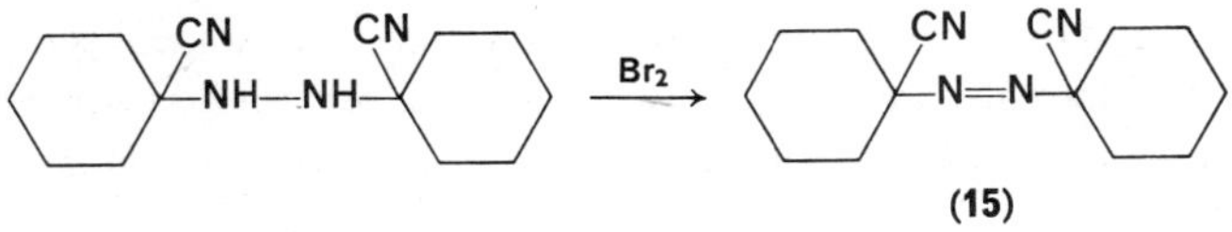

(6)

(equation 6). Compound **16** was similarly prepared and in the same yield[35].

Symmetrical aliphatic azo compounds have also been synthesized from hydrazines via oxidation with lead tetra-acetate, but in some cases, particularly with regard to *N,N′*-dicarbonylazo derivatives, the products are thermally unstable[14]. On the other hand, monoacylazo compounds are more stable and isolable by this method[14]. However, very recently Fantazier and Herweh[36] prepared *N,N,N′,N′*-tetramethylazobisformamide (**17**) by oxidation of the hydrazobisformamide at carefully controlled temperature (20 ± 2°C) with lead tetra-acetate suspended in dichloromethane.

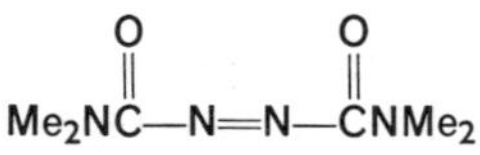

(17)

An example of a symmetrical unsaturated aliphatic azo compound recently obtained by oxidation of the hydrazo precursor is azodiallyl, $CH_2{=}CHCH_2{-}N{=}N{-}CH_2CH{=}CH_2$; the oxidant in this case being mercuric oxide[37].

B. *Unsymmetrically Substituted Hydrazines to Azo Compounds*

In recent years there has been considerable interest in studies on the thermal, photolytic and radiolytic decomposition of unsymmetrical aliphatic azo compounds as convenient sources of free radicals, which in turn has stimulated work on the oxidation of hydrazines for the synthesis of the necessary starting materials. Investigations of the oxidation of 1,2-dialkylhydrazines as a route to unsymmetrical azo-alkanes have been carried out by Spialter and colleagues[38] using oxidants such as potassium chromate, copper(II) salts, hydrogen peroxide, iodine, lead oxides (PbO, Pb_3O_4 and PbO_2) and silver oxide. However, they found that these reagents were not satisfactory, for instance the lead oxides were not powerful enough and silver oxide caused catalytic decomposition with evolution of nitrogen. They finally selected mercuric oxide as the preferred oxidant and used it in aqueous suspension at room temperature to prepare a series of azo compounds of general formula R—N=N—R^1, where R is —Et or —Pr and R^1 is —Pr, i-Pr, —Bu or i-Bu. The yields obtained were however low (less than 60 %) and this was attributed to a number of factors such as incomplete oxidation, side-reactions, decomposition of the sensitive azo product, and adsorption or complexing of starting material and/or the azo-alkane by mercury salts.

Mercuric oxide has been used in quite recent work for the preparation of unsymmetrical aliphatic azo derivatives containing alkyl and alkenyl groups. For instance, Takagi and Crawford[37, 39] oxidized 1-allyl-2-(1-propyl)hydrazine (**18**) in ether solution with a slurry of red mercuric oxide and sodium sulphate to obtain 1-propylazo-3′-propene (**19**) (equation 7).

$$\underset{(\mathbf{18})}{CH_3CH_2CH_2—NH—NH—CH_2CH{=}CH_2} \xrightarrow{HgO} \underset{(\mathbf{19})}{CH_3CH_2CH_2—N{=}N—CH_2CH{=}CH_2} \quad (7)$$

Similarly synthesized were methylazo-3-propene and t-butylazo-3-propene (in 75 % yield), the azo compounds prepared in this study being carefully purified by preparative gas chromatography. Slow treatment with yellow mercuric oxide has been employed to produce the optically active (methylazo-α-phenylethane in good yield from (–)-*N*-methyl-*N*′-α-phenylethylhydrazine in n-pentane, the solution being kept below 25°C[40].

Cohen and Wang[16] have made methylazo-2,2,2-triphenylethane (**20**) by subjecting the hydrazine to the action of 30 % hydrogen peroxide in aqueous

$$Ph_3CCH_2—NH—NH—Me \xrightarrow{H_2O_2} \underset{(\mathbf{20})}{Ph_3CCH_2—N{=}N—Me} \quad (8)$$

sodium bicarbonate (equation 8). On the other hand, these workers found that air bubbled through a solution of 2,2,2-triphenylethylhydrazine in boiling benzene for 8 h caused mainly conversion to benzophenone rather than the azoalkane.

Low temperature oxidations of unsymmetrical hydrazines with lead tetra-acetate have also been shown to be a useful source of azo compounds. It has been reported[14] that oxidation of *N*,*N*-dialkyl-, diaryl-, and diaralkylhydrazines with this reagent at –60°C affords the tetrazenes (**21**) in good yield; the reaction mechanism, in the absence of a trapping agent, being an intermolecular process involving the *N*-nitrene intermediate (**22**) (equation 9). Lynch and MacLachlan[41] have recently synthesized 1-benzoylazo-1-

$$R_2N{-}NH_2 \xrightarrow{Pb(OAc)_4} \left[R_2N{-}\ddot{N} \longleftrightarrow R_2\overset{+}{N}{=}\overset{-}{N}\right] \xrightarrow{R_2NNH_2} R_2N{-}N{=}N{-}NR_2 \quad (9)$$

(**22**) (**21**)

R = alkyl, aryl or aralkyl group

cyclohexanenitrile (**23**) in 91% yield by oxidizing the hydrazo precursor with lead tetra-acetate in methylene chloride at –20°C.

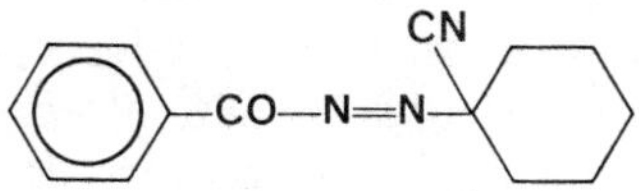

(**23**)

Bock and colleagues have used *N*-bromosuccinimide as an oxidizing agent under a variety of conditions for the preparation of some interesting unsymmetrical acylazo compounds[42–44]. For example, benzoylazodiphenylphosphine oxide (**24**) was obtained from the hydrazine in this

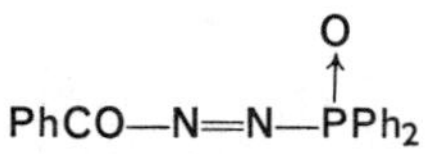

(**24**)

manner[33]. Campbell and co-workers[31] have also synthesized unsymmetrical aliphatic acylazo compounds for Diels–Alder reactions involving cyclopentadiene, by oxidation of appropriate dicarbonylhydrazines with *N*-bromosuccinimide in carbon tetrachloride. Substances of general formula RCO—N=N—COR^1, where R is —Ph or —CMe$_3$, and R^1 is —Ph or —Me, were thus obtained.

C. Cyclic Hydrazines to Azo Compounds

The oxidation of hydrazines has become one of the most important synthetic routes to cyclic azo compounds[5,6], various oxidizing agents, e.g. mercuric oxide, nitric acid, silver oxide, hypochlorous acid, copper(II) chloride, and molecular oxygen, together with a wide range of solvents, having been used.

A good deal of research has been performed with mercuric oxide, which appears to have become a preferred oxidant for the preparation of cyclic azo derivatives. Lombardino and co-workers[45] discovered that immediate oxidation of freshly made hydrazines of general formula (**25**) with this reagent gave rise to the corresponding azo compounds (**26**) (equation 10).

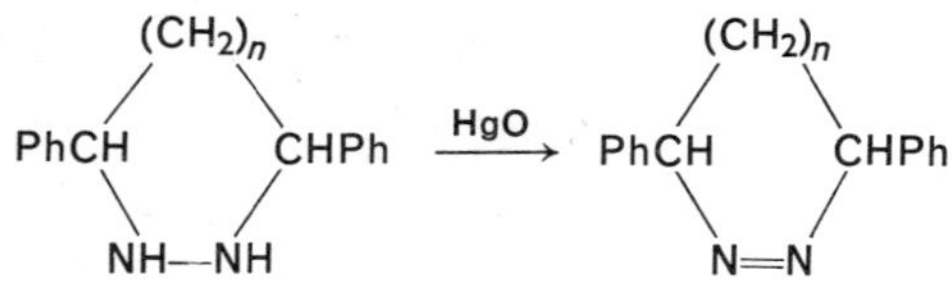

(10)

(**25**) (**26**)

The compound having $n = 3$, which was obtained in 30% yield and slowly isomerized to the hydrazone, was also produced by air oxidation. The eight-membered ring azo derivative ($n = 4$), afforded in 57% yield, on the other hand did not exhibit such isomerization. Crawford and colleagues[46] have reported that oxidation of alkyl derivatives of pyrazolidine (**27**) in pentane with red mercuric oxide in the presence of magnesium sulphate is

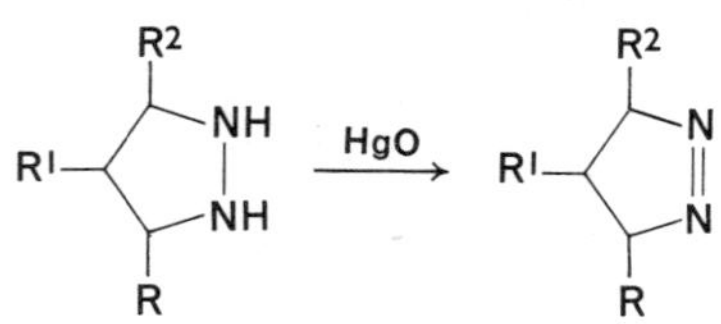

(11)

(**27**) (**28**)

a preferred route to 1-pyrazolines (**28**) (equation 11). This method is rapid and leads to good yields (50–80%) when carried out as near neutrality as possible so as to avoid acid or base catalysed tautomerism to the 2-pyrazolines. Similar oxidations using silver oxide in methanol gave faster reaction and produced yields of about the same order of magnitude. Treatment of the parent substance, pyrazolidine, with molecular oxygen using methanol as solvent and copper(II) acetate as catalyst effected slow conversion to 1-pyrazoline (71%)[46].

Oxidations of some cyclic hydrazines with mercuric oxide have led to the formation of unsaturated hydrocarbons. Thus, exposure of hydrazine (**29**) to this reagent has afforded (**30**), with rapid evolution of nitrogen; the reaction probably proceeding via the unstable azo intermediate, which could lose nitrogen rapidly in a concerted cycloreversion to give **30** directly[47] (equation 12). In contrast, similar treatment of hydrazine (**31**) produced the corresponding azo compound (**32**), which proved to be stable (equation 13). On the other hand, Berson and Olin[48] also found that oxidation of the

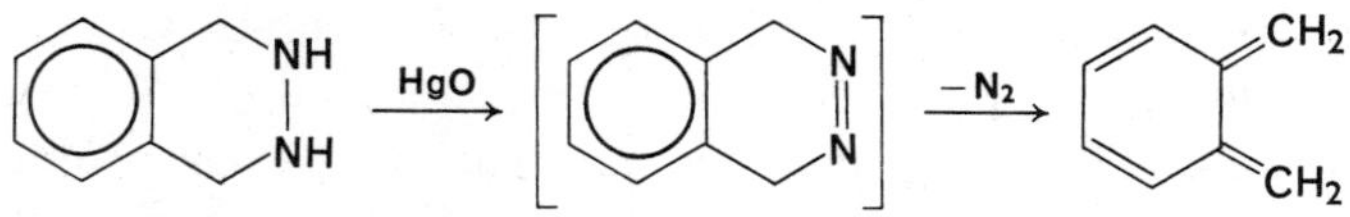

(29) (30) (12)

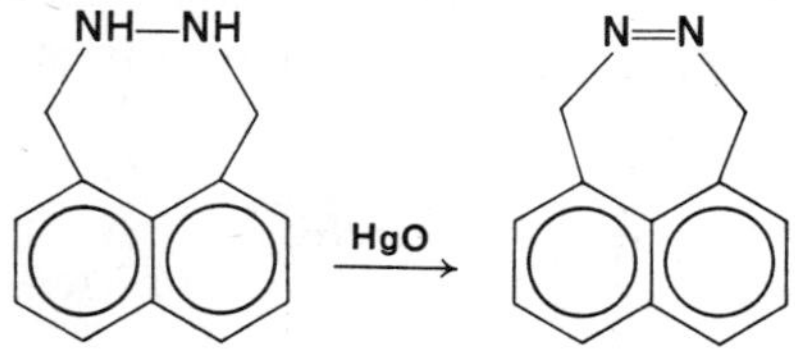

(31) (32) (13)

cis-hydrazo compound (**33**) with air, yellow mercuric oxide, or manganese dioxide at 25°C gave rise to an approximately quantitative yield of nitrogen and *trans-trans*-2,4-hexadiene (**34**), presumably via the unstable azo inter-

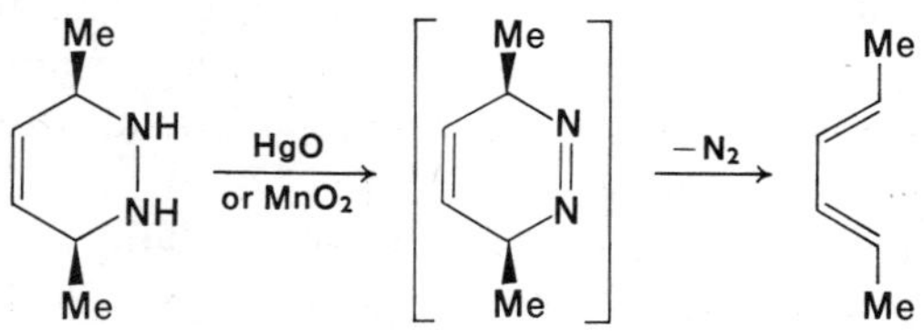

(33) (34) (14)

mediate (equation 14). Similar oxidation of the *trans*-hydrazine gave nitrogen and *cis-trans*-2,4-hexadiene. However, ultraviolet spectroscopic evidence for the presence of the azo intermediates presumed to be involved in both of these reactions could not be obtained, even at −50°C.

More recently, Overberger and co-workers[49, 50] have employed mercuric oxide to prepare eight- to ten-membered cyclic azo compounds from appropriate hydrazines. For instance, they oxidized a mixture of *cis*- and *trans*-3,8-dimethyl-1,2-diazacyclo-octane, which after separation by chromatography yielded *cis*-3,8-dimethyl-*cis*-1,2-diaza-1-cyclo-octene (**35**) and the *trans-trans* isomer. Similarly synthesized were *trans*-1,2-diaza-1-cyclononene and *trans*-1,2-diaza-1-cyclodecene; these being the first

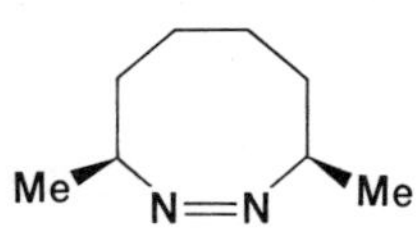

(35)

reported preparations of medium-sized cyclic azo derivatives having trans-azo linkages.

In comparison to acyclic hydrazines, in general the cyclic analogues are very sensitive to air oxidation, and this has been confirmed by several researches carried out in recent years[51–54]. For example, Overberger and Lapkin[51] showed that the cyclic bishydrazo derivative (**36**) could not be isolated even in an inert atmosphere due to its extreme ease of oxidation to

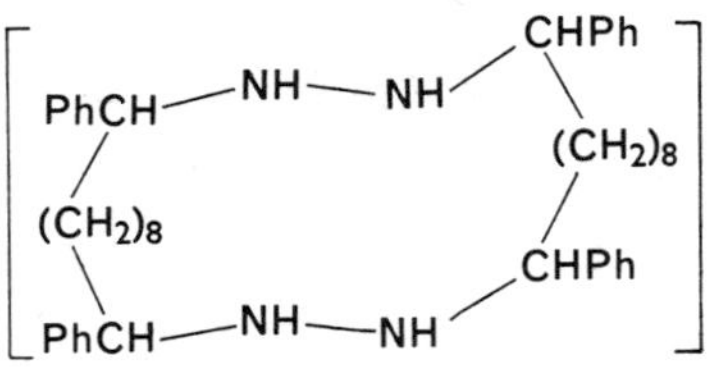

(36)

the corresponding 24-membered bisazo compound (**37**). Furthermore, although catalytic hydrogenation of **37** over 10% palladium on charcoal took place quantitatively, oxidation occurred during attempts to isolate **36**, the only recoverable product being **37**. Similar results were also obtained with the 28-membered analogues.

Recently, Snyder[55] in reference to the extreme lability to air oxidation of cyclic hydrazines, undertook semi-empirical m.o. calculations for the smallest cyclic hydrazine, diaziridine (**38**) and its conjugate acid, in an attempt to 'understand the influence of geometry on the reducing capacity

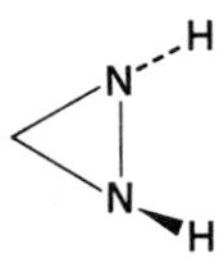

(38)

of hydrazines'. The lowest energy conformation of diaziridine is predicted to be the *trans*-isomer (**38**); the *cis*-, half-planar, and planar forms being computed to be less stable. In the corresponding azo derivative, diazirine (**39**), the nitrogen lone electron pairs strongly interact and are considerably delocalized throughout the σ framework of the molecule. It was concluded

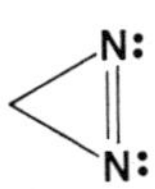

(39)

that the 'differential capacity for oxidation of cyclic versus acyclic hydrazines' is due to electronic structure following the geometries of the systems in question; lone pair interaction being responsible for the short lifetimes of cyclic hydrazines in the presence of very mild oxidants; and it was shown that removal of such interaction by protonation should inhibit facile oxidation[55].

I. Acylhydrazines

Some interesting work has been done lately with regard to the preparation of cyclic α-carbonylazo compounds (some of which are very unstable) from appropriate hydrazines. Chapman and Dominianni[56] have oxidized phthalylhydrazide *in situ* with lead tetra-acetate to the unstable phthalazine-

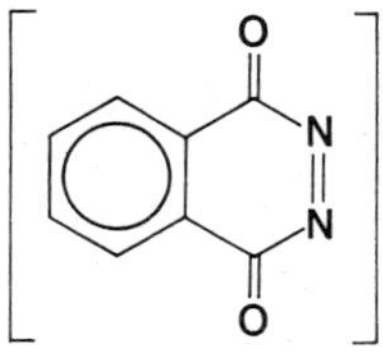

(40)

1,4-dione (**40**). Other workers[57,58] have also used lead tetra-acetate to oxidize similar substrates, e.g. urazoles.

Another reagent employed for the oxidation of cyclic acylhydrazines is t-butyl hypochlorite[59,60]. For example, treatment of 4-phenylurazole (**41**) in dry acetone with this reagent at –50 to –78°C gives 4-phenyl-1,2,4-triazolin-3,5-dione (**42**) in 80% yield[59] (equation 15).

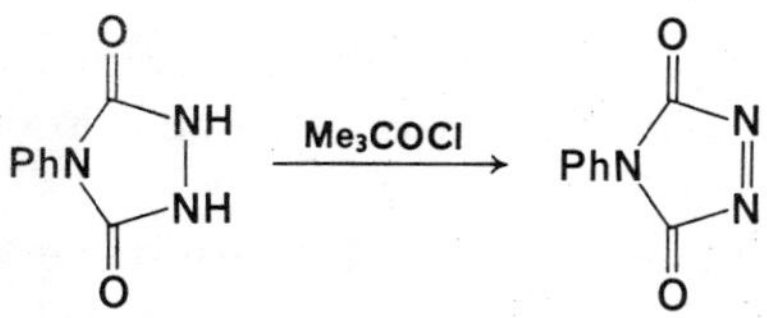

(41) **(42)** (15)

Very recently[61], it was reported that 1,2,4-triazoline-3,5-dione (**43**), which had not been previously isolated and characterized, was synthesized 'in solution' by addition of dinitrogen tetroxide in dichloromethane to a suspension of urazole (**44**) in the same solvent at 0°C (equation 16). Under these conditions, there was gradual formation of a deep pink solution, which was decolourized immediately on addition of cyclopentadiene at

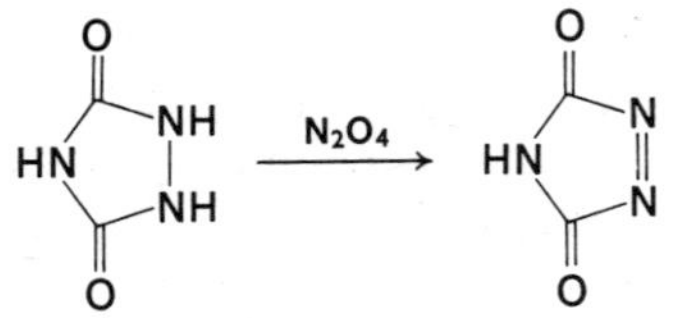

(44) (43)

(16)

0°C; the product then being a Diels–Alder adduct (84% yield based on **44** consumed). Similar results were obtained when using lead tetra-acetate as oxidant. The pink colourations observed for the solutions from these reactions at 0°C lasted for several hours, thereby suggesting that **43** is more stable than has previously been supposed.

2. Bicyclo and polycyclohydrazines

A number of interesting azo derivatives have been recently synthesized via oxidation of bicyclo- and polycyclo-hydrazines. For instance, Allred and Johnson[62] have oxidized hydrazine (**45**) with copper(II) chloride to obtain the corresponding azo derivative as the copper(I) chloride complex, which on treatment with aqueous ammonia at –20°C liberated the free azo

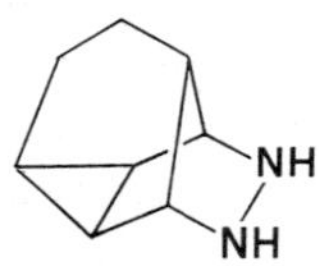

(45)

compound. Copper(II) chloride has been used quite widely for oxidation of bicyclo-hydrazines, the azo compounds usually being produced as complexes with copper(I) chloride and the free azo product being obtained from addition of suitable bases.

The bicyclohydrazine (**46**) was found to oxidize very easily in air in the presence of copper(I) chloride (instead of copper(II) chloride) to afford, via the corresponding complex, 2,3-diazabicyclo[2.2.1]-2-heptene (**47**)[63] (equation 17). Compound **47** was similarly prepared by using mercuric oxide[63].

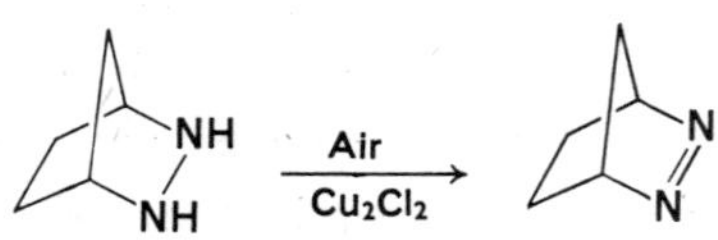

(46) (47)

(17)

Rieber and colleagues[64] have prepared azo compounds (**48**) and (**49**) in the form of their copper(I) complexes by the action of copper(II) chloride

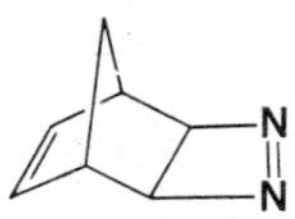

(48)

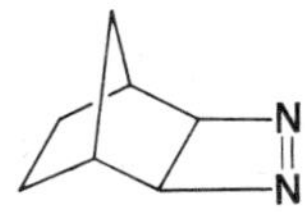

(49)

on the hydrazines. They also synthesized 2,3-diazabicyclo[2.2.2]octa-2,5-diene (**50**) from the bicyclohydrazine by treatment with t-butyl hypochlorite at −78°C. However, **50** decomposes rapidly at room temperature, in common with other 1,4-bis-allylic azo compounds.

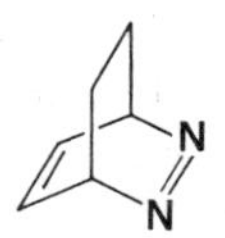

(50)

Lay and co-workers[65] have obtained azo compound (**51**) for use in fragmentation reactions, by oxidizing the hydrazo precursor in toluene with red mercuric oxide (equation 18).

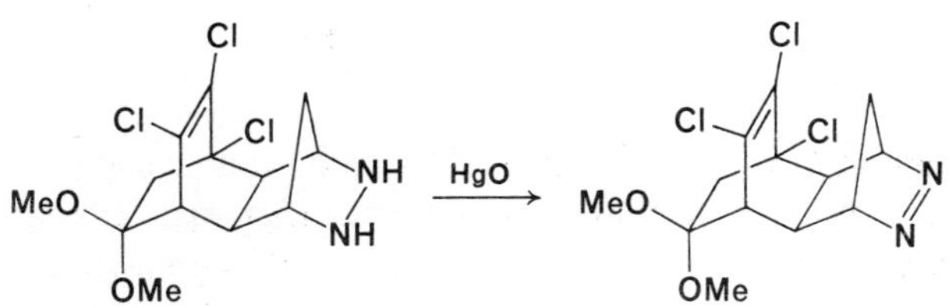

(51) (18)

Berson and Olin[48, 66] have carried out interesting studies on the oxidation of bicyclohydrazines which have led to conclusions regarding the mechanism involved. They oxidized the *trans*-hydrazine (**52**) and its two *cis*-isomers at 25°C, the products being nitrogen and 2,5-heptadiene; and concluded that the results obtained were more consistent with a concerted retrograde homo Diels–Alder mechanism than a two-step sequence involving an intermediate in which loss of nitrogen is more rapid than bond rotation[48]. They also prepared ethereal solutions of the *cis-anti* azo compound (**53**) and its *trans*- and *cis-syn* forms by treating the hydrazo precursors with manganese dioxide at –50°C; these solutions being found to be stable for several days at –70°C[66].

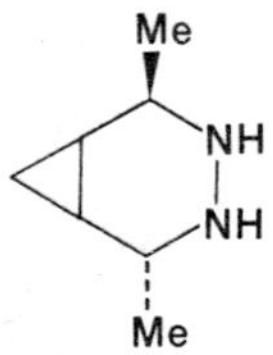

(52)

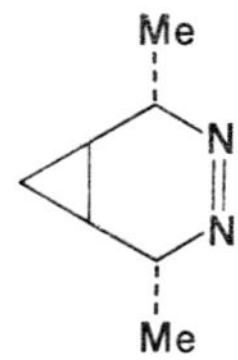

(53)

Several bicyclic *cis*-azoalkanes, which are of use as precursors for the synthesis of other organic species, have been produced very recently by controlled oxidation of hydrazines[67]. Thus, compounds of general formula (**54**), where $n = 1$–3, were conveniently oxidized to the azoalkanes (**55**) in good yield, by bubbling oxygen into solutions of **54** in the presence of catalytic suspensions of 5% palladium on charcoal in methanol (equation 19).

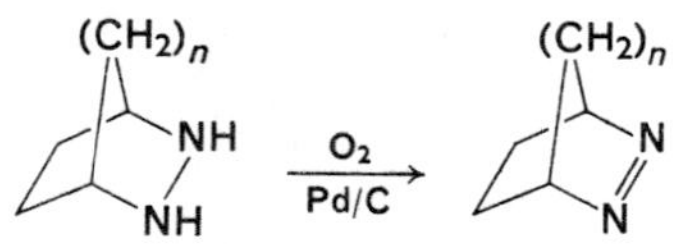

(54) **(55)** (19)

Other recent studies have concerned the oxidation of cyclic bishydrazo compounds. For example, Shen[68] discovered that exposure of the bis-hydrazine (**56**) to yellow mercuric oxide or air caused immediate oxidation with the formation of nitrogen and *cis*-9,10-dihydronaphthalene (**57**); and postulated that under the reaction conditions employed oxidation of **56** gives the corresponding unstable bisazo compound, which undergoes fragmentation. Paquette[69] has also reported that oxidation of deuterated

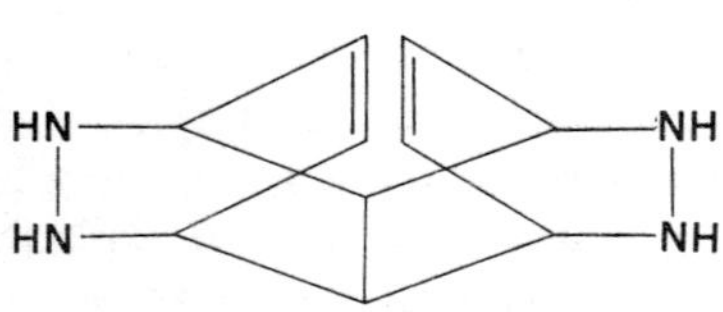

(56)

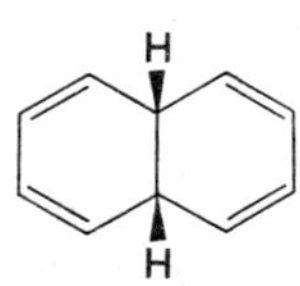

(57)

56 in like manner and under the same conditions leads to the isolation of compound (**58**).

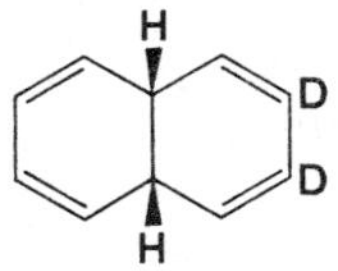

(58)

D. Hydrazines to Azoxy Derivatives

A survey of the literature reveals that oxidations of acyclic or cyclic hydrazines directly to azoxy derivatives have not been reported, presumably because oxidations of these substrates with mild reagents generally stop at the azo stage or lead to fragmentation reactions depending upon the stability of the azo product. However, it is possible that azoxy compounds in some cases are formed along with azo derivatives during oxidations of aliphatic hydrazines, and the extent of such formation of azoxy compounds would be expected to be dependent upon the conditions used and the nature of the oxidant. It has been recently pointed out that the problem of the possible formation of azoxy derivatives during oxidations of hydrazines to azo compounds has been neglected, and that the products of these reactions should be subjected to chromatographic analysis[11].

III. OXIDATION OF ALIPHATIC AZO COMPOUNDS

The oxidation of aliphatic azo compounds has become a preferred route to the corresponding azoxy derivatives, and consequently this type of conversion has received increasing attention lately. Di-imide (HN=NH), the prototype azo compound, which has attracted the interest of chemists for many years, was until very recently considered to be extremely unstable and at normal temperatures to have only transient existence (e.g. a half-life of a few seconds in the gas phase)[70]. However, Willis and Back[71] have carried out a detailed study of the chemistry of di-imide (generated by a microwave discharge in hydrazine vapour in a flow system) and found the lifetime in the vapour phase to be 'surprisingly long' (2·4 min at 30°C). The vapour showed a weak 'azo' absorption with considerable bond structure at 320–420 nm, and decomposed to give hydrogen, nitrogen, hydrazine and probably ammonia. These studies also showed that solid di-imide is quite stable, no decomposition occurring up to the m.p. of solid NH_3—N_2H_2 mixture (–65°C) and that it exists for many minutes in liquid mixture with ammonia at –65 to –30°C. Over the past few years, di-imide has been generated (but not isolated or detected) *in situ* in aqueous media, and because of its easy oxidation has been used to a considerable extent as a reducing agent[70] for various types of substrates, including azoxy and azo compounds (see following chapter).

Both *trans*- and *cis*-aliphatic azo compounds may be converted to the corresponding azoxy derivatives by the application of appropriate oxidants, but since they are prone to acid-catalysed isomerization to hydrazones and are sensitive to decomposition, the reaction conditions have to be controlled to achieve successful conversions. The preferred reagents for these transformations are the relatively weak peracids, such as perbenzoic and peracetic acid, since their use in conjunction with suitable solvents diminishes the possibility of isomerization[11]. Although a number of successful oxidative syntheses of aliphatic azoxy compounds have been reported in the recent literature, it should be pointed out here that the chemistry of aliphatic azoxy derivatives had in general received little attention before 1967.

A. Symmetrical Azo to Azoxy Compounds

Langley and colleagues[72] have prepared some symmetrical aliphatic azoxy compounds by oxidizing the corresponding azo derivatives with perbenzoic acid at 0°C in dichloromethane. In this manner, 1-azoxypropane and 2-azoxypropane (**59**) were obtained in good yield. However, attempts to oxidize 1-azopropane and 2-azopropane with hydrogen peroxide in acetic acid or with nitric acid were not successful, apparently because acid-catalysed isomerization had occurred before oxidation. These workers also showed that perbenzoic acid converted azocyclohexane to the azoxy derivative in quantitative yield.

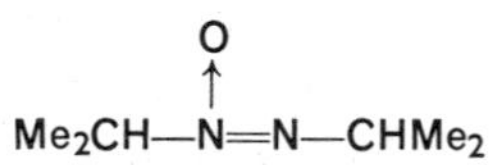

(59)

In later work, Freeman[73] used perbenzoic acid at 0–5°C to convert *trans*-ω-azotoluene (**60**) to *trans*-ω-azoxytoluene (**61**), the latter product

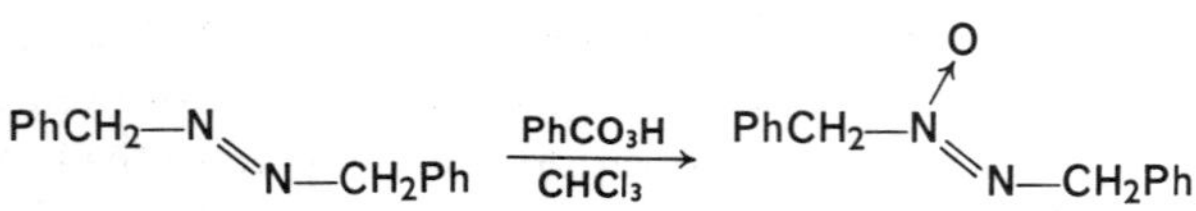

(20)

(60) **(61)**

being produced in 67% yield (equation 20). Oxidation of azoisobutane (**62**) with 40% peracetic acid at 5–10°C gives rise to the corresponding azoxy derivative (**63**)[73] (equation 21).

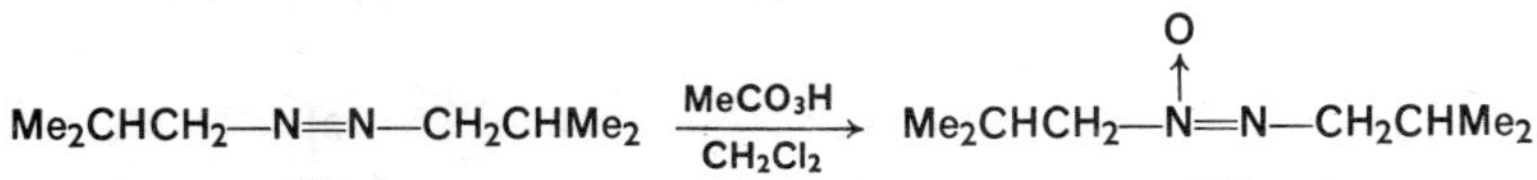

(**62**) (**63**) (21)

Azoxyalkanes, including symmetrical derivatives, are presently being looked at with more interest because some of them have been found to be carcinogenic, and others have possible applications as fungicides. Azoxyalkanes have also been used as starting materials for the synthesis of oxadiaziridines. These developments have stimulated interest in the synthesis of aliphatic azoxy compounds. Another aspect that has attracted the attention of chemists is the preparation of acyclic aliphatic *cis*-azoxy compounds, because as recently as 1969 no genuine derivatives of this type had been reported in the literature[74]. However, since that time some *cis*-azoxyalkanes have been obtained by careful oxidation of the *cis*-azo precursors. For example, *cis*-azoxyisopropane (**64**) has been synthesized in 72% yield by peracid oxidation of *cis*-azoisopropane (**65**)[24] (equation 22).

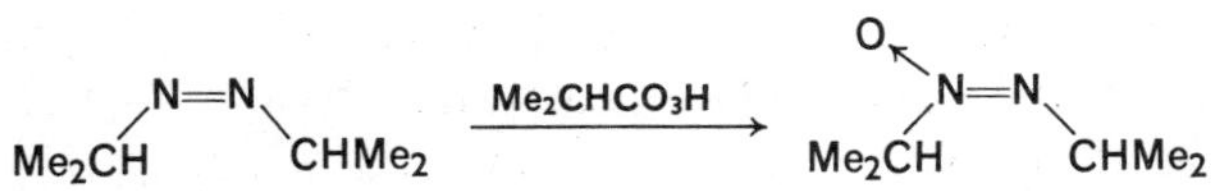

(**65**) (**64**) (22)

B. *Unsymmetrical Azo to Azoxy Compounds*

In the case of the oxidation of unsymmetrical aliphatic azo compounds, R—N=N—R[1], the entering oxygen may in theory attack either of the nitrogen atoms, hence giving rise to the possibility of position isomerism and the formation of mixtures of azoxy derivatives. Despite these possibilities, for azomethanes (R = Me), it had until recently been believed that the incoming oxygen atom invariably attacks the nitrogen atom linked to the methyl group[11], i.e. the less sterically hindered position. However, work by Gillis and Hagarty[75] on the oxidation of α,β-unsaturated azomethanes has shown that at least in some cases the reverse is true, attack taking place on the more hindered nitrogen. In fact, the position of the oxygen atom in the azoxy group of the products obtained was found to be dependent upon alkyl substitution. When an anhydrous ether solution of 2-(methylazo)-

propene (**66**) was treated with 1 equivalent of 40% peracetic acid conversion to 2-(methyl-NNO-azoxy)propene (**67**) in 44% yield resulted (equation 23). The results obtained in this study for oxidations of other azomethanes are summarized in Table 1. The determination of the position of the oxygen

TABLE 1. Oxidations of α,β-unsaturated azomethanes[75], R—N=N—Me

R	Oxidant	Azoxymethane	
		Structure	Yield, (%)
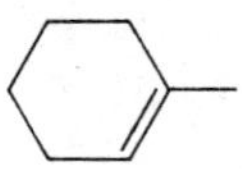	$MeCO_3H$	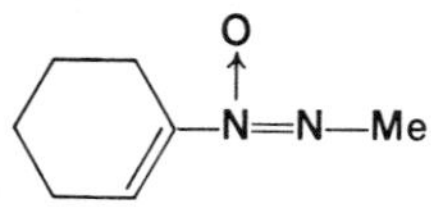	66
$Me_2C{=}CH—$	$MeCO_3H$	$Me_2C{=}CH—N{=}N(\rightarrow O)—Me$	34
$Me_2C{=}C(Me)—$	*m*-$ClC_6H_4CO_3H$	$Me_2C{=}C(Me)—N{=}N(\rightarrow O)—Me$	42

atom in unsymmetrical azoxy compounds by chemical means has always been fraught with difficulties, but fortunately the advent of nuclear magnetic resonance spectroscopy has gone a long way towards solving this problem. Gillis and Hagarty used this technique, together with infrared and ultraviolet spectroscopy, for the assignment of structures to the azoxymethanes they prepared.

$$\underset{(\mathbf{66})}{H_2C{=}C(Me)—N{=}N—Me} \xrightarrow{MeCO_3H} \underset{(\mathbf{67})}{H_2C{=}C(Me)—N(\rightarrow O){=}N—Me} \quad (23)$$

These workers also reported that treatment of 2-(methylazo)-3-methyl-2-butene (**68**) with peracetic acid did not lead to an azoxy compound, but rather the ester 3-acetoxy-3-methyl-2-butanone, presumably formed via the epoxide or by acetolysis of the azoxy derivative. On the other hand, exposure of **68** to *m*-chloroperbenzoic acid in anhydrous ether at room temperature gave rise to a non-separable mixture, which was shown from spectral evidence to contain 2-(methyl-NON-azoxy)-3-methyl-2-butene.

In view of the results obtained in the above-mentioned study, it is evident that more extensive and systematic investigations on the orientation of the incoming oxygen atom in oxidations of unsymmetrical aliphatic azo compounds are needed to achieve a greater understanding of the factors involved.

Another problem concerning the synthesis of unsymmetrical azoxyalkanes has been the lack of a suitable general method for this purpose. Thus, as recently as 1972 Moss and co-workers[76] pointed out that no general synthesis for these substances was available that combined the following attributes: (*i*) flexibility (applicable to a wide variety of unsymmetrical azoxyalkanes), (*ii*) direction (capable of giving a single 'structurally predictable' azoxyalkane), (*iii*) allows the use of easily available starting materials, and (*iv*) accommodates constraints on chirality at the α and α' carbon atoms. They also drew attention to the fact that preparations involving ultimate precursors or intermediates having linked nitrogen atoms of identical oxidation state, as is the case for the oxidation of azoalkanes, do not generally satisfy the flexibility and direction requirements. These researchers reported a new method of preparation for unsymmetrical azoxyalkanes consisting of alkylation of alkane diazotates, and purified their products by means of gas chromatography[76, 77]. This general synthesis for azoxyalkanes approaches the four criteria listed above and clearly offers interesting possibilities which might be advantageously investigated.

Unsymmetrical mixed (i.e. aliphatic–aromatic) azoxy compounds have been synthesized via oxidation, which exhibit both position and stereoisomerism. For instance, Brough and colleagues[78] have prepared three of the four possible isomers of representative azoxy derivatives of the type $ArCH_2(N_2O)Ar^1$, the two trans isomers being synthesized by perbenzoic acid oxidation of the azo precursor, $ArCH_2—N{=}N—Ar^1$. Oxidation of *trans-p*-chlorobenzylazobenzene (**69**) with this reagent gave the *trans*-azoxy

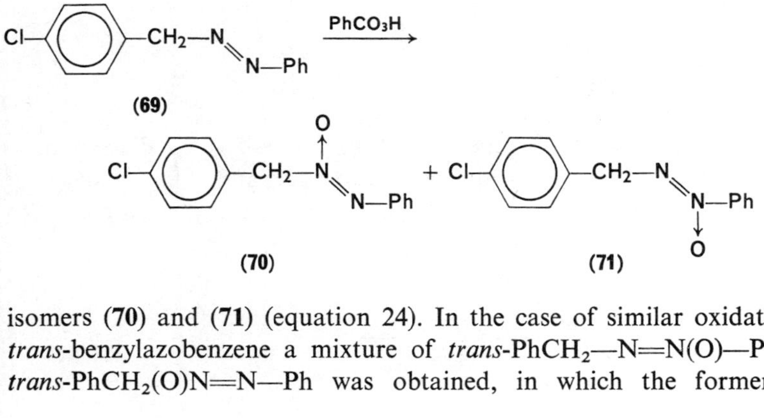

(**69**) (**70**) (**71**) (24)

isomers (**70**) and (**71**) (equation 24). In the case of similar oxidation of *trans*-benzylazobenzene a mixture of *trans*-$PhCH_2—N{=}N(O)—Ph$ and *trans*-$PhCH_2(O)N{=}N—Ph$ was obtained, in which the former predominated. However, the corresponding *cis*-azoxy isomers could not be prepared by careful oxidation of the azo derivatives[78]. Other unsymmetrical mixed azo compounds, such as phenylazomethane, have been oxidized with perbenzoic acid[73].

Some work has also been done on the oxidation of mixed bis-azo compounds. For example, Woodward and Wintner[79] oxidized the bis-azo derivative (**72**) with perbenzoic acid and obtained bis-azoxy compound (**73**) (equation 25).

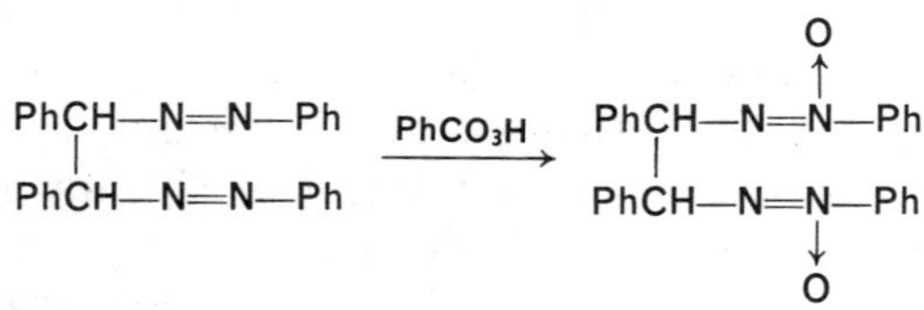

(25)

(**72**) (**73**)

C. Cyclic Azo to Azoxy Derivatives

Cyclic *cis*-azoxyalkanes are still relatively rare since up to about ten years ago none of these compounds had been reported in the literature, and in the meantime only a limited number of these derivatives have been synthesized. One of the earliest reported preparations was that of 3-benzoyloxy-3,5,5-trimethyl-Δ'-pyrazoline-1-oxide (**74**), which was unexpectedly obtained from oxidation of 3,5,5-trimethylpyrazoline (**75**) with perbenzoic acid in dichloromethane[80] (equation 26); this transformation presumably taking place via the benzoyloxyazo precursor.

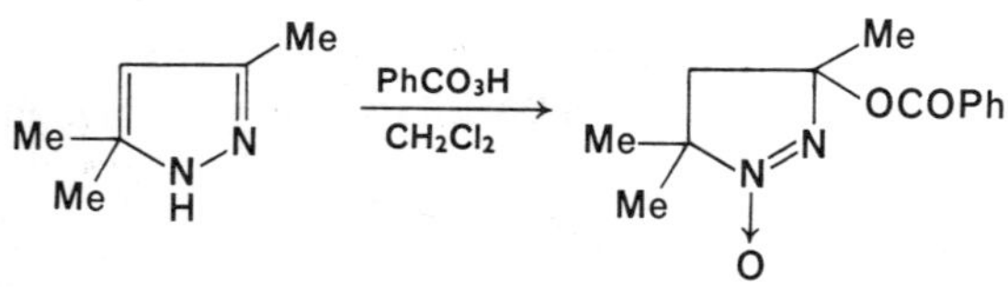

(26)

(**75**) (**74**)

In more recent work, Greene and Hecht[74] oxidized 2,3-diazabicyclo-[2.2.1]-2-heptene (**47**) with *m*-chloroperbenzoic acid and obtained an azoxy compound in 57% yield, to which the *cis* configuration was assigned (**76**)

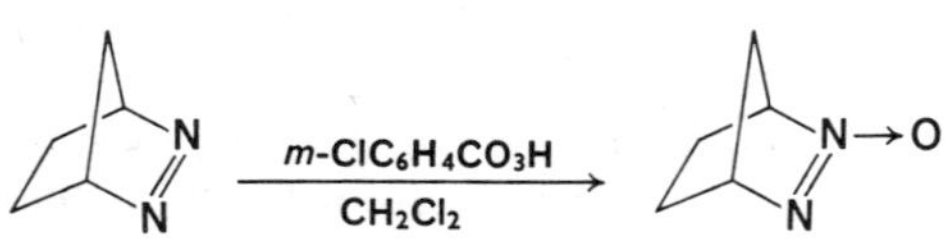

(27)

(**47**) (**76**)

after consideration of n.m.r. spectral evidence (equation 27). They also synthesized optically active **76** having $[\alpha]_D^{20} = 1\cdot 2°$ (CH_2Cl_2) by oxidation of **47** with (+)-peroxycamphoric acid. In addition, these workers oxidized *trans*-3,5-diphenylpyrazoline (**77**) to the corresponding *cis*-azoxy derivative (**78**) in 70% yield by treatment with *m*-chloroperbenzoic acid in dichloro-

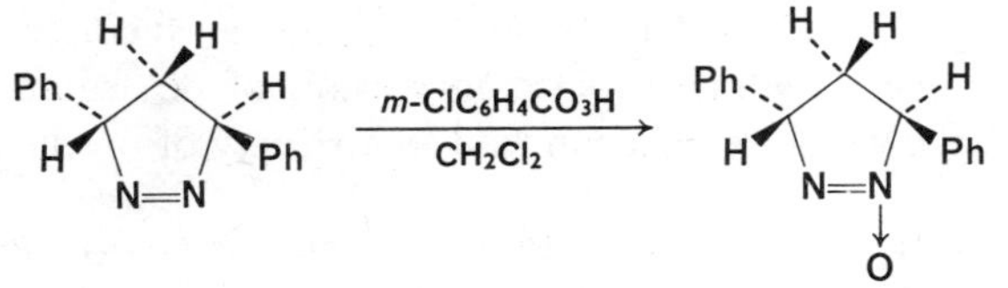

(28)

(77) (78)

methane, (equation 28). A series of bicyclo *cis*-azoxy derivatives has been prepared from compounds of general formula (**79**), where R is -Ph or -Me and $n = 1$–4, by hydrolysis (refluxing with excess potassium hydroxide in ethylene glycol–water under nitrogen for 1–24 h), followed by oxidation (refluxing with excess 30% aqueous hydrogen peroxide for 2 h)[81] (equation

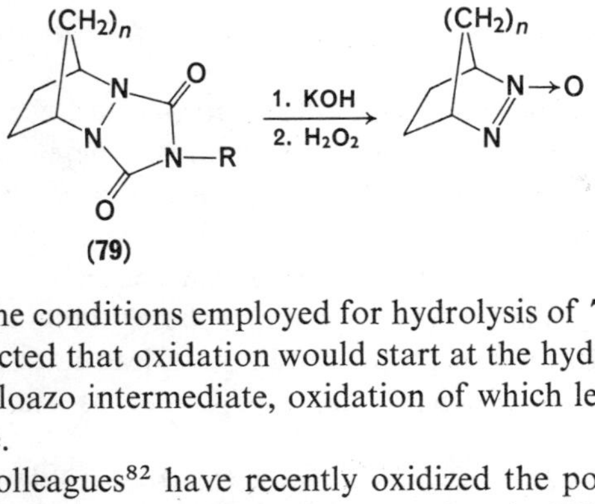

(29)

(79)

29). In view of the conditions employed for hydrolysis of **79** in this study, it would be expected that oxidation would start at the hydrazo stage giving rise to the bicycloazo intermediate, oxidation of which leading to the *cis*-azoxy derivative.

Snyder and colleagues[82] have recently oxidized the polycycloazo compound, diazabasketene (1,1′-diazabishomocubene) (**80**) to the corresponding pentacyclo *cis*-azoxy derivative (**81**), obtained in 70% yield, by the action of 85% *m*-chloroperbenzoic acid in chloroform at room temperature for 2 h (equation 30).

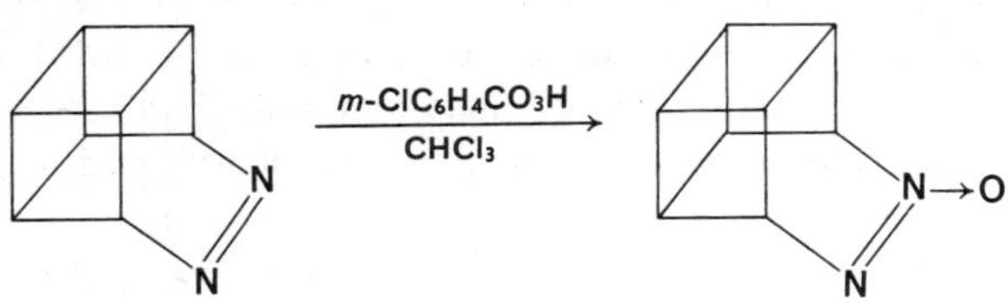

(30)

(80) (81)

IV. OXIDATION OF AROMATIC HYDRAZO COMPOUNDS

A. Arylhydrazines to Azo Derivatives

The oxidation of arylhydrazines has undergone considerable study in recent years, and has often been shown to involve either di-imides or free radicals as reactive intermediates[83,84]. Thus, phenyldi-imide (**82**) has frequently been considered to be an unstable intermediate in oxidation reactions of phenylhydrazine. Decomposition of **82** would give rise to phenyl radicals, which in turn can lead to a variety of products depending

$$\mathrm{PhNH{-}NH_2} \longrightarrow \underset{(\mathbf{82})}{[\mathrm{Ph{-}N{=}N{-}H}]} \longrightarrow \mathrm{Ph}\cdot \longrightarrow \text{Products} \qquad (31)$$

upon the substrates present (equation 31). The non-detection of **82** in oxidation reactions using chemical agents has been attributed to its facile oxidation[85]. However, electrochemical oxidation of phenylhydrazine at controlled potential in anhydrous organic medium under argon and on a large platinum anode has recently resulted in the obtention of **82**[85]. The results of this work are summarized in equation 32.

$$3\,\mathrm{PhNH{-}NH_2} \longrightarrow 2\,e^- + \mathrm{Ph{-}N{=}N{-}H} + 2\,\mathrm{PhNH{-}NH_3^+} \qquad (32)$$

Arylhydrazines are oxidized by lead tetra-acetate via aryldi-imides to give aryldiazonium ions at low temperatures, whereas aryl radicals are formed at room temperature[86]. Addition of monoaroylhydrazines to two equivalents of this reagent leads to the formation of aroyldi-imides as intermediates, which may be converted (via triacetate complexes) to the corresponding aromatic acids, the latter being obtained in high yield (equation 33). On the other hand, addition of lead tetra-acetate to monoaroylhydrazines affords *N,N'*-diacylhydrazines as the main products.

$$\mathrm{ArCONH{-}NH_2} \xrightarrow{\mathrm{Pb(OAc)_4}} \mathrm{ArCO{-}N{=}N{-}H} \xrightarrow{\mathrm{H_2O}} \mathrm{ArCO_2H} \qquad (33)$$

Oxidation of *N*-aryl-*N,N'*-diacylhydrazines with lead tetra-acetate in protic solvents causes mainly conversion to acylazobenzenes, whereas similar treatment in aprotic media favours transformation to azoacylals[87].

Some work has been done on the oxidation of 1-substituted 2-arylhydrazines as a route to unsymmetrical aryl-azo compounds. Cohen and Wang[88] oxidized 1-benzhydryl-2-phenylhydrazine (**83**) to phenylazodiphenylmethane (**84**) in 54% yield by the action of 30% hydrogen peroxide

$$\underset{(\mathbf{83})}{\mathrm{Ph_2CH{-}NH{-}NH{-}Ph}} \xrightarrow[\mathrm{EtOH/Et_2O}]{\mathrm{H_2O_2}} \underset{(\mathbf{84})}{\mathrm{Ph_2CH{-}N{=}N{-}Ph}} \qquad (34)$$

in ethanol/ether at room temperature for 2 days (equation 34). However, attempts to obtain similarly methyl- and benzylazodiphenylmethane were unsuccessful, the products being oils, which were apparently mixtures of hydrazones. Bellamy and Guthrie[89] have oxidized several 1-alkyl-2-phenylhydrazines to the phenylazoalkanes, Ph—N=N—R, by prolonged treatment under nitrogen with yellow mercuric oxide at room temperature. The results achieved in this work are listed in Table 2. The phenylazoalkanes

TABLE 2. Oxidations of 1-Alkyl-2-phenylhydrazines Ph—NH—NH—R with HgO[89]

R	Phenylazoalkane (%)
Cyclohexyl	89
Ethyl	87
Propyl	88
Butyl	86

made in this manner were pure, mobile yellow liquids, which were quite stable. For instance, phenylazocyclohexane was not affected by exposure to oxygen in dry hexane for 2 h.

Recently, 3-oxo-1-phenylazo-1-cyclohexene (**86**) was prepared in 90% yield by oxidation of the hydrazine (**85**) in methanol or glacial acetic acid

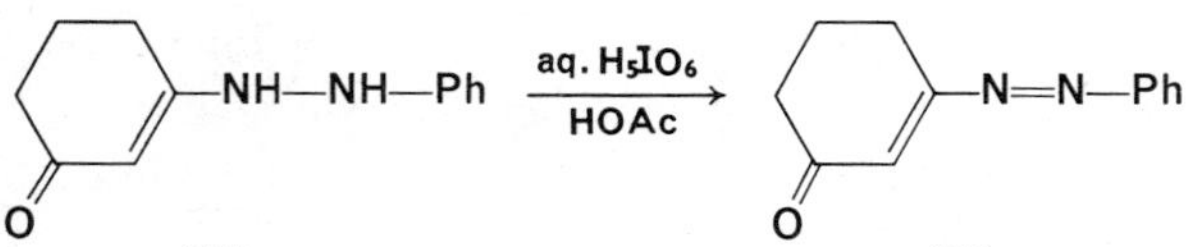

(35)

(85) (86)

with aqueous periodic acid at room temperature[90] (equation 35). However, when lead tetra-acetate in acetic acid or sodium periodate were used as oxidizing agents the yields of **86** decreased to 55 and 70%, respectively. The mechanism of the reaction of periodic acid (a two-electron oxidant) with **85** may be thought of as a simultaneous attack of an electrophilic and nucleophilic species (present, for example, in aqueous acetic acid–periodic acid) on the vinyl and phenylhydrazino groups, respectively. Pryor and Smith[91] have shown that *p*-nitrophenylazotriphenylmethane (**87**) may be synthesized by oxidation of the hydrazo precursor in ether with a two-fold excess of isoamyl nitrite under reflux conditions for 1 h.

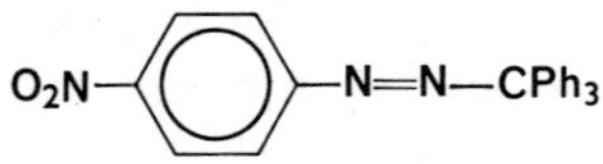

(87)

Carpino and co-workers[32] have prepared a series of substituted t-butyl phenylazoformates in good yield from the appropriate phenylhydrazines by the action of *N*-bromosuccinimide (NBS) in pyridine at room temperature. The compounds so prepared and their yields are given in Table 3.

TABLE 3. Oxidation of substituted t-butyl phenylhydrazoformates with NBS[32]

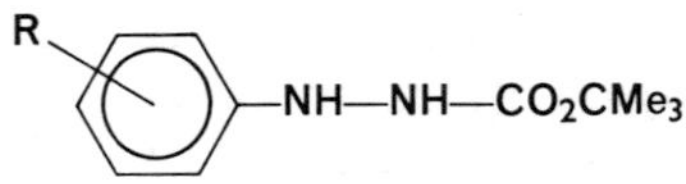

R	t-Butyl phenylazoformate (%)
H	94
p-Bromo-	86
p-Nitro-	91
o-Methoxy-	100

Attempts were also made to synthesize in an analogous manner t-butyl benzoyl-, *p*-nitrobenzoyl-, and acetylazoformate, but only the benzoyl compound (obtained in 70% yield) was stable enough to be isolated under the conditions employed.

Bock and colleagues[33] have also used the NBS method to oxidize phenylhydrazines to numerous derivatives of phenylazocarboxylic acid (**88**),

$Ph—N{=}N—CO_2H$ **(88)** $Ph—N{=}N—SO_3H$ **(89)** $Ph—N{=}N—PO_3H_2$ **(90)**

phenylazosulphonic acid (**89**), and phenylazophosphonic acid (**90**). In addition, NBS was found to be an excellent oxidant for phenylhydrazocarboxylic acid derivatives of type (**91**), where X is —Ph, alkyl, alkoxy, $—NR_2$ or phenoxy group, giving rise to good yields of the corresponding azo compounds (**92**) (equation 36). Phenylazosulphonic acid derivatives of

$$\text{Ph—NH—NH—COX} \xrightarrow{\text{NBS}} \text{Ph—N=N—COX} \quad (36)$$

(91) (92)

general formula Ph—N=N—SO_2X, where X is —Ph or alkyl group, were similarly prepared. Phenylazodiphenylphosphine oxide (**93**) has also been made in excellent yield (90%) using the same reagent[44].

$$\text{Ph—N=N—}\overset{\overset{\text{O}}{\uparrow}}{\text{P}}\text{Ph}_2$$

(93)

B. Hydrazobenzenes to Azobenzenes

I. Hydrazobenzene

Over the past ten years or so, various oxidants have been applied to the parent substance, hydrazobenzene (**94**), for its conversion to azobenzene (**95**) (equation 37), and one notable mechanistic investigation has been

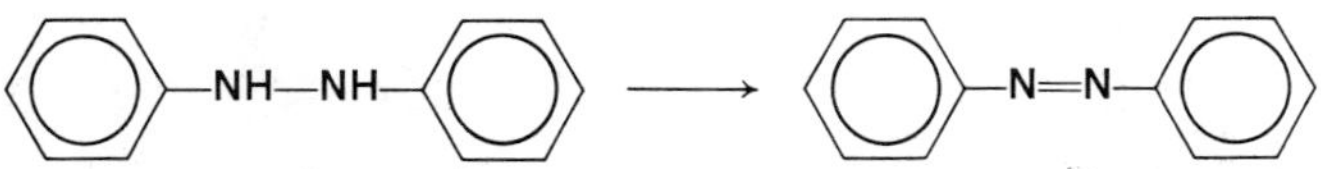

(37)

(94) (95)

carried out on this transformation. Thus, hydrazobenzene undergoes easy oxidation on treatment with iodine pentafluoride in dichloromethane at low temperature to give *trans*-azobenzene in 74% yield[92]. It also reacts smoothly (via hydrogen abstraction) with diethyl azodicarboxylate under mild conditions to produce azobenzene, the oxidant being hydrogenated to diethyl hydrazodicarboxylate. This reaction, which takes place remarkably fast even in cold benzene, affords **95** in almost quantitative yield, and no further reaction occurs after the abstraction of hydrogen[93]. The electrochemical oxidation of hydrazobenzene has been reported by Wawzonek and McIntyre[94].

The kinetics of the oxidation of hydrazobenzene to *trans*-azobenzene by iodine in aqueous ethanolic solution have been studied by May and Halpern[95] to attempt to decide whether the oxidation takes place by a one-step mechanism or by a multi-step route involving the α,β-diphenylhydrazyl radical as an intermediate. Although this reaction was found to be very rapid, a much slower parallel oxidation path involving tri-iodide ion was also discovered. Both of these reactions exhibited second-order kinetics, inverse dependence of the rate constants on the ethanol concentration being observed; and the addition of quinone and various oxidizing and reducing metal ions did not cause significant effects. These results were interpreted in terms of a simple bimolecular route for both reactions (equation 38) and a very low activation energy, notably for the reaction with iodine, was explained on the basis of a cyclic activated complex (**96**), stabilized by the formation of relatively strong H–I bonds and by 'a substantial portion of the resonance energy of the incipient azobenzene

(94) (95) (38)

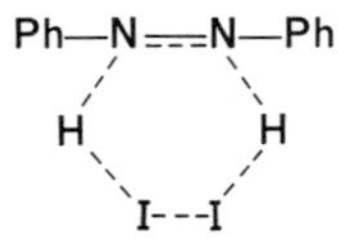

(96)

molecule'. An alternative hydride transfer mechanism leading to initial formation of HI, I^- and the conjugate acid of azobenzene was considered to be less likely due to the absence of evidence for a highly polar transition state[95].

Treatment of an *N*-acyl derivative of hydrazobenzene, namely *N*-benzoyl-*N,N'*-diphenylhydrazine (**97**) with lead tetra-acetate has been recently shown to give rise to both *cis*- and *trans*-azobenzene (equation 39). In dichloromethane–acetic acid, the product was essentially *trans*, whereas reaction under nitrogen in dry benzene with lead tetra-acetate (free from acetic acid) gave over 20% of the *cis* isomer, 'even though no precautions

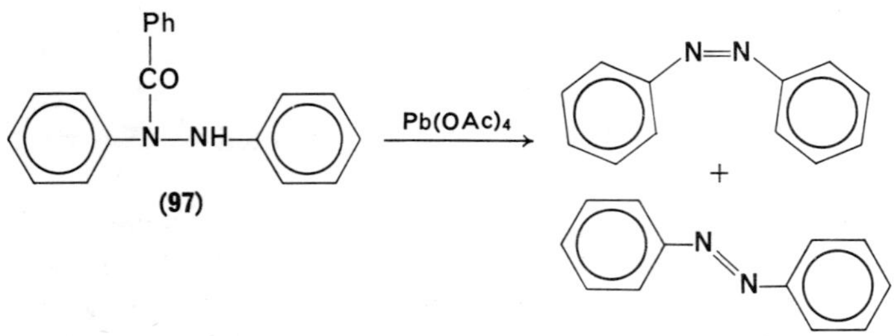

(97) (39)

were taken to prevent equilibration of the two forms during isolation'[87]. These findings are in accord with reaction via routes (i) and (ii), the latter pathway being important only in the absence of an external nucleophile (Scheme 1).

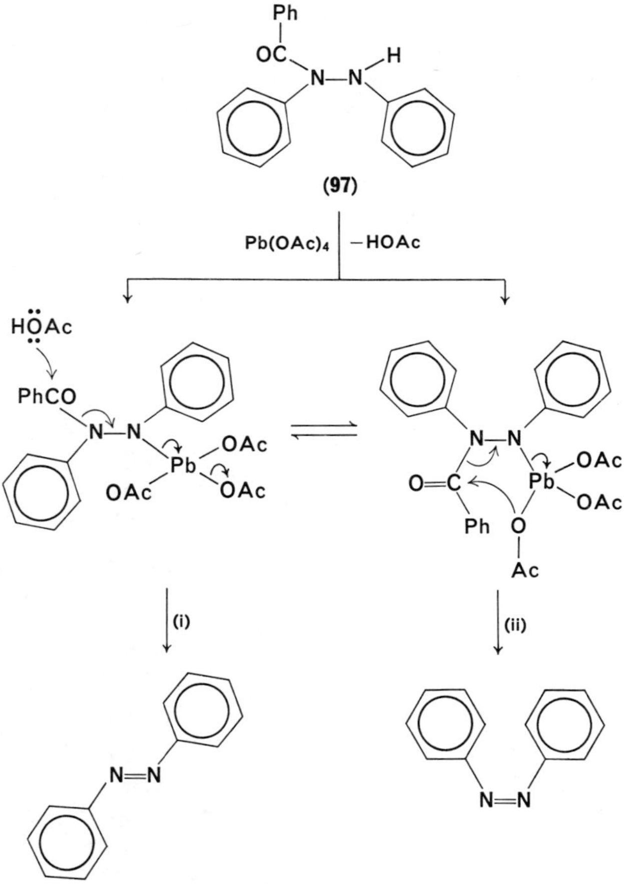

SCHEME 1

2. Substituted hydrazobenzenes

It has long been known that hydrazobenzenes, although more stable than the aliphatic hydrazines, are quite sensitive to air oxidation, usually taking on a yellow to red colouration due to partial or complete conversion to the appropriate azo derivative. Recently[96] the air oxidation of a series of 2,2′-, 3,3′-, and 4,4′-dihalohydrazobenzenes was studied spectrophotometrically and the rates of reaction to the azobenzenes were interpreted in terms of electronic and steric effects. In another study[97], 2,2′- (**98**) and 4,4′-diethylhydrazobenzene (**99**) were shown to have quite different sensitivities to air oxidation. For example, when an ethanolic solution of **98** was boiled in air to the point of crystallization, the product redissolved, and this treatment repeated several times, there was only partial conversion to the azo derivative. In contrast, similar treatment of **99** caused rapid and complete conversion to 4,4′-diethylazobenzene (**100**) (equation 40). Furthermore, essentially the same results were obtained when **98** and **99** were allowed to stand in air at room temperature. These differences may be mainly attributed to steric hindrance around the hydrazo linkage in **98** by the bulky ethyl groups[97]. 3,3′-Dimethylhydrazobenzene has also been found to be suscept-

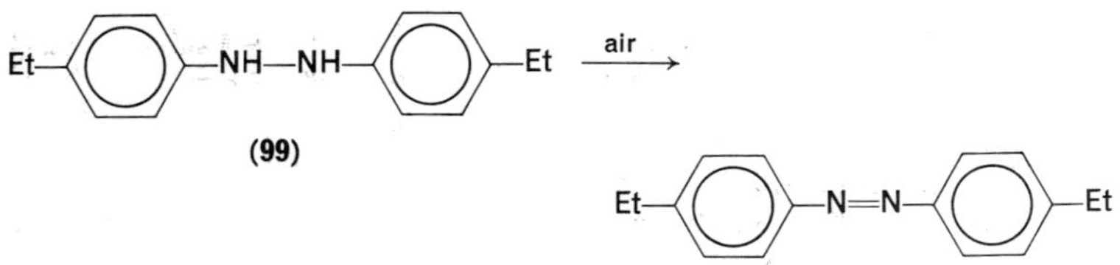

(99)

(100)

(40)

ible to oxidation to the azobenzene on exposure to air[98]. Grammaticakis[99] has reported the oxidation of a series of dichlorohydrazobenzenes to the corresponding azobenzenes by means of heating in air or potassium ferricyanide.

The electrochemical oxidation of *p*-substituted anilines has been found to give rise to the appropriate azobenzenes in rather low yield, the reaction proceeding via the 4,4′-disubstituted hydrazobenzenes[94]. The electrochemical oxidation of 4,4′-dichlorohydrazobenzene using a platinum anode has been reported[94].

In earlier work, a variety of oxidizing agents, including for example iodine, lead peroxide in acetone, and potassium dichromate and acetic acid, was used for the oxidation of hydrazobenzenes to azo compounds. However, in more recent years, other reagents such as 30% hydrogen peroxide in glacial acetic acid, and sodium hypohalites, have been employed in this connection. For instance, 2,2′,5,5′-tetra-t-butylhydrazobenzene, which is virtually unaffected by boiling its ethanolic solution in air, is oxidized to the azobenzene (75% yield) by refluxing in ethanol in the presence of excess 30% hydrogen peroxide and glacial acetic acid[100]; these rather drastic conditions being necessary to overcome steric hindrance[101]. Similar treatment of 2,2′,3,3′-tetrachlorohydrazobenzene affords 2,2′,3,3′-tetrachloroazobenzene[102].

Studies on the benzidine rearrangement of substituted hydrazobenzenes under the necessary acidic conditions have also revealed that azo derivatives are frequently formed under said conditions, and that fission to amines takes place as well. A recent study in this regard by Banthorpe and Winter[103] has thrown some light on the mechanism of oxidation of hydrazobenzenes to azo compounds in this type of reaction. These workers reported that acid treatment of 3,3′,5,5′-tetrabromohydrazobenzene caused benzidine rearrangement (78%), and disproportionation to 3,3′,5,5′-tetrabromoazobenzene together with fission to 3,5-dibromoaniline (22%). Furthermore, they showed that disproportionation also took place with 4,4′-dichlorohydrazobenzene, and used tracer experiments to show that in this case the azo product was formed by oxidation rather than through intermolecular recombination of anilino or nitrene fragments, the cross product being absent when using both the normal and tetradeuteriated substrate.

It has recently been reported[104] that decafluorohydrazobenzene (**101**) is more stable to oxidation than hydrazobenzene, but still decomposes in air, and apparently to the azo compound. Both **101** and 4*H*,4′*H*-octafluorohydrazobenzene (**102**) decompose to the azo derivatives (and probably to the anilines as well) on heating to 170°C; a reaction which also occurs with

$$2\ RC_6F_4{-}NH{-}NH{-}C_6F_4R \xrightarrow{170^\circ C} RC_6F_4{-}N{=}N{-}C_6F_4R + 2\ RC_6F_4NH_2 \qquad (41)$$

where $R = p$-F or p-H

hydrazobenzene at lower temperature (equation 41). The greater resistance of **101** and also **102** to oxidation has been attributed to the electronegative fluorine atoms which render electron-abstraction more difficult. However, **101** and **102** are quite readily converted to the azobenzenes (yields of 90 and 98%, respectively) on exposure to aqueous sodium hypochlorite for 2 h at room temperature[104] (equation 42).

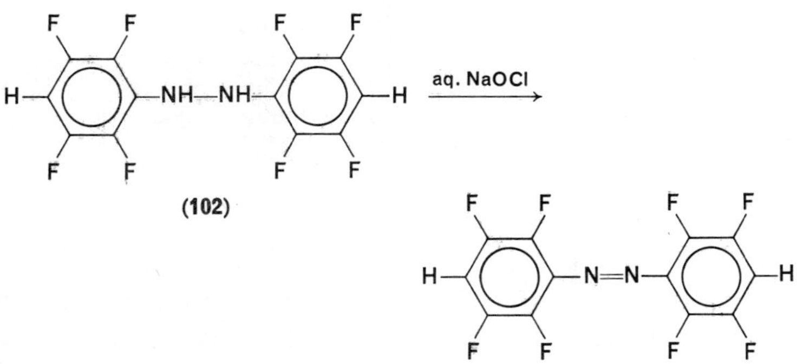

(42)

As far as the author is aware, no conversions of unsymmetrically substituted hydrazobenzenes to azo compounds have been reported in recent years.

C. Hydrazobenzenes to Azoxy Derivatives

The preparation of azoxybenzenes from hydrazobenzenes has received little attention to date, presumably because these substances (particularly symmetrically substituted compounds) may be more conveniently synthesized by the oxidation of azobenzenes or via reduction of appropriate nitrobenzenes. However, a number of symmetrical dichloro- and dibromohydrazobenzenes have been oxidized with 30% hydrogen peroxide in glacial acetic acid to give the dihalogenated azoxybenzenes in very good yield[101]. The reaction conditions were found to be important here. Thus, the yields of azoxybenzenes varied considerably depending upon the reactant proportions employed, as well as temperature and time. It is interesting to note here that in the case of 2,2′-dibromohydrazobenzene, where steric hindrance of the hydrazo group is expected, only a 3% yield of the azoxybenzene resulted even after prolonged treatment at 55–60°C. Some of the results obtained in this investigation are given in Table 4.

TABLE 4. Oxidations of substituted hydrazobenzenes with hydrogen peroxide in glacial acetic acid[101]

Hydrazobenzenes		30% H_2O_2 (ml)	HOAc (ml)	Temp. (°C)	Time (h)	Azoxybenzene (%)
Substituents	Amount (g)					
2,2′-Dichloro-	1·15	50	150	60–70	48	79
3,3′-Dichloro-	0·44	10	50	65–70	18	100
4,4′-Dichloro-	0·79	15	100	70–85	2·5	88
3,3′-Dibromo-	0·45	10	50	65–75	21	66
4,4′-Dibromo-	0·39	10	100	80–85	2	88
2,2′,3,3′-Tetrachloro-	0·10	20	50	Reflux	3	71
2,2′,4,4′-Tetrachloro-	0·10	40	50	Reflux	24	15
2,2′,5,5′-Tetrachloro-	0·11	40	50	Reflux	24	35
3,3′,5,5′-Tetrachloro-	0·11	10	50	Reflux	1	85
2,2′-Dimethyl-3,3′-dichloro-	0·29	25	100	62–80	6	89

In the same study, a series of symmetrically tetrasubstituted hydrazobenzenes was also oxidized with the above-mentioned reagent, and in several cases improved yields of the corresponding azoxy compounds were achieved. For instance, 3,3′,5,5′-tetrachlorohydrazobenzene (**103**) when so oxidized at 55–80°C for 24 h afforded the azoxy derivative (**104**) in 85% yield (equation 43), together with traces of the azobenzene. In contrast, similar treatment of 2,2′,5,5′-tetrachlorohydrazobenzene (**105**) at higher

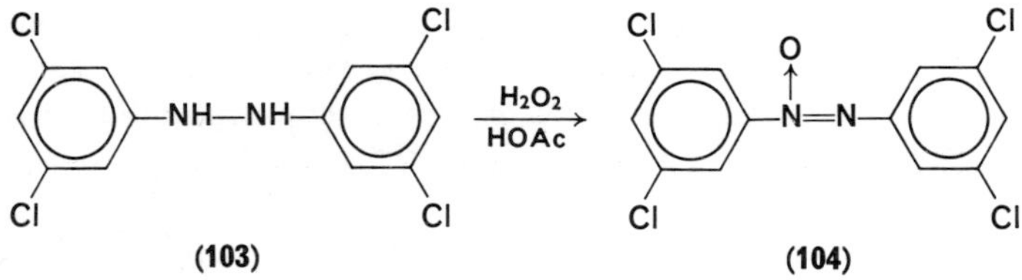

(43)

temperature (85–90°C) and for longer reaction time (72 h) did not lead to the azoxybenzene; the oxidation stopping with the formation of 2,2′,5,5′-tetrachloroazobenzene (**106**), obtained in 77% yield. Some of the oxidations of tetrasubstituted substrates are summarized in Table 4. In general, quite drastic conditions were needed to oxidize hydrazobenzenes having substituents in the 2 and 2′ positions, to the azoxy stage. 2,2′,4,4′-Tetrachlorohydrazobenzene and **105** were also oxidized to the azoxy derivatives with 30% hydrogen peroxide in acetic anhydride. This study showed that the substituted hydrazobenzenes used were remarkably stable in the oxidizing media selected, little or no decomposition having taken place in most instances[101].

V. OXIDATION OF AROMATIC AZO COMPOUNDS

A. Azobenzenes to Azoxybenzenes

The oxidation of azobenzenes has been studied extensively using a multitude of reagents, this type of reaction having proved to be an excellent and important route to azoxybenzenes, and particularly symmetrically substituted compounds[1]. Azobenzenes are generally much more stable than aliphatic azo derivatives, and this factor has permitted and indeed in many cases necessitated the use of strong oxidants and drastic conditions to bring about desired conversions to azoxybenzenes. An early review by Bigelow[105] has covered the chemistry of aromatic azoxy compounds up to about 1930, and gives information on the first conversions of azobenzenes to azoxybenzenes. The oxidation of these substances by means of organic peracids has been a subject of interest for many years. Swern[106] has listed all the azoxybenzenes prepared in this way up to 1949 and mentioned that such oxidations frequently led to quantitative yields.

It has been shown that the oxidation of substituted azobenzenes with perbenzoic acid exhibits second-order kinetics and that substituents have a considerable effect on the rate of reaction and the energy of activation[107]. Electron-releasing groups increase the rate whereas electron-attracting substituents decrease the rate of reaction. The effects of *meta* and *para* substituents follow Hammett's free energy relation. The extent of conjugation between the azo-linkage and the benzene ring is an important factor in *trans*-azobenzenes. In general, the greater the conjugation, the less available the lone pair electrons on the nitrogen atoms, and hence the more difficult the oxidation[107].

Most of the recent work on oxidations of azobenzenes has been carried out using hydrogen peroxide, peracetic acid, or perbenzoic acid[11]. Oxidations usually lead to the formation of the more stable *trans*-azoxy isomers.

I. Azobenzene

Badger and co-workers[108] have determined the rate of reaction of azobenzene (**95**) with perbenzoic acid in chloroform, and brought to light some interesting differences in the reactivities of the *cis*- and *trans*-isomers. The oxidation of *cis*-azobenzene at room temperature and in the dark gave *cis*-azoxybenzene (**107**), and that of *trans*-azobenzene afforded *trans*-azoxybenzene (**108**) in almost quantitative yield, when carried out in the

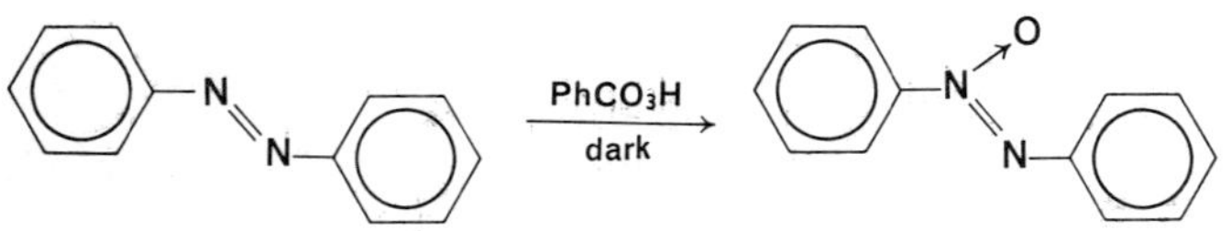

(44)

(95) (108)

dark (equation 44). Oxidation of *trans*-azobenzene in sunlight gave **108** and some **107**. The kinetic study revealed that *cis*-azobenzene reacted much more rapidly than the *trans*-isomer, thus indicating that the 'electron density and electron availability' at the azo linkage in the former is considerably greater than that in the latter. These workers also found that much higher reaction rates resulted when perbenzoic acid that had undergone a considerable amount of spontaneous decomposition was used, and showed that this was due to a catalytic effect caused by the presence of benzoic acid.

Although Badger and colleagues reported that the exclusion of light was necessary for conversion of *cis*-azobenzene to **107**, it has been pointed out by Sandler and Karo[11] that since *cis*-azoxybenzenes were later prepared by oxidation while warming with a heating lamp, this requirement may not be absolutely essential. An explanation of the formation of both **107** and **108** from the oxidation of *trans*-azobenzene was subsequently offered by Webb and Jaffé[109], who suggested two possible mechanisms, namely: oxidation to **108** followed by reversible isomerization to **107**; and reversible isomerization to *cis*-azobenzene and then oxidation to **107**, the latter pathway being favoured.

More recently, Pentimalli[110] oxidized azobenzene by means of perbenzoic acid; and peracetic acid[111], peracetic acid/98% hydrogen peroxide in chloroform[112], and 30% hydrogen peroxide in glacial acetic acid[101, 113], have also been applied to the parent substance. When azobenzene was exposed to the latter reagent under various conditions it was found that increasing the temperature together with a shortening of the reaction time led to a reduction of 20% in the yield of azoxybenzene[101]. The electrochemical oxidation of azobenzene has also been reported[94].

On the other hand, azobenzene is not attacked by some oxidants. For example, when it is dissolved in pure iodine pentafluoride and the solution warmed to 100°C, no reaction takes place, the substrate being recuperated after hydrolysis of the oxidant[92]. It has also been shown to be 'unexpectedly resistant' to ozonolysis in acetic acid or chloroform after being subjected to the action of 1–8 molar equivalents of ozone in oxygen mixtures at –40 to 34°C. Under these conditions no formation of azoxybenzene could be detected and the azobenzene was recovered almost quantitatively[114].

2. Symmetrically substituted azobenzenes

a. Disubstituted azobenzenes. A number of studies on the oxidation of symmetrically disubstituted azobenzenes have been reported. Thus, Badger and co-workers[108, 115] measured the rates of oxidation of a series of 4,4′-disubstituted azobenzenes, 2,2′- and 3,3′-dimethyl- and 3,3′-dichloroazobenzene, with perbenzoic acid; and also succeeded in converting *cis*-3,3′- and 4,4′-dimethylazobenzene (**109**) to the *cis*-azoxy derivatives with the same reagent in the dark. Gore and Wheeler[111] have prepared 3,3′-difluoro- and 4,4′-dinitro-, dibromo- and difluoroazoxybenzene by peracetic acid oxidation of the azo precursors. Webb and Jaffé[112] have used quite concentrated solutions of hydrogen peroxide to obtain *cis*-azoxy derivatives from the *cis*-azobenzenes. For example, they oxidized **109** to the *cis*-azoxybenzene using 98% hydrogen peroxide.

A series of dihalogenated azobenzenes has been oxidized to the corresponding azoxy compounds by means of 30% hydrogen peroxide in glacial acetic acid, the reactions being performed at about 60–70°C for 24 h[113]. In general, the yields of the azoxybenzenes from substrates having substituents in the 3 and 3′ or 4 and 4′ positions were very good, whereas those obtained from azobenzenes with groups in the 2 and 2′ positions were lower. This study revealed that conditions had an important bearing on the course of reaction. Thus, the way in which the hydrogen peroxide was added was found to be important, the best results being achieved by regular addition of small portions of reagent during the first 2 h. The azobenzenes most soluble in acetic acid were the most readily oxidized. The reaction time was also important, low yields resulting after heating for only a few hours, hence the heating period was extended to 24 h. Some of the results obtained are given in Table 5. 2,2′-Dichloroazobenzene (**110**) was quite stable to oxidation

TABLE 5. Oxidations of disubstituted azobenzenes with hydrogen peroxide in glacial acetic acid

Azobenzenes		30% H_2O_2 (ml)	HOAc (ml)	Temp. (°C)	Time (h)	Azoxy-benzene (%)	Ref.
Substituents	Amount (g)						
2,2′-Dichloro-	2·0	41	150	60–70	24	28[a]	113
3,3′-Dichloro-	2·0	41	200	62–75	24	93	113
4,4′-Dichloro-	0·5	31	200	60–82	24	89	113
2,2′-Dibromo-	0·2	10	50	63–66	24	13[b]	113
3,3′-Dibromo-	0·3	10	50	64–65	24	100	113
4,4′-Dibromo-	0·2	16	100	65–68	24	100	113
3,3′-Di-iodo-	0·2	10	100	62–67	24	96	113
2,2′-Dimethyl-	0·5	25	50	65–75	17·5	70	97
3,3′-Dimethyl-	0·4	25	50	61–65	3	87	97
4,4′-Dimethyl-	0·4	15	50	65–70	1	89	97
2,2′-Diethyl-	0·4	25	50	55–65	4	40	97
4,4′-Diethyl-	0·2	5	50	65–70	1·5	52	97
4,4′-Di-isopropyl-	0·1	5	25	Reflux	0·1	87	100
4,4′-Di-t-amyl-	0·3	30	100	Reflux	0·5	75	117

[a] 47% of substrate recuperated.
[b] No azo derivative recovered.

under the above conditions (only 28% conversion to the azoxy derivative); and an even lower yield resulted with 2,2′-dibromoazobenzene, which underwent considerable decomposition. In view of its stability, later on further attempts were made to oxidize **110** using stronger oxidants, such as chromic acid in acetic acid, and nitric acid, but these failed, the substrate being recovered essentially unaffected; however when 30% hydrogen peroxide in glacial acetic acid was applied to **110** under reflux conditions very good conversion (91%) to the desired 2,2′-dichloroazoxybenzene was achieved[101].

The latter study[101] showed that the refluxing of stable azobenzenes with 30% hydrogen peroxide in glacial acetic acid for a relatively short time constitutes a good method for the preparation of azoxybenzenes; the advantages of this technique being that the azobenzene is more soluble under these conditions, the higher temperature leads to effective oxidation, and the reaction time is shortened thus reducing decomposition. However, this procedure is not suitable for the oxidation of 2,2′-di-iodoazobenzene, which although largely unaffected by 30% hydrogen peroxide in glacial acetic acid at 70–80°C, is completely decomposed on treatment with an excess of this reagent at higher temperatures[116]. It would appear that steric hindrance of the azo linkage by the bulky iodine atoms impedes oxidation to the azoxy compound in this case.

A number of dialkylazoxybenzenes have been recently synthesized via the oxidation route. Thus, the three isomeric symmetrical dimethylazoxybenzenes, 2,2′- and 4,4′-diethylazoxybenzene[97], and 4,4′-di-isopropyl-[100] and di-t-amylazoxybenzene[117] have been prepared from the azobenzenes by the use of 30% hydrogen peroxide in glacial acetic acid (Table 5). The 2,2′-dialkylazobenzenes proved difficult to oxidize, a large excess of reagent being needed to produce 2,2′-dimethylazoxybenzene in high yield (90%); and in order to prepare the 2,2′-diethyl analogue in over 60% yield, either an extended reaction period or elevated temperature was necessary as well[97]. Here again steric effects would be expected to be in operation. It has been previously reported[115] that 2,2′-dimethylazobenzene reacts more slowly than azobenzene and much more slowly than 4,4′-dimethylazobenzene with perbenzoic acid, and that steric hindrance is a factor here, the entropy of activation being somewhat greater than that for the 4,4′-analogue.

Dewar and Goldberg[118] have reported the synthesis of 4,4′-dialkoxyazoxybenzenes in good overall yield (67–76%) by oxidation of the crude azo precursors (obtained via lithium aluminium hydride reduction of the appropriate 4-alkoxynitrobenzenes) with 30% hydrogen peroxide in acetic

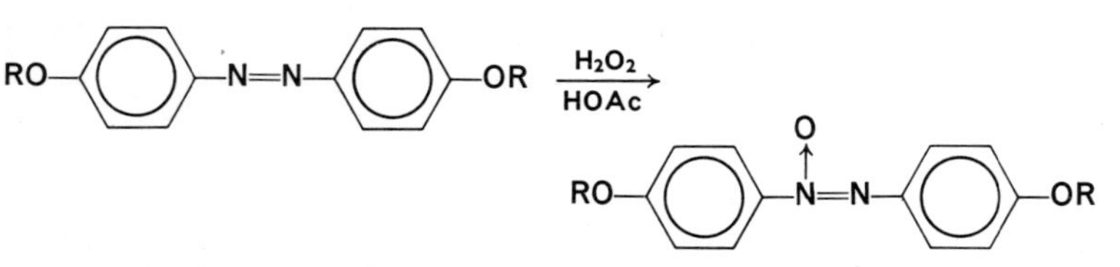

(45)

R = *n*-hexyl, ethyl, or methyl group

acid at 65°C for 36 h (equation 45). The azoxybenzenes so obtained were purified by chromatography on alumina, followed where necessary by repeated treatment with the oxidant to remove small quantities of unreacted azo compound.

The use of 30% hydrogen peroxide in glacial acetic acid as a preferred oxidant for the conversion of azobenzenes to azoxy compounds has continued to the present time. For instance, 4,4′-diacetylazobenzene has been oxidized to the azoxy derivative with this reagent[119], and very recently[120], 2,2′-di-isopropyl- and dibenzylazoxybenzene, and 2,2′-azoxybiphenyl have been prepared from the corresponding azo compounds by treatment with this oxidant at 60–70°C for 9 h.

b. Tetrasubstituted azobenzenes. In recent years the preferred oxidant for the conversion of symmetrically tetrasubstituted azobenzenes to the corresponding azoxy compounds has been hydrogen peroxide in acetic acid. Gagnon and Newbold[113] have prepared some tetrachloroazoxybenzenes in varying yield by treating the azo compounds with 30% hydrogen peroxide in glacial acetic acid at about 60–70°C, and showed that the reaction conditions and the manner in which the oxidant was added were of importance. In this work 2,2′,3,3′-, 3,3′,4,4′-, and 3,3′,5,5′-tetrachloroazoxybenzene (**104**) were obtained in good yield (above 70%), whereas 2,2′,4,4′- (**111**) and 2,2′,5,5′-tetrachloroazoxybenzene (**112**) could be produced in only about 25% yield, three-quarters of the substrate being recovered in each case. However, the yield of **111** by this route was subsequently increased to 88% by performing the oxidation under reflux conditions[101], and that of **112** was more than doubled in like manner[102]. The resistance of 2,2′,5,5′-tetrachloroazobenzene (**106**) to oxidation is illustrated by the fact that it is not converted to **112** by sodium dichromate in acid medium at room temperature[101]. Attempts have been made to oxidize 2,2′,4,4′-tetrachloroazobenzene and **106** with 30% hydrogen peroxide in acetic anhydride under reflux conditions, but although rapid reaction took place, decomposition also occurred resulting in low yields of **111** and **112**, respectively.

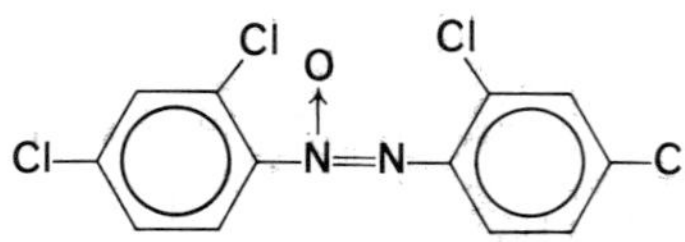

(111)

Relatively little work on the oxidation of tetra-alkylated azobenzenes has been reported. Recently, 2,2′,4,4′-tetraethylazobenzene was oxidized by refluxing with 30% hydrogen peroxide in glacial acetic acid to give rise to the azoxy compound in 96% yield[100], and 2,2′,5,5′-tetramethyl- and tetraethylazoxybenzene have been similarly prepared[100,121]. Treatment of

2,2′,6,6′-tetramethylazobenzene with the same reagent at 60°C affords the tetramethylazoxybenzene in quantitative yield[120].

A number of dihalodialkyl- and dihalodialkoxyazoxybenzenes have been synthesized by oxidation of the appropriate azo compounds. For example, 3,3′-dichloro-2,2′-dimethyl- and 5,5′-dichloro-2,2′-dimethoxyazoxybenzene were made in good yield by heating the azo precursors at 65–80°C with 30% hydrogen peroxide in glacial acetic acid[101]. Recently, series of dihalodimethylazobenzenes, which had not been previously oxidized, were similarly converted to the azoxy derivatives, and the ease of oxidation was found to depend upon the structure of the azo compound[121,122]. Prolonged treatment under drastic conditions was needed to transform dihalo-2,2′-dimethyl- and 2,2′-dihalodimethylazobenzenes to the azoxybenzenes, and this was attributed to steric effects. For instance, conversion to 5,5′-di-iodo-2,2′-dimethylazoxybenzene could only be effected after prolonged refluxing with excess oxidant, whereas 4,4′-difluoro-3,3′-dimethylazobenzene, in

TABLE 6. Oxidations of dihalodimethyl- and dihalodialkoxyazobenzenes with hydrogen peroxide in glacial acetic acid[a] [121, 123]

Azobenzenes: Substituents	Azobenzenes: Amount (g)	30% H_2O_2 (ml)	Time (h)	Azoxybenzene (%)
4,4′-Difluoro-3,3′-dimethyl-	0·2	20	0·4	86
5,5′-Difluoro-2,2′-dimethyl-	0·2	25	0·5	87
2,2′-Dichloro-3,3′-dimethyl-	0·2	30	0·5	85
2,2′-Dichloro-5,5′-dimethyl-	0·2	30	0·6	84
3,3′-Dibromo-4,4′-dimethyl-	0·2	30	0·7	86
5,5′-Dibromo-2,2′-dimethyl-	0·2	30	0·8	85
5,5′-Dichloro-2,2′-dimethoxy-	1·1	25	1	85
2,2′-Dichloro-5,5′-dimethoxy-	1·1	25	1	88
4,4′-Dichloro-2,2′-dimethoxy-	1·0	25	0·8	85
4,4′-Dichloro-3,3′-dimethoxy-	1·0	20	0·5	91
3,3′-Dichloro-4,4′-dimethoxy-	1·0	15	0·5	93
2,2′-Dichloro-5,5′-diethoxy-	0·2	30	2·4	69
2,2′-Dibromo-5,5′-diethoxy-	0·2	30	2	61[b]
2,2′-Dibromo-5,5′-dimethoxy-	0·1	30	2	55[c]
3,3′-Dibromo-4,4′-dimethoxy-	0·2	20	18	87
4,4′-Dibromo-3,3′-dimethoxy-	0·2	20	18	89

[a] All oxidations were carried out in 100 ml glacial acetic acid and under reflux conditions, except the last two which were performed on a steam bath.
[b] Traces of substrate detected.
[c] 23% of substrate recovered.

which the azo group is unhindered, underwent ready oxidation to the azoxy derivative under mild conditions. In some cases, oxidation of the dihalodimethylazobenzene was a valuable alternative route to the azoxy compound when the latter could not be synthesized by reduction of the appropriate halonitrotoluene, for example, this being the case for 5,5′-difluoro-2,2′-dimethylazoxybenzene[121]. Some of the results from these studies are given in Table 6.

In another investigation[123], ten dihalodialkoxyazobenzenes were similarly oxidized to the azoxy compounds and in most cases high yields were achieved, as may be seen from Table 6. The hindered 2,2′-dibromodialkoxyazobenzenes, as expected, proved difficult to oxidize. On the other hand, attempts to oxidize a number of tetra-alkoxyazobenzenes in this way were not successful, complete decomposition taking place when these substrates were heated under reflux with varying amounts of hydrogen peroxide[123].

Nitric acid has been used as an oxidant for some tetrasubstituted azobenzenes, but this reagent can also simultaneously bring about nitration. Singh and co-workers[124] have reported that treatment of 3,3′-dichloro-2,2′-dimethylazobenzene (**113**) with a mixture of fuming nitric acid and glacial acetic acid causes both oxidation and nitration, the product of the reaction being 3,3′-dichloro-2,2′-dimethyl-4,4′-dinitroazoxybenzene (**114**) (equation 46).

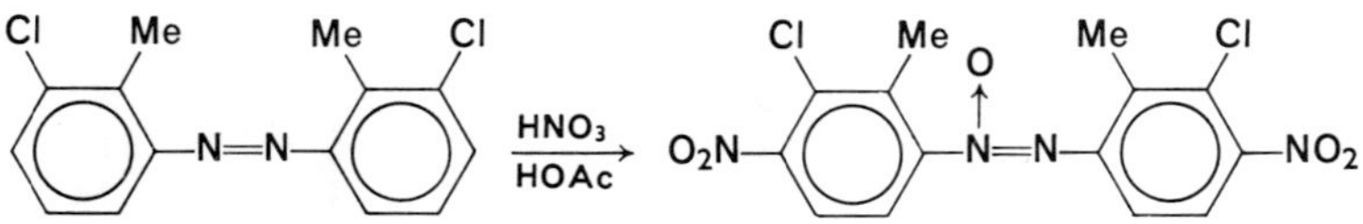

(**113**) (**114**) (46)

c. *Polysubstituted azobenzenes.* Recent work on the oxidation of polysubstituted azobenzenes has been mainly concerned with halogenated substrates. Peracids and other reagents have been used to oxidize these systems to the azoxybenzenes, excellent yields being achieved in a number of instances.

Burdon and colleagues[125] prepared 4*H*,4′*H*-octafluoroazoxybenzene (**115**) from the azo compound (**116**) in dichloromethane by prolonged refluxing (50 h) with a mixture of 89% hydrogen peroxide and trifluoroacetic anhydride (in effect, trifluoroperacetic acid), but even under these severe conditions considerable amounts of **116** were recovered (equation 47). Decafluoroazobenzene (**117**) also proved to be very difficult to oxidize, being virtually unaffected after refluxing in dichloromethane with the above

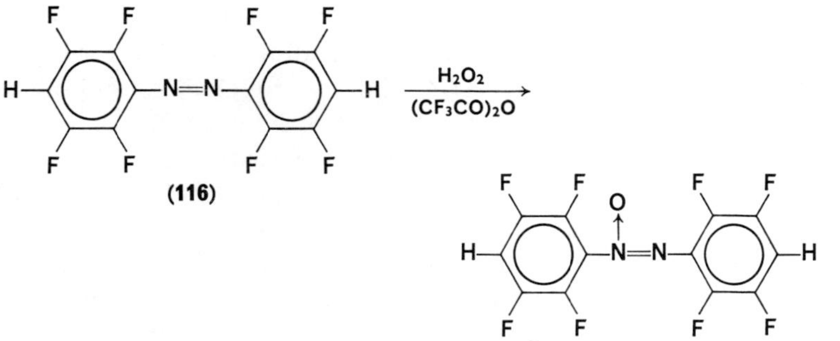

(116)

(115)

(47)

oxidant for 3 days or with performic acid for 4 days. It was also reported that **117** was not oxidized after exposure to either peracetic acid or fuming nitric acid at room temperature for 5 days. The much greater resistance to oxidation of **117** compared to that of azobenzene was attributed to the presence of electronegative fluorine substituents which should render the electron-abstraction necessary for oxidation more difficult. In contrast, these workers found that octafluoro-4,4′-dimethoxyazoxybenzene was formed in 82% yield by oxidation of the azo compound with trifluoroperacetic acid, the easier oxidation in this case evidently being due to the electron-releasing capacity of the methoxy groups.

In later work[126], decafluoroazoxybenzene was synthesized in very high yield (94%) by treating **117** with aqueous 85% hydrogen peroxide and trifluoroacetic anhydride in dichloromethane under reflux conditions. Furthermore, **115** was obtained in 98% yield by similar treatment of **116**; and almost pure 2*H*,2′*H*-octafluoroazoxybenzene (**118**) was likewise

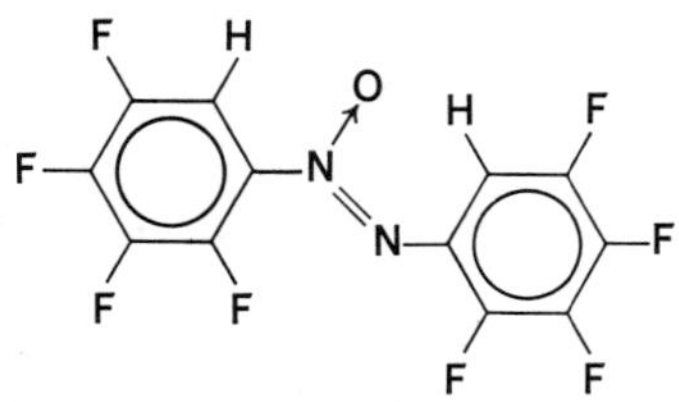

(118)

quantitatively prepared from the azobenzene. It is important to note here that the concentrations of the reagents determined the success of these reactions, and especially that of **117**, the oxidations proceeding readily in homogeneous solutions. The use of too much hydrogen peroxide or too dilute a solution of oxidant led to separation of an aqueous phase possessing most of the oxidizing power needed for successful reaction. It will be recalled that the importance of reaction conditions for oxidations of other azobenzenes was mentioned earlier (see Section V, A, 2a and b).

The electrochemical oxidation of *sym*-hexachloroazobenzene using a platinum anode has been reported[94]. Decachloroazoxybenzene has been synthesized in 97% yield by refluxing decachloroazobenzene (**119**) in a mixture of chloroform, 98% hydrogen peroxide, and trifluoroacetic anhydride for 5 h. It was also found that **119**, in which the azo linkage is more sterically hindered than that in **117**, was not affected by either performic or peracetic acid[127].

In very recent work[127], the sterically crowded 2,2′,4,4′,6,6′-hexamethylazoxybenzene (**120**) was prepared in 84% yield by refluxing *sym*-hexamethylazobenzene (**121**) in chloroform with 85% *m*-chloroperbenzoic acid for 4 h (equation 48).

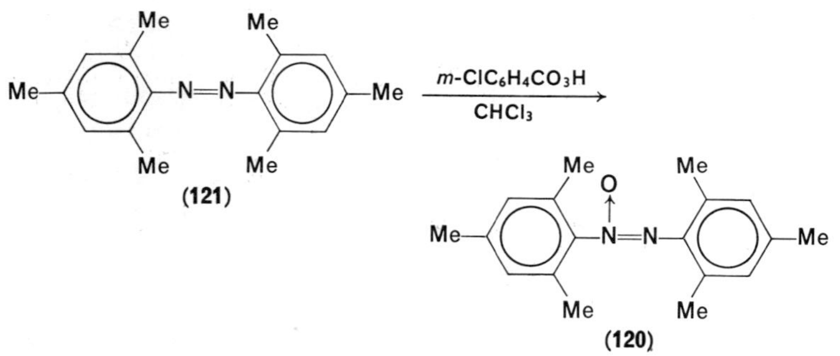

(48)

3. Unsymmetrically substituted azobenzenes

Unsymmetrically substituted azoxybenzenes, besides exhibiting *cis–trans* isomerism, can also exist in two other isomeric forms, depending upon the position of the oxygen atom in the azoxy group, i.e. the ONN- and NNO-isomers. One would therefore expect that mixtures of ONN- and NNO-isomers would invariably result from the oxidation of unsymmetrically substituted azobenzenes. However, it was recently pointed out that this has rarely been observed, perhaps due to difficulties with structural studies or inadequate separation methods[11]. The separation of individual isomers of azoxybenzenes from mixtures is a formidable challenge. Fractional recrystallization has been used, but is tedious and does not always give reliable results. In recent years modern separation techniques have been employed, but difficulties were encountered in attempts to separate isomers by column or gas chromatography. The identification of structure of

ONN- and NNO-isomers of azoxybenzenes is not easy either, relatively few methods being available for distinguishing between such isomers[107,128,129]. Spectroscopic analysis has been used in this regard but the ultraviolet, visible, infrared and n.m.r. spectra are very similar. Thus, chemical degradation is still generally required for definite structural determination of the compounds in question. However, very recently a rapid, direct and accurate analytical method has been developed for the analysis of mixtures of isomeric azoxybenzenes using a lanthanide p.m.r. shifting agent[130].

Although peracids have been frequently used over many years as oxidants for azo compounds, including unsymmetrically substituted derivatives, surprisingly little is known of the mechanisms involved. Thus, as recently as 1970 it was pointed out by Mitsuhashi and co-workers[131] that 'although the oxidation of *trans*-azobenzenes with peroxyacids has been investigated extensively, no consideration has yet been given to the mechanism'. These workers cited two possible types of interaction between the peracid (e.g. perbenzoic acid) and the azo linkage, one involving competitive attack on either of the lone pairs of electrons of the nitrogen atoms (equation 49); and the other on the —N=N— bond where the transition state is symmetrical with respect to each benzene ring, leading to the rate-determining formation of an oxadiaziridine-type intermediate, which immediately forms

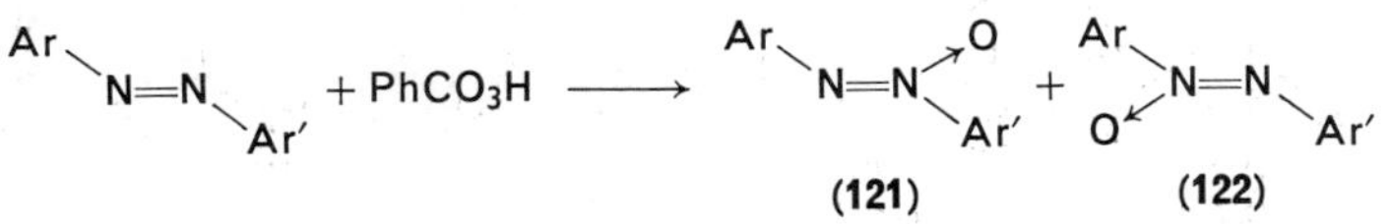

(49)

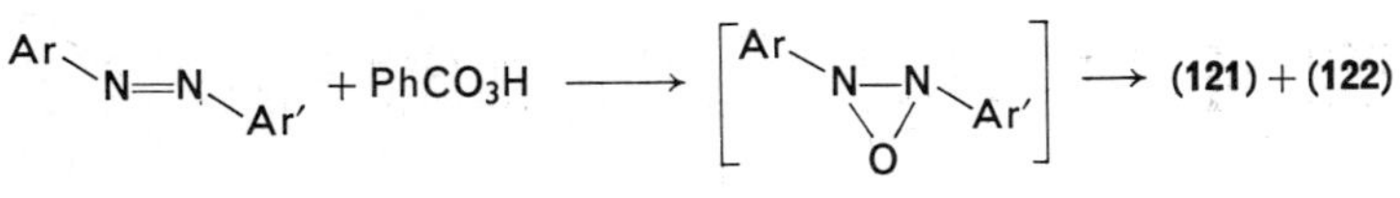

(50)

isomeric azoxy derivatives (equation 50). To distinguish between these two mechanisms, these investigators determined the reaction rates and the isomer ratios for the oxidation of a series of unsymmetrically and symmetrically substituted *trans*-azobenzenes and *trans*-azonaphthalenes with perbenzoic acid in benzene at 25°C. They calculated second-order rate constants from rate of consumption of peracid and obtained isomer ratios for the azoxy derivatives from 4-methyl- and 4-methoxy substituted azo compounds by comparing the intensities of the signals due to the 4-methyl or 4-methoxy protons in the n.m.r. spectra. The experimental data recorded were found to be in agreement with a simple one-step mechanism which does not necessarily involve an oxadiaziridine intermediate. Another interesting finding was that the 2-methyl or α-naphthyl (Ar) group enormously increased the formation of **121** relative to **122**, an effect attributable to steric hindrance caused by substitution at the *ortho* position.

a. Monosubstituted azobenzenes. Badger and Lewis[115] have determined the rates of oxidation of series of monosubstituted azobenzenes (3- and 4-methyl, methoxy, nitro, carbethoxy and halo derivatives) with perbenzoic acid at three temperatures, and calculated the energies of activation. They found that all the compounds in these series had the same entropy of activation, that electron-releasing groups decreased and electron-attracting substituents increased the heat of activation and free energy of activation; and that there was a linear relationship between the rate constants and the Hammett σ constants of the substituents.

Gore and Wheeler[111] have reported that the oxidation of 4-dimethylaminoazobenzene with peracetic acid yields 4-dimethylamino-ONN-azoxybenzene. On the other hand, Newbold[101] found that oxidation of *trans*-3-nitroazobenzene with 30% hydrogen peroxide in glacial acetic acid at temperatures varying from about 60 to 90°C gave mixtures of 3-nitro-NNO- (**123**) and -ONN-azoxybenzene in yields of up to 90%, the isomers being separated by fractional recrystallization. Behr and co-workers also obtained mixtures of these isomeric azoxybenzenes from treatment of 3-nitroazobenzene with peracetic acid, with **123** predominating[129]. These workers separated the isomers by column chromatography, and also prepared them by unambiguous independent synthesis definitely to establish their structures.

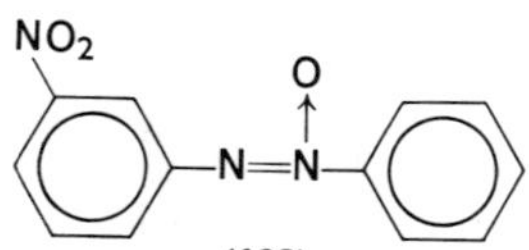

(123)

Hahn and Jaffé[132] have prepared series of *trans*-4- and 4′-monosubstituted azoxybenzenes of general formula (**124**), where R is Me, Cl, Br or OMe, and R^1 is H; and where R is H, and R^1 is Me, Br, OMe, OEt or NO_2, by oxidizing the appropriate azobenzenes with peracetic acid. It was recently

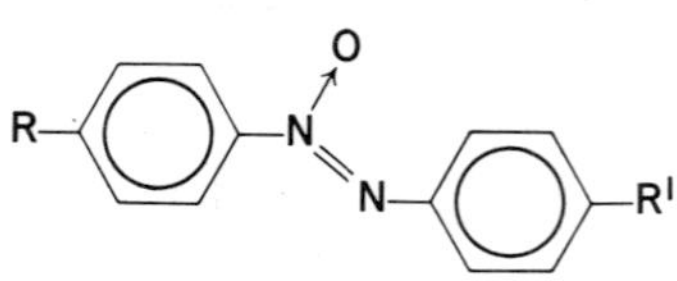

(124)

reported[133] that 2-methylazobenzene is oxidized to 2-methylazoxybenzene when heated with perhydrol in glacial acetic acid on a water bath, but apparently no attempt was made to separate the possible isomers.

Some *cis* monosubstituted azoxybenzenes have been synthesized in low yield by the oxidation of mixtures of the *cis*- and *trans*-azo compounds with peracetic acid and 98% hydrogen peroxide in chloroform[112]. Nuclear magnetic resonance spectroscopy was used to establish the *cis* configuration of the azoxy group in 4-bromo-, chloro-, and methylazoxybenzene, obtained in this manner, and the existence of the ONN- and NNO-isomers of these products was also shown.

b. Disubstituted azobenzenes. Oxidations of unsymmetrically disubstituted azobenzenes may lead to two types of unsymmetrically disubstituted azoxybenzenes, i.e. those having both substituents on the same benzene ring, and those having one group on each ring. A recent example of the preparation of an azoxy product of the first type was the conversion of 4-nitroazobenzene-2-carboxylic acid to the corresponding azoxybenzene by means of perhydrol in glacial acetic acid at 100°C[133]. A very recent case is the oxidation of 2,6-dimethylazobenzene (**125**) with 30% hydrogen peroxide in acetic acid at 60°C, which gives almost exclusively 2,6-dimethyl-NNO-azoxybenzene (**126**) because of steric hindrance[120] (equation 51).

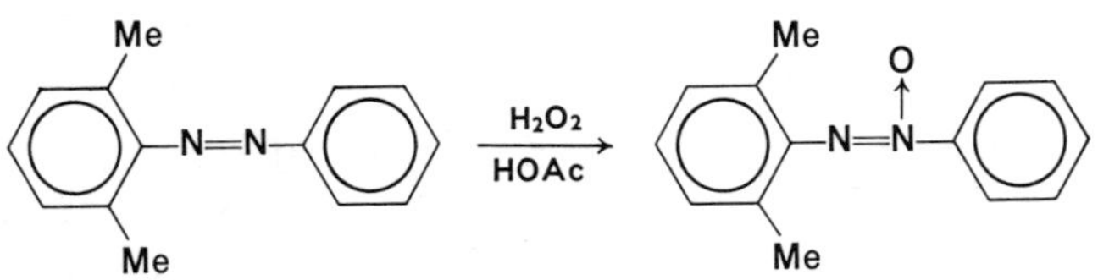

(**125**) (**126**) (51)

Over the past few years, some interesting research has been carried out on substrates having one substituent on each ring. Thus, *trans*-4-methyl-4′-bromo- and 4-bromo-4′-methylazoxybenzene have been synthesized by peracetic acid oxidation of the *trans*-azo compounds[132]; and 3-methyl-3′-nitro- (**127**) and 3-chloro-3′-methylazobenzene (**128**) have been oxidized with 30% hydrogen peroxide in glacial acetic acid at 85–90°C to give azoxybenzenes, whose structures were ascertained by bromination and reduction of the bromo derivatives to the corresponding amines[134]. For the product from **127**, the oxygen atom of the azoxy group was found to be adjacent to the benzene ring having the methyl group, this being accounted for by the methyl group exerting its electron-releasing effect to increase the availability of the lone pair of electrons on the nitrogen atom nearest to it, and the electron-attracting effect of the nitro substituent which would deactivate the adjacent nitrogen atom with regard to attack by electrophilic oxygen. Furthermore, with **128** oxidation to **129** took place, this product being obtained in 99% yield.

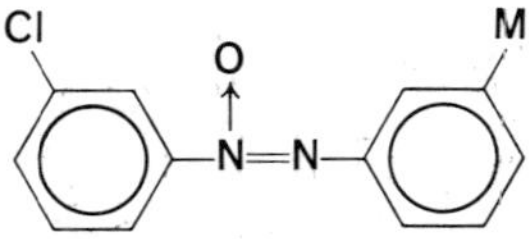

(129)

Haug and collaborators[135] have synthesized unsymmetrically 4,4′-disubstituted azoxybenzenes by oxidation of the *trans*-azo compounds with 30% hydrogen peroxide in acetic acid, and showed that the products had the *trans* configuration by means of ultraviolet spectroscopic and dipole moment studies. For example, such oxidation of *trans*-4-chloro-4′-ethoxy-azobenzene gave a mixture of the *trans*-ONN- (**130**) and -NNO- (**131**)

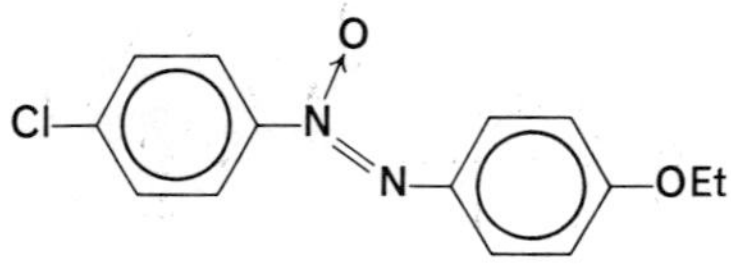

(130)

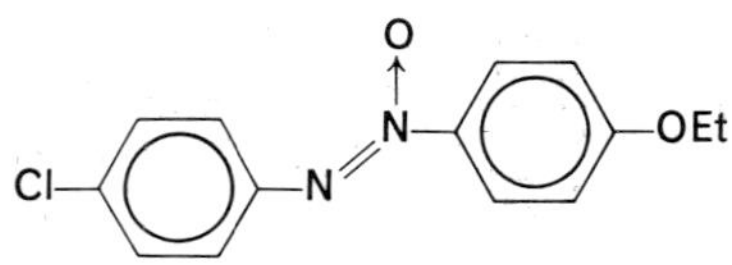

(131)

isomers of the azoxybenzene. Steinsträsser and Pohl[136] have also prepared series of 4-n-alkyl-4′-n-alkoxy- and n-acyloxyazoxybenzenes by oxidizing the azo precursors with 30% hydrogen peroxide in acetic acid at 60°C (equation 52); and in each case obtained mixtures of the NNO- and ONN-

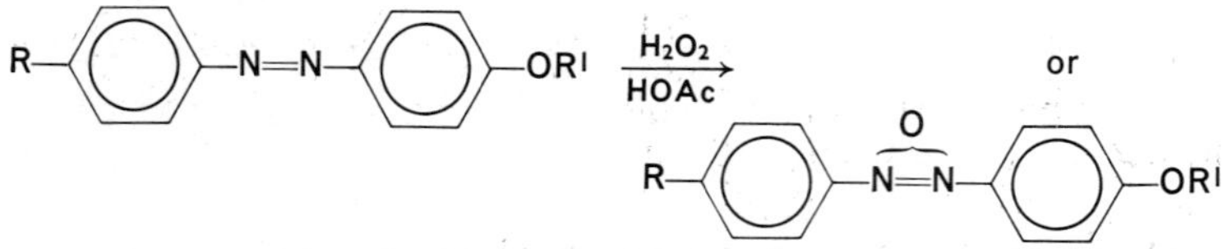

(52)

R is Et, Pr, Bu, Am or hexyl group ,and R^{I} is Me, Et, Pr, Bu, Am or hexyl group or CO-hexyl, CO-Bu or CO-heptyl group.

isomers. Some of the products made in this work, which are of interest for liquid crystal studies, were examined by p.m.r. spectroscopy and shown to consist of mixtures containing 59–62% $RC_6H_4—N{=}\overset{\overset{O}{\uparrow}}{N}—C_6H_4OMe$ and

38–41 % $RC_6H_4{-}\overset{\overset{O}{\uparrow}}{N}{=}N{-}C_6H_4OMe$. For instance, the eutectic mixture of the isomers of 4-n-butyl-4′-methoxyazoxybenzene was composed of 66 % NNO-isomer and 34 % ONN-isomer.

Until quite recently, little quantitative work had been done to determine the effect of *ortho* substituents upon the oxidation of azobenzenes, presumably because of the lack of reliable analytical methods[137]. However, in 1972, Berwick and Rondeau[130] reported the direct examination of the ratios of isomeric azoxybenzenes, produced via peracetic acid oxidation of a series of *o*-substituted azobenzenes, by using tris(1,1,1,2,2,3,3-heptafluoro-7,7-dimethyl-4,6-octanedionato) europium to separate isomeric p.m.r. signals. Before analysis with this shifting agent phenols and acids were converted to acetates and esters. The reactions were carried out in acetic acid at 25°C using 90 % hydrogen peroxide and a catalytic amount of sulphuric acid, the yields of azoxybenzenes being in general >85 %. The crude products were purified without changing the isomeric contributions, and the isomer ratios were found to be stable for >20 h under the conditions employed. The magnitude and direction of the regioselectivity of the oxidations were measured, and some of the results from this work are given in Table 7.

TABLE 7. Oxidations of *ortho*-substituted azobenzenes with 90 % hydrogen peroxide in glacial acetic acid[130]

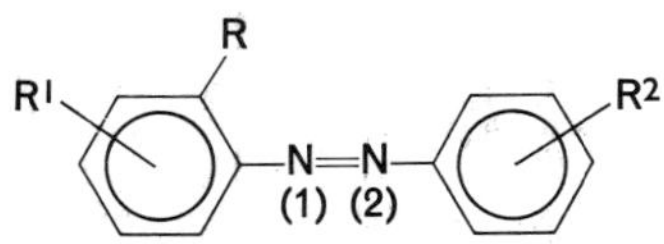

Substituents			Nitrogen atom oxidized
R	R[1]	R[2]	(1):(2)
Me	H	2-CO_2Me	1:3·2
Cl	H	4-Me	1:24
OH	5-Me	H	2·6:1
CO_2Me	H	4-Cl	1:5·0
CO_2H	H	4-Cl	1:20
CO_2Me	H	4-OMe	1:3·8
OC(=O)Me	5-Me	H	1:9·3

With the exception of the hydroxyl group, which favoured oxidation of the nitrogen atom next to the *ortho*-substituted benzene ring, all the *ortho* substituents directed oxidation to the other nitrogen atom, i.e. that furthest away; and this was attributed to mainly inductive and steric effects, resonance effects appearing to be of little consequence. The relative magnitudes of this N(2) selectivity also gave rise to an interesting finding for the case of N(2) selectivity due to the *ortho* carboxylic acid group which proved to be 2–3 times as great as that caused by the carbomethoxy group, in spite of the fact that the latter is the larger substituent. This difference was explained in terms of the role played by internal hydrogen bonding rather than increased steric N(2) selectivity due to solvent co-ordination.

This investigation has also provided an attractive alternative regiospecific route to azoxybenzenes, involving firstly oxidation of azobenzene-2-carboxylic acids to azoxybenzene-2-carboxylic acid mixtures (>98% isomerically pure) with a 14–20:1 preference for oxidation of N(2); and then decarboxylation, as shown in Scheme 2.

c. Trisubstituted azobenzenes. The oxidation of the two categories of unsymmetrically trisubstituted azobenzenes has received relatively little

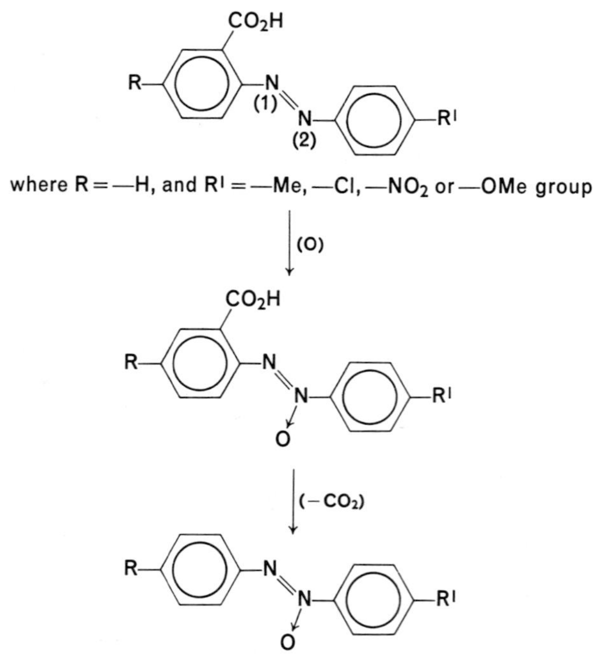

SCHEME 2

attention lately, but in the studies that have been reported hydrogen peroxide in acetic acid was once again the preferred oxidant. Thus, Newbold[101] oxidized 4,4′-dichloro-2-nitroazobenzene with 30% hydrogen peroxide in glacial acetic acid at 67–75°C, and obtained only 4,4′-dichloro-2-nitro-NNO-azoxybenzene (**132**), in 79% yield, which was considered to have the *trans* configuration. The exclusive production of **132** here would seem to be due to both steric hindrance by the nitro group and the inductive effect of the latter, which would deactivate the adjacent nitrogen atom towards oxidation.

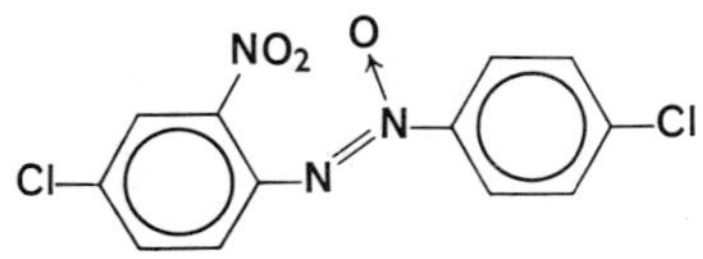

(132)

A similar case to that mentioned above is the very recent[120] oxidation of 2,4,6-trimethylazobenzene with the same reagent, which leads to the exclusive formation of 2,4,6-trimethyl-NNO-azoxybenzene (**133**). This result is to be expected after consideration of steric effects and in analogy with the like finding for the oxidation of 2,6-dimethylazobenzene (**125**), which was described earlier (see Section V, A, 3b).

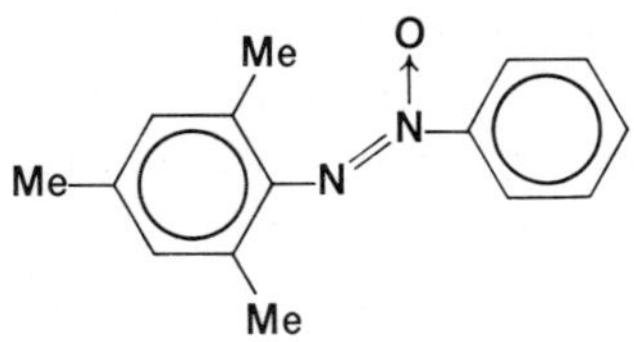

(133)

d. Polysubstituted azobenzenes. Recent work on the synthesis of unsymmetrically polysubstituted azoxybenzenes by oxidation of appropriate azo compounds has been concerned with highly chlorinated products. Thus, 2,3,4,5,6-pentachloroazoxybenzenes of general formula (**134**), where

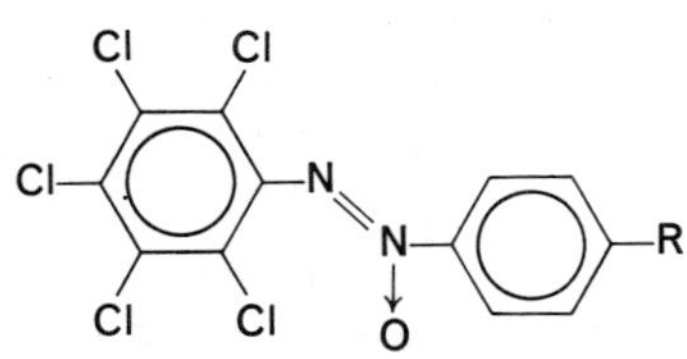

(134)

R = —H, —Br, —Cl, —Me or —OMe, have been prepared in almost quantitative yield (>95%) from the corresponding azobenzenes by treatment with 98% hydrogen peroxide and trifluoroacetic anhydride in chloroform under reflux conditions; and the n.m.r. spectra of the azoxybenzenes were determined[126]. In this study, model azoxybenzenes having one completely deuterated benzene ring were also synthesized for further n.m.r. studies, by oxidation of pentadeuterioazobenzenes with 98% hydrogen peroxide and glacial acetic acid in chloroform, mixtures of the ONN- (**135**) and NNO- (**136**) isomers resulting, where were separated by

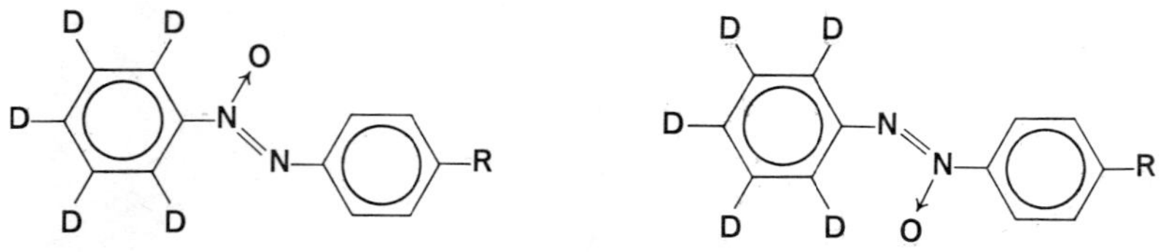

R = —Br, —Me or —OMe

(135) **(136)**

column chromatography. Examination of the n.m.r. spectra of the pentadeuterioazo- and azoxybenzenes revealed that the trifluoroperacetic acid oxidations of the pentachloroazobenzenes had afforded exclusively the NNO- isomers of the azoxy derivatives. Furthermore, this finding was confirmed by mass spectroscopy. This preferential formation of the NNO-isomer could be mainly accounted for on the basis of steric effects. The above investigation also showed that the pentachloroazobenzenes selected for study resisted oxidation by other peracids, such as peracetic and perbenzoic acid, reaction being incomplete even after several days of refluxing with these oxidants.

B. *Azonaphthalenes to Azoxynaphthalenes*

I. Symmetrical azonaphthalenes

Badger and Lewis[107] have determined the rates of oxidation of the two symmetrical azonaphthalenes with perbenzoic acid at four temperatures, and calculated the energies of activation; interpreting the results obtained in terms of the extent of conjugation of the azo linkage with the ring system, and of steric hindrance. They reported that 2,2′-azonaphthalene (**137**) was oxidized faster than azobenzene to give the 2,2′-azoxy derivative, and that 1,1′-azonaphthalene (**138**) oxidized more slowly to afford 1,1′-azoxynaphthalene (**139**). The peri-hydrogens cause steric interference with the oxygen atom in **139**[138], thus it was concluded that steric hindrance must be a factor in all oxidations occurring at an azo-nitrogen atom next to the 1-position in naphthalene. These workers also showed that **138** has a much larger heat of activation than its isomers, and the steric factor was confirmed by a significant increase in the entropy of activation compared with that of either azobenzene or **137**.

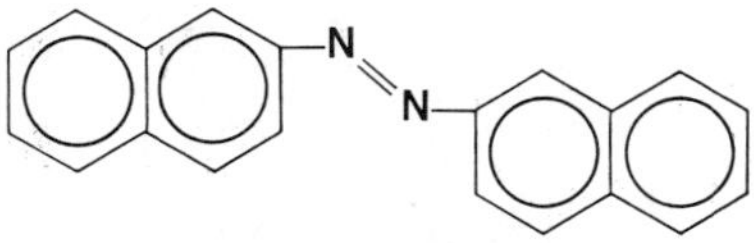

(137)

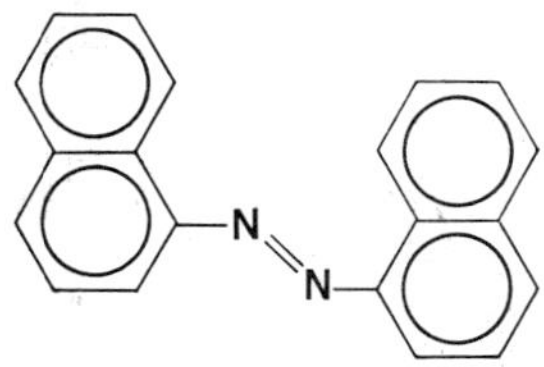

(138)

Recently[139], 2,2′-dimethyl-1,1′-azoxynaphthalene was synthesized in 46% yield by treatment of the azonaphthalene in glacial acetic acid with 30% hydrogen peroxide at 60–70°C; and 2,2′-dimethoxy-1,1′-azonaphthalene was converted to the azoxy derivative by prolonged oxidation with peracetic acid in chloroform at room temperature.

2. Unsymmetrical azonaphthalenes

The rate of oxidation of 1,2′-azonaphthalene with perbenzoic acid has been determined and it was found that this oxidant gave only one azoxy product, which from spectral and other evidence was considered to have structure (**140**)[107]. Since the azo-nitrogen atom adjacent to the 1-position is sterically hindered, exclusive formation of isomer **140** is not surprising in this case.

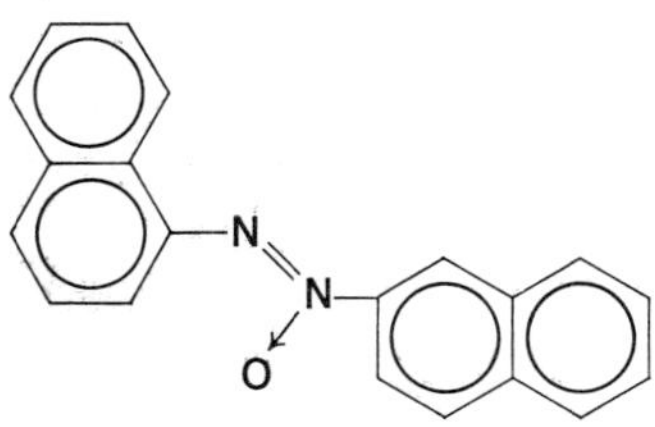

(140)

Similar studies were carried out on the two phenylazonaphthalenes using perbenzoic acid[107]. Thus, 1-phenylazonaphthalene on treatment with this oxidant gave only one product, which was shown by spectral studies to be azoxy isomer (**141**), as expected. On the other hand, similar oxidation of

2-phenylazonaphthalene gave a mixture of the NNO- (**142**) and ONN- (**143**) isomers of the corresponding azoxynaphthalene, with an isomer ratio of about 41 : 59, which was almost the same as the ratio of the reaction rates of azobenzene and **137** with perbenzoic acid.

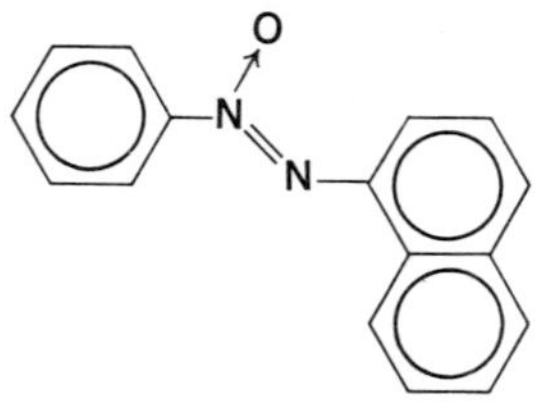

(141)

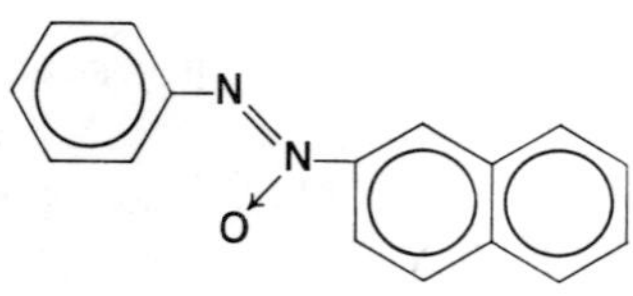

(142)

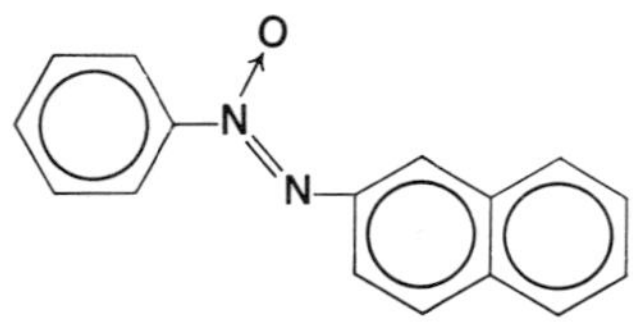

(143)

The rate of reaction of *trans*-1-(*p*-tolylazo)-naphthalene (**144**) with perbenzoic acid has been recorded, together with the isomer ratio for the mixture of *trans*-azoxy products thus obtained[131].

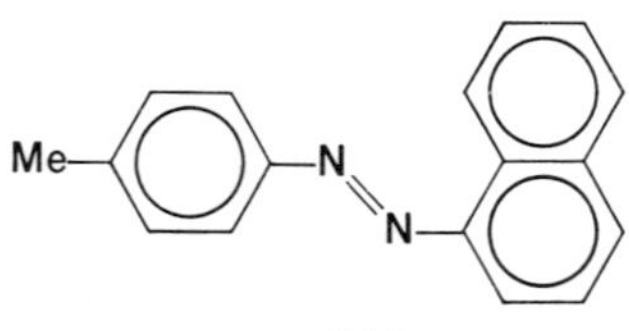

(144)

In recent work[140], 1-(2,6-dimethylphenylazo)-2-methoxynaphthalene (**145**) was oxidized with 30% hydrogen peroxide in glacial acetic acid at 50–60°C, and research is presently under way to determine if the product obtained is (2,6-dimethylphenyl)-NNO-1-azoxy(2-methoxynaphthalene), the -ONN- isomer, or a mixture thereof. This is an interesting case, in that in **145** both azo-nitrogen atoms are sterically hindered, and hence the formation of a mixture of the isomeric phenylazoxynaphthalenes would be expected to result from a successful oxidation.

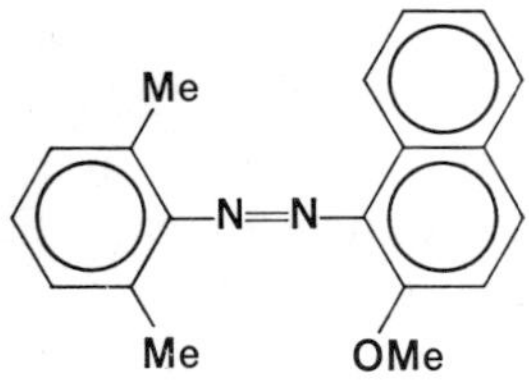

(145)

VI. CONCLUSIONS

A good deal of work has been done recently on the oxidation of aliphatic hydrazines to azo compounds, with mercuric oxide becoming a preferred oxidant for this purpose, although copper(II) chloride has also been widely used for the synthesis of alicyclic azo systems. The oxidation of azo derivatives has proved to be an important route to aliphatic azoxy compounds. Oxidations of acyclic or cyclic hydrazines to azoxy compounds do not appear to have been reported, whereas this type of transformation was accomplished with some hydrazobenzenes. A significant advance has been the synthesis, via oxidation, of interesting new aliphatic cis-azo and azoxy compounds, including polycycloazo systems.

Although some research has been carried out on the oxidation of hydrazobenzenes to azobenzenes, much more attention has been paid to conversions of azobenzenes to azoxybenzenes; the most popular oxidants for these purposes being hydrogen peroxide and various peracids. In most cases, the *trans* products were apparently obtained, but a number of *cis*-azoxybenzenes have also been synthesized by careful oxidation of the *cis*-azo precursors. From the studies on the oxidation of azobenzenes and azonaphthalenes it is clear that steric effects may have a great influence on both the rate and course of reaction, and that in general electron-attracting groups render oxidation of the azo linkage difficult; however, there is a need for more investigations on the effects of substituents in order to have a good understanding of these aspects. In contrast to the mild reaction conditions generally used for oxidations of aliphatic hydrazines and azo derivatives, quite drastic conditions were advantageously employed for the oxidation of some stable hydrazo and azobenzenes.

The reports on the oxidation of unsymmetrically substituted aliphatic and aromatic azo compounds have brought to light some of the factors that determine which of the nitrogen atoms of the azo group will be oxidized, but it is evident that a lot more research is necessary in this area. The mechanistic aspect of the oxidation of both aliphatic and aromatic hydrazo and azo compounds was until quite recently almost completely neglected, hence much remains to be done in that field as well.

Two of the most striking and fruitful developments in recent years have been the increasing use of modern techniques for the separation and purification of products from oxidations of hydrazo and azo compounds; and the successful application of nuclear magnetic resonance spectroscopy for determining the structures of these products, and particularly those of isomeric azoxy derivatives.

VII. ACKNOWLEDGMENTS

The author thanks the National Research Council of Canada for several grants, and Mr. Guy Cormier for help with part of the literature survey. He is also grateful to Drs. E. Buncel and N. J. Bunce for having provided unpublished information.

VIII. REFERENCES

1. H. Zollinger, *Azo and Diazo Chemistry—Aliphatic and Aromatic Compounds*, Interscience Publishers, Inc., New York, 1961.
2. R. Stewart, *Oxidation Mechanisms—Applications to Organic Chemistry*, W. A. Benjamin Inc., New York, 1964.
3. W. A. Waters, *Mechanisms of Oxidation of Organic Compounds*, Methuen and Co., Ltd., London, 1964.
4. K. B. Wiberg (Ed.), *Oxidation in Organic Chemistry*, Academic Press Inc., New York, 1965.
5. C. G. Overberger, J. P. Anselme and J. G. Lombardino, *Organic Compounds with Nitrogen–Nitrogen Bonds*, Ronald Press, New York, 1966.
6. P. A. S. Smith, *Open Chain Nitrogen Compounds*, Vols. 1 and 2, W. A. Benjamin Inc., New York, 1966.
7. N. V. Sidgwick, *The Organic Chemistry of Nitrogen*, 3rd ed., Clarendon Press, Oxford, 1966.
8. I. T. Miller and H. D. Springall, *A Shorter Sidgwick's Organic Chemistry of Nitrogen*, Clarendon Press, Oxford, 1969.
9. R. L. Augustine, (Ed.), *Oxidation*, Marcel Dekker, Inc., New York, 1969.
10. K. L. Rinehart, *Oxidation and Reduction of Organic Compounds*, Prentice-Hall, Inc., Englewood Cliffs, N. J., 1973.
11. S. R. Sandler and W. Karo, in *Organic Functional Group Preparations*, Vol. II, (Ed. A. T. Blomquist), Academic Press, New York, 1971.
12. P. Rys and H. Zollinger, *Fundamentals of the Chemistry and Application of Dyes*, Wiley, London, 1972.
13. B. Sklarz, *Quart. Revs.*, **21,** 3 (1967).
14. J. B. Alward, *Quart. Revs.*, **25,** 407 (1971).
15. N. L. Weinberg and H. R. Weinberg, *Chem. Revs.*, **68,** 449 (1968).
16. S. G. Cohen and C. H. Wang, *J. Am. Chem. Soc.*, **77,** 2457 (1955).
17. C. G. Overberger and H. Gainer, *J. Am. Chem. Soc.*, **80,** 4556 (1958).

18. S. Seltzer, *J. Am. Chem. Soc.*, **83,** 2625 (1961).
19. S. Seltzer and E. J. Hamilton, *J. Am. Chem. Soc.*, **88,** 3775 (1966).
20. F. Gerson, E. Heilbronner, A. van Veen and B. M. Wepster, *Helv. Chim. Acta*, **43,** 1889 (1960).
21. D. Y. Curtin and T. C. Miller, *J. Org. Chem.*, **25,** 885 (1960).
22. J. C. Kauer, *Org. Syn. Coll.*, **4,** 411 (1963).
23. S. E. Scheppele and S. Seltzer, *J. Am. Chem. Soc.*, **90,** 358 (1968).
24. J. Swigert and K. G. Taylor, *J. Am. Chem. Soc.*, **93,** 7337 (1971).
25. M. J. Gibian and R. C. Corley, *J. Am. Chem. Soc.*, **94,** 4178 (1972).
26. J. R. Shelton and C. K. Liang, *Synthesis*, 204 (1971).
27. J. R. Shelton and C. K. Liang, *J. Org. Chem.*, **38,** 2301 (1973).
28. E. Fahr and H. Lind, *Angew. Chem. Int. Ed. Engl.*, **5,** 372 (1966).
29. S. G. Cohen, R. Zand and C. Steel, *J. Am. Chem. Soc.*, **83,** 2895 (1961).
30. M. C. Chaco and N. Rabjohn, *J. Org. Chem.*, **27,** 2765 (1962).
31. J. A. Campbell, D. Mackay and T. D. Sauer, *Can. J. Chem.*, **50,** 371 (1972).
32. L. A. Carpino, P. H. Terry and P. J. Crowley, *J. Org. Chem.*, **26,** 4336 (1961).
33. H. Bock, *Angew. Chem. Int. Ed. Engl.*, **4,** 457 (1965).
34. H. Bock and J. Kroner, *Chem. Ber.*, **99,** 2039 (1966).
35. C. G. Overberger, P. T. Huang and M. B. Berenbaum, *Org. Syn. Coll.*, **4,** 66 (1963).
36. R. M. Fantazier and J. E. Herweh, *J. Org. Chem.*, **38,** 2560 (1973).
37. K. Tagaki and R. J. Crawford, *J. Am. Chem. Soc.*, **93,** 5910 (1971).
38. L. Spialter, D. H. O'Brien, G. L. Untereiner and W. A. Rush, *J. Org. Chem.*, **30,** 3278 (1965).
39. R. J. Crawford and K. Tagaki, *J. Am. Chem. Soc.*, **94,** 7406 (1972).
40. A. Tsolis, S. G. Mylonakis, M. T. Nieh and S. Seltzer, *J. Am. Chem. Soc.*, **94,** 829 (1972).
41. T. R. Lynch and F. N. MacLachlan, *Can. J. Chem.*, **50,** 2143 (1972).
42. H. Bock and G. Rudolph, *Chem. Ber.*, **98,** 2273 (1965).
43. H. Bock and E. Baltin, *Chem. Ber.*, **98,** 2844 (1965).
44. H. Bock, G. Rudolph and E. Baltin, *Chem. Ber.*, **98,** 2054 (1965).
45. C. G. Overberger, J. G. Lombardino, I. Tashlick and R. G. Hiskey. *J. Am. Chem. Soc.*, **79,** 2662 (1957).
46. R. J. Crawford, A. Mishka and R. J. Dummel, *J. Am. Chem. Soc.*, **88,** 3959 (1966).
47. L. A. Carpino, *J. Am. Chem. Soc.*, **85,** 2144 (1963).
48. J. A. Berson and S. S. Olin, *J. Am. Chem. Soc.*, **91,** 777 (1969).
49. C. G. Overberger and J. W. Stoddard, *J. Am. Chem. Soc.*, **92,** 4922 (1970).
50. C. G. Overberger, M. S. Chi, D. G. Pucci and J. A. Barry, *Tetrahedron Lett.*, 4565 (1972).
51. C. G. Overberger and M. Lapkin, *J. Am. Chem. Soc.*, **77,** 4651 (1955).
52. P. G. Gassman, K. T. Mansfield, G. N. Taylor and K. B. Wiberg, *Org. Synth.*, **49,** 1 (1969).
53. E. L. Allred, J. G. Hinshaw and A. L. Johnson, *J. Am. Chem. Soc.*, **91,** 3382 (1969).
54. E. L. Allred and J. G. Hinshaw, *Chem. Commun.*, 1021 (1969).
55. J. P. Snyder, *Tetrahedron Lett.*, 2451 (1972).
56. O. L. Chapman and S. J. Dominianni, *J. Org. Chem.*, **31,** 3862 (1966).
57. R. A. Clement, *J. Org. Chem.*, **25,** 1724 (1960).

58. J. C. Stickler and W. H. Pirkle, *J. Org. Chem.*, **31,** 3444 (1966).
59. R. C. Cookson, S. S. H. Gilani and I. D. R. Stevens, *Tetrahedron Lett.*, 615 (1962).
60. T. J. Kealy, *J. Am. Chem. Soc.*, **84,** 966 (1962).
61. J. E. Herwen and R. M. Fantazier, *Tetrahedron Lett.*, 2101 (1973).
62. E. L. Allred and A. L. Johnson, *J. Am. Chem. Soc.*, **93,** 1300 (1971).
63. R. M. Moriarty, C.-L. Yeh and N. Ishibi, *J. Am. Chem. Soc.*, **93,** 3085 (1971).
64. N. Rieber, J. Alberts, J. A. Lipsky and D. M. Lemal, *J. Am. Chem. Soc.*, **91,** 5668 (1969).
65. W. P. Lay, K. Mackenzie and J. R. Telford, *J. Chem. Soc.(C)*, 3199 (1971).
66. J. A. Berson and S. S. Olin, *J. Am. Chem. Soc.*, **92,** 1086 (1970).
67. M. L. Heyman and J. P. Snyder, *Tetrahedron Lett.*, 2859 (1973).
68. K. Shen, *Chem. Commun.*, 391 (1971).
69. L. A. Paquette, *J. Am. Chem. Soc.*, **93,** 7110 (1971).
70. S. Hünig, H. R. Müller and W. Thier, *Angew. Chem.*, **77,** 368 (1965).
71. C. Willis and R. A. Back, *Nature*, **241,** 43 (1973).
72. B. W. Langley, B. Lythgoe and L. S. Rayner, *J. Chem. Soc.*, 4191 (1952).
73. J. P. Freeman, *J. Org. Chem.*, **28,** 2508 (1963).
74. F. D. Greene and S. S. Hecht, *Tetrahedron Lett.*, 575 (1969).
75. B. T. Gillis and J. D. Hagarty, *J. Org. Chem.*, **32,** 95 (1967).
76. R. A. Moss, M. J. Landon, K. M. Luchter and A. Mamantov, *J. Am. Chem. Soc.*, **94,** 4392 (1972).
77. R. A. Moss and M. J. Landon, *Tetrahedron Lett.*, 3897 (1969).
78. J. N. Brough, B. Lythgoe and P. Waterhouse, *J. Chem. Soc.*, 4069 (1954).
79. R. B. Woodward and C. Wintner, *Tetrahedron Lett.*, 2693 (1969).
80. J. P. Freeman, *J. Org. Chem.*, **27,** 2881 (1962).
81. V. T. Bandurco and J. P. Snyder, *Tetrahedron Lett.*, 4643 (1969).
82. J. P. Snyder, L. Lee and D. G. Farnum, *J. Am. Chem. Soc.*, **93,** 3816 (1971).
83. P. C. Huang and E. M. Kosower, *J. Am. Chem. Soc.*, **89,** 3911 (1967).
84. F. L. Scott and J. A. Barry, *Tetrahedron Lett.*, 2461 (1968).
85. G. Cauquis and M. Genies, *Tetrahedron Lett.*, 3537 (1968).
86. J. B. Aylward, *J. Chem. Soc.* (*C*), 1633 (1969).
87. W. A. F. Gladstone, *J. Chem. Soc.* (*C*), 1571 (1969).
88. S. G. Cohen and C. H. Wang, *J. Am. Chem. Soc.*, **77,** 3628 (1955).
89. A. J. Bellamy and R. D. Guthrie, *J. Chem. Soc.*, 2788 (1965).
90. A. J. Fatiadi, *J. Org. Chem.*, **35,** 831 (1970).
91. W. A. Pryor and K. Smith, *J. Am. Chem. Soc.*, **92,** 5403 (1970).
92. T. E. Stevens, *J. Org. Chem.*, **26,** 2531 (1961).
93. F. Yoneda, K. Suzuki and Y. Nitta, *J. Am. Chem. Soc.*, **88,** 2328 (1966).
94. S. Wawzonek and T. W. McIntyre, *J. Electrochem. Soc.*, **114,** 1025 (1967).
95. J. W. May and J. Halpern, *Can. J. Chem.*, **39,** 1378 (1961).
96. L. J. Saulnier and B. T. Newbold, unpublished work.
97. B. T. Newbold and D. Y. Tong, *Can. J. Chem.*, **42,** 836 (1964).
98. D. J. Byron, G. W. Gray and B. M. Worrall, *J. Chem. Soc.*, 3706 (1965).
99. M. P. Grammaticakis, *Bull. Soc. Chim. Fr.*, 951 (1951).
100. L. J. Saulnier, *M.Sc. Thesis*, Université de Moncton (1966).
101. B. T. Newbold, *J. Org. Chem.*, **27,** 3919 (1962).
102. M. H. Akhtar, *M.Sc. Thesis*, Université de Moncton (1966).
103. D. V. Banthorpe and J. G. Wintner, *J. Chem. Soc., Perkin II*, 868 (1972).

104. J. Burdon, C. J. Morton and D. F. Thomas, *J. Chem. Soc.*, 2621 (1965).
105. H. E. Bigelow, *Chem. Revs.*, **9,** 117 (1931).
106. D. Swern, *Chem. Revs.*, **45,** 1 (1949).
107. G. M. Badger and G. E. Lewis, *J. Chem. Soc.*, 2151 (1953).
108. G. M. Badger, R. G. Buttery and G. E. Lewis, *J. Chem. Soc.*, 2143 (1953).
109. D. L. Webb and H. H. Jaffé, *Tetrahedron Lett.*, 1875 (1964).
110. L. Pentimalli, *Tetrahedron*, **5,** 27 (1959).
111. P. H. Gore and O. H. Wheeler, *J. Am. Chem. Soc.*, **78,** 2160 (1956).
112. D. L. Webb and H. H. Jaffé, *J. Am. Chem. Soc.*, **86,** 2419 (1964).
113. P. E. Gagnon and B. T. Newbold, *Can. J. Chem.*, **37,** 366 (1959).
114. R. E. Miller, *J. Org. Chem.*, **26,** 2327 (1961).
115. G. M. Badger and G. E. Lewis, *J. Chem. Soc.*, 2147 (1953).
116. B. T. Newbold, *J. Chem. Soc.*, 6972 (1965).
117. G. Cormier and B. T. Newbold, unpublished work.
118. M. J. S. Dewar and R. S. Goldberg, *Tetrahedron Lett.*, 2717 (1966).
119. E. C. Blossey and A. Corley, *Chem. Commun.*, 895 (1972).
120. D. J. W. Goon, N. G. Murray, J.-P. Schoch and N. J. Bunce, to be published.
121. M. N. Islam, *M.Sc. Thesis*, Université de Moncton, (1970).
122. P. E. Belliveau, *M.Sc. Thesis*, Université de Moncton, (1966).
123. R. A. Rao, *M.Sc. Thesis*, Université de Moncton, (1968).
124. J. Singh, P. Singh, J. L. Boivin and P. E. Gagnon, *Can. J. Chem.*, **40,** 1921 (1962).
125. J. M. Birchall, R. N. Haszeldine and J. E. G. Kemp, *J. Chem. Soc.* (*C*), 1519 (1970).
126. E. T. McBee, G. W. Calundann, C. J. Morton, T. Hodgins and E. P. Wesseler, *J. Org. Chem.*, **37,** 3140 (1972).
127. E. Buncel and R. A. Cox, *Chem. Commun.*, 1259 (1972); E. Buncel and R. A. Cox, private communication.
128. L. C. Behr, *J. Am. Chem. Soc.*, **76,** 3672 (1954).
129. L. C. Behr, E. G. Alley and O. Levand, *J. Org. Chem.*, **27,** 65 (1962).
130. M. A. Berwick and R. E. Rondeau, *J. Org. Chem.*, **37,** 2409 (1972).
131. T. Mitsuhashi, O. Simamura and Y. Tezuka, *Chem. Commun.*, 1300 (1970).
132. C.-S. Hahn and H. H. Jaffé, *J. Am. Chem. Soc.*, **84,** 949 (1962).
133. H. Rembges and F. Kröhnke, *Chem. Ber.*, **103,** 3437 (1970).
134. J. Singh, P. Singh, J. L. Boivin and P. E. Gagnon, *Can. J. Chem.*, **41,** 499 (1963).
135. U. Haug, J. Pelz and H. Usbeck, *Die Pharmazie*, **24,** (8), 442 (1969).
136. R. Steinsträsser and L. Pohl, *Tetrahedron Lett.*, 1921 (1971).
137. V. M. Dzimoko and K. A. Dunaevskaya, *Zh. Obshch. Khim.*, **31,** 3385 (1961).
138. G. M. Badger and R. G. Buttery, *J. Chem. Soc.*, 2156 (1953).
139. N. J. Bunce, private communication.
140. D. J. W. Goon and N. J. Bunce, work in progress.